The Evolution of Communication

The Evolution of Communication

Marc D. Hauser

A Bradford Book
The MIT Press
Cambridge, Massachusetts
London, England

This book was set in Times Roman by Achorn Graphic Services and was printed and bound in the
United States of America.

Library of Congress Cataloging-in-Publication Data

Hauser, Marc D.
     The evolution of communication / Marc D. Hauser.
         p.   cm.
     "A Bradford book."
     Includes bibliographical references (p.   ) and index.
     ISBN 0-262-08250-0
     1. Animal communication. 2. Communication. I. Title.
QL776.H38   1996
591.59—dc20
                                    95-52290
                                      CIP

I dedicate this book to the people
who have brought so much to my life:
My mentors, Dorothy Cheney, Peter Marler, and Robert Seyfarth
My family, Alberta, Alexandra, Dana, and Jacques Hauser
and
My closest friends, Kim Beeman, Carolee Caffrey, & Matthew Grober

Contents

Preface

Everybody who writes is interested in living inside of themselves. That is why writers have two countries, the one where they belong and the one in which they live really. The second one is romantic, it is separate from themselves, it is not real but it is really there.
—Gertrude Stein (1940), Paris, France

I have always been curious about the factors that shape each person's thoughts—especially those thoughts that culminate in a piece of scientific writing or work of art. Knowing about the history underlying the work is useful because it provides one with a framework for evaluating the creator's perspective. In music, for example, Alban Berg's work becomes all the more interesting and beautiful when one places it in its historical path, preceded by some of the giants of classical music—Beethoven, Wagner, and Strauss.

My own intellectual development has been influenced by a number of scientific giants—at least they are giants from my own Lilliputian perspective. I want to briefly recreate this development here because I think it will serve the reader well in evaluating some of the arguments and claims that are raised throughout the book.

As an undergraduate at Bucknell University concentrating in animal behavior, I worked under the supervision of Douglas Candland, a first-rate comparative psychologist whose intellectual breadth and lecturing style were enough to convince any undergraduate that animal behavior was the career of choice. My first course in animal behavior centered around Ed Wilson's towering oeuvre *Sociobiology*. Thus started my first encounter with adaptationists' accounts of behavior, including human behavior. I pursued my interests at UCLA, and in 1982 started the path to a PhD. My advisers were Robert Seyfarth and Dorothy Cheney. Now here was an interesting blend. Both Robert and Dorothy received PhDs from Robert Hinde in the Sub-Department of Animal Behaviour at Cambridge University and thus were firmly entrenched in the ethological tradition. However, due to their postdoctoral work with Peter Marler, they were gradually emerging as pioneers in what has become known as cognitive ethology. In essence, Robert and Dorothy initiated a research program on wild nonhuman primates that looked at how different sorts of cognitive mechanisms had been designed to solve ecologically relevant problems. My thesis work took me to East Africa, where I studied the behavioral ecology of vervet monkeys, focusing in particular on ontogenetic problems. Robert and Dorothy's influence can be seen at two levels.[1] First, they

---

1. The word *influence* is loaded here. Some of what I learned from Robert and Dorothy was through direct instruction, whereas other aspects were through osmosis or observational learning. They perhaps don't even realize what I have gathered over the years—parasitized may be a more appropriate term!

taught me how to find and test interesting questions, and second, they encouraged my disposition to seek out problems that cut across several disciplines. Upon finishing my thesis, I considered myself a behavioral ecologist with interests in the cognitive sciences. In 1987, I joined Richard Wrangham on a new project of forest chimpanzees. The chimpanzees lived in the Kibale Forest, Uganda, and my goals were to provide a comprehensive understanding of their vocal repertoire and, more specifically, to attempt to decode the meaning of their calls. Although it turned out to be exceedingly difficult to study chimpanzee vocal behavior, my time in the field with Richard turned out to be exceptionally important. Not only did I learn a great deal from him about forest ecology and wildlife, but I also sharpened my appreciation for the frustrations and limitations of field research. The latter led me to a change in projects. From Uganda and chimpanzees I went to Puerto Rico and rhesus monkeys, and from Richard Wrangham to Peter Marler. Working with Peter gave birth to a long series of changes in my intellectual interests. Specifically, I became increasingly interested in the anatomical and psychological mechanisms underlying communication and, in particular, in using nonhuman primates to test current theoretical issues in speech production and perception. Peter continued to promote my interests, giving me considerable latitude with regard to the questions I addressed. Peter's influence on me is perhaps most profound of all because I was ready[2] to be shaped.

Peter Marler completes the truly pedagogical period of influence. To complete the tour, let me briefly mention a handful of people in both closely and more distantly related disciplines who have had a significant influence on my thoughts. My background in linguistics was initiated at the UCLA Phonetics Lab, where I had the good fortune to be guided by Peter Ladefoged, Ian Maddieson, and Pat Keating. From UCLA, we move to the Haskins Lab, where I encountered Carol Fowler and Michael Studdert-Kennedy. Finally, we shift to UC-Davis and Harvard, where I have continued to be influenced by three of the broadest thinkers around, Sarah Blaffer Hrdy, John Locke, and Steve Pinker. Throughout this period, my knowledge of acoustics has been fine-tuned by Kim Beeman, bioacoustician extraordinaire. Most recently, my thoughts on cognitive science have been guided by Susan Carey, Richard Herrnstein, and Steve Kosslyn.

I hope this historical digression provides a useful angle on why I have some of the thoughts that I do. I don't blame history for the quirks in my thinking, but I will acknowledge its influence. The present is waiting. Onward.

---

2. I emphasize "ready" because during th Seyfarth-Cheney-Wrangham era, I was in the rambunctious teenager phase, rebelling against all. One learns quickly, however. I hope S-C-W forgive this phase.

Acknowledgments

Parents are usually acknowledged last. I must acknowledge them first. Both my mother and father have always encouraged me to explore my passions. This is perhaps the greatest gift one could receive, and I am certainly grateful.

Continuing on an emotional plane, I would like to warmly thank Alison Newsome for a beautifully crafted book cover and Wanda Reindorf for her sparkling smile, kindness, and support.

My intellectual development has been influenced by many, and I have developed the course of this historical path in the preceding preface. Here, I must thank all those who have encouraged me, criticized me, and made me think harder than I ever have about evolution, neurobiology, cognition, and behavior: Evan Balaban, Kim Beeman, Halle Brown, Susan Carey, Tim Caro, Dorothy Cheney, Jeff Cynx, Irv DeVore, Christopher Evans, Don Griffin, Sandy Harcourt, Richard Herrnstein, Sarah Blaffer Hrdy, James Intrilligator, Jerome Kagan, Nancy Kanwisher, Mark Konishi, Steve Kosslyn, Jerald Kralik, Peter Ladefoged, John Locke, Peter Marler, Dave Moody, Cindy Moss, Doug Nelson, Steve Nowicki, Steve Pinker, Josef Rauschecker, Mike Ryan, Robert Seyfarth, Elizabeth Spelke, Bill Stebbins, Kelly Stewart, Peter Tyack, and Richard Wrangham.

I would also like to thank my home institution, Harvard University, for providing me with an extremely comfortable and enjoyable environment for thinking. Most importantly, I would like to thank my colleagues in the Departments of Anthropology and Psychology. They are not only supportive, but friendly and intellectually inspiring as well.

For providing audio recordings or illustrations, I thank Andy Bass, Daniel Blumstein, Susan Boinski, Borror Laboratory of Bioacoustics–Ohio State University, Chris Clark, Cornell Laboratory of Ornithology, Richard Coss, Steven Dear, Paul Ekman, Christopher Evans, Carl Gerhardt, Robert Gibson, Fred Harrington, Steve Insley, Mark Konishi, Don Kroodsma, Joseph Macedonia, Peter Marler, Brenda McGowan, Matthew Rowe, Andrea Megal-Simmons, Michael Ryan, Jim Simmons, Peter Tyack, Haven Wiley, David Yager, and Ken Yasukawa.

The MIT Press, and especially Betty Stanton and Jerry Weinstein, have given me a wonderfully supportive feeling throughout this process, in addition to considerable flexibility. This support has made the entire venture both relaxing and rewarding.

Special thanks to Mark Konishi, Peter Marler, and Steven Pinker for reading through the entire first draft of the manuscript and for providing me with page after page of insights. As a result of their truly Herculean efforts, this book has been transformed into what I believe is a more thoughtful and balanced treatment of the literature.

The Evolution of Communication

1 Synopsis of the Argument

1.1 General Comments

Nothing would work in the absence of communication. Flowers must communicate with bees in order for pollination to be successful. Male songbirds must communicate with females if they are to mate and rear young. Lions on a cooperative hunt must communicate with each other about how they will attack their prey. A human infant must communicate with its parents so that the needs of both are met. Great orators, such as Jesse Jackson, must use their communicative skills to captivate and manipulate the emotions of their audience. Computer programmers must design software to communicate with their hardware. And who knows, perhaps there are extraterrestrials trying to communicate with us at this very moment, although perhaps more successfully with some of us mere mortals than with others.

But why do birds sing rather than speak Mandarin? Why don't human infants scream and cry when they are content as peas in a pod, but coo and gurgle when they are angry, annoyed, or in pain? And why doesn't Jesse Jackson simply convey his expressive skills by blinking his eyebrows, Morse code style? These are questions about design, and what is true of all communication systems that work is that they have specific design features. The design features of a communication system are the result of a complex interaction between the constraints of the system and the demands of the job required. Consider the traffic light. There are two sorts that I am aware of. The old-fashioned ones consist of red, yellow, and green lights that can be seen at any distance, but have the disadvantage that they are difficult to detect in bright sunlight. The newfangled ones often look as though they are dead when seen at a distance, but come to life when approached, regardless of lighting conditions. Thus both morphs have been designed to convey specific information about the flow of traffic, and the advantages and disadvantages of each relate to the constraints of the technology. For natural communication systems, such as those observed in the plant and animal kingdoms, constraints can be seen at several levels including neurobiological, physiological, and psychological. These constraints are important, for they determine the relative success of the organism in responding to socioecologically relevant stimuli in the environment.

This book is about the evolution of communication. For all organisms, including humans, communication provides a vehicle for conveying information and for expressing to others what has been perceived. But organisms differ with regard to what they can convey and what they can perceive. Consequently, there are a diversity of communication systems in the natural world, and I hope to shed light on the causal factors that have been responsible for the evolution of such diversity. To accomplish this goal, we require a comparative perspective that tackles the variation in communication forms head-on. In this book, therefore, I follow the ethologist–Nobel laureate Nikolaas Tinbergen (1952) and explore the diversity of animal communication systems from four perspectives. These perspectives, I believe, provide the only fully encompassing and explanatory approach to communication in the animal kingdom, including human language:

1. *Mechanistic* Understanding the mechanisms (e.g., neural, physiological, psychological) underlying the expression of a trait.

2. *Ontogenetic* Determining the genetic and environmental factors that guide the development of a trait.

3. *Functional* Looking at a trait in terms of its effects on survival and reproduction (i.e., its fitness consequences).

4. *Phylogenetic* Unraveling the evolutionary history of the species so that the structure of the trait can be evaluated in light of ancestral features.

These perspectives or problems are not hierarchically structured. They represent a coherent theoretical and methodological framework for both studying and explaining communication. Although researchers may prefer to work on some problems and not others,[1] variation can only be fully explained when comprehensive answers to all four issues have been addressed (for a recent discussion of the interaction between causal and functional analyses, see Alcock and Sherman 1994; Curio 1994).

Perhaps the most exquisite testimony to the success of Tinbergen's framework comes from the study of birdsong, where observations and experiments have addressed questions of mechanism, ontogeny, function, and phylogeny (Konishi et al. 1989).[2] In brief, detailed neurobiological studies have revealed dedicated neural pathways underlying song production and perception, and in many species such pathways and their associated nuclei are sensitive to changes in circulating

---

1. When the fields of sociobiology and behavioral ecology emerged in the 1970s, many researchers restricted their analyses to problems of function or ultimate causation. Although this restriction continues to characterize a good deal of work in behavioral ecology, an emerging trend in studies of animal behavior is to revert to the Tinbergian prescription, integrating studies of function with studies of mechanistic causation, ontogeny, and phylogeny. An elegant example of such integration, discussed in greater detail further on in the book, is Michael Ryan's studies of vocal communication in frogs.

2. The discussion here is intentionally brief and general. A more complete account of within- and between-species variation in birdsong is provided in chapters 4 to 7.

hormones. Ontogenetically, song develops gradually over time, constrained in part by species-specific auditory templates and shaped by acoustic features of the environment, both social and ecological. Studies of wild birds indicate that song serves at least two important functions—territorial defense and mate attraction—and females commonly use song to guide their mating decisions. In terms of phylogeny, not all birds sing, and among those that do, there is considerable variation in structure even among closely related species. In sum, therefore, current understanding of birdsong far exceeds many other areas of animal behavior because the question "Why?" has been addressed from the four Tinbergian perspectives.

A close look at what we know about human language suggests that our understanding of mechanism and ontogeny is sophisticated (see chapters 4, 5, and 7), whereas our understanding of function and phylogeny is relatively poor.[3] In fact, rarely have researchers asked whether language should be considered an adaptation[4] (for important exceptions, see Lieberman 1984, 1991; Pinker and Bloom 1990; Pinker 1994a) and, if so, what its fitness consequences are for members of the species (Cheney and Seyfarth 1990; Dunbar 1993). There are at least four reasons for this omission, which I develop only briefly here (see chapter 2). First, for some linguists, the analytical lens focuses exclusively on the problem of syntax. Those working on this aspect of language show little interest in the fact that syntactic structure provides humans with an extraordinary communicative tool and, as a result, show little interest in the evolutionary pressures that may have led to this form of communication rather than some other form. Second, language is sometimes conceived as a trait that is detached from our biology—most of the interesting features of language are the result of cultural processes. Clearly, anything that is not part of our biology cannot be an adaptation, *sensu strictu*. Third, interest in the evolutionary origins of language has concentrated

3. By the phylogeny of language, I am referring to the evolutionary changes in communicative structure and function preceding the emergence of language in modern humans (i.e., *Homo sapiens sapiens*), rather than historical changes in linguistic form following the emergence of modern humans (for a recent, nontechnical review, see Ruhlen 1994).

4. I am using the term *adaptation* to mean a trait that evolved in response to a particular socioecological problem, where the design features of the trait can be shown to be causally related to the individual's probability of survival and reproductive success; such features represent the best solution to the problem out of a number of possible solutions. In this sense, my use of *adaptation* is similar to the one proposed by Reeve and Sherman (1993), and different from those who consider trait history to be important (S. J. Gould and Vrba 1982). For those interested in phenotypes and their fitness consequences, history is important (Lauder, Leroi, and Rose 1993), but current design features and functional utility are more relevant to the problem at hand. These ideas are more fully developed in chapter 3.

on syntax and the computational machinery that it appears to require. In some accounts, such as Chomsky's (1986), the machinery is considered a unique feature of modern humans. Without precursors, it makes little sense to talk about gradual evolution by natural selection. Last, evolutionary thinking has generally played a minor role in the social sciences. This statement is certainly true of most researchers working on human language, with a few notable exceptions (e.g., Fernald 1992b; Kuhl 1989; Lieberman 1991; Locke 1993; MacNeilage 1991, in press; Pinker 1994a; Pinker and Bloom 1990; Studdert-Kennedy 1981, 1991). In sum, the study of human language requires a good dose of Tinbergen's medicine, and especially, medicine flavored with functional and phylogenetic ingredients.

To set up the argument I will develop in this book, let me start with a few distinctions that are critical to a sophisticated comparative analysis.[5] First, it is important to distinguish between phylogenetic analyses that consider the evolutionary history of a trait, and adaptationists' analyses that consider how a trait contributes to an individual's survival and reproductive success. Thus, if one were interested in the phylogenetic origins of mating signals, one might approach this problem by looking within a narrow range of organisms (e.g., comparing song structure among Passerine species; see Read and Weary 1992) or from within a relatively more expansive range, comparing distantly related species (e.g., contrasting the displays used by insects and birds in the context of lek mating; Bradbury 1985). The goal here is to quantify the structure of a signal (e.g., its defining features) and then determine when, evolutionarily, the signal first appeared within a group of organisms. For this kind of analysis to succeed, however, a phylogeny is required, based on nonvocal characteristics such as genetic sequence data (nuclear or mitochondrial) and morphological measurements of relevant anatomical traits. In addition to addressing the phylogenetic problem, one might further examine whether a particular mating signal is an adaptation by documenting how it contributes to an individual's genetic fitness. Thus one might examine whether variation in mating success can be accounted for by variation in signal structure. For example, in a species where females choose mates on the basis of their advertisement calls, do males with low-pitched vocalizations obtain more matings than males with high-pitched vocalizations?

Research on communication has tended to adopt a relatively narrow comparative approach with regard to the range of species explored (for some important

5. For a general discussion of issues in comparative biology, see Ridley's (1983) introductory text. More specific and technical details are provided in Harvey and Pagel (1991). I develop some of these issues in section 1.2.2.

exceptions, see Marler 1955, 1967, 1977; W. J. Smith 1977), and I believe this narrow range has hindered our understanding of the evolution of communication systems. In the absence of broad taxonomic comparisons, it is difficult to assess both the adaptive significance and phylogenetic history of a trait, either in terms of homology or homoplasy.[6]

My approach in this book, therefore, is to provide a framework for studying communication from a broad comparative perspective, using both closely and distantly related species to inspect how adaptive solutions to specific problems have evolved. This project requires a synthesis of existing data and theory together with an attempt to provide general methodological tools that can be used with a wide diversity of species. Needless to say, I cannot hope to provide a complete solution. I believe, however, that the synthetic approach adopted here will lead to new insights, that broad-spectrum methodological techniques are available, and that, with a bit of effort, these can be readily implemented into the study of a taxonomically diverse group of organisms.

The empirical work reviewed is restricted to natural communication systems that use auditory, visual, or audiovisual signals. There are two reasons for this restriction. First, by focusing on taxonomic groups that use either auditory or visual signals, we open the opportunity to explore homologous as well as homoplasic traits for a substantial set of species, thereby enriching the power of the comparative approach. Although a considerable amount of work has been conducted on other signaling modalities, including chemical, tactile, and electrical, adding on these modalities would cause some taxonomic groups to be overrepresented and others to be underrepresented. This is especially the case for electrical signaling, where studies of fish far exceed all other animal groups. Additionally, because human language is communicated via auditory and visual channels,[7] the design features of our own communication system can be directly compared to the design features of other animals without the complications of different sensory

6. For those less familiar with this distinction, a simple and succinct discussion can be found in Hodos and Campbell (1990). In brief, homoplasies represent traits that are similar and have evolved independently in two distantly related taxonomic groups. Homoplasies commonly arise from convergent evolution, a process that results from the fact that when two species confront similar ecological problems, selection typically provides similar solutions. In contrast, when a trait is similar in two taxonomic groups as a result of shared evolutionary ancestry, then the trait is considered homologous. In general, traits that are functionally significant with regard to survival and reproduction tend to show up in the evolutionary record in a wide variety of distantly related taxonomic groups. As a result, such traits tend to be convergent or homoplastic. See discussion in section 1.2.2.

7. Although other sensory modalities may provide additional communicative information (e.g., chemosignals that influence the menstrual cycle; McClintock 1971) and tactile signals used by the blind and deaf (e.g., the Tadoma method; Vivian 1966), the auditory and visual channels are primary.

modalities. Second, I focus on natural communication systems, excluding research on nonhumans who have been taught a novel or artificial system. What we have learned from the latter studies is both impressive and clearly of importance in any discussion of comparative cognition. In this book, however, I am primarily concerned with communication systems that evolved in response to naturally occurring problems in the species-typical environment.

1.2 Some Background Information

1.2.1 Communication and Information

Almost every author who has written on the topic of communication has provided a working definition, and as several analyses reveal, it is not an easy task (Mellor 1990). Definitions are interesting for at least two reasons. They provide information on the particular theoretical slant of the author and typically serve as guiding (constraining?) forces in empirical investigations. Table 1.1 provides a sample of definitions of communication from researchers in sociobiology, behavioral ecology, sensory ecology, neuropsychology, cognitive psychology, and linguistics— disciplines with a major representation in this book.

The concepts of *information* and *signal* form integral components of most definitions of communication, including the ones listed in Table 1.1. Both concepts are associated with long lists of operational definitions. Because this is a book about communication between organisms, my bias is to couch information and signal in terms of their functional design features. Thus information is a feature *of* an interaction (i.e., not an abstraction that can be discussed in the absence of some specific context; see Box 1.1) between sender and perceiver. Signals carry certain kinds of informational content, which can be manipulated by the sender and differentially acted upon by the perceiver. Signals have been designed to serve particular functions, and the functions they serve must be evaluated in light of both production and perception constraints. Thus the stomatopod *Gonodactylus bredeni,* a small marine shrimp, defends its nest cavity with a visual display that involves either a simple extension of its powerful chiliped (i.e., claw) or an extension and ground strike (Caldwell and Dingle 1975). Potential competitors can either accept this aggressive signal at face value and retreat, or challenge the cavity owner by intruding further.[8] Constraints on the system include the ener-

---

8. As discussed in chapters 6 and 7, Caldwell's (1986) elegant experiments on this stomatopod reveal that this system is extremely plastic, and even suggest the possibility of cavity owners sending dishonest information in the form of a functional ''bluff.''

Table 1.1
Definitions of Communication: A Sampler

Authors (discipline)	Definition
Wilson (1975) Sociobiology	"Communication occurs when the action of or cue given by one organism is perceived by and thus alters the probability pattern of behavior in another organism in a fashion adaptive to either one or both of the participants" (p. 111).
Hailman (1977) Ethology	"Communication is the transfer of information via signals sent in a channel between a sender and a receiver. The occurrence of communication is recognized by a difference in the behavior of the reputed receiver in two situations that differ only in the presence or absence of the reputed signal. . . . the effect of a signal may be to prevent a change in the receiver's output, or to maintain a specific internal behavioral state of readiness" (p. 52).
Dusenbery (1992) Sensory ecology	"The term 'true communication' is restricted to cases in which the transmitting organism engages in behavior that is adaptive principally because it generates a signal and the interaction mediated by the signal is adaptive to the receiving organism as well" (p. 37).
Krebs and Davies (1993) Behavioral ecology	"The process in which actors use specially designed signals or displays to modify the behaviour of reactors" (p. 349).
Kimura (1993) Neuropsychology	"The term is used here in a narrower sense, to refer to the behaviors by which one member of a species conveys information to another member of the species" (p. 3).
Johnson-Laird (1990) Cognitive psychology	"Communication is a matter of causal influence. . . . the communicator [must] construct an internal representation of the external world, and then . . . carry out some symbolic behaviour that conveys the content of that representation. The recipient must first perceive the symbolic behaviour, i.e. construct its internal representation, and then from it recover a further internal representation of the state that it signifies. This final step depends on access to the arbitrary conventions governing the interpretation of the symbolic behaviour" (pp. 2–4).
Lindblom (1990) Linguistics	"Human communication . . . includes forms of verbal communication such as speech, written language and sign language. It comprises nonverbal modes that do not invoke language proper, but that nevertheless constitute extremely important aspects of how we communicate. As we interact, we make various gestures—some vocal and audible, others nonvocal like patterns of eye contact and movements of the face and the body. Whether intentional or not, these behaviors carry a great deal of communicative significance" (p. 220).

Box 1.1
The Statistical Theory of Information

Although the computer generation has dramatically improved the flow of information, there is still no generally accepted definition of *information*. Nonetheless, a commonly used framework, proposed by Shannon and Weaver (1949), is the statistical or mathematical theory of information; in addition to engineers, the implications and applications of this approach have been discussed by biologists (e.g., Fagen 1978), psychologists (Attneave 1959), and philosophers (Dretske 1981). In simple language, Shannon and Weaver proposed that the function of information is to reduce the observer's uncertainty about a particular event. For example, imagine a boxing match between two heavyweight contenders; let's call them George and Michael. Before the match, each knows a bit about the other's boxing talents based on fighting records and, perhaps, observed fights. In this sense, George and Michael have information, and their uncertainty about the outcome of the fight is clearly nonzero. When George and Michael enter the ring and take their respective corners, they gain additional information. George is older than Michael and looks tired, but is acting cocky. Michael refuses to make eye contact, looks cool and collected, and is in great physical shape. The bell rings, and George comes flying out and swings wildly at Michael, missing. More information. At each step of the interaction, therefore, there is some reduction or change in uncertainty, and this can be formalized by the following equation:

$$H(X) = - \sum_{i=1}^{k} p_i \log_2 p_i$$

Here, X designates a set or field of events x_i to x_k (e.g., walking into the ring, throwing a punch, etc.), and p_i stands for the probability that the ith event of set X will occur. This formalization can be used to assess (1) the extent to which a receiver's uncertainty is reduced by the transmission of the sender's signal and (2) in cases of two or more interacting individuals, whether communication has occurred. Regarding the latter, the goal of the Shannon-Weaver approach is to determine whether the type of response selected by the receiver appears to be causally related to the type of signal selected by the sender. More complete treatments of these issues, as they relate to natural communication systems, are provided by Markl (1985), Dusenbery (1992), and Allen and Hauser (1993).

getic costs associated with generating the chiliped display and the ability of the intruder to properly perceive the display given its visual system[9] and the channel in which the signal was conveyed.

We tend to think of biological signals as conveying or carrying information. In general, this characterization is accurate. For purposes of clarity, however, I will

---

9. Interestingly, stomatopods have highly mobile eyes that look like translucent crystal balls sitting upon rotating stalks (Cronin and Marshall 1989). This mobility has presumably been designed to provide cavity owners with the potential to scan in all directions without having to move outside of the cavity. In addition to their mobility, recent neurobiological studies have revealed that the stomatopod visual system consists of eight cones, thereby providing a potentially exquisite system for detecting subtle difference in color among conspecifics. Such fine-grained discriminations may be extremely important given the potentially lethal consequences of aggressive interactions—a single strike with the claw can result in immediate death.

draw a technical distinction between *cues* and *signals*. Cues, like signals, represent potential sources of information. Cues, however, differ from signals in two important ways.[10] First, cues tend to be permanently ON, whereas signals are more plastic and can be in an ON and OFF state. As a result, signals, but not cues, are produced in response to socioecologically relevant and temporally varying changes in the environment. Second, cues typically correspond to an individual's or species' phenotype, and their expression carries no immediate extra cost. Signals, although individual- and species-specific, are associated with significant costs of expression. Thus, for example, a number of poisonous species have distinctive warning colors (Guilford 1989a, 1990; see chapter 6). These colors are cues. They are permanently ON, they are part of the individual's phenotype, they require no extra cost to produce, and they have been designed to be informative. In contrast, several avian and mammalian species produce warning calls in response to predators. Such calls can be turned ON and OFF (e.g., whether or not an animal calls is mediated by changes in the social environment; Marler, Karakashian, and Gyger 1991), they are costly to produce, and they have been designed to be informative (Marler 1955).

A second distinction is necessary between signals and cues on the one hand, and what I will call *signs*.[11] Cues, such as sexual ornaments and warning colors, have been selected to be informative, and such information can be extracted directly from the individual. For instance, in several avian species, females appear to use tail length as an accurate predictor of male fitness (e.g., Andersson 1982; Møller 1988b, 1989, 1993). Tail length,[12] as a cue, is likely to be an accurate predictor because maintaining this trait is costly. Signs, in contrast to cues, are either temporally or spatially displaced from the individual and are informative to perceivers[13] as a result of their association with biologically significant features of the environment. For example, several species construct nests for raising their young (e.g., birds) or for rest (e.g., chimpanzees). Among birds, parasitic species (e.g., cowbirds) require mechanisms for recognizing potential hosts, and it is possible that the species-specific architecture of the nest is a relevant piece of

10. I recognize that some exceptions to this generalization may exist. I leave the subtlety of the distinction between cues and signals for later chapters.

11. The use of the word "sign" here has little to do with its implementation in semiotics.

12. As discussed more thoroughly in chapter 6, tail length is commonly viewed as a sexually selected trait, exaggerated in males relative to females. In some species, females preferentially mate with males who have long tails.

13. What may be a sign to a perceiver could very well be a signal for a sender. Thus advertisement calls produced by male frogs would be signals to female frogs, but from the perspective of a predatory bat, they are signs.

information. Similarly, forest monkeys could use chimpanzee nests to avoid predation, though this hypothesis has never been formally tested. In these situations,
nests become informative (i.e., they acquire predictive power) because of specific
environmentally relevant associations. They have not, however, been designed
to be informative in this specific way. To hammer the distinction home, consider
the tracks left by predatory species such as lions and pythons. As a result of
their locomotory patterns and their frequent travels through dusty soils, these
species leave traces of their presence. A "smart" prey species might learn that
particular traces are associated with danger whereas other traces are not. Clearly,
lions and pythons do not intend their traces to be informative. From the prey's
perspective, however, they would represent informational bonanzas, signs of potential danger.

The distinction between signals and cues is not novel (see, for example, Markl
1985; Seeley 1989; Dusenbery 1992). However, my own characterization of the
distinction differs from previous authors in one crucial way. To illustrate, consider Seeley's (1989) succinct account (p. 547): "Signals are stimuli that convey
information and have been molded by natural selection to do so; cues are stimuli
that contain information but have not been shaped by natural selection specifically
to convey information." In contrast to Seeley, I suggest, and will empirically
substantiate further on, that both signals and cues have been designed to convey
information, but signs have not. Most of the communication systems discussed
in this book involve signals, rather than cues or signs.

1.2.2 The Comparative Method: Which Species to Compare and What to Conclude?

In addition to his theory of natural selection, the comparative method is what
made Darwin great. If you don't believe this claim, look at any of his major
works. They are packed with interspecific comparisons based on detailed studies
and anecdotal observations. As previously mentioned, however, studies of communication have generally compared closely related species. What is sorely
needed is for communication to be treated on a broader phylogenetic scale. For
this effort to succeed, however, we require methods that will generate comparable
data and statistical techniques that will allow us to properly interpret evolutionary
relationships. At present, there are few methodological tools that can be treated
as species-independent. That is, most observational and experimental techniques
have been designed with a particular species in mind. One of a handful of exceptions is the habituation-dishabituation paradigm, used to quantify an individual's
ability to discriminate between two stimuli (see chapter 3). Thus developmental

psycholinguists (e.g., Eimas et al. 1971) have used this paradigm in studies designed to assess the human newborn's ability to make fine-grained phonemic distinctions, as well as its knowledge of the physical world (Baillargeon 1994; Spelke 1991, 1994). Recently, ornithologists have used this technique to look at the problem of categorical perception (Nelson and Marler 1989), whereas primatologists (Cheney and Seyfarth 1988) have used it to decode the meaning of vocalizations. Throughout the book, I will point to techniques that have been employed with different species and whether they yield results that can be directly compared. I will also discuss techniques that have been restricted to a particular species, but might be usefully implemented by researchers studying other species.

Within the past few years, statistical tools for phylogenetic comparisons have increased in sophistication and have gradually invaded behavioral ecological studies. This change is due in part to the efforts of Paul Harvey and his colleagues (reviewed in Harvey and Pagel 1991). The infusion of phylogenetic considerations has at least three important implications for the study of behavior. First, phylogenetic analyses can help us discern when a trait evolved, in how many lineages, and how long it survived over evolutionary time. Second, documenting phylogenetic relationships allows one to avoid statistical comparisons among nonindependent taxonomic units. Thus, for example, in studies of the evolution of brain size, species within genera often cluster, and thus species should not be considered as independent points. In this case, it is more appropriate to analyze the data at the generic level. Finally, comparative analyses often shed light on correlated traits. Identifying correlated traits is important because regression analyses typically require one or more variables to be partialed out so that one can directly assess the independent effects of one variable on the other. In the brain size example, we speak of dolphins, chimpanzees, and humans as having large brains *relative* to their body weights. We do so because brain size and body weight are, generally, positively correlated. In research on the neural structures subserving song production, we find that species with large repertoires have relatively larger song nuclei than species with small repertoires (DeVoogd et al. 1993).[14]

The comparative method is not without problems. The problems are most acute when phylogenetic relationships are poorly documented (e.g., ambiguities in the evolutionary tree due to incomplete molecular or morphological data) and when variables or traits are coded dichotomously as *present* versus *absent*. The latter

14. A few words of caution here. Some (Konishi and Ball, personal communication, unpublished data) have argued that the volume of song nuclei is more directly related to song output—how often birds sing—rather than to repertoire size. I give this issue a more thorough treatment in chapter 4.

problem is most relevant to the issues discussed in this book. The simplest way to illustrate this point is by referring to the heated debates in animal-learning psychology over the question of animal intelligence. Briefly, a great deal of research carried out from 1950 to 1980 focused on the possibility of general learning rules and cognitive abilities. Opponents of this view (e.g., Macphail 1987a, 1987b) argued that because interspecific comparisons were often made on the basis of a single experiment, it was not possible to distinguish between differences in performance and differences in ability. In terms of understanding the evolution of intelligence, ability is the most relevant measure.

The presence-absence distinction is at the heart of most discussions of language evolution. Although there is nothing theoretically misguided about this distinction, it must be used with caution. This is particularly the case in comparative studies of communication, where the units of comparison commonly differ between species (see chapter 3) and knowledge of the communication system is typically asymmetric for the species being compared. For example, it is often stated that a fundamental difference between human language and nonhuman animal communication is that the former has access to an infinite set of meaningful utterances whereas the latter is more restricted. The difference in expressiveness lies in the fact that language makes use of the combinatorial power of syntax whereas other communicative forms lack syntax. Although this distinction is currently accurate, it is perhaps premature. Only a handful of studies (see chapters 2, 3, and 7) have even broached the possibility of a nonhuman animal form of syntactical structure (Hailman and Ficken 1987; Hailman, Ficken, and Ficken 1985, 1987; Robinson 1984). There are two (related?) reasons for this sparsity. First, the unit of analysis for most studies of nonhuman animal communication is *the call* or *the song*. Strings of calls and song sequences are certainly produced by animals, and often such strings are comprised of heterogeneous call units. However, quantitative analyses of call sequences require complex statistics and large sample sizes (see chapter 3 for a more detailed discussion). The latter are often lacking. Second, there can be no formal study of syntax without first studying call meaning.[15] The study of meaning is only in its infancy with respect to

15. In discussing the structure of nonhuman animal communication, Peter Marler (1977) has suggested that we distinguish between two types of syntax, *phonological* and *lexical*. Phonological syntax requires the combination of two or more acoustically distinct signals to create a new signal, the meaning of which is unrelated to the meaning of the components, if any. For instance, the word *came* can be arranged to create the word *mace*. There is some evidence that nonhuman primates, such as the capuchin monkey (Robinson 1984), cotton-top tamarin (Cleveland and Snowdon 1981), and gibbon (Mitani and Marler 1989) may exhibit phonological syntax. Lexical syntax is analogous to what most

nonhuman animal signals (see chapter 7). Until these two problems have been remedied, the evolution of syntactical abilities will remain unclear.

The last point I would like to make in this section is in reference to the distinction between homology (C. B. G. Campbell 1988) and homoplasy (Hodos 1988). As highlighted previously, if two species with a common phyletic history exhibit structurally similar traits, then we call such traits homologous. In contrast, if two species lack a common phyletic history but exhibit structurally similar traits, then such traits are considered homoplasies. Homoplasies typically arise because even distantly related species will confront environmental problems with a limited set of adaptive solutions. The distinction is important and relevant to comparative studies of communication because different evolutionary forces can be responsible for generating homologous and homoplasic similarities. For example, as described in greater detail in chapter 5, songbirds tend to go through a phase of vocal development known as subsong. During this phase, the young bird produces long, structurally diverse song monologues. The compositions of these songs are, in general, quite different from those produced in adulthood and yet contain similar elements. Like songbirds, human infants also exhibit a phase of vocal development where long, structurally diverse monologues are produced and such monologues consist of speechlike elements. In humans, this phase has been called *babbling*. Researchers working on avian song development (Marler 1970a, 1984; Marler and Pickert 1984) have suggested that subsong is like babbling. In this case, babbling and subsong represent homoplasies because humans and birds do not share a common ancestor who babbled or had subsong. In contrast to birds, there is only suggestive evidence that nonhuman primates go through a babbling phase (Snowdon 1982, 1990). If further work strengthens the claim, then babbling in human and nonhuman primates would be considered homologous, since both primates are likely to have shared a common ancestor who babbled.

1.3 Outline of the Book

This book was conceived as a purely scholarly scientific piece aimed at experts in the neurosciences, evolutionary biology, cognitive-developmental psychology, linguistics, and anthropology. As I began to research the various topics to be covered and talk to colleagues about the issues to be discussed, I realized that

---

linguists consider to be syntax *sensu stricto* and occurs when the meaning of a compound signal, whether consisting of parts of a word or several words, derives from the meanings of its components. Thus far, there is no convincing evidence of lexical syntax in the animal kingdom.

several courses were groping for a text that discussed issues in comparative communication. Consequently, I have attempted to write a book that is aimed primarily at the expert while being useful to those wishing to pick out pieces of the book for undergraduate and graduate instruction. The latter has been addressed at two levels. First, for each major topic I work through a small set of cases that I believe provide elegant and comprehensive studies of a particular process. For instance, in discussing the neurobiology of communication, I review some of the exciting new developments emerging from studies of birds and frogs where the finely choreographed interplay between perception and production systems has been extensively documented. By focusing on a relatively small subset of studies, students can learn to appreciate "the best of" from a given research area and hopefully understand how a problem is solved in the best of all worlds. In general, the case studies exhibit a taxonomic bias toward vertebrates, and more specifically, anurans, birds, and primates. The reason for this bias is that by sticking with these groups it is possible to maintain a level of conceptual continuity throughout the book that would not be possible with several other equally interesting organisms such as insects, fish, and reptiles. The case studies also exhibit a modality bias, restricted to communication in the auditory and visual domains. As a result, some of the fascinating studies on olfactory communication in insects and electric communication in fishes fail to make a significant appearance in the book. Although some of these organisms undoubtedly provide a better illustration of a particular phenomenon, the restriction I have imposed allows for a richer comparison across levels of analysis. Second, key concepts and methodological issues are flagged by providing compact discussions within "boxes," such as Box 1.1, "The Statistical Theory of Information." In contrast to the main text, each box provides a more clearly defined set of operational definitions, description of methodological techniques, and references to important review or primary articles.

Chapter 2 provides a historical overview of some of the different approaches to studying the evolution of communication systems, including the evolution of language. As highlighted in some of the previous sections, I find the history of work in this area particularly important because there is a tendency for the new generation to either ignore or misinterpret the older generation of work and, in some situations at least, for previous work to block our path toward a strong comparative research program. Additionally, some of the views espoused have led to highly testable predictions, many of which remain to be explored. Chapter 3 provides a basic introduction to conceptual issues in the study of communication, including a general discussion of the ecology of communication, an analysis of problems associated with the definition and classification of fundamental units of communication, and a preliminary peek at how Tinbergen's research program can be worked out. Chapters 4 to 7 place comparative communication into the structure of Tinbergen's four causal questions, with the following modifications and subdivisions. Under the problem of mechanistic explanations of communication, I have created two chapters, one on neurobiology (4) and one on psychology (7). In essence, the topic of mechanism is divided into relatively low- and high-level neural processes. For chapters 4, 5, and 6, I have organized the case studies into three socioecological contexts—mating, survival, and socialization—focusing on both auditory and visual signals and cues. My motivation for organizing the examples into these three contexts is so that questions of design can be evaluated in light of evolutionarily significant problems. Thus, under mating, I describe the neurobiological specializations underlying anuran perception of advertisement calls and then discuss how sensory mechanisms coevolve with production mechanisms, leading to significant effects on the fitness of males and females. Chapter 7 also focuses on auditory and visual communication systems, but rather than looking at the contexts of mating, survival, and socialization, I concentrate instead on more general theoretical problems such as categorical perception and deception. Though there is no explicit chapter on phylogeny, I have attempted to address this gap by adopting a phylogenetic perspective throughout the book and, wherever it is relevant, referring directly to known phylogenetic relationships and historical constraints.

As a warning to the reader, let me note that the three contexts I have selected for discussing the design features of a communication system overlap considerably, and for some systems it is difficult to select a particular context. Thus signals and cues used during dominance-related interactions are clearly involved in socialization, but also serve an important function in survival and, ultimately, reproduction. The problem of contextual overlap becomes all the more challenging in discussing human communication. For purposes of organizational clarity, therefore, I have decided to cover human language under the socialization heading. Throughout the book, however, I will consider the possibility that the structure and function of language were shaped by problems associated with mating and survival, something that Darwin (1871) certainly believed.

Chapter 4 examines the neurobiological mechanisms underlying the production and perception of communication systems, focusing in particular on auditory signals. At present, little is known about the neurobiology of visual signals and cues, with the exception of primate facial expressions and human sign language. Chapter 5 explores the ontogeny of communication, using observational and experimental data to assess the relative contributions of genetic and experiential factors to the developmental process. Chapter 6 discusses current understanding of the function or adaptive significance of different communication systems, looking at the question of design in light of how the production and perception of signals and cues influence individual fitness. Chapter 7 concentrates on the psychological mechanisms guiding the production and perception of communicative signals and cues. Specifically, I discuss the importance of feature extraction during categorization of visual and auditory stimuli, the "amount" of affective and referential information conveyed, and the extent to which individuals voluntarily communicate honest as opposed to dishonest information. The final chapter summarizes the general issues raised in the book, synthesizes the patterns uncovered, and provides a few specific suggestions regarding future research in comparative communication.

Enough warm-up. To work!

2 The Evolution of Communication: Historical Overview

2.1 Introduction

This chapter provides a topographical road map to some of the most dominant theoretical perspectives on the evolution and design of communication systems. The historical perspective is important for at least two reasons. First, some of the most luminary scholars of our time have dedicated their mental powers to speculations about the evolution of communication, and the ideas that have emerged provide explicit predictions about the form and function of communication systems. Interestingly, many of these predictions remain untested. In this historical treatment, therefore, I hope to lay bare some of the reasons for such conceptual dormancy and, along the way, to inspire more empirical work. Second, to avoid endlessly reinventing the communicative wheel, I develop the different theoretical perspectives in some detail so that current findings can be placed in a relatively unambiguous relationship to their historical origins. I begin in section 2.2 with a discussion of how biologists, dating back to Charles Darwin, have thought about the general design of communication systems. These ideas are rich, for they attempt to account for the structure and function of communication in all taxonomic groups, using a unified theory. In sections 2.3 and 2.4, I discuss a cluster of theoretical perspectives with a more narrow taxonomic focus. Specifically, these two sections concentrate on how linguists and biologists have attempted to explain the evolution of human language as a species-specific form of communication. Although narrow taxonomic analyses are not uncommon in biology, and this observation certainly applies to analyses of communication, human language evolution has been singled out here because of the considerable amount of critical discussion and controversy it has generated relative to other species and their communication systems. Needless to say, no study of language evolution has ever ignored the possibility of evolutionary precursors, visible within extant species. However, a range of views exist with regard to the importance of precursors, as well as to what "counts" as one. These different perspectives are relevant because they can guide our thinking about the significant design features of language, and what they do for the individual.

Section 2.5 is an attempt to reconcile some of the conflicting perspectives by arguing that a theoretical summation or averaging of the different views comes closer to a complete explanation of how communication systems evolved than any particular account viewed in isolation. In essence, the historical path we will travel in this chapter will lead us to an exciting position: by uniting different perspectives and levels of explanation, we can develop a coherent, synthetic treatment of the evolution of communication that attempts to encompass all organisms. Although some linguists (Lieberman 1984, 1991; Pinker 1994a; Pinker and Bloom 1990) and some biologists (Cheney and Seyfarth 1990; Endler 1993b; S. Green and Marler 1979; Marler 1961, 1967, 1977, 1992b) have pushed this line of thinking, I don't think that it has been pushed far enough.

2.2   The Design of Natural Communication Systems

Interest in communication loomed large in much of Darwin's writings. For example, in his treatise on sexual selection (Darwin 1871), he not only focused considerable attention on the signals used by males and females during courtship rituals, but also considered the possibility that the diversity of human languages was the result of a process akin to speciation:

The formation of different languages and of distinct species, and the proofs that both have been developed through a gradual process, are curiously the same. . . . We find in distinct languages striking homologies due to community of descent, and analogies due to a similar process of formation. The manner in which certain letters or sounds change when others change is very like correlated growth. . . . Languages, like organic beings, can be classed either naturally according to descent, or artificially by other characters. Dominant languages and dialects spread widely and lead to gradual extinction of other tongues. A language, like a species, when once extinct, never, as Sir C. Lyell remarks, reappears. (Darwin 1871, 59–60)

Several historical linguists have followed up on Darwin's intuitions, and much current work in this area involves an assessment of language change and stability over time (for a general overview of this area, see Ruhlen 1994).

One of Darwin's most significant theoretical statements on communication emerged from his consideration of animal emotions and the vehicles for their expression. In 1872 he set out to provide a general account of both acoustic and visual expressions of emotions in humans and nonhumans. As in all of his earlier discussions, he attempted to document and account for similarities and differences in expression across the animal kingdom by considering their design features and function. The following quote captures the relevant details of Darwin's "bottom line":

No doubt as long as man and all other animals are viewed as independent creations, an effectual stop is put to our natural desire to investigate as far as possible the causes of Expression. By this doctrine, anything and everything can be equally well explained; and it has proved pernicious with respect to Expression as to every other branch of natural history. With mankind some expressions, such as the bristling of the hair under the influence of extreme terror, or the uncovering of the teeth under that of furious rage, can hardly be understood, except on the belief that man once existed in a much lower and animal-like condition. The community of certain expressions in distinct though allied species, as in the movements of the same facial muscles during laughter by man and by various monkeys, is rendered somewhat more intelligible, if we believe in their descent from a common progenitor. He who admits on general grounds that the structure and habits of all animals have been gradually evolved, will look at the whole subject of Expression in a new and interesting light.   (Darwin 1872, 12)

Though this quote is restricted to visual expressions, Darwin adopted the same approach (descent with modification) to vocal expressions, and this included his explanation of function. Specifically, he argued that expressions were designed to convey information about the signaler's emotional or motivational state, with some signals reflecting an underlying ambiguity or conflict between different emotional states (his theory of "antithesis"; see chapter 7) such as fear and aggression. To give one example of his thinking, consider the vocalizations used by animals during aggressive encounters. Based on a set of observations, Darwin suggested that aggressive animals typically produce low-pitched growls whereas fearful animals tend to produce high-pitched squeals or screams. As described more completely in chapter 7, one reason why Darwin expected this relationship to be universal was that in most species, aggressive animals are larger and more dominant than fearful animals. Because size is likely to show a negative correlation with pitch, aggressive vocalizations produced by dominants will tend to be lower in pitch than fearful vocalizations produced by subordinates. This kind of thinking led Darwin to a series of predictions and conclusions regarding both the universality of expressions in the animal kingdom and the recognition that species differences were prevalent and fundamentally related to their ecology and, at least for humans, to their mental capacity (see section 2.3).

In the late 1940s to early 1950s, ethology emerged as a powerful new discipline whose conceptual and methodological framework differed fundamentally from the competing field of comparative psychology. In very general terms, comparative psychologists studied animal behavior under controlled laboratory conditions using rigorous experimental methods. They paid little attention to the organism's natural history and the impact of ecological pressures on species-specific behavior, and were dedicated to uncovering general laws of learning. Ethologists, in contrast, relied heavily on observations collected in the species-typical habitat, placed significant emphasis on understanding the diversity of behavioral adaptations, and were primarily concerned with innate mechanisms guiding and constraining behavior. Concerning communication, the grandfathers of ethology—Lorenz, Tinbergen, and von Frisch—largely accepted Darwin's treatment, especially the idea that signals were designed to communicate information about the signaler's motivational or emotional state. Thus, Lorenz (1966) explored the communicative exchanges between mothers and their young in a variety of avian species, focusing in particular on the signals and cues used during imprinting. For Lorenz, signals used during imprinting would insure attachment and maternal response to offspring in distress. Tinbergen (1952, 1953) looked at some of the key releasers during aggressive interactions between stickleback fish, in addition to exploring the variety of displays used by gulls during competitive interactions and courtship. Last, von Frisch (1950, 1967b) provided in-depth analyses of the honeybee's communication system, concentrating in particular on signals used to convey information about the location and quality of food.

Before moving on to the ethologist's conceptualization of communication, let me briefly digress to an important historical point. Research during the early days of ethology was largely influenced by a conceptual framework that saw the forces of natural selection operating at the level of the group. Thus the main function of communicative signals, as with other behaviors, was to insure survival of the group or species, rather than the individual. For example, when a squirrel gives an alarm call to a predator, it does so in order to protect its group. The primary challenge to this perspective emerged in the 1960s, based on the theoretical arguments of G. C. Williams (1966), W. D. Hamilton (1963, 1964), and John Maynard Smith (1964), and most significantly, I believe, the empirical work of the ornithologist David Lack (1966, 1968). To capture the essence of this challenge, let us return to our alarm-calling squirrel. The puzzle in this example is to account for the apparently altruistic act: given that alarm calling is costly (e.g., increases the caller's probability of being detected and eaten by the predator), what is the benefit? As stated, a group-selection argument claims that the caller incurs a cost so that other group members have a better chance of escape—the benefit. However, thinking in terms of the individual, or more appropriately, its genes, generates a very different interpretation. In a population where individuals always produce alarm calls when they detect predators, selection would readily favor a mutation that caused an individual to withhold the alarm call, run for cover, and save its own skin. Individuals who call tend to die, whereas those who do not call tend to live and thus reproduce. And in genetic terms, reproducing means passing on one's genes to subsequent generations. This interpretation, with selection operating at the level of the individual or gene, has become the dominant perspective in the study of animal behavior and was largely responsible for changing how biologists think about communication and design. I am getting ahead of myself, so let us return to the ethological framework for studying communication.

The early ethologists focused primarily on the evolutionary origins of signals, their ontogeny, and their informational content (reviewed in Halliday and Slater 1983; W. J. Smith 1977). The general view (see Figure 2.1) was that signals emerged from nonsignals, what Tinbergen (1952) called *derived activities*, and what many ethologists considered intention movements—behaviors that appeared goal-oriented. Once nonsignals gain functionality (i.e., begin to predict environmental processes that influence the probability of survival and mating success), they become ritualized, emerging as communicative signals. For example, a male might place his hand on a female's head prior to mating in order to gain his balance. Initially, hand placement is not a signal. However, it might gradually serve this purpose, informing the female that the male is moving into the copulatory position. This process, in turn, might lead to further elaboration, such that the male repeatedly pats the female on the head as he mounts. Thus head patting becomes a ritualized signal in the context of mating. What distinguished ritualized signals[1] from their ancestral precursors was that they were highly repetitive, stereotyped, and exaggerated. As Krebs (1987) states, "An ancestral dither may have evolved into a dance akin to that of a whirling dervish" (p. 164).

Once a ritualized signal evolved, its ultimate form was believed to be designed for maximizing information transfer. Selection therefore operated on the sender to provide recipients with signals conveying unambiguous information. As a result, signals within the repertoire were distinctive and readily distinguishable from others in the repertoire. Rare signals and large repertoires were selected against because they would lead to a slowed-down response in recipients. In other words, as repertoire size increased, recipients would need to store more information and this growth would ultimately increase the probability of responding inappropriately—confusing one signal for another. Given that the number of signals in each species' repertoire is both small and quite conservative across taxonomic groups, repertoire evolution would arise as a result of new signals replacing old ones in a cyclical pattern (Moynihan 1970). With this logic in mind, several ethologists (Andersson 1971; Stokes 1962) set out to provide rich analyses of communicative sequelae, documenting the probabilities of, for

1. The general view of ritualization was that the change occurring between precursor to stereotyped display was a genetic one, involving organic/biological evolution. However, because many ethologists recognized the possibility that environmental factors might also influence the path from precursor to stereotyped display, as well as the ultimate form of the display, the concept of *conventionalization* was also invoked. Remember that at the time of this earliest formalization, ethologists, in contrast to comparative psychologists, were pushing a heavily nativist position.

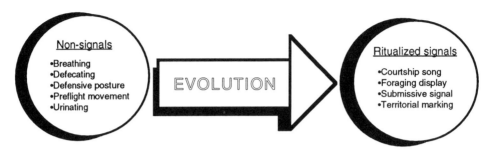

Figure 2.1
A schematic representation of how the early ethologists depicted the evolutionary origins and subsequent development of animal signals.

example, an aggressive display posture in a dominant leading to a submissive posture in a subordinate.[2]

A final point to make about the ethological position on communication is both theoretical and methodological. In general, much of the early work conceptualized the communication arena as being guided by *innate releasing mechanisms* that, in turn, led to *fixed action patterns*. The releasing mechanism was believed to be innate because certain kinds of responses, such as the stickleback's aggressive display and the chick's begging behavior, were elicited soon after birth and apparently in the absence of relevant experience. Such responses typically followed a stereotyped pattern, and as a result they were originally described as fixed action patterns.[3]

Though much of the conceptual luggage associated with the terms *innate releasing mechanism* and *fixed action pattern* have been dropped, some of the important methodological approaches associated with such concepts have been preserved. For example, an important attribute of this early research was a keen sensitivity to the salient stimulus features of the communication environment. Thus, when Tinbergen and some of his students discovered the importance of the stickleback's red belly in aggressive contexts, they soon began to dissect the necessary and sufficient features for eliciting an aggressive response. Thus, was it the red belly on the stickleback, just redness, or redness on something that looked like a stickleback, and if so, then what does "looking like a stickleback"

2. These analyses relied heavily on such mathematical formulations as Shannon-Weaver's information theory (Box 1.1) and Markovian analyses (chapter 3).

3. Few of the responses considered were actually *fixed,* but rather represented something like the *modal* pattern of response to a particular situation (Barlow 1977).

mean, and so on? The same kind of experimental approach was used by Lorenz to map out the featural topography of the imprinting process, and more recent researchers such as Bateson, Horn, and their colleagues (Bateson 1966, 1991; Horn 1991) have provided details of how the brain both guides and constrains this process.

The first challenge to the ethological perspective on communication arrived in the mid-1970s, when several ethologically trained researchers—carrying the disciplinary flag of sociobiology or behavioral ecology—began adopting a more functional perspective, with a strong emphasis on gene-centered thinking. The essence of this challenge was to take the traditional view of signal ritualization and flip it on its head: in contrast to the ambiguity-reduction view, where ritualized signals were considered to be highly informative, the new view claimed that ritualized signals were foils, designed to increase ambiguity by concealing the signaler's "true" motivations or intentions. The logic underlying this argument—what I will call the behavioral ecologist's perspective—can be distilled to the following points:

1. Competitive interactions over valued resources (e.g., mates, food) are guided by individuals attempting to maximize benefits relative to costs. The ultimate goal is to maximize the number of genes contributed to subsequent generations.

2. Such interactions are mediated by species-typical displays or signals.

3. Contra the traditional ethological view: selection should operate against individuals using displays that are highly predictive of their subsequent behavior (i.e., displays that are highly informative with regard to motivational state or intent).[4]

4. Rather, signals used in competitive interactions are like bidding wars in poker: keep a straight face, bluff for as long as you can, and only show your hand when the stakes are too high and you are about to lose big time.

5. Given the poker metaphor, stereotyped signals will be favored because they increase signal ambiguity by concealing the individual's actual emotional or

4. As Hauser and Nelson (1991) have pointed out, there has been a confusion in the literature with regard to the notion of intentional signaling. Those working within a behavioral ecology framework use the term to refer to the predictability of a signal. Thus, somewhat in line with the old idea of "intention movements," an individual signaling his intent to be aggressive can be translated as an individual producing a signal that reliably predicts the probability that aggression will ensue. In contrast, those working within an animal cognition framework use the notion of intentionality in the rich philosophical sense, of aboutness and conveying beliefs, desires, and so on. The first sense of intentional signaling is discussed in greater detail in chapter 6, and the second sense of the term is discussed in chapter 7.

motivational state (e.g., readiness to fight or flee). By definition, stereotypical signals show little structural variation and thus are limited with regard to the information they can encode about an individual's competitive state.

6. Animals engaged in aggressive competition will therefore be guided by two factors. On the one hand, each individual automatically offers information to its opponent that is unfakable and associated with fighting ability. This component, known as resource holding potential (RHP), includes such features as body size and weaponry (antlers, canines). In many contests, individuals are asymmetrically matched on RHP, and this fact alone can decide the outcome of an encounter (big guy beats up little guy). When RHP is more symmetrical, however, then the second factor—communicative signals—will play a more significant role. Such signals do not, however, provide reliable information about the intent to fight or flee. Consequently, the outcome of a competitive interaction must be decided by a volley of signals, with each individual attempting to extract the most useful information with regard to the relative probability of winning or losing a fight.

The perspective sketched here makes use of two conceptual tools: selection operating at the level of the individual or gene, and evolutionary game theory. So that we can move swiftly into the more detailed aspects of the behavioral ecologist's framework, let me briefly make a few comments on the units of selection problem and game theory; all of the points raised here are developed more fully in chapter 6. The idea that selection favors gene propagation was a fundamental advance in the history of evolutionary biology, for it helped resolve the paradox of altruism—cases where individuals appeared to either forgo reproduction completely or engage in activities that would potentially reduce the prospects of future reproduction. The conceptual kernel of this line of thinking, generally credited to W. D. Hamilton (1963, 1964), was to consider how behavior is guided by an individual's genetic relationship to other individuals in the group. Specifically, Hamilton argued that individuals have been selected to maximize their *inclusive fitness,* a term that refers to the number of genes one passes on to subsequent generations as a result of direct reproduction (i.e., the number of offspring you produce who survive and reproduce) and indirect reproduction (i.e., the number of individuals you help survive and reproduce as a function of your degree of genetic relatedness to them).[5] Thus, for example, in several species of

---

5. The definition I provide here glosses over some distinctions that were made by Hamilton in his original treatment of inclusive fitness. These distinctions are not important for our discussion here, but are relevant to those concerned with keeping track of gene propagation. For the reader interested in the details of inclusive fitness theory, see Alan Grafen's (1992) important paper.

insect (bees, wasps) and at least one mammal (naked mole rats), some individuals completely forgo their own reproduction in order to help closely related group members raise young. In other species (scrub jays, jackals), where the options to breed in a given year may be poor (e.g., high-quality territories are occupied), individuals help rear their siblings in one year, and then may attempt to reproduce on their own in the following year. In each of these cases, individuals are investing in their genes—that is, in the probability that their genes will be passed on from generation to generation. Inclusive fitness theory, then, can readily account for the many seemingly costly behaviors observed in the animal kingdom (Wilson 1975).

Game theory, as many readers will know, was developed by economists to explain the dynamics of competitive interactions where specific resources are at stake and alternative strategies for obtaining the resources are possible. Maynard Smith (1974, 1979, 1982) applied the logic and mathematical tools of economic game theory to problems in biology in an attempt to assess whether evolutionarily stable strategies—or ESSs—existed. In a standard game (see Box 6.4), individuals compete over a resource and have access to one of two strategies: fight or flee. In such games, each strategy is associated with a payoff, based on a cost-benefit ratio, and the strategies are frequency dependent (i.e., the strategy selected depends, in part, on what the rest of the population is doing). The goal of evolutionary game theory is to determine whether, over evolutionary time (i.e., practically, iterations in a computer simulation or an algebraically derived optimality solution), a stable solution can be obtained. Evidence for an ESS then comes from a set of equations and conditions (specified by the details of the payoff matrix) which indicate that no mutational strategy can invade a population with $X\%$ of individuals playing strategy 1 and $Y\%$ playing strategy 2. To make this relationship concrete, imagine a population in which all individuals first signal "fight" when competing over a resource and then, if their opponent mirrors with a "fight" signal, they respond with an "escape" or "flee" signal. This simple decision rule might not be an ESS if a mutant strategy can invade, allowing individuals to signal fight on two consecutive exchanges, regardless of the competitor's response. This mutant strategy would disrupt the stability of the game if individuals gained a disproportionate share of the resources.

Let us now return to our discussion of communication. To summarize, the behavioral ecology perspective suggests that during competitive interactions, animal signals will be designed to conceal information about intent and motivational state. Moreover, ritualized signals do not evolve by means of selection for motivationally transparent signals. Rather, they evolve by means of a competitive arms race, with signalers attempting to manipulate susceptible receivers and receivers resisting such manipulations—the old sales pitch and resistance routine. With the theoretical groundwork laid down, research efforts soon focused on establishing the relative predictability of signals and displays in terms of conveying information about the signaler's subsequent behavior. Thus, for example, Caryl (1979) went back and reanalyzed some of the original data sets published by Stokes (1962) and Andersson (1971) on aggressive displays in birds and found that the probability of a particular display being followed by the putatively intended behavior (e.g., a wing stroke display followed by an aggressive attack) was relatively low. Consequently, a submissive individual watching the aggressive display of a dominant would not be able to reliably predict whether or not the dominant would attack if the subordinate failed to retreat. These reanalyses provided the first piece of evidence that signals do not always convey accurate information about the signaler's motivational state, as suggested by the early ethologists (although see Hinde 1981, for an important comment about Caryl's misinterpretation of the "traditional ethological position").

As the game theory perspective gained momentum among behavioral ecologists, two more explicit proposals were brought forward, both emphasizing individual-selection thinking and fitness-maximizing strategies. Zahavi (1975, 1977) developed the first of these theoretical accounts of signal design. The crux of his argument was that signals are honest if and only if they are costly to produce and maintain. Although this perspective was intended to provide a general framework for understanding communication, many of Zahavi's examples centered on the mating arena. Here is the logic of his argument, laid out in skeleton form, and applied to the mating context:

1. In a variety of species, females attempt to mate with high-quality males and are able to exert mate choice.

2. To maximize their genetic fitness, females must be able to reliably distinguish (i.e., discriminate) between high- and low-quality males.

3. Males must express traits (e.g., cues or signals) that reliably covary with their quality, and females must be able to recognize this covariation.

4. The correlation between trait reliability and male quality is established by natural selection on the basis of costs to the male (e.g., reduced survivorship) of carrying the trait. In the absence of significant costs, the trait becomes unreliable, because any individual in the population can express the trait.

5. Thus honest traits are ones that are costly to produce and maintain.

6. The costs paid are, however, calibrated to the individual's current condition or quality. Thus males in good condition will pay a relatively smaller cost for the trait than males in poor condition.

Using this logic, Zahavi proposed that some of the ridiculously exaggerated traits seen in the animal kingdom (immense candelabra-type antlers, high-volume boom-box vocalizations) could be readily accounted for by his theory, and more particularly, by what he called the *handicap principle*. In brief, exaggerated traits not only appeared costly in terms of their production and maintenance, but also appeared deleterious with regard to survival. For example, males with absurdly long tails or shockingly bright colors would surely be more vulnerable to predation than males with short tails or dull colors. Natural selection should therefore eliminate the showy males and favor the more cryptic ones. But individuals who sport the exaggerated traits and live to tell the tale must be truly extraordinary males, with genotypes that can readily tolerate the survival costs of the trait. Consequently, Zahavi argued, females should pick males with these "handicaps" because they have made it through a survival filter. The end product of this complicated process is a conflict between the pressures of natural selection and sexual selection: Natural selection operating against the exaggerated trait and sexual selection, by means of female choice, operating in its favor.

Zahavi's arguments were initially rejected by evolutionary biologists (Kirkpatrick 1986; Maynard Smith 1976), primarily on the basis of the fact that population genetic models failed to provide evidence that a handicap system could evolve and remain evolutionarily stable. However, more recent work has provided not only formal theoretical support for the handicap principle (Grafen 1990; Grafen and Johnstone 1993; Johnstone and Grafen 1992, 1993), but some empirical support as well (Andersson 1982; FitzGibbon and Fanshawe 1988; Hauser 1993b; Møller 1988b). In summary, and as developed in greater detail in chapter 6, the notion of cost appears to be fundamental to the generation and expression of honest signals. What remains unclear is (1) whether honest signals are more or less likely to appear in certain social situations than others and (2) how much cost is either necessary or sufficient to both generate and maintain an honest signal (Maynard Smith 1991, 1994).

The most significant alternative to the "traditional ethological view" of animal communication was Dawkins and Krebs' (1978) theoretical framework. I would like to develop this framework in some detail here because from a functional perspective, it is the only comprehensive account available, and one that has some important implications for how we think about design from the three other Tinbergian perspectives: mechanistic, ontogenetic, and phylogenetic. Using the logic of evolutionary game theory, Dawkins and Krebs also championed the view that selection will act against conveying accurate information and in favor of concealment. The novel twist, however, was the notion that in competitive situations, signals and signalers would be designed to manipulate the behavior of receivers. For example, fledgling birds have begging calls that are designed to obtain added care from their parents, parasitic cuckoos have been designed to manipulate the parenting abilities of their hosts, and aggressive vocalizations are designed to intimidate wimpier competitors. How does this manipulative account add to our understanding of signal design? Recall that in Maynard Smith's game theory account, individuals who conceal information about their intentions or motivational states are favored by natural selection. Thus individuals either abstain from signaling or produce signals that are, in some sense, uninformative. In contrast, the Dawkins and Krebs position makes predictions about the specific morphology of the signal and how such morphological features translate into fitness benefits for the signaler and fitness costs for the receiver. For example, in some primates, individuals bristle their hair when they are aggressive. The apparent function of this display is to make the signaling animal look larger and thus, presumably, more intimidating. If an individual spots another with its hair bristled, it can move off in the opposite direction and thus avoid the impending attack. Once the association between hair bristling in signalers and retreat in receivers is established, an evolutionary option becomes available: signalers can bypass the more costly attacks and simply hair bristle in order to cause others to retreat.

Dawkins and Krebs' manipulative account of communication was readily received by the behavioral ecologists. Nonetheless, as researchers began to work through the theoretical and empirical implications of this framework, significant problems emerged (Hinde 1981; Markl 1985; van Rhijn and Vodegel 1980).[6] The criticisms can be summarized as follows. The original manipulative account

---

6. Markl's (1985) discussion of communication is of particular interest for its remarkable wit and use of political innuendos. For example, in discussing the reciprocal roles of sender and receiver, he states, "Communicating animals have to be Leninist: trust is good, control is better, but that holds for *both parties,* if the degeneration of manipulation into exploitation is to be avoided" (p. 165). Further on, in discussing the use of resource holding power in aggressive contests, Markl points out, "Although the loser at least avoids being damaged in such an asymmetrical fight, he certainly cannot gain resources by giving in, so to call this mutualism would really be at best a Brezhnev doctrine of cooperation" (p. 166).

1. was signaler centered and paid far too little attention to the active role of the receiver in both guiding the communicative process and constraining the form of the signal generated.

2. failed to take into account that in several social, long-lived species, where cooperative interactions are common, there will be strong selection for signals that convey honest information about the signaler's intentions and motivations.

Krebs and Dawkins (1984) responded to the criticisms of their position by giving the receiver a more prominent role in the communication process and by providing some intriguing ideas on signal evolution in both competitive and cooperative situations. Specifically, they argued that receivers will be selected to be good *mind readers,* sifting out the accuracies from the inaccuracies of the signal conveyed—metaphorically, receivers should be resistant to the actor's sales pitch. Thus, countering the manipulative intent of the signaler, is a receiver who is attempting to decipher the true intent and qualities of the signaler. For example, if Joe Fish sends an aggressive threat display to Fred Fish, there will be strong selection on Fred to establish whether Joe is likely to be a tough competitor or a pushover, and Fred's assessment will be based on his past experience with Joe and what can be perceived (extracted) on the basis of Joe's current physical condition—for example, although Joe may have been a formidable opponent last week, he may now be in poor condition as a result of repeated challenges from other competitors. These dynamically changing parameters provide the fuel for what Krebs and Dawkins consider the natural outcome of competitive social interactions: an evolutionary arms race with selection for increasingly manipulative signalers and sales-resistant receivers.

The theoretical suggestions that we have reviewed thus far account for several puzzling observations and provide a unifying framework for thinking about the design of communication systems across the animal kingdom. But, the entire discussion has focused on competitive interactions where the interests of the participants are in conflict because of the resources at stake. Can the behavioral ecologist's perspective also account for the structure and design of signals used in cooperative interactions? For Krebs and Dawkins, the answer is an unambiguous yes, and the predictions that are generated about signal design and function are beautifully simple and, to date, have been insufficiently tested. To develop these ideas, it is important to recognize one of the primary differences between a competitive and a cooperative situation: in the former, one individual needs to be convinced that a resource should be forfeited, whereas in the latter, both individuals should already be convinced that by helping each other, there is a higher probability of successfully commandeering the resource. Empirically, one tends to find that during competitive interactions over valued resources (e.g., food, mates), signals are loud and exaggerated, and consequently costly to produce. In contrast, Krebs and Dawkins have suggested that cooperative signals should be quiet, subtle, produced with minimal cost, and responded to with high sensitivity (i.e., the perceiver's threshold for responding should be low). As a result, signals used during cooperation should take the form of "conspiratorial whispers." The intuition seems right here: just imagine how unconvincing the Reverend Jesse Jackson would be if he whispered all of his orations about government policy into a loudspeaker system!

In summary, a more balanced view of communication that considers both senders and perceivers has been emerging, gradually. Interestingly, although classical ethologists and neuroethologists have long been aware of the importance of considering senders and receivers (Capranica 1965; Konishi 1970; Marler 1955, 1967; E. S. Morton 1975), those behavioral ecologists interested in the evolution of communication were primarily concerned with the signaling end of the process and only more recently have developed an interest in how the perceptual mechanisms of potential perceivers can constrain both the type and structure of the signal emitted (reviewed in Endler 1993b; Guilford and Dawkins 1991). In section 2.5, I will return to the interface between functional accounts of signal design and more mechanistic accounts, and attempt to set up a marriage between the two that is synthetic.

2.3   Language Evolution: Linguists Take a Look

The power of communication between members of the same tribe by means of language has been of paramount importance in the development of man; and the force of language is much aided by the expressive movements of the face and body. We perceive this at once when we converse on an important subject with any person whose face is concealed. Nevertheless there are no grounds, as far as I can discover, for believing that any muscle has been developed or even modified exclusively for the sake of expression. The vocal and other sound producing organs, by which various expressive noises are produced, seem to form a partial exception; but I have elsewhere attempted to show that these organs were first developed for sexual purposes, in order that one sex might call or charm the other.   (Darwin 1872, 354)

2.3.1   Uniqueness

Let's face it. We have, and probably always will have, an obsession about our uniqueness. The history of our species shows that every time a discovery has been made that challenges our domination of the animal kingdom, we are disbelieving at first and, once convinced, unleash all of our intellectual horsepower and search for something else that will set *us* apart from *them*. The list of something elses is quite long, and includes such things as tool use, pedagogy, and cooperative hunting. As anthropologists, biologists, and psychologists have explored these topics in detail (Boesch and Boesch 1992; Caro and Hauser 1992; McGrew 1992; Teleki 1973), several versions of the uniqueness claim have fallen by the wayside. For example, although chimpanzee tool use is certainly not in the same league as human tool use (consider, for starters, the mundane Swiss army knife, which no chimpanzee has yet crafted), it clearly shares a number of structural and functional similarities. Given the phylogenetic proximity of humans and chimpanzees (Begun 1992; Lewin 1993; Pilbeam 1984; Ruvolo 1991), this similarity would appear to represent a convincing case of homology.

I find the search for human uniqueness particularly intriguing. It is clear that we differ in several important ways from other animals, and as a biologist, I would find it puzzling if this were not the case. After all, we evolved under ecological conditions that created specific selection pressures to solve problems in new and adaptive ways. But we tend to be much more possessive about some traits than others, and this attitude has, I believe, blinded our ability to pursue some of the more interesting conceptual issues that emerge from the discovery of phylogenetic differences in the expression of a genetic, morphological, behavioral, or cognitive trait. Specifically, after discovering that a trait is unique, we should ask: Why did it evolve? What does this trait currently do for the individual and what might it have done for the individual when it originally evolved? I will return to this problem in a moment, but for now consider the following sample of features, unique to humans, but that if discovered in other animals would most likely fail to elicit any kind of emotionally charged rescue attempt, although they might elicit a chuckle:

Only humans

1. have large pendulous breasts.

2. dye their hair with such colors as shocking pink and lagoon blue.

3. file their fingernails.

4. wear clothes.

5. dive off cliffs in Acapulco, for fun!

The truly threatening cases are examples of animals whose apparent thought processes—or their overt expression—come dangerously close to our own. No challenge has been more intense and more vehemently defended than human language. Before embarking on the history of this debate, however, consider the quotes in Table 2.1 taken from a diversity of scholars in evolutionary biology, neurobiology, and linguistics.

As these quotes illustrate, scholars in a variety of disciplines have long thought that language represents the crowning achievement of our cognitive evolution. It has not always been fashionable, however, to work on the problem of language evolution. In the early part of the nineteenth century, for example, interest in language origins was squelched by the Société de Linguistique de Paris, primarily because the disagreements were so acrimonious and, apparently, unproductive. As a result of the intellectual "verboten" sign, scientific treatment of language evolution was markedly absent from the published literature in linguistics before the late 1950s and 1960s (Hockett 1960b; Lenneberg 1967; Lieberman 1968; Lieberman, Klatt, and Wilson 1969).

What reinvigorated the quest for the evolutionary origins of human language? Although this question has yet to receive careful treatment by historians of science, it is my impression that at least two factors were critical. First, an important consequence of the modern synthesis in evolutionary biology in the 1940s (J. Huxley 1942; Mayr and Provine 1980) was the conclusion that evolutionary change is, in general, a gradual process, and the result of recombination and selection on genetic variation. As a result of this view, the structure and function of a particular trait was examined in light of ancestral, and most importantly, intermediate traits. Language, as good a trait as any, would therefore be viewed as a communicative form that evolved from earlier forms. It thus became fashionable to ask about the structure of those earlier forms. Second, the 1950s and 1960s saw a relative explosion of research on the biological foundations of language, including technically sophisticated analyses of the physiological and anatomical processes underlying the perception and production of speech (Cooper et al. 1952; Kimura 1961; Klatt et al. 1968; Lenneberg 1967; Liberman et al. 1967; Stetson 1951). As knowledge in this area increased, at least some researchers started to broach comparative questions in an attempt to uncover what, if anything, was unique about the human speech system.[7] With this prelude, let us now turn to a discussion of how linguists have explored the origin and subsequent evolution of human language. Though the various players in this story could be evaluated in light of their differences, I prefer to focus on the strengths of their contributions, and how they have shaped our overall understanding of a su-

---

7. It is worth noting here that it is only within the past twenty or so years that the linguistics community has come to acknowledge the similarities between spoken and signed language. As a result, most discussions of language evolution have focused on speech. For further discussion, see Petitto (1993).

Table 2.1
A List of Quotes on the Uniqueness of Human Language

Author	Quote
T. H. Huxley (1863) Biologist	"Believing, as I do . . . , that the possession of articulate speech is the grand distinctive character of man . . . , I find it very easy to comprehend that some . . . inconspicuous structural differences may have been the primary cause of the immeasurable and practically infinite divergence of the Human form from the Simian strips" (p. 63).
Piatelli-Palmarini (1989) Linguist	"Since language and cognition probably represent the most salient and the most novel biological traits of our species . . . it is now important to show that they may well have arisen from totally extra-adaptive mechanisms" (p. 23).
Lieberman (1991) Speech and hearing scientist	"Research over the past fifty years shows that two unique aspects of human language—speech and syntax—enhance the speed of communication. . . . If we lacked human speech we would be limited by the constraints of the mammalian auditory system that make it impossible to keep track of rapid sequences of sound . . . to paraphrase René Descartes, we are because we talk" (pp. 3–4).
Mayeux and Kandel (1991) Neurobiologists	"Language is a uniquely human capability. . . . From a biological standpoint, language is not a single capability but a family of capabilities, two of which, comprehension and expression, can be separated by distinctive functional sites in the brain" (p. 850).
Denes and Pinson (1993) Speech and hearing scientists	"Speech, in fact, is one of the few basic abilities—tool making is another—that set us apart from other animals and are closely connected with our ability to think abstractly" (p. 1).
Pinker (1994a) Psycholinguist	"Language is obviously as different from other animals' communication systems as the elephant's trunk is different from other animals' nostrils. Nonhuman communication systems are based on one of three designs: a finite repertory of calls (one for warnings of predators, one for claims to territory, and so on), a continuous analog signal that registers the magnitude of some state (the livelier the dance of the bee, the richer the food source that it is telling its hivemates about), or a series of random variations on a theme (a birdsong repeated with a new twist each time: Charlie Parker with feathers)" (p. 334).

premely difficult problem. I consider this to be a more productive approach, for it highlights an often-neglected feature of scientific progress: the conceptual development of thought based on a continuously developing set of empirical building blocks.

2.3.2   Noam Chomsky

Individuals in a speech community have developed essentially the same language. This fact can be explained only on the assumption that these individuals employ highly restrictive principles that guide the construction of grammar. (Chomsky 1975, 11)

In the 1950s, Noam Chomsky (1957) initiated a research program that was designed to understand the constraints on human language variability—specifically, the reasons why certain sounds, sound sequences, and word orders never appeared in the languages of the world. Although his writings on language evolution have not been extensive, his detailed description (Chomsky 1957, 1965, 1986) of some of the key features of language is so profound that it has shaped the way in which both linguists and nonlinguists think about the problem. In the following sketch, therefore, I attempt to distill some of the most important ingredients of his theory, especially those that are most relevant to problems of communication, design, and evolution.

1. Humans, but not other species, have a module in the brain—the *language organ*—that was designed to carry out combinatorial calculations. The module is autonomous, "encapsulated" (Fodor 1983).

2. This combinatorial machinery provides algorithms for specifying the details of our communicative utterances. The fact that we use this machinery *for* communication is, however, accidental.

3. In its most powerful formulation, Chomsky's theory provides a single set of rules from which all possible grammatical sentences can be derived for a given language.

4. The rules that make up what Chomsky calls universal grammar are composed of abstract constraints that determine the kinds of categories implemented in the language.

5. Universal grammar is innately specified.

6. All humans share a universal grammar; nonhuman animals lack universal grammar.

7. Points 1–6 constrain the range of variability in human languages.

This sketch raises several issues that need to be fleshed out in greater detail. Consider point 1, a clearly articulated claim regarding comparative neuroanatomy. If humans have a dedicated language organ that no other species has, then how did it get there and why? The "how" part of this question of course refers to the problem of mechanism, specifically evolutionary mechanisms such as random genetic drift, migration, and selection on both beneficial mutations and preexisting structures. The "why" part of the question, in contrast, refers to the possible adaptive significance of the language organ. In general, Chomsky has not taken a firm stance on either piece of this problem. There appear to be two reasons for his choosing not to do so. On the one hand, he considers our lack of knowledge about developmental neurobiology and constraints on the brain to significantly limit our understanding of how the language organ evolved. Although a considerable amount of research has recently been dedicated to the study of neurodevelopmental constraints and their role in brain evolution (e.g., Deacon 1990a, 1990b, 1991; Finlay and Darlington 1995; Rakic 1995), relatively little comparative work has focused on the microcircuitry of the language organ. Second, Chomsky considers the theory of natural selection to be lacking as an account of either the design or function of the language organ. I disagree with this claim, and so have other linguists (see discussion of Lieberman and Pinker in sections 2.3.4 and 2.3.6). But before we can examine Chomsky's insights within the framework of natural selection theory, let us look more closely at the comparative neuroanatomical claim.

Chomsky's view that only modern humans have a language organ can be examined in two ways. Taken from one angle, his claim does not exclude the possibility that the language organ evolved from one of our more immediate ancestors. Recall that selection will tinker with existing structures, and the ones that would have been most similar to our own would have been present in some of the early *Homo* species, or perhaps even earlier, in the australopithecines. Although several investigators have used endocasts to provide information on brain volume, and even coarse-grained assessments of cerebral structures (Falk 1991; Holloway 1983, 1995) and blood supply (Saban 1995), such analyses fall short of providing us with the requisite material for evaluating whether our hominid ancestors had the relevant neural gear for processing language. In the end, as neurological investigations of brain-damaged patients have revealed (Caplan 1987, 1992; Damasio and Damasio 1992; Maratsos and Matheny 1994), brain size and structural composition tell us far less than knowledge of the underlying circuitry and its functionality. Put simply, even if *Australopithecus afarensis*—the famous ''Lucy''—had a language organ, we would not be in the position to definitively assess its role in communication or, for that matter, any other behavioral expression. This point should be swallowed completely, especially in light of Chomsky's (1986, 1990) own belief that our use of the language organ *for* communication was quite accidental (see point 2).

The second approach to Chomsky's comparative point is to examine the brains of our closest living relatives, to assess whether homologous structures exist on an anatomical basis, and, if so, to determine whether such structures are functionally homologous. This two-pronged approach is important because similarity in structure need not imply similarity in function. A quote from Darwin (1871) makes the point cleanly:

The fact of the higher apes not using their vocal organs for speech, no doubt depends on their intelligence not having been sufficiently advanced. The possession by them of vocal organs, which with long-continued practice might have been used for speech, although not thus used, is paralleled by the case of many birds which possess organs fitted for singing, though they never sing. Thus, the nightingale and crow have vocal organs similarly constructed, these being used by the former for diversified song, and by the latter merely for croaking.   (p. 59)

As I will review in greater detail in chapter 4, studies have revealed that in both New and Old World monkey species, anatomical homologues to Broca's and Wernicke's areas exist. However, when these areas are experimentally lesioned, no significant effects on vocal production are observed (reviewed in Jürgens 1990). These results have led to the general conclusion that the locus of control for primate vocalizations appears to be subcortical, in particular, the limbic system. In humans, limbic structures are involved in our more emotive utterances, such as laughter, screaming, and crying. In contrast to the production data, single-unit recordings from rhesus macaques suggest that the homologue to Wernicke's area may be critically involved in the perception of species-specific calls (Rauschecker, Tian, and Hauser 1995). Unfortunately, these kinds of studies are only in their infancy and must mature considerably if they are to provide the requisite tests of Chomsky's comparative anatomical claim.

If we accept Chomsky's position that the language organ is a species-specific neuroanatomical trait, we open another round of evolutionary problems, such as the following:

1. What kind of advantages might have resulted from the language organ at the time of its evolutionary birth?

2. If the language organ arose due to mutations, followed by selection, what kinds of population structures and selection pressures would have favored its maintenance *in* the population?

3. Given the current design features of our language organ, which ones would have given us a selective advantage while walking around the savanna? Which ones provide us with an advantage now?

Answers to these questions will be difficult to obtain. But a rich comparative approach that makes use of a smorgasborg of evidence, including fossils, neurobi-

ology, and behavior, is the only coherent approach to the problem, and the empirical chapters of this book represent one attempt to fuel this approach.

A final issue to discuss is Chomsky's claim about universality and the absence
of grammar in nonhumans. A difficulty with this claim is that it forces a clear
answer to the question, What are the crucial properties of universal grammar,
and do all of the 4,000-plus languages of the world have each and every one of
these properties? (For a recent discussion of the details underlying this issue, see
Chomsky 1990; Pinker 1994a.) Assume, for the sake of discussion, that one can
precisely describe the features of our universal grammar. What about grammar
in nonhumans? The general answer to this question has been a resounding: No!
I would like, however, to ask the reader to indulge my request to fend off the
temptation to side with the general consensus and to consider some alternatives.
The reason why you should agree to my proposal (laid out in greater detail in
chapters 3, 7, and 8) is that it provides an avenue for future research, an avenue
that is likely to provide some of the critical empirical data to assess the evolutionary uniqueness of human grammar. Let me summarize the conceptual argument
and then give a possible example.[8] Briefly, we can discuss the properties of human
grammar in detail because we have an exquisite understanding of the units that
comprise our grammar. Specifically, we know about phonemes, morphemes,
words, phrases, and so forth. And we know how each of these elements can be
strung together into a potentially unique and meaningful utterance. But the bottom line for nonhuman animals is this: we don't fully understand what the relevant
units of communication are, and thus we are crippled in our ability to say, one
way or the other, whether grammatical structure underlies their utterances. This
kind of agnostic response cannot, however, be used indefinitely, and at some
point the linguists will rightly yell: "Okay, you ethologists. Time's up. Do you
or don't you have evidence of grammar?" So that we can answer this question,
we need to be looking in the right place, and the following represents a set of
observations that may be ripe for the picking:

1. In a variety of nonhuman primate species, individuals vocalize before they
move off into a new area. Somehow, and this is the mysterious part, such vocalizations appear to serve the function of group coordination and movement. The
information content of these vocalizations has yet to be described in detail, but
the social situation seems primed for a grammatically structured utterance, where

8. I would like to thank Steve Pinker for pushing me to think about the kind of evidence one might
use to substantiate this claim.

something like word order is important (e.g., "I am heading out toward the big sleeping tree and will see you in the bush below" as opposed to "I am heading out toward the big sleeping tree and will see you up in the canopy").

2. Several primate species (e.g., chimpanzees, mangabeys) produce long-distance calls that consist of multiple units, some of which are repeated. The ordering of the primary units in these calls is consistent. Thus, for example, the chimpanzee's pant-hoot tends[9] to start with a series of relatively soft hoots, which then moves into a series of climactic screams and then, occasionally, terminates with an additional series of hoots. Thus there is structure within a call type, suggesting the possibility of something akin to phonemes within words.

3. In black-capped chickadees and capuchin monkeys, different call types within the repertoire are strung together in sequences based on ordering rules (e.g., A before B and C but never after B or C). Missing from this analysis, however, is a clear description of the meaning or semantics of each call type.

With these observations in mind, therefore, I hope the reader will hold his or her vote on nonhuman animal grammar[10] until the end of the book.

Before moving on to our next historical piece, let me reiterate what I consider to be Chomsky's three most important contributions to the problem of language evolution and communication. First, the structure of language is determined by a specialized brain module. Second, all humans have this module, but nonhuman animals do not. Third, the language module is responsible for a set of universal rules or grammar that provides humans with the capacity to generate a limitless range of meaningful utterances.

2.3.3   Derek Bickerton

Derek Bickerton, like Chomsky, has spent a long career looking at the fine structure of language, especially the fascinating cases of language change that have occurred as a result of different cultures coming together. These cases are interesting because they provide fundamental insights into the historical reconstruction of our languages and how they have changed over time. For example, when

---

9. Caution should be used in interpreting the precise order of chimpanzee pant-hoots because there is at least suggestive evidence that in different populations different units are used (Mitani et al. 1992).

10. Reminder: I am talking about the possibility of a *natural grammar*, one that underlies the communication used by animals in their species-typical environment, rather than the kind of grammar that has been claimed to underlie the artificial communication systems used by apes and dolphins (Herman, Pack, and Palmer 1993; Herman, Richards, and Wolz 1984; Holder, Herman, and Kuczaj 1993; Savage-Rumbaugh 1986; Savage-Rumbaugh et al. 1993).

two communities lacking a common language are confronted with a situation that requires communication, we see, as Bickerton (1981) has so carefully documented, the emergence of a kind of protolanguage, or what is known as a *pidgin*. Relative to natural languages, the structure of a pidgin is quite simple, often consisting of short word strings and only a few grammatical items. Over time, especially with the subsequent generation of offspring, we see a refinement in the structure and usage of the language—what is known as a *creole*. This process of change from pidgin to creole shows how innate mechanisms guide the original rules for language production and subsequently constrain language structure and usage.

Armed with the insights from his studies of pidgins and creoles, Bickerton (1990) set out to explain the evolution of human language. He starts with what he considers the central paradox: "[L]anguage must have evolved out of some prior system, and yet there does not seem to be any such system out of which it could have evolved" (p. 8). This claim is, on a coarse-grained level, very much like Chomsky's. It differs, however, in at least two important ways. First, for Bickerton, the crucial difference between human language and all other communication forms lies in the structure and content of the representation. Whereas animal signals use *primary representations* of whole situations (e.g., food, a predator), human language is a *secondary representational system* that consists of parts, each with its own goals—verbs referring to action, adjectives to properties of objects, and so on. The significance of grammar, then, is that it provides us with an unlimited variety of utterances and therefore an infinite pool of expressions for representing our thoughts, including thoughts about things we have experienced or even contemplated experiencing. The reason why Bickerton must call on representational aspects of language is that in looking at the structural and functional aspects of nonhuman animal signals (e.g., the number of discrete sound categories, the entities that such signals refer to in the world), he sees nothing like human language. Of course, the claim that we do not see similarities in structure or function between humans and extant nonhumans (even our closest living relatives, the monkeys and apes) need not represent a paradox, though we may be deeply disappointed and puzzled by the absence of similarity. The reason for the apparent lack of similarity is that evolutionary continuity, in terms of structural and functional properties of a system, is most likely to be seen in our most recent ancestors, where the tinkerings of natural selection are likely to be most apparent. Evolution within the hominid line (i.e., the one that includes modern humans) has occurred over some 4 to 5 million years, ample time for natural selection to do its thing with regard to changes in communication. This argument does not,

of course, invalidate the search for commonalities among all of the living species—if it did, I wouldn't be writing this book! But, as discussed in section 2.2,
we must be careful to distinguish between similarities that arise due to convergent
evolution and similarities that arise due to common descent. And within the latter,
a distinction is necessary between similarities that have arisen due to evolution
of a trait that is shared with a recent common ancestor as opposed to a more
remote, but nonetheless common, ancestor.

The second difference between Chomsky and Bickerton lies in how the continuity paradox is resolved. For Bickerton, one needs to explain the difference in
representational capacities between those that lack language and those that have
it. But the starting point is what Bickerton calls a protolanguage, a system of
communication that has the rudimentary structure of full-blown language; for
example, protolanguages use secondary representations but lack some of the
significant formal properties of language such as the mapping between word order
and meaning, and the necessity of grammatical morphemes for structuring the
utterance. The privileged users of protololanguage are the signing chimpanzees,
children under the age of about two years, feral children, and first-generation
speakers of a pidgin. Bickerton considers this approach to provide a handsome
payoff with regard to the conceptual turf of the language origins problem:

If there indeed exists a more primitive variety of language alongside fully developed human
language, then the task of accounting for the origins of language is made much easier. No
longer do we have to hypothesize some gargantuan leap from speechlessness to full language, a leap so vast and abrupt that evolutionary theory would be hard put to account
for it. We can legitimately assume that the more primitive linguistic faculty evolved first,
and that contemporary language represents a development of the original faculty. Granted,
this assumption does not smooth the path, for the gulf between protolanguage and language
remains an enormous one. But at least it makes the task possible, especially since the
level of representational systems achieved by some social mammals amounts to a stage
of readiness, if not for language, at least for some intermediate system such as protolanguage. (Bickerton 1990, 128)

Having laid out the conceptual argument, Bickerton then moves on to the empirical evidence for protolanguages, followed by a discussion of how one gets from
the protolanguage to language. Here is the argument, and some of the logic that
underlies it:

1. Fundamental genetic changes were responsible for the emergence of the first
protolanguages. We see some evidence of a protolanguage in the ape "language"
studies. Bickerton does not provide any details about the evolutionary emergence
of the first protolanguage, nor which evolutionary mechanisms were responsible
for its appearance.

2. Protolanguage represents the intermediate between no language and language. But humans do not go through any intermediary stages between protolanguage and language.

3. Thus the evolutionary transition between protolanguage and language resulted from catastrophic processes. Specifically, a single macromutation (one gene), which appeared in the ancestral Eve, was responsible for the transition from protolanguage to full-blown language.

4. This single gene was responsible for considerable structure-function changes. For example, comparing signing chimpanzees with protolanguage to normal human adults with language, this gene must have minimally (a) rewired the neural circuitry to support syntactic structure and (b) modified the structure of the vocal tract so that the specific sounds of speech could be produced.

5. The gene responsible for the transition was fixed in the population by means of natural selection.

Bickerton's thoughts on the transition from protolanguage to full-blown language are fascinating, especially the commonalities among signing chimpanzees, two-year-old children, feral humans, and first-generation speakers of a pidgin. However, there are at least two problems with his suggestion that a single macromutation was responsible for the critical evolutionary transition. First, as Bickerton and others have clearly articulated, language recruits a number of significant cognitive modules and peripheral organs, all of which require extensive coordination. The probability that such complexity of design is the result of a single gene is extremely low. Even something *relatively* more simple, like the vertebral column, is coded by hundreds of genes, responsible for segmentation and structural arrangement. As an interesting historical aside, even Bickerton once took a hard line on this issue, arguing in his 1981 book *Roots of Language* that it makes little sense to think that language emerged "in its entirety from Jove's brow by some beneficient and unprecedented mutation" (p. 215), because "evolution has advanced not by leaps and bounds but by infinitesimal gradations" (p. 221). Second, Bickerton justifies his claim about a single mutational step by citing recent evidence from evolutionary studies of mitochondrial DNA. Specifically, in 1987, Cann, Stoneking, and Wilson published a paper indicating that the root of all modern humans could be traced back to a single female ancestor, popularly called Eve. To understand this claim, a few facts first. The bulk of the genetic material (50,000 to 100,000 genes) in the human genome (and this is generally true of many other vertebrates) is nuclear, that is, located within the nucleus of a cell. This nuclear DNA, or nDNA, consists of 22 autosomes and two sex chromosomes, the X and Y. In addition to the nDNA, however, is mitochondrial DNA, or mtDNA. In humans, there are 37 mitochondrial genes, and they have a unique characteristic: unlike nDNA which recombines (genetic mixing from the maternal and paternal lines), mtDNA is only transmitted (inherited) through the maternal line. Because of this pattern of transmission, differences between parent and offspring mtDNA can only be due to mutation, and several studies have shown that the rate of mutation in mtDNA is, on average, about 10 times what has been observed for functionally similar classes of nDNA.

Now comes the interesting part. Given the apparently rapid rate of change in mtDNA, the evolutionary research crowd was astonished to learn that humans exhibited only 20–50% of the intraspecific variability observed in several other mammalian species. This pattern led to the proposal that the section of mtDNA examined for modern humans could be traced back approximately 200,000 years to a single female, popularly labeled Eve.

The problem in Bickerton's discussion is twofold. On the one hand, he interprets the molecular data as evidence that modern humans appeared on the scene with all of their species-typical traits perfectly in place, and that the change from premodern to modern was due to a catastrophic mutational event. But there is no evidence for this interpretation, and in fact, Cann and colleagues (1987) explicitly speak against this kind of conclusion. Specifically, although their analyses indicate that mtDNA diversity can be traced back to a common ancestor living 140,000 to 280,000 years ago, this finding "need not imply that the transformation to anatomically modern *Homo sapiens* occurred in Africa at this time. The mtDNA data tell us nothing of the contributions to this transformation by the genetic and cultural traits of males and females whose mtDNA became extinct" (p. 35). Second, Bickerton considers Eve to be the source of all the important traits we have acquired as modern humans, including the suite of characteristics supporting our language. This theory is problematic because it excludes the importance of nDNA and confuses the effects of mtDNA transmission on patterns of variation. To make this point clear, I draw on a succinct treatment of these issues by Cann (1995):

If we try to trace the person's nuclear alleles, we have to consider the probability of transmission in each generation, and the probability that any one ancestor might be polymorphic at the locus of interest. We must also account for the possibility of genetic recombination. The result is some complicated mathematics for nuclear gene transmission. Yet, if we consider the person's mtDNA only, there is one and only one ancestor in this family pedigree. That is the maternal great-great-grandmother. There may be a mitochondrial "Eve" in this family, but she is not the only woman alive in her generation. So, the sense in which "Eve" matches her biblical model is rather remote. I think this helps us to understand some of the confusion generated in the popular press about the mitochondrial mother of us all and the playful misnomer "Eve." The biological truth is that she may represent the only distinct common ancestor that we all share.   (p. 129)

Although the particular evolutionary scenario invoked by Bickerton has problems, I find his proposition about language as a vehicle for transmitting complex representations to be of considerable interest, and one that can serve to structure future research on nonhuman animals. As I will review in chapter 7, current work on animal representation is only in its infancy. Whereas some studies suggest, in agreement with Bickerton, that nonhuman animals lack abstract relational concepts (Herrnstein 1991), other studies suggest that animal representations are complex and, at least in the case of nonhuman primates, may underlie the intricate social relationships and political power struggles that have been described (Allen and Hauser 1991; Cheney and Seyfarth 1990; de Waal 1982, 1989).

2.3.4   Philip Lieberman

With regard to both theoretical and empirical contributions to the study of language evolution, Philip Lieberman (Lieberman 1968, 1984, 1991; Lieberman, Klatt, and Wilson 1969) must be credited with the longest-running and most active research program. As with Chomsky's voluminous writings, I cannot provide a detailed review of all the relevant findings from Lieberman's work, but rather will attempt to give the reader a flavor of some of the more significant points.

Lieberman's perspective[11] on the origins and evolution of language starts with a series of observations. First, modern adult humans have a supralaryngeal vocal tract that is fundamentally different from all other animal vocal tracts (Crelin 1987; Negus 1929, 1949). A coarse-grained summary of the key featural distinctions is provided in Table 2.2.

The distinctions made in Table 2.2 led Lieberman (1968; Lieberman, Klatt, and Wilson 1969) to a series of predictions and subsequent tests about the range of sounds and sound contrasts that would be permissible given constraints imposed by vocal tract morphology. Using formal models in acoustics (Fant 1960; Flanagan 1963), anatomical data on vocal tract length and structure, and a small set of sound recordings from rhesus monkeys and chimpanzees, Lieberman argued that nonhuman primates

---

11. In contrast to the other linguists who have approached this problem, Lieberman is trained in both engineering and linguistics. This fact may partially explain why he has pursued some of the more mechanical aspects of human language and their evolutionary precursors.

Table 2.2
A Featural Comparison between Human and Nonhuman Supralaryngeal Vocal Tracts (adopted from
Lieberman 1984 and Negus 1949)

Feature	Humans	Nonhumans
Position of larynx in throat	Low	High
Length of mandible	Long	Short
Size and shape of tongue	Large and rounded	Short and flat
Angle between pharyngeal and oral cavities	Approximately 90° bend	No bend or very slight
Number of resonating cavities*	Two-plus	One

*The two tubes are the pharyngeal and oral cavities. The "plus" comes from the added possibility
of using the nasal cavity as a resonator. I return to these issues in chapter 4.

1. cannot produce the point vowels[12] of human speech (/a/, /i/, /u/).

2. use a limited set of articulatory gestures to modify the resonances of the
supralaryngeal vocal tract.

3. produce an unstable pitch period[13] due to relatively weak laryngeal control.

As with Chomsky and Bickerton, Lieberman also highlights specific differences
between humans and nonhumans, but instead of universal grammar or the complexity of representation, he emphasizes sound production, and the uniqueness
of speech as an acoustic signal. The significance of point 1 emerges from the
observation that because most human languages use the three point vowels, they
are fundamental attributes of spoken language. The phylogenetic claim that nonhuman animals are incapable of producing the point vowels is generally uncontro-
versial (it is an empirical claim that has not been sufficiently tested but has
received some empirical support), but its significance can only be determined in
light of the relative importance of these vowels in speech perception, and language more generally; the same can be said for the acoustic significance of vocal
tract resonances and the stability of the pitch period. Work on the articulatory
and perceptual significance of the point vowels in human speech (e.g., Chiba and
Kajiyama 1941; Stevens 1972) indicates that they provide important definitional
boundaries for a language's acoustic space. They also provide the means by

12. The point vowels, which tend to be observed in all human languages, define an acoustic space
within which speakers differentiate other vowels within the language. When the average formant
frequencies are plotted for these vowels, they provide an indication of the range of acoustic variation
for a speaker. As such, they form the basis for vocal tract normalization.

13. The pitch of the voice results from oscillations of the vocal folds. Such oscillations generate an
acoustic signal known as the fundamental frequency which, in normal humans, tends to approximate
a saw-toothed waveform.

which listeners can extract information relevant to vocal tract length (vocal tract normalization) and consequently the age and sex of the speaker. An interesting question emerging from this claim is, Although monkeys and apes may be incapable of producing the point vowels, do they produce sounds that are functionally equivalent? That is, are there sounds within the repertoire of nonhuman primates that in some sense anchor the acoustic space and provide reliable information about the speaker's age and sex? To my knowledge, this question has yet to be addressed in detail for any species, but there are empirical hints, and we pick these up in chapter 4.

Using the logic from the points listed, Lieberman has made three further suggestions regarding the significance of vocal tract anatomy for understanding language evolution. First, the supralaryngeal vocal tract of a human neonate is more like that of a nonhuman primate than like that of an adult human. At approximately three months, the larynx begins to drop, thereby creating the bend in the oropharyngeal juncture. At about six months, the shape of the human infant's tract is approximately adultlike. Lieberman then makes the interesting observation that prior to six months, the infant's tract is designed for maximum digestive efficiency, whereas after six months, digestive efficiency drops and vocal efficiency rises. As discussed in chapter 5, some of the first signs of speech emerge at six months, a stage often called early babbling. Lieberman's second point is that if one were to place a modern human vocal tract into a Neanderthal, it wouldn't fit properly. Specifically, a direct insertion of the modern tract would place the larynx inside the Neanderthal's chest, an impossible location. Alternatively, one would need to modify certain components of the supralaryngeal vocal tract,[14] but as Lieberman (1984) has argued, doing so would significantly compromise the individual's ability to produce the sounds and sound contrasts of human speech. As a result of these anatomical considerations, Lieberman claims that human speech emerged with the evolution of a modern vocal tract—some time after the Neanderthals. Last, Lieberman (1991) has argued that the peripheral machinery for vocal production is driven by a brain that has been designed to maximize the efficiency of "high-speed vocal communication" (p. 286), and in particular, computation of rule-governed syntax. To quote: "The evolution of human speech, complex syntax, creative thought, and some aspects

14. Houghton (1993) has recently attempted to address the possibility of Neanderthal speech by recreating the supralaryngeal vocal tract of the La Chapelle-aux-Saints Neanderthal. Houghton's conclusion is that Neanderthals were capable of human speech. Unfortunately, and as Lieberman (1994) rightly points out, Houghton's assumption of tongue size violates what is known about human tongues, thereby negating the validity of his conclusions.

of morality is linked—and . . . the driving force that produced modern human beings in the last 200,000 years or so was the evolution of speech adapted for rapid communication" (p. 2).

Although there has been some disagreement in the literature about Lieberman's assertions regarding comparative vocal tract and brain anatomy (see, e.g., Crelin 1987; Falk 1991; also, see comments in section 4.4), the significance of his research program, like Chomsky's and Bickerton's, is that it provides a set of testable hypotheses. But unlike Chomsky and Bickerton, Lieberman's views fall clearly within a Darwinian framework, emphasizing the adaptive significance of particular features of the communication system:

1. The structure of the human vocal tract provides a selective advantage over other configurations because it can produce sounds that are not nasalized; nasalization causes utterances to be less readily identified by listeners.

2. The human vocal tract produces quantal sounds (e.g., /a/, /u/, /i/), defined by distinct spectral peaks, and resulting in fewer perceptual errors; these sounds "are better suited for communication than other sounds" (Lieberman 1984, 58).

3. Speech provides a significantly greater rate of data transmission than other communication systems.

We find, once again, a set of hypotheses that relate directly to problems in comparative biology, in particular functional anatomy. A difficulty with some of these hypotheses, however, is that they fail to make clear which taxonomic groups are being compared. Consider the claim about nasalization. Although the nonhuman vocal tract may produce more nasalized sounds than the human vocal tract, the perceptual system of nonhuman primates may be designed to handle the acoustic consequences of nasalization. In other words, the perceptual difficulty of nasalization (point 1) only makes sense in light of the design features of the perceptual system discussed. As detailed in chapters 4 and 5, though some rhesus monkeys produce perceptually nasal vocalizations, we have no indication that such individuals are less readily understood than individuals producing less nasalized vocalizations (Hauser 1992a). The same kind of argument applies to quantal sounds. Nonhuman animals may not be able to produce such sounds, but their perceptual systems have coevolved with their vocal tracts and thus exhibit design features that have been selected to minimize perceptual errors.

Summarizing thus far, our walk along the path of history has provided at least three important considerations: Chomsky's language organ and universal grammar, Bickerton's view that language is the most powerful vehicle in the animal kingdom for representing our complex thoughts, and now, Lieberman's position that the evolution of the human vocal tract was a fundamental step in the development of a communication system that was designed to maximize information transmission and minimize perceptual errors. I now take the final turn in our historical path and discuss the views of two linguists who have attempted to capture the most significant design features of human language: Charles Hockett and Steven Pinker.

2.3.5   Charles Hockett

Charles Hockett (1960a, 1960b) was one of the first linguists to suggest that language evolved gradually over time and that continuity with nonhuman animal systems could be documented. Hockett's view is notable for it provides one of the few attempts to break language down into its constituent parts and evaluate whether, and to what extent, other organisms incorporate different components into their own communication systems. Table 2.3 summarizes the primary components or design features of language as discussed in Hockett's original treatise.

Early attempts to place nonhuman animal communication systems within this framework indicated that only two of the thirteen features—traditional transmission and duality of patterning—were not observed in any nonhuman species, and one feature—displacement—was only observed in the honeybee's "dance language" (von Frisch 1967a). As indicated further on in the book, however, recent research efforts have uncovered suggestive evidence that cultural transmission underlies changes in the vocal signals of some birds and whales (Balaban 1988b; Balaban 1988c; Marler and Tamura 1962; K. B. Payne and R. S. Payne 1985; R. B. Payne, L. L. Payne, and Doehlert 1988) and that in some songbirds and primates discrete call types are often combined by means of a rudimentary syntax (Balaban 1988a; Hailman and Ficken 1987; Hailman, Ficken, and Ficken 1987; Mitani and Marler 1989; Snowdon 1990). Importantly, however, no nonhuman species has yet been able to amass all 13 features within its repertoire. Thus the featural set provided by Hockett represents a unique characterization of only one communication system: human language.

Although Hockett's framework is appealing from a comparative perspective, in that it provides criteria upon which to contrast different communication systems, it has generally failed to motivate theoretical discussion or empirical tests. I can see at least two reasons why Hockett's analysis may have failed in this way. First, it does not provide any insights into the functional significance of each of the thirteen design features. For example, what are the fitness consequences of displacement, and what socioecological pressures would favor its evolution?

Table 2.3
Hockett's Design Features of Communication

Feature	Definition
1. Auditory-vocal channel	Sound is transmitted from the mouth to the ear.
2. Broadcast transmission and directional reception	An auditory signal can be detected by any perceiver within hearing range, and the perceiver's ears are used to localize the signal.
3. Rapid fading	In contrast to some visual and olfactory signals, auditory signals are transitory.
4. Interchangeability	Competent users of a language can produce any signal that they can comprehend.
5. Total feedback	All signals produced by an individual can be reflected upon.
6. Specialization	The only function of the acoustic waveforms of speech is to convey meaning.
7. Semanticity	A signal conveys meaning through its association with objects and events in the environment.
8. Arbitrariness	The speech signal itself bears no relationship with the object or event that it is associated with.
9. Discreteness	Speech is comprised of a small set of acoustically distinct units or elements.
10. Displacement	Speech signals can refer to objects and events that are removed from the present in both space and time.
11. Productivity	Speech allows for the expression of an infinite variety of meaningful utterances as a result of combining discrete elements into new sentences.
12. Traditional transmission	Language structure and usage is passed on from one generation to the next via pedagogy and learning.
13. Duality of patterning	The particular sound elements of language have no intrinsic meaning, but combine to form structures (e.g., words, phrases) that do have meaning.

Second, it does not provide a weighting of the importance of each feature, where the weighting of a feature is related to the species-typical habitat. Without understanding something about the importance of the feature set, it is difficult to assess what is necessary for a functional language and what is sufficient. I address this last issue in chapter 8 while attempting to grapple with the problem of building a communicating organism.

2.3.6   Steven Pinker

The most recent attempt to reconcile current theories of language structure and function with Darwinian evolution is Steven Pinker's (Pinker and Bloom 1990; Pinker 1994a) theoretical analysis. Pinker's argument is somewhat tricky because he is Chomskyan in his analysis of language (i.e., universal grammar is central), but Darwinian in his view of evolution (natural selection is the only evolutionary mechanism that can possibly account for the complex design features of language). As the quote in Table 2.1 reveals, Pinker considers the gulf between human language and nonhuman animal communication to be immense. However, what appears to be a non-Darwinian discontinuity for Chomsky and Bickerton is no problem for Pinker, or for a Darwinian account. The following is a distillation of his perspective, which is powerful and lucid, and fits beautifully with the conceptual goals of this book, focusing on design features and Darwinian processes of evolution:

1. Language, like the eye or the sonar system of bats, has complex design features.

2. Natural selection is the only possible mechanism that can account for a trait with complex design features.

3. A trait with complex design features can only evolve by means of gradual changes.

4. The fact that our closest living relatives lack language is not a problem for the theory because these species do not represent our *direct* ancestors. Our direct ancestors (*Australopithecus* spp. and archaic *Homo*) lived several million years ago, and it is within this ancestral line that earlier forms of language—protolanguages—evolved.

5. The capacity for language is not the accidental by-product of having a relatively big brain. Rather, selection favored the capacity because of its adaptive significance.

6. Three observations provide evidence of genetic variation, sufficient for selection to act upon: (a) variation between normal humans in grammatical competence; (b) in those families with a preponderance of left-handers, right-handed individuals rely more on lexical than syntactical analyses; (c) potential evidence of an inherited deficit in grammatical ability within a family (M. Gopnik 1990).

7. All humans are born with a universal grammar—the "language instinct." This grammar constrains the range of potential variation, allowing for both linguistic diversity and universals (Comrie 1981; Hawkins 1988).

8. The language instinct is located in the brain, lateralized to the left hemisphere. Although a modular view of the neural substrate for language is favored, evidence of a nonmodular, multipurpose system would not cause problems for the Darwinian account; selection can favor a structure with more than one function.

Pinker's primary contribution to the language origins problem then is that he has persuasively argued, contra Chomsky and Bickerton, that Darwin's theory of natural selection is the only theory that can account for both the structure and function of human language. What is interesting, from a historical perspective, is that several other scholars who have also read Darwin have somehow missed some of his crucial take-home messages. In particular, they have missed points 1, 2, 3, and 4. Point 4, which is Darwin's descent by modification, is particularly important because failure to appreciate the implications of this point leads to the following kind of faulty logic:

Observation 1: Humans and nonhuman primates are closely related.

Observation 2: Trait X in humans is similar to trait X' in nonhuman primates.

Deduction 1: Trait X in humans evolved from trait X' in nonhuman primates.

The logic is flawed because the living nonhuman primates, even chimpanzees, are not our direct ancestors. Consequently, and as mentioned in my discussion of Bickerton's perspective, the lack of language or languagelike design features in the communication systems of our nearest relatives does not represent a paradox, nor does it speak against the Darwinian view of gradual evolution guided by natural selection.

2.3.7   Summary

In summary, the linguistic perspective has offered a suite of claims and predictions concerning the evolution of language. Many of the claims open the door for comparative approaches to studying communication systems generally, and the design features of human language more specifically. If we are interested in the evolutionary forces that have shaped the structure of a trait such as language, then comparative approaches are critical. To recap, here is what we have picked up along the historical path:

1. Chomsky emphasizes the importance of the language organ, a computational system that provides the mental grammar subserving our species-specific communication system. Natural selection theory cannot account for the design features of human language.

2. Bickerton suggests that human language evolved from a protolanguage. The primary advantage of full-blown language over protolanguage is in its capacity to represent complex facets of the observable and unobservable events in our lives.

3. Lieberman emphasizes the unique design of the human vocal tract. Over the course of human evolution, selection favored a vocal tract that was optimized for rapid transmission of information relative to digestive efficiency.

4. Hockett provided a list of design features that characterizes all human languages. Although other species exhibit some of these features, no single species exhibits them all.

5. Pinker shows how a Darwinian account of language evolution is the only possible account. This is because natural selection is the only mechanism that can account for the complex design features of a trait such as language.

Let us now turn to a discussion of how biologists have approached the problem of language evolution and, in particular, how they have attempted to test some of the ideas developed by linguists.

2.4 Language Evolution: Biologists Take a Look

Articulate language is, however, peculiar to man; but he uses in common with the lower animals inarticulate cries to express his meaning, aided by gestures and the movements of the muscles of the face. This especially holds good with the more simple and vivid feelings, which are but little connected with our higher intelligence. Our cries of pain, fear, surprise, anger, together with their appropriate actions, and the murmur of a mother to her beloved child, are more expressive than any words. It is not the mere power of articulation that distinguishes man from other animals, for as everyone knows, parrots can talk; but it is his large power of connecting definite sounds with definite ideas; and this obviously depends on the development of the mental faculties.   (Darwin 1871, 54)

2.4.1 General Comments

Charles Darwin, who collected beetles and studied barnacles, was forced, over and over again, to use his theory of natural selection to account for human behavior, and this included both human language and facial expression. In *The Expression of the Emotions in Man and Animals,* Darwin (1872) confronted the nonhuman-human comparison head-on, moving elegantly between observations of behavior and insights into some of the driving forces underlying adaptive variation. And for Darwin, there was no question that both human gestural expressions and language were derived from an animal ancestor, but that natural selection resulted in significant changes in the design features of the communication system:

I cannot doubt that language owes its origin to the imitation and modification, aided by signs and gestures, of various natural sounds, the voices of other animals, and man's own distinctive cries. When we treat of sexual selection we shall see that primeval man, or rather some early progenitor of man, probably used his voice largely, as does one of the gibbon-apes at the present day, in producing true musical cadences, that is in singing; we may conclude from a widely-spread analogy that this power would have been especially exerted during the courtship of the sexes, serving to express various emotions, as love, jealousy, triumph, and serving as a challenge to their rivals. The imitation by articulate sounds of musical cries might have given rise to words expressive of various complex emotions.   (Darwin 1871, 56)

We thus learn two important lessons from Darwin's treatment of human language evolution. First, the structure and function of human language can be accounted for by natural selection, and second, the most impressive link between human and nonhuman-animal forms of communication lies in the ability to express emotional state.

The founders of ethology, Lorenz, Tinbergen, and von Frisch, generally picked up where Darwin left off, focusing their research efforts on nonhuman animals, but drawing comparisons with humans. This tradition has continued steadily ever since (Hinde 1987), through all of the various disciplinary mutations (e.g., from ethology to sociobiology to behavioral ecology and so on). Numerous debates, however, have emerged over the relevance, and consequently the importance, of nonhuman animal data for understanding human behavior. These debates can be avoided by making two distinctions explicit. First, by keeping the phylogenetic tree firmly ensconced in one's mind, one avoids blurring the difference between homology and homoplasy. As previously discussed, similarity between two organisms can evolve because of phylogenetic ancestry or because of a limited solution set to significant ecological problems or pressures. Second, the behavior of nonhuman animals is often described with terms that are used for human behavior. Sometimes, an author actually *means* that term X is being used in the same way for humans and nonhumans. In other situations when term X is applied to nonhumans, the author is using a shorthand for something like "the behavior is *functionally X*" or "the animal behaves *as if X*." The distinction is perhaps most important when term X has something to do with the organism's cognitive capacity (see chapter 7), as when ethologists describe animals as having "language" (Roitblat, Herman, and Nachtigall 1993), "culture" (McGrew 1992), "theories of mind" (Cheney and Seyfarth 1990; Povinelli 1993; Premack and Woodruff 1978), and so on.

To illustrate the topography of comparative analyses, as well as the cautionary distinctions raised here, consider von Frisch's (1967a) work on the honeybee's communication system. When an individual returns to the hive after a foraging trip, it will often perform a dance display that conveys information about the location and quality of food (see especially chapter 7). Von Frisch described this system of communication as the honeybee's *dance language*. His attribution of the term *language* was meant as an analogy, based on the function of the signal, not, obviously, its structure. Since honeybees do not share a common ancestor with humans, whose signaling system represented something like an intermediate, the similarity between the dance language and human language represents a potential case of convergence; it is almost certainly not a homology. And of course, the type of similarity involved falls under the category of *is functionally like X*. Unlike current knowledge of human language, we know relatively little about the cognitive processes underlying the honeybee's dance, although some, such as J. L. Gould (1990), have argued that both the transmission and reception of information is guided by a mental map. Let us now leave von Frisch and the honeybees and turn to more recent and explicit treatments of the similarities between human language and nonhuman animal communication.

   Marler (1955, 1961, 1967) and W. J. Smith (1969) were the first ethologists to provide a formal characterization of animal communication systems and, from their characterization, draw parallels to human language (Marler 1970a, 1975; W. J. Smith 1977). Though both ethologists studied birds, Marler also conducted field studies of East African primates, including apes (Marler 1969a; Marler and Tenaza 1977) and monkeys (Marler 1970b, 1973). As is so often the case in animal behavior, there is a significant, bidirectional interaction between the type of organism studied and the theoretical and methodological approach adopted. This effect may partially explain why Marler and Smith's views differed and continue to differ to this day. Let us look at the details of their claims.

2.4.2   Peter Marler

Peter Marler's contribution can be seen in two general domains, the first having to do with some of the common structural features of animal signals (S. Green and Marler 1979; Marler 1977), especially auditory ones, the second having to do with the similarities between animal signals and human language (Marler 1961, 1975; Marler, Evans, and Hauser 1992). To illustrate the first point, consider his analysis of structure-function relationships in avian alarm calls (Marler 1957). At the time Marler was thinking about such issues, the general view, promoted by Lorenz and Tinbergen, was that animal signals were arbitrary and, as such, provided significant insights into the phylogeny of the trait, free from the problems of evolutionary convergence. Marler's observations of avian vocalizations produced in the context of predator detection led him to a different perspective, and one that argued quite strongly for nonarbitrary acoustic features, designed to maximize either silent predator evasion or vociferous predator attack. As Figure 2.2 illustrates, birds use structurally different calls when they are mobbing a

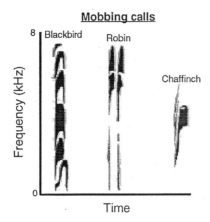

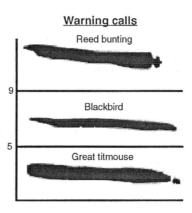

Figure 2.2
Structural differences in the acoustic morphology of avian mobbing and warning calls. (Redrawn from Marler 1957).

predator as opposed to when they are warning group members about the presence of a predator. In the former case, commonly called mobbing calls, the function of the signal is to recruit help in chasing the predator away. Consequently, the acoustic features of the system are designed to maximize localizability. In contrast, the function of warning calls is to alert group members to danger without alerting the predator. As a result, the design features of the warning call make localization by the predator difficult. Studies conducted with birds of prey have demonstrated the relative difficulty of localizing mobbing as opposed to warning calls (C. H. Brown 1982; Klump and Shalter 1984).

In addition to his functional analyses, Marler also made a number of insightful observations with regard to interspecific differences in the structural features of animal signals. One of the more significant suggestions was his proposal that communication systems could be distinguished on the basis of whether signals within the repertoire were graded or discrete. A repertoire is graded if signal variation is continuous, lacking clear acoustic boundaries for demarcating one signal type from another. Such acoustic continuity, however, can be cut up by the perceptual system, creating categorically distinctive classes of sound (reviewed in Harnad 1987). Discrete repertoires consist of featurally bounded signals with no intermediates between signal types. Viewed in this way, graded repertoires associated with perceptual systems that can partition the acoustic stream into functionally meaningful categories would have the potential to encode considerably greater amounts of information than discrete repertoires. Table 2.4 summa-

Table 2.4
Summary of the Social Organization, Territoriality, and Vocal Behavior of Some African and Asian Primates (modified from Marler 1975, 1976)

Species	Group Size	Terrestrial	Territorial	Communication Emphasis*	Sexual Dimorphism	Vocal Repertoire
Blue monkey	Small	No	Yes	bg	Yes	Discrete
Red-tailed monkey	Small	No	Yes	bg	Yes	Discrete
Vervet monkey	Small	Yes	Yes	bg + wg	Yes	Discrete
Black and white colobus	Small	No	Yes	bg + wg	Yes	Mixed
Red colobus	Large	No	No	wg	No	Graded
Japanese macaque	Large	Yes	No	wg	No	Graded
Chimpanzee	Large	Yes	No	wg	No	Mixed

*bg = between group; wg = within group.

rizes some of Marler's (1975, 1976) early classifications of nonhuman primates with respect to the graded-discrete distinction.

Although the discrete-graded distinction was laid out dichotomously, Marler actually viewed it as a continuum, allowing for the possibility of meaningful within-category variation. In fact, such within-category variation was the focus of one of his student's investigations of vocal communication among wild Japanese macaques (*Macaca fuscata*). S. Green (1975a) showed (Figure 2.3) that within the category of calls known as coos, there were at least seven acoustically distinct variants, each variant associated with a different social context. Subsequent experimental studies have revealed that such subtle acoustic differences are perceptually meaningful to the animals (reviewed in Stebbins and Sommers 1992).

There are some problems with the discrete-graded distinction and, in particular, with Marler's classification of primates into these categorical bins. The problems are due to unsatisfactory methods for testing his hypothesis and, more importantly, a general lack of quantitative data on both the acoustic properties of primate vocal repertoires and the perceptual salience of critical acoustic features *defining* call types or classes (see discussion in chapter 3). For example, without a clear understanding of the most fundamental (lowest) unit of communication within a species repertoire it is difficult to assess whether the repertoire is graded, discrete, or some combination of the two. Obtaining such information will depend on three types of data: (1) broad sampling of the vocal repertoire (i.e., different socioecological contexts and age-sex classes of callers), thereby facilitating quantification of acoustic variation; (2) quantitative acoustic analyses to define

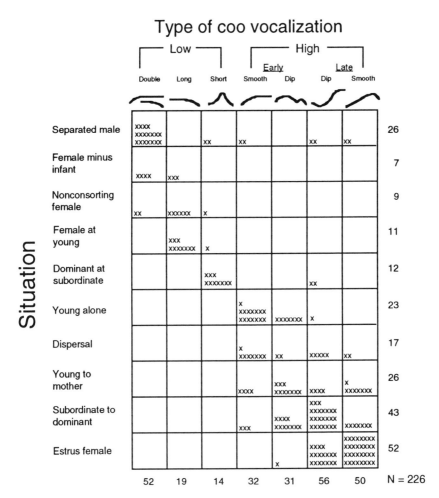

Figure 2.3
The relationship between social context or situation and the acoustic morphology of the coo's pitch contour (modified from S. Green 1975a). Each cell represents the number of cases (i.e., *x*'s) where a coo of a particular contour was produced in a particular situation.

some of the statistically significant properties of repertoire morphology; and (3) data collected from types 1 and 2 that will facilitate the implementation of experiments designed to determine what the organism considers perceptually meaningful.

Two sets of data have influenced Marler's thinking on the relationship between human language and nonhuman animal communication: developmental studies of birdsong and the meaning of avian and primate calls. In a landmark paper, Marler (1970a) pinpointed seven parallels between birdsong and human speech (these themes are developed in greater detail in chapters 4, 5, and 7):

1. Young learn the species-typical repertoire from adult models.

2. Dialects are formed as a result of learning.

3. Experientially guided learning is most significant during a critical period.[15]

4. To develop a normal vocal repertoire, young must be able to hear sounds from their species-typical repertoire and to hear themselves reproduce such sounds.

5. Like human infants, young birds also go through a series of developmental stages, including a subsong phase that resembles babbling.

6. Vocal imitation, in and of itself, may be self-reinforcing.

7. The left hemisphere is dominant for the control of sound production.

These interspecific similarities are obviously homoplasies given the phylogenetic distance between birds and humans. What is interesting about these findings, from a historical perspective, is that they are one of the few data sets on nonhuman animal communication that appear to have had a significant effect on the way in which some researchers studying human language think about their problems. For example, Marler's ideas on the "instinct to learn" and "sensitive periods" (Marler 1989) have influenced developmental psycholinguists such as Kuhl (1989) and Newport (J. S. Johnson and Newport 1991; Newport 1991). This impact is important, for it has provided the avenue for productive interactions between biologists, psychologists, and linguists (Marler and Terrace 1984).

Marler's interest in the problem of call meaning can be traced back to the early 1960s. In one of the first formal treatments of animal communication, Marler (1961) proposed that zoologists had much to learn from linguists and semioticians, and suggested that the frameworks laid out by researchers such as Cherry (1957),

---

15. Although Marler referred to the developmental window of opportunity as a critical period, his later writings reflect a greater appreciation for the plasticity of this window, and consequently we shift conceptually from critical to "sensitive" periods. These ideas are picked up in chapter 5.

Ogden and Richards (1923), and Morris (1946) on the nature of signs and pragmatics were ripe for the picking. What was encouraging about these approaches is that they ensured "from the outset that studies of communication systems shall not be restricted to the languages of man" (Marler 1961, 295). Marler saw, at the time, two general areas of inquiry: (1) to assess whether the communicative signals of animals were purposive[16] and (2) to determine the informational content of the signal expressed. In modern parlance, do nonhuman animals intend to signal to others, and what is the semantic content of their signals?

To address these problems, Marler and his students have conducted observations and experiments on wild primates and, most recently, domestic chickens (C. S. Evans, L. Evans, and Marler 1994; Marler, Evans, and Hauser 1992). The earliest work on this topic was inspired by Struhsaker's (1967) observations of East African vervet monkeys (*Cercopithecus aethiops*), in particular their alarm call system. Questioning the commonly held view[17] that animal signals merely reflect changes in affective state (emotional or motivational state), Marler (1978) suggested that the vervet monkey's alarm call system represented a potential case of symbolic signaling. His initial reasoning was based on the fact that acoustically distinctive alarm calls were produced in response to different classes of predator. Seeing a predator or hearing a particular alarm call elicited a specific escape response, and one that appeared to be designed to maximize the probability of escape given the predator's hunting strategy. For example, vervets ran under bushes upon seeing an eagle, whereas they ran up into trees upon seeing a leopard; eagles can take vervet monkeys out of trees, but leopards cannot. The definitive test of this hypothesis was carried out several years later when Seyfarth, Cheney, and Marler (1980a, 1980b) conducted playback experiments showing that the acoustic features of each alarm call type were sufficient to elicit the behaviorally appropriate escape response. These results, Marler (1985) argued, provided evidence that vervet alarm calls conveyed referential information about the external environment (e.g., the type of predator encountered).

What is perhaps most interesting about Marler's thoughts on the meaning of nonhuman animal vocalizations is how they have changed over the past thirty

16. In this early analysis, we can already see that Marler had begun to think about how the notion of purposiveness, or intentionality, could be examined in animals by looking at the relationship between signal production and social context or audience. This theme is one of the major foci of his current research, and one that I explore in greater detail in chapter 7.

17. Interestingly, Struhsaker's work on vervets, as well as many of Marler's suggestions of referential signaling, emerged at approximately the same time that the Gardners (Gardner, Gardner, and Van Cantfort 1989) began discussing the referential and symbolic abilities of their chimpanzee Washoe.

years. Thus, in his early writings (Marler 1961, 1982; Marler, Karakashian, and Gyger 1991), animal signals were considered to reflect sophisticated cognitive processes, including the capacity for representation (Marler 1985) and deception (Gyger and Marler 1988). Thus we find the following kinds of statements (italicized words have been added) made in connection with discussion of vervet monkey alarm calls and rhesus monkey screams (Gouzoules, Gouzoules, and Marler 1984):

They *represent* the kinds of precautions you would expect them to take if they understood the nature of the threat implied by that which each call type *symbolizes*.   (Marler 1982, 88)

Each call appears to *represent* to other listening monkeys not only the severity of the fight but also the class of opponent present, as defined by rank and kin relationships to the caller. We conclude that the calls of non-human primates reflect both affective and *cognitive* processes. The *signs* of these two kinds of process are often intermingled and conveyed by different acoustic vehicles, *as in human speech*.   (Marler 1985, 211)

More recently, Marler's experimental approach and writings on the topic of signal meaning have moved away from high-level cognitive interpretations and toward more low-level perceptual mechanisms linked to the particular stimulus features of the environment (C. S. Evans, L. Evans, and Marler 1994; Marler, Evans, and Hauser 1992). Specifically, we see a shift from looking at representations to looking at the precise stimulus features that elicit alarm and food calling, and from more loosely controlled experimental conditions carried out with chickens in seminaturalistic enclosures (Gyger and Marler 1988; Marler, Dufty, and Pickert 1986a, 1986b; Marler, Karakashian, and Gyger 1991) to more controlled laboratory conditions where live chickens have been supplanted by digital video playbacks of recorded chickens (C. S. Evans, L. Evans, and Marler 1994; C. S. Evans and Marler 1992, 1994).

In summary, therefore, Marler's views on the evolution of human language are relevant to the problems of homoplasy and homology. Regarding the former, it is clear that the developmental mechanisms leading to the acquisition of an adult, species-typical repertoire are similar for passerine songbirds and human infants: both species control vocalization by auditory feedback, critical for vocal learning. What is less clear is why, from a functional perspective, such mechanisms have evolved in these species and not closely related ones (e.g., nonpasserines or nonhuman primates). In terms of homology, Marler would argue that nonhuman primates produce signals that have rudimentary referential properties and that in some species one sees a hint of syntactic potential. Unlike his research on birdsong, which has had a significant effect on other disciplines such as developmental psychology and psycholinguistics, his research on call meaning has, thus far, made a smaller impact. For most linguists, it appears that claims about meaning and reference in nonhumans are so weak relative to the expressive power of human language that the data are not pertinent to studies of language evolution or use (Bickerton 1990; Lieberman 1991; Pinker 1994a). The critical question is, What would constitute convincing evidence of evolutionary precursors to human language in a species that can at least be conceived as a reasonable approximation to our direct ancestor? Chapters 4 to 7 provide some of the relevant data, and chapter 8 provides a potential framework for answering the question more definitively.

2.4.3   W. John Smith

Like Marler, W. John Smith has focused considerable attention on the structure-function relationship of animal signals. An important distinction in his earliest writings was between *message* and *meaning*. Concerning the former, Smith (1977) stated:

Most display messages make the behavior of the communicator to some degree more predictable by the recipient of the message. . . . When we speak of the messages of a display we mean the information available to an individual as a result of having received just the display; all other sources of information are considered contextual. . . . Because most messages indicate some selection within the behavioral repertoire of the communicator, an investigator recognizes each message by the particular class of behavior consistently correlated with the displays that encode it.   (pp. 145–146)

The meaning of a signal, for Smith, was closer to the notion of function and intimately tied to the receiver's response (W. J. Smith 1977, 262):

Because ethologists study behavior, they can take the position that a display has meaning to a recipient in some event to the extent that they see him base his selection of a response on the display and on information from other sources contextual to it.

Based on this theoretical perspective, Smith proposed an analysis of the structure of nonhuman animal communication that differed from Marler's (1955, 1961). He suggested that animal communication systems could be characterized by seven features:

1. Repertoires are limited to a small number of distinctive displays (15 to 45).
2. These displays break down into a small number of unique messages.
3. Understanding the meaning of a message depends critically on its context.

4. There are twelve messages that refer to identification, probability (i.e., the probability of a behavioral act occurring), general set (i.e., displays that are associated with a number of different activities), locomotion, attack, escape, non-agonistic subset, association, bond-limited subset, play, copulation, frustation.

5. Because the set of messages is limited, each one carries a significant burden in terms of conveying information about current events.

6. Most species are expected to show a significant positive correlation between the number of types of social interaction and the diversity of displays within the repertoire with different messages.

7. Most of the information conveyed concerns the signaler's motivational or emotional state. Signals do not refer to objects and events in the external environment.

Smith's analysis raises several important issues relevant to current discussions of language evolution and, more generally, the design features of animal communication systems. The first—one that was glossed by some behavioral ecologists[18]—is that communication is an interactional process involving sender and perceiver. As Smith and several other ethologists have pointed out, it is important to recognize the design features of the signal in light of the design features of the perceptual mechanism for receiving the signal. This coevolutionary process has been a focus of interest in animal communication studies, as well as in studies of human language (Liberman 1992; Liberman and Mattingly 1985), and has received considerable attention over the past ten years (Mattingly and Studdert-Kennedy 1991; Ryan and Wilczynski 1988) because of theoretical and methodological advances within evolutionary biology (especially techniques for constructing phylogenies) and neurobiology.

Smith's perspective not only influenced some of the earliest studies of animal communication, but also influenced how others outside of ethology thought about the structure of animal communication. In particular, based on Smith's synthetic writings, many psychologists and linguists such as Premack, Lieberman,

18. Due to criticisms, Krebs and Dawkins (1984) revised their earlier views on animal communication so as to incorporate the role of the receiver more fully. Nonetheless, several recent papers by evolutionary biologists (e.g., Guilford and Dawkins 1991) seem to pass over the *historical* fact that work on human and nonhuman animal communication over the past thirty years has focused a great deal of attention on the perceptual mechanisms guiding and constraining the signaling system (e.g., C. H. Brown and Waser 1988; C. H. Brown 1989a, 1989b; C. H. Brown and Gomez 1992; Gerhardt 1992a; Liberman and Mattingly 1985; Marler 1955; Marler 1961; Mattingly and Studdert-Kennedy 1991; Morton 1975, 1982).

Bickerton, and Pinker have continued to pass on the view that, for nonhuman animals, signals merely reflect changes in affective state, and repertoire size is limited to a few (<25) functionally distinctive signals (reviewed in Marler 1992a). As research by Marler and many others (Cheney and Seyfarth 1990; C. S. Evans, L. Evans, and Marler 1994; C. S. Evans and Marler 1994; Hauser 1993a; Snowdon 1990) has revealed, there is a dire need to revamp Smith's position, especially the way others interpret this position. In brief, though nonhuman animals clearly convey important information about their affective states, they also appear to convey information about objects and events in the external environment. Additionally, and because of the ability to convey information about changes in the affective state and properties of their external environment, repertoire size is unlikely to be as limited as Smith proposed. However, to repeat the broken record of the past few pages, the size of the repertoire can only be quantified when we have a better sense of the perceptually meaningful units within the vocal repertoire.

2.5 Synthesis

Historical treatments of science can provide both an accounting of facts and a record of the causal agents of change. In this chapter I have provided a discussion of each of these topics with the hope that interested readers who are new to the problem will find a useful introduction to the primary ideas from the past and present. For the expert, I hope that my historical treatment has provided a fair account of the key players, their theoretical views, and what I consider to be some of the strengths and weaknesses of the different perspectives.

Although this book focuses on the evolution of communication, I have contrasted historical ideas related exclusively to human language with those dealing with communication systems more generally. The reason for treating these separately is both historical and theoretical. Historically, researchers in linguistics, psychology, and anthropology have long been interested in the evolution of language. Biologists have not. Rather, biologists have tended to focus on communication in nonhuman animals, treating human language as a special case. What I have argued in this chapter, and will develop more formally here, is that human communication, like other communication systems, has its own suite of design features (Pinker 1994a, 1995). The interesting problem lies in understanding the function of each system's design features and the selective pressures that led to such functional and structural organization. The theoretical point stems from the distinction that linguists wish to make between ''language'' as a formal symbolic structure organized around syntax and "language" as a communicative tool. For those interested in the syntactic structure of language, evolutionary issues are typically moot. For those interested in language as communication, evolutionary issues are essential. In this summary, I extract the essence of the historical developments and attempt to reconfigure the various perspectives into a more coherent theoretical picture.

A starting claim of all theoretical treatments of language evolution is that humans possess a system of communication that is fundamentally different from that of all other communication systems in the natural world. This statement, though true, is less interesting in my opinion than the question of *why* human communication exhibits design features that differ from those of other animals. Needless to say, communication systems can be characterized by an enormous set of potential features, and depending upon one's theoretical bent, one will find some features more interesting and important than others. However, a point that I stress in this book is that a rich description of a communication system, including human language, will only be obtained when we have addressed problems related to phylogeny, ontogeny, proximate causation, and ultimate consequences—Tinbergen's four questions. Unfortunately, these different problems have often been addressed as if they could be ranked in terms of their relative importance. This approach has not only created extensive rifts between researchers, but has also fostered a sense of false security with regard to the depth of our understanding of communication systems and their evolution. Thus, although we can delight in the fact that studies of insect communication have been able to precisely determine the neural circuitry underlying acoustic communication (Hoy 1992; Michelsen 1992; Römer 1992), we know somewhat less about whether selection designed the system for maximizing the reproductive opportunities of males and females during mating interactions, or whether the acoustic system was designed as a general-purpose machine that can solve multiple problems in the auditory domain (Bailey 1991). Similarly, though we know a vast amount about the factors guiding the production and perception of human speech, it is not yet clear why, over evolutionary time, selection favored quite dramatic alterations in the morphology subserving speech production, but favored extraordinary conservatism with regard to the mechanisms underlying speech perception. Taking an integrative approach to the study of communication will greatly enhance our ability to address such issues and many more.

For linguists following in the Chomskyan tradition, communication and language must be discussed separately. Thus language represents the output of a specific neural module, designed to carry out combinatorial manipulations of fundamental syntactic units. Communication, on the other hand, includes what we do with our syntactically structured utterances, in addition to what we do with the pitch of our voice, the position of our body, the expressions of our face, and so on. The information that I have reviewed in this chapter, as well as the observations that I will discuss in the following chapters, are irrelevant to the formal study of language that has been, and continues to be, of utmost importance to Chomsky, his students, and critics. In contrast, the material covered in this book *is* relevant to those who have been interested in the evolution of human communication, and here the syntactic structure of human communication is important. Thus, in the remainder of this book, I will focus on the communicative properties of human language, since these are directly relevant to problems in evolutionary biology.

Before I attempt to reconcile and synthesize some of the different perspectives sketched in this chapter, let me confront a problem that at least some readers may be asking themselves about evolutionary theory and its importance to studies of human behavior. First, for those who are simply interested in what humans do, how they do it, and why they do it, it is possible to make the argument that evolutionary issues are moot. At some level, I buy this argument. If one wants to know, for example, why a child's ability to discriminate nonnative speech contrasts disappears at approximately one year, then one can address the question without even thinking about the evolution of this perceptual mechanism. However, I personally feel that this is a narrow perspective, and one that can be greatly enriched by thinking about the history of evolutionary pressures and the capacities of other organisms. Second, if one is interested in evolution, then punctuating one's argument with a conclusion such as "Humans have syntax and nonhumans don't" simply leaves open the more interesting question of why such differences exist, why humans took the evolutionary path they did, and what advantages they reap from having a communicative architecture that other species lack. Third, although studies of genetic and anatomical evolution allow for fairly robust claims regarding the presence or absence of a particular trait in a given taxonomic group, similar claims are more difficult to substantiate for behavioral or cognitive traits. This difficulty should not, however, completely stifle our research efforts. Unlike genes and anatomy, comparative studies of behavior and cognition are often forced into the position of stating that a given trait is absent under a certain set of conditions. Such a position leaves the critical reader with the feeling that there is always an easy out in terms of defending against claims of absence: one can state, with confidence, that given the *right* conditions, the trait will be expressed. The defense is weak, but the point has some validity, especially in light of current attempts to understand cognitive evolution. As discussed in chapter 1, comparative studies of behavior and cognition require a battery of methodological approaches, because analyses that rely exclusively on single techniques are vulnerable to the criticism that apparent failures in ability are more properly interpreted as failures in performance. Several techniques *are*, I believe, available. The path is thus paved for studying cognitive evolution and for creating a phylogeny of cognitive and communicative abilities.

Returning to language and the linguistic approach, some researchers have taken an explicitly evolutionary perspective, detailing both how certain features of language evolved and what sorts of evolutionary pressures may have been responsible for such changes. Lieberman, for instance, has emphasized the importance of evolutionary changes in the vocal tract, suggesting that selection has favored a morphological alteration that enhances communicative efficiency at the cost of digestive efficiency. Pinker, in contrast, focuses on the structure of language, especially syntax, and argues that natural selection is the only mechanism that could have led to its complex design features. The views espoused by Lieberman and Pinker, though Darwinian in form, represent two opposing positions within current studies of human behavior. Specifically, those who consider themselves Darwinian anthropologists (e.g., Richard Alexander, Laura Betzig, Monique Burgerhoff-Mulder) argue that evolutionary studies of human behavior require an explicit demonstration that traits considered adaptive yield fitness payoffs (e.g., behavior A leads to more offspring than behavior B). The other perspective, associated with the Darwinian or evolutionary psychologists (e.g., Leda Cosmides, Donald Symons, John Tooby), suggests that traits should be considered adaptive if selection favored the particular design features of the trait in the environment of evolutionary adaptedness, or EEA. Under this distinction, Lieberman is a Darwinian anthropologist, and Pinker is a Darwinian psychologist. Although these positions are often cast as alternatives, they strike me as perspectives addressing different problems and requiring different sorts of evidence. Thus, for Lieberman's argument to work, we need evidence that the payoffs associated with a communicatively efficient vocal tract outweigh the costs associated with decreased digestive efficiency. Additionally, and as pointed out more explicitly in chapter 4, we would require evidence that the structure of speech depends critically on the structure of the vocal tract. Although there is reasonable evidence for the communicative-digestive efficiency trade-off, it is less clear that speech *requires* the vocal tract of modern humans. That is, if humans had a vocal tract with, for example, a smaller tongue, less curvature in the oropharyngeal cavity, and a larynx placed higher in the neck, the sounds produced would certainly be different, but might nonetheless be sufficient for communication, especially linguistically structured communication. This proposal is based, in part, on the observation that there is generally an arbitrary association between sound structure and meaning. The sounds that humans use to construct their lexicon, though constrained by a species-typical vocal apparatus, gain their meaning through the process of learning. For example, whereas American children of the sixties and seventies called things they liked *groovy* or *cool,* children of the nineties have landed on terms such as *fresh, phat,* and *macdaddy.* The driving force behind such rich expression is the human brain, capable of generating a rich representational tapestry, output in the form of words as well as signs. I develop the ideas underlying the problem of sound-meaning associations more fully in later chapters.

Pinker's argument can be looked at in two ways. On the one hand, we have his clearly articulated position that natural selection is the only evolutionary mechanism that can account for the complex design features of human language. This claim is completely consistent with Darwin and modern evolutionary theory. On the other hand, we have the less explicitly articulated claim that the current design of human language reflects earlier selective pressures on our most direct and common ancestors, the early hominids. Specifically, at some point in our evolutionary past, there were socioecological pressures that pushed a fundamental change in the design features of hominid communication (or the neural substrate guiding such communication), and such changes were selected for because of their effects on survival and mating success. Darwin's (1871) thoughts are relevant here:

When we treat of sexual selection, we shall see that primeval man, or rather some early progenitor of man, probably used his voice largely, as does one of the gibbon-apes at the present day, in producing true musical cadences, that is in singing; we may conclude from a widely spread analogy that this power would have been especially exerted during the courtship of the sexes, serving to express various emotions, as love, jealousy, triumph, and serving a challenge to their rivals.   (p. 56)

Though I am sympathetic to Pinker's argument that selection seems to be the only reasonable mechanism for explaining the design of human language, what is not yet clear is what the precise selective pressures were like during the EEA that led to the expression of language. It is my hunch, and one that I will pursue in this book, that we may gain insights into this problem by establishing how the socioecology of extant primates and extinct hominids differed and how such differences led to alternative evolutionary pathways for communicating with conspecifics. For example, is it the case, as Dunbar (1993) has argued, that increases in group size within the Primates led to strong selection pressure for a mechanism that would facilitate communicative coordination and interaction among a large number of individuals whose personalities, interests, ranks, and abilities change dynamically over time? Or, were there specific ecological changes that led to selection for an explosive lexicon, designed to describe the dynamically changing features of the environment? Or, were social *and* ecological pressures involved, resulting in a selective advantage for those individuals who were able to coordinate social interactions and acquire greater access to rich resources? These questions may be difficult to answer, but my guess is that comparative studies of communication are likely to provide the richest empirical avenue.

Ethologists and behavioral ecologists who have thought about language evolution have generally focused on the extent to which the signals produced by nonhuman animals are *like* human signals, particularly words and sentences. Concerning words, such researchers as Marler, Cheney, and Seyfarth have looked both at the referential properties of nonhuman animal vocalizations in addition to the affective (e.g., emotions) and psychological (e.g., intentional states) factors leading to their production. The literature suggests (reviewed in chapter 7) that in some nonhuman primates and some birds, individuals produce vocalizations that are, at least rudimentarily, like human words. These results are exciting, for they advance the sophistication of nonhuman animal systems from mere expressions of affective state to at least coarse-grained expressions of external events. Nonetheless, current work in this area is limited in three significant ways. First, we know extremely little about the psychological constraints on generating referential signals. Studies of child language acquisition tell us that sophisticated word use emerges along with sophisticated development of other cognitive abilities, including the ability to recognize mental states in others and to take advantage of rich conceptual representations. To address these sorts of issues in nonhuman animals, we will require studies that clarify their cognitive limitations and abilities, together with studies that detail the precise meaning of communicative signals. Second, it remains unclear why selection would favor a more specific referential system than currently exists in nonhuman animals. What advantage would obtain from the ability to succinctly describe events in the world, both those currently experienced and those experienced in the past and stored in memory? Honeybees, for example, clearly profit from the ability to first store complex representations of food location and quality and then subsequently produce signals that convey such information, displaced in space and time. But do animals living in cohesive social groups, where individual knowledge is, commonly, group knowledge, require this ability? And even if reproductive and survivorship benefits could be gained from the ability to communicate about displaced events, the appropriate selection pressures may not have been present in the evolutionary history of the organism. Determining the conditions under which an explosive and referentially precise lexicon are favored will ultimately shed light on the evolution of our own communication system. Third, understanding the meaning of nonhuman signals will require a deeper understanding of the fundamental units of their communication system. As previously discussed, researchers have often made the assumption that the "call" or "song" represents the most significant unit within the repertoire. But such assumptions are generally not substantiated by the relevant perceptual tests, designed to assess what individuals perceive as meaningful units. Until such experiments have been conducted, however, we will remain in the dark on a number of interesting problems, especially the problem of syntactically governed organization of elements within the repertoire.

In summary, ethologists and behavioral ecologists need to devote as much energy to thinking about the similarities between human and nonhuman animal communication as about the causes underlying observed differences. For those interested in language evolution, there has been a tendency to search for similarities and then, upon finding differences, bow the head, put the tail between the legs, and terminate the research. And yet, those who are fundamentally interested in evolution should be as delighted about similarities among taxonomic groups as they are about differences. The process of evolution leads to a branching of possibilities, and the causes of branching as well as the lengths of the limbs (e.g., how long a trait remains unchanged) are of considerable importance. By facing this research challenge head-on, we will be able to construct a phylogeny of communication systems in much the same way that we can create a phylogeny of hands or blood types.

Ethologists and behavioral ecologists interested in the design features of communication have taken a variety of approaches, extending from mechanistic analyses of the nervous system (e.g., neuroethological investigations into the releasing mechanisms of signal generation and perception) to studies of adaptation (e.g., the relationship between signal structure and measures of survival and reproductive success). In the 1950s and 1960s researchers focused on the origins and subsequent evolution of animal signals, arguing that vocalizations and visual displays were designed to convey information to relevant perceivers. In the mid-to-late 1970s, this view was challenged on several counts. Some, like Dawkins, Krebs, and Maynard Smith, argued that selection should favor individuals who play poker, concealing their true intentional states and manipulating perceivers to their own reproductive benefits. In contrast to the early ethologists, therefore, the behavioral ecologists promoted a perspective that placed dishonest manipulative signaling at its core. At about the same time, Zahavi championed a different framework, but one that was also driven by an interest in function and adaptation. Specifically, Zahavi claimed that honest and dishonest signals were both possible, but that only individuals in relatively good condition would be able to tolerate the costs of producing some signals. As such, honest signals would be relatively costly to produce and perceivers would be selected to recognize the relationship between the signaler's current condition and the costs associated with generating a signal.

At present, there seems to be evidence for all three perspectives, and I see at least two reasons for such evidence to exist. First, the relative honesty and informational content of a signal will depend critically on the socioecology of the species investigated, in addition to the socioecology of the particular interaction under study. Thus, under some conditions, providing honest information will be favored even when the actual costs of production are relatively low. For example, in species where vocalizations are used to form defensive coalitions or to reconcile following an aggressive interaction, the vocalization itself may not be costly to produce, though the payoffs of honest signaling of intent are high. In contrast, where the stakes are potentially high, as occurs during fights over valued resources, there may be strong selection for hiding the truth, or at least stretching it a bit. Additionally, the particular details of the signaling environment will be influenced by the social structure of the group which, in turn, will influence the frequency and quality of subsequent interactions with the same and different individuals. The dynamics of future interactions are important to consider, especially in situations where individuals have the capacity to store information about the details of prior interactions and use this to guide subsequent interactions. Second, for all three theoretical perspectives, the primary issue is to determine the conditions that favor information exchange. Thus early ethological research examined how simple intention movements were shaped (ritualized) into signals carrying quite explicit information. Such ritualized signals were designed to facilitate conflict resolution between competitors fighting over resources. Most researchers working at this time recognized the fact that individuals would often be placed in a state of motivational conflict, and this might lead to informationally ambiguous signals. Behavioral ecologists, such as Krebs and Dawkins, described several conditions under which signals of intent would be selected against and manipulative signaling favored. Using a comparable evolutionary framework, Zahavi generated some predictions for when honest signals would be favored, focusing in particular on signal costs and signaler condition. In sum, therefore, the history of research on animal communication leaves us with an important message: the structure and function of a signal will be influenced by the communicative context, the physical condition of the signaler and intended signal recipient, the costs of signal production, and the payoffs associated with particular outcomes from the communicative interaction.

In my opinion, the different views espoused by linguists, psycholinguists, ethologists, and behavioral ecologists all provide important ingredients for the general picture that I wish to paint in this book. Put simply, all of the views have something to offer if we are interested in the evolution of communication, and the most elegant account will come from an integration of perspectives. But how? For brevity, and to encourage the reader to move on, let me give a cookbook version of how I envisage the research program, saving the meat of the argument for the later chapters. First, if we are interested in the design features of a species' communication system, we should start by providing a clear characterization of the structure and function of each feature. Such characterizations are important for both within- and between-species comparisons. Second, in examining the structure of a communicative feature, we should establish not only how it develops, but also under what circumstances (e.g., what are the necessary and sufficient environmental parameters for proper development?). Third, to assess the function of a particular feature, we need to consider the selective advantage to the signaler as well as to the perceiver, in addition to the possibility that what is biologically beneficial to the signaler may be biologically costly to the perceiver (an arms race of competing selection pressures). Furthermore, we should engage in these problems in the light of current thinking in evolutionary biology, especially how the feature contributes to the genetic fitness of the individual. Fourth, in considering how selection fine-tunes the featural set, we should develop an understanding of the kinds of constraints that might influence the pattern of evolutionary change. The constraints I am envisaging emerge from both phylogenetic considerations and more immediate issues related to underlying neural machinery. The next chapter provides some of the relevant conceptual tools for working through these problems.

3 Conceptual Issues in the Study of Communication

The goal of this chapter is to provide a discussion of some conceptual issues that are relevant to how we think about and study communication. The presentation is intentionally quite general so that the importance of each idea can be kept at the fore of our discussion, separate from the technical details of particular experimental or observational techniques. The breakdown of the chapter is as follows. In section 3.1, I look at issues of signal design in light of the ecology of the communicative channel. Here we see how environmental factors shape the design features of a species' communicative repertoire, both in terms of signal production and perception. Section 3.2 addresses the problems of similarity and classification, focusing especially on the fundamental units of communication. This is a problem of utmost importance, both because it sheds light on the design constraints of a particular communication system and because it represents one of the most significant methodological barriers to current progress in comparative studies of communication. Section 3.3 concludes the chapter and provides a glimpse of how an integration of Tinbergen's fundamental questions about causality can be brought to bear on the problem of communication, using nonhuman primates as a test case.

3.1 Signals Designed for a Complex Environment

3.1.1 The Ecology of Signal Transmission

The diversity of communication systems treated in this book are associated with a diversity of ecological conditions for communication. Thus honeybees use both visual and auditory signals to convey information about the location and quality of food, and in general, most of this information is conveyed and received in the darkness of the hive. During the mating season, frogs produce advertisement calls within densely populated leks situated, typically, in moist and heavily vegetated areas. Birds sing on perched areas to advertise their territorial possessions to neighbors and attract potential mates from afar. Bats, during prey pursuit, produce high-amplitude echolocation pulses and use the returning echo to both detect the prey species and maximize the probability of capture. For some bat species, prey are found in the air whereas for others they are found under the water's surface. And humans, sitting in a cafe perhaps, or within the sardine-like crowds of a Tokyo subway, attempt to channel private whispers to their friends amid a sometimes deafening backdrop of conversations from conspecific competitors.

To dissect the relevant details of an ecologically rooted analysis, let us start

with the most basic level—a communication system that consists of three fundamental features: (1) the source of the signal, (2) the channel for signal transmission, and (3) the potential pool of perceivers (Figure 3.1).

In this general formulation, although the sender may direct his or her signal at
a designated perceiver, rarely do auditory or visual signals and cues lend themselves to completely private channels of communication. Auditory and visual
signals and cues are readily parasitized by uninvited perceivers, both at the intra-
and interspecific level. For example, a frog may produce his advertisement call
to deter other male competitors and attract potential mates. Predatory bats care
little about what the frog intends to communicate and simply make use of the
auditory environment as a reliable source of information about their next meal
(Ryan 1985; Tuttle and Ryan 1981).

The morphology of a signal at its source will never be identical to the morphology perceived, because the transmission path from sender to potential perceiver
is associated with at least three general factors known to influence signal structure. As illustrated in Figure 3.2, the signal is altered at the moment of liberation
from the source as a result of spreading. Thus even highly directional sounds

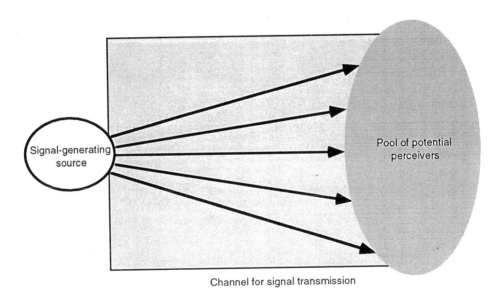

Figure 3.1
Three fundamental features of a communication system: the signal-generating source, the channel for
transmission, and the pool of potential perceivers.

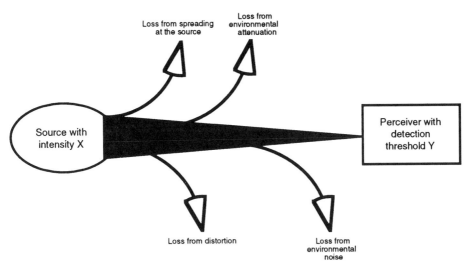

Figure 3.2
Factors causing signal loss during transmission from source to perceiver. Signal structure is influenced in three ways (indicated by curved arrows): (1) general spreading effects due to lack of complete directionality, (2) attenuation due to environmental influences (e.g., vegetation, climate), and (3) reduction in strength of signal due to background noise. The actual signal received will depend on these three sources of loss, in addition to the starting intensity of the signal (X) and the sensory threshold (Y) of the perceiving organism.

such as bat and dolphin sonar will experience diffusion effects (Au 1993). Once the signal has been launched, it is at the mercy[1] of the environment. In general, the vegetative and climatic components of the environment will cause significant alteration of both spectral and temporal properties of the signal, whereas biotic noise will cause an overall decrease in the signal-to-noise ratio (see Table 3.1). It is not uncommon to think of these effects as being most significant in the auditory domain where, for example, characteristics of the habitat can result in sound reverberation and attenuation (see below). Signals and cues produced in the visual domain, however, are also subject to environmental influences, and such factors have recently been the subject of studies by evolutionary biologists

1. Studies reviewed further on in this chapter, as well as in other chapters of the book, strongly suggest that selection has favored signals with particular design features, matched to achieve optimum transmission in the species-typical environment. Thus in C. H. Brown and Waser's (1984, 1988) elegant studies of nonhuman primate vocalizations, experimental results indicate that calls that function in intergroup interactions and require long-distance transmission are produced within a spectral range (i.e., the *sound window*) that minimizes attenuation.

working on visual communication systems. For example, in field and laboratory research by Endler and colleagues (Endler 1987; Endler 1991a; Endler 1993b; Houde and Endler 1990) on guppies (*Poecilia reticulata*), it has been demonstrated that male coloration represents a trade-off between conspicuousness to females (sexual selection pressure) and inconspicuousness to predators (natural selection pressure). Females prefer to mate with more conspicuously colored males, but such males are more likely to be seen by visually guided predators. From the perceiver's perspective, however, color is not a trait with absolutely fixed properties because changes of light color in the environment will affect such features as hue, brightness, and saturation. As a result, what may look like vivid red under one condition may appear to be pale brown under another condition. The scenario depicted in Figure 3.2 is therefore applicable to both auditory and visual signals and cues.

In any communication system, the sender will have some control over the intensity of the signal generated. Thus, during territorial encounters, a howler monkey might roar softly or at the maximum amplitude, augmented by the resonances of his oversized vocal sack (Schön Ybarra 1988; Whitehead 1987). Similarly, a red-winged blackbird might reveal all or only a small portion of his red epaulet when an intruder is detected (Metz and Weatherhead 1992). The range in signal intensity must, of course, correspond at some level with the perceptual abilities of the pool of perceivers; this is depicted in Figure 3.2 as the perceiver's detection threshold (Y). The threshold for detection will be influenced by the mechanics of the sensory system (e.g., what colors or frequencies can be detected, discriminated), the current motivational and attentional state of the organism, and the properties of the environment. Each of these factors is important. An individual may have the neurophysiological substrate required to discriminate small differences in frequency, but because of attentional distractors in the environment, may completely miss the signal conveyed. This possibility raises a general problem, discussed in greater detail further on. Studies of communication require knowledge of what *can* be discriminated under ideal conditions and what *is* discriminated and acted upon under natural conditions. The first analysis will typically take place under controlled laboratory conditions and will generate data on the upper and lower limits of the sensory system. The latter piece of information demands highly sensitive response assays to insure that the signal is, in fact, detected by the sensory system. The importance of a response assay is all the more important when behavioral measures are involved because one needs to distinguish between an individual who fails to respond because the signal was

Table 3.1
Biotic and Physical Factors Influencing the Process and Evolution of Communication
(modified from Endler 1993b, 217)

Generation and Emission	Environmental Transmission	Reception and Processing	Response Decision Based on Reception
Biophysical limits to form and intensity	Background noise	Biophysical and bio-chemical limits	Other signals
Energetic limits	Interfering signals	Sensory adaptive state and attentiveness	Time wasted during choice
Biochemical limits	Attenuation and distortion	Need to be attentive	Reasons and need for choice
Energy storage	Blocking	Need for alerting signals	Predator risk
Timing and location (predation, seasonality, spatial location, time of day)	Reflection and refraction	Constraints on reception time	Parasite risk
Environmental constraints	Spectral and temporal properties	Noise	Quality of signal and its components
Physiological state	Information density	Jamming	Physiological state
Information content versus quality	Temperature	Signal reception rate	Signal channel use
Genetic variation	Self-interference	Pattern recognition needs	Genetic variation
Cultural patterns (history, regional dialects)	Timing and location		Cultural patterns (history, regional dialects)

never detected and an individual who detects the signal but chooses[2] to refrain from responding.

The previous discussion provided some general comments on how the environment sets up constraints on communication. But studies of environmental acoustics[3] have actually taken such issues much further, providing a suite of tools for

---

2. The word *chooses* is used here in two ways (see brief discussion in chapter 2). The first skirts the issue of high-level cognitive decision making and simply means that an individual has available to it a set of alternative actions given current environmental contingencies, and has selected one action over another. The second use of the term, most safely applied to normal human adults, is that the individual makes an intentional decision to respond in a particular way based on beliefs, desires, and conceptual representations. Intentional signaling in humans and nonhumans is covered in greater detail in chapter 7.

3. Some of the same principles apply to visual signals, though with the exception of work by Endler (1987, 1991b, 1993a), extremely little research has been carried out.

understanding how habitat structure and climatic factors force specific design limitations on signal structure. The following synopsis is therefore provided so that the reader can be aware of the complexity of ecological factors involved in communicative signaling. More complete treatments are provided in Dusenbery (1992), Gerhardt (1983), Wiley (1994), and Wiley and Richards (1982).

To set the stage, imagine two crickets, a male and female, sitting atop some vegetation (Figure 3.3). The male produces a courtship signal, directed at the female. Several questions arise: What vegetative factors will influence the structure of the signal from the point of emission to the point of reception? Can the signaler compensate in any way for vegetative influences, producing signals whose design features maximize the probability of preserving signal structure? What is the ideal height above the substrate for communicating? Does the perceiver evaluate the signal received relative to some template representing an unperturbed signal (McGregor and Krebs 1984b; E. S. Morton 1986)? Does the perceiver approach the communicative arena with certain kinds of expectations about environmental influences, as has been shown for humans and their expectations about echo effects (Clifton et al. 1994)? How is signal structure influenced by nonvegetative factors, such as climatic changes and the interaction with temporally varying calls of other species? To assess these questions, bioacousticians have performed experiments in the laboratory and under natural conditions (Bregman 1990; Brenowitz 1982; C. H. Brown 1989a; Marten, Quine, and Marler 1977; E. S. Morton 1975; Wasser and Brown 1984; Yost and Gourevitch 1987), and the yield is a rich description of how signal structure is transformed along the way to a potential perceiver (Table 3.2).

A common methodological approach in environmental acoustics involves the following three steps: First, record a signal under relatively ideal conditions or generate a computer-synthesized signal. Second, play the signal back under different ecological conditions. Third, record the signal played back and compare the change in acoustic morphology with the originally emitted signal. Signal comparison in the time domain can be achieved by using a cross-correlation function between the original and transmitted signal. If there is no signal distortion, then the value returned will be 1.0; any value less than 1.0 indicates distortion due to environmental effects. Comparisons in the frequency domain involve subtracting the Fourier transforms (a mathematical dissection of a complex signal into a set of simpler signals) of each signal's waveform (adjusted for amplitude differences). The result is a difference spectrum whose area can be derived and compared with that obtained for the original signal. These spectrotemporal techniques are described in detail by C. H. Brown and Gomez (1992), who have used this ap-

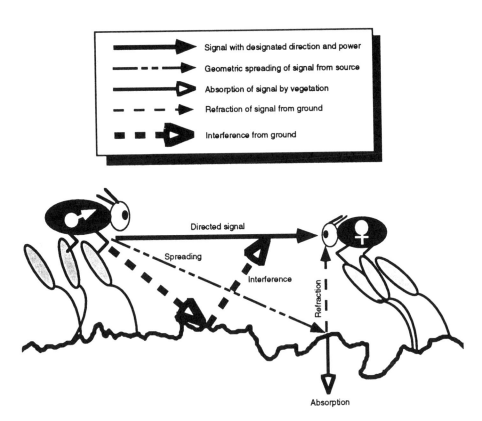

Figure 3.3
An illustration of the complex biotic factors influencing signal transmission in the auditory domain.
Here, the large male insect on the left has sent a signal to the smaller female insect on the right. As
described in the text, the structure of the signal received will be greatly altered from the structure
originally emitted. Such alterations, including spreading, absorption, and refraction, are due to properties of the vegetation and ground surface.

proach to investigate how forest and savanna environments influence the design
features of primate vocalizations.

Now imagine that you are riding the sound pressure wave of the signal emitted
from the male cricket to the female cricket, observing the changes in structure
along the way. At the point of emission, the signal experiences spreading effects,
and the primary consequence of spreading is loss of signal energy; you are now
riding on a much narrower wave. As a result, signal emission must attain an
energy threshold in order to insure that at least some form of the signal arrives at the perceiver. Unlike the effects of spreading, which always occur and

Table 3.2
Some Factors Influencing the Transmission of Auditory Signals

Factor	Definition
Absorption	An audio signal loses energy as a result of contact with a medium; the medium contacted may convert the signal's energy into another form (e.g., heat).
Attenuation	A decrease in the amplitude or intensity of an audio signal due to absorption, scattering, or distance from the source.
Diffraction	A part of the audio signal is redirected because of contact with an absorbing or reflecting medium.
Geometric spreading	Signals radiate in several directions from the source. That is, they are not perfectly directional from source to receiver. Omnidirectional radiation is associated with energy loss.
Interference	Signals are reflected from the substrate, to later interact with the originally transmitted signal.
Reflection	An audio signal is bounced back in the direction of the emitting structure as a result of contact with a reflective medium.
Refraction	The direction and speed of an audio signal are altered/perturbed by contact with a given medium or changes in climatic conditions (e.g., temperature gradients).
Reverberation	As a result of multiple scattering events, there is a temporal delay in the arrival of the signal, which is often perceived as an echo and appears as a blurring of pulses in a time-amplitude waveform.
Scattering	Because of contact with an object, the audio signal undergoes a complex, multidirectional change in the original transmission direction.

presumably can be compensated for to some extent by the sender in terms of signal design, the other kinds of transmission factors are less predictable (see Table 3.2). The lack of predictability means that it is more difficult for senders to optimize signal design, especially over relatively long distances. Thus, as you ride the wave to your intended listener, you will be riding a wave with somewhat uncertain physical properties. Nonetheless, data on insects (Römer and Bailey 1986), frogs (Littlejohn 1977), birds (Brenowitz 1982), and primates (C. H. Brown 1989a) have provided convincing evidence that the structural properties of the environment have a direct impact on the spectrotemporal design features of acoustic signals (for a thorough discussion of changes in signal acoustics as a function of environmental effects, see Piercy and Daigle 1991). The focus of most of this research has been to determine the active space or effective range of particular signals given ecological and climatic factors, and to determine whether the specific structure of signals within the repertoire of a species exhibits the kinds of design features that would maximize short- as opposed to long-range communication. In general, high-frequency signals tend to be attenuated more significantly by the environment than low-frequency signals (Marten, Quine, and Marler 1977; Wiley and Richards 1978). Apparently as a result, we find, for example, that many avian and mammalian species produce long-distance signals consisting of significant low-frequency components, whereas close-range signals are generally characterized by high-frequency components (Ryan and Brenowitz 1985). Some extraordinary examples of this phenomenon are the low-frequency signals of whales and elephants, used to coordinate group movement over miles and miles of water and land, respectively.

In summary, therefore, any consideration of signal design must consider the ecological context in which such signals evolved and the problems that they were designed to solve. Consequently, research on the structure of communicative signals must always be rooted in the natural world, and if possible, the species-typical environment. Once the research is so rooted, however, it is possible, and highly desirable, to take the problem into the laboratory, in order to fine-tune the analytical techniques and determine how the physical composition of the signaling channel alters the morphology of the call. This work in turn will set up the empirical foundation for assessing the range and causes of acoustic variation, and consequently the potential problems for perceivers listening in. I now turn to some issues associated with the ecology of perception.

3.1.2   The Ecology of Signal Detection

In section 3.1.1, I pointed out that perception of a signal depends on the perceiver's ability to detect and discriminate a signal, as well as his or her current attentional status. This relationship between discriminability and attention lies at the heart of signal detection theory. There are several excellent treatments of this theory for the nonexpert, especially as they relate to issues in this book (Dusenbery 1992; see Box 3.1), and thus I provide only a brief discussion here.

Imagine that you are walking through a tropical rain forest in Southeast Asia. The forest has a towering canopy that allows only small bursts of light to shine through to the floor. There are vines draped all over and a virtual kaleidoscope's worth of color from the trees that are in fruit, the avian community that decorates the area, and the reflections that emerge from the small rivers that meander throughout. In addition, there are buzzing insects, melodious songbirds, and various grunts and roars from the primate community. But what you see and hear will depend, to a large extent, on what you are looking for, your experience in detecting visual and auditory stimuli in such an environment, and the design features of your sensory system. If you are an ornithologist or bird enthusiast looking to ''tick'' one more species for your life list, then your ability to locate

Box 3.1
Signal Detection Theory

Much of signal detection theory (Poor 1988) was developed with the military in mind, and consequently the terminology typically connotes some aspect of warfare. A simple way to understand the issues involved and their relationship to communication is to imagine a species such as a gazelle that must avoid falling prey to a predatory cat such as a cheetah. To avoid being eaten, the gazelle has evolved an alarm call system (FitzGibbon 1989, 1990; FitzGibbon and Fanshawe 1988) that includes both auditory signals (e.g., snorting) and visual signals (e.g., stotting). From a signal detection perspective, the issue becomes: What is an appropriate level of vigilance for detecting predatory cats? The gazelle's vigilance system must be set so that it can remain alert to detect the cat directly or tap into the alarms of conspecifics. Time spent vigilant represents time away from other activities that are also critical to survival, such as feeding. The situation for gazelles and cheetah is illustrated in the following matrix:

<table>
<tr><td rowspan="2"></td><td rowspan="2"></td><td colspan="2" align="center">Cheetah/alarm is</td></tr>
<tr><td align="center">Present</td><td align="center">Absent</td></tr>
<tr><td rowspan="2" align="center">Gazelle's
response</td><td align="center">Escape</td><td align="center">*Hit*
(appropriate response)
Gazelle flees</td><td align="center">*False Alarm*
(inappropriate response)
Gazelle flees</td></tr>
<tr><td align="center">No escape</td><td align="center">*Miss*
(inappropriate response)
Gazelle continues eating</td><td align="center">*Correct Rejection*
(appropriate response)
Gazelle continues eating</td></tr>
</table>

In the case depicted, there are two appropriate responses (i.e., Hit and Correct Rejection) and two inappropriate responses (i.e., False Alarm and Miss). The mathematics of signal detection theory generates a series of probability curves that reveal the trade-off between false alarms and misses, given the signal-to-noise ratio in the environment. Using the terminology of signal detection, each curve reflects the relationship between the subject's criterion for detection (β) and his or her sensitivity (d).

In the present example, a miss is far more costly to the gazelle than a false alarm. A miss can lead to death, whereas a false alarm only leads to unnecessary energy expenditure and an additional time out from feeding. Thus the gazelle's optimum strategy requires a level of vigilance (i.e., sampling in statistical terms) that will maximize hits and minimize misses. False alarms are to be expected at some level, but should also be minimized because they represent an energetic cost.

The subject's sensitivity to the situation will tend to vary considerably. Thus, if a hungry cheetah has recently been seen moving within the home range area of a group of gazelles, each gazelle is likely to be more vigilant (i.e., more sensitive) than if a cheetah hasn't been seen in weeks. In general, if the level of sensitivity is high and the costs associated with a miss are significant, then false alarms may be expected to increase in frequency in the interest of efficiency (e.g., anything even remotely resembling a cheetah will elicit flight).

and discriminate a broad range of birds is highly developed, and you can most probably accomplish this detection task in both the auditory and visual domains. But if you are a tourist who has spent most of his life in New York City, your senses may be just as acute, but you are likely to miss a lot.

The fundamental problem in signal detection theory is to quantify a subject's optimum strategy for detecting a stimulus in the environment, focusing especially on detection under marginal conditions, where it is sometimes difficult to ascertain whether the signal is there or not. In essence, this determination involves an estimation of (1) the probability distribution of a signal being present given the level of environmental noise (i.e., the signal-to-noise ratio) and (2) the relative costs associated with the failure to detect a signal that is present as opposed to mistakenly concluding that a signal is present when it is not. These ideas are developed more completely in Box 3.1.

Signal detection theory has been used successfully to understand the psychophysics involved in categorization of auditory and visual signals. In laboratory psychophysics, an important goal is to map out the discriminability of stimuli within a defined stimulus space. In the auditory domain this task might entail comparisons of pure tones that differ in frequency or duration, whereas tests in the visual domain might involve comparisons between objects that vary along a brightness or size continuum. In each test, the experimenter attempts to determine the limits of discrimination by asking *how different* two stimuli must be in order for the subject to detect a difference. The quantitative measure obtained is known as the *just noticeable difference* or *JND*. For example, studies of humans demonstrate that frequency differences of only 0.1% can be discriminated for pure tones (i.e., signals with only one frequency) with a bandwidth less than 5 kilohertz (Moore 1988); above 5 kilohertz, discrimination requires larger differences in frequency (i.e., the JND increases in magnitude). Comparable data are presented by Fay (1992) in a thorough review of vertebrate sound discrimination.

A useful concept developed by Nelson and Marler (1990) based on their naturalistic experiments of birdsong is that of the *just meaningful difference* or *JMD*. The distinction between JND and JMD is simply the difference between what the perceptual system can resolve under ideal conditions—JNDs—and what the perceptual system chooses to recognize as a biologically significant difference—JMDs. And what the system responds to will depend critically upon the individual's current motivational and attentional state. Thus, under some conditions, information about individual identity may be important (e.g., whether a song is from a territorial neighbor or intruder), whereas under different conditions, information about identity might be irrelevant (e.g., alarm call to a predator).

Although the JMD concept originated from studies in the auditory domain (e.g., Nelson 1988; Gerhardt 1992a), it, like the JND, is modality independent and thus equally applicable to visual signals.

To clarify the JND-JMD distinction, imagine a passerine species whose song falls within the 1000–1500 Hz frequency range. Under laboratory conditions, members of this hypothetical species can discriminate songs that differ by 10 Hertz or more, even if the frequency of the song falls outside the species-typical range (i.e., >1500 Hz). For this species, then, 10 Hz represents the JND. Results from field playback experiments with songbirds, for which song frequency is an important factor in species identification, might indicate that males fail to produce a species-typical response (e.g., countersinging, aggressive attack) to a song if the frequency deviates from the natural norm for that species by 50 Hz or more. In this case, the JMD would be 50 Hz and would be bounded by the frequency range between 1000 and 1500 Hertz.

In general, to return to the issue raised at the end of section 3.1.1, studies of communication must design experiments to determine not only what is discriminable, but also what the organism considers to be meaningful discriminations to make and, especially, under what conditions. The laboratory is ideally suited for the first problem, for one can isolate the most critical variables and push the perceptual system to its most efficient state. The second problem can be conducted under both laboratory and field conditions. For example, laboratory studies can more readily control speaker-subject distances and thereby determine how sound amplitude influences response strength. In contrast, field studies can test the relative importance of spectral changes in song structure when an individual is on or off of its territory, an ecologically relevant situation.

3.1.3 Adaptation and Signal Design

In the previous subsections we discussed how ecological factors affect the design features of signal production and perception. The notion of design is, of course, importantly related to the notion of adaptation. Because these ideas surface repeatedly throughout the book, I would like to take a few pages to lay out some of the relevant issues. For many biologists, the study of adaptation or adaptive significance represents the most fundamental problem in biology (Mayr 1982). However, the notion of adaptation remains controversial, not only for biologists (S. J. Gould and Lewontin 1979), but for scholars in other disciplines as well (Kitcher 1986; see reviews in Lauder, Leroi, and Rose 1993; Reeve and Sherman 1993). The aim of this section is not to resolve the controversy but to highlight a few key problems and domains of disagreement.

As a general gloss, there have been two approaches to the problem of adaptation: historical and nonhistorical. For biologists interested in historical processes (e.g., paleontologists) three distinctive definitional criteria have been suggested:

1. A trait is an adaptation if and only if its current structure and function are similar to those that were originally selected for when the trait evolved. If the trait currently serves a different function, then it should be considered an exaptation (S. J. Gould and Vrba 1982).

2. A trait is an adaptation if it demonstrates shared-derived features from a recent ancestor that evolved under significant selection pressure (e.g., Harvey and Pagel 1991).

3. A trait is adaptive if it exhibits complex functional design features that reflect the tinkering of natural selection over a long period of time (e.g., G. C. Williams 1966; Dawkins 1986).

As Reeve and Sherman (1993) have pointed out, what is problematical about these historically laden definitions is that they depend on somewhat arbitrary determinations of where, in the evolutionary record, to look for the original function, and they require evidence of the history of selection pressures that, for some traits such as behavior, are impossible to obtain. To clarify, consider the illustration in Figure 3.4. In the top row, we have a scenario that fits perfectly with our first criterion, and the one championed by Gould and Vrba (1982). Thus our fictitious beast has two forward-reaching, straight gray tusks (A) at T0 that are maintained in their original form, and we will assume, with their original function, until time T1. Rows two and three represent relatively similar evolutionary patterns, but with slightly different temporal events. Under our first criterion, the organism's curved gray tusks (A1) at T1 are like those exhibited in the present, and thus we say that the trait is an adaptation, again assuming we know the original function and selection pressures. However, if we push the record back to T0, we find the straight gray tusks. Now, although the straight and curved tusks are structurally similar, they have different functions, and thus the curved tusks can no longer be considered an adaptation—*even if* they provide a selective advantage to those individuals that express this trait. The same is true for the scenario in row three, but with a more shallow temporal record. In row four, a structurally and functionally different trait evolves at T1—a corkscrew-type tusk emerging from the center of the upper lip (B)—and is preserved into the present. This new tusk is an adaptation if we go back as far as T1, but not if we move further to T0. And again, we must know something about the history of selection pressures.

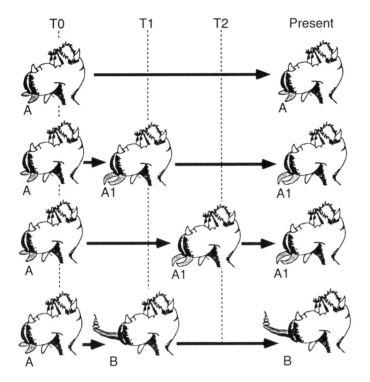

Figure 3.4
Different evolutionary scenarios for how a trait might change over time and how knowledge of the fossil record will influence our understanding of adaptive significance. In its present state, the hypothetical creature depicted has the original trait, two small emergent gray tusks from underneath the upper lip (A), or a corkscrewlike tusk emerging from the middle of the upper lip (B). A1 represents a slightly modified state from A, but with a new function; unlike B, however, A1 is not considered a new trait.

Although these comments are not meant to diminish the importance of historical processes or of considering traits in their proper phylogenetic perspective, they do underscore the difficulties associated with a definition of adaptation that requires identification of historical processes. In response to such problems, Reeve and Sherman (1993) propose an alternative definition that is more suitable for researchers interested in current phenotypes, why some do better than others and how selection pressures work in the present. Thus they state, ''An adaptation is a phenotypic variant that results in the highest fitness among a specified set of variants in a given environment'' (p. 9). Given the issues we will cover in the subsequent chapters of the book, this definition has three important implications. First, it states that we should be looking at the current design features of a communication system relative to other possible (theoretical or observed) features. More specifically, because natural selection nonrandomly sorts among phenotypes, the current phenotypic set will be represented by those individuals contributing the most genes to subsequent generations. Second, it allows for the possibility of identifying adaptations independently of the historical factors or mechanisms that led to its current expression. Third, although it does not deny the importance of complex design in identifying adaptations, it does not require complexity of design for classifying a trait as adaptive.

In summary, although I will primarily adhere to the Reeve and Sherman perspective in this book, studies of adaptive significance would ideally obtain the following information for a given trait: (1) a phylogenetic characterization of the group, derived from behavioral, morphological, and molecular data; (2) a thorough assessment of the phylogenetic tree so that homologies can be distinguished from homoplasies, and further, so that primitive as opposed to shared derived traits can be contrasted; (3) experimental and observational data on the relationship between different traits of interest to assess whether one trait is merely correlated with another or functionally independent; (4) documentation of the fitness consequences of the trait under different socioecological conditions.

3.2 Problems of Similarity and Classification

To respond to things appropriately, animals must identify them as belonging to various categories: potential mates, potential food items, potential predators, and so on. All sentient animals therefore simplify the world's diversity by imposing their own categorical distinctions upon it. People make more numerous, complex, and finely subdivided distinctions than other animals do because the languages we use allow us to do so. Language also turns many of our verbalized conceptual categories into shared social constructs, which we can consciously deliberate, criticize, and reformulate at will. (Cartmill 1994, 115)

Similarity plays a fundamental role in theories of knowledge and behavior. It serves as an organizing principle by which individuals classify objects, form concepts, and make generalizations. Indeed, the concept of similarity is ubiquitous in psychological theory. It underlies the accounts of stimulus and response generalization in learning, it is employed to explain errors of memory and pattern recognition, and it is central to the analysis of connotative meaning. (Tversky 1977, 327)

3.2.1 The Concept of Similarity

In philosophical studies of mind and language, the concept of *similarity* has played an extensive role, especially in terms of the work it does for a theory of concepts (Fodor 1994; Peacocke 1992). However, as Fodor has articulated, the concept of similarity is intimately tied up in the notion or concept of *identity*. Thus, we find discussions of Twin Earth, where water is made up of H_2O on one planet and XYZ on the other; of the morning star and evening star; of a fictitious land where an emerald is "grue" at time t, and "bleen" at time $t + 1$, forcing the inductive problem of assessing whether the emerald is the same or a different stone, whether it is like a green or a blue emerald, and so on; and lastly, of concerns over whether a ship where all the old planks have been replaced by new planks is, nonetheless, the same ship.

Central to these philosophical discussions is heated disagreement over whether one can actually get at an agreed-upon understanding of similarity (of the proposition or hypothesis of what counts as alike) given its apparently subjective flavor. It depends—doesn't it?—on qualia, and as such, depends on knowing what it is like to be the individual having such qualia. And this brings us to a fairly dominant position in philosophy, that there is no such thing that it is like to be another organism—human or nonhuman—because, quite simply, there is no such thing that it is like to have someone else's subjective experiences (Nagel 1974).

Although it may not be possible to specify what it is like to be someone else, to be precisely aware of what they are aware of, it is nonetheless possible to assess how others respond to various situations and as such, determine what they consider to be functionally similar entities.[4] And we absolutely need to tackle the problem of similarity for it takes on considerable importance in almost every discipline that we touch upon in this book: neurobiology, evolutionary biology, cognitive and developmental psychology, linguistics, and anthropology. Thus in neurobiology one presents visual or auditory stimuli and obtains, for example, firing rate data from single cells or "maps" of cortical activation from imaging tools such as PET and fMRI. Based on this output, one makes a statistical judgment about whether or not the stimuli are the same or different, and uses this judgment to determine how the brain processes similar stimuli in the world. In evolutionary biology, data from gene sequencing, functional morphology, and

---

4. This is a somewhat behavioristic notion of similarity (Herrnstein 1991) and of concepts in general, placing emphasis as it does on the ability to sort things. Thus having a concept of similarity, or more generally, of X, means being able to sort X's from non-X's. But, as Fodor (1994) has properly articulated, two organisms might sort things into similar piles (X's on one side, non-X's on the other) and yet may have different organizing concepts (i.e., inferences about how to sort). The classic example here is of one person who uses the concept "triangle" to sort geometric figures as opposed to another person who uses the concept "trilateral." The former will look for things with a particular set of angles, the latter, for things with a particular set of sides, and yet both people will end up with similar geometric objects in their respective X and non-X piles. Thus sorting behavior does not really get at concepts (their mental, representational instantiation), nor does it really get at the psychology of similarity in a fully satisfying fashion.

behavior are used to construct phylogenies, and as such, are used to reconstruct the evolutionary processes that have led to either homologies (similarity by common descent) or homoplasies (similarity by convergent or parallel evolution in distantly related taxa). In cognitive psychology, assessments of categorical perception and concept formation attempt to determine how tokens in the world are placed within functionally coherent categories based on the features they share with a prototypical exemplar. Linguists interested in language acquisition attempt to determine how interspeaker variability is normalized so that different-sounding tokens of the same word are perceived as referring to the same object. And in anthropology, especially anthropology with a cognitive bent, cross-cultural studies attempt to determine universals, as in the case of universally similar facial expressions or folk taxonomies for classifying colors, plants, and animals.

As this synopsis suggests, the notion of similarity is theoretically critical and represents empirically one of the most difficult problems we will face in this book. It is difficult because, for the problem of communication, we not only wish to know how the organism's perceptual system determines whether or not two signals are similar, but whether the similarities or differences are meaningful in the given context—yes, the problem, once again, of just noticeable differences as opposed to just meaningful differences. Let me now turn to some specifics that link the notion of similarity up with problems of classification schemes in communication.

3.2.2   Similarity and Classification

For the researcher interested in understanding a species communication system, whether human or nonhuman, there is the daunting task of cataloging representative exemplars into what are putatively meaningful categories—that is, of determining which exemplars are similar and thus fall naturally into a specified category. Thus, as a simple cut, one might observe a mammalian species and find that all harsh-sounding vocalizations are given by animals in an aggressive state, whereas all tonal high-pitched vocalizations are given by animals who are fearful. Similarly, someone interested in the cross-cultural patterns of prosody in human language might find that in cases where approval is communicated, a rapid frequency modulation is imposed on the pitch contour, whereas signals of negation are conveyed by brief frequency downsweeps. In each case, the researcher has determined, presumably by acoustic analyses, that there are characteristic features associated with particular contexts. And in reference to quantitative analyses, *characteristic* means something like "most of the exemplars recorded have features $X_1 \ldots X_n$." Two critical questions emerge from such a "similarity" analysis: first, are some features more significantly weighted than others, and second, what does the organism consider to be the relative importance of each feature? The first question generates significant problems for analytical treatments of communicative signals, and I pick these up in section 3.2.3. The second question, which focuses on perceptual salience, raises some interesting problems for experimental research. I would like to just briefly mention one technique that has commonly been used to address this general issue: the habituation-dishabituation procedure. It is a procedure that will be repeatedly mentioned throughout the book, and for the discussion that follows, serves to highlight some central issues associated with analyses of similarity, identity relationships, and concepts.

The habituation-dishabituation procedure has been used under both laboratory and field conditions to establish how a sensory system categorizes stimuli—what it treats as the same or different. In response to repeated exposure to a stimulus, organisms from aplysia to humans evidence habituation—a diminution or arrest in response that cannot be explained by muscle or sensory fatigue. To simplify (also see Box 3.2), imagine that a small nagging four-year-old is trying to obtain her mother's attention because she wants ice cream and has been repeatedly saying "Mommy . . . Mommy . . . Mommy . . ." for the past five minutes, sometimes loudly, sometimes softly, sometimes whining, and sometimes shrieking. Needless to say, most mothers will rapidly ignore such broken-record requests, not because they are tired and are incapable of orienting toward their child, but because repeated exposure to the class of stimuli has made it uninformative. If the child suddenly turns around and sees her brother and shouts "Tommy," the mother is likely to orient both to her daughter and to her son. To the human ear, and the mother in our scenario in particular, "Tommy" sounds different from "Mommy," is importantly informative, and leads to a response. That is, the sensory system perceives an acoustically significant difference between "Tommy" and "Mommy," and therefore a renewed level of response is observed, or what is known as dishabituation.

The habituation-dishabituation paradigm has been used in several areas of research as an assay to uncover perceptual categorization of auditory and visual stimuli. In studies of communication, one of its most useful applications has been in cases where the experimental subject is incapable of providing a verbal report of similarity—that is, prelinguistic human infants and nonhuman animals.[5] For

5. Actually, not all nonhuman animals. Researchers working with captive animals have managed to train their subjects to "report" perceptual differences, either through the use of visual symbols as in the studies by Savage-Rumbaugh and colleagues (1993) with bonobos and Schusterman and colleagues

Box 3.2
The Habituation-Dishabituation Paradigm: A Methodological Sampler

The typical design of a habituation-dishabituation experiment is as follows. Assume that A1 and A2 represent acoustic stimuli from the same category (e.g., different speakers saying the vowel sound /u/ as in *you*), whereas B1 comes from a different category (e.g., the vowel sound /o/ as in *boat*). The response assay or measure might be the amount of time subjects spend looking in the direction of the speaker following audio playback. In the following figure (left panel), subjects first habituate to repeated presentation of A1 and then show a small rebound in response, or response revival (rr) to A2; the presentation of A2 is considered the dishabituation phase. The panel on the right illustrates the same experimental approach, but contrasts stimuli from different categories (i.e., A1 with B1). Once again, subjects habituate to repeated presentation of A1 but show greater dishabituation to B1 than to A2.

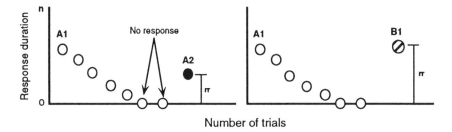

The difference between the two data sets is that the A1-B1 comparison elicits a stronger dishabituation than the A1-A2 comparison. In this scenario, therefore, the strength or magnitude of the dishabituation can be used as a quantitative measure of the difference between two stimuli; when the dishabituation response is small (i.e., approaches zero), the two stimuli are considered to be perceptually similar.

example, pioneering studies by Eimas and colleagues (1971) tested the human newborn's ability to discriminate and categorize English phonemes. Exemplars from different phoneme classes were played while infants sucked from a nonnutritive nipple. As infants heard a particular phoneme repeated over and over, they would suck at a decreasing rate. When a perceptually distinct phoneme was introduced, sucking rate increased. These studies and several others that followed provided evidence that prior to producing human language, human infants are sensitive to many important phonetic distinctions and, like human adults, judge

(1993) with sea lions, or through spoken English as in the work by Pepperberg (1987a, 1987b, 1991) with an African gray parrot. In the latter, for example, tests have been designed to show that a particularly brilliant parrot (''Alex'') can use the concept of *same* to distinguish between colors and textures.

exemplars of a particular phonetic category based on a suite of shared acoustic features.

This quite general technique has recently been implemented in an interesting way to study the meaning of nonhuman primate vocalizations (see chapter 7 and critical discussion in chapter 8). Specifically, Cheney and Seyfarth (1988, 1990) carried out field playback experiments to determine whether vervet monkeys use differences in the call's acoustic structure or differences in the call's meaning to guide their behavioral responses. In one experiment, for example, they compared individuals' responses (in general, looking time) to two acoustically distinct alarm calls, one given to eagles and the other given to leopards. Trials were conducted by repeatedly playing back one alarm call type (e.g., eagle alarms) and then, following habituation (i.e., either failure to look toward the speaker or a greatly diminished response), playing back a single presentation of the other call type (e.g., leopard alarms). Results showed that habituation to one call type was followed by dishabituation (i.e., subjects turned and looked toward the speaker) to the other call type, and this pattern was observed independently of the caller's identity. That is, habituation to "Sam's" eagle alarm call was followed by dishabituation to both Sam's leopard alarm call and "Fred's" leopard alarm call. In contrast, when Cheney and Seyfarth used the same protocol to compare two other vocalizations from the repertoire ("wrrs" and "chutters"), they found something quite different. Wrrs and chutters are produced in the same general context (i.e., aggressive intergroup encounters), but are acoustically different. Results showed that following habituation to Sam's wrr, presentation of Sam's chutter failed to elicit a significant change in response; however, following habituation to Sam's wrr, individuals responded (i.e., dishabituation) to presentation of Fred's chutter. In summary, results suggested that subjects were responding to differences in call meaning rather than to differences in acoustic structure and that for some contexts caller identity represents an important secondary source of information. Specifically, for wrrs and chutters, similarity is judged on the basis of meaning, and caller identity matters. For alarm calls, similarity also appears to be judged on the basis of meaning, but caller identity seems to matter less. A more complete discussion of these results is provided in chapters 7 and 8.

A common expectation in studies using the habituation-dishabituation procedure is to expect relatively symmetrical patterns of response. Thus, in the Cheney and Seyfarth experiment described in the preceding paragraph, one expects that if chutters and wrrs have similar meaning, and if meaning is what vervets are using in this task, then the pattern of habituation from chutters to wrrs should be the same as from wrrs to chutters. It may, however, be a mistake to expect symmetry, a point that has been clearly articulated by Tversky (1977) in his classic paper on the features of similarity (see also Medin, Goldstone, and Gertner 1993). To use one of his examples, in response to the question "Is Cuba like Russia?" most people are likely to answer yes, whereas in response to the question "Is Russia like Cuba?" many people would be expected to answer no. The reason is that in terms of political ideology, Cuba is like Russia, and political ideology most likely dominates in our conception of Cuba, especially when it is set up first in the proposition. However, Russia is associated with considerably more than political ideology (Borscht and blintzes for food, Tolstoy for literature, Tarkovsky for film, and so on). Consequently, Cuba only shares one or a small number of features with Russia, and would thus be considered different. Returning to our example of vervet wrrs and chutters, though both calls tend to be given in the context of intergroup encounters, wrrs are commonly associated with the initial stages of the encounter, but as the level of aggressiveness escalates, chutters are produced. If the motivational state of the caller is important to listeners—a reasonable assumption—then one might expect that the magnitude of dishabituation would differ following habituation to wrrs as opposed to chutters.

These kinds of asymmetries raise several important questions for the study of similarity and classification. Specifically, some entities in the world will not stand in symmetrical relationship to each other with regard to the concept of similarity, even though they may share several featural similarities. Additionally, what will often determine our own assessment of similarity, or that judged by another species, is a particular weighting of features. In the psychological literature, the notion of a prototype was proposed (Rosch 1975) in an attempt to formalize the suite of features that appeared most important or dominant within a category. Prototypes were invoked to account for the fact that, for example, a large number of people could agree about the tokens that fall within a broad array of concepts. Thus, although we may have very different experiences with birds, we all tend to agree that some things (features) are relevant to being a bird (e.g., having feathers, being able to fly) and some things are not (e.g., having antlers, being able to play the fiddle). And although definitional theories of concepts or prototypes are problematical, there is nonetheless a sense in which features play a role in how organisms figure out what is similar and what is different. Let us now turn to a discussion of the direct impact that notions of similarity and classification have on how we determine the relevant units of analysis in a communication system.

3.2.3 Units of Analysis and Their Classification in Communication

A critical issue in the study of communication is to determine the unit or units of analysis (e.g., Chomsky and Halle 1968; S. Green and Marler 1979; Slater 1973, 1983). Although researchers studying animal communication often determine the relevant units on the basis of production features (e.g., bouts of calling, acoustic structure of the signal), units of analysis must ultimately be derived from the perceiver's perspective. This boils down to a problem of categorization and an attempt to determine how relatively continuous streams of auditory and visual information are dissected by the perceptual system into functionally meaningful classes or categories. This is not simply a technical problem, but a conceptual one that is faced, for example, by the developing human child attempting to parse units such as "words" from the ongoing stream of continuous speech. And, although one would think that the "word" represents a well-defined entity, even here we find considerable disagreement among specialists (e.g., Christophe et al. 1994; Macnamara 1982; Markman and Hutchinson 1984; Premack 1990; Vihman and McCune 1994). In this section, I review some of the common approaches to studying the units of communication. This section will not attempt to provide any explicit solutions to the problems associated with understanding fundamental units, but will pave the way for achieving solutions in the later chapters.

Communication is a type of behavior. As a result, we can use general ethological analyses (S. Green and Marler 1979; P. Martin and Bateson 1993; W. J. Smith 1977) to frame our discussion of the appropriate units of communication. In general, there are two approaches,[6] involving either a description of the behavior's structure or a description of the behavior's apparent function. Thus, in studying an animal's vocal repertoire, we might describe a signal in terms of its duration, loudness, and pitch. Alternatively, we might attempt to provide a more complete description of the shape or contour of the signal or signals using a combination of pitch-extraction techniques (Fernald and McRoberts 1994) and sound similarity algorithms (Beeman 1992; Buck and Tyack 1993; C. Clark, Marler, and Beeman 1987). This would be a structural approach. In contrast, we might label the signal as an *alarm call*[7] based on the fact that the signal is given

---

6. These two approaches can be mapped onto a distinction within anatomy between studies that describe the form of an anatomical structure (e.g., the larynx consists of a set of vocal folds, cartilages, and muscles situated between the trachea and oropharynx) and studies that describe its function (e.g., the larynx is the initial source of acoustic output).

7. P. Martin and Bateson (1993) recommend that neutral terms be used to describe behavior so that function is not prematurely presumed. Thus, *peep* is a more appropriate descriptor than *distress call.*

during confrontations with predators and apparently serves the function of alerting group members.

In structural studies of communication, as in any structural analysis, the descriptive level can dissolve into an infinite regress, whereby increasingly fine-grained details are measured. Thus, in an audio signal such as the one presented in Figure 3.5, it is common to see researchers use such measures as signal duration, number of pulses, and interpulse duration; the same kind of approach holds for analyses of visual signals such as the facial expression of a rhesus monkey (Figure 3.6). However, the number of *possible* parameters measured is almost infinite. We therefore require an objective procedure to limit the number of parameters measured. Such limitations will depend on our understanding of the perceptual salience of signal features to both sender and perceiver, as well as on the amount of time required to analyze each feature. The first issue has to do with biology, the second with technology.

Experiments are required to assess which features of a signal are perceptually salient. A general approach is to take a relatively complex signal, decompose it into simpler units, and then ask the subject to provide a relative weighting of each feature's importance when other features are either excluded, distorted, or held constant. Thus a face can be divided into the eyes, nose, and mouth, and particular details can be manipulated, such as the distance between the eyes, the position of the mouth, the flaring of the nostrils, and so on (Ekman 1982; Ekman and Friesen 1978). In a task involving the recognition of particular facial expressions, one might then ask whether the identification of an "open mouth threat" requires protrusion of the lips *and* eyebrow raising, or whether lip protrusion is sufficient. Image manipulation software, available on several microcomputer platforms (e.g., Adobe Photoshop and Morph for the Macintosh), provides a simple technique for altering visual signals (e.g., see Etcoff and Magee 1992, for a nice example of how a graded series of stimuli can be created for a continuum of facial expressions), and for creating composite faces (e.g., Perrett, May, and Yoshikawa 1994). Similar software is available for manipulating auditory signals (e.g., Beeman 1992; Charif, Mitchell, and Clark 1993).

I agree with Martin and Bateson on this, even though researchers, including myself, often misbehave. I think that when researchers use a label such as *peep*, it simply stands for (i.e., is shorthand for) its function (e.g., "that call given by distressed individuals in the context of alarm"). This assertion is somewhat like the claim made by Lakatos and other philosophers about the Popperian view of science: although in theory scientists may look like they are attempting to falsify their hypotheses, in reality they are often attempting to confirm their deeply rooted beliefs about the way the world works.

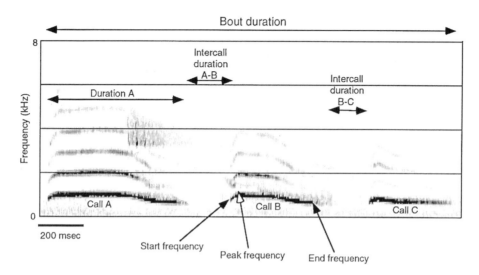

Figure 3.5
Some features used to quantify the structure of an auditory signal (a spectrogram of a Madagascar harrier hawk call). The *x*-axis is time in milliseconds, and the *y*-axis is frequency in kilohertz.

Studies of human language are exemplary in that the units of analysis are generally defined at the outset and are typically derived from perceptual studies. Thus phoneticians have developed an entire symbol system to facilitate identification and quantification of phonemes (International Phonetic Association 1949; Chomsky and Halle 1968; Lindblom and Maddieson 1988; Maddieson 1984). Although the table of phonetic symbols has been revised and continues to undergo revision, it represents a widely agreed upon convention, thereby providing uniformity among studies that focus on the phonetic level. Similarly, researchers studying sign language have also provided an objective set of measures for defining functional units in the visual domain (Lane, Boyes-Braem, and Bellugi 1976; Loomis et al. 1983; Poizner et al. 1986).[8] In such analyses, a three-dimensional movement-monitoring system is used to track the speed, path, and configuration of particular signs and sign combinations. The system uses multiple optoelectronic cameras designed to pick up light-emitting diodes placed on the hands and arms. When signs are produced, the computer takes the input from the diode

8. Rigorous measures for identifying facial expressions in humans have also been developed by Ekman and colleagues (Ekman 1982; Ekman and Friesen 1969, 1978), and some of these are discussed in chapters 4 and 7.

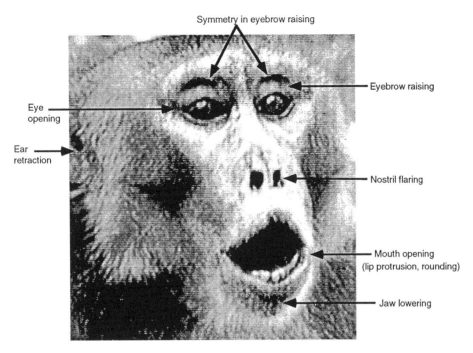

Figure 3.6
Some features used to quantify the structure of a visual signal (rhesus monkey open mouth threat expression).

positions in space and reconstructs a three-dimensional representation of the movement.

In contrast to studies of human spoken and signed language, studies of nonhuman animal communication have used a variety of approaches to defining communication units, and consequently it is difficult to make comparisons across studies, both within and between species. In the auditory domain, the most fundamental unit has generally been the *call* or *song,* and each of these descriptors has typically been defined in terms of both structure and function. Thus songs tend to be longer in duration than calls and are most commonly given in the context of competition for resources (mates or food) (Horn 1992; Kroodsma 1982). As techniques in digital signal processing have improved over the past ten years, it has become increasingly easier to collect data on a large set of acoustic parameters. Although such technological power has led to an exceptionally rich description of call structure, the logic behind the selection of features is often unclear and in

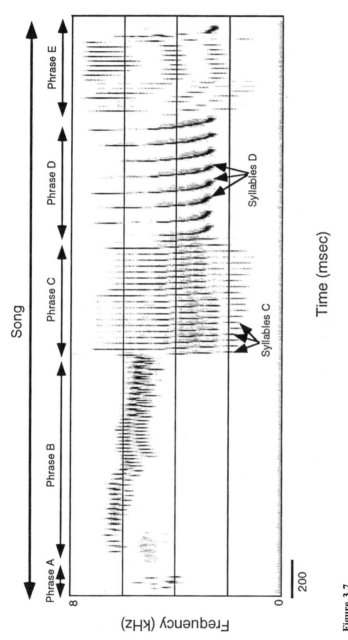

Figure 3.7
Sound spectrogram of a canary's song, showing some of the relevant features for analysis.

general seems to bear little relationship to the question of perceptual salience raised earlier.

A second unit-of-analysis problem for studies of nonhuman animal communication is that although individuals produce discrete calls, songs, and facial expressions, it is quite common for socially interacting individuals to produce a string of vocalizations or a sequence of visual displays. And, such strings or sequences often consist of different types of vocalizations or different kinds of visual displays. For example, studies of rhesus monkeys indicate that the *girney* vocalization is used in the context of affiliative social interactions (Hauser 1992b). This call is most frequently given in bouts of two to three calls, each call separated by relatively short periods of silence. Sometimes, however, girneys are also given *with* other calls in the repertoire such as the *coo, gecker,* and *grunt.* Analytically, such sequences present considerable problems (see Box 3.3), and at present few studies have explored the possibility that strings of calls mean something different from single-call utterances (Ficken and Popp 1992; Hailman and Ficken 1987; Hailman, Ficken, and Ficken 1987; Marler 1977; Robinson 1984; Snowdon 1982). And yet, if we are to compare, for example, the syntactical structure of human language (both spoken and signed) with the structure of nonhuman animal communication, then we need to move beyond analyses of single calls or displays (see the excellent discussion of this and related points by Kuczaj and Kirkpatrick 1993).

The timing of displays within a sequence represents yet another problem associated with analyses of communication. This is particularly important for studies concerned with vocal exchanges (Harcourt, Stewart, and Hauser 1993), or what have commonly been called *duets* (Farabaugh 1982) or *conversations* (Hauser 1992b; Snowdon 1988; Snowdon and Cleveland 1984). The analytical challenge lies in understanding whether the signal delivered represents a directed response to a prior signal or the initiation of a new bout. In general, studies of vocal exchange have used temporal rules to assess whether an individual has initiated a communicative bout or is responding to one. For example, in studies of the squirrel monkey by Symmes, Biben, and Masataka (Biben 1993; Biben, Masataka, and Symmes 1986; Masataka and Biben 1987), results indicate that *responses* to "chuck" vocalizations typically (i.e., with a high statistical probability) occur within a 0.5-second period; vocalizations occurring outside of this time window are more likely to reflect the initiation of a new vocal bout. Symmes and colleagues (reviewed in Symmes and Biben 1988) suggest that these vocal interactions provide evidence that squirrel monkeys engage in

Box 3.3
Analytic Approaches to Behavioral Sequences

Visual and auditory displays often consist of action sequences. Thus, during a threat display, monkeys often raise their eyebrows, protrude their lips, and bristle their hair. Sometimes, these actions occur independently, sometimes sequentially, and sometimes simultaneously. What we require are quantitative techniques to establish action dependencies—that is, the probability of a given action being repeated or preceded or followed by another action. Markov analyses, which use transition matrices, represent a common solution to this problem (Bakeman and Gottman 1986; Hailman 1977; P. Martin and Bateson 1993). Under the general problem of Markovian analysis, there are several ways in which sequential actions can be cached out. Consider a relatively simple situation in which there are two actions (i.e., displays) A and B. If A is given, then either B follows (assuming for the moment, a relatively short time interval) or it does not. Conversely, if B is given, then either A follows or it does not. Here, the sequence is determinant and can be described in terms of a 1-0 matrix where a value of 1 designates the occurrence of an action, and 0 its absence. In this matrix, whenever A is produced, B follows, and whenever B is produced, A follows. Neither A nor B is repeated sequentially (i.e., one never observes AA or BB).

		Previous output	
		A	B
Subsequent output	A	0	1
	B	1	0

The preceding matrix might be characteristic of a very stereotyped display, what ethologists traditionally called a fixed action pattern. As recognized by more recent studies of behavior (Barlow 1977), fixed action patterns are quite rare. A more common observation is that sequences of behavior are probabilistic and depend upon complex changes in the individual's motivational state together with changes in the perceiver's state. For this situation, the transition matrix would consist of conditional probabilities rather than fixed values. For example, consider a situation where A is a mild threat display and B is an intense threat display with two individuals engaged in an aggressive volley of displays:

		Previous output	
		A	B
Subsequent output	A	0.1	0.9
	B	0.9	0.1

What this matrix indicates is that the probabilities of B following A and A following B are high, whereas the probability of either A or B being repeated is low.

To date, sequential analyses have played a minor role in studies of communication. One reason for this neglect is that matrices often become unmanageable because of the relatively large number of displays or signals employed in a given interaction, in addition to the need for observers to obtain continuous records of interaction. Although the latter is possible given recent advances in computer technology (Noldus, van de Loo, and Timmers 1989) and video technology, such analyses are time-consuming and require large sample sizes.

question-answer sessions designed to provide information about the nature of affiliative relationships.

One problem with such analyses is that they are based on the assumption that responses to signals must occur within restricted periods of time, otherwise they are classified as independent. To clarify this point, consider a discussion between two roommates about what they want to eat for dinner:

"Bert, let's have some chicken for dinner?"
"That sounds good," says Ernie. "Let's have some mushrooms as well and a salad."
Bert and Ernie enter the kitchen and begin cooking. Five minutes later, Bert speaks to Ernie without making eye contact.
"I think I will make a vinaigrette dressing to go along with the salad."

Now, in the last utterance, Bert has potentially started a new conversation, but he is also responding to the prior conversation initiated by Ernie. Thus, long pauses need not reflect bout termination, as commonly assumed by studies of nonhuman animal vocal exchange. More to the point, the temporal pattern underlying the production of signals within a group of individuals is likely to be a misleading parameter in the analysis of signal exchanges or conversations until we have a more precise understanding of the signal's meaning (see chapter 7).

A second approach to the study of vocal exchange has been to explore some of the physiological constraints on signal production. For example, research by Heiligenberg (1986, 1991) on the jamming avoidance response in electric fish and by Narins (1982, 1992) on advertisement calls in frogs indicates that there are neurophysiological processes that determine *when* signals can be produced. For example, in the Neotropical tree frog *Eleutherodactylus coqui*, experimental results indicate that after call production, there is a behavioral refractory period in which calls are neurophysiologically suppressed. As a result of this physiologically imposed pause in production, neighboring individuals are less likely to produce mating calls at the same time and thus are more likely to avoid call overlap (i.e., acoustic interference).

In a study of free-ranging vervet monkeys and rhesus macaques, results suggest that physiological changes associated with the control of vocal pitch may guide the pattern of vocal exchange (Cooper and Sorensen 1981; Hauser and Fowler 1991; Hauser 1992b). Specifically, over the course of a call bout (i.e., a string of two to three calls of the same type), the pitch of an individual's voice (i.e., the fundamental frequency) consistently declines in these species. Analyses of vocal exchanges suggest that group members use this spectral change as a cue for initiating a vocal response during a period when call overlap is minimized. Based on studies of human phonation (Titze, 1976, 1989), declination in the fundamental is due to at least two physiological changes: (1) a decrease in subglottal pressure that causes a decrease in the rate of vocal fold vibration and (2) a relaxation of laryngeal muscle (e.g., cricothyroid) tension. In general, human and nonhuman primates exhibit similar mechanisms of phonation (Jürgens 1990; Lieberman 1985; Negus 1949), though there are clearly a number of subtle differences in the anatomy and function of the vocal folds and accompanying musculature (Schön Ybarra in press). Observations of vervets and rhesus therefore suggest that the timing of vocal exchanges may be shaped by physiological constraints, which, in turn, may determine what counts as a communicative unit.

Although the preceding discussion has focused on vocal exchanges, the problems raised are equally applicable to interactions involving the exchange of visual signals (van Hooff 1982; West and Zimmerman 1982). In fact, the problems may well be more complicated for visual signals because they lack clearly demarcated start and end points.

Intimately related to the issues discussed thus far is the problem of *discrete* versus *graded* signals, initially raised by Marler (S. Green and Marler 1979; Marler 1975; see section 2.4.2) in the context of nonhuman animal vocal signals, but equally applicable to visual signals. A signaling system is considered discrete if precise boundaries can be demarcated for each signal.[9] Such demarcation must be considered from both the production end (i.e., generating distinct signals) and the perception end (i.e., perceiving distinct signals). For example, imagine a hypothetical repertoire that includes four acoustic signals A, B, C, and D. If the repertoire is discrete, it should be possible to distinguish each signal from the other based on specific acoustic features. In addition, if the signal classes are to have different meanings, perceptual experiments ought to reveal that such differences in acoustic morphology are discriminable and elicit behaviorally appropriate responses (e.g., if A is associated with aggressive threat and B with fear from threat, then upon hearing A, individuals might be expected to flee, with no intermediate or mixed responses). In contrast, if A, B, C, and D are graded, then acoustic boundaries between signals cannot be demarcated, and perceptual experiments should reveal more graded responses. Concerning the latter, repertoires that are acoustically graded may nonetheless be treated as perceptually discrete. Thus human speech and elements of other species' repertoires represent systems that exhibit gradedness on the production end, but discreteness on the

---

9. For the moment, I am leaving aside the possibility of mixed repertoires, including both discrete and graded components.

perceptual end. This description highlights the point, discussed in greater detail in chapter 7, that a common design feature of perceptual systems is to serve as efficient categorization machines, requiring algorithms for placing exemplars into functionally meaningful classes (Herrnstein 1991). The degree to which within-category distinctions are made will ultimately depend upon their importance to survival and reproduction. Although "a rose is a rose is a rose . . ." may be true under some circumstances, if red roses are unpalatable and pink roses are delectable, then one would expect such fine-grained classifications to be selected for.

In summary, cross-species comparisons of communication are currently hindered by the lack of standardized descriptive units. Important progress will be made when we have a better understanding of which features are most salient with regard to signal recognition (i.e., the problem of classifying exemplars into functionally meaningful categories; Dooling 1992; Gerhardt 1992b; Nelson and Marler 1990) and extend our level of analysis from single tokens of a display to display sequences. The latter is obviously most relevant to studies of nonhuman animal communication, and in this light, research on both spoken and signed human language can serve as important sources of methodological insight.

3.3 Potential Fruits of Tinbergen's Research Design

In chapter 1, I laid out the Tinbergian research program, one that is based on the view that biological traits can only be completely explained by addressing questions about the mechanisms, ontogenetic processes, adaptive significance, and phylogeny underlying the expression of a trait. In this section, I wish to give the reader some idea of what this program can do for our intellectual curiosities by running through a sample case study. My primary motivation here is not to give a fully detailed account, but rather to whet the reader's appetite by showing what is possible when different sorts of conceptual and empirical perspectives are integrated. An auxiliary reason for doing this here is that the critical empirical chapters of the book are divided into Tinbergen's different causal questions, and integration occurs by cross-referencing. An alternative approach to writing this book might have been to divide the chapters by taxonomic group (e.g., insects, fish, anurans, birds, mammals) and within each, integrate the four causal questions. Although this would have made for a very good read and represents the ultimate goal of the proposed research program, there are problems, they are unfortunate ones, and the rest of this book is designed to help uncover how we can solve them. Specifically, though we tend to know a great deal about one or two of Tinbergen's questions for a given taxonomic group, we know relatively little about the other questions. It is therefore the goal of this section to inspire more integrative work, and to punch this point home, I return to it yet again in the final chapter, armed with considerably more knowledge.

The case I wish to focus on here is vocal communication in rhesus monkeys. I concentrate on this species, not just because it happens to be one of the organisms I work on, but because some interesting issues can be raised under each of the four main topics of Tinbergen's framework, including the subdivision between neurobiological and psychological (cognitive) problems (Figure 3.8). I start with a general summary of macaques, including some of the early work on their vocal communication. I then turn to a synthesis of what we now know about the neurophysiological mechanisms underlying vocal production and perception, the ontogeny of the repertoire, the functional consequences of specific vocalizations, and the psychology underlying their vocal behavior. Throughout these brief summaries, I also provide some comments on relevant phylogenetic issues.

The macaques, genus *Macaca,* are catarrhines or Old World monkeys. Within this genus there are approximately 16–19 species (Fooden 1980; Melnick and Pearl 1987), and they are distributed in woodland and rain-forest habitats throughout Asia and Africa. In general, all species, including rhesus, live in multimale multifemale social groups, structured on the basis of fairly rigid dominance hierarchies; males generally outrank females, and females tend to inherit the rank of their mothers. Females enter a period of estrus, with breeding typically restricted to a few months out of the year. Although some species exhibit paternal care, most of the burden falls on females. Like other mammals, females stay in their natal groups for life, whereas males move out and into neighboring groups once they have reached reproductive maturity.

One of the first studies of vocal communication in nonhuman primates was conducted on rhesus monkeys (Rowell and Hinde 1962) and set the stage with regard to some of the acoustic and contextual variants within the repertoire. Specifically, Rowell and Hinde demonstrated that rhesus monkeys in captivity produced both tonal and atonal vocalizations, and used these to convey information about their affective states, particularly aggression and fear.

Several years later, Philip Lieberman, D. H. Klatt, and W. H. Wilson (1969) embarked on a study that was aimed at assessing whether the nonhuman primate vocal tract was capable of producing some of the critical sounds and sound contrasts of human speech. Analyses of a few sounds from a rhesus monkey, together with a vocal tract model, suggested not. Specifically, though the rhesus vocal tract (considered to be a reasonable model for other monkeys) could produce some consonants, it was not capable of articulating the point vowels (see chapters

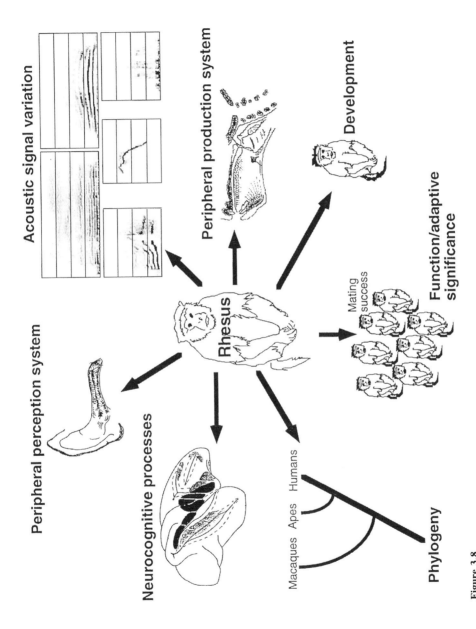

Figure 3.8
Some of the ingredients involved in a Tinbergian research program on rhesus monkey vocal communication.

2 and 4). Moreover, Lieberman argued, most of the significant energy in their vocalizations was restricted to the fundamental frequency, with little evidence of supralaryngeal filtering (i.e., resonances or formants). Based on these conclusions, subsequent physiological studies by Larson, Ortega, and DeRosier (1988) focused on analyses of the laryngeal musculature, showing, for example, that stimulations of the cricothyroid led to changes in vocal pitch. Moreover, electrostimulation studies revealed that the periaqueductal gray area was significantly involved in vocal production, whereas higher cortical structures seemed to be unimportant in either production or perception (Aitken 1981; Pandya, Seltzer, and Barbas 1988).

With the advent of sophisticated computer techniques for sound analysis and more rigorous experimental methods for sampling vocal behavior under natural conditions, the picture painted above has changed quite a bit with regard to the range of acoustic variation in the repertoire, its ontogeny and associated neurophysiology, the functional consequences of vocalizing in different socioecological contexts, and some of the higher-level cognitive processes underlying signal production and perception. I now turn to some of these developments.

Based on research from a free-ranging colony of rhesus monkeys living on the island of Cayo Santiago, Puerto Rico, we now know that this species produces somewhere between 25 and 30 discrete call types (Gouzoules, Gouzoules, and Marler 1984; Hauser and Marler 1993a); recall from the previous discussion that this value represents an estimate given that our understanding of relevant units of analysis is quite poor. Although the Cayo population is not confronted by predation, it is generally assumed that most of the major socioecological contexts occur on the island (e.g., finding food, dominance interactions, intergroup encounters, affiliative behaviors) and thus that most of the vocalizations in the species-typical repertoire have been recorded. In terms of spectrotemporal variation, it is clear that changes in the fundamental frequency are critically involved in encoding meaningful differences both within and between call types, as suggested by Lieberman's analyses. Thus differences in the fundamental frequency contour of the affiliative coo vocalization are associated with differences in individual identity, with some evidence that members of the same matriline produce structurally similar coos (Hauser 1990). In addition, as pointed out in the previous section, rhesus monkeys appear to use changes in the fundamental frequency (i.e., a decrease in pitch across a string of calls) to encode information about bout termination, and this information may be used by listeners in conversational interactions (Hauser and Fowler 1991).

In addition to changes in the fundamental frequency, recent evidence also points to significant changes in spectral energy above the fundamental, apparently associated with changes in the configuration of the supralaryngeal vocal tract. There are three pieces of evidence. First, an observational study of rhesus monkeys by Hauser, Evans, and Marler (1993) showed that during vocal production, there are significant changes in the length and shape of the vocal tract as a result of changes in the position of the lips, jaw, and teeth. In a detailed kinematic analysis of the coo vocalization, results showed that changes in the position of the jaw were statistically correlated with changes in the first resonance frequency, but not with changes in the fundamental frequency. This finding provided correlational support for the idea that, like human speech production (Fant 1960), source (larynx) and filter (the supralaryngeal tract) operate independently during rhesus vocal production. In a second observational study (Hauser 1992b), analyses of the coo vocalization revealed that some individuals produced this vocalization in a perceptually nasal fashion whereas other individuals produced them in a nonnasal mode. The acoustic consequence of what human listeners perceived as perceptual nasality was a reduction in spectral amplitude at the center of the call's frequency range and an apparent insertion of spectral energy (resonances) between the primary harmonics. Although it was not possible to confirm that the nasal cavity was being engaged for this vocalization, it seems likely given that nasal productions in human speech exhibit similar acoustic changes. Additionally, and of interest for our discussion of development, those individuals producing perceptually nasal coos were all females and all members of a single matriline. These observations suggest that the ability to produce nasal coos may represent a learned mode of vocal production.

The third piece of evidence was obtained from an experimental study (Hauser and Schön Ybarra 1994) involving the use of xylocaine (a form of novocaine) to perturb the articulatory system. The aim was to obtain more causal information on the relationship between changes in vocal tract configuration and acoustic morphology. When xylocaine is administered to a subject's lips, the subject is temporarily incapable of either protruding or retracting them. Two call types were analyzed: (1) a coo vocalization that, under natural conditions, is produced by protruding the lips and (2) a scream vocalization that is produced by retracting the lips. For subjects treated with xylocaine, therefore, coos would be produced with a relatively shorter vocal tract, whereas screams would be produced with a relatively longer vocal tract. Results revealed that for coos, the first two formant frequencies were significantly higher under xylocaine treatment than under normal conditions; no other acoustic differences were detected. This result is precisely what one would expect if rhesus fail to compensate for the perturbation and produce the coo with a relatively shorter vocal tract (i.e., a shorter vocal tract is associated with higher resonances). In contrast, screams produced under xylocaine treatment were acoustically indistinguishable from those produced under normal conditions. This finding either suggests that rhesus can compensate for the perturbation during scream production or that the xylocaine treatment failed to impose a sufficiently significant alteration to affect (statistically) call structure. In summary, results from observations and experiments suggest that both laryngeal and supralaryngeal maneuvers play a role in configuring the acoustic morphology of the rhesus monkey's vocal repertoire.

Given the observed acoustic variation, what do we know about perceptual salience? After all, what we as human observers consider to be salient acoustic features of the repertoire may turn out to be irrelevant from the rhesus monkey's perspective. In contrast to the detailed psychophysical analyses conducted on Japanese macaques (reviewed in Moody, Stebbins, and May 1990; Stebbins and Sommers 1992), which emphasize this species' exceptional ability to perceive quite subtle spectrotemporal events (especially frequency modulation), relatively little work has been conducted on rhesus monkeys, and most of the studies have used artificial stimuli to determine auditory frequency profiles or sound localization capabilities. Thus we know from the work of Stebbins, Brown, and colleagues that rhesus monkeys and humans are comparable with regard to hearing range (audibility functions), as are their abilities to localize sound in the horizontal and vertical planes. More specifically, however, C. H. Brown, Beecher, et al. (1979) showed that tonal vocalizations such as the coo are less accurately localized than atonal vocalizations such as barks, and for both vocalizations, signals presented in the horizontal plane are more accurately localized than those presented in the vertical plane (C. H. Brown, Schessler, et al. 1983). The latter appears to fit with the general ecology of the rhesus, which is one dominated by a terrestrial existence.

Turning from the peripheral processing system to the central nervous system, recent work by Deacon (1991) has demonstrated cytoarchitectural homologues to Broca's and Wernicke's areas, though at present no functional work has been done to determine how these areas guide or constrain vocal output. However, based on tracer analyses, Deacon suggests that these higher cortical structures are unlikely to play a role in mediating vocal communication. In contrast, auditory neurophysiological work by Rauschecker, Tian, and Hauser (1995; Rauschecker, et al. in press) shows that the lateral belt surrounding the primary auditory cortex is fundamentally involved in the perception of conspecific calls and may represent an important evolutionary precursor to areas involved in human speech perception.

To address questions regarding phylogenetic patterns in brain evolution, Hauser and Andersson (1994) conducted a study to determine whether the robust left hemisphere bias for processing human speech (reviewed in Hellige 1993; Davidson and Hugdahl 1995) is based on an evolutionarily ancient neural mechanism, perhaps dating back at least as far as the Old World monkeys. Studies of Japanese macaques had already demonstrated a significant left hemisphere bias for processing a single species-typical vocalization (M. R. Petersen et al. 1978). Using a field playback technique, results showed that adults consistently evidenced a left-hemisphere bias for processing conspecific calls, but showed a right-hemisphere bias for a heterospecific's call. In addition, infants under the age of one year failed to show a hemisphere bias for either conspecific or heterospecific calls, suggesting that the brain asymmetry observed in adults develops gradually over time, presumably influenced by both maturation of the brain and experience with the vocal repertoire.

Two observations have already alluded to developmental changes in vocal communication among rhesus macaques: the production of nasal coos and hemispheric biases in perception of conspecific calls. However, two experimental studies, using a cross-fostering procedure, have looked more carefully at the ontogeny of this species' vocal repertoire. Specifically, rhesus macaque infants were reared by Japanese macaque mothers (and vice versa) to determine whether such infants would modify their vocal output to meet the structural requirements of their cross-fostered parent's repertoire, or whether they would continue to produce their own species-typical vocal output. Unfortunately, the results are ambiguous. In one study (Masataka and Fujita 1989), rhesus infants produced food coos that more closely resembled the adult form of Japanese macaque food coos than the adult form of rhesus macaques. In contrast, Owren and colleagues (1992, 1993) found no evidence of such vocal modification, with cross-fostered rhesus infants continuing to produce the species-typical coo structure throughout development; the same pattern of results were observed for cross-fostered Japanese macaque infants. At present, therefore, there is little evidence that acoustical experience plays a significant role in the ontogeny of vocal morphology (as opposed to call usage and comprehension; see chapter 5). This conclusion is generally consistent with what has been observed for other nonhuman primate species.

Functional analyses of the rhesus monkey's vocal repertoire have focused on calls produced in the context of food discovery and mating. Based on observations and experiments, Hauser and Marler (Hauser 1992c; Hauser and Marler

1993a, 1993b) have found that when rhesus monkeys find food, they give one or more of five acoustically distinctive vocalizations. Some of these calls are given to low-quality, common food items, whereas the others are given to high-quality, rare food items. When individuals call, other group members come running. Observational data reveal that males call significantly less often in the context of food than females, and females with large matrilines call more often than females with small matrilines. These results suggest that kin selection may exert significant pressures on this calling system, though at present there is no evidence that kin selectively benefit (e.g., gain more food) from calling by their relatives. Under experimental conditions, results reveal that individuals who find food and call obtain more food and receive less aggression than those who remain silent. In fact, silent discoverers who are caught with food are severely beaten up, and this is true for high- and low-ranking individuals. These results suggest that although there may be significant benefits to remaining silent when discovering a food source (i.e., sometimes you get away with it and obtain a lot of food), the potential costs of aggression are high and may ultimately cause a significant decrease in survivorship.

In the context of mating, males produce extremely loud and individually distinctive calls during mating (i.e., during thrusts with or without ejaculation). Observations of a large number of males and females revealed that males who called during copulation obtained significantly more matings than males who were silent. However, vocal males also received significantly more aggression than did silent males. These results suggest that there are potential survival costs to calling (natural selection pressure), but potential reproductive benefits (sexual selection pressure). At least two important factors remain to be examined before the adaptive significance of this calling system can be firmly established. First, we do not know whether males in good condition are more likely to call than males in poor condition. Second, it is unclear whether females use the male's vocal behavior as an important parameter in female mate choice.

The data on food calling in rhesus have also been used to address two questions regarding the psychological or cognitive basis of communication in this species. First, when rhesus monkeys find food and call, do such vocalizations provide referential information about the specifics of the discovery (e.g., what kind of food, how much, its location, etc.)? And second, when they find food and fail to call, are they intentionally suppressing information from group members? In terms of cognitive mechanisms, the first question is critically related to the second. Observational and experimental research on referential signaling in this species (Hauser and Marler 1993a, 1993b) has demonstrated that when rhesus are hungry, they call at higher rates than when they are relatively satiated. Hunger level, however, does not appear to be related to the *type* of call used. Here, animals appear to select different call types to convey information about the *type* of food item encountered, with some calls given to high-quality, rare food items and other calls given to lower-quality, common food items. Thus the spectral properties of the call would appear to convey information about an external referent, specifically the relative quality and rarity of the food discovered.

Concerning the second question, there is some evidence that the failure to call is, at least partially, a calculated decision. First, when individuals find food, their first response is to scan the area, presumably because they are looking for enemies that might beat them up or allies that might help them defend or share in the food source. Thus the discovery of food does not elicit a reflexive response to call. Second, the population is not divided into individuals who call and individuals who do not, though males call less often than females. Third, animals who live as residents within a social group appear to implement different calling algorithms than animals (i.e., males) who live peripherally, waiting to join a social group. Specifically, when peripheral males find food under experimental conditions, they never call. Interestingly, when such silent discoverers are found by other peripheral males or members of a social group, they never receive targeted aggression. This lack of response stands in striking contrast to the aggressive response received by silent discoverers who *are* members of a social group. These subtleties suggest the possibility, though they by no means definitively prove, that rhesus monkeys may intentionally withhold information about their food discoveries under certain conditions. These results have important implications for the study of mind in nonhuman animals, and even such topics as moralistic aggression.

With a finale like the last sentence, let me end this chapter and move on to the empirical chapters of the book. I hope this section has given the reader a sense of the power of integrating different levels of analysis and shown how Tinbergen's framework is the only workable one for the study of communication.

4 Neurobiological Design and Communication

4.1 Introduction

This chapter addresses some of the neuroanatomical and neurophysiological processes underlying the production and perception of communicative signals. Over the past fifteen years fundamental principles in neuroscience have been uncovered through the integration of research on behavior and neurobiology (Capranica 1965; Heiligenberg 1991; Hoy 1992; Konishi 1989; Marler 1991b, 1992b; Nottebohm 1981, 1989; Suga 1988). The type of behavior that has yielded some of the deepest neurobiological insights is communication. For example, there is now evidence from research on birds of dedicated neural circuitry for song production and perception that undergoes organizational changes during development and in adulthood, mediated by complex socioendocrinological processes. And studies of bats have uncovered exquisitely fine-tuned neurons arranged in a tonotopic map that enable this flying mammal to navigate in the dark and localize prey with extraordinary accuracy. At each investigatory step, studies of insects, fish, frogs, birds, bats, and primates have allowed for a natural interchange between behavioral and neurobiological findings and insights.

An evolutionary and ecologically motivated neuroscience is relatively new but shows great promise for at least two reasons. First, by looking at problems that are of direct relevance to a species' survival and reproduction, it is possible to generate quite specific predictions about the design of the nervous system. It is in this area that the marriage between behavior and neurobiology becomes powerful. Second, early studies in neurobiology focused on relatively simple behaviors (e.g., locomotory) and used artificial stimuli[1] to determine the specific operating properties of the neural system. Hubel and Wiesel's (1959; reviewed in Hubel 1988) pioneering work on the visual system of the cat is testimony to the power of this approach. Moreover, there were good reasons to look at simple systems with highly controlled techniques. They worked, and they produced results that undoubtedly will remain as textbook examples in the neurosciences. However, advances in auditory and visual signal processing techniques within the past fifteen years have enabled researchers to shift from artificial stimuli to biologically natural and meaningful stimuli (Capranica 1965; Heiligenberg 1986; Konishi 1985; Marler and Nelson 1992; Nottebohm 1989; Perrett and Mistlin 1990; Suga 1988). This shift allows us to look under the cortical hood and determine how the

---

1. Studies of vision often use lines, colored geometric shapes, and gradients, whereas studies of audition typically employ presentations of pure tones or white noise, varying in frequency and amplitude.

nervous system generates and perceives signals that it was designed to process. As in any interdisciplinary venture, a good deal of intellectual retooling is required. Thus those approaching the problem from an ethological perspective have a lot of neurobiology to learn. In contrast, those approaching from neurobiology need to gain a better understanding of natural behavior and what it means for a nervous system to be designed to solve particular socioecological problems.

Over evolutionary time, a combination of both random and nonrandom processes have led to changes in brain structure and size. Regarding communication, it is important to examine such changes in detail because they are likely to shed important light on the design features of both signal production systems and signal perception systems. To take a well-known biological observation, large animals tend to have larger brains than small animals (Jerison 1973; Jerison and Jerison 1988). When allometric relationships are taken into account, including such factors as body size and growth, we find that humans have considerably larger brains than one would expect. We also find that species such as chimpanzees, capuchin monkeys, dolphins, and parrots have bigger brains than other representatives of their families. It is tempting to conclude with the equation

BIGGER BRAINS = GREATER INTELLIGENCE

But this conclusion would be mistaken on two levels. First, it simply begs the question of what a big brain *does* (i.e., computationally) that a little brain cannot. And second, it implies that the notion of intelligence can be readily defined and, more importantly, compared across species. Thus far, at least, no one has managed to delineate the key symptoms of intelligence in humans, let alone operationalize the concept for cross-species comparisons. The main point of this chapter is to understand how the socioecology of a species has created problems that demand a specialized brain, that is, a brain with computational abilities designed to solve the problems that it confronts in its species-typical environment. In this sense, we are looking for highly specialized neural architectures (i.e., not measures of general intelligence) that appear to be customized by natural selection for the task at hand. By looking at brain evolution in this way, we are likely to uncover significantly more interesting patterns—in particular, species differences that are telling with regard to the design of the communication system. As one example, consider the modern human brain. There is no question that it is relatively bigger (approximately 1400 cc) than in all other species, even closely related living species such as the apes and monkeys (< 450 cc). But what is most interesting in terms of thinking about the neural substrate subserving human language, both signed and spoken, is the fact that certain parts of the brain are much much

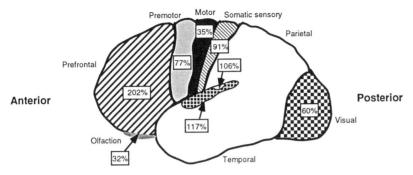

Figure 4.1
A schematic illustration of the human brain, showing some of the critical cortical areas and how they have changed in size relative to other primates. Specifically, the percentages represent the predicted size of an area relative to what would be expected for a primate with a comparably sized human brain (data from Deacon 1992).

larger than what one would expect based on analyses of other species (Figure 4.1; see recent analyses by Finlay and Darlington 1995). For example, the prefrontal cortex of the human brain is 202% larger than would be expected for a primate with a similarly sized brain (Deacon 1992, in press). Although the prefrontal cortex is not considered a "language area" per se, it appears to play a critical role in language processing: it is the primary area for working memory (i.e., storing representations in mind for immediate use), a computation that is crucially involved in both producing sentences and comprehending their meaning.

Understanding the neurobiology of a communication system requires a description of the mechanisms underlying the production and perception of signals. One could choose to discuss these issues in any order. I have, however, decided to begin with production mechanisms because communication must start with a signal. Nonetheless, it should be clear from the outset that the production system is ultimately constrained by and coevolves with the perception system. Although production systems can, in theory, produce an infinite range of variation, the biological relevance of such variation is limited unless the perceptual system has the ability to decode it. And in the context of communication, decoding is not synonymous with discrimination. The range of variation must be both discriminable and meaningful. Discriminability depends on the psychophysical features of the sensory system. A signal's meaning comes from its association with fitness-related events in the environment.

In most of the case studies that will be described in this chapter, we know relatively more about the neurobiology of signal perception than we do about the neurobiology of signal production. This imbalance is largely due to the fact that at present one can readily test an animal's response to a biologically meaningful signal (perception), but there are significant difficulties associated with eliciting one (production). In addition to the asymmetry in our knowledge about the neurobiology of perception and production, we also know considerably more about auditory signals than we do about visual ones. As a result, most of the case studies come from the auditory domain.

A second organizational decision concerns the sequence in which peripheral as opposed to central mechanisms are detailed. Because stimulation of a peripheral sensory system is often the first step in communication, I begin each case study by discussing the structure and function of such systems and then turn to a description of the underlying brain circuitry. Pedagogically, I also think that it is easier for the reader to first digest information about systems that can be readily observed with the naked eye and are rather less mysterious than what goes on inside the phylogenetically diverse set of *black boxes* in the animal kingdom.

4.2 Mating Signals: Frogs and Birds

4.2.1 Anuran Advertisement Calls

The anurans (i.e., frogs and toads), like most other organisms that communicate in the auditory domain, produce a variety of vocalizations (Rand 1988). In contrast to other vertebrates (e.g., songbirds and primates), however, most anuran vocalizations are structurally quite simple and stereotyped. By far the most conspicuous auditory signal in the repertoire is the advertisement call (Figure 4.2), used by males to both attract potential mates and to notify other male competitors. There are at least three reasons why the advertisement call has been at the center of neurobiological investigations. First, because of its conspicuousness, high rate of delivery,[2] and relative acoustic simplicity, the call can be readily recorded under natural conditions and then manipulated for use in laboratory or field playback experiments (Gerhardt 1988a, 1992a; Littlejohn 1977). Second, because the production of such calls is closely associated with the mating period, there are likely to be important neural links between the auditory and endocrine systems

2. Under both laboratory and field conditions, it has been demonstrated that several anuran species (e.g., *Physalaemus pustulosus, Hyla versicolor*) call continuously for periods of up to seven hours. Based on energetic studies, such calling appears to elevate oxygen consumption by approximately four to five times that observed in resting animals (Bucher, Ryan, and Bartholomew 1982; Ryan, Bartholomew, and Rand 1983).

(Aitken and Capranica 1984; Capranica 1965, 1966; Urano and Gorman 1981; Wilczynski, Allison, and Marler 1993). Third, all frogs produce advertisement calls. Consequently, the neural substrate underlying call production and perception can be examined phylogenetically (Ryan 1988; Schneider 1988; Walkowiak 1988).

When one thinks of a calling frog, one typically imagines an individual with an inflated air sac. Although the air sac (which receives its air from expiratory forces generated by the lungs) plays a role in sound amplification (Capranica and Moffat 1983; W. F. Martin 1972), the laryngeal cavity (muscles, vocal cords, cartilages), or sound source, is sufficient for call production; experimentally puncturing the sac does not perturb the spectral properties of the call, but does reduce the sound-pressure level. The anuran larynx is similar in structure to that of reptiles, comprised of a cricotracheal cartilage attached to the arytenoid cartilages. It differs from the mammalian larynx in that it lacks a thyroid cartilage. Among anurans, the size of the cricotracheal cartilage decreases and the arytenoids increase as one moves phylogenetically from the more primitive species within the genus *Bombina* to the more advanced species within the Ranidae. In all species that have been examined, the male's larynx is both larger and heavier than the female's larynx (Schneider 1988), even though in several species, females weigh more than males; the same pattern of sexual dimorphism has, interestingly, also been found in humans (a homoplasy), where adult males, whose body weight may be less than adult females, typically have a larger larynx (and prominent adam's apple) and a lower voice. There is also considerable variation in laryngeal weight among males of a given species, and there is strong selection on such

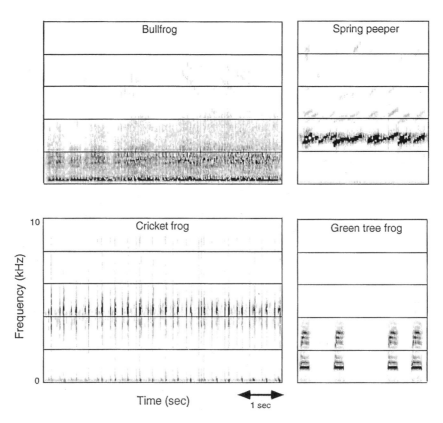

Figure 4.2
Sound spectrograms of a sample of frog advertisement calls. Time (seconds) is plotted along the
x-axis, frequency (kilohertz) along the *y*-axis.

variation as a result of both male-male competition and female choice (see chapter
6 for a more complete discussion of these ideas). Specifically, because one often
finds a significant negative correlation between male body size and vocal pitch[3]
(Davies and Halliday 1978; Ryan 1985), females can and do use differences in
pitch to guide their mating decisions, tending to mate preferentially with larger
males—those with deeper croaks.

3. In much of this work, researchers assume that there is a positive correlation between overall body
weight and the weight of the larynx. Although a study by Nevo and Schneider (1976) of green toads
(*Bufo viridis*) indicates a negative correlation between the size of the larynx and the fundamental
frequency, extrapolation to other species is not entirely warranted.

The vocal cords are largely responsible for the tonal quality of the signal generated. Among anuran species, there are three general vocal cord types (Schmid 1978). Several species exhibit the T-type, reflecting the shape of the cords. As with any morphological classification, however, there is variation within each type. Although we know relatively little about the causal connection between vocal cord structure and call structure, observations of *Bombina* and *Alytes* suggest that the smoothly curved features of the cords are probably responsible for the harmonically structured calls produced by these species (Schmid 1978).

Using data on laryngeal movements obtained by direct observations through the opening of the skull, Schmidt (1965) showed that during the production of mating calls in *Hyla cinerea* and *H. versicolor,* vocal cord operation is controlled by the laryngeal muscles and the central nervous system. This configuration stands in contrast to most other vertebrate systems, including other anurans (e.g., *Bufo valliceps;* W. F. Martin 1971), where acoustic signals are produced by the generation of air from the lungs that passively causes the vocal cords to separate and vibrate.

More causal information on the relationship between laryngeal morphology and call structure was obtained in an experiment performed by E. Weber (1976) on the mating call of *Hyla savignyi*. The laryngeal muscles were first ablated, then serum gonadoptropin injected to stimulate production of mating calls. The overall effect was to reduce the distinctive pulsatile property of the call, yielding instead a call with varying degrees of noise bridging each pulse in the sequence. Experiments with other species and other calls also indicate that ablation of the laryngeal musculature has its most disruptive effects in the temporal domain.

The central mechanisms underlying call production were first studied in the genus *Rana,* using electrical stimulation and focal lesions (Arnnson and Noble 1945; Schmidt 1966, 1968, 1969). Subsequently, other species were examined (e.g., *Xenopus*) using autoradiography to localize and trace the pathways between functional areas (Kelley 1980). In general, all of these studies indicate that the preoptic nucleus is the primary center for the production of advertisement calls and that males show a significantly stronger neuronal response in this area than do females. In addition, although females rarely call under natural conditions, hormone therapy (i.e., administration of either testosterone propionate or dihydrotestosterone) causes a detectable increase in call rate.

Before turning to the peripheral and central mechanisms underlying anuran perception, it is important for the reader to have a slightly more complete picture of the acoustic environment in which advertisement calls are produced and perceived. During the mating season, most males cluster within an area and wait for females to arrive. While waiting, some males call and some are silent. The density of calling males in a given area is likely to vary, and thus the level of competing vocalizations will differ for each male based on his location within the environment. The flip side, of course, is that females will be exposed to different densities of calling males and different sound-pressure levels as a result of distance to the chorus. In addition to the acoustic competition created by conspecifics, additional noise will be imposed based on auditory signals from other animals, including other anurans.

Given the acoustic environment for anuran advertisement calls, with males competing among each other for access to receptive females and females moving into a calling arena to find an appropriate mate, sexual selection theory[4] would predict that males and females will differ with regard to the design features of their perceptual systems. Thus, to increase the probability that females will hear, approach, and mate with them, anuran males must be able to detect relatively noise-free opportunities for calling. This ability requires a perceptual system that can properly assess background noise in three domains: overall intensity, signal duration, and signal bandwidth (frequency range). In contrast, the female frog's perceptual system must first localize potentially relevant signals within the general cacophony of the forest, second, discriminate between conspecifics and heterospecifics, and third, use the advertisement call of males to find a male of high genetic quality. The anuran perceptual system must therefore operate like an exquisite feature detector, but at multiple levels of analysis (Gerhardt 1992b).

From a methodological perspective, anurans are ideally suited for experimentation on auditory processing (Gerhardt 1988a; Littlejohn 1977; Walkowiak 1988). When male frogs hear natural or synthetic playbacks of advertisement calls, they either respond by calling or remain silent. In addition, perception of a mating call elicits significant changes in testicular volume as well as a detectable electrodermal skin response. Female frogs respond with phonotaxis (i.e., movement toward the sound) to playbacks of male advertisement calls, with response strength mediated by such factors as temperature and the spectral features of the call (see especially the discussion on pages 121–129, and also chapters 6 and 7).

Anurans do not have an external ear.[5] Most anurans have two middle-ear systems, a tympanic middle ear that includes a tympanum, middle-ear cavity, and

4. Darwin's (1871) theory of sexual selection, described in greater detail in chapter 6, consisted of two key components: intrasexual competition and intersexual competition. In most species, intrasexual competition occurs among males for access to females, whereas intersexual competition occurs in the form of female mate choice. For each species, the pattern of intra- and intersexual competition will be largely determined by the type of limiting resource.

5. The tympanum is present in most anurans, but is often located on the surface of the head in the form of a disc.

Table 4.1
Significant Features of the Amphibian and Basilar Papillae (modified from Zakon and Wilczynski 1988)

Feature	Amphibian Papilla	Basilar Papilla
Frequency range (kHz)	0.1–1.4, some species-dependence	1.0–4.0, species-dependent
Tonotopic organization	Yes	For most species, only tuned to one frequency
Hair cell density	High (@1000)	Low (@100)
Best excitatory frequency is affected by temperature	Yes	No
Size or sexual dimorphism	Rare	Common
Tuning mechanism	Mechanical/electrical	Primarily mechanical

stapes, and an opercularis system (Hetherington 1994). These middle-ear[6] systems connect directly with the oral cavity, a connection that represents a fundamental difference in design features from the mammalian system and leads to a situation where auditory signals can make contact with the eardrum from both sides. As a result, the anuran auditory system is specialized for highly directional sound localization (Eggermont 1988). The inner ear of anurans differs from all other vertebrates in that it consists of two spatially separated hearing organs,[7] the amphibian papilla and the basilar papilla (E. R. Lewis and Lombard 1988; Zakon and Wilczynski 1988). Based on cellular recordings obtained from a number of species, each papilla shows the characteristic V-shaped tuning curve of vertebrate nerves (see Box 4.1), but the best excitatory frequency is lower in the amphibian papilla (100–1400 Hz) than it is in the basilar papilla (> 1000 Hz). There are other important differences between these structures (see Table 4.1). In the amphibian papilla, length correlates positively with the range of best excitatory frequencies, whereas in the basilar papilla, the weight of the species correlates negatively with the range of best excitatory frequencies. Moreover, whereas the tuning properties of the amphibian papilla are comparable across species, the basilar papilla shows species-specific design features with variation at the intraspecific level mediated by size, sex, and geographic distribution. Finally, the

6. Not all anurans have a middle-ear structure. In a phylogenetic analysis by Duellman and Trueb (1986), results indicated that in 11 out of 21 families, at least one species lacked a middle-ear structure. Such "earless" frogs are interesting because they appear to have quite highly developed localization capabilities in the absence of structures believed to contribute to localization (e.g., pressure gradient created on the basis of a connection between the two middle ear cavities).

7. Evidence of spatial separation comes from single-unit recordings and horseradish peroxidase dye fills, indicating that the neural fibers from each papilla are distinct.

Box 4.1
Single-Unit Recordings and Tuning Curves

Technological advances within the past 30 to 35 years have made it possible to record activity levels (i.e., changes in the rate of action potentials) in a single neuron from either an anesthetized or an awake individual. The output of such recordings can provide insights into how the nervous system processes auditory and visual stimuli, and as a result, can provide researchers with important information about how cells and cell ensembles classify potentially communicative events.

The technique works by inserting a microelectrode probe into a target neuron to measure changes in firing rate. (More recent advances also permit simultaneous recordings from multiple cells.) Following baseline recordings, the subject is presented with either artificial stimuli (e.g., pure tones varying in frequency or spots of light varying in intensity) or natural signals (e.g., a vocalization or face); comparisons are then made between firing rates obtained during baseline and during test conditions. For example, one might determine the set of frequencies in an auditory signal or the orientations of a face that elicit the strongest firing rates.

Two important pieces of information are derived from single-unit recordings. The tuning curve represents the range of values that elicit a response (i.e., increase in firing rate over baseline) from the neuron being recorded. Thus, in the following figure, there is one frequency that elicits a maximum response—known as the *best excitatory frequency (BEF)*—and other frequencies that elicit weaker responses for a given sound-pressure level. In the tuning curve on the left, BEF is centered at 100 Hz for a sound-pressure level of 25 dB. In order for other frequencies to elicit a comparably strong response, the sound-pressure level would have to increase.

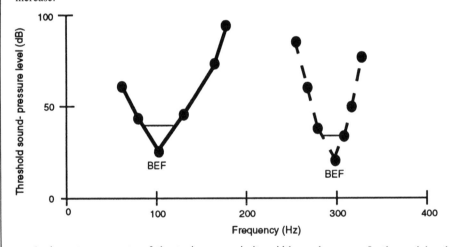

An important property of the tuning curve is its width or sharpness. In the peripheral auditory area, tuning is contingent upon the metabolic state of the inner ear. As a result, for ectothermic species such as frogs and toads, changes in temperature have a significant effect on auditory processing. The sharper the tuning curve, the more precisely stimuli can be discriminated. One way to quantify the tuning characteristics of a neuron is to calculate its Q_{10dB} value. This value is obtained by dividing BEF by the frequency width of the curve at 10 dB above the BEF (the horizontal line in each tuning curve). For example, the tuning curve on the left has a BEF of 100 and a frequency width of 25 Hz at 10 dB above the BEF. The Q_{10dB} value would then be 100 divided by 25, or 4.

Although tuning curves have played a central role in studies of visual and auditory perception, Capranica (1992) has argued that their importance should not be overrated. He makes three points. First, tuning curves do not provide the necessary information for assessing how specific neural areas process complex auditory signals. Second, with the exception of neurons in the bat auditory cortex dedicated to echolocation (Simmons 1989; Suga 1988; see section 4.3), most vertebrate auditory tuning curves are quite broad and lack the kind of tuning-specificity that is often invoked or assumed. For example, recent research on the green tree frog *Hyla versicolor* (Lim 1990) produced broad tuning curves with relatively low Q_{10dB} values. However, based on firing-rate data obtained in response to pure tones varying in frequency, there was no evidence of frequency selectivity, and responses to particular frequencies were significantly influenced by changes in intensity: as intensity increased, the neuron's most sensitive frequency (i.e., the frequency eliciting the highest firing rate) decreased. Third, because of an emphasis on tuning curves and Q_{10dB} values, there has been a tendency for studies of auditory neurophysiology to focus on spectral rather than temporal information. Although the assertion is controversial, Capranica suggests that because sound-production systems have been designed to produce biologically salient information in the temporal domain, perceptual systems will have been selected to be most efficient at decoding temporal information, not spectral information. Capranica thus makes a plea for us to concentrate more explicitly on time-based measures of responsiveness to changes in acoustic morphology. Several anuran researchers, especially Narins and his colleagues, have already begun to map out how temporal information is processed by the auditory pathways involved in communication (Dunia and Narins 1989a, 1989b; see discussion that follows). These issues are picked up in greater detail in this chapter, as well as in chapters 6 and 7.

amphibian papilla contains approximately 1,000 hair cells, whereas the basilar papilla contains approximately 100 hair cells.

An elegant example of sex differences in the neural basis of perception is illustrated by experiments with the Neotropical tree frog, *Eleutherodactylus coqui* (Narins and Capranica 1976). In this species, males produce a two-note advertisement call, the "coqui." The "co" component, used in male-male competition, is followed 100–150 msec later by the higher frequency "qui" note, used to attract potential mates. Although the best excitatory frequencies of the amphibian papilla are the same for males and females, males have a greater number of cells tuned to the "co" note than do females. And while females have a larger number of cells within the basilar papilla that are tuned to the "qui" note, overall, the best excitatory frequency of the basilar papilla is tuned to a higher frequency in males than it is in females.

Because anurans are ectothermic, temperature has a direct effect on their behavior. This is most clearly illustrated by temperature-sensitive changes in responsiveness by the two papillae. In both the American toad (Moffat and Capranica 1976) and spring peeper (Wilczynski, Zakon, and Brenowitz 1984), decreases in temperature from 22°C to 11°C result in a decrease in best excitatory frequencies for the amphibian papilla, but no change for the basilar papilla. From a functional perspective, such sensitivity to temperature changes has important consequences for studies designed to look at the fitness consequences of different mating strategies (Ryan 1988; see chapter 6). For example, a male's ability to attract females with an advertisement call will depend upon the female's body temperature (Gerhardt and Mudry 1980). Therefore, from a neural perspective, a male cannot count on the ability to consistently ''match'' the female's best excitatory frequency.

A critical step in perception is to localize the sound source. In general, this task can be accomplished by an auditory system that is capable of calculating interaural differences in the signal's arrival time, intensity, and spectrum. In a typical localization study, such as those performed on the barn owl by Konishi and his colleagues (Knudsen, Blasdel, and Konishi 1979; Knudsen, Esterly, and Knudsen 1984; reviewed in Konishi 1993), individuals are tested on only one trial, and playback stimuli are presented for brief periods of time. Single test trials are conducted to avoid the possibility that the response elicited is guided by location memory (i.e., remembering where the sound source is located). Brief sound presentations are implemented into the experimental procedure so that the onset of orientation and phonotactic movement occurs after stimulus offset. Thus far, most studies of sound localization in frogs (reviewed in Rheinländer and Klump 1988) have opted for a more naturalistic design whereby stimuli are broadcast repeatedly from a fixed location (though see recent experiments by Klump and Gerhardt 1989). As a result, we know little about the frog's ability to localize different angles of sound in the acoustic environment and know relatively more about the phonotactic zigzag response (Figure 4.3) used to localize sounds in the horizontal plane.

Interaural distance in frogs and toads is approximately 1–2 cm. Based on experiments with several species within the Hylidae and Ranidae, results indicate that mating calls can be localized to within 15° (Fay and Feng 1987; Rheinländer and Klump 1988). In general, the underlying mechanism for sound localization in anurans appears to be a pressure gradient, whereby interaural differences in pressure provide the requisite information about directionality (Eggermont 1988).

A majority of the results summarized thus far were collected under closed-field conditions (e.g., isolated chambers or enclosures). As several researchers have pointed out (Megela-Simmons, Moss, and Daniel 1985; Narins and Zelick 1988; Zakon and Wilczynski 1988), such conditions underestimate the complexity of the acoustic environment, both from the production and perception end. Thus the difficulty of finding periods of silence in which to produce signals (i.e., periods

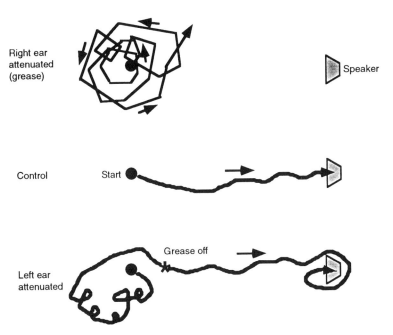

Figure 4.3
Results from a phonotactic experiment with the barking tree frog, *Hyla gratiosa*. Experiments involved playbacks of the male's advertisement call to females with the right ear attenuated (top), neither attenuated (middle), or the left ear attenuated (bottom). Ear attenuation was achieved by placing a thin coat of silicone grease on the eardrum. The arrows refer to the direction of the phonotactic path. The asterisk (*) in the bottom panel marks the location where the grease fell off (redrawn from Feng, Gerhardt, and Capranica 1976).

where acoustic overlap is minimal) is considerable given the large size of anuran choruses (hundreds of calling males). Similarly, attempting to localize a particular signal within such choruses is far more difficult than the conditions established in the laboratory, where individuals are typically presented with, at most, a two-choice situation. More recent studies have attempted to broaden the perspective by increasing the complexity of the stimulus environment and by obtaining direct neurophysiological measures of neural selectivity (Xu, Gooler, and Feng 1994). For example, when Gerhardt (1982) set up a more complicated experimental paradigm, involving four speakers rather than two, female *Hyla cinerea* performed less well in a frequency discrimination test. These data serve a cautionary note for those wishing to use neurophysiological and psychophysical data to assess selection pressures on particular traits observed under free-field conditions: perceptual abilities described under controlled laboratory conditions may not be

Bullfrog auditory nuclei

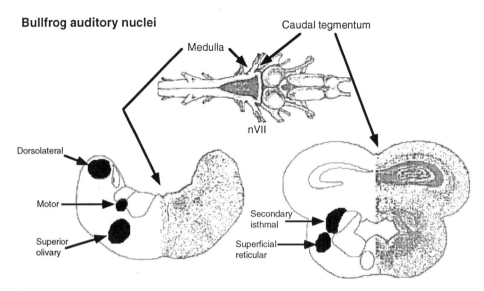

Figure 4.4
Schematic (left half) and Nissl-stained (right half) illustration of some of the more important nuclei in the bullfrog's auditory pathway (redrawn from Wilczynski 1988).

directly relevant (i.e., may not generalize) to the perceptual problems confronted in the species-typical environment.

As with all vertebrates, the anuran inner ear is connected to the brain stem via the VIIIth cranial nerve (Wilczynski 1988; Will and Fritsch 1988). In general, the structural properties of the advertisement call (e.g., intensity, time, frequency) are encoded at the level of the VIIIth nerve, whereas the central nervous system is responsible for decoding the call and recognizing it *as* a conspecific mating call. Information on VIIIth nerve projections have been obtained for a diversity of anuran species[8] using radioactive labeling techniques (e.g., horseradish peroxidase). Afferents from the VIIIth nerve terminate in nuclei of the ventral zone, the dorsolateral nucleus, the reticular formation, and the cerebellum. Concerning the two papillae, fibers project into the brain through the dorsal root of the VIIIth nerve, with amphibian papilla projections dorsal to basilar papilla projections. Fibers from the two papillae then send their primary projections to different regions of the dorsolateral nucleus (see Figure 4.5 for a description of the cir-

8. As pointed out by Will and Fritsch (1988), there are important interspecific differences with regard to the specific projections observed between nuclei within the auditory pathway.

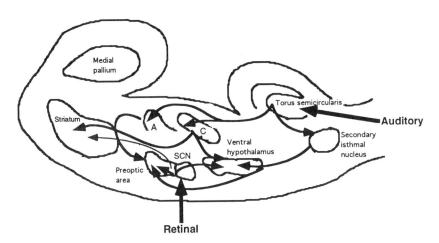

Figure 4.5
Schematic representation of the primary auditory pathways in the anuran brain, including connections that influence the feedback loop between acoustic stimulation, hormonal changes, and reproductive behavior. SCN: suprachiasmatic nucleus; A: anterior thalamic nucleus; C: central thalamic nucleus (redrawn from Wilczynski, Allison, and C. Marler 1993).

cuitry), thereby creating a tonotopic map of frequencies.[9] The sacculus, a secondary structure which is responsive to low-frequency signals, has terminal projections in the ventrolateral region of the dorsolateral nucleus. In addition to these projections, there are at least four other medullary nuclei involved in auditory processing, indicating that complex, multichannel computations occur at this level.

In summarizing current knowledge of the anuran brain stem auditory pathway, Wilczynski (1988) highlights five main points. First, all of the primary auditory nuclei exhibit a tonotopic structure. Although the tonotopic organization of the mammalian auditory system is relatively more sophisticated than what has been reported for anurans, this difference may lie in the fact that anurans, in contrast to mammals, must integrate auditory input from two hearing organs (i.e., the amphibian and basilar papillae) rather than one. Integration may represent computational challenges that necessitate decreased tonotopic efficiency.[10] Second,

9. There are extensive commissural interconnections between the two dorsolateral nuclei, and tuning curves indicate that the best excitatory frequencies are the same for cells on each side. Such symmetry serves to maintain the tonotopic frequency representation. Note, however, that in contrast to many birds and most bats, tonotopy in frogs is weak because each neuron is broadly tuned.

10. Although Wilczynski's suggestion is reasonable here, it is also possible to think of this neural division of labor as facilitating integration, in much the same way that the domain-specific modulariza-

each brain stem center in the auditory pathway shows bilateral connections, and as in mammals, such connections are likely to facilitate sound localization. Third, information flow occurs over several parallel pathways, and thus projections from one nucleus can be traced to all nuclei that lie anteriorly, Fourth, several brain stem centers have descending projections that appear to serve a feedback function. Last, there is some evidence, though not as strong as reported for the barn owl or moustached bat, that the auditory nuclei are arranged hierarchically such that ascent through the auditory pathway is associated with an increase in the complexity of the signal to which each station responds.

The final pathways for the auditory system lie between the torus semicircularis and the forebrain (Neary 1988). The torus semicircularis sends out two primary projections, one to the secondary isthmal nucleus and one to the ventromedial area of the central thalamic nuclei. From these two nuclei, a considerable number of fibers project to the ventral portion of the hypothalamus, a primary neuroendo-crinological center. As previously mentioned, the connection between the auditory and endocrine system may lie at the heart of the system that links vocal and reproductive behavior in anurans.

How do the different pathways work? That is, are the nuclei and their respective projections and pathways responsible for encoding and decoding different properties of the calls perceived? To answer these questions, we turn to data from single-unit tuning curves—the primary source of information for this problem. At the spectral level, results indicate that the auditory midbrain represents the first major processing station. Tuning curves from neurons in the dorsal lateral nucleus are V-shaped and are comparable to those obtained from the auditory periphery. They respond to both tonal and noisy signals, with neurons sensitive to low-frequency signals exhibiting the lowest excitatory thresholds. In parallel with the response characteristics of the VIIth nerve, neurons in the dorsal lateral nucleus show a positive correlation or dependence between the intensity of the excitatory stimulus and inhibitory threshold.

Moving to the next step in the auditory pathway, neurons in the superior olivary nucleus, like those in the dorsal lateral nucleus, are tonotopically represented, with low-frequency neurons located in the dorsolateral portion and high frequencies in the ventromedial portion. The primary differences between these two nuclei are that for the superior olivary nucleus, inhibitory thresholds are lower, inhibitory effects of two-tone signals are greater, and some neurons are actually

tion of the human brain, together with hemispheric differentiation, has been looked at in terms of facilitating task demands.

inhibited by the presence of broadband noise. The end result is that the superior olivary nucleus represents the first stage in the auditory pathway where spectral information is processed.

Neurons in the torus semicircularis lack tonotopic structure and exhibit a significantly broader and more complex range of spectral responses than do neurons in the other two nuclei discussed thus far. For example, recordings have revealed that some neurons respond to single or combination tones within a restricted frequency range, whereas other neurons are responsive to all frequencies within the audible range of the species.

The pretectum and caudal thalamus represent the final stations for auditory processing. Both areas receive projections from the superior olivary nucleus, and their response characteristics are similar. Specifically, there is evidence of considerable selectivity for signals with bimodal frequency distributions, and especially spectral distributions that "match" the species-typical mating call. For example, in *Rana pipiens*, the best excitatory frequencies match those typically generated by males giving an advertisement call. Neural activity or responsiveness is suppressed when an additional frequency component is added to the signal. Based on the complexity of this region's neural selectivity, it has been suggested (Fuzessery 1988) that neurons in the pretectum and caudal thalamus may function as mating call detectors.

In addition to spectral sensitivity, neurons in the anuran auditory system are also responsive to temporal information. Temporal information—in particular, amplitude modulation—plays an important role in anuran signal recognition. For example, two species of toad, *Bufo americanus americanus* and *B. woodhousii fowleri,* produce spectrally similar advertisement calls, but with significant differences in pulse repetition rates. The primary role of the auditory periphery appears to be as a filter, centered at about 100–250 Hz. Neural responses are guided by changes in phase, but not periodicity. It is not until the dorsal lateral nucleus and the superior olivary nucleus, however, that one begins to observe neurons that are responsive to temporal information such as rate.

Research on the neotropical tree frog, mentioned earlier, illustrates the importance of temporal information in anuran communication systems. Data from single-unit recordings (Dunia and Narins 1989a, 1989b) show that the auditory nerve is capable of temporal integration (i.e., the time period over which a neuron integrates information), varying as a function of characteristic frequency. Fibers with characteristic frequencies below 500 Hz had the longest integration times, fibers with frequencies between 500–1300 Hz showed the shortest integration times, and fibers with the highest frequencies (greater than 1300 Hz), showed intermediate integration times. The functional significance of this frog's sensitivity to temporal features, and in particular, brief periods of silence between the "co" and "qui" note, is that males can minimize acoustic jamming (call overlap), thereby maximizing the probability of sending a relatively undegraded signal to females. From the female's perspective, the behavioral and neurophysiological adaptations that minimize acoustic jamming are important, for they provide her with an opportunity to more readily evaluate the signals of individual males.

Narins, Ehret, and Tautz (1988) have further revealed that in addition to the traditional sound-receiving mechanism, *E. coqui* also possesses a sound-receiving system on the portion of the body surrounding the lung cavity. This area is heterogeneous with regard to its response to ambient sound. However, the area overlying the lung cavity is most sensitive to signals in the 1300–2600 Hz frequency range. This range corresponds to measured frequencies of the "qui" note and to the best excitatory frequency of the basilar papilla. As a result, it appears that the sensitivity of the lateral portion of the body to sound vibration may provide this frog species with a highly refined mechanism for sound localization, critical to females in their attempt to choose an appropriate mate.

An interesting consequence of the temperature-sensitive changes in anuran behavior is that temporal features of the advertisement call, and their perception, are influenced by fluctuations in temperature. For example, in the European tree frog, *Hyla arborea,* Schneider (1977) has demonstrated that pulse repetition rates

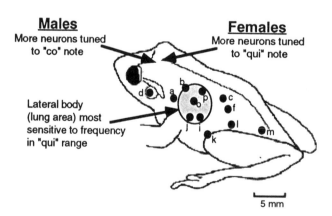

Figure 4.6
Schematic illustration of the Puerto Rican tree frog (*Eleutherodactylus coqui*) highlighting the areas of the overlying lungs that are sensitive to sound. The shaded area is most sensitive to sounds in the frequency range corresponding to the "qui" portion of the advertisement call (from Narins, Ehret, and Tautz 1988).

increase as a function of increases in male body temperature. A comparative study by Stiebler and Narins (1990) of the Pacific tree frog (*Hyla regilla*)—a species living in a highly variable thermal environment—and the neotropical tree frog (*E. coqui*)—a species living in a more constant thermal environment—revealed significant auditory response changes to experimentally induced changes in temperature. Specifically, tuning curve data revealed that for both species, center frequencies did not change within the basilar papilla, but shifted to a higher frequency range for recordings obtained from the amphibian papilla. Maximum Q_{10dB} values were obtained at 22°C for *H. regilla* and at 24°C for *E. coqui*. In playback experiments, Gerhardt (1978b) has shown that female gray tree frogs (*H. versicolor*) choose potential mates on the basis of temporal information that matches the species-typical structure. What is most intriguing, however, is that the perceptual system of the female appears capable of compensating for temperature fluctuations that influence the structure of the male's calls. Studies suggest that the ability of neurons in the auditory midbrain to shift their firing rates as a function of changes in temperature may underly the ability of females to compensate for changes in pulse rate (Brenowitz, Rose, and Capranica 1985; Rose, Brenowitz, and Capranica 1985).

Perceptual systems ultimately constrain the form of the signal generated. Nowhere has this fact been more elegantly demonstrated in the animal kingdom than among anurans. Although studies of several anuran species have documented an intricate interplay between the features of the auditory system (e.g., characteristic frequency and sharpness of auditory nerves) and the spectral contour of the advertisement call, I focus here on two of the most carefully detailed cases, carried out on the Túngara frog, *Physalaemus pustulosus,* and the cricket frog, *Acris crepitans* (Keddy-Hector, Wilczynski, and Ryan 1992; Ryan and Drewes 1990; Ryan et al. 1990b; Ryan, Perrill, and Wilczynski 1992; Ryan and Wilczynski 1988, 1991; Wilczynski, Keddy-Hector, and Ryan 1992). These studies are exemplary because they have explored (1) the neural basis of perception, (2) the anatomical structure of the vocal apparatus, (3) the acoustic features of the advertisement call that most significantly affect mate choice, (4) population variation in call structure and nerve fiber tuning, and (5) the selective pressures that have led to the particular design features of the system observed. Some of the implications of this work for current problems in sexual selection theory are discussed in greater detail in chapter 6.

In the first of a series of studies, Ryan and Wilczynski (1988) demonstrated that auditory neurons in the basilar papilla of female cricket frogs are tuned lower (i.e., the best excitatory frequency) than the dominant frequency of the male

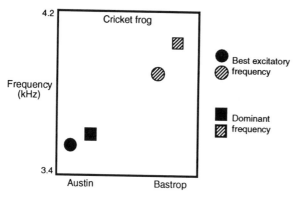

Figure 4.7
The relationship between the dominant frequency of the cricket frog's advertisement call (circles) and the best excitatory frequency of the female's basilar papilla (squares). The Austin population is indicated by the black symbols, the Bastrop population by the striped symbols. Each data point represents the mean, adjusted for differences in body weight (redrawn from Ryan and Wilczynski 1988).

advertisement call. Moreover, recordings from two geographically separated populations indicated that dominant frequencies were significantly different, as were the best excitatory frequencies. Figure 4.7 illustrates both the correspondence between production and perception, and interpopulation variation.

Given the results already discussed, as well as more recent analyses of a larger range of populations (Wilczynski, Keddy-Hector, and Ryan 1992), Ryan, Perrill, and Wilczynski (1992) set out to test the hypothesis that female mate choice is guided by sensory biases that lead to selection for males whose calls "match" the auditory system's best excitatory frequency. As stated, this hypothesis claims that female choice is the driving force (the causal agent) underlying variation in the expression of the male's trait. Using a phonotactic procedure (two-speaker choice test, assessing female approach behavior) involving playbacks of calls from three different populations, results indicated that where females exhibited a preference (i.e., approached one playback speaker more often than the other), they preferred the lower-frequency call; preference for the lower-frequency call emerged even when the choice was between the female's own and a foreign population. Moreover, within populations, relatively large females tended to prefer lower-frequency calls whereas smaller females preferred higher-frequency calls. These results, together with those published for the Túngara frog, illustrate the importance of considering the coevolution of sender-perceiver systems and the extent to which sensory biases can exert strong pressures on the pattern of selection observed both within and between populations (see section 6.2).

Because of the close association between calling and mating, one would expect to find a number of significant pathways between auditory processing areas and hormonal centers in the brain that mediate physiological state. Wilczynski, Allison, and C. Marler (1993) have found that the suprachiasmatic nucleus (involved in the control of circadian rhythms in vertebrates) sends projections to the preoptic area and ventral hypothalamus (these two areas represent the control centers for the release of gonadotropin releasing hormone, GnRH). Thus the pathway between the suprachiasmatic nucleus and GnRH control regions provides a mechanism by which social signals are coupled to environmental changes relevant to reproduction.

In a study designed to look directly at how vocal activity influences hormonal profiles, advertisement calls were played repeatedly to male *Rana temporaria* before and after the breeding season (Brzoska and Obert 1989). Results indicated that the normal time course of change in testes volume (i.e., shrinkage) was retarded as a result of exposure to conspecific calls.

In summary, neurobiological studies of anuran vocal communication reveal both highly specialized design features, apparently reflecting specific selection pressures, and quite general design features, shared in common with other closely and distantly related species. Thus, in contrast with other well-studied vertebrates, anurans have two hearing organs, each designed to solve different acoustic problems. This specialization, together with division of neural labor, results in an enhanced ability to process the dominant vocalization in this group's repertoire: the advertisement call, produced by males in intra- and intersexual competition, and perceived by females as a source of information about the location of an appropriate mate. Specialization at this level, however, appears to result in a compromise on a different level: reduced tonotopic organization and more broadly tuned neurons relative to some other vertebrates. In addition, because changes in temperature have a direct effect on both the production and perception of vocalizations, the anuran brain has evolved compensatory mechanisms to enable females, for example, to properly evaluate male survival quality based on call characteristics. Shared with other species, as revealed in greater detail in sections 4.2.2 and 4.3, is the observation that the anuran brain has been built to discriminate conspecific calls from heterospecific calls and to solve this discrimination problem by perceptually "finding" critical acoustic features within a complex environment. In addition, as in many avian species (see next section), the neural structures involved in acoustic processing are intimately connected to the neural structures involved in regulating the individual's hormonal profile, especially those hormones involved in reproductive behavior.

4.2.2   Avian Song

Those of us who were born and raised on the east coast of the United States are intimately familiar with the relatively demarcated changes associated with the four seasons. Announcing the spring are the aesthetically appealing melodies of the songbirds. But why spring? Why, in most species, do males sing, but not females? Why are some species, like the mockingbird, capable of producing an almost infinite variety of song types whereas other species, such as the white-crowned sparrow, are limited to a single song type? At one level, these questions demand an understanding of the neurophysiological substrate underlying avian song production and perception. And at another level, they demand an understanding of the evolutionary function of such communicative processes and interspecific patterns. The following section provides a synthesis of a voluminous literature on how birds produce their songs, why males rather than females are generally responsible for the *job* of singing, and how the perceptual system has been designed to detect subtle variations in song structure. The discussion will concentrate primarily on the neuroanatomical and neurophysiological mecha-

nisms guiding adult song, and will leave the exquisite empirical edifice documenting ontogenetic changes in song production and perception for chapter 5. Chapter
6 provides information on the adaptive significance of avian song.

In terms of sound production, birds are impressive vocalists, with signals covering a broad frequency range and a diversity of acoustically varied elements
including buzzes, trills, and pure-tone whistles. For example, the brown-headed
cowbird (*Molothrus ater ater*) produces a song that starts with a 200 Hz segment
and ends with a segment that approaches 11,000 Hz (King and West 1983). In a
small sample of species, such as the oilbird (*Steatornis caripensis*) and grey swiftlet (*Collocalia spodiopygia*), individuals use relatively low-frequency[11] clicks for
echolocation, thereby facilitating navigation through dark caves (Suthers and
Hector 1982, 1985). And in the songs of most avian species there are overlapping
elements that are spectrally unrelated, leading to the suggestion that some birds
are capable of recruiting two independent voices (Greenewalt 1968; Nowicki and
Marler 1988; Stein 1968; Suthers 1990). Our knowledge of the neurobiology of
avian communication is, however, primarily restricted to song, with relatively
little understanding of the factors mediating *call* production and perception.[12]
Most authors distinguish between calls and songs on the basis of their duration
and spectral structure: songs are longer and more tonal (either pure-tone sinusoids
or harmonically structured) than calls. As indicated in the section heading, the
following discussion is restricted to song, a signal that plays an important function
in avian mating systems. The empirical beauty of avian song, as discussed for
the anuran advertisement call, is that it provides the opportunity to examine the
neural interface between the auditory and endocrine pathways.

A general sound production system consists of three components: an air-
generating mechanism, a vibrating mechanism, and a resonating cavity. Birds
employ a three-component system and, like all other vertebrates, use their
respiratory system to set the sound production machine in motion. Because

11. In contrast to the high-frequency echolocation pulses of bats and dolphins, which can reach values
of close to 100 kHz, echolocation pulses in swiftlets and oilbirds are generally less than 10 kHz.

12. Studies of acoustic signaling in frogs, birds, bats, and dolphins—taxonomic groups for which our
understanding of the proximate causes of variation are significant—have tended to focus on a highly
limited and specialized portion of the repertoire. Thus studies of frogs and birds have concentrated
on their mating signals, whereas for both bats and dolphins, an almost exclusive emphasis has been
placed on echolocation. As a result, our understanding of the mechanisms of production and perception in these species may not generalize to other signals in the repertoire. In contrast, research on
human (Fant 1960; Sundberg 1987) and nonhuman primate communication (Jürgens 1979, 1990) has
focused on a much more comprehensive set of signal types within the repertoire. Generalizations
about the neurophysiological substrate underlying vocal communication can thus be more readily
asserted.

birds often produce long and quite complex sounds, flow rate through the vocal tract is often substantial. Although many sounds can be produced from a single inspiration-expiration cycle, recent research (Hartley and Suthers 1989) on canaries (*Serinus canaria*)—using data from tracheal airflow and air sac pressure—indicates that individuals use minibreaths between the silent pauses of their song. In contrast to a normal inspiratory cycle, such minibreaths enable canaries to produce songs of relatively longer duration.

Birds possess an anatomically unique sound-generating structure known as the syrinx (Figure 4.8). Like its functional analogue in anurans and mammals, the larynx, there is considerable interspecific variation in the structure of the syrinx. For example, the syrinx of suboscines is considerably less complex in terms of muscle control than in the syrinx of oscines, or true songbirds (Feduccia 1980). The following characterization of syringeal structure and function thus focuses on the oscines, but provides details of interspecific differences as they relate to specialized design features of the communication system.

Among the oscines, the syrinx is located between the primary bronchi and the heart, enclosed inside the interclavicular air sac. The trachea joins with the top of the syrinx, whereas the bronchi (which are connected to the lungs) join with the base of the syrinx. Below the connection between the syrinx and the bronchi are a series of C-shaped cartilaginous rings. Across these rings, and on each bronchus, are thin elastic membranes known as the medial tympaniform membranes. It has been postulated that these membranes are critically involved in vibration, and thus far they have played a central role in models of phonation (Casey and Gaunt 1985; Greenewalt 1968).

Each side of the syrinx represents an anatomical duplicate of the other, with separate innervation from the right and left twelfth hypoglossal nerve. This apparent structural symmetry, together with spectrographic data indicating two different sound sources, has led to several hypotheses about the mechanical operation of the syrinx during song production. Greenewalt (1968) was the first to propose a comprehensive quantitative model of syringeal function, suggesting that the medial tympaniform membranes of each bronchus represent the source of sound production, that each side of the syrinx operates independently of the other, and that there are no significant filtering effects on the source (i.e., the suprasyringeal cavity has no effect on the spectral properties of the signal). Suthers (1990) and Suthers, Goller, and Hartley (1994) provided the first piece of experimental support for Greenewalt's "two-voice" theory based on direct measures of acoustic output and syringeal function in two songbirds, gray catbirds (*Dumetella carolinensis*) and brown thrashers (*Toxostoma refum*). Suthers used thermistors to

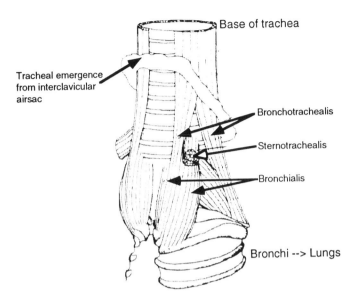

Base of trachea

Tracheal emergence
from interclavicular
airsac

Bronchotrachealis

Sternotrachealis

Bronchialis

Bronchi --> Lungs

Figure 4.8
Schematic illustration of the avian syrinx, including some of the more important muscles involved in
sound production (redrawn from Nowicki and Marler 1988).

record flow rate from the trachea and a piezoresistive pressure transducer to
obtain pressure changes from the air sac—a true technological *tour de force*.
Results indicated that in both species components of the song could be produced
by either side of the syrinx alone, by synchronous contributions from right and
left side, or by rapid alternation between sides. When the right and left side of
the syrinx operated simultaneously, both sides were capable of producing either
the same or different sounds. The latter provides the most convincing evidence
of two independent sources.

Research on canaries has provided conflicting data with regard to the relative
contribution of each side of the syrinx to sound output. In the first study, Nottebohm and Nottebohm (1976) found that song structure was negatively affected
(i.e., fewer syllables were produced) by sectioning the left tracheosyringeal
branch of the hypoglossal nerve, but only minor effects were observed when the
right side was sectioned. In contrast, McCashland (1987) found no evidence of
asymmetry in syringeal function based on data from canaries whose right or left
bronchi were plugged. More recently, Hartley and Suthers (1991) have demonstrated, using bronchus plugs and then ipsilateral denervation of the syringeal

muscles, that the left side of the syrinx is primarily responsible for generating syllables.

In addition to the studies mentioned, a suite of new studies adds further confusion to the kinds of generalizations that can be made with regard to syringeal function in songbirds. In particular, H. Williams and colleagues (1992) have demonstrated that in the zebra finch there is right-side dominance with regard to the production of song components. Allan and Suthers (1994) have demonstrated that in the brown-headed cowbird the left side of the syrinx is generally responsible for producing low-frequency introductory notes, whereas the right side produces high-frequency introductory notes. More significant asymmetries are observed when other components of the song are examined. Thus, in contrast to the left side of the syrinx, the right side produces notes that are higher in frequency, more varied, and more significantly frequency modulated. In this sense, therefore, brown-headed cowbirds are also right-side dominant. Last, Goller and Suthers (1995) present data on brown thrashers showing that lateralization only occurs for muscles involved in regulating airflow through each side of the syrinx. In contrast, syringeal muscles that control the precise acoustic morphology of song notes (i.e., the phonetic structure of the song) are involved bilaterally.

Based on studies of other species and a reevaluation of the vibrational properties of the medial tympaniform membranes, Nottebohm (1976) and then Gaunt and colleagues (Casey and Gaunt 1985; Gaunt and Gaunt 1985) provided an alternative to Greenewalt's model of syringeal function. This model suggests that the syrinx, like a hole-tone whistle, generates sound by producing stable vortices. Although this suggestion may explain syringeal function in some species, it does not account for the results obtained by either Suthers and colleagues (Suthers 1990; Goller and Suthers 1995) for catbirds and thrashers, nor for Nowicki's (1987; Nowicki and Capranica 1986) data on several other oscines. In fact, Nowicki's recent experiments suggest at least two important emendations to the current framework for understanding avian song production. First, following unilateral denervation of syringeal nerves in the black-capped chickadee, it is evident that the linear sum of the left and right postoperative signals does not equal the normal signal produced preoperatively. Moreover, unlike the results obtained for catbirds and thrashers, there is no evidence that each side simultaneously contributes different elements. Consequently, data on chickadees suggest that the two sides of the syrinx are not necessarily independent, but can be coupled, thereby generating nonlinear interactions. Second, Nowicki (1987) employed a technique—originating with Hersch's (1966) work on Stellar's Jays (*Cyanocitta stelleri*) and Japanese quail (*Coturnix coturnix*)—that involved placing individuals

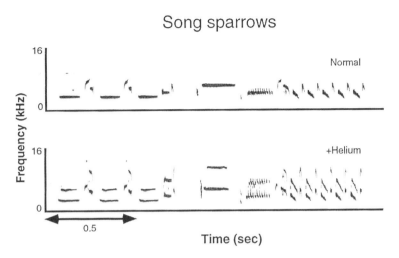

Song sparrows

Figure 4.9
The effects of helium on song production in song sparrows. The top spectrogram is from an individual singing in normal air, the lower spectrogram is of the same individual singing in helium. Notice how the second harmonic for each note is present in the lower spectrogram but not in the upper one (redrawn from Nowicki 1987).

from nine oscine species in a helium environment[13] and recording their song. Because helium-generated songs contained harmonic overtones that were absent in the pure-tone signals generated under normal conditions, Nowicki concluded that the suprasyringeal cavity imposes a significant filter function on the source (Figure 4.9). Moreover, based on kinematic studies of songbirds (discussed in Nowicki, Westneat, and Hoese 1992; Westneat et al. 1993) and spectrographic analyses of echolocation signals produced by oilbirds (Suthers and Hector 1988), it is now clear that several avian species are capable of modifying the source function by means of suprasyringeal modifications (i.e., articulation and anatomical asymmetries in vocal tract shape).

In summary, the sound-producing apparatus of songbirds is remarkably versatile, varying in the mode of operation from species to species. In some species, it appears, as Greenewalt suggested more than twenty years ago, that each side of the syrinx acts independently and is capable of producing spectrally distinct

13. Helium causes a change in the propagating velocity of sound and thus has a direct impact on resonance frequencies. In studies of human speech (Beil 1962), for example, results indicate that the fundamental frequency is generally unaffected by helium whereas the formant frequencies are shifted upward. This shift is what causes Mickey Mouse–esque voices to emerge after you have swallowed helium from a balloon!

notes at the same time, as well as spectrally identical notes at the same and different times. In other species, the right and left sides of the syrinx may be coupled, thereby imposing nonlinear effects on the spectral function. And although there is now convincing evidence that some oscine species actively manipulate the filtering properties of their vocal tract to alter spectral contours, most songbirds produce pure-tone signals, suggesting that pitch-related information is perhaps of paramount importance, and is certainly so in song recognition (Marler 1969b; Nowicki et al. 1989). This emphasis on the fundamental frequency stands in striking contrast to data on humans and some nonhuman primates (discussed further on in this chapter and in chapter 7), where harmonic structure, resonance frequencies, and bursts of broadband noise carry significant weight with regard to call structure, and certainly in the case of human speech, call function.

Before turning to the neural substrates underlying song production, I first provide a brief summary of how the peripheral auditory system works during song perception (for a review, see Hulse 1989; Saunders and Henry 1989). This shift from production to perception is particularly relevant to the oscine song system because in all species auditory feedback plays a critical role in the acquisition and maintenance of species-typical songs (Konishi 1965a, 1965b, 1989; Marler 1987; Marler and Waser 1977; Nottebohm 1968). And as I will document, the neural pathways for song production are intimately connected with the pathways for song perception (Bottjer, Mieser, and Arnold 1984; Cynx and Nottebohm 1992a, 1992b; Williams and Nottebohm 1985; reviewed in Konishi 1985, 1989, 1994a, 1994b).

The peripheral auditory system of birds is apparently quite ancient, sharing a number of structural similarities with the auditory system of the Crocodylia (Manley and Gleich 1992). The mechanical operation of the avian middle ear structure (Figure 4.10) has been described by Manley and his colleagues as a second-order-lever system (Manley 1990). What this entails is a series of interconnecting processes between the tympanic membrane and the oval window. As the tympanic membrane moves, the inferior process of the extracolumella moves in and out, creating a lever motion. The most significant limitation of this type of hearing system is that its efficiency is considerably diminished at frequencies above 4.0 kHz. Consequently, the upper limit of hearing in birds is approximately 12.0 kHz, though most songbirds are limited to the 6.0–9.0 kHz range.

The hair cells in the avian auditory system have typically been classified into four categories, based primarily on size and degree of specialization. The number of hair cells varies between species. Most songbirds have a short basilar papilla (approximately 3–5 mm) with several thousand hair cells, whereas the barn owl—representing the opposite extreme—has a longer papilla (approximately 11 mm)

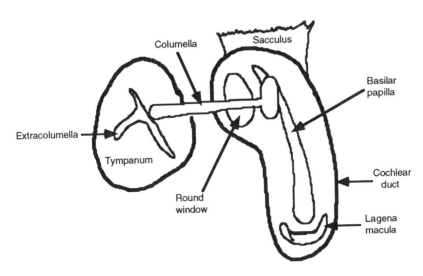

Figure 4.10
Schematic illustration of the middle ear of a bird (redrawn from Manley and Gleich 1992).

and over 16,000 hair cells (Fischer, Köppl, and Manley 1988). As with other vertebrate species, tuning curves obtained from VIIIth nerve afferents in birds exhibit a best excitatory frequency that ranges from approximately 100 Hz to approximately 6.0 kHz. In general, Q_{10dB} values indicate that within the 100–6000 Hz range, tuning curves are relatively sharp, though variably so. Additionally, of those species that have been carefully studied—chicken and starling—there is a systematic distribution of nerve fiber response frequencies as a function of location on the papilla (Figure 4.11). Perhaps most intriguing from a comparative perspective is the fact that although songbirds exhibit extensive variation in song structure, they represent a relatively homogeneous group with regard to hearing sensitivity and bandwidth (Dooling 1992; Fay 1992). This pattern—conservative hearing systems and variably designed acoustic production systems—is one that we shall be revisiting throughout this book.

At least one study of an avian species suggests that the excitation pattern (i.e., the distribution of neural excitation as a function of the fiber's characteristic frequency) of cochlear nerves is different from that of mammals. Specifically, work on starlings (*Sturnus vulgaris;* Gleich 1994) has demonstrated that in response to a broad range of frequencies (0.125 to 2.0 kHz) and sound-pressure levels (20 to 90 dB SPL), there is a significant asymmetry in the excitation pattern, with a significantly steeper slope on the high- than low-frequency side. Moreover,

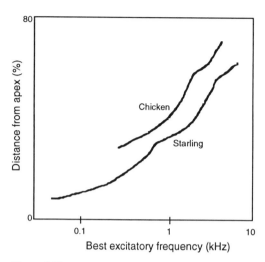

Figure 4.11
Tonotopic arrangement in the avian basilar papilla, showing best excitatory frequencies as a function of placement/location (redrawn from Manley and Gleich 1992).

whereas mammals typically show excitation patterns that covary with changes in sound-pressure level, changes in sound-pressure level did not have a significant effect in starlings.

Although the gross anatomy of the avian papilla is similar across species, there are some important differences at a more microanatomical level that provide interesting insights into the functional aspects of the communication system and environment. Thus, as mentioned previously, there are differences in the density and orientation of the hair cells, and in the length and thickness of the papilla. In the barn owl, for example, the density of hair cells is relatively high, the orientation pattern of the cells differs from that of the starling and chicken, and there is a significant thickening of the basal portion of the basilar membrane. It is believed that these species-specific features provide the barn owl with acute sensitivity to high-frequency sounds, necessary for accurate localization (C. A. Smith, Konishi, and Schull 1985). Last, although several features of the avian basilar papilla are similar to those that have been described for other vertebrates, recent studies of the chicken, pigeon, and starling suggest an important difference: whereas most species exhibit either electrical or mechanical tuning of the hair cells, at least some avian species evidence a combination, with electrical tuning at the apex and mechanical tuning at the basal end. What is not yet known is why, from a functional perspective, selection should have favored such mechanisms in birds but not other vertebrates.

How does the brain orchestrate all of the extraordinary sensorimotor functions that underly the avian song system? The first critical piece to this puzzle emerged in the mid-1970s when Nottebohm and colleagues (Nottebohm, Kelley, and Paton 1982; Nottebohm, Stokes, and Leonard 1976) uncovered several discrete nuclei within the canary brain that appeared to be dedicated to song control and that were generally either absent or greatly reduced in size in females relative to males; this gender difference represents a key insight given the virtual absence of singing in female passerines.

As illustrated in Figure 4.12, the neural pathway for song production originates in the higher vocal center (HVc)[14] and then makes its way to the syrinx via three other nuclei: the robustus archistriatum (RA), the dorsomedial nucleus (DM) of the intercollicular complex, and the tracheosyringeal portion of the hypoglossal nerve (nXIIts). These primary nuclei are located bilaterally, and their projections tend to be unilateral.[15] Lesioning any step along this pathway either completely eliminates song output or causes significant distortion of song structure.

Nottebohm's pioneering studies paved the way for research in five major areas: (1) describing additional components of the song pathway, especially the interconnection between nuclei responsible for production and those involved in perception, (2) assessing interspecific differences in the song control regions based on differences in ecology and mating behavior, (3) determining the neural basis for sex differences in singing behavior, including the interface between hormonal and auditory control regions, (4) documenting both the ontogenetic and seasonal changes in neural structure and function, and (5) study of the neural basis for vocal learning. I now turn to an overview of some of the findings in these areas, leaving the majority of the observations on ontogeny for chapter 5.

The primary pathway for song production starts with projections leading from HVc to RA. The dorsal area of RA then sends projections to DM, whereas the ventral portion of RA sends projections directly to nXIIts. DM projects to the caudal portion of nXIIts. The final step in this pathway consists of nXIIts projections, responsible for innervating the syringeal muscles, together with projections from RA to two nuclei (nucleus ambiguus and nucleus retroambigualis) that

---

14. In the early birdsong literature, this nucleus was inappropriately labeled the hyperstriatum ventrale, pars caudalis.

15. Recent research by Wild (1993, 1994) indicates that both RA and DM send relatively weak projections to the contralateral hypoglossus and respiratory nuclei.

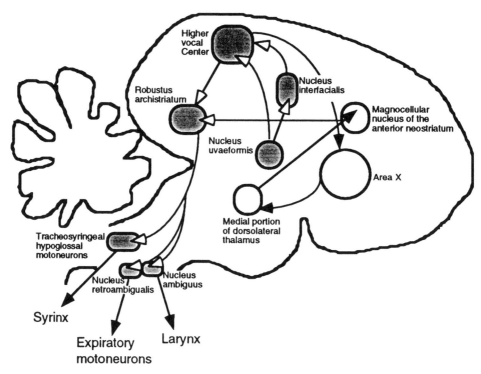

Figure 4.12
A schematic illustration of the primary nuclei involved in the song production system of passerine birds. Arrows indicate primary projections between nuclei. The dark-shaded nuclei are primarily involved in song output, whereas the stippled nuclei are more directly involved in song development and are described in greater detail in chapter 5 (circuitry details from Arnold 1992; Bottjer 1991; and Konishi 1994b).

control breathing and the timing of signals (Wild 1993, 1994). In addition to the song production circuitry, recent studies have revealed significant interactions between the song production and perception nuclei (Williams and Nottebohm 1985). Thus, HVc projects to area X (located in the lobus parolfactorius), which projects to the lateral dorsal thalamus (DLM), which projects to the lateral portion of the magnocellular nucleus of the anterior neostriatum (I-MAN), which sends terminal projections to RA. In contrast to the effects of lesions placed in or between the pathways connecting HVc, RA, DM, or nXIIts, lesions located in the pathways between X-DLM-MAN show little disruption of the song system in adult birds. However, the X-DLM-MAN loop appears to play a critical role in song development. This conclusion is based on the observation that if lesions are placed within the loop before or during song acquisition, individuals either fail to learn a song or show considerable delays (Bottjer, Mieser, and Arnold 1984; Scharff and Nottebohm 1989; Sohrabji, Nordeen, and Nordeen 1990).

Ongoing neuroanatomical studies reveal that several other nuclei send either direct or indirect projections to HVc and RA, including input from the forebrain auditory area, field L (Kelley and Nottebohm 1979). Based on observations that HVc and the hypoglossal motor nerve respond to auditory stimuli, and that lesioning either HVc or RA eliminates such responses, the current view is that HVc represents a critical juncture for the interaction of auditory and motor input necessary for song production (Arnold 1992; DeVoogd 1994).

Once relatively discrete nuclei had been identified, several questions emerged. Given changes in song during the year—changes that correspond with shifts in hormonal profile—were there any detectable changes in brain structure? Because most songbirds are characterized by male songsters and silent females, are such behavioral differences due to differences in the brain? Do females simply lack the structures or internal wiring that males possess, or do features of their socio-ecology fail to elicit activity in these structures (i.e., neural absence *versus* neural presence plus behavioral dormancy)? And last, are the neural structures subserving song consistent across species, or do species that renew their songs seasonally or imitate quite freely have brains that differ fundamentally from those species that acquire only one song and sing it for the rest of their lives?

One of the first attempts to address these questions came from Nottebohm and Arnold's (Arnold 1980; Arnold, Nottebohm, and Pfaff 1976; Nottebohm 1989; Nottebohm and Arnold 1976) work on canaries and zebra finches (*Taeniopygia guttata*), aimed at understanding the relationship between testosterone levels and changes in brain and song structure. Male canaries undergo considerable changes in testosterone level during a year, peaking just before the mating season. Note acquisition occurs primarily before the surge in testosterone and afterward as well. More generally, most songbird species that have been examined show high levels of androgens during singing (Wingfield and Marler 1988), and singing rates can be elevated by increasing day length and by experimentally administering testosterone. These generalizations are checked by some important differences between species, and between males and females within species; I discuss such variation later in this section.

Given the association between testosterone and note acquisition, what corresponding changes in the brain have been observed? Nottebohm and colleagues discovered that in the male canary, HVc and RA both undergo volume increases of 99% and 77%, respectively, as a function of the seasonal change from winter into spring. In addition to these volumetric changes, seasonal differences in synaptic morphology have also been identified, including relative increases in the size of pre- and postsynaptic profiles in RA, the number of transmitter vesicles per synapse, and the proportion of neuropil in nXIIts (Clower, Nixdorf, and DeVoogd 1989; DeVoogd, Nixdorf, and Nottebohm 1985).

What makes songbirds so fascinating as a taxonomic group for brain-behavior analyses is the extensive variation between species in song complexity and change over time, overlaid by significant differences in singing performance between the sexes (Brenowitz 1991). Concerning the former, one of the most striking contrasts exists between species that acquire a single song at the end of their first year and then use this song throughout life (close-ended), as opposed to species that modify the structure and number of songs from season to season throughout their adult life (open-ended). This difference in song plasticity raised the possibility that adult birds undergo structural changes in the brain associated with modification of song morphology, a neurobiological process that at least in mammalian species is generally restricted to young developing individuals. When Nottebohm and colleagues (Alvarez-Buylla, Theelen, and Nottebohm 1988; Alvarez-Buylla, Kim, and Nottebohm 1990; Goldman and Nottebohm 1983; Nottebohm 1989) looked at adult canaries, they found significant evidence of neurogenesis and migration to many parts of the brain, including the song system. One function of these changes is the birth and functional integration of new neural circuits. Most importantly, many of the new neurons project into RA and, although new neurons were found within HVc, the total number within HVc did not change seasonally, suggesting that the new neurons replace the ones that die. The hypothesis then, is that cell birth and death are associated with song learning and song forgetting, respectively.

Following up on these general findings, a comparative investigation of neurogenesis in canaries and zebra finches was undertaken (Alvarez-Buylla, Kim, and Nottebohm 1990). The technique involves injecting individuals with [3H]thymidine for a period of 14 days, thereby providing a marker of DNA synthesis; cells labeled with [3H]thymidine emerge within the first 60 minutes following injection, with labeling arrested after 90 minutes. After 128 days, fluorogold is injected into the right and left RA. New cells and their projections can then be counted and traced. As Figure 4.13 indicates, both canaries and zebra finches of all ages showed new neurons and new projections. There were, however, significant age, season, and species differences. Among canaries, year-old birds in the spring exhibited significantly less neurogenesis than either year-old or four-year-old birds in the fall; year-old fall birds showed higher neurogenesis in total and in

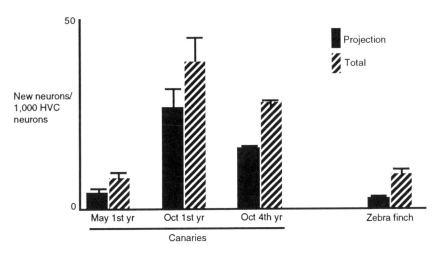

Figure 4.13
Changes in the proportion of HVc and RA-projecting HVc neurons as a function of age, season, and species. The striped bars refer to all neurons, whereas the black bars refer to the number of projection neurons. Values above the black bars represent the proportion of labeled neurons projecting to RA. Canaries: year-old May birds ($n = 4$) represent individuals captured during the breeding season, when song is stable. Year-old ($n = 6$) and four-year-old October birds ($n = 2$) represent individuals bred and raised in captivity, outside the breeding season, when song undergoes structural changes. Zebra finch: all year-old birds ($n = 6$) (redrawn from Alvarez-Buylla, Kim, and Nottebohm 1990).

terms of RA-projecting neurons than four-year-old fall birds. Last, all canary groups exhibited greater neurogenesis than zebra finches. These results show that neurogenesis occurs in close-ended learners, but in contrast to open-ended learners, this process is generally arrested after the first year of life.

The studies reviewed here have revealed seasonal changes in the size of song nuclei and significant inter-specific differences in the relationship between the volume of these nuclei and song complexity or repertoire size (reviewed in Arnold 1992; DeVoogd 1991, 1994). There are, however, important caveats to keep in mind when evaluating such findings. First, volumetric changes can result from a wide variety of neural processes, including synaptic shrinkage or expansion, deletion or insertion of large synapses, increase or decrease in cell density, and changes in synaptic connections from dendritic spines to dendritic shafts. Needless to say, the particular process of change is crucially relevant to understanding how song production and perception vary seasonally. Second, studies by Gahr (1990) suggest that the traditional method of estimating the volume of a neural region by Nissl staining may, under some circumstances, lead to misleading conclusions. Working with male canaries, Gahr measured seasonal changes in HVc using Nissl staining and an estrogen antibody labeling technique. Both approaches yielded similar volumes in the breeding season. In the nonbreeding season, however, the Nissl stain revealed a relative decrease in HVc, whereas the estrogen antibody revealed no change. A strong implication of these results is that the estrogen-antibody staining procedure is more accurate than the Nissl staining procedure under conditions where seasonal changes in song nuclei are coupled with changes in the morphology of relevant neurons, rather than in the density of neurons. Consequently, Nissl staining may be an inappropriate technique for quantifying the volume of nuclear regions when cellular processes are mediated by such factors as photoperiod.

To evaluate Gahr's (1990) criticism of the Nissl staining technique, and therefore the validity of the claim that song nuclei such as HVc change seasonally (i.e., with changes in testosterone), Johnson and Bottjer (1993) castrated male canaries, maintained them on short fall-like days, and administered either testosterone, antisteroid drugs, or nothing as the control. Treatment with testosterone was associated with an increase in HVc volume, the density and size of cells within HVc, and the absolute number and percentage of androgen target cells within HVc. Testosterone did not, however, affect the absolute number of cells in HVc. This finding stands in contrast to results obtained from female canaries where testosterone causes an increase in the absolute number of cells in HVc (Brenowitz and Arnold 1990). Johnson and Bottjer's results further suggest that the effect of testosterone, or perhaps its metabolites, is to regulate the overall size of HVc by altering the distribution of projection neurons and androgen target cells, neurons that are specifically connected to RA. In relation to Gahr's (1990) findings, therefore, Johnson and Bottjer suggest that seasonal changes should not be equated with changes in testosterone. Instead, some process, as yet unidentified, occurs during the photoperiodic changes associated with fall and spring.

The sex differences in singing behavior observed among many oscine species are strongly correlated with both macro- and microanatomical differences in brain structure (reviewed in DeVoogd 1991, 1994). Based on analyses of almost twenty species, the following features have been shown to be greater (i.e., more numerous, larger, longer) in males than in females: (1) volume of HVc and RA; (2) number of neurons in HVc, RA, and MAN; (3) size of cell bodies in HVc, RA, MAN, DM, and nXIIts; (4) length of dendrites in HVc and RA; (5) number of dendritic spines for principal neuron type in RA; (6) size of neuronal somata; and (7) distribution of androgen target cells.

The documented sex differences in brain structure are all the more fascinating when one contrasts the "typical" temperate oscine species with other species in which females sing as much as males, and under conditions where normally silent females can be induced to sing by means of hormonal therapy. Concerning the first, studies by Brenowitz and colleagues (Brenowitz and Arnold 1986, 1989; Brenowitz, Arnold, and Levin 1985) have revealed that as the degree of sexual dimorphism in song behavior decreases, so does the degree of sexual dimorphism in the structure of the song regions. Thus, in duetting song birds such as some of the tropical wrens, females have the same song nuclei as males, and their macro- and microanatomical features are comparable.

Though the hormonal profiles of male and female songbirds differ, including brain receptors, areas of activation, and so forth, a great deal of research in this area has focused on the effects of hormonal manipulations on female singing behavior (reviewed in Arnold 1992; DeVoogd 1991, 1994). For example, to determine the effects of testosterone on female canaries, S. D. Brown and Bottjer (1993) treated some females with testosterone for four weeks, then removed the testosterone and assessed changes in HVc, RA, and area X. The female control groups were (1) testosterone implants maintained for greater than four weeks, and (2) empty implants. All groups showed efferent projections from HVc to RA and area X, showing that testosterone is not *necessary* for creating this portion of the song circuit. Rather, the formation of the circuit may be a prerequisite for hormone-induced singing. In all testosterone-treated individuals, RA was larger. Individuals who had testosterone implanted and then removed showed RA volumes comparable in size to controls without testosterone. This finding shows that although the administration of testosterone initially caused an increase in RA, its removal was associated with regression. Area X showed no change in response to testosterone administration, a result that contrasts with a previous study by Bottjer, Schoonmaker, and Arnold (1986) where area X increased in response to testosterone. One important difference is that in the S. D. Brown and Bottjer (1993) study, females were kept on a short, fall-like photoperiodic schedule, whereas in the earlier study (Bottjer, Schoonmaker, and Arnold 1986), they were on a springlike schedule for North America. Males also did not show a significant change in area X in response to testosterone administration. Thus, for both male and female canaries maintained on a short-day photoperiod, changes in area X are not regulated by testosterone alone—changes in area X seem to be the result of an interaction between changes in testosterone and photoperiodic schedules.

A final result from the Brown and Bottjer study is that there was no significant correlation between the volume of RA or area X and the number of syllables produced by testosterone-treated females. Although some studies have found a positive correlation between the volume of particular song-control nuclei and repertoire size (Brenowitz and Arnold 1986; Canady, Kroodsma, and Nottebohm 1984), other studies have failed to find such a correlation (Kirn et al. 1989; Simpson and Vicario 1991). Nonetheless, the most recent and comprehensive analysis of this problem, involving measurements of 41 oscine species, has revealed a significant positive correlation between the relative volume of HVc and the number of song types typically found in the repertoire (DeVoogd et al. 1993). One interpretation for these kinds of correlations is that offered by Arnold (1992), who suggests that the capacity to sing requires large HVc to RA projecting neurons, whereas song complexity is related to the number of neurons in the vocal network. This suggestion is based on the result that the size of neurons in RA is not sexually dimorphic in species in which both species sing, even for species where song is less complex in females. A second interpretation (Konishi and G. Ball, personal communication) is that the size of the nucleus is a reflection of effort or song output, with repertoire diversity a secondary correlate. Thus, in species that sing more, song nuclei are larger. More formal tests of these hypotheses are needed, both at the inter- and intraspecific level.

Thus far, we have primarily focused on the production end of the song system. There are, however, significant neural connections between production and perception, as might be expected in an organism that commonly relies on auditory feedback for guiding vocal output. An important investigation of these connections was initiated by Williams and Nottebohm (1985), basing their experiments on the motor theory of speech (Lieberman and Mattingly 1985; Liberman et al. 1967; see section 4.3 and Box 4.4). The motor theory claims that when speech is perceived, we hear it as a sequence of articulatory gestures rather than as a specific sound sequence. Recording from HVc and the tracheosyringeal branch of the hypoglossal nerve of male zebra finch, results indicated that in response to pure-tone sound bursts, neural activity started in HVc and was followed 12–18 msec later by activity in nXIIts. Moreover, lesioning HVc or RA completely eliminated the response in nXIIts. Thus motor commands are neurally "entwined" with the perceptual event. Last, Williams and Nottebohm found that motor neurons in the nXIIts were differentially responsive to different types of syllables in the zebra finch song (Figure 4.14). Specifically, frequency-modulated signals elicited more significant excitatory responses in neurons located in the posterior portion of the nXIIts, whereas unmodulated high frequency signals had an inhibitory effect. Conversely, frequency-modulated syllables had an inhibitory effect in the anterior portion of the nXIIts, whereas unmodulated high-frequency syllables had an excitatory effect. The implication here, then, is that in order for birds to perceive the proper acoustic features of a song syllable, the percept must

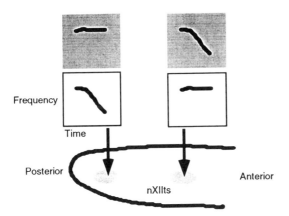

Figure 4.14
Syllable-specific responses in the tracheosyringeal neuron (nXIIts) of the zebra finch. Results indicate
that frequency-modulated syllables have a significant excitatory effect (black spectrographic tracing
on white) on neurons in the posterior portion of the nXIIts, whereas unmodulated high-frequency
syllables have an inhibitory effect (black spectrographic tracing on gray). In contrast, frequency-
modulated syllables have an inhibitory effect in the anterior portion of the nXIIts, whereas unmodulated high-frequency syllables have an excitatory effect (redrawn from Williams and Nottebohm
1985).

be converted into a series of motor actions required to produce the sound. As
such, these results are considered the avian equivalent (homoplasy) to the motor
theory of human speech perception.

Complimenting the findings of Williams and Nottebohm are more detailed studies of neural specificity in the white-crowned sparrow. Specifically, research by
Vicario (1991) demonstrated that nXIIts consists of a myotopic map of the syrinx.
Thus stimulation at one locus results in contraction of one muscle. As mentioned
earlier on, there are pathways from HVc down to nXIIts, and within HVc there
are cells that are strictly responsive to the bird's own song. Thus, as Margoliash
(1983, 1987) has demonstrated using single-unit recording techniques, once song
has crystallized toward the end of the first year, there are cells that only fire to
playbacks of the individual's crystallized song. Reduced firing rates are, however,
observed in response to songs from the same dialect (relative to songs from
outside of the dialect) and to songs within the dialect that share structural elements (e.g., some of the same notes or syllables). Most impressively, cells that
are highly responsive to the bird's own song are less responsive to the same
song played backward. This is a striking finding because it demonstrates that cell
specificity is not based on overall spectral properties of the song, but rather

on the temporal sequence of spectral events (syllables) that are unique to each individual's song. Furthermore, the specificity shows that there are fundamental units within the song and that these must be produced in a particular sequence in order for the song to be recognized, neurally, as "own" versus "other."

Further evidence for organization at the level of the syllable is provided by both behavioral and neurobiological studies of the zebra finch. Cynx (1990) showed that when zebra finches sing, visual disruptions caused by presenting stroboscopic flashes result in individuals halting song output after syllable strings. In other words, there is not a simple tape loop that, once engaged, runs itself out. Rather, individuals can stop singing before they complete an entire song. Additionally, Vu, Mazurek, and Kuo (1994) have demonstrated that in zebra finch, HVc, its afferents, or both in unison, may be responsible for the timing of phrases (syllable strings) within a song sequence. This conclusion was derived from pathway lesions that disrupted the temporal structure of their song.

To further explore the neurobiological mechanisms underlying perception of avian song, Cynx, Williams, and Nottebohm (1992) have conducted a series of experiments to determine whether there are brain asymmetries involved in the perceptual analysis of zebra finch song. These experiments were theoretically inspired by research on humans, indicating left hemisphere dominance for processing the formal, structural properties of language (i.e., semantics, syntax), whereas the right hemisphere tends to be dominant for processing the emotive properties of language, conveyed, for example, by its melody (see section 4.4.2). The prediction for birds, then, is that if song, like human language, is processed by specialized neural circuitry, and if selection has favored hemispheric differentiation of function, then the left hemisphere should be dominant with respect to the species-typical features of song; this prediction is based, of course, on the assumption that birds and humans might confront comparable problems leading to convergent evolution and thus to homoplasies in brain-behavior relationships.

One set of birds received a lesion in the left nucleus ovoidalis (thalamic auditory relay nucleus), whereas the other set received a lesion in the right nucleus ovoidalis. These lesions eliminated, unilaterally, auditory input to either the right or left hemisphere. Subsequently, individuals were trained in an operant procedure to discriminate between their own song and a cage mate's, or between two unfamiliar zebra finch songs that differed in terms of the harmonic structure of one syllable. Results showed that individuals with left-side lesions took longer to discriminate between own and cage mate's song than individuals with right-side lesions. In the task involving two unfamiliar songs, however, individuals with right-side lesions reached criterion for discrimination more slowly than individuals with left-side lesions. Thus the two hemispheres of the zebra finch brain appear to play different roles in processing conspecific song, with familiarity and acoustical complexity of the perceptual task contributing significantly. These results are difficult to compare with those obtained for human language because it is unclear how familiarity and acoustical complexity can be translated into speechlike features that are processed asymmetrically in the human brain. Nonetheless, results provide clear evidence of hemispheric biases with regard to processing acoustically meaningful stimuli.

As previously mentioned, females do not sing in most oscine species. Females do, however, often exhibit brain structures associated with song production—albeit usually atrophied by comparison with males—and clearly possess the perceptual apparatus to respond selectively to song (Catchpole, Leisler, and Dittami 1984; Cynx and Nottebohm 1992a, 1992b; Searcy 1992; Searcy and Brenowitz 1988; Searcy and Yasukawa 1990). To assess the importance of song-related brain nuclei in perception, Brenowitz (1991) lesioned HVc in female canaries and then tested the selectivity of their response to conspecific song and the song of a heterospecific, the white-crowned sparrow. Results showed that intact female canaries only respond with courtship to conspecific song, whereas HVc-lesioned females produced courtship displays to both conspecific and heterospecific song. Therefore, HVc appears critical for song discrimination in female canaries.[16] Moreover, results suggest that in contrast to males (Williams and Nottebohm 1985), females develop the ability to discriminate among song types without having produced song.[17] The mechanism responsible for the ontogeny of this ability has not yet been detailed.

In summary, the songbirds, like the anurans, exhibit evidence of both specialized neurophysiological design and conservative maintenance of mechanisms shared with other vertebrates. To make the contrast most meaningful, and to bring back the problem of design again, recall that the primary socioecological problem for a male songbird is to use its song as a means of establishing and maintaining a territory, in addition to broadcasting information to females about mate availability and, potentially, male quality in terms of an ability to raise

16. The results from canaries are unlikely to generalize across songbirds. For example, female zebra finches without HVc recognize conspecific song, and in intact females there are no connections between HVc and RA.

17. Note that male white-crowned sparrows are also capable of discriminating between conspecific and heterospecific song prior to the onset of song production (Nelson and Marler 1993).

offspring, survive, and defend a territory. From the female's perspective, the perceptual problem involves sorting conspecific song from heterospecific song, and within the category of conspecifics, attempting to discriminate between mates that will survive and contribute either directly or indirectly to reproductive success. In terms of environmental acoustics, both the production and perception components of the songbird's advertisement signal are fundamentally different from what we observed in frogs: whereas frogs call within a chorus, there are rarely more than a few neighboring songbirds calling at the same time.

Concerning signaling specializations, the syrinx has evolved as a unique vertebrate adaptation, providing some songbirds with the ability to produce two independent acoustic signals. And in species that have this ability, a brain has been designed to coordinate the peripheral output system and to perceive the complexity of its structure. Thus, in striking contrast to the advertisement call of many anuran species, the song morphology of most passerines is extremely complex, both in terms of the types of notes or syllables produced and in terms of the temporal organization of such notes and syllables within a song.

Because of the temporal association of the advertisement signal with the mating season, there are fundamental neural connections between the auditory and endocrinological regions of the brain in both anurans and songbirds, and in both groups males are the primary vocalizers. Where these two taxonomic groups differ, however, is the extent of neural plasticity at the species level and between the sexes. Thus, in some songbirds, neurogenesis waxes and wanes in apparent coordination with seasonal changes in the acquisition and deletion of new song material, though as pointed out, there is no firm evidence for a causal connection between such neurobehavioral changes. Additionally, whereas anuran males are solely responsible for producing the advertisement call, in some birds males and females sing. As a result, one finds that as the degree of sexual dimorphism in singing behavior decreases, so does the degree of sexual dimorphism in brain architecture.

4.3   Survival Signals: Bats

A human hand is quite different from, for instance, a batwing in shape and function. However, they share the same set of bones. Like the set of bones, the neural mechanisms [for auditory processing] . . . are probably shared by humans and other mammals. . . . One important problem is to determine the extent to which speech-sound processing is based upon the specialization of shared mechanisms, or whether it involves mechanisms that are so unique that it is not shared at all by other animals.   (Suga 1988, 716–717)

This section describes the auditory neurophysiology of a group of predatory species—bats—whose perceptual system has been fine-tuned for sound localization and prey recognition. The auditory emission and processing to be discussed, however, generally falls outside of the concept of communication developed in chapter 1. In bats, most research has focused on the echolocation system[18] rather than on auditory signals used in intraspecific communication (though see Kanwal et al. 1994 for recent analyses of communicative elements of the mustached bat's vocal repertoire and Wilkinson, in press, for a discussion of communication calls used by several bat species during group foraging and recruitment to feeding sites). When a bat echolocates, it is both the sender and the perceiver. The function of echolocation is to provide the signaler with information about prey type, distance, velocity, and spatial location. As such, bat echolocation is comparable to the noncommunicative tapping used by aye-ayes (a prosimian primate) to extract insects from hollowed branches. Nonetheless, it is one of the most elegant stories about the coevolution of neural systems underlying sound

18. A set of observations (Leonard and Fenton 1984) and playback experiments (Balcombe and Fenton 1988) suggest that echolocation calls may also serve a true communicative function (*sensu* chapter 1). In at least some Vespertilionid bats, individuals appear to respond selectively to echolocation calls of conspecifics as a source of information regarding prey availability. Although such selective responsiveness has been interpreted as evidence that echolocation sounds are communicative, there is no evidence that individuals produce echolocation *in order to* communicate information about prey availability (i.e., that the sounds have been designed to be communicative).

production and perception, and it is one of the few stories we have in the context of survival systems. It is therefore a story that I can't resist telling.

4.3.1 Bat Echolocation: The Problem

What special problems do bats confront in their environment that might have selected for echolocation—a biological sonar system (Galambos 1942; Galambos and Griffin 1942; Griffin 1944; Griffin and Galambos 1941)? Most bats, especially the microchiropterans, live in caves and are primarily nocturnal. Thus individuals hunt for prey at night and navigate through dark caves. Their vision is relatively poor (though see discussion of visual acuity in the Microchiroptera by Pettigrew 1988), whereas their auditory system is highly developed.[19] These conditions, along with the sensory processes that have emerged, suggest that bats have been designed to dissect the salient properties of their world via auditory mechanisms rather than visual ones.

Though a considerable amount of work has been conducted on both the social behavior (Bradbury and Vehrencamp 1976; McCracken 1987; Wilkinson 1987) and echolocation system of bats (Fay and Popper 1995; Griffin 1958; Pollak and Casseday 1989; Simmons 1989; Suga 1988), the species investigated have generally differed for these two topics of research. Nonetheless, some general phylogenetic patterns have emerged with regard to the structure of the echolocation pulse, and these differences shed light on a number of important socioecological parameters, including habitat structure, prey density, and predatory behavior.

Studies of the psychoacoustics and neurobiology of bat echolocation have generally focused on six species (Figure 4.15). Most researchers divide bats into two types based on the structure of their echolocation sounds: frequency-modulated (FM) and constant frequency/frequency-modulated (CF/FM) bats. The echolocation pulse of FM bats consists of a relatively rapid frequency downsweep (i.e., the frequency of the signal starts high and ends low). In contrast, CF/FM bats produce an echolocation pulse that starts with a long unmodulated frequency component and terminates with a brief frequency downsweep. Typically, FM pulses last between 0.5 and 5.0 msec, whereas CF/FM pulses last 30–60 msec, and up to 200 msec. Both echolocation types are exceedingly loud, with sound-pressure level (measure of relative intensity) recordings ranging from 80 to 120 dB

---

19. Again emphasizing trade-offs between different sensory modalities—no species has evolved extraordinary sensory efficiency in all modalities—showing that selection for particular sensory specializations represents a compromise with regard to other modalities (constraints on overall brain size, but not on how particular regions of the brain evolve toward increasing specialization, matched to environmental demands).

SPL. In general, FM bats tend to hunt their prey in relatively open environments, whereas CF/FM bats tend to hunt in thick canopied forests.[20]

Though the general structures of these two sonar types are distinctive, there is significant within-type variation, and much of this variation results from changes in the kind of echo feedback the bat receives during prey pursuit. For example, long unmodulated pulses tend to be emitted at the start of a pursuit (Figure 4.16). As the bat moves closer to its prey, however, the pulses become shorter and more significantly modulated. In addition, there is a general decrease in pulse amplitude as prey are approached, thereby maintaining a relatively constant echo-amplitude profile during capture.

Before we turn to a discussion of how the vocal tract generates echolocation pulses and how the hearing system dissects spectrotemporal properties of the returning echo, I would like to summarize the key problem and design solution for bats by quoting directly from Dawkins (1986), who states the issue eloquently:

Bats . . . have an engineering problem: how to find their way and find their prey in the absence of light. . . . Natural selection working on bats [provided a solution], their "radar,"

---

20. A recent exception to this pattern has been reported for two of the world's smallest bats, *Craseonycteris thonglongyai* and *Myotis siligorensis* (Surlykke et al. 1993). These phylogenetically distantly related bats hunt for small insects in open areas, and both species produce CF/FM echolocation pulses. Though high-frequency sounds are attenuated in the open air (see chapter 3), the size of these two species constrains them to relatively small insect prey; such small prey only reflect echoes at high frequencies.

Common Name	Latin Name	Biosonar Type	Spectrogram
Big brown bat	*Eptesicus fuscus*	Loud FM	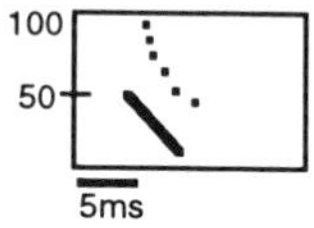
Mexican free-tailed bat	*Tadarida brasiliensis*	Loud FM	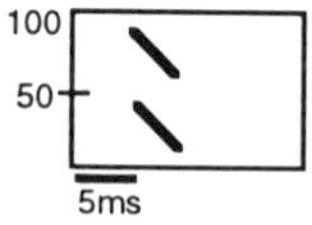
No common name	*Molossus ater*	Loud FM	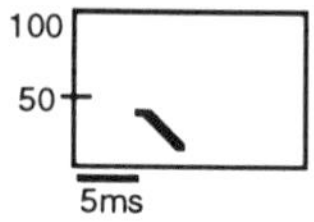
Greater horseshoe bat	*Rhinolophus ferrumequinum*	Long CF/FM	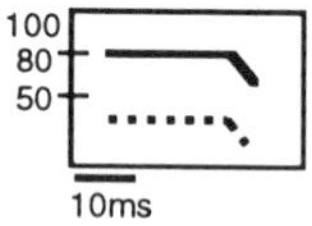
Rufous horseshoe bat	*Rhinolophus rouxi*	Long CF/FM	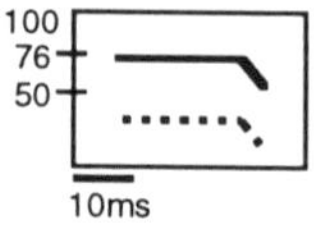
Mustached bat	*Pteronotus parnelli*	Long CF/FM	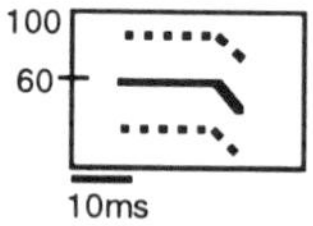

Figure 4.15
A list of bat species as function of biosonar type. These species have been most extensively studied. The solid tracings represent the dominant frequencies, whereas the dashed tracings represent harmonics. The y-axis is frequency (kilohertz), and the x-axis is time (milliseconds) (redrawn from Pollak and Casseday 1989).

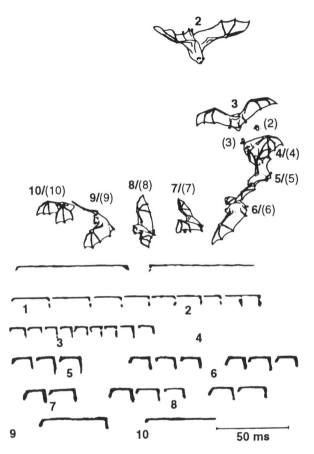

Figure 4.16
Changes in the horseshoe bat's position and echolocation pulses during prey pursuit. The numbers in bold next to the bat refer to a frame of movement. The numbers in parentheses refer to the prey's position. Spectrograms of the bat's echolocation pulses are shown below, the number reflecting the bat's position in the sequence above (redrawn from Neuweiler, Bruns, and Schuller 1980).

that achieves feats of detection and navigation that would strike an engineer dumb with admiration. . . . It is misleading to speak of bats as though they were all the same. It is as though we were to speak of dogs, lions, weasels, bears, hyenas, pandas and otters in one breath, just because they are all carnivores. Different groups of bats use sonar in radically different ways, and they seem to have "invented" it separately and independently, just as the British, Germans, and Americans all independently developed radar. . . . The smaller bats appear to be technically highly advanced echo-machines. They have tiny eyes which, in most cases, probably can't see much. They live in a world of echoes, and probably their brains can use echoes to do something akin to "seeing" images, although it is next to impossible for us to "visualize" what those images might be like. . . . These bats are like miniature spy planes, bristling with sophisticated instrumentation. Their brains are delicately tuned packages of miniaturized electronic wizardry, programmed with the elaborate software necessary to decode a world of echoes in real time. Their faces are often distorted into gargoyle shapes that appear hideous to us until we see them for what they are, exquisitely fashioned instruments for beaming ultrasound in desired directions. (pp. 22–24)

The Production of Biosonar   A primary difference between the larynx of microchiropterans and other mammals is the presence of a matching set of thin membranes that lack muscle fibers and lie along the edge of the vocal folds. To change the tension of the vocal membranes, the bat must alter the activity of the cricothyroid muscles, thereby imposing configurational changes (i.e., flexing the joint) on the cricothyroid joint (Suthers and Fattu 1973). These vocal membranes are considered to play the most important role in the generation of ultrasonic vocalizations. Evidence for a causal connection between the vocal membranes, the cricothyroid muscle ensemble, and ultrasound production comes from a set of experimental studies with the big brown bat (*Eptesicus fuscus*) and the greater horseshoe bat (*Rhinolophus ferrumequinum*). In the big brown bat, following transverse cuts to both vocal membranes, individuals are rendered virtually silent. Sectioning both sides of the cricothyroid muscles, but keeping the vocal membranes intact, causes a shift from the characteristic FM pulse to an abnormal CF pulse.

Because bats need to produce sonar pulses at high rates and high intensity during both normal flight and capture sequences, the respiratory system must operate under formidable energetic constraints (Thomas and Suthers 1972). The greater the respiratory volume, the greater the number of pulses per respiratory cycle, or the greater the length of each individual pulse (Suthers 1988). Increases in pulse intensity are associated with increases in target range (i.e., the distance at which targets can be detected). Pulse intensity depends on the rate of airflow through the glottis, the subglottic pressure and resistance, the filtering properties of the supralaryngeal tract, and the degree of signal radiation (see chapter 3) from the oral cavity and lips (Fattu and Suthers 1981; Suthers 1988).

The key evolutionary adaptation within the Microchiroptera seems to be an ability to produce relatively intense sonar pulses within the expiratory portion of the respiration cycle, in the absence of excessive interference to the overall respiratory needs of the individual. In the mustached bat (*Pteronotus parnelli*), pulses lasting 15–20 msec can be produced at repetition rates of up to 30 pulses/ second within a single exhalation. Among FM bats, such as the big brown bat, a single pulse per respiratory cycle is produced at low repetition rates (e.g., 10 pulses/sec). As repetition rates increase, as occurs during the final stages of a predatory attack (Figure 4.16), individuals will produce ten or more pulses per expiration and achieve a repetition rate of 100/second. In contrast, the greater horseshoe bat produces some of the longest pulses among bat species and, conse- quently, utilizes most of the expiratory capacity of the lungs. Species using shorter pulses appear to use the maximum expiratory capacity by producing the maximum number of pulses without inspiratory interruption.

Because of the positive correlation between high-intensity sonar pulses and prey-detection range, there is strong selection on bats to produce high-intensity sonar pulses. High-intensity pulses require high subglottic pressure, and both Hildebrandt's horseshoe bat (*Rhinolophus hildebrandti*) and the mustached bat reach subglottic pressures of approximately 40 cm H_2O. In contrast, subglottic pressure for even maximum-intensity shouting in human speech only attains a maximum subglottic pressure of 30 cm H_2O (Isshiki 1964). In producing high- intensity pulses, frequently exceeding 100 dB SPL, bats efficiently convert sub- glottic pressure to radiated power with minimal loss. Repeating the point raised earlier, efficient vocal output is of major significance to bats given the trade-offs between the energetic costs associated with flight (Thomas 1987) and sonar gener- ation, both critically tied to successful prey capture.

Thus far we have discussed the relationship between respiration and the time- amplitude profile of the sonar signal. We now turn to the mechanisms responsible for generating the highly specialized spectral contours produced by both FM and CF/FM bats. The control of these contours is spectacular, as evidenced by the fact that in some species individuals can maintain acoustically distinctive "sig- nature" frequencies when moving in small groups (Habersetzer 1981), other spe- cies vary the slope of the FM contour as a function of their distance to target prey (Simmons 1989; Simmons, Howell, and Suga 1975), and some species are capable of shifting the absolute frequency of the sonar pulse as a function of the relationship between their own movement and the movement of targets in the environment (i.e., Doppler shift compensation; see Box 4.2; reviewed in Pollak and Casseday 1989).

The cricothyroid muscles are largely responsible for the changes in spectral structure observed in the bat's echolocation system. For example, when the frequency drops in an FM signal, there is a general decrease in the tension of the cricothyroid muscles on the vocal folds. More causally, sectioning the cricothyroid muscles results in a relatively flat spectral contour with a greatly reduced absolute frequency (Suthers and Fattu 1982). A second mechanism for controlling frequency (phonation) is via a gating mechanism, whereby vocal fold vibration is turned on and off during different portions of the cycle—a burstlike mechanism. This is a difficult mechanism to coordinate (i.e., coordination between muscle action and membrane tension), especially as pulse repetition rates speed up during the final portions of prey pursuit.

As previously described for some songbirds (Nowicki 1987; Westneat et al. 1993), several bat species are known to make use of their supralaryngeal resonance cavities (Figure 4.17) to alter the spectral properties of the echolocation signal. In fact, some of the most detailed quantitative research on vocal tract filtering in a nonhuman animal comes from studies by Suthers and his colleagues (Hartley and Suthers 1988; Suthers, Hartley, and Wenstrup 1988) of the horseshoe bat. In one study, the transfer function (see Box 4.3) was derived by comparing the spectral properties of the pulses produced under normal conditions and in a helium environment; recall that helium changes some of the critical sound transmission properties of the tract. Results showed that the filtering properties of the supralaryngeal cavity caused a relative diminution in the fundamental and the third harmonic, and a relative increase in intensity of the second harmonic. In a second study, the effects of the nasal cavity and the tracheal pouches were explored by selectively filling these areas. Results indicated that when the tracheal pouches were blocked during pulse emission, the only significant effect was a 15–19 dB increase in the fundamental frequency. In contrast, when the nasal cavity was filled, viable effects were obtained, including increases of 10–12 dB for the fundamental, overall decrease in spectral energy, and selective increase and attenuation of different harmonics.

In summary, the echolocation pulses of bats are under exquisite control by the peripheral organs. This control permits bats to fine-tune the acoustic output, necessary for obtaining an auditory image of the prey environment. But in an engineering sense, and this is the problem for the next section, the high-intensity signal coming out of the bat's mouth must be processed by a hearing system that

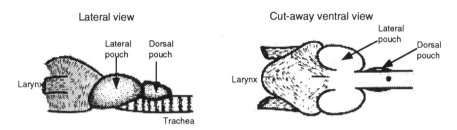

Figure 4.17
Schematic illustration of the larynx, tracheal pouches, and trachea of *Rhinolophus hildebrandti* (redrawn from Suthers 1988).

is sensitive to relatively quiet returning echoes. To capture the beauty of this design problem, let me again steal from Dawkins (1986):

Now here is the problem that would strike the engineer trying to design a bat-like machine. If the microphone, or ear, is as sensitive as all that, it is in grave danger of being seriously damaged by its own enormously loud outgoing pulse of sound. It is no good trying to combat the problem by making the sounds quieter, for then the echoes would be too quiet to hear. And it is no good trying to combat *that* by making the microphone (''ear'') more sensitive, since this would only make it more vulnerable to being damaged by the, albeit now slightly quieter, outgoing sounds! It is a dilemma inherent in the dramatic difference in intensity between outgoing sound and returning echo, a difference that is inexorably imposed by the laws of physics. . . . Bats developed ''send/receive'' switching technology long long ago, probably millions of years before our ancestors came down from the trees. . . . The bat is able to . . . switch its ears off temporarily. The muscles [of the ear] contract immediately before the bat emits each outgoing pulse, thereby switching the ears off so that they are not damaged by the loud pulse. Then they relax so that the ear returns to maximal sensitivity just in time for the returning echo.   (pp. 27–28)

The Perception of Biosonar   Once bats have emitted an echolocation pulse, their hearing system is, crudely put, in a temporary holding phase, waiting for the echo to return for processing. But as the quote from Dawkins reveals, the hearing system goes on hold both to prevent damage from the loud emitted pulse and because it must wait for acoustic information from the boomeranged echoes. The problems we will address in this section are twofold. First, what kind of information is available in an echo, and second, what can the bat's perceptual system extract and use in order to maximize prey detection and capture? On a purely behavioral level, it was known early on that bats could discriminate between prey and nonprey items (Griffin 1967; Webster and Durlach 1963). These observations set the stage for a series of experiments designed to show how bats use their biosonar system to discern the shape, texture, size, distance, velocity, and location of their prey.

A crucial inroad to understanding the perceptual mechanisms underlying echolocation is the recognition that bats and their prey are both moving through the environment during an attempted capture sequence.[21] Thus, when a biosonar signal reaches an insect, the returning echo will be influenced by the orientation and movement of the insect, relative to the bat's position (Figure 4.18; Box 4.2). The periodic wing-beat frequency of the insect will impose periodic frequency modulations on the echo, and this is especially the case for the CF component (creating spectral peaks and valleys amid a flat plateau). In addition, as the insect's wings cycle through their oscillations, they will present different surface orientations, and these will create amplitude modulations, or *glints,* in the echo.

The returning echo not only informs the bat that an insect has been detected, but actually provides information on the species of insect detected (Figure 4.19); this is especially the case for CF/FM bats. Schnitzler and colleagues (1983) played tonal signals at eight different insect species that were tethered but flying. Results indicated that each insect species sent back an acoustic "signature," information of direct use to the bat in deciphering what has been encountered (Emde and Menne 1989; Emde and Schnitzler 1990; Feng, Condon, and White 1994; L. A. Miller 1988; Moss and Zagaeski 1994). Further evidence for the importance of wing-beat frequency, as well as the level of sensitivity to subtle frequency changes, comes from the observation that bats will actually bypass stationary insects. In addition, it has been demonstrated in horseshoe bats that differences of ± 12 Hz around the fundamental frequency of 83 kHz can be detected, whereas most of the insects that are preyed upon provide frequency changes of ± 1000 Hz (Schnitzler and Flieger 1983). Thus bats clearly have the perceptual gear to detect subtle changes in prey movement,[22] thereby increasing the probability of capture.

---

21. As pointed out in the section on the neurobiology of anuran communication, it is important to recognize that most studies of the psychophysics and neurobiology of bat echolocation have been conducted under conditions that fail to capture the complexity of the natural context for prey capture. However, several investigators have attempted to ameliorate this imbalance by embedding echoes within noise (Roverud and Grinnell 1985) or by introducing a variety of potentially disruptive signals, including clicks of the arctiid moth while attempting to evade capture by bats (Surlykke and Miller 1985; reviewed in Fay and Popper 1995).

22. A recent report by Schnitzler and colleagues (1994) has extended the range of prey from insects to fish, showing that the greater bulldog bat (*Noctillo leporinus*) uses echoes from body movements of fish together with changes in the physical properties of the water to fine-tune its capture maneuvers.

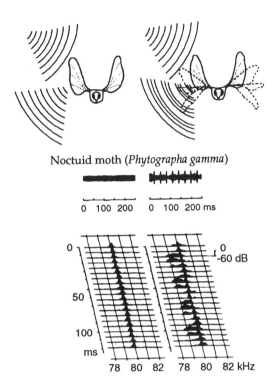

Figure 4.18
Changes in the spectral properties of an echo as a function of prey movement. The upper portion of
the figure provides a schematic illustration of the waveform as it strikes the insect, stationary on the
left, flying on the right. In the middle, typical waveforms of the echo striking a noctuid moth (*Phyto-
grapha gamma*); an 80-kHz tone was emitted from a loudspeaker in the direction of a tethered moth
positioned at 30 degrees. At the bottom, real-time spectra of the echo (redrawn from Pollak and
Casseday 1989).

A more striking set of psychophysical perceptual results have been obtained
with *Eptesicus fuscus* in studies designed to examine the importance of target
jitter (period-to-period variation in the fundamental frequency) in range determination. Using an analog system to play back an echo from the bat's original sonar
emission, Simmons (1979) first showed that a 1–2-microsecond level of jitter
could be detected in the echo delay. These results, together with subsequent
reports of nanosecond-level discrimination (Simmons et al. 1990, 1994), were
treated skeptically by a number of bat researchers. The reason for skepticism
was both theoretical and methodological. On theoretical grounds, the results suggested that *Eptesicus* was capable of discriminating echo differences that far

Box 4.2
Doppler Shift Compensation

Some bat species, such as the horseshoe and mustached bats, exhibit what is known as Doppler shift compensation (reviewed in Pye 1983). A Doppler shift occurs when there is an increase in perceived frequency (pitch) as a function of changes in the individual's motion, as occurs when you are sitting in a moving train listening to another train pass by. When a bat produces a biosonar pulse while remaining still, there is only a minor change in frequency of 50–100 Hz from pulse to pulse. But when the bat echolocates during flight, echoes from objects in the environment are shifted up in frequency relative to the bat's movement. Bats can compensate for this change by a Doppler shift (Schnitzler 1970a, 1970b), where the frequency of the pulse is dropped. In fact, studies have shown that the frequency change is approximately equal to the amount increased by Doppler shifting. This effect is illustrated in the following figure, using the horseshoe bat's echolocation pulse; modified from Pollak and Casseday (1989).

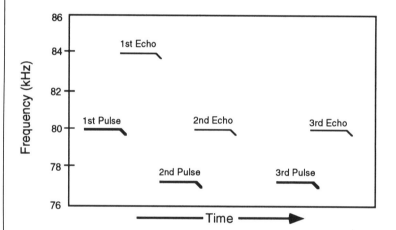

The primary consequence of Doppler shift compensation is that the bat maintains the echo CF component within a tight frequency window where a large proportion of neurons are finely tuned, thereby facilitating spectral discrimination. This ability enables bats, especially those living in densely cluttered environments, to tease apart differences that are due to insect prey from differences due to nonprey items in the habitat. As Pollak puts it: "In vision, changing the direction of gaze keeps images of interest fixated on the fovea. Similarly, Doppler compensation ensures that acoustic signals of interest are processed by a region of the sensory surface innervated by a large number of primary fibers having exceptionally narrow tuning curves. The resonant segment of the bat's cochlea, then, is essentially an 'acoustic fovea'" (Pollak 1992, 774).

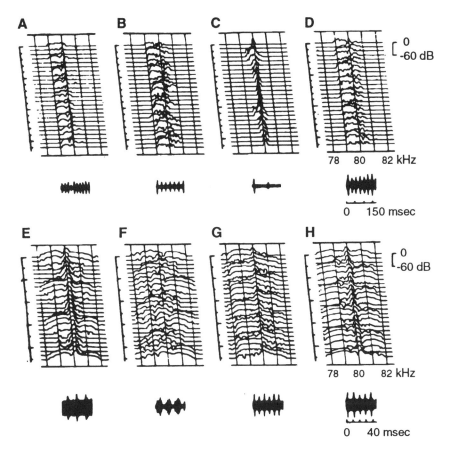

Figure 4.19
Acoustic signatures produced in the echo returned from eight tethered but flying insect species:
(*A*) *Ichneumonidae*, (*B*) *Spilarctica lupricipeda L.*, (*C*) *Campaea margaritata L.*, (*D*) *Axylia putris*,
(*E*) *Tipula* sp., (*F*) *Cantharidae*, (*G*) *Tachinidae*, (*H*) *Dolichovespula saxonica*. Echos were obtained
from an 80-kHz tone. All insects were tethered at a 60-degree angle from the speaker. The top portion
shows the real-time spectrum; the lower, the waveform (redrawn from Schnitzler et al. 1983).

exceeded what was biologically necessary for prey capture. More importantly, perhaps, this apparent perceptual ability surpassed what was imaginable for a nervous system that must also take into account variation introduced by the individual's motor actions (Moss and Schnitzler 1995). On methodological grounds, the experiments were considered problematic (for a more complete discussion, see Pollak 1993) because of the possibility of introduced spectral artifacts from playback echoes and clutter echoes (i.e., overlapping signals). Moreover, in contrast with digital systems, the analog device used by Simmons lacked the ability to precisely control the timing and structure of echo feedback. Nonetheless, some researchers, such as Moss and Schnitzler (1995) are not convinced that spectral artifacts are a possible confound. They argue that spectral changes introduced by echo overlap are outside of the bat's hearing range. Moreover, in studies by Simmons et al. (1990), spectral artifacts were explicitly ruled out as a possible explanation of jitter discrimination.

Although the precision of the bat's perceptual system would seem to give it an unfair advantage in prey capture, selection has also managed to provide prey with their own bag of defensive tricks. Specifically, insect prey have evolved counter-measures that include specialized hearing organs designed to increase sonar detection (Figure 4.20) and behavioral responses such as altering wing-beat frequency and the pattern of movement (Hoy 1992; Yaeger and Hoy 1986). The bat-insect coevolutionary arms race[23] thus represents the outcome of competing selection pressures on sensory systems, designed to pick up fine details of the acoustic environment.

Given the bat's ability to produce biosonar to compensate for various changes in the environment and to receive a wealth of information from the returning echo (Table 4.2), how much of this information is discriminable and used for fine-tuning the prey-capture sequence? Most bats that have been studied have quite large external ears. In some of these species (especially CF/FM bats), such as the rhinolophids, the pinnae (external parts of the ear) can be rotated through wide angles and moved in alternation (i.e., one pinna moves forward as the other moves back), and movement is often correlated with biosonar pulse emission (Griffin et al. 1962; Webster and Durlach 1963). In species with movable pinnae, changes in their position will create interaural intensity differences known to be important for sound localization in the vertical plane (for a thorough review, see

---

23. Many other arms races have been described where significant changes in the sensory apparatus provides one species with a slight fitness edge. For example, though most flies lack a hearing organ, a few species of parasitoid tachinid flies have evolved a hearing system that is finely tuned to the frequency of the cricket's chirp, their primary host (Robert and Hoy 1994).

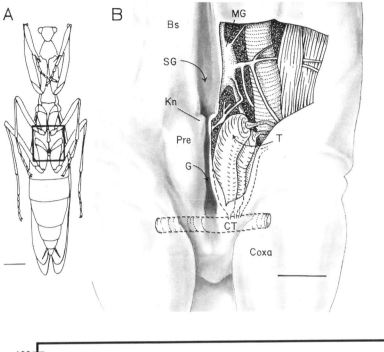

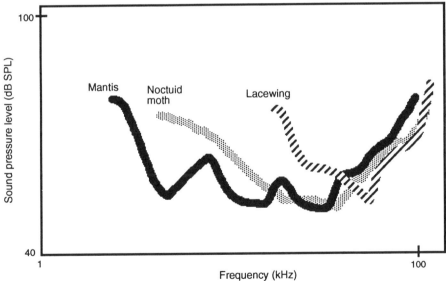

Figure 4.20
The upper panel shows the specialized hearing organ of the praying mantis. The lower panel shows tuning curves of three insect prey, including the praying mantis. Note that for all three species, there is increased sensitivity (lowest point in each tuning curve) at the frequency of several echolocating predatory bats (redrawn from Hoy 1992; Yaeger and Hoy 1986).

Table 4.2
Properties of the Echo as a Function of Target Features (derived from studies of the mustached bat, Suga 1988).

Echo Features	Target Features
Frequency modulation	Distance
Doppler shift	Velocity
DC component	Relative velocity
AC component	Flutter
Amplitude	Subtended angle
Delay	Range
Amplitude + delay	Size
Amplitude spectrum	Fine characteristics
Binaural cues	Azimuth
"Pinna-tragus" cue	Elevation

Moss and Schnitzler 1995). In species that lack the ability to move their pinnae, vertical sound localization results from spectral changes derived from the echo reflecting off of the pinna and tragus (Lawrence and Simmons 1982).

To test for the importance of the pinnae in sound localization, researchers have performed experiments that immobilize their movement. For example, Mogdans, Oswald, and Schnitzler (1988) trained the greater horseshoe bat to fly through a complex environment involving an arrangement of vertical or horizontal wires. In contrast to individuals with normally functioning pinnae, those with surgically immobilized pinnae (i.e., severing the motor nerves and ear muscles) exhibited a significant decrease in flight performance with regard to the vertical plane, but not in terms of the horizontal plane. From these results, the authors conclude that horseshoe bats, and probably other CF/FM bats as well, use pinna movement in conjunction with scanning in order to enhance interaural intensity differences of the CF echo component.

For effective prey capture, the bat's brain must accomplish three tasks: (1) assess target range, (2) determine target location with respect to vertical and horizontal coordinates, and (3) identify the type of target detected (i.e., whether it is prey, and if so, what species). In several species, range appears to be determined by an assessment of the temporal interval separating the emitted pulse and its echo component. This perceptual calculation has been elegantly demonstrated for the big brown bat where delay differences of 1.0 millisecond can be accurately resolved (Simmons 1971, 1973; see review in Moss and Schnitzler 1995). Localizing sound is achieved by processing interaural disparities, either interaural intensity differences or differences in arrival time. Given the highly directional properties of their outer ear, interaural intensity differences appear to be the most important, creating differences in intensity of 30–40 dB and enabling them to resolve target differences of 1.5–4 degrees (Simmons et al. 1983). Target detection or recognition is primarily accomplished by assessing the difference in spectral structure of the emitted sonar pulse and the returning echo. In sum, therefore, the bat requires a neural system that can store a representation of the emitted sonar for subsequent comparison with the echo, and the comparison must operate in the time, frequency, and amplitude domains.

From the outer ear we move to the inner ear and, in particular, the cochlea. Most of the research in this area has been conducted on the horseshoe bat (Bruns 1976a, 1976b) and mustached bat (Kössl and Vater 1985a, 1985b; Zook and Leake 1989). Thus the following synopsis may not generalize to other species, especially the FM bats. In the horseshoe bat, cochlear thickness is greatest at the base and then, after a significant decrease, remains constant up to the apex. The width, however, is quite narrow at the base and then increases dramatically at the apex. The mustached bat, in contrast, has a much more uniform cochlea, with relatively minor changes in thickness and width.

Studies using cochlear microphonics in the mustached bat have revealed a region of high sensitivity around 60 kHz (Henson, Schuller, and Vater 1985), corresponding precisely to the dominant frequency of the biosonar pulse. Specifically, audiograms indicate that the sound-pressure level required to elicit a response at 60 kHz is approximately 40 dB less than at other frequencies. Close to 35% of the basilar membrane corresponds to the 62–45 kHz range, which correlates precisely with the CF and FM components of the dominant frequency (i.e., the second harmonic of the pulse). Moreover, Suga and colleagues (1987) have shown that as the bat matures, the particular dominant frequency that characterizes each individual's echolocation signal—its signature—becomes disproportionately represented. Though current understanding of cochlear mechanics is relatively poor, researchers generally agree that the bat system is an exaggerated or "hypertrophied" form of the general mammalian blueprint (Pollak and Casseday 1989; Pollak 1992).

As mentioned earlier on, Capranica (1992) has raised some concerns with regard to the validity and interpretation of tuning curves (Box 4.1). Specifically, he argued that for most vertebrates, neurons are only weakly turned to particular frequencies. Such criticisms, however, do not apply to a majority of tuning curves generated for echolocating bats (Figure 4.21). In fact, if there is anything that distinguishes the bat's auditory system from that of all other mammals, it is the

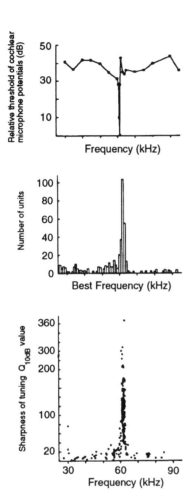

Figure 4.21
Sharpness of neural response in the mustached bat. The top panel shows an audiogram from cochlear microphonic recordings. The middle panel shows a distribution of best frequency responses obtained from cells located in the inferior colliculus. The lower panel provides Q_{10dB} values for cells in the inferior colliculus. Note that all three response measures indicate selectivity at approximately 60 kHz, the primary frequency of the mustached bat's CF component. Other cells do not exhibit such sharp tuning (redrawn from Pollak 1992).

overrepresentation of neural tissue for highly specific frequencies, together with an ensemble of neurons that respond selectively to a narrow frequency band. In mustached bats, a majority of neurons are tuned to 60 kHz, whereas in horseshoe bats a significant proportion of cells are tuned to 80 kHz, and for both species, Q_{10dB} values are relatively high. These are precisely the frequencies of the CF component of each species' sonar pulse.[24]

A more microscopic look at the peripheral neurons in the cochlear nucleus and in the auditory cortex indicates that in certain species there are cells tuned to a single frequency-amplitude cluster and others to a homogeneous or heterogeneous class of frequencies and amplitudes. For example, Suga and his colleagues (reviewed in Suga 1988) have found that in the mustached bat some cells are tuned to a single CF or FM component of the echo whereas other cells are tuned to frequency combinations (i.e., combination-sensitive neurons) from the sonar and echo, such as FM_a-FM_b, CF_a-CF_b, and CF_a-FM_b (the subscripts a and b refer to the number of the harmonic in the signal). The latter are particularly interesting in light of signal-processing complexity at the level of the single cell. Specifically, in areas dedicated to CF-CF processing, cells only respond to particular frequency combinations, and such responses are relatively immune to fluctuations in amplitude. The immunity to amplitude changes appears to result from inhibitory action from neighboring cell populations.

There are several important auditory nuclei and pathways in the bat's brain (Figures 4.22 and 4.23), and as documented for other species, different components of this system are responsible for analyzing particular properties of the acoustic environment. In general, three patterns emerge from studies of the auditory processing system of the bat: (1) nuclei are tonotopically structured, with projections to higher nuclei; (2) the lower brain stem is characterized by a set of parallel hierarchical pathways; and (3) a large proportion of these pathways are specifically designed to process different features of the auditory signal. One of the primary pathways of interest starts in the cochlear nucleus (receiving neural inputs from the cochlea) and then projects directly to the superior olivary complex, followed by projections to the inferior colliculus and up into the auditory cortex. Within this pathway, a great deal of work has focused on the inferior colliculus, primarily because there are quite striking differences between FM and

24. In contrast to the mustached bat, cochlear microphonic potentials obtained from the horseshoe bat are not as significant, even though a clear resonant peak is observed at 80 kHz (Henson, Schuller, and Vater 1985), the dominant frequency of the CF component of the sonar emission. Moreover, tuning curves recorded from the periphery of FM bats are much more comparable to those obtained for nonecholocating mammals.

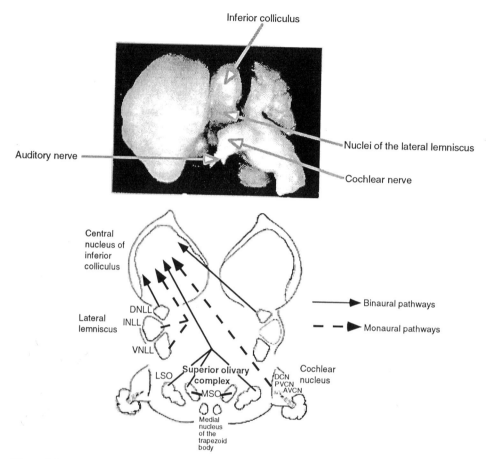

Figure 4.22
The brain of the mustached bat, focusing in particular on key auditory areas and pathways. The upper portion shows the left side of the brain (cerebrum at left, brain stem at right) with a portion of the cerebellum removed. The bottom portion of the figure provides a schematic representation of the brain stem (transverse section), detailing the main auditory nuclei, and ascending projections that are either binaural (solid line) or monaural (dashed line). AVCN: anteroventral cochlear nucleus; DCN: dorsal cochlear nucleus; DNLL: dorsal nucleus of the lateral lemniscus; INLL: intermediate nucleus of the lateral lemniscus; LSO: lateral superior olive; MSO: medial superior olive; PVCN: posteroventral cochlear nucleus; VNLL: ventral nucleus of the lateral lemniscus (redrawn from Pollak and Casseday 1989; and Pollak 1992).

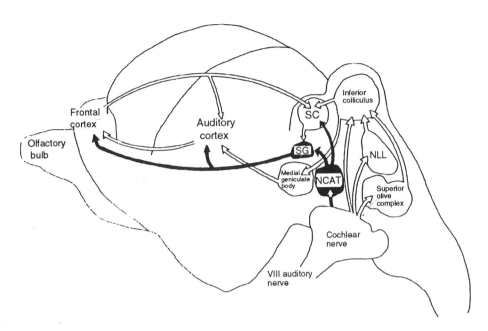

Figure 4.23
Some of the major auditory pathways in the bat, with connections to frontal cortex. SC: superior colliculus; SG: suprageniculate nucleus; NCAT: nucleus of the central acoustic tract; NLL: nuclei of the lateral lemniscus (redrawn from Pollak and Casseday 1989).

CF/FM bats. Specifically, whereas FM bats show a tonotopic arrangement that is comparable to the general mammalian pattern—each frequency is associated with a neural sheet that receives projections from a comparable frequency region located within a lower auditory brain-stem nucleus—CF/FM bats, and the mustached bat in particular, have a spectacular overrepresentation of neural tissue dedicated to processing 60 kHz sound (Ross, Pollak, and Zook 1988).

Within the inferior colliculus, populations of neurons are responsible for processing either monaural or binaural information. In contrast to the mammalian visual system, where input from both eyes is first combined and processed in the cortex, binaural processing in the bat starts within the superior olivary nuclei. Monaural cells typically obtain excitatory stimulation from the contralateral ear. Binaural cells, in contrast, either receive excitatory input from both ears or excitation from the contralateral ear and inhibitory input from the ipsilateral ear. As indicated in Figure 4.21, there is an extensive topographic arrangement of binaural and monaural cells within the inferior colliculus.

A somewhat surprising finding with regard to the cytoarchitecture of the mustached bat's auditory system, and a finding that emphasizes the special design features of this species, is the identification of the medial superior olive (Zook and Casseday 1982), a structure that is typically present in large mammals but absent in smaller species, with sizes approximating those of the Microchiroptera. The presence of this structure appears to be correlated with the relative importance of detecting low-frequency signals and using binaural time cues for localization. In species that detect high-frequency signals and use interaural intensity differences, the lateral superior olive is present and large, but the medial superior olive is lacking. At present, however, the function of the medial superior olive in the mustached bat is somewhat unclear, though several recent papers (Kanwal et al. 1994; Ohlemiller et al. 1994) suggest that it is used in detecting more communicative signals within the repertoire that are relatively lower in frequency than the sonar emission, and that can be localized via binaural time cues.

If we move from the coarse-grained level of the nuclei involved in processing acoustic signals, to the more fine-grained level of the neurons or neuronal ensembles, we find a suite of design features that greatly facilitate the bat's ability to localize a sound and parse it into its component parts. In the mustached bat, for example, work by Fuzessery, Pollak, and others (Fuzessery and Pollak 1984; Fuzessery, Wenstrup, and Pollak 1990) indicates that for 60-kHz signals, excitatory-inhibitory neurons are responsible for detecting sounds in the azimuthal plane, whereas excitatory-excitatory neurons are most responsive to sounds detected in the vertical midline. The perception of range is determined by echo delay-tuned neurons. In the big brown bat, for example, single-unit recordings from the auditory cortex indicate that cells calculate the sequential arrival times of echoes, thereby providing an ''auditory scene'' (Bregman 1990) of a number of objects at different distances from the subject (Dear, Simmons, and Fritz 1993). Turning to the detection of prey type, we find neurons that respond with phase-locked discharges to the spectral modulation within the echo (G. Schuller 1979a), and such discharge patterns change, on-line, as the bat actively pursues its prey (G. Schuller 1979b).

In summary, the bat nervous system is beautifully designed for picking up the salient features of biologically meaningful signals—echoes returned from potential prey. Although we should not accept bat sonar as a communicative signal, our knowledge of the mechanisms underlying the production and perception of this auditory event is superlative and serves us well in thinking about the possible tools or approaches that can be used to investigate other, more communicative systems, including human speech. In fact, as the following quote indicates, Suga has argued that bats are more appropriate than nonhuman primates for studying the special neural features of human speech (Suga 1988):

> If monkey acoustic communication were similar to human speech, the physiological correlations between monkey and human auditory centers would also be high. This is, however, not the case. Human speech is unique and highly specialized. There is an enormous difference in both quantity and quality between human speech and monkey acoustic communication, although both species belong to the same taxonomic order. Therefore, studying the monkey's auditory cortex to understand the functional organization of the human speech areas is not appropriate. (p. 685)

There are, I believe, two problems with this claim. First, it confuses the kinds of evolutionary similarities (homologies versus homoplasies; synapomorphies versus symplesiomorphies) one can observe in the animal kingdom. Thus the similarity we see in auditory specializations of the bat and human are likely to be homoplasies, due, perhaps, to parallel evolution. In contrast, similarities we observe among primates (monkeys, apes, and humans) are more likely homologies, though depending upon how far back in the evolutionary tree we go, we may be looking at either synapomorphies (shared, recently evolved traits) or symplesiomorphies (shared primitive traits). The next section explores both similarities and differences in neural architecture among primates, both human and nonhuman. As argued earlier in the book, an evolutionary perspective tells us that similarities *and* differences are important for understanding the history of selection pressures on a group of organisms. Second, just as it would be misleading to talk about the bat's auditory system due to the extensive variation between bat species, it is also misleading to talk about the monkey's auditory cortex. To put it crudely, not all monkeys were designed equally, and this includes the design of their auditory systems.

4.4   Social Signals: Nonhuman and Human Primates

4.4.1   Nonhuman Primate Vocalizations: General

Unlike studies of insects, anurans, birds, and bats, which have tended to focus on highly specialized vocalizations such as advertisement calls and echolocation pulses, studies of nonhuman primates have explored the mechanisms underlying the perception and production of a wide range of calls within the repertoire. One reason for this focus is that most nonhuman primates, perhaps with the exception of gibbons and siamangs (Chivers and MacKinnon 1977; Mitani and Marler 1989; Mitani 1985), lack highly specialized vocalizations either in terms of the frequency

of their production or in terms of dedicated neural machinery. As a result of this diversity within the repertoire, current knowledge of the proximate mechanisms guiding nonhuman primate vocal communication is less detailed than for other species, but is perhaps more representative of general auditory processing mechanisms.

Nonhuman primates vocalize in a wide range of contexts, and for many species a large number of calls within the repertoire are used during social interactions. As described in greater detail in subsequent chapters, there is often significant variation in acoustic morphology within a specific social context; within-context variation is typically associated with subtle changes in the dynamics of the interaction, including the relative rank and kinship of the participants. Thus we find that in species such as the rhesus macaque (*Macaca mulatta*) and pigtail macaque (*Macaca nemestrina*) subordinates produce one type of scream to call for help when they are being physically attacked by an unrelated dominant and produce a second type of scream when they are being threatened in the absence of physical contact by a genetic relative who is dominant (Gouzoules, Gouzoules, and Marler 1984; Gouzoules and Gouzoules 1989). For nonhuman primates, therefore, selection appears to have shaped a vocal tract system with design features set up to produce a broad range of acoustic variation in a broad range of socioecological contexts. In this section, I begin by providing an overview of how the peripheral organs guide the production and perception of conspecific vocalizations, and then discuss current knowledge of the brain mechanisms responsible for orchestrating the communication system.

As one might expect, a great deal of research on nonhuman primate communication has been guided by research on human speech. Although comparative approaches are powerful and are emphasized throughout the book, interest in demonstrating similarities and differences between human and nonhuman primates has led to a somewhat more narrow focus than studies of other nonhumans. What is unfortunate about this approach is that it has resulted in a rather piecemeal and unconnected data set concerning the central and peripheral mechanisms underlying communication. Specifically, in contrast to studies of avian song, neurobiological studies of nonhuman primate communication did not emerge from careful analyses of repertoire variation; in fact, to this day, such descriptive accounts are lacking for the majority of species. Instead, researchers jumped into the brain and onto the peripheral organs without a complete understanding of what those systems were designed to drive. Thus there have been attempts to find whether nonhuman primates have homologues to some of the important neural regions for human language such as Broca's and Wernicke's areas. Although structural homologues appear to exist (Aitken 1981; Deacon 1991; Jürgens 1990), the function of these areas remains somewhat obscure because knowledge of the vocal repertoire, in particular its fundamental acoustical units, has yet to be described in sufficient detail.

Historical criticisms aside, research on primate acoustical communication has come into its own within the past ten or fifteen years. One of the reasons for this change is the advent of sophisticated computer technology. Using digital signal-processing algorithms and hardware that permits rapid acquisition and processing of a large sample of signals, primatologists have begun to provide increasingly more subtle and quantitative descriptions of within- and between-species acoustical variation (Cleveland and Snowdon 1981; Gouzoules and Gouzoules 1989; Hauser 1991; Macedonia 1990, 1991; Mitani and Marler 1989; Seyfarth and Cheney 1984). What these data sets provide are important insights into the range of spectrotemporal variation and, consequently, some hints as to how laryngeal and supralaryngeal maneuvers may function during vocal production.

Sound Production Mechanisms   Nonhuman primates tend to vocalize during expiration, generating the necessary airflow by filling and then depleting portions of their lungs. The air emerging from the lungs causes vocal fold vibrations, resulting in phonation. The waveform produced at the glottis—the fundamental frequency—is saw-toothed, representing the periodicity of vocal fold vibration. Looking at the vocal repertoires of different species, it is clear that there is considerable within- and between-species variation in the fundamental frequency, and much of this variation can be accounted for by differences in body weight

(Hauser 1993a). Thus large species such as gorillas (*Gorilla gorilla*) and chimpanzees (*Pan troglodytes*) tend to produce calls with relatively low fundamental frequencies, whereas small species such as the marmosets and tamarins (Callitrichidae) tend to produce calls with relatively high fundamental frequencies. There are, however, exceptions to these general trends, and recent research by Schön Ybarra (in press) suggests that some of the unexplained variation in fundamental frequency range and stability among primate species may be due to morphological specializations of the vocal folds and laryngeal muscles. For example, in many nonhuman primates, the edge of the vocal fold lip is generally unrestricted (i.e., relatively free to move). As a result, exerting high muscle tension on this lip will lead to high-pitched vocalizations and, in some species, ultrasonic calls (Masters 1991; Zimmerman 1981). On the other hand, because the lip is unrestricted, period-to-period variation is likely, creating what Lieberman (1968) described as the instability of primate phonation. In sum, these morphological features indicate that in contrast to the human larynx, the nonhuman primate larynx allows for a greater range in vocal pitch as well as greater instability.

Because nonhuman primates tend to produce vocalizations during expiration, one might expect to find a relationship between decreasing air pressure, the rate of vocal fold vibration, and the fundamental frequency. Specifically, for relatively long calls, lung deflation should be associated with a decrease in the rate of vocal fold vibration and, consequently, a decrease in the fundamental frequency (see section 4.4.2 for a similar discussion of human phonation). To assess this possibility, Hauser and Fowler (1991) examined changes in the fundamental frequency for calls produced by free-ranging vervet monkeys during intergroup encounters (''wrrs'') and by rhesus monkeys during intragroup affiliative interactions (''girneys''). Results showed that during bouts (a series of 2–3 consecutive calls) of wrrs and girneys, there was a statistically significant decline in the fundamental frequency. Moreover, for vervet monkeys, the decline in frequency appeared to be used as a cue for conversational interactions: the frequency drop was highly correlated with bout termination, and individuals rarely interrupted each other. The only exception to this pattern was for juveniles and infants who showed no consistent pattern of frequency modulation across a bout of calls and, apparently as a result, were interrupted more often than adults.

The supralaryngeal vocal tract of nonhuman primates is structurally similar to that of other mammals, consisting of a relatively straight tube (Negus 1949). This configuration stands in contrast to the human supralaryngeal tract (see Figure 4.24) which has been characterized as a two-plus tube because of the characteristic 90-degree bend between the laryngeal and oropharyngeal cavities. However,

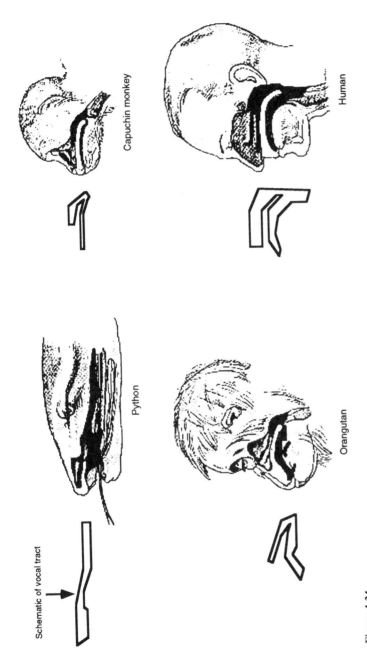

Figure 4.24
Comparative vocal tract anatomy (Negus 1949). Blackened portions highlight primary cavities, illustrating the fundamental contrast in configuration between humans and other animals; note especially the difference in the degree of curvature between laryngeal and oropharyngeal cavity. To the left of each illustration are schematic drawings of the vocal tract, including the nasal cavity.

as will be revealed in greater detail, nonhuman primates can impose configurational changes on the supralaryngeal tract (Schön Ybarra, in press) by means of articulatory maneuvers of the lips, mandible, and possibly the tongue.

In contrast to current understanding of sound production in insects (Bailey 1991), frogs (Capranica and Moffat 1983), birds (Greenewalt 1968; Nowicki, Westneat, and Hoese 1992; Suthers 1990), bats (Suthers 1988), and humans (Borden and Harris 1984; Lieberman and Blumstein 1988), relatively little is known about the kinematics of vocal production in nonhuman primates. Although several early studies provided intriguing speculations about nonhuman primate sound production based on morphological descriptions, Lieberman, Klatt, and Wilson (1969) were the first to use information about vocal tract structure to quantitatively model sound performance. Although the primary motivation for this work was to assess whether nonhuman primates could produce some of the critical sounds of human speech, the end product was of more general interest and provided two empirically testable predictions. Specifically, Lieberman and colleagues (Lieberman 1984, 1985, 1991; Lieberman, Klatt, and Wilson 1969) suggested that (1) the fundamental frequency of phonation would be highly unstable and (2) the range and plasticity of articulatory routines would be limited, partly because of a relatively short supralaryngeal vocal tract and a tongue with restricted mobility.

Over the past ten to twenty years, primatologists have provided general observations and experimental results that permit more careful assessment of Lieberman's predictions. Regarding phonation, it appears that several nonhuman primate species produce vocalizations with relatively unmodulated fundamental frequency contours (see Figure 4.25 for examples). However, nonhuman primates also produce calls with highly modulated frequency contours and considerable period-to-period variation. Although structural properties of the glottis may be responsible for some of this variation (Schön Ybarra, in press), a number of studies now indicate that factors such as ontogenetic and emotional changes, or selection for particular design features (C. H. Brown and Gomez 1992), also play a significant role in configuring the fundamental frequency contour. For example, Hauser (1989) has shown that in the intergroup wrr of vervet monkeys, the stability of the fundamental frequency increases gradually over a developmental period of two to three years. Goedeking (1988) has shown that in common marmosets the stability of the fundamental changes as a function of the caller's motivation to play, and thus variation in frequency appears to covary with an affective state. Last, as pointed out by C. H. Brown and Waser (1988), changes in the frequency contour of primate vocalizations, like the changes observed in avian alarm and mobbing calls, are likely to represent structural changes designed to maximize propagation through the environment (see chapter 3).

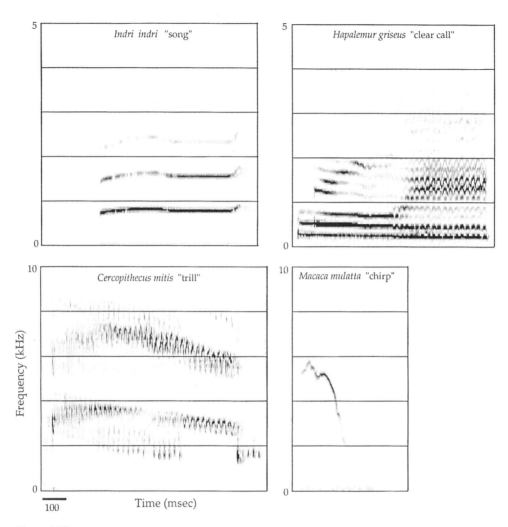

Figure 4.25
A sample of spectrograms of nonhuman primate vocalizations, illustrating variation in the fundamental frequency. Some vocalizations are heavily modulated (*Cercopithecus mitis*) whereas others exhibit a relatively flat frequency contour (e.g., *Indri indri*).

Based on Marler and Tenaza's (1977) qualitative assessment of the correspondence between articulation and sound structure in chimpanzees (Figure 4.26),
Bauer (1987) used 16-mm footage to provide the first quantitative analysis of
changes in lip configuration of a male chimpanzee during the production of submissive screams and aggressive barks. The theoretical motivation underlying this
work was to test the notion of an orofacial code (Darwin 1872; Morton 1977;
Ohala 1983, 1984), the idea that particular gestures are coupled to particular
frequencies, reflecting rather distinctive motivational states. Using frame-by-
frame analyses, Bauer found that during the production of submissive screams,
the lips are retracted, and the fundamental frequency is relatively high. In contrast, during the production of aggressive barks, the lips are rounded and pro-
truded, and the fundamental frequency is relatively lower. Thus, as predicted by
Darwin and Morton, states of submission appear to be associated with high-
pitched calls, whereas states of aggression appear to be associated with low-
pitched calls (August and Anderson 1987; Hauser 1993a).

There are two potential problems with Bauer's analyses. First, he did not
provide an independent measure of motivational state—that is, facial expressions
were used to define the individual's motivational state. Second, the magnitude
of change in the fundamental frequency is unlikely to be due to changes in lip
configuration alone. In fact, given current models of speech production (Fant
1960; Lieberman and Blumstein 1988) that are most applicable to nonhuman primates (Fitch and Hauser, in press), lip configuration is likely to have only a small
effect on the laryngeal output (Bickley and Stevens 1987), but a quite significant
effect on the resonance frequencies of the call. This difference exists because
changes in lip configuration will tend to cause an alteration in the filtering properties of the tract, but are unlikely to cause major changes in laryngeal muscle
tension or glottal movement (see Box 4.3).

To assess the possibility of supralaryngeal filtering during vocal production,
Hauser, Evans, and Marler (1993) examined the role of articulation in free-ranging
rhesus monkeys. Audio-video records of vocally interacting individuals revealed
three coarse-grained articulatory gestures: (1) mandibular lowering and raising,
(2) lip protrusion and retraction, and (3) the separation between the upper and
lower teeth. Thus individuals place their lips in an O-shaped configuration during
the production of aggressive barks, retract their lips during submissive screams,
and initiate repeated, rapid lip protrusion-retraction oscillations during the production of affiliative girney vocalizations.

A more quantitative analysis of the filtering properties of the supralaryngeal
tract of rhesus monkeys was obtained using frame-by-frame measurements of

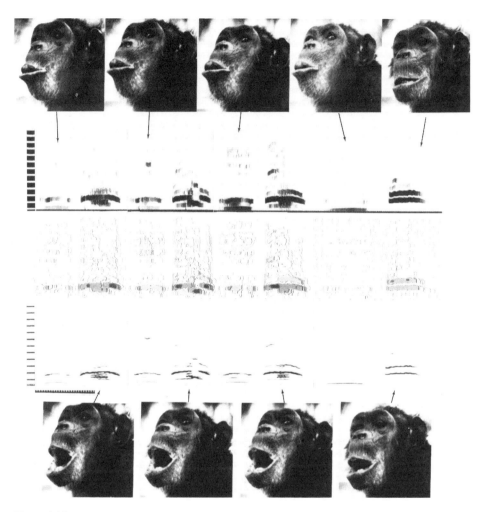

Figure 4.26
A frame-by-frame illustration of a chimpanzee pant-hoot, illustrating the correspondence between articulation and sound structure (from Marler and Tenaza 1977).

Box 4.3
Source-Filter Systems in Acoustics

Current knowledge of the acoustics of sound production systems comes primarily from studies of human speech, where a great deal is known about the mechanisms of phonation and the articulatory gestures involved in filtering the sound source (Baken 1987; Kent and Read 1992; Lieberman and Blumstein 1988). A dominant model in this area is Fant's (1960) source-filter theory of speech production. This is a nonfeedback model that starts with the claim that source (i.e., glottal vibrations) and filter (i.e., configuration of the supralaryngeal cavity) are independent. The most likely context for such independence occurs when the frequency of the lowest resonance is significantly higher than the fundamental frequency. In contrast, source-filter dependence or feedback (also known as nonlinear systems; Baken 1990; Kent and Read 1992) occurs when the resonance and fundamental frequencies overlap. Nonfeedback systems appear most appropriate for human speech, whereas feedback systems are appropriate for most wind instruments, and most likely, the singing human voice (Sundberg 1987); species such as the indri and gibbon, who produce highly melodic songs, may be more properly characterized by nonfeedback models.

   To apply the models of speech acoustics to nonhuman animals, details about phonation, articulation, and vocal tract morphology are sorely needed, and this is especially the case for nonhuman primates. However, even in the absence of such information, we can begin to assess the relevance of feedback versus nonfeedback systems in nonhuman animal communication by looking at spectrographic data in relation to simple equations generated from acoustic principles. For example, starting with a straight tube closed at one end (e.g., glottis) and open at the other (e.g., lips), we can predict the wavelength of a sound by the following equation:

$$I = c/f$$

where I = wavelength of a sound, c = speed of sound propagation in air (335 m/sec), and f = frequency of the sound. Using this equation, we can see that the wavelength of a 400-Hz rhesus monkey coo call is 0.84 m. Once the wave propagates down the tube, some of its frequencies will be emphasized, and others will be deemphasized; tube length and shape determine which frequencies are emphasized. To calculate the resonance frequencies of a cavity, acousticians have developed a number of equations; parameters of the equation vary as a function of tube structure (e.g., flared end, points of constriction). The human vocal tract, for example, is modeled as a quarter-wavelength resonator, and the following equation is used to derive each resonance:

$$f = (2n + 1)c/4L$$

where n = the number of the resonance frequency to be calculated and L = the length of the vocal tract from the point of closure or constriction (i.e., glottis) to the point of opening (i.e., mouth opening).

mandibular position during the production of affiliative coos. For each frame, values of mandibular position were examined relative to two spectral components of the call: (1) fundamental frequency (i.e., laryngeal source) and (2) first resonance frequency. If source and filter are independent (*sensu* Fant's (1960) nonfeedback model; see Box 4.3) during the production of coos, then changes in mandibular position should have no effect on the fundamental frequency and a significant effect on the resonance frequencies. Results indicated that during coo production, the mandible was lowered, reaching a maximum about two-thirds of the way through. Although the fundamental frequency contour was virtually flat, there were significant changes in the first resonance frequency, and these were highly correlated with changes in mandibular position. These results suggest, therefore, that changes in the mandible are associated with changes in the filtering properties of the tract, but not with changes in laryngeal function.

To explore the possibility that changes in articulation play a more causal role in configuring call structure, Hauser and Schön Ybarra (1994) adopted a technique developed by speech scientists (reviewed in Borden 1979) to temporarily *silence* a key articulator during sound production. The technique involved injection of xylocaine into the lips. Xylocaine, like novocaine obtained at the dentist's office, is a local anesthetic that suppresses efferent and afferent impulse transmission. Consequently, when rhesus monkeys were treated with xylocaine, lip movement was significantly diminished for approximately one hour. The importance of this manipulation, therefore, was to cause individuals to produce some calls with a shorter vocal tract (i.e., block lip protrusion) and other calls with a relatively longer vocal tract length (i.e., block lip retraction).

Although individuals treated with xylocaine were not trained to produce particular types of vocalizations, one individual produced a large number of coos and the other produced a large number of noisy screams. The configurational features of these two calls represent opposite ends of a continuum from extensive lip protrusion (coos) to extensive lip retraction (screams). The xylocaine manipulation therefore forced individuals to produce coos with a shorter vocal tract and to produce screams with a longer vocal tract. Based on Fant's source-filter theory, the following predictions were generated. If individuals can compensate for the perturbations imposed (MacNeilage and DeClerk 1969), then xylocaine should have no effect on the acoustic morphology of the call produced. If, on the other hand, compensation is not possible, then the xylocaine manipulation should cause changes in the resonance frequencies, but not in the fundamental frequency. Results indicated that for coos, there were no significant differences in the fundamental frequency or duration of the calls, but significant changes in the two formant frequencies. Specifically, formant frequencies were higher for coos produced with xylocaine relative to coos produced in its absence. This is exactly what one would predict based on acoustic theory and an articulatory system that fails to compensate for the imposed perturbation: shorten cavity length, and formant frequencies increase.

In contrast to coos, the acoustic morphology of screams produced under xylocaine treatment did not differ from those produced without xylocaine. The lack of difference was evident for acoustic features associated with laryngeal and supralaryngeal function. Moreover, the first few calls produced under xylocaine were no different from those produced later on or during control conditions, suggesting that auditory feedback did not play an important role in attaining the acoustic target.

Results from Hauser and Schön Ybarra's experiments suggest that for coos, blocking lip production causes a significant change in formant structure. What remains to be demonstrated is that such acoustic changes are perceptually meaningful to the animals[25] and that perhaps with additional experience, animals can learn to compensate for the imposed perturbation.

For rhesus screams, xylocaine did not cause a significant change in call structure. There are two possible explanations for this result. First, the change in lip position from normal (i.e., retracted) to xylocaine-treated (i.e., approximating the resting position) represents only a minor change in vocal tract length. Consequently, the experimental perturbation failed to create a significant articulatory challenge. Second, xylocaine imposed a significant change in vocal tract length, but the articulatory system is sufficiently flexible to compensate for the change and attain the intended acoustic target. One possible articulatory avenue for compensation would be to shift the position of the larynx, a maneuver that appears at least possible on anatomical grounds (Schön Ybarra, in press; T. Deacon, personal communication). Distinguishing between these two alternative explanations will require further experiments, employing such techniques as cineradiography to directly observe laryngeal movement.

---

25. To date, several studies of nonhuman primate communication have demonstrated significant within-species acoustic variation, but have made little progress in showing how members of the species respond to such variation. In Hauser and Schön Ybarra (1994), the possibility remains that differences in formant structure, though statistically significant, are not perceptually meaningful with regard to classifying exemplars into call types. Thus, although coos produced with xylocaine are acoustically different from those produced without xylocaine, they may nonetheless be classified from a perceptual perspective as a perfectly good coo. This observation once again raises the distinction between just noticeable differences and just meaningful differences.

Sound Perception Mechanisms   Studies of primate perception have focused primarily on the more tonal components of the repertoire, especially the lower frequencies (i.e., fundamental). As discussed in the previous subsection, however, several nonhuman primate species produce complexly structured atonal calls with considerable spectral information in the upper frequencies. Whether such information is perceptually salient is currently unclear (though see Owren and Bernacki 1988; Owren 1990). Much of the discussion that follows is therefore based on perceptual experiments conducted with more tonal signals in the repertoire (emphasizing changes in the fundamental frequency) or with nonbiological sounds (e.g., time-varying sinusoids with and without amplitude modulation) that appear to incorporate some of the significant features of conspecific vocalizations.

To capture the type of perceptual problem confronted by some nonhuman primates, let us start with a quick summary (see chapter 7 for detailed discussion) of one of the most comprehensive studies to date. Based on Green's (1975a, 1975b) field study of the Japanese macaque (see chapter 2), which yielded the first account of an entire macaque vocal repertoire, Stebbins and his colleagues (reviewed in Stebbins and Sommers 1992) initiated a series of experiments to determine the mechanisms underlying perceptual discrimination of one acoustically variable class of calls in the repertoire, the coo. The general aims of this work were to determine whether there are species-specific perceptual mechanisms and to assess which features of the call are most significant with regard to acoustical categorization. The first study revealed that whereas Japanese

macaques readily learned a discrimination task involving what appeared to be a species-specific marker—the temporal location of the peak frequency—closely related species learned this task more slowly but performed equally well with a task based on a more general feature, namely the fundamental frequency (Zoloth et al. 1979). Follow-up studies showed that two variants of the coo were processed in a categorical fashion (May, Moody, and Stebbens 1989). Specifically, depending upon the location of the peak frequency, exemplars were either classified as smooth early high (SEH) coos or smooth late high (SLH) coos.[26] Subsequent psychophysical experiments (May, Moody, and Stebbins 1988; Moody and Stebbins 1989), involving selective manipulation of spectrotemporal parameters, revealed that frequency modulation was by far the most important acoustical feature for distinguishing between these call types.

How do the peripheral organs of the nonhuman primate auditory system process biologically meaningful signals, and what mechanical constraints limit the individual's ability to resolve the acoustical details of a signal? As with studies of production mechanisms, most of our understanding of nonhuman primate hearing comes from studies of macaques. I first provide descriptive information on the anatomy of the primate ear, and then discuss its analytic abilities; because the gross-level anatomy of the primate ear (Figure 4.27) is relatively homogeneous across species,[27] the discussion provided here will also be relevant to the section on human hearing. Specific perceptual specializations will be pointed out where relevant. I focus my discussion on studies aimed at understanding basic perceptual problems, rather than those that have used primates to test explicit hypotheses about speech perception (Hienz and Brady 1988; Kuhl and Padden 1982, 1983; Sinnott 1989a, 1989b), the latter are treated in chapter 7, where I discuss the topic of categorical perception.

Figure 4.28 captures the general process by which sound in the environment is perceived by the primate auditory system. The pressure sound wave (vocalization, speech) arrives at the outer ear, or pinna, and then moves through the external auditory meatus or channel. The pinna is both a receiving dish for sound and a protective device, shielding the sensitive components of the middle ear from injury. Moreover, the pinna functions as a filter, providing the relevant cues for localizing sound. Just like the vocal tract, the meatus or channel leading from

---

26. A recent study by Hopp and colleagues (1992), using a slightly different paradigm from the one used by May, Moody, and Stebbins, failed to find evidence for categorical perception using the same species and coo exemplars. Possible differences between these studies are examined in chapter 7.

27. In contrast to the comparative work on primate vocal tract anatomy, relatively few species have been investigated with regard to specializations for hearing.

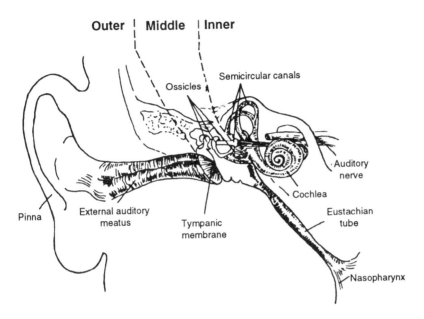

Figure 4.27
The human ear, comparable, on a gross-anatomy level, to that of other primates (redrawn from Borden and Harris 1984).

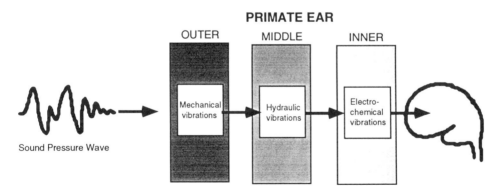

Figure 4.28
Schematic illustration of changes in the transmission process during the perception of sound by primates. The chain of events starts on the left with a sound pressure wave and moves through the ear to the brain on the far right.

the pinna to the middle ear serves as a resonator, boosting the amplitude of some frequencies and decreasing the amplitude of other frequencies. The pressure sound wave moves from the outer to the middle ear by means of mechanical vibrations, first passing the tympanic membrane, and then moving on to the ossicles. The ossicles serve as the primary bridging mechanism between the tympanic membrane and the oval window, a membrane that leads to the cochlea of the inner ear. The primary function of the middle ear is in boosting the amplitude of the signal that moves into the inner ear. The middle ear thus acts as an exquisite transformer.

The inner ear is filled with a fluid known as perilymph. Within this fluid lies the vestibular system, involved in maintaining proper body orientation, and the cochlea, which is the primary hearing structure. At the base of the cochlear duct is the basilar membrane, and on top of this membrane is the organ of Corti, a set of hair cells whose primary function is auditory reception. Pressure waves within the cochlea cause the basilar membrane to move, in turn causing the hair cells to move. The analysis of a complex sound (a Fourier transform) is thus based on such movements, in particular the place and timing of stimulation along the basilar membrane. As discussed earlier for bats, the basilar membrane is segmented according to frequency, with high frequencies represented at the thinner and stiffer base, and low frequencies represented at the more mobile and wider tip. The cells within the cochlea lead to a bundle of nerve fibers that collectively form the eighth cranial nerve, the primary neural avenue to the proximally located auditory cortex, situated within the temporal lobe.

Although we still know relatively little about the diversity of anatomical and psychophysical properties of nonhuman primate auditory systems, it seems likely that interspecific differences will emerge given variation in repertoire composition across species. For example, whereas most of the gorilla's repertoire falls below 5.0 kHz (Harcourt, Stewart, and Hauser 1993), vocalizations produced by several small New World monkeys and prosimians exceed 20 kHz (Pola and Snowdon 1975; Zimmerman 1981, 1985). Let us now explore some of the specific details of nonhuman primate perception, using the macaque as a model species. It is a model species methodologically and empirically, though it is not necessarily representative of other species, confronted with different ecological pressures for hearing (e.g., small New World monkeys living atop the rain-forest canopy).

To understand the perceptual capacities of macaques (primarily, the Japanese macaque, *M. fuscata;* rhesus, *M. mulatta;* and pigtailed macaque, *M. nemestrina*), Stebbins, Moody and colleagues (reviewed in Moody 1994; Moody, Stebbins, and May 1990; Stebbins and Sommers 1992) have conducted a series of experiments using psychophysical procedures. In the typical paradigm, a chair-restrained subject is trained on a discrimination task that requires depressing a lever when a particular exemplar of a sound class is presented. In one of the earliest studies (Serafin, Moody, and Stebbins 1982), involving pigtailed macaques, psychophysical tuning curves[28] were derived based on presentations of pure-tone frequencies that were either embedded in noise (simultaneous masking) or immediately followed a burst of noise (forward masking). Although both masking procedures yielded sharply tuned curves at the lower frequencies and a gradual broadening of the response at higher frequencies, there were also significant differences. In the simultaneous masking experiments, frequency selectivity increased up to 16 kHz, whereas the peak in selectivity for the forward masking experiments occurred at 1–4 kHz. Moreover, tuning curves were sharper in the forward masking tests. In general, studies of humans indicate that forward masking procedures generate sharper tuning curves. Data on macaques therefore suggest that tuning curve differences are largely due to the details of the masking procedure.

To further assess the perceptual capacities of macaques, especially specializations for processing species-typical vocalizations, additional experiments have been conducted with the Japanese macaque. Results indicate extensive intersubject variability in terms of frequency and intensity discriminations (Prosen et al. 1990; Sinnott, Petersen, and Hopp 1985), variability that decreases with training (i.e., experience with the task). Regarding frequency, subjects can discriminate signals in the range of 7–33 Hz for a carrier frequency of 1000 Hz; the variation in sensitivity is due to both interindividual differences and testing procedures. Difference limens for intensity are in the 1.5–5 dB range for signals in the 500–1000 Hz range. Macaque-human comparisons indicate that frequency sensitivity is higher in humans, whereas measures of sensitivity to intensity differences are comparable.

Two important acoustic components of the macaque vocal repertoire, especially the more tonal elements (e.g., affiliative coo vocalizations), are amplitude and frequency modulation. Moody and colleagues (Moody 1994; Moody et al. 1986; Sommers et al. 1992) have shown by means of standard psychophysical tests and multidimensional scaling[29] that macaques are extremely sensitive to

---

28. Psychophysical tuning curves are comparable to neural tuning curves (Box 4.1), in that they provide information on the specificity of the perceptual system's response to a range of frequency values, relative to intensity. In both types of tuning curves, Q_{10dB} values are used to describe the sharpness of the response.

29. A statistical technique that provides a quantitative visual and numerical assessment of how exemplars within a population cluster into groups, related by some metric of similarity.

changes in frequency and amplitude modulation, in terms of both the fundamental frequency and the formant frequencies. Thus they are not only sensitive and able to discriminate independent changes in frequency and amplitude, but are also perceptually skilled at discriminating relationships between frequency and amplitude—the intensity of the spectral envelope. These perceptual experiments suggest that future studies of communication in macaques, which have tended to focus on simple univariate acoustic dimensions (e.g., duration, peak frequency), will require attention to complex interactions between a suite of spectral features, including the perceptual salience of resonance frequencies that result from vocal tract filtering (Hauser, Evans, and Marler 1993).

Given the perceptual abilities described, how does the primate auditory system localize sound? The first experimental test[30] of sound localization was conducted on wild mangabeys (*Cercocebus albigena*), a forest-living monkey. During intergroup interactions and territorial defense, this species produces a relatively loud vocalization known as the "whoopgobble" (Waser 1977b). This call is temporally segmented into a frequency-modulated introductory "whoop" followed by silence and then a series of low frequency "gobble" notes. The whoop appears to attract the attention of other group members (an alerting function), whereas the gobbles facilitate localization. The data presented in Figure 4.29 illustrate the path taken by one mangabey tracking the sound emitted from the speaker, and the histograms below show, based on a number of playback trials, the degree of orientation error. In general, localization errors were greater when the speaker was far (mean = 6 degrees, range = 0–21 degrees) than when it was near (mean = 3 degrees, range = 0–9 degrees), presumably reflecting the effects of both distance and habitat effects on signal structure and propagation.

More quantitative research on sound localization in primates has been conducted in the laboratory, using techniques that were previously described for anurans and barn owls. Specifically, C. H. Brown and colleagues (1979) have demonstrated that in macaques, harmonically structured clear calls (e.g., coos) are less accurately localized (Figure 4.30) than broadband noisy calls (e.g., barks).

Moreover, tests (C. H. Brown et al. 1983) using the same class of vocalizations revealed that macaques were more accurate when localization occurred in the horizontal plane than when it occurred in the vertical plane (Figure 4.31). Given the fact that the two species tested are generally considered terrestrial primates, this difference in localization accuracy makes good evolutionary sense.

---

30. In fact, Waser (1977a) was the first to conduct field playback experiments with nonhuman primates.

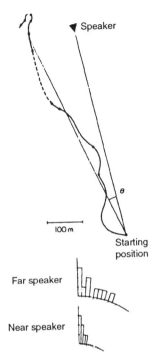

Figure 4.29
Localization by wild mangebeys as evidenced by tracking behavior during playbacks of the whoopgobble, a call used during intergroup interactions. The upper portion of the figure shows the approach
behavior of a male, relative to the position of the speaker. The lower portion of the figure shows the
distribution of angular errors following playbacks where speaker subject distances were short (near
speaker) and long (far speaker). The y-axis represents the number of playbacks, the x-axis is the
error in degrees (redrawn from Waser 1977b).

The studies reported on in this subsection and in chapter 3, suggest that selection has favored primate signals that are designed to propagate through the environment with minimum degradation and, further, are readily localizable. There
are three primary constraints: the vocal apparatus, the perceptual system, and
the specific structure of the habitat. To examine the interaction between these
factors, C. H. Brown and Waser (Brown and Waser 1988; Brown 1989a, 1989b;
Brown and Gomez 1992; Brown and May 1990; Brown and Waser 1984) have
conducted a series of field and laboratory playback experiments with blue monkeys and mangabeys that map out the propagation properties of the calls and the
discriminative abilities of their perceptual systems. To cite one example of this
work (Brown and Waser 1984), results presented in Figure 4.32 show that for

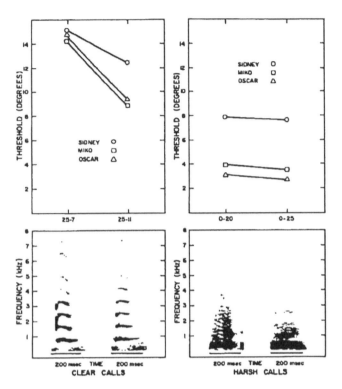

Figure 4.30
Sound localization angles by macaques (*M. mulatta* and *nemestrina*) for two clear calls (left column) and two harsh call (right column) (from C. H. Brown et al. 1979).

both blue monkeys (*Cercopithecus mitis*) and humans, there is relatively high sensitivity at 1–4 kHz, a frequency range that is intensively used in both species' vocal repertoire. Interestingly, however, when relative audibility functions (i.e., hearing functions normalized to 0 dB for humans) are plotted, we find that blue monkeys are more sensitive than humans and rhesus monkeys in the low-frequency (125–250 Hz) range. A plausible explanation for the difference is that one of the more significant calls in the blue monkey's repertoire, the "boom" (Marler 1973), has a majority of its energy centered at 125 Hz. Another possibility is that, because maximal sensitivity is largely determined by the resonant frequencies of the ear canal, blue monkeys have a relatively longer ear canal than either rhesus or humans. Independently of these explanations, however, observations suggest that relative to humans and rhesus, there appears to have been stronger selection on blue monkeys for detecting low-frequency signals.

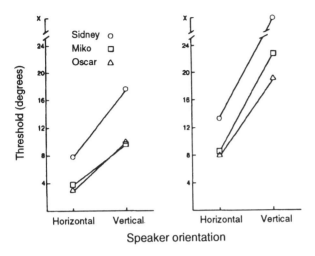

Figure 4.31
Localization accuracy (minimum audible angles) by macaques during tests involving playbacks of harsh calls (0–20) and clear calls (25–12) presented in the horizontal and vertical planes (from Brown et al. 1983).

A

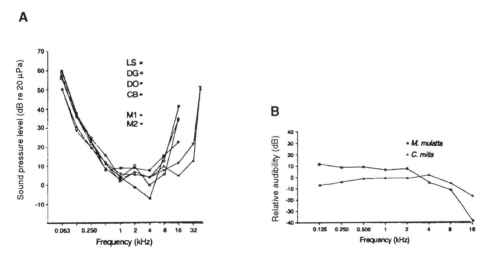

Figure 4.32
(*A*) Audiograms derived from standard laboratory psychophysical techniques. Data from four human subjects (LS, DG, DO, CB) and two blue monkeys (M1, M2). (*B*) Relative audibility functions for blue monkeys (*C. mitis*) and rhesus macaques (*M. mulatta*), normalized to human hearing analyses conducted under the same conditions. Points above 0 dB indicate that subjects are less sensitive than humans, whereas points below indicate greater sensitivity (redrawn from Brown and Waser 1984).

Paralleling work on birdsong, deafening experiments have been conducted with both adult and infant squirrel monkeys (Talmage-Riggs et al. 1972) to determine the role of auditory feedback on vocal production. Surprisingly, results (see chapter 5 for a more complete discussion) indicate that in the absence of auditory feedback, young continue to develop a normal vocal repertoire and adults continue to produce species-typical calls on both a short- and long-term basis. As will be discussed in the next subsection, the lack of effect is surprising because there appear to be direct neural connections between the vocal production and perception systems of the squirrel monkey brain.

In summary, as we discovered for frogs, birds, and bats, nonhuman primates have coevolved finely tuned sender-receiver systems. And like these other organisms, selection has apparently resulted in greater interspecific variation in the sound-producing organs than in the sound-receiving organs. Although a relatively impoverished comparative data set warrants cautious interpretation, it appears that primates may have evolved more specialized systems for vocal tract filtering than other species, both mammals and nonmammals. If so, then this finding would suggest specialized central nervous system mechanisms for decoding the information produced at the periphery. As we will soon learn, humans have evolved such mechanisms.

Brain Mechanisms As we will discover in section 4.4.2, human language, both spoken and signed, is primarily controlled by higher cortical functions. In contrast, subcortical areas such as the limbic system appear to be responsible for the production of more emotive human utterances (e.g., laughter, crying, screaming). The story that will unfold in this section reveals that for at least two intensively studied nonhuman primates, squirrel monkeys and macaques, the production of vocal communication is controlled by the midbrain, diencephalon, and limbic system (Deacon 1991, in press; Jürgens 1990, 1992; Larson, Ortega, and DeRosier 1988). Higher cortical structures may only be involved in the perception of vocal signals (Rauschecker, Tian, and Hauser 1995; Rauschecker et al., in press).

Though a number of early studies attempted to explore the relationship between vocal communication and brain function in nonhuman primates (Dusser de Barenne, Garol, and McCulloch 1941; Leyton and Sherrington 1917), the first detailed investigation of this relationship was initiated by a team of German biologists, focusing on the squirrel monkey (reviewed in Jürgens 1990, 1992). The squirrel monkey was an excellent choice for neurobiological studies (Jürgens et al. 1967) because analyses of its vocal repertoire had already been published

(Ploog, Hopf, and Winter 1967; Winter 1969; Winter, Ploog, and Latta 1966). The first series of studies revealed that most of the vocal repertoire of captive squirrel monkeys could be elicited by electrically stimulating areas within the midbrain and limbic system (Figure 4.33), and in general, stimulation sites were associated with single call types.[31] The latency to elicit a vocalization varied between sites, as did the number of different vocalizations elicited. In the periaqueductal gray area, for example, where the greatest diversity of call types was produced, latencies were short (50–60 msec). In contrast, latencies in the cingulate cortex and supplementary motor cortex were relatively long (220–300 msec). Associated with electrical stimulation of these brain areas is not only the production of species-typical vocalizations, but in addition, emotionally consistent behaviors (e.g., submissive behaviors associated with calls produced in contexts of submission or fear) and the simultaneous production of several autonomic responses such as changes in heart rate, salivation, and licking. These data, together with tracer studies of efferent projections, indicate that the periaqueductal gray area is involved in the orchestration of both arousal and motor control of facial, laryngeal, and respiratory activities involved in vocal production.

   In addition to the brain areas revealed through single-unit recording procedures, tracer studies (Jürgens and Pratt 1979; Müller-Preuss and Jürgens 1976) have demonstrated that there are direct projections from the anterior cingulate to the midbrain periaqueductal gray area. Lesioning these areas, however, neither eliminates vocal output nor significantly alters call morphology. Consequently, though the cingulate-to-midbrain pathway is clearly involved in vocal production, it is not the final motor pathway. What is more likely, and supported by comparable research on macaques (Sutton 1979), is that this area of the brain somehow facilitates or inhibits emotively based vocalizations (Deacon 1991)—that is, it is involved in the voluntary[32] control of vocal output. Last, Jürgens and colleagues (1990, 1992; Kirzinger and Jürgens 1982) have used lesioning techniques to draw functional neuroanatomical parallels between the control centers for human speech and squirrel monkey communication. They show that whereas humans suffer significantly from lesions to Broca's area (i.e., motor aphasias; see section

---

31. These data do not, of course, show that the stimulated areas are *singly* responsible for controlling vocal output.

32. Note that the notion of "voluntary control" should be treated cautiously here, based on the fact that nonhuman primate vocalizations are difficult to condition (i.e., shape individuals to alter the structure of their vocalizations, or even the rate of delivery; Peirce 1985). Moreover, the idea of voluntarily controlling vocal output veers dangerously close to the notion of "intentional signaling" for which there is very little evidence (see chapter 7).

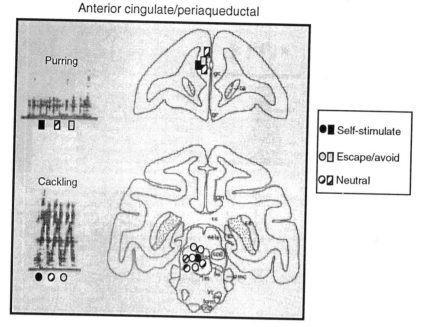

Figure 4.33
Results from electrical stimulation experiments in the squirrel monkey, emphasizing two areas that are significantly involved in the production of species-typical vocalizations—the anterior cingulate and periaqueductal gray area. In the upper portion of the figure, stimulation cites elicited an affiliative call known as purring, generally associated with self-stimulation of neutral behavior. In the lower portion of the figure, stimulation cites elicited a dominance-related call known as cackling, typically associated with escape/avoidance behavior or a neutral response (redrawn from Jürgens 1979).

4.4.2), squirrel monkeys do not—the gross structure of their vocalizations is similar pre- and postoperatively (Figure 4.34). These results and the conclusions drawn from them should be treated cautiously. Specifically, the acoustical analyses that have been carried out to compare pre- and postoperative vocal output were performed in a highly coarse-grained fashion—qualitative inspection of sound spectrograms. Hence, it is possible that more fine-grained analyses would reveal significant changes in call structure. If such differences should emerge in subsequent tests, it will be important to assess whether they represent perceptually meaningful differences to the monkeys.

Studies of squirrel monkeys have also examined the neural substrate underlying perception of species-typical calls. Lesion studies (Hupfer, Jürgens and Ploog 1977) have revealed that if the primary and secondary auditory cortex are com-

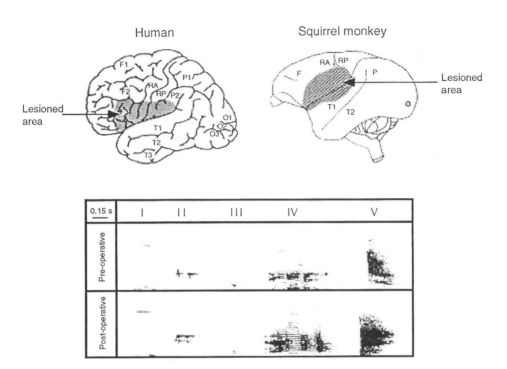

Figure 4.34
Upper left: human brain with lesion leading to motor aphasia. F, F1–F3: frontal cortex; O, O1–O3: occipital cortex; P, P1–P3: parietal cortex; RA, RP: anterior and posterior Rolandic cortex; T1–T3: temporal cortex. *Upper right:* squirrel monkey brain with lesion leading to the production of similar vocalizations compared with the preoperative period. Exemplars of five different call types are presented in the lower portion of the figure (redrawn from Jürgens 1990).

pletely removed, individuals are incapable of discriminating sounds (e.g., conspecific versus heterospecific), but can nonetheless hear them (i.e., the brain detects an acoustic signal). In contrast, if the temporo-parieto-preoccipital association cortex (an area that is cytoarchitectually the homologue to Wernicke's areas) is lesioned, discrimination of species-typical vocalizations is unaffected.

Single-unit recording procedures, comparable to those used in studies of insects, frogs, birds, and bats, have been employed by several research teams interested in finding cells in the auditory cortex that respond selectively to conspecific calls as opposed to calls from heterospecifics or artificial signals such as pure tones. Thus far, there is no evidence of call specificity in the squirrel monkey. For example, Newman and Wollberg (1973a, 1973b) found that although 90% of cells within the auditory cortex respond to species-specific calls, they respond to several different call types and, more importantly with regard to the specificity issue, also respond to pure tones that appear to carry no biological significance.[33] In addition, some cells were more responsive to different call types than to structurally similar variants within a call type, and a number of cells showed temporally varying response intensities with the same call type. In summary, therefore, featural decoding of species-specific calls does not occur within the auditory cortex of squirrel monkeys.

In the previous subsection on perception, I pointed out that when squirrel monkeys are deafened, in infancy or adulthood, their vocal repertoires appear to remain intact. Though quantitative acoustic analyses were not performed, these results nonetheless suggest that auditory feedback plays a relatively insignificant role in squirrel monkey communication. From a neurobiological perspective, this result is puzzling, because several recent studies have revealed direct connections between the production systems and perception systems (Müller-Preuss 1988). For example, there are pathways from the anterior cingulate and periaqueductal gray area to the auditory association cortex, and from the medial geniculate body and inferior colliculus to the periaqueductal gray (recall that these are key areas in the bat's system as well). These observations suggest that at some level there must be a convergence of vocal motor behavior and auditory perception (Jürgens 1990). Additional research is sorely needed.

To contrast with Jürgens' work on the squirrel monkey, let us now briefly review what is known about the neurobiology of communication in the macaque.[34] The contrast is important because behavioral and acoustical analyses to be discussed in some of the subsequent chapters indicate that there are fundamental differences in the vocal behavior of these two species. For example, whereas the entire vocal repertoire of the squirrel monkey appears fully formed at birth or soon after (Newman 1985; Newman and Symmes 1982; Winter et al. 1973), there is evidence from several macaque species that their repertoire unfolds gradually, with experience playing a significant role in the ontogeny of call usage and comprehension (Gouzoules and Gouzoules 1989, 1990; Masataka and Fujita 1989; Owren et al. 1992, 1993). Moreover, whereas squirrel monkey vocalizations only convey information about the caller's affective state (emotional/motivational

---

33. As with all studies aimed at understanding neural response specificity, artificial signals that appear to carry little in the way of biologically meaningful information may nonetheless elicit cellular responses because they in fact capture important features of the natural signal.

34. Though there is considerable variation in repertoire structure across macaque species, most of the neurobiological work in this area has ignored such variation. Most of the research has focused on three species: *Macaca fascicularis, M. nemestrina,* and especially *M. mulatta.*

changes), analyses of rhesus and pigtail macaque calls suggest that information is conveyed about affective state, as well as objects (e.g., food) and events (e.g., dominance-related interactions) in the external environment (Gouzoules and Gouzoules 1989; Gouzoules, Gouzoules, and Marler 1984; Hauser and Marler 1993a, 1993b), suggesting the possibility of a semantically rich vocal repertoire. Differences in repertoire structure and function are likely to be driven by details of each species' neural architecture.

Like studies of squirrel monkeys, research on macaques has used neuroanatomical tracers, single-unit recordings,[35] and experimental lesions to investigate the neural substrate underlying vocal production and perception (Deacon 1984, 1991; Larson and Kistler 1986; Larson, Ortega, and DeRosier 1988; Sutton 1979; Figure 4.35). Because early lesion work, involving removal of higher cortical areas apparently homologous to Broca's and Wernicke's, revealed insignificant effects on vocal production (Aitken 1981), experiments within the past ten or fifteen years have concentrated on lower brain areas, especially the periaqueductal gray, anterior cingulate, and several limbic structures (Figure 4.35).

Tracer studies reveal fibers descending from the periaqueductal gray area to several motor control regions. Single-unit recordings from the periaqueductal gray area, together with EMG data, revealed that most of the cells increased their firing rates prior to muscle action and vocal output (Figure 4.36), whereas a smaller portion showed a decrease in firing rates. This finding suggests that cells within the periaqueductal gray do not respond to sensory stimuli involved with vocalization, but rather are involved in the planning of vocal communication. In addition, there appeared to be some evidence of cell ensemble selectivity for certain types of vocalizations and muscle groups. Thus, as illustrated in Figure 4.36, some cells and muscle complexes were active during the production of barks, but not shrieks. Though some cell specificity was observed, there was no evidence of a topographical arrangement of cells within the periaqueductal gray, either in terms of cell-to-call type relationships or in terms of cell-to-muscle complex relationships.

Tracer studies of macaques and squirrel monkeys have yet to reveal efferent projections from the periaqueductal gray area to the hypoglossal nucleus, a nucleus that consists of neurons involved in the control of the tongue. As we will

---

35. Interestingly, Larson and colleagues have managed to condition individuals to vocalize for a fruit juice reward and consequently, obtain neural recordings from awake subjects. All subsequent attempts to condition nonhuman primates to vocalize have failed (Peirce 1985), leading to the conclusion that their vocal behavior is not under voluntary control and cannot be modified by experience; see chapters 5 and 7 for a more thorough discussion.

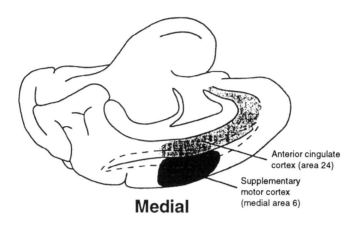

Anterior cingulate
cortex (area 24)

Supplementary
motor cortex
(medial area 6)

Medial

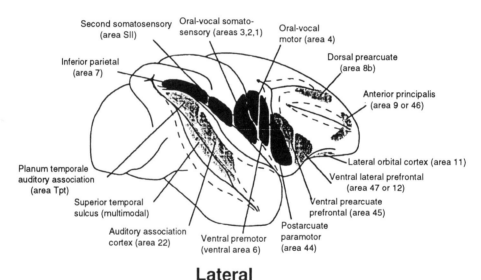

Second somatosensory
(area SII)

Oral-vocal somato-
sensory (areas 3,2,1)

Oral-vocal
motor (area 4)

Inferior parietal
(area 7)

Dorsal prearcuate
(area 8b)

Anterior principalis
(area 9 or 46)

Planum temporale
auditory association
(area Tpt)

Lateral orbital cortex (area 11)

Ventral lateral prefrontal
(area 47 or 12)

Superior temporal
sulcus (multimodal)

Ventral prearcuate
prefrontal (area 45)

Auditory association
cortex (area 22)

Ventral premotor
(ventral area 6)

Postarcuate
paramotor
(area 44)

Lateral

Figure 4.35
Medial and lateral view of the macaque cortex, indicating some of the proposed homologues to
cortical language areas in humans. Shaded areas are interconnected with the prearcuate (light gray)
or postarcuate (dark gray) region of the ventral frontal lobe (redrawn from Deacon 1991).

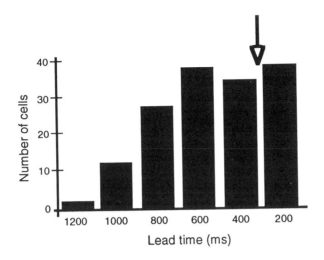

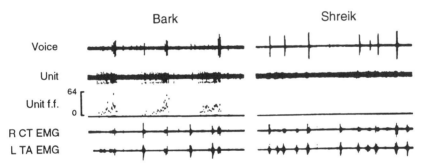

Figure 4.36
Upper panel shows a frequency distribution of the timing of single-unit activity relative to the onset of vocalization (lead time measured in milliseconds) and EMG prior to vocalization (arrow). The lower panel shows unit activity for an aggressive bark vocalization and a submissive shriek vocalization. From top to bottom: time-amplitude waveform of the voice, unit potentials, unit firing frequency (f.f.), EMGs of right cricothyroid (R CT) and left thyroarytenoid (L TA) muscles (redrawn from Larson and Kistler 1986).

learn in section 4.4.2, the tongue is critically involved in the articulation of human speech. Though quantitative analyses of primate vocal production have not been conducted, preliminary observations by M. D. Hauser and C. H. Brown (unpublished data) of rhesus macaques, blue monkeys, and mangabeys suggest that for several vocalizations, the position of the tongue is altered during vocalization. For example, as rhesus monkeys produce their primary alarm call ("shrill bark"; Hauser and Marler 1993a), they first slide the tongue back and then arch it toward the upper palate. Consequently, there appears to be control over tongue movement, but neural sites other than the periaqueductal gray must be involved. Moreover, it is not yet clear whether such articulatory movements are causally related to call structure.

Early work on the neurobiology of acoustic perception in macaques (rhesus and stump-tailed) was limited with regard to issues in communication because much of the neural mapping was carried out with pure tones and on anesthetized animals (Merzenich and Brugge 1973; Mesulam and Pandya 1973; Pandya, Seltzer, and Barbas 1988; Pfingst and O'Connor 1981). However, this research revealed a tonotopic arrangement (low to high frequencies represented rostocaudally) within the primary auditory cortex (A1) and also what appeared to be the neural homologue to ocular dominance columns in the visual system—binaural interaction columns (Brugge and Merzenich 1973); the function of such columnar organization in the auditory cortex has yet to be detailed for macaques or other primate species, but is most likely comparable to what we find in the visual domain (Hubel 1988; Imig et al. 1977).

The basal portion of the cochlea, which codes for relatively higher frequencies, has a disproportionate representation within primary auditory cortex. In addition, an auditorily responsive cortical region surrounding primary auditory cortex has also been identified (Figure 4.37), but its specific function in processing acoustic stimuli has yet to be established. Intriguingly, some studies find significant individual differences in the relative amount of neural space dedicated to particular frequencies. For example, Aitkin, Kudo, and Irvine (1988) reported that for the common marmoset, one individual had an area of primary auditory cortex that was 2 mm in diameter and dedicated to sounds in the 14-kHz region. For others, such areas mapped onto sounds extending continuously up to 30 kHz. It has been proposed that these differences are due to neural plasticity that occurs in response to either damage or the selective use of certain frequencies in vocal behavior (Merzenich and Schreiner 1992; Weinberger and Diamond 1988). Similar arguments have been made for interspecific differences in somatosensory maps in primates (Merzenich et al. 1987, 1989).

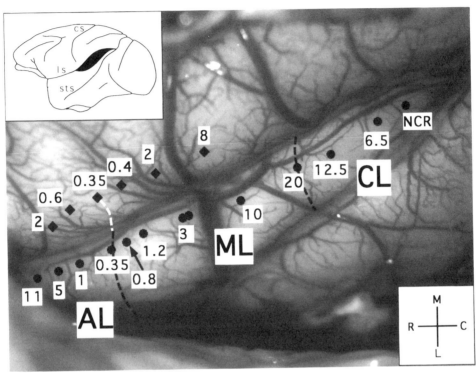

B06   Best Center Frequencies (kHz)

Figure 4.37
A view of the rhesus macaque brain, emphasizing stimulation sites and the approximate locations of auditory areas: anterolateral area (AL), middle lateral area (ML), candolateral area (CL), superior temporal sulcus (STS), central sulcus (CS), and lateral sulcus (LS) (figure provided by J. Rauschecker).

Recently, Rauschecker and colleagues (Rauschecker, Tian, and Hauser 1995; Rauschecker et al., in press) have begun to explore auditory processing in the rhesus macaque using more complex acoustic stimuli, including broadband noise with different center frequencies as well as vocalizations from the rhesus repertoire. Recording from the lateral belt surrounding primary auditory cortex, they have found cells whose response is facilitated by the presentation of spectrally complex sounds relative to the presentation of pure tones. Moreover, when calls from the repertoire were played back, there was evidence of linear and nonlinear summation effects based on particular pieces or components of the call (e.g., the upper or lower set of harmonics within the coo call). Further experiments are under way to explore how awake individuals process specific features of their calls and to determine which areas of the auditory belt surrounding primary auditory cortex are critically involved in acoustic perception, and whether there is a hierarchical system of processing.

Studies of songbirds and humans (see section 4.4.2) have revealed significant anatomical and functional asymmetries in brain regions involved in the perception and production of species-typical vocalizations (Bradshaw and Rodgers 1993; Cynx, Williams, and Nottebohm 1992; Galaburda 1995; Hellige 1993; Kimura 1993; Nottebohm and Nottebohm 1976). Given the phylogenetic distance between birds and humans, this functional and anatomical similarity in hemispheric processing would appear to be an elegant example of convergence. In contrast to these taxonomic groups, relatively less is known about functional asymmetries for communication in nonhuman primate species. The emerging picture suggests, however, that among Old World monkeys and apes, at least the direction of the asymmetry is consistent with that shown in humans.

The first test of functional asymmetries for communicatively meaningful stimuli was initiated by Petersen and colleagues (1978) using a psychophysical approach and a reaction-time measure. The experiment was designed on the basis of Green's (1975a) field research with wild Japanese macaques, reviewed in the previous subsection. Petersen and colleagues trained Japanese macaques and closely related species to discriminate between coos where the frequency peak was located at the start of the call ("smooth early highs") with those where the frequency peak occurred at the end of the call ("smooth late highs"); these two contours appear to map on to ecologically significant contexts (chapter 2). Calls were selectively presented through headphones to either the right or left ear and reaction time scored. Results showed that for Japanese macaques, but not closely related species, reaction time for discrimination was significantly faster when calls were played to the right ear than when they were played to the left ear. Consequently, Petersen and colleagues concluded that like humans, Japanese macaques evidence a left-hemisphere bias for processing species-typical vocalizations.

To provide more direct evidence of asymmetries in neural processing, Heffner and Heffner (1984, 1990) repeated the Petersen et al. experiments, but in addition to testing normals, selectively lesioned areas of the brain believed to be critical in processing complex auditory stimuli. Specifically, subjects received one of four lesions: (1) left auditory cortex, (2) right auditory cortex, (3) right and left auditory cortex, and (4) dorsolateral region. Heffner and Heffner predicted that lesions in the left hemisphere would cause more significant effects on discrimination than

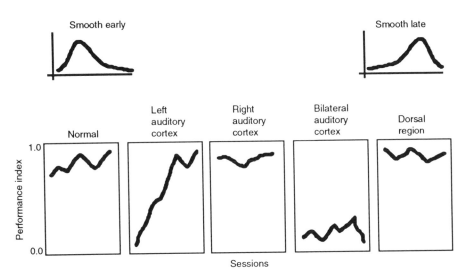

Figure 4.38
Results from psychophysical experiments using Japanese macaques on a discrimination task. The task involved discriminating one of two coo types (illustrated in the upper portion of the figure), a smooth early high and a smooth late high. There were five treatment groups: (1) Normals compared with individuals who experienced lesions to the (2) left auditory cortex, (3) right auditory cortex, (4) right and left auditory cortex, and (5) dorsal region. The data have been smoothed to show the relative performance on the discrimination task (y-axis) over a series of experimental sessions (x-axis) (redrawn from Heffner and Heffner 1984).

right-hemisphere lesions. As Figure 4.38 reveals, performance was significantly worse for subjects experiencing left-auditory-cortex lesions than for subjects experiencing lesions to either the right auditory cortex or the dorsolateral region. However, the performance of individuals with lesions to the left auditory cortex showed quite dramatic recovery over time, and subjects with bilateral lesions never recovered. These additional results suggest both the possibility of plasticity in brain function and the importance of both right and left auditory cortex in processing species-typical vocalizations. In other words, though the left hemisphere appears dominant for vocal perception in Japanese macaques, the extent of this hemispheric asymmetry seems less than what has been documented for humans processing language (see section 4.4.2).

   The studies by Petersen et al. (1978; Petersen and Jusczyk 1984) and Heffner and Heffner (1984, 1990) suggest that in Japanese macaques, the left hemisphere plays a dominant role in processing species-specific coo vocalizations. What is unclear is whether the demonstrated asymmetry holds for other call types within the repertoire and for other species.[36] Moreover, an assumption underlying the claim that macaques process coos in the same way as humans process language is that both types of vocalization carry *more* semantic than emotional content. However, the informational content (see section 7.2) of Japanese macaque coos has not been explicitly addressed. If coos convey primarily emotive information, then a left-hemisphere bias would represent the opposite pattern from humans, where the right hemisphere is dominant for the paralinguistic or melodic structure of language.

To extend the laboratory work on Japanese macaques, and to increase our understanding of how hemispheric biases in perception may function under natural conditions, Hauser and Andersson (1994) conducted a series of field playback experiments with rhesus monkeys living on the island of Cayo Santiago, Puerto Rico. The experiments were conducted by first placing a speaker 180° behind a subject and then playing back a vocalization. The primary prediction was that the subject's head-orienting response would reveal a bias in hemispheric processing. Thus turning to listen with the right ear would indicate a left-hemisphere bias, whereas turning to listen with the left ear would indicate a right-hemisphere bias.

Records of the head-orienting response of rhesus were obtained from 80 adults and 52 infants (4–12 mos) based on playbacks of 51 vocalizations, including 3–5 exemplars from 12 call types (Hauser and Marler 1993a); aggressive, fearful, and affiliative call exemplars were used. As a control, the alarm call (4 different exemplars) of the ruddy turnstone (*Arenaria interpres*) was played back. The turnstone is a seabird that lives on Cayo Santiago throughout the year. Its alarm call is frequently heard and is typically given in response to human presence. Although human observers do not interact with rhesus, humans evoke alarm during the annual trapping season and when infants are approached. As a result, turnstone alarms represent potentially useful information for rhesus monkeys.

Results (Figure 4.39) showed that for adults, 76.3% turned their right ear toward the speaker to listen to conspecific calls, whereas 86.7% turned their left ear for the turnstone's call. Moreover, the right-ear bias for conspecific calls was consistent across the different call types. Infants, in contrast to adults, showed no asymmetry in response to either conspecific calls (54.1% turned their right ear), or heterospecific calls (60.0% turned the right ear). In addition, the relatively symmetric response by the infants was consistent across all three call types.

36. Note that Ehret (1987a) has also demonstrated a left-hemisphere bias for acoustic perception in mice. This bias was observed in tests involving species-typical ultrasonic contact calls.

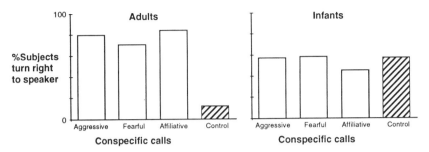

Figure 4.39
Results from playback experiments of conspecific calls and one heterospecific control call (ruddy turnstone alarm call) to adult and infant (4–12 mos) rhesus macaques. The y-axis indicates the proportion of subjects turning their right ear toward the speaker. The x-axis represents the category of call types played back, including the three conspecific calls and the control (redrawn from Hauser and Andersson 1994).

Hauser and Andersson's (1994) results suggest that in adult rhesus monkeys, as in Japanese macaques and humans, the left hemisphere plays a dominant role in processing species-specific vocalizations. This asymmetry does not, however, appear to be present at birth. The development of hemispheric dominance for processing conspecific calls may depend upon a fully mature brain, as well as sufficient exposure to and understanding of the vocal repertoire. As will be discussed in greater detail in chapter 5, studies of nonhuman primate vocal development (reviewed in Hauser and Marler 1992; Snowdon 1982, 1990; Snowdon and Elowson 1992) have shown that an infant's ability to produce a call in the appropriate socioecological context and to comprehend its meaning requires considerable experience over the course of one to two years.

In a recent study by Berntson, Boysen, and Torello (1993), an event-related potential (ERP) approach, involving the recording of brain activity from the scalp, was used to examine perception of auditory stimuli in a juvenile chimpanzee. Three categories of auditory stimuli were presented: simple tones, human speech (i.e., caregiver saying the chimpanzee's name "Sheba"), and chimpanzee vocalizations; the authors did not provide any details about which vocalizations were used, but they did not play back the subject's own vocalizations. In response to all three signal categories, late positive components of the ERP trace were primarily in the frontocentral region of the scalp. The structure and temporal location of the late positive component appears analogous to the P300 peak obtained when human subjects perceive an auditory event and is therefore likely to be associated with attentional state. In addition, responses to both human speech and chimpanzee calls were stronger (i.e., higher amplitude peaks in the waveform) and lateralized to the right hemisphere when compared to the weaker and symmetric response to the tonal stimuli. Although these results are promising in terms of using ERPs with a nonhuman animal, the pattern of response observed is difficult to interpret in light of current observations of hemispheric asymmetries for speech in humans (what does a human utterance of the chimpanzee's name mean to the chimpanzee?). Tests conducted in the future would profit from using a broader range of acoustic stimuli, especially conspecific vocalizations that are likely to be of greater biological significance than human speech.

In summary, studies of the neurobiological basis of vocal communication in nonhuman primates have revealed that subcortical structures are primarily involved in call production, whereas both subcortical and cortical structures appear to play a role in call perception. Unlike songbirds and frogs, no pathways have yet been uncovered between the auditory and endocrinological control regions of the brain. Although physiological changes are clearly involved in both the production and perception of nonhuman primate vocalizations, primate vocal repertoires are not dominated by a sexual advertisement signal, and this is the case even in species with highly restricted periods of mating.

The lack of similarity between birds and frogs on the one hand, and nonhuman primates on the other is, of course, not surprising, given their phylogenetic distance. Perhaps more surprising, however, is the fundamental difference in the neural substrate underlying vocal production and perception in bats and nonhuman primates—both mammals. Yet, once again, we find that salient environmental pressures lead to selection pressures that have designed brains and communication systems that are custom fitted to the problem at hand. For bats, navigating through the dark and finding prey are critical to survival. The solution: a brain engineered for the task and a signal that can do both—biosonar. For primates, most are diurnal, foraging and navigating during the day. Finding food is thus a visual problem, even though some signals have evolved to announce the discovery of food (see chapters 6 and 7). For primates, no repertoire seems to be dominated by a single call type or by a single functional context for communication. Apparently as a result, there is not, as we see in bats, a chunk of the brain that is overwhelmingly dedicated to, for example, a single dominant frequency. What we do find, however, is a vocal tract (and accompanying facial musculature) that may be more directly suited for encoding rich spectral structure into the signal than is the case for other species. Consequently, and this has yet to be detailed, we may find that the nonhuman primate brain has been designed to more effectively decode spectrally complex vocal signals, where spectral complexity maps onto semantic complexity. This is not an unreasonable proposition, especially since the human brain, vocal tract, and language appear to have co-evolved for this purpose (Lieberman 1984, 1991; Deacon 1991).

4.4.2   Human Language

Why the organs now used for speech should have been originally perfected for this purpose, rather than any other organs, it is not difficult to see. Ants have considerable powers of intercommunication by means of their antennae, as shewn by Huber, who devotes a whole chapter to their language. We might have used our fingers as efficient instruments, for a person with practice can report to a deaf man every word of a speech rapidly delivered at a public meeting; but the loss of our hands, whilst thus employed, would have been a serious inconvenience. As all the higher mammals possess vocal organs constructed on the same general plan with ours, and which are used as a means of communication, it was obviously probable, if the power of communication had to be improved, that these same organs would have been still further developed; and this has been effected by the aid of adjoining well-adapted parts, namely the tongue and lips. The fact of the higher apes using their vocal organs for speech, no doubt depends on their intelligence not having been sufficiently advanced.   (Darwin 1871, 58)

Speech is not produced as beads are put on a string, one phone after another. The sounds overlap and flow into one continuously changing stream of sound, further bonded by slowly changing modifications overlaid upon it. These overlaid changes are the prosody, the rhythm and music of speech.   (Borden and Harris 1984, 131)

In section 4.4.1, I stated that nonhuman primates, relative to other vertebrates, vocalize in a much wider variety of contexts. What represents a variety of contexts for nonhuman primates is, of course, only an infinitesimally small subset of the contexts that are associated with the production of human language, spoken or signed. In fact, whereas most nonhuman animal signals are effectively *tied* to a given context, human signals are contextually liberated, exhibiting the powerful property of displacement (Hockett 1960a, 1960b). This feature, which allows humans to talk about objects and events in the past, present, and future, together with an almost infinite set of signals and signal combinations for describing a particular object or event, would seem to give humans a considerable communicative advantage. Although we will return to some of the functional implications of human communication systems in chapters 6 and 8, it is important to consider these issues here because they raise a number of problems for those interested in the neural processes subserving such communicative power. How, for example, does the brain enable an individual to first recall an event in the past, then retrieve the requisite vocabulary to describe that event? In what way does this retrieval → description process differ from one involving, first, some thought about a future event, followed by description of this event? And in each case, how are the units of communication retrieved from the brain's storage system? These are some of the questions that will be addressed in this section. I start with spoken language, and then turn to a discussion of signed languages. For both languages, I first describe how the requisite peripheral organs do their work and then review some of the relevant neuroanatomy. Because the literature on the neurobiology of human language is enormous, all that I can hope to do in this section is provide a flavor of some of the more significant issues and exciting new developments. Where it is possible, I focus on topics that are relevant to interspecific comparisons and our attempt to understand the evolution of brain mechanisms underlying communication.

Like frogs, bats, monkeys, and apes, humans communicate by sending air from the lungs to the larynx, and from the larynx out through the supralaryngeal cavity.[37] Unlike these other species, however, the human vocal apparatus appears[38] to access a much wider range of articulatory configurations, and this fact is due, in part, to its specialized design features. Specifically, and as the comparative anatomist Negus (1949) pointed out almost fifty years ago, the human vocal tract consists of at least two tubes or cavities (oropharynx and laryngopharynx), connected at a ninety-degree angle. In contrast, all other species are characterized by a single-tube system with little curvature (see Figure 4.24). This configurational difference arises from the fact that during hominid evolution, the larynx descended,[39] thereby eliminating the velo-epiglottal seal that facilitates simultaneous breathing and eating in nonhumans. Consequently, the human vocal tract is exquisitely well adapted for sound production, but less so for eating (Lieberman 1984). As previously pointed out, the lack of a multiple-cavity system in nonhumans need not imply lack of tract filtering by supralaryngeal maneuvers. However, the configuration of the human vocal tract, in concert with a relatively small and mobile tongue, permits a wider range of articulatory gestures, gestures that are extensively used to create the many sound contrasts of human language. In this section, I describe some of the mechanics underlying human speech production,

37. Much of the description that follows is particular to the vocal apparatus of normal human adults. As studies have revealed, however, the vocal system of human infants and even young children differs considerably from that of adults, and thus models of vocal production are unlikely to be valid across the human life span.

38. A cautionary flag is raised here because nonhuman animal studies of supralaryngeal maneuvers are only in their infancy.

39. As will be discussed in greater detail in chapter 5, newborn human infants have a raised larynx until approximately 3–4 months of age, and thus the configuration of their vocal tract is much more like that of the nonhuman primate.

including the role of laryngeal and supralaryngeal action on the spectrotemporal properties of the acoustic waveform.

When adult humans speak, they tend to produce their utterances during the expiratory phase of the respiratory cycle.[40] Consequently, the termination of an utterance tends to be associated with a decrease in lung pressure and airflow at the lips; differences in the general pattern of airflow emission can arise when words within a sentence are stressed or modified to convey affective information, or to emphasize conversational endpoints (Cooper and Sorensen 1981; Gelfer 1987; Gelfer, Harris, and Baer 1987; Pierrehumbert 1979). A feedback control system allows humans to regulate the muscles involved in respiration (Borden and Harris 1984; Draper, Ladefoged, and Whitteridge 1959; Hixon 1973), and a number of observations point to the conclusion that the suite of muscles involved in speech production are programmed to meet the linguistic demands of the sentence. For example, the amount of air brought into the lungs prior to speaking is proportional to the length of the sentence to be uttered, as well as to the specific compositional structure of the sentence (M. R. Lieberman and P. Lieberman 1973).

Once air passes from the lungs, through the trachea, and up to the larynx, a group of muscles is then responsible for determining the amount of air passed further on to the supralaryngeal cavity (Figure 4.40). These muscles, primarily the cricothyroid, thyroarytenoid, and cricoarytenoid, regulate the action and tension of the vocal cords. Such muscle activity, together with the level of air pressure, determines the fundamental frequency of phonation or what listeners perceive as the pitch of the human voice.[41] It should be noted, however, that depending upon the structure of the sentence, changes in the fundamental frequency can be primarily accounted for by changes in subglottal air pressure, changes in muscle tension, or some combination of the two (Figure 4.41).

In addition to the changes in muscle tension and air pressure that accompany phonation, several other factors influence the acoustic waveform emitted and, consequently, the sound perceived. For example, the weight of the glottis is typically lighter in children than it is in adults, and lighter in women than it is in men. In general, the lighter the glottis, the higher the pitch. Body weight and glottis weight are not, however, statistically correlated, and though children under

---

40. Many vocalizations produced by human infants, such as crying, occur during repeated expiratory-inspiratory cycles.

41. Though much of speech involves voiced sounds from the larynx, there are also voiceless utterances, sounds that are generated without periodic phonation (e.g., the /s/ in *soft* or the /h/ in *hurry*).

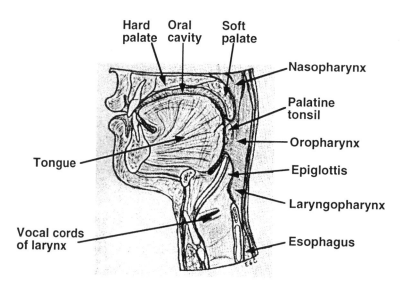

Figure 4.40
The modern human vocal tract, highlighting the most significant features (redrawn from Crelin 1987).

the age of three years tend to produce higher fundamental frequencies (250–540 Hz) than adults (150–250 Hz), young children can produce a fundamental as low as 30 Hz (Keating and Buhr 1978; Robb and Saxman 1985) albeit with very little energy. Last, though we tend to assume that gender differences in the voice can be completely accounted for by differences in pitch, a significant proportion of the variation between males and females can be explained by differences in the length of the supralaryngeal vocal tract and its configuration. Thus vocal tract length is longer in men than it is in women, and consequently formant frequencies are generally lower in men than they are in women. In young, prepubescent boys and girls, there are no significant sex differences in vocal tract length or glottis weight. However, one can readily discriminate between the voices of girls and boys. Sachs and colleagues (cited in Lieberman and Blumstein 1988, 131) observed that this difference arises from the fact that boys round their lips more than girls, and this tendency lowers their formants.

Added on to the effects of glottis weight are differences in what phoneticians call phonation registers (reviewed in Baken 1987; Lieberman and Blumstein 1988). Phonation registers are created by shifting the position of the laryngeal muscles and cartilages. Familiar to many readers will be the falsetto register of operatic singers, associated with an increase in the fundamental frequency rela-

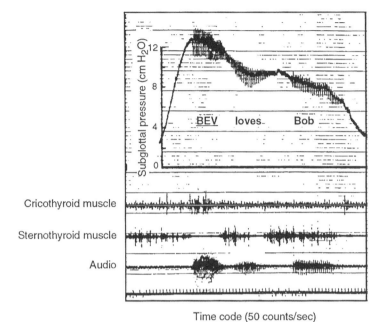

Time code (50 counts/sec)

Figure 4.41
Changes in subglottal air pressure and muscle activity as a function of the acoustic output produced during speech (redrawn from Lieberman and Blumstein 1988).

tive to the chest register. A less common phonation mode is the fry register, associated with weak muscle tension on the vocal folds. As a result, the fundamental frequency is not only quite low (approximately 60 Hz), but it is also highly variable; such period-to-period variation is used by clinicians to diagnose malfunctioning larynges (Baken 1987).

The next step in the speech-production process involves the transmission of acoustically structured air from the larynx to the supralaryngeal tract. In general, several discussions of speech production make the assumption, following Fant (1960), that the output from the larynx is unaffected by the supralaryngeal tract— that source and filter are independent (see Box 4.3). Though this model generally works, and will be emphasized throughout the following discussion, the reader should be aware that the configuration of the supralaryngeal tract does have an effect on the glottal waveform (Bickley and Stevens 1987). The interaction between source and filter arises from a combination of factors including aerodynamic coupling, changes in muscle activity and posture, and aerostatic effects that arise from alterations in the transglottal air pressure. Though interactions with the supralaryngeal cavity cause only small changes in the glottal waveform (10–20 Hz), such modifications are significant for those concerned with the more prosodic aspects of speech (melody, stress, intonation; see, for example, the discussion of parent-child speech in chapter 5).

To initiate our discussion of speech articulation, close your eyes and slowly say "scooby-doo." To produce this utterance, you start with an unvoiced /s/, which creates a kind of whistling noise and recruits significant movement of the tongue. You then protrude your lips and raise your mandible for the /u/, and then retract and bring your lips together for the /b/. For the "-doo" you bring the tip of your tongue up to the base of the hard palate, just beneath the teeth, protrude your lips, and raise your mandible. For all of these movements, specific muscles are recruited into action. For example, control over lip configuration is generally achieved by three muscles, the orbicularis oris, depressor labii inferior, and levator labii superior. In contrast, changes in the position of the tongue arise from a much larger set of muscles.[42] Though we can describe the changes in muscular activity that arise for each of the speech sounds noted, there are at least two reasons why it would be inappropriate to conclude that there is a one-to-one correspondence between particular muscle activities and particular sound contrasts. First, there is considerable within- and between-individual variability in the types of muscles recruited and the extent to which particular muscles are activated throughout the production of a speech sound. Second, though a given muscle may be consistently used to produce a specific sound, it may also be used to generate other sound contrasts.

If there is variation in the muscles recruited to produce speech, then presumably there should be considerable variation in the output, and this would surely challenge the efficacy of speech comprehension. However, research on motor production strongly suggests that individuals "shoot" for an acoustically appropriate acoustic target and this is achieved via the coordinated activity of several functionally related muscles—muscles that are employed to carry out automatized, but flexible, motor programs (reviewed in Perkell and Klatt 1986; schematically illustrated in Figure 4.42). Moreover, as will be discussed in the following subsection, the human perceptual system is extremely talented with regard to pulling out the intended meaning from a relatively variable speech stream. That

---

42. For a more complete listing of the muscles involved in speech production, see Lieberman and Blumstein (1988), Borden and Harris (1984).

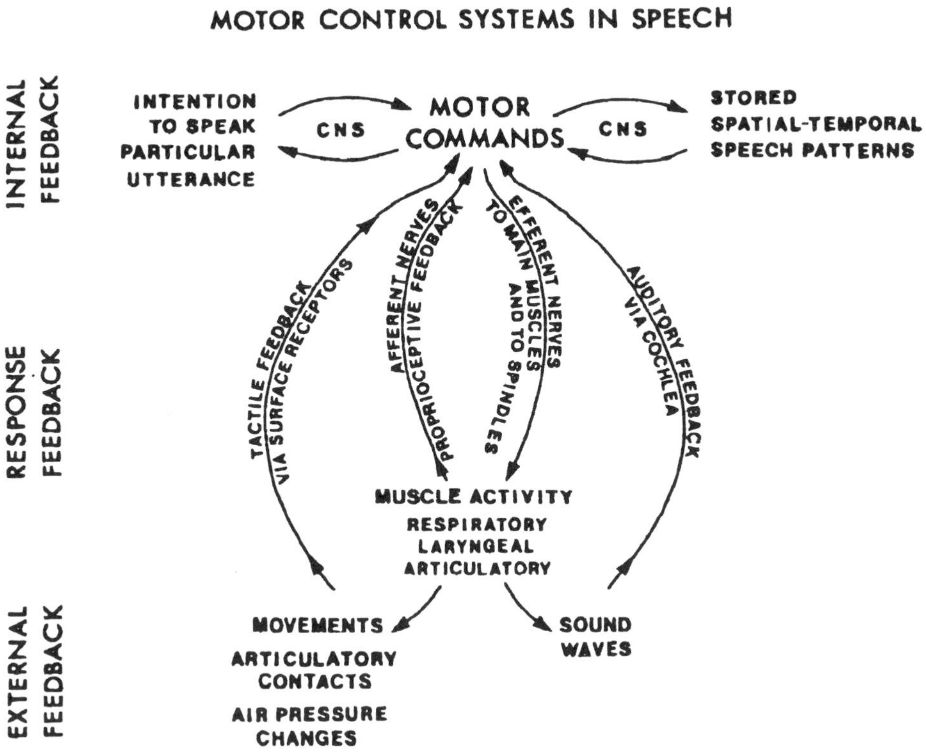

MOTOR CONTROL SYSTEMS IN SPEECH

Figure 4.42
Overview of possible motor control systems for speech production, emphasizing the role of different feedback mechanisms (from Borden and Harris 1984).

is, variance in production is typically converted by the auditory system into perceptual invariance (Abbs 1986; Perkell and Klatt 1986; Stevens 1972).

Support for the view that individuals intend to achieve a specific acoustic target and tend to store segments of an utterance in memory prior to production comes from experiments where a speaker's vocal tract is perturbed during an utterance and also from natural observations of speech errors, known as spoonerisms. Consider first observations from bite block experiments (Folkins and Zimmerman 1981; Lindblom, Lubker, and Gay 1979). Bite blocks are pieces of wood or solid plaster that can be clenched between a subject's teeth. While they are inserted, the subject is told what to say. Simultaneously, the articulatory event is filmed and EMG data recorded from the appropriate muscles. Results indicate that subjects employ different articulators and articulatory gestures to achieve the intended acoustic target. In addition, though the muscles used to control mandibular position could not be used in the bite block condition, the muscles subserving such articulatory changes were nonetheless activated. Together, these observations indicate that the human vocal tract compensates for perturbations and attains acoustic targets by means of automatized motor programs (Abbs and Gracco 1984; Folkins 1985; Folkins and Zimmerman 1981; Gracco 1988; Lindblom, Lubker, and Gay 1979). Additionally, experience plays a role in shaping the particulars of the articulatory routine, as evidenced (Lubker and Gay 1982) by differences between American and Swedish speakers in the timing of lip rounding and protrusion for the vowel /u/. As discussed in chapters 5 and 7, these articulatory differences have a significant impact on child language acquisition, both in terms of purely perceptual phenomena (Kuhl et al. 1992) and in terms of cross-modal processes (Kuhl and Meltzoff 1982, 1988). This is the kind of evidence that lends support to the notion of a linguistic face (Locke 1993).

Spoonerisms are interesting because they show that although errors of speech occur, they are constrained by rules of production. For example, the famous English clergyman William Spooner was well-known for his grammatically correct but semantically confused phrases such as "Work is the curse of the drinking class." Other examples from Fromkin's (1973) studies indicate that certain vowel and consonant substitutions are possible (e.g., "The optimal number" → "The moptimal number") but not others (e.g., out of thousands of speech errors, no one has reported "The ngoptimal number," because in English, "ng" cannot occur as an initial syllable).

A fascinating set of studies has been conducted by Graves and colleagues on the motor asymmetries that arise during speech articulation. Specifically, based on the finding that the left hemisphere is dominant during speech processing and that the left hemisphere controls the motor activity of the lower half of the right side of the face, one would predict right-side-of-the-face dominance during speech production. The first test of this hypothesis was conducted with normal human subjects (Graves, Goodglass, and Landis 1982). Asymmetries in lip opening were scored by calculating differences between left and right sides of the mouth from a video frame and from the relative movement of tiny lights placed on the lips. Results from both sampling procedures indicated a significant right-side bias in articulation. More recent work (Graves and Landis 1985, 1990; Graves, Strauss, and Potter 1990; Graves and Potter 1988) extends this pattern of articulatory asymmetry, showing that singing with words is associated with a right-side bias, whereas singing without words, emotional expression, and prosodic expression are associated with symmetry. Moreover, when one side of the mouth was temporarily anesthetized and thus removed from the articulatory process, the quality of the articulation as judged by perceptual tests was higher on the right than on the left side of the mouth. Together, these results have been used to further strengthen the claim that the left hemisphere is dominant for language processing, whereas the right hemisphere is dominant with regard to the melodic and emotional features of language that are not structurally "hooked" to grammar (e.g., tone in tone languages such as Mandarin).

Perception Systems for Spoken Language   The gross anatomy of the human auditory system is very much like that of nonhuman primates, and it was described in the subsection "Sound Perception Mechanisms." In this section we turn our attention to some of the factors guiding human speech perception, and the kinds of specialization that have evolved to process speech as opposed to other acoustic signals;[43] because the general discriminative abilities of the human auditory system are relevant, however, I provide a brief discussion of basic psychoacoustical results (for excellent reviews of this literature, see Bregman 1990; Green 1988; Gulick 1989). The story that will unfold highlights a recurrent theme in this chapter, and those that follow: production and perception systems for communication need not coevolve in perfect synchrony. For human speech, it appears that although the vocal tract has undergone significant changes in morphology over evolutionary time, our peripheral auditory system has remained in a relative state of stasis. An unresolved problem, then, is what kinds of selective pressures might have favored change as opposed to stasis.

A common aim of research on communication in insects, fish, frogs, birds, bats, and nonhuman primates is to uncover those acoustic features that allow individuals to recognize members of their species, to distinguish between mates and rivals, and to discriminate between functionally distinctive calls within the repertoire. Studies of human speech generally do not ask about species-typical acoustic features, though a considerable amount of effort has been invested in the notion of speech as a special acoustical signal (Mullenix and Pisoni 1989). In my opinion, these two issues are different. In general, those interested in the

43. Although an extraordinary amount of work has been conducted in basic human psychoacoustics (Bregman 1990; Espinoza-Varas and Watson 1989; Green 1988), I focus primarily on studies that have made explicit use of speech stimuli, at the level of either phonemes or whole words. Where relevant, I also discuss experiments on the role of prosodic or paralinguistic features of speech. I intentionally restrict this discussion, but acknowledge that much research on speech perception recognizes the fact that under certain conditions, the perceptual apparatus responds similarly to speech and nonspeech acoustic stimuli. Such studies, some of which are discussed in chapter 7, have formed the core of the "speech is special" debate; see, for example, discussions in Harnad (1987), Lieberman (1991, 1992), and Studdert-Kennedy (1986a).

special properties of speech look for particular neural structures or perceptual processes that occur with speech, but not nonspeech acoustical signals. The nonspeech signals used, however, are typically biologically and ecologically meaningless (e.g., sinusoids), although occasionally, they attempt to *mimic* the salient acoustic features of speech (e.g., sine-wave speech). For example, no study, to my knowledge at least, has attempted to assess at what age human infants show a preference for listening to conspecific as opposed to heterospecific sounds; this kind of experiment would mirror those that have been conducted with human faces (e.g., preferences for maternal versus nonmaternal; see section 5.4). Nor have studies asked whether phenomena such as cross-modal perception, discussed in chapter 7, would occur with the voices and faces of nonhuman animals. Where studies of speech perception excel, however, is with regard to identifying the primary units of perceptual analysis and the features associated with such units. A general goal of speech perception research, then, is to assess how the listener decodes the continuous speech stream into functionally significant units (e.g., phonemes, words) and why some variation in acoustic form is meaningful and some meaningless.

As we learned in the section on speech production, the larynx is generally responsible for producing acoustic features that are paralinguistically meaningful (e.g., pitch contours that identify emotional state), whereas changes in the configuration of the supralaryngeal tract result in linguistically meaningful consequences.[44] Concerning the latter, the relevant acoustic features are the resonance or formant frequencies (F# = the number of the formant frequency). For example, formant transitions (especially F1 and F2) in voiced stop-vowel pairs (Figure 4.43) represent the most significant acoustic change, in terms of both production and perception. This claim is based on experiments that involve synthesis and manipulation of acoustic parameters, followed by perceptual tests requiring both discrimination and identification of phonemic contrasts.

Because current research on speech perception is aimed at testing between several competing theories, I have sketched the essence of some of these theories in Box 4.4. In general, these different perspectives are primarily concerned with understanding how linguistic meaning is derived from a signal that lacks acoustic-phonetic invariance, clear segmentation of units, and consistency from one speaker to the next.

---

44. As a reminder, recall that a large number of factors enter into the particular structure of the sound waveform, including variation in the precise configuration of the tract, voice register, speaking rate, and intensity.

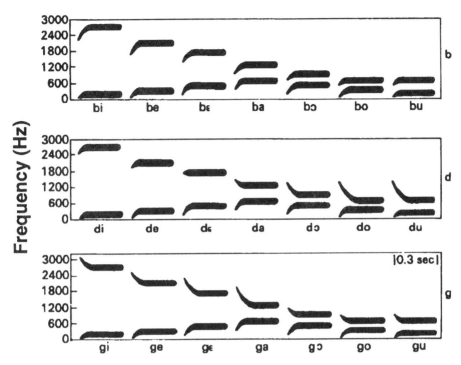

Figure 4.43
Synthesized consonant-vowel strings showing changes in the first two formants, F1 and F2 (redrawn from Lieberman and Blumstein 1988).

Given the problems of invariance and segmentation, how are perceptual units derived? To start, consider the problem of perceptual normalization and, in particular, the recognition of vowels produced by different speakers. Studies by Peterson and Barney (1952) revealed that men, women, and children produce the same vowel with different absolute formant frequencies. Intuitively, one might therefore suppose that listeners could use a simple formula to normalize for differences in vocal tract size. This formula would fail because there are not only differences in vocal tract length among age-sex classes, but differences in vocal tract shape as well. For example, women have an oral cavity that is, on average, approximately 1.25 cm shorter than it is in men, whereas their pharynx is approximately 2 cm shorter. To further complicate matters, various aspects of conversational style, such as speaker rate and stress, will also influence formant frequencies. To account for listener recognition of vowels, most theoretical treatments suggest that individuals either refer to some stored representation of the

Box 4.4
Some Theories Relevant to Human Speech Production and Perception

Over the past twenty or more years, linguists and psycholinguists have developed a number of theories to explain both the form of human speech and the mechanisms that allow for perception, or decoding of the message. The trick has been to explain how we understand the message on the basis of so few invariant acoustic cues. Following Borden and Harris (1984), the major theories can be divided into two categories, those that view perception as an active process and those that view it as a passive process. To facilitate discussion of some of the empirical literature in this area, I briefly describe the key points associated with each theory. More complete treatments can be found in Borden and Harris (1984), Perkell and Klatt (1986), and Lieberman and Blumstein (1988).

1. *Motor theory* (Liberman 1992; Liberman and Mattingly 1985; Liberman et al. 1967): The main tenet of this position is that the relationship between acoustic signal and phoneme is derived from the listener's knowledge of articulation (i.e., the vocal tract configuration required to produce the particular phoneme). Underlying this claim is the subsidiary point that speech is a special type of acoustic signal that is processed by a special perceptual apparatus. In sum, therefore, the motor theory postulates that humans have quite explicit knowledge of the relationship between articulatory gestures and linguistic units.

2. *Analysis-by-synthesis theory* (Stevens and Halle 1967): Like the motor theory, analysis by synthesis also makes a connection between production and perception. In contrast to motor theory, however, the connection has more to do with the specific acoustic morphology of the signal than with the articulatory gesture used to produce the signal. In essence, listeners decode the message by breaking down the incoming signal into acoustic units and matching these units to some internal representation.

3. *General auditory/phonetic theory* (Fant 1967; Pisoni and Sawusch 1975; Sawusch 1986): Decoding occurs primarily at the level of the auditory system, which is set up with auditory templates (similar to those described in the section on birdsong) and feature detectors. The production system is not invoked or called up during the perceptual processing of speech. Rather, processing depends on a match between innate brain structures and the features of an incoming utterance.

4. *Quantal theory* (Stevens 1972): Like motor theory and analysis-by-synthesis, quantal theory also relates production to perception but places more emphasis on how the vocal tract constrains the range of variation. Specifically, in some areas of the vocal tract, extensive variation in the articulatory routine results in only minor changes in the acoustic output, whereas in other areas of the tract, minor variations in articulation result in extensive variations in the acoustic output. Languages appear to take advantage of these relationships to produce highly distinctive or quantal sounds, especially consonants.

5. *Action theory* (Fowler et al. 1980): The starting point of this perspective is the Gibsonian (J. J. Gibson 1950, 1966) view of ecological perception and, more explicitly, the claim that perception is based on action. There are two crucial relationships: one between the speaker's actions and the environment, and one between the act of speaking and speech perception. Like motor theory, action theory also posits that the message is recovered by recreating the requisite articulatory information. In contrast to motor theory, however, action theory works at the level of the entire speech act rather than the phoneme.

vowel for purposes of calibration or use contextual information provided by neighboring consonants (Ladefoged and Broadbent 1957; Lindblom and Studdert-Kennedy 1967). Moreover, both vowel and consonant-vowel recognition are significantly influenced by higher-order units such as words, or words within sentences. In fact, several studies have shown that degradation of the phonetic unit can be tolerated (i.e., recognition accuracy can be maintained) when specific semantic or syntactic constraints occur (Connine and Clifton 1987; Salasoo and Pisoni 1985).

Studies of consonant and consonant-vowel identification reveal that both temporal and spectral information are important for identification, and as with vowels, contextual parameters are significant. For example, the voiced (*b, d, g*) and voiceless (*p, t, k*) stop consonants are distinguished by an asynchrony between the burst of expiratory energy and voicing. Moreover, as one progresses temporally from consonant to steady-state vowel production, the formant transition tends to be relatively brief. Verification of the importance of such parameters comes from tests using naturally produced consonant-vowel strings and synthetic exemplars that represent qualitative changes in the duration of formant transitions and in the presence or absence of energy bursts (Delattre, Liberman, and Cooper 1955; Liberman, Delattre, and Cooper 1958).

Summarizing a vast literature, acoustic cues for recognition of speech segments follow a kind of decision tree (Borden and Harris 1984). For example, if there is harmonic structure, no noise, and relatively low frequency energy, then the sound is a vowel. If the signal is nonperiodic and has relatively high frequency energy, then it is a consonant. For consonants, if there is a sharp noise onset, then it is an affricate (e.g., *ch* as in *chair* and *j* as in *jar*), and if the duration of the noise segment is long, then it's a fricative (e.g., *f* as in *fat* and *v* as in *vampire*). Needless to say, these are generalizations, and as highlighted previously, perceptual recognition is dependent on both local features of the utterance (i.e., in terms of phonetic-acoustic effects) and its global features (i.e., in terms of semantics and syntax).

Because many of the most recent and exciting issues in speech perception are directly tied to problems in development and psychology, I leave the discussion of such issues to subsequent chapters. Specifically, chapter 5 uses data on child language acquisition to assess the special nature and properties of speech, focusing in particular on the kinds of linguistic knowledge available to the preverbal child, how such knowledge is used to segment the acoustic speech stream, and how ontogenetic changes in experience shape the perceptual process. Chapter 7 continues with the issue of speech as a special acoustic signal and discusses the problems of categorical perception and cross-modal perception. Before concluding this section, however, I turn to a brief discussion of sound localization, a topic that provides phylogenetic continuity with the previous sections.

As with other mammals, studies of sound localization in humans[45] have generally revealed that interaural intensity differences are dominant when frequencies are high, whereas interaural time or phase differences are dominant when frequencies are low (Gourevitch 1987; Yost 1980). Other factors have also been suggested to be important in sound localization, including the role of the pinnae in sound filtering, complex interactions between spectrotemporal features of the signal, visual cues, and body posture. For example, several studies (reviewed by Middlebrooks and Green 1991; Yost 1980; Yost and Gourevitch 1987) have revealed that when high-frequency signals contain low-frequency repetitions, they can be readily localized by means of interaural time differences. From a comparative perspective, directional hearing in humans is far more accurate than it is in most other terrestrial mammals, especially nonhuman primates. Thus average interaural intensity difference thresholds have been measured at approximately 1.0 dB (125–8000 Hz), localization thresholds at 2–4 degrees (250–8000 Hz), and interaural time differences of 15–35 microseconds (250–1250 Hz). These differences, which in some cases are orders of magnitude higher than in closely related primate species (C. H. Brown and May 1990), are apparently not due to interspecific differences in head size or the mobility of the pinnae. What is clear, however, is that within the primates, phylogenetically more ancient species such as the prosimians and monkeys have broader audibility functions (i.e., hear a broader range of frequencies) and reduced frequency resolution when compared with the apes and humans.

Studies of brain imaging (reviewed in Posner and Raichle 1994) have been extensively used within the past five years to look at speech processing (see next section), in particular those aspects of speech that involve higher cortical processing (e.g., naming, syntax). In contrast, imaging techniques have been infrequently used to assess more basic auditory processing, or even the perception of phonemes (for a review and a discussion of possible avenues for future research, see Elliott 1994). Yet, of those studies conducted, there are several interesting findings that warrant further exploration, especially in light of debates concerning the special nature of speech. For example, Zattore and colleagues

45. It should be noted that a majority of work on human sound localization has been conducted using signals presented via headphones, rather than in a free-field environment. When headphones are used, researchers refer to *lateralization,* whereas when free-field presentations are employed, the term *localization* is used (Yost and Hafter 1987).

(1992) used positron emission tomography (PET; regional cerebral blood flow) to understand which cortical areas are associated with phonetic, as opposed to pitch, discrimination. In both tasks, the same signal was presented. In the phonetic task, subjects were required to determine whether the final consonants of two consonant-vowel-consonant strings were the same or different. In the pitch discrimination task, subjects had to respond when the second consonant-vowel-consonant stimulus had a higher pitch than the first. Phonetic discrimination was associated with primary activation in the left hemisphere, in particular Broca's area. In contrast, pitch discrimination was associated with activation in the right prefrontal cortex. Similar studies, using PET as well as functional magnetic resonance imaging (fMRI), can be conducted to ask other fundamental questions in psychoacoustics.

Brain Mechanisms for Spoken Language

The intimate connection between the brain, as it is shewn by those curious cases of brain-disease, in which speech is specially affected, as when the power to remember substantives is lost, whilst other words can be correctly used.   (Darwin 1871, 58)

Human language has effectively colonized an alien brain in the course of the last two million years. Evolution makes do with what it has at hand. The structures which language recruited to its new tasks came to serve under protest, so to speak. They were previously adapted for neural calculations in different realms and just happened to exhibit enough overlap with the demands of language processing so as to make ''retraining'' and ''reorganization'' minimally costly in terms of some as yet unknown evolutionary accounting. Many of the structural peculiarities of language, its quasi-universals, and the way that it is organized within the brain no doubt reflect this preexisting scaffolding.   (Deacon 1991, 164)

During primate evolution, several lines of evidence point to important changes in the structure and size of the brain, and at some level these changes appear to be related to modifications in the structure and function of human and nonhuman primate communication systems, and the environments in which they evolved. As highlighted by the preceding quote from Deacon and mentioned in previous sections of this chapter, a number of studies now provide evidence that language processing is based on a set of dedicated neural systems. A significant proportion of the data on humans come from individuals with brain damage—so to speak, natural experiments. More recently, however, neurologists, neuropsychologists, and cognitive neuroscientists interested in the neurobiological bases of language have taken advantage of new developments in brain imaging (PET, fMRI, ERP) in addition to cortical stimulation of patients with epileptic seizures. These approaches are extremely powerful and, together with the voluminous literature on brain-damaged patients, have made it possible for a new view of the neural substrates underlying language production and perception to emerge. I present a selective review in this subsection. More complete technical treatments can be found in Caplan (1987, 1992), Kimura (1993), Maratsos and Matheny (1994), and Dronkers and Pinker (in press), with more general discussions in Calvin and Ojemann (1994), Damasio and Damasio (1992), and Pinker (1994a).

The earliest evidence for language-related brain structures came from the findings of Broca (1861) and Wernicke (1874), who discovered patients with specific lesions and quite specific linguistic deficits—what are now commonly known as language aphasias. Moreover, many of the early neurological accounts and subsequent anatomical investigations reported significant asymmetries in language functioning, with the left hemisphere playing a more dominant role than the right (Figure 4.44). As a result of these findings, it has generally been concluded that the aphasias can be placed into two neat classes, Broca's and Wernicke's. In general, Broca's aphasics appeared to suffer from problems of production such as poor articulation and lack of fluency, but their comprehension was intact. In contrast, the typical Wernicke's aphasic had fluent production, but with quite bizarre content, and comprehension was generally poor.

As in all general accounts of biological phenomena, the story on aphasias was too simple. And the new story, which I will try to briefly develop here, is by no means complete, with controversies raging, both because of inconsistent definitions across research teams and the use of qualitative as opposed to quantitative techniques for evaluating the locus and consequences of brain damage (see critical review in Dronkers and Pinker, in press).

There is considerable variation in language deficits, and much of this is due to variability in the locus of damage. Thus individuals classified as Broca's and Wernicke's aphasics can exhibit widely ranging lesions (Figure 4.45) and highly specific linguistic deficits. For example, depending on how much of the motor strip has been damaged, Broca's aphasics will exhibit varying degrees of articulatory problems. Consequently, those working on the neurobiology of language have been forced into the role of splitters, rather than lumpers, providing increasingly more specific labels for the kinds of aphasias observed. Table 4.3 provides a glimpse of some of this variation, though even here, experts in the field will find much to argue about given that no concensus has been reached on the kinds of linguistic deficit associated with particular regions of damage.

In addition to the population-level variability in aphasias, Kimura (1993) has recently summarized data on sex differences in the incidence of aphasias, as well

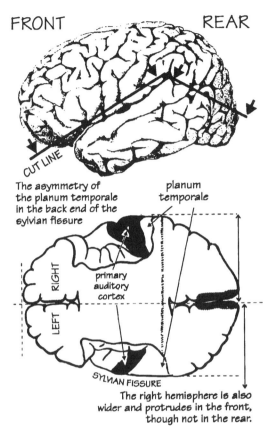

FRONT   REAR

CUT LINE

The asymmetry of
the planum temporale
in the back end of the
sylvian fissure

planum
temporale

RIGHT

primary
auditory
cortex

LEFT

SYLVIAN FISSURE

The right hemisphere is also
wider and protrudes in the front,
though not in the rear.

Figure 4.44
Some of the more significant language-related anatomical asymmetries of the human brain. This schematic figure highlights the planum temporale, located near the posterior end of the sylvian fissure. Also illustrated is the right hemisphere's anterior protrusion (from Calvin and Ojemann 1994).

Broca's aphasia

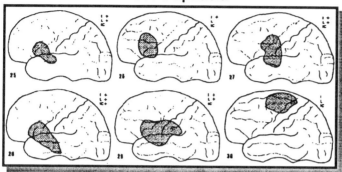

Wernicke's aphasia

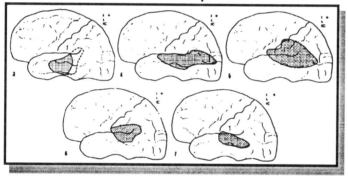

Figure 4.45
Variation in the locus of damage for human subjects exhibiting the symptoms of acute Broca's (top) and Wernicke's (bottom) aphasia; data obtained from CT scans (redrawn from Mazzocchi and Vignolo 1979).

Table 4.3
A Sample of Language Aphasias, Their Putative Locus of Damage, and Apparent Functional Consequences

Type of Aphasia	Locus of Damage	Description
Global aphasia	Left perisylvian region	Propositional speech is either completely or almost completely absent, including oral production, oral comprehension, reading, and writing.
Mixed transcortical aphasia	A border surrounding the perisylvian	Comprehension in the auditory and visual speech domains is lost, though subjects can repeat what they hear.
Transcortical sensory aphasia	Temporo-parietal-occipital junction, posterior to Wernicke's	Comprehension is impaired, but speech repetition and production are unaffected.
Transcortical motor aphasia	Prefrontal area rostral and dorsal to Broca's	Speech production is impaired, but speech repetition and comprehension are unaffected.
Broca's aphasia	Broca's area (posterior regions of the second and third divisions of the inferior frontal gyrus)	Spontaneous speech is often poorly articulated, nonfluent, effortful, and can be characterized by loss of grammatical words. Speech repetition is poor, and naming can also be impaired. Comprehension is typically unaffected.
Wernicke's aphasia	Wernicke's area	No articulatory impairments for spontaneous speech, but there often is difficulty with selection of appropriately meaningful words. Repetition and naming are poor. Comprehension of both spoken and written words is significantly impaired.
Conduction aphasia	Arcuate fasciculus	Occasional difficulties with articulation. Repetition is heavily impaired. Naming and comprehension in the auditory domain show slight impairments.

Information on aphasias obtained from reviews by Caplan (1987, 1992), Demonet, Wise, Frackowiak (1993), Dronkers and Pinker (in press), Gordon (1990), and Maratsos and Matheny (1994).

as in their neuroanatomical topography (Figure 4.46). These findings not only are inherently interesting, but in addition are directly relevant to the previous sections of this chapter indicating sex differences in the neural substrate for communication, and the effects of early hormonal influences on the structure of the brain.

Neuroanatomical work indicates that the splenium—a part of the corpus callosum, connecting the two cerebral hemispheres—is wider in females than it is in males, and this sex difference is observed in the fetus (deLacoste, Holloway, and Woodward 1986; deLacoste-Utamsing and Holloway 1982). Such differences suggest that interhemispheric communication (Hoptman and Davidson 1994) may

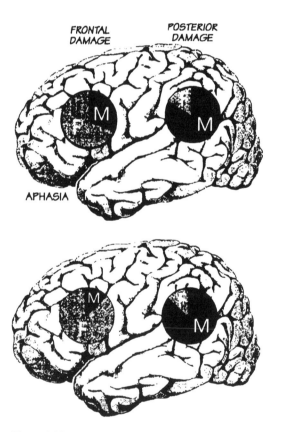

Figure 4.46
Differences in the location and prevalence of language aphasias and motor apraxias in adult men and women (data summarized from Kimura 1992; figure from Calvin and Ojemann 1994).

differ between the sexes, that such differences are set up early in life, and as a result, asymmetries in cognitive functioning may show up in infancy and subsequently be shaped by significant experiences.

Kimura and her colleagues (Kimura 1993; Kimura and Hampson 1994) have conducted a number of studies to explore sex differences in cognitive functioning. Relevant to the current discussion are analyses of both speech and motor aphasias. Based on tests of a large sample of patients ($n = 314$) suffering from left-hemisphere damage, results indicated that more men (48.5%) exhibited signs of aphasia than did women (30%). But this coarse-grained analysis covers up some interesting anatomical subtleties with significant functional consequences (Figure 4.46). Specifically, when the left anterior portion of the frontal cortex is damaged, more women than men suffer from aphasia. In contrast, when damage is left posterior, more men than women suffer from aphasia (Cappa and Vignolo 1988; Kimura 1993). The population-level effect therefore appears to be due to the fact that when restricted damage to the vascular system occurs, it is more likely to occur in posterior than in the anterior regions of the brain. As a result, males are more vulnerable to aphasias than are females. Some of the specifics of these effects are illustrated in Figure 4.47, showing that for five out of six speech-related tests, females with anterior damage are more impaired than males. For posterior damage, males generally show greater impairment than females, but sample sizes are small.

As highlighted in the introduction to this section, new developments in brain imaging have opened the door to studies aimed at revealing how the normal human brain processes language. Such research is important in and of itself, but also serves the function of checking the validity of the view that studying the damaged brain provides fundamental insights into the function of particular neural structures. Put more succinctly, I now turn to a discussion of how the normal brain processes language and whether current findings correspond with, or differ from, studies of the damaged brain.

In PET and fMRI research on language processing, the general approach has been the subtraction technique where, for example, subjects are asked to (1) look passively at a monitor showing written words, (2) listen passively to words, (3) repeat words that are presented auditorily or visually on a monitor, and (4) generate an appropriate verb for a noun presented either auditorily or visually (e.g., when asked to produce the appropriate verb for the noun ''bike,'' subjects are expected to produce the word ''ride''). Summary data from PET-rCBF studies (Petersen et al. 1989, 1990; Posner and Carr 1992) are presented in Figure 4.48.

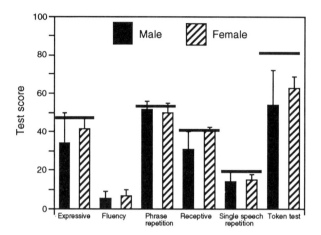

Figure 4.47
Sex differences in linguistic tasks for patients with left anterior pathology. Values are means and standard deviations. Thick horizontal bars represent the maximum scores for each particular test (data from Kimura 1993).

Results showed that passively viewing words was associated with significant activation of the extrastriate area located on the inner surface of the left hemisphere. When hearing words, the auditory cortex was activated, in particular Wernicke's area (left hemisphere). For both the visual and auditory task, words and pseudowords showed similar activation patterns. Moving from passive to more active linguistic tasks,[46] we find that during the production of words, either heard or read, four major areas are activated: primary motor cortex, supplementary motor cortex, cerebellum, and insular cortex. Each of these areas is involved in motor control. Interestingly, however, neither Broca's nor Wernicke's area was activated during this speech production task. As Posner and Raichle (1994) point out,

Skeptics, of course, questioned the ability of PET to reliably detect all changes. This charge is always difficult to refute, but, as it turned out, the charge was incorrect. Fortunately, the roles of Broca's and Wernicke's areas were clarified when subjects were asked to generate a use for the nouns presented to them.   (p. 119)

Moreover, other PET studies have indicated that Broca's area is active during speech production (see Maratsos and Matheny 1994; Poeppel, in press).

---

46. It should be noted that some linguists have been highly critical of current imaging work on language, arguing that the tasks used to assess areas of activation are artificial and fail to capture psychologically relevant processes (e.g., Pinker 1995; Poeppel, in press).

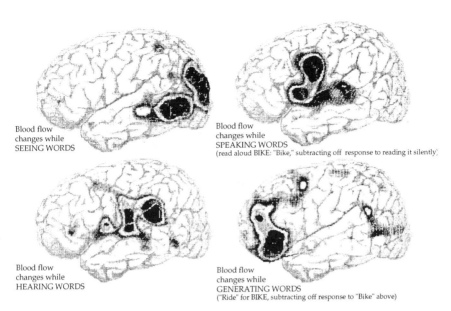

Blood flow
changes while
SEEING WORDS

Blood flow
changes while
SPEAKING WORDS
(read aloud BIKE: "Bike," subtracting off response to reading it silently)

Blood flow
changes while
HEARING WORDS

Blood flow
changes while
GENERATING WORDS
("Ride" for BIKE, subtracting off response to "Bike" above)

Figure 4.48
Summary of data on cerebral blood flow changes when normal subjects are seeing, speaking, or
hearing words. ''Generating words'' differs from ''speaking words'' in that the former is recorded
when subjects are asked to produce, e.g., the appropriate verb for someone on a bicycle, whereas
the latter is produced in response to someone saying, ''Say BIKE'' (gray-scale images derived from
color PET scans presented in Raichle 1992; schematized here by Calvin and Ojemann 1994).

In the verb-generation task, there was strong activation in the left frontal cortex, including Broca's area. This suggested, based in part on memory research
by Goldman-Rakic and her colleagues (reviewed in Goldman-Rakic 1987), that
linguistic tasks that involve the frontal cortex rely on access to stored representations, rather than external cues or well-rehearsed information. In addition to the
left frontal cortex, the anterior cingulate cortex, the left posterior temporal cortex, and the right cerebellum were also active. The anterior cingulate appears to
play a role in attention and, more specifically, in selecting among alternative
choices such as the appropriate verb-noun pairing. Wernicke's area, located
in the posterior temporal cortex, was surprisingly inactive during the verb-
generation task, suggesting that other areas subserve this particular type of
auditory processing. The final surprise came from the activation of the right
cerebellum. Though this area has been clearly implicated in motor control, it
nonetheless remained active when the speech component of the task was sub-
tracted out. This finding suggested the possibility, which will be discussed more

fully, that the cerebellum plays a role in the acquisition of new information and, perhaps especially, linguistic information. Based on these results, Posner, Raichle, and colleagues have suggested that the verb-generation task may actually call up two different neural pathways, one involving automatic processing and one involving nonautomatic processing; the latter would involve the right cerebellum.

Most of the research using PET and MRI has focused on broken speech, both at the production and perception end. Recently, however, Mazoyer and colleagues (1993) have conducted a study designed to reveal how the brain parses the units of speech and, more specifically, whether there are specific regions of the brain that function differently in terms of the acoustical, phonological, lexical, prosodic, grammatical, and conceptual features of speech. The subjects were native French speakers who were presented with five different listening tasks: (1) listening to a story spoken in Tamil, an unfamiliar language to French speakers, (2) a story spoken in French, (3) French sentences where content words were replaced by pseudowords, (4) French sentences where content words were replaced by semantically anomalous words, and (5) lists of French words. Results from normalized cerebral blood flow and PET revealed that the superior temporal gyri (right and left sides) were the only active areas for all five tasks, suggesting that this area is responsive to auditory stimuli, but somewhat indiscriminately responsive with regard to acoustic structure and content. However, even with such a coarse-grained response, the left superior temporal gyrus was more active than the right when French was spoken than when Tamil was spoken. This result suggests at least some level of specialization for the native tongue. When the French story was heard, the left side of the brain showed activation of four other areas: the middle temporal gyrus, the inferior frontal gyrus, temporal region, and Brodmann's area 8. In contrast, only the superior temporal gyrus and temporal region were activated on the right side.

The left middle temporal gyrus appeared to be an important area for phonological decomposition, based on the fact that this area was not active for stories in Tamil but was active for stories in French. Comparable evidence for the role of the middle temporal gyrus has been reported by Zatorre et al. (1992).

At the lexical level of analysis, results point to the inferior frontal gyrus, because this area was active during the presentation of French word lists. Although studies have implied that the inferior frontal gyrus is also important in word production, as evidenced by observations of patients with Broca's aphasia who show neurological damage to this region, results reported here show that it is also important in comprehension. However, this conclusion must be treated somewhat cautiously because during the presentation of pseudowords, the inferior frontal region was also activated, but asymmetrically so, with the right side exhibiting greater activation than the left. Last, because meaningful utterances showed different areas and asymmetries of activation from either broken speech, or continuous but meaningless speech, certain areas such as the superior and middle temporal gyri and area 8 may play a more important role in the conceptual analysis of speech.

A recent critique by Poeppel (in press) of imaging studies of language processing suggests that we should be cautious in our inferences about brain localization. Several recent experiments using PET have provided data on areas of activity during language processing. Unfortunately, as depicted in Figure 4.49, there is not much agreement between studies as to the precise relationship between structure and function: although all of the studies cited in the following paragraphs focused on phonological processing, there was extraordinary variation in the areas of activation.

Some of the problems raised by Poeppel's critique, and others' concerns as well (Maratsos and Matheny 1994), may soon be resolved as researchers whose

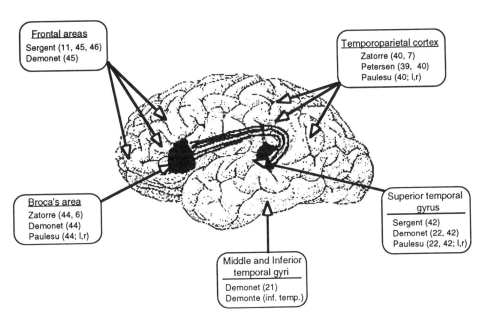

Figure 4.49
Brodmann areas (in parentheses following author names) uncovered by PET in various studies of language processing (redrawn from Poeppel, in press).

expertise lies in the technical side of brain imaging combine forces with those whose expertise lies in the technical aspects of linguistics and psycholinguistics (Pinker 1994b; Raichle 1994). In addition, a growing number of studies are combining data from PET and fMRI with data from ERP and cortical stimulation (Figure 4.50). This kind of multimethod approach is by far the most powerful because it shows not only areas of significant activation and function, but also the potential time course of activation.

Before leaving experimental studies of language processing via PET or fMRI, let me turn to a recent report of significant sex differences in language perception in normal human adults. Using a modified form of fMRI known as echo-planar functional magnetic resonance imaging, Shaywitz and colleagues (1995) obtained activation data from two primary cortical regions: extrastriate area and inferior frontal gyrus (approximately Brodmann's 44/45). Four same-different judgment tasks were run. The first involved a comparison between two sets of lines with particular orientations. The second task required subjects to determine the similarity between two sets of consonant strings, potentially differing in the pattern of case alteration. Task three involved a rhyme judgment for nonsense words. In the final task, subjects were asked to determine whether two words had membership in the same semantic category. The logic of these tasks was to isolate orthographic processing (subtracting the case from the line task) and phonological processing (subtracting rhyming from case).

Results revealed that males exhibited much greater activation in the left hemisphere than the right, whereas females were relatively symmetrical. Breaking these general results down into cortical region, the case task showed greater activation of extrastriate areas, whereas the rhyming task elicited greater activation of the inferior frontal gyrus. Most importantly, during the rhyming task males showed significantly greater activation of the left inferior frontal gyrus than the right, but females showed bilateral activation; the extrastriate region was bilaterally active for both males and females. These data provide evidence of gender differences in the neural basis of language and support previous findings from brain-damaged patients and from normals using more indirect tests of brain function (see Kimura 1993).

Though most discussions of the neural mechanisms subserving language have focused on higher cortical processes, several recent studies have attempted to strengthen the claim that other neuroanatomical structures, commonly believed to function in motor control alone, are also involved in language. The two most important candidates are the basal ganglia (Lieberman 1991) and the cerebellum

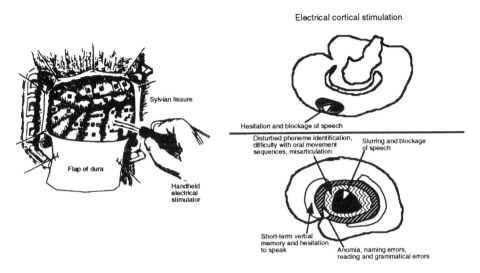

Figure 4.50
On the left, schematic illustration of cortical stimulation technique (Calvin and Ojemann 1994). On the right, summary of data on the language-related functions based on cortical stimulation (from Deacon 1991).

(Leiner, Leiner, and Dow 1993; Middleton and Strick 1994). The basal ganglia are of interest because individuals with Parkinson's disease not only show generalized motor malfunction, but also exhibit specific deficits in the production of syntactic structure (Lieberman 1992).

The cerebellum has long been viewed as one of the major centers for motor control, but new research from neuroimaging in particular has suggested that during language processing, and other cognitive tasks as well, there is significant cerebellar activity in the absence of motor output. The primary area of activity is the dentate nucleus, located within the lateral portion of the cerebellar cortex (Figure 4.51).

In contrast to the rest of the cerebellum, this lateral region became relatively larger over the course of evolution. Moreover, the neodentate nucleus is not completely differentiated in monkeys, and in apes it is nowhere near as large as it is in humans. These neuroanatomical changes, together with evidence of connectivity between the cerebellar and cerebral cortices (e.g., direct connections between the dentate nucleus and the inferior prefrontal cortex—Brodmann's areas 44 and 45, or Broca's area), raise the interesting possibility that the dentate nucleus subserves cognitive functions underlying language processing and

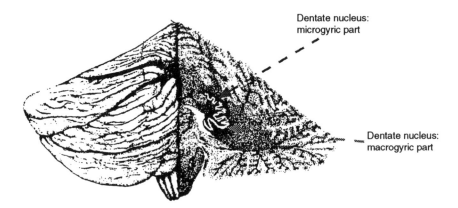

Dentate nucleus:
microgyric part

Dentate nucleus:
macrogyric part

Figure 4.51
The right side of this schematic illustration of the cerebellum represents a coronal slice, revealing the
dentate nucleus (redrawn from Leiner, Leiner, and Dow 1993).

production. More specifically, Leiner and colleagues have gone so far as to suggest that the evolution of the human cerebellar system was a prerequisite for the
evolution of human language. Though imaging studies have revealed activation of
the lateral portion of the cerebellum during language-processing tasks (discussed
earlier) and patients with damage to the cerebellum often exhibit a host of cognitive deficits (e.g., attention shifting, decision making and planning), this view of
the cerebellum is controversial (Bloedel 1993; Glickstein 1993). The controversy
stems from both conflicting neuroimaging results and different perspectives on
the relative strength of cerebellar-cerebral connections.

Signed Language   In the previous sections on spoken language, we discovered
that the left hemisphere is dominant with regard to the structural complexities of
language and that specific linguistic components such as grammar and the lexicon
can be selectively impaired. The common view among linguists is that signed
language exhibits all the formal properties of spoken language, including its expressive power. Consequently, we can ask whether the neural substrate subserving this visual language is the same as or different from that subserving an
auditory language. If it is the same, then this fact would suggest that the brain
has been designed to process language with either auditory or visual input. And
this suggestion would indeed be fascinating from a comparative and phylogenetic
perspective, because to my knowledge at least, no other species houses a brain
that is sufficiently flexible to generate and perceive a communicative sequence in

more than a single sensory modality.[47] Before discussing the neural architecture underlying signed language, let us take a brief look at some of its structural properties. More detailed information is provided in chapters 5 and 7.

Within linguistics, the changing status of sign language from some primitive and derivative communication system to a fully expressive natural language emerged from two broad sets of findings (Bellugi and Studdert-Kennedy 1980; Klima and Bellugi 1979). First, detailed studies of sign languages throughout the world revealed that they were based on the same kinds of mechanisms as were spoken languages, including syntactic regularities, a sublexical level of organization, inflectional patterns, and rules for compounding. Second, the linguistic mechanisms underlying each sign language were completely unrelated to the spoken language of the culture (e.g., American English and American Sign Language [ASL]) and were often fundamentally different from other sign languages from cultures sharing similar spoken languages. (For example, ASL and British Sign Language [BSL] have little in common, though American and British spoken English are similar; Figure 4.52 shows finger spelling in ASL and BSL.) Moreover, whereas different sign languages often share similar hand forms, meanings can differ (Figure 4.53). With these similarities uncovered, it was possible to consider "the internal structure of ASL's[48] lexical units [its signs] and of the grammatical scaffolding underlying its sentences [thereby providing] new perspectives on brain organization for language" (Bellugi, Poizner, and Klima 1990, 522).

Though signed and spoken languages share a number of organizational features, the visual-gestural mode has a significant impact on grammatical form. Specifically, whereas spoken languages order their elements sequentially and in a linear fashion, signed languages use a multidimensional space to simultaneously articulate the different elements. In fact, the grammar of sign languages explicitly makes use of space to provide structure for its elements. Consequently, since normal humans tend to show right-hemisphere dominance for processing spatial information, one might expect comparable hemispheric asymmetries in processing

---

47. This is not to say that other species lack the capacity to communicate in more than one modality. But, as pointed out for birds for example, the neural substrate for song is in fact quite different from that subserving the production and perception of calls. Thus, if one lesions some of the critical song pathways, song deteriorates, and no other communication system emerges to replace it, especially not a system in a nonauditory modality. However, no one has created a population of songless individuals and watched to see what happens, over the long run, to their social interactions and communication system.

48. This perspective applies to all sign languages. Most work in this area, however, has focused on ASL.

British Sign Language   American Sign Language

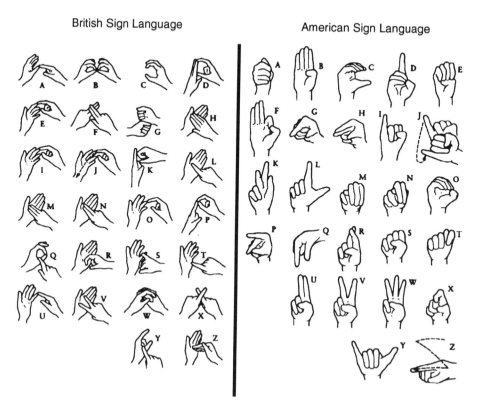

Figure 4.52
Differences in finger spelling between British Sign Language and American Sign Language (redrawn from Crystal 1987).

signed languages. This hypothesis would be falsified if the spatial information was strictly linguistic, therefore processed preferentially by the left hemisphere—the language-dominant hemisphere. To examine these competing predictions, Bellugi and colleagues (Bellugi, Poizner, and Klima 1983; Klima, Bellugi, and Poizner 1988) examined the consequences of right- and left-hemisphere damage on signing.

   A battery of language tests and nonlanguage tests (particularly, spatial tasks) were administered to normal deaf subjects, and to deaf subjects suffering from either right- or left-hemisphere damage (Figure 4.54). On a coarse-grained level, all deaf subjects with left-hemisphere damage showed significant aphasias, whereas none of the subjects with right-hemisphere damage did so (Figure 4.55).

Similar sign forms,
different meanings

Sign forms appearing
in CSL, but not in ASL

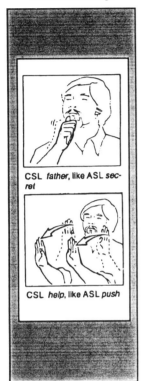
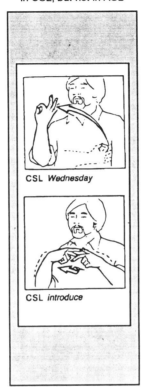

Figure 4.53
Similarities and differences between Chinese Sign Language (CSL) and American Sign Language
(ASL). The panel on the left shows cases where the form of the sign is similar in CSL and ASL, but
the meanings differ.The panel on the right shows cases where the sign in CSL is not possible in ASL
(redrawn from Klima and Bellugi 1979).

Left hemisphere damaged signers

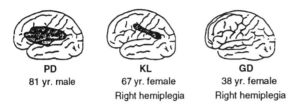

PD	KL	GD
81 yr. male	67 yr. female	38 yr. female
	Right hemiplegia	Right hemiplegia

Right hemisphere damaged signers

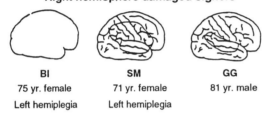

BI	SM	GG
75 yr. female	71 yr. female	81 yr. male
Left hemiplegia	Left hemiplegia	

Figure 4.54
Patient population for assessing the relationship between location of brain damage and signing deficits
(from Bellugi, Poizner, and Klima 1990).

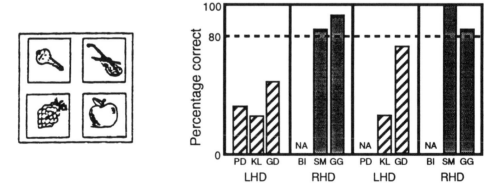

Figure 4.55
Differences in signing deficits for subjects with right- and left-hemisphere damage. Subjects were
asked to pick out the two pictures associated with the ASL equivalent of rhyming (i.e., using two of
the three primary sign parameters—handshape, movement, and articulation). LHD = left hemisphere
damage; RHD = right hemisphere damage. Refer to Figure 4.54 for locus of damage by subject
(x-axis).

Rather, those individuals with right-hemisphere damage were impaired on nonlinguistic spatial tasks. Moreover, production signing deficits could not be explained
as a simple consequence of motor disfunction because such subjects performed
well on tests involving nonlinguistic motor gestures. On a more fine-grained level,
there was considerable variation in the severity of the aphasias for patients with
left-hemisphere damage, with some showing agrammatism and others exhibiting
discourse errors. Thus, in parallel with cases of spoken language aphasia, there
is considerable heterogeneity in the severity of signed language aphasias, and
presumably the type of impairment exhibited is closely related to the locus of
neural damage. And in both signed and spoken aphasias, damage to the left
hemisphere leads to more significant linguistic loss than damage to the right hemisphere, both in terms of production and perception.

Tapping into the cortical stimulation technique employed by Calvin and Ojemann for spoken language (see Figure 4.50), Haglund and colleagues (1993) carried out both cortical stimulation and single-unit recordings on a hearing patient
who was fluent in ASL but was suffering from intractable seizures in the left
anterior frontotemporal region. Intracarotid amobarbital perfusion tests[49] revealed involvement of both hemispheres in language, but a significant left-
hemisphere bias, especially for reading. Results from cortical stimulation mapping
and single-unit recordings showed that some sites were involved in spoken and
signed naming as well as in the interpretation of signs. However, there were also
sites with clear dissociations between spoken and signed languages. For example,
cortical stimulation revealed that areas involved in sign language were located in
the anterior temporal region. Single-unit recordings from superior temporal gyrus
indicated a higher level of activity in response to spoken naming as compared to
signed naming. Interestingly, stimulation within Broca's area showed deficits in
spoken and signed language. As Haglund and colleagues argue, this result suggests that "Broca's area . . . should be viewed as important to language motor
output, regardless of whether orofacial or limb movements are involved in output" (pp. 24–25); this quote emphasizes the views espoused in Broca's early
writings on this topic. Once again, we find strong support for the position that
the dominant language hemisphere is, in general, bimodal, but that there are
significant dissociations that relate directly to the structural differences between
signed and spoken languages.

---

49. A simple technique used to functionally "cool" or deactivate one hemisphere, leaving the other
operative.

In summary, studies of the neurobiology of human language, both signed and spoken, have revealed specialized structures underlying production and perception—so specialized in fact, that focal damage can result in the loss of names for vegetables or animals, or the loss of syntax (Caramazza et al. 1994). At the periphery, the modern human vocal tract represents a unique evolutionary adaptation, and one that appears to maximize the efficiency of sound production while proving somewhat costly with regard to feeding efficiency. Where sound efficiency comes in is in terms of the rate of information transmission that is possible with a two-plus tube system and, of course, a perceptual system that is capable of decoding such information. And needless to say, the transmission speed is not only faster in the vocal than in the visual mode (sign language), but is also more efficient, especially in cluttered visual environments or when one's hands are preoccupied.

In contrast with the production system, the human auditory system is structurally and functionally similar at the periphery to what we observe in other mammals (e.g., nonhuman primates and nonprimate mammals are capable of discriminating a number of significant sounds that appear within human languages). Where humans appear to differ fundamentally from other animals—especially their close relatives, the nonhuman primates—is in the neural substrate driving production and perception of species-typical signals. For example, whereas subcortical structures appear to play the most significant role in nonhuman primate communication, both subcortical and cortical structures are heavily involved in human communication. One potential reason for this difference, detailed more completely in chapter 5, is that whereas learning is critically involved in human language acquisition, it plays either no role or a very limited role in the ontogeny of nonhuman primate communication. The importance of learning is likely to place significant pressures on brain areas that subserve such functions, and these are likely to be higher cortical areas. Whether higher cortical structures evolved to subserve noncommunicative functions first, and then were coopted for communication—or the reverse—has yet to be determined.

As demonstrated for frogs and birds, humans also exhibit sexual dimorphism with regard to the neural architecture subserving communication. However, unlike frogs and birds, such sex differences do not translate into differences in the kinds of signals produced but rather in the way signals are processed. Thus it is not the case that males produce words or intonations that females are incapable of producing, or vice versa. Rather, males and females appear to differ with regard to how they process language, and this is at least partly due to the fact that in men there is greater neural asymmetry than in women. Moreover, due to apparent anatomical differences between men and women, damage to the brain often results in cognitively more problematic deficits in men than in women.

To conclude, perhaps one of the most striking design features of the human brain with regard to communication is its apparent ability to accept information from either the auditory or visual modalities as relevant input. Although some would argue that this facility indicates that the human brain is *amodal* in terms of language processing, it seems more likely that selection has favored a brain that is *bimodal*. The semantic difference is important. Specifically, amodality suggests that either modality will do, whereas bimodality implies that although either modality will be accepted, one is dominant. Thus normal humans will always use speech as their dominant mode of communication. And this approach works because the developing brain sits in a waiting state, prepared to accept the most dominant sensory input. In normals, this will be acoustic stimulation. In contrast, if an individual is born with damage to the auditory system, then the second most dominant sensory system will click into operation, and this will typically be the visual system. As a result, sign language emerges, carrying all of the structural properties of spoken language.

4.4.3 Facial Expression and Perception in Primates

The functional architecture of the brain is designed to accommodate the cognitive demands of the multitude of operations that have to be performed to adapt to the environment. With respect to face processing, the stringent perceptual demands, made necessary by the common format of all faces that differ in subtle variations in their configurations, call for structures that are capable of performing quick and reliable differentiations; the need to access relevant personal information about these faces for their recognition also requires the involvement of structures able to relate facial representations to their pertinent memories. (Sergent and Signoret 1992, 61)

A ubiquitous feature of all organisms is the capacity to produce and perceive signals or cues that identify them as members of a particular class of individuals such as conspecifics, kin, and mate (see section 7.2). The level of detail and significance of within-class variation will depend on both the resolving power of the sensory system and the selective pressures favoring a particular level of resolution. Thus, in the case of eusocial insects such as sweat bees, selection has favored the expression of olfactory cues specific to the hive as well as more fine-grained cues that allow individuals to assess degrees of genetic relatedness. In passerine birds, individuals are readily identified on the basis of their song, and for a large number of species, memorized information about individual songs is used to assess whether an intruder is a familiar neighbor or stranger, from the same dialect or different dialect (Beecher 1982; Beecher, Campbell, and Burt 1994; McGregor and Krebs 1984a, 1984b, 1989; Weary, Lemon, and Perreault 1992; Williams and Slater 1990). Such information can even be stored and retrieved over several mating seasons (Godard 1991).

Among nonhuman and human primates, the face is perhaps the most critical source of information about individual identity,[50] and the work summarized in this section will show how the primate brain has been designed to detect not only faces, but particular features of the face and how they are configured to produce communicatively meaningful expressions (i.e., signals) of emotion. As with so many other issues in biology, and especially those discussed in this book, Darwin (1872) was one of the first to identify the commonalities between nonhuman and human primate faces and their capacity to be communicatively expressive (Ekman 1973; Fridlund 1994). Although the latter issue is discussed in greater detail in chapter 7, it is worth pointing out here that it is in *The Expression of the Emotions in Man and Animals* that we see the first signs of Darwin using a nonadaptationist perspective to explain both the origins and current utility of facial expressions (reviewed in Fridlund 1994). For Darwin, facial expressions were simply reflexive responses to changes in affective state. It is not entirely clear why he shied away from using his own evolutionary tool kit to interpret faces (see the interesting discussion by Fridlund 1994), but this is a matter for later chapters. For now, let us turn to a discussion of primate facial expressions, starting with nonhumans and ending with humans. For both primates, I begin with a brief synopsis of the peripheral organs involved in production, then turn to a more detailed discussion of the neural circuitry underlying production and perception.

Nonhuman Primates   The production of facial expressions requires the manipulation of facial muscles. From a phylogenetic perspective, primate species can be distinguished on the basis of the degree to which they can control facial musculature (Huber 1931). In general, the great apes and humans have a far more intricate set of facial muscles than do the monkeys and prosimians (Figure 4.56). However, the kinds of muscles that are available for configuring the face into an expression

---

50. Several studies have demonstrated, based on acoustic analyses and statistical procedures for quantitatively assessing clustering of data points (e.g., discriminant function analysis), that individual identity is also encoded in primate vocalizations (Hauser 1990, 1994; Lieblich et al. 1980; Macedonia 1986). Most of these studies are, however, limited to a relatively small sample of individuals and few experiments. For example, Cheney and Seyfarth (1980) have assessed whether the acoustic features that appear to "define" an individual—to give it an acoustic signature—are perceptually salient and sufficient for individual recognition in the absence of other sources of information.

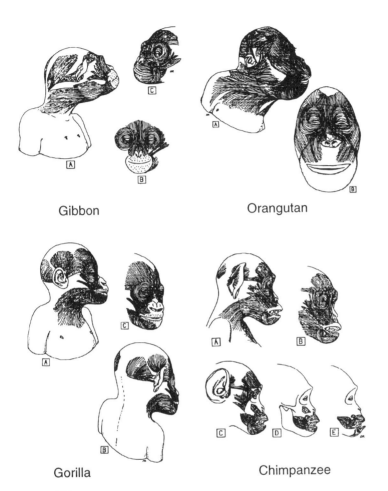

Figure 4.56
Comparative facial musculature of the great apes (redrawn from Huber 1930b).

of emotional state fail to provide a complete characterization of species' differences. For example, although gorillas (*Gorilla gorilla*) and chimpanzees (*Pan troglodytes*) both live in social groups, engage in complex social relationships (e.g., coalitions and alliances used in dynamically changing dominance interactions), and have similar facial musculature (Figure 4.56), gorillas display a relatively impoverished repertoire of facial expressions (Andrew 1962, 1963; Harcourt and Stewart, personal communication). A thorough understanding of the causes of interspecific variation in primate facial repertoires will require a more complete understanding of their psychologies and how they are used to negotiate social interactions. These issues are picked up in chapter 7.

Although virtually no work has been done on the kinematics of nonhuman primate facial expressions (Chevalier-Skolnikoff 1973; Hauser 1993c), Huber (1931) has provided a detailed comparative account of facial musculature,[51] and Izard (1971) has conducted a series of experiments to determine the importance of particular facial muscles in social communication. Concerning anatomy, one general trend within the primates has been toward increasing control of the midfacial region (primarily the lips) and a decrease in control over the ears. Thus, we find that in lemurs, the relative number and control of the muscles surrounding the lips is reduced, whereas in macaques and chimpanzees, a fine network of

---

51. It is unfortunate, given recent technological advances in quantifying motor actions and control, that the most detailed and comprehensive accounts of facial musculature and vocal tract structure were both published more than fifty years ago (Huber 1930a, 1930b, 1931; Negus 1929, 1949).

muscles (particularly the orbicularis oris) provides these species with considerable control over the configuration of the lips. In addition to changes in the midfacial region, there has also been significant modification of the eye region, with humans demonstrating the most extensive level of muscle control among primates.

To assess more causally the relation between muscle control, facial expression, and social communication, Izard conducted an experiment that involved first removing an individual rhesus monkey from its social group, then sectioning the eighth cranial nerve, followed by replacement into the colony. This lesioning technique completely eliminates control over facial expression, perhaps with the exception of what can be communicated by means of the eyes alone. Results indicated quite significant effects on the quality and quantity of social interactions. Following experimental loss of facial control, individuals tended to drop in dominance rank and experience an increase in aggressive interactions; experimental females experienced a decrease in the strength of the mother-infant bond. These results suggest that in the absence of facial expressions, body postures and vocal signals are insufficient to maintain the quality of particular types of social interaction.

Turning to more direct measures of brain activity, little is known about the relative contribution of the right and left hemispheres to the production of nonhuman primate facial expressions. Ifune, Vermeire, and Hamilton (1984) presented split-brain rhesus monkeys with videotaped sequences of human and nonhuman primates, in addition to sequences of other animals and scenes. Significantly more facial expressions, both submissive and aggressive, were elicited during right-hemisphere stimulation than during left-hemisphere stimulation. Although the results presented by Ifune, Vermeire, and Hamilton (1984) are clear, the use of split-brain subjects makes it difficult to assert with confidence that intact rhesus show right-hemisphere dominance for the production of facial emotion.

Recently, Hauser (1993e) has presented data from free-living, intact rhesus monkeys which show that during the production of facial expression, the left side of the face begins moving into the expression earlier and is more expressive than the right side (see Table 4.4). This face asymmetry, which implicates right-hemisphere control, was documented for four different facial expressions, spontaneously produced during naturally occurring social interactions. Three expressions were clearly associated with social interactions that were likely to result in negative/withdrawal emotion, and one, the copulation grimace, is possibly associated with positive/approach emotion. The emotional valence distinction made here has most recently been championed by Davidson (1992, 1995), who

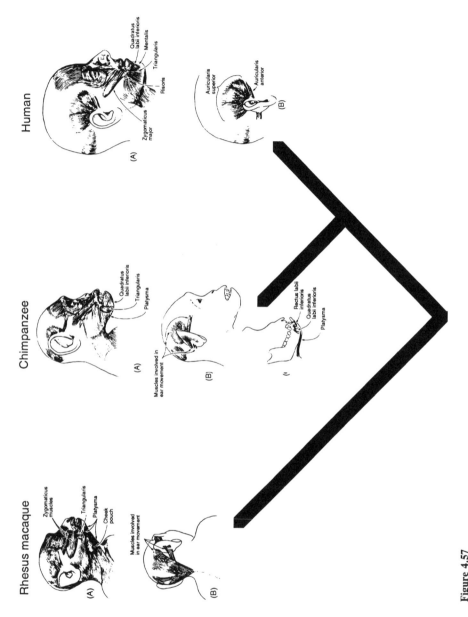

Figure 4.57
Changes in facial musculature from rhesus to chimpanzees and, last, to humans (redrawn from Huber 1930b).

Table 4.4
Asymmetries in Production of Rhesus Monkey Facial Expressions (from Hauser 1993e)

Expression	Social Situation	Variable	Sample Size	Face Expressions (number)	Asymmetrical Expressions (percent)	Left-Bias Expressions (percent)
Fear grimace	Produced by subordinate attacking or threatened by a dominant	Timing	17	60	76.5*	92.3†
Fear grimace		Number of skin folds	19	63	78.9†	86.7†
Fear grimace		Lip retraction	12	36	91.7†	90.9†
Fear grimace to resting position		Timing	4	8	75.0	66.7
Copulation grimace	Produced by a male during copulation	Timing	8	35	50.0	50.0
Copulation grimace		Number of skin folds	7	30	85.7	83.3
Copulation grimace		Lip retraction	4	21	100.0	75.0
Copulation grimace to resting position		Timing	8	17	100.0*	87.5*
Open mouth threat	Produced by a dominant attacking or threatening a subordinate	Timing	10	21	60.0	100.0*
Ear flap threat	Produced by a dominant attacking or threatening a subordinate	Timing	4	9	100.0	100.0

*$p < 0.05$
†$p < 0.01$

argues, based on EEGs, that in human adults, the right hemisphere is more active during negative/withdrawal emotion, whereas the left hemisphere is more dominant for positive/approach emotion. Although all facial expressions examined thus far for the rhesus monkey exhibit a left-side-of-the-face bias (right hemisphere), the general pattern of asymmetry for emotional expression is consistent with the pattern documented for humans (see next subsection), as well as for rats and chicks (reviewed in Bradshaw and Rodgers 1993; Hiscock and Kinsbourne 1995).

Single-unit recording studies of face recognition in nonhuman primates have been carried out for over fifteen years (Gross, Rocha-Miranda, and Bender 1972; for a recent review, see Tovee and Cohen-Tovee 1993). To properly evaluate this research, it is important to keep three issues distinctly in mind. First, we must bring back the problem of neural tuning raised earlier for frogs and bats (Box 4.1), and, in particular, ask whether it really makes sense to talk about cells being tuned to faces, of being "face cells" or "face-selective cells"? Second, we must explore not only how neurons or neuronal ensembles recognize faces as faces, but also how they manage to recognize facial expressions and individual faces. This task represents, once again, the problem of categorization, and more precisely, the problem of establishing the resolution of the perceptual system. Third, research on the neural substrate guiding face processing in nonhuman primates is restricted to rhesus macaques. As a result, taxonomic generalizations are limited. In fact, given interspecific variation in the relative importance of faces (both in terms of identity and as a source of communication) within the primates, it is highly likely that interspecific differences in neural processing will be uncovered. For example, though forest-dwelling primates must identify conspecifics and heterospecifics, and presumably use the face as one of several relevant cues, facial expressions typically play a less significant role than auditory signals; this modality difference is further emphasized by the fact that in contrast to terrestrial species (e.g., macaques, baboons, gorillas, chimpanzees), where the face tends to be naked, forest-dwelling primates (e.g., colobus, gibbons, orangutans) tend to have more facial hair. Thus the kinds of neural specializations that have been suggested for rhesus monkeys (Hasselmo, Rolls, and Baylis 1989; Heywood and Cowey 1992), a ground-dwelling species that tends to live in relatively open habitats, may not generalize to species living in different ecologies.

One of the strong claims to emerge from early research on face recognition was that there are "face cells," comparable conceptually to the famous grandmother or gnostic cells that Lettvin (cited in Gross 1992), Konorski (1967), and other early neurobiologists speculated about. The proposed evidence for such cells (reviewed in Gross 1992; Tovee 1995; Tovee and Cohen-Tovee 1993) was derived from the observation that neurons, particularly those located in the inferior temporal cortex (IT; Figure 4.58), are far more responsive (i.e., have higher spike rates) to faces than to other stimuli (visual and nonvisual), and the specificity of the response to faces is maintained across transformations in color, size, or orientation. Moreover, it seemed intuitively reasonable to argue for the presence and significance of cells dedicated to processing faces, given their importance in communicative interactions. As Gross (1992) puts it: "It is more crucial for a monkey to differentiate among faces than among any other categories of stimuli," and, "Faces are more similar to each other in their overall organization and fine detail than any other stimuli that a monkey must discriminate among"

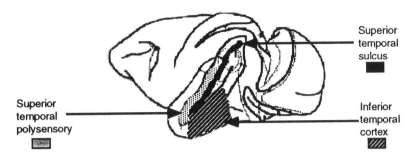

Superior
temporal
sulcus

Superior
temporal
polysensory

Inferior
temporal
cortex

Figure 4.58
Schematic illustration of some of the major areas of the rhesus macaque brain that are responsive to faces; data obtained from single-unit recordings.

(p. 6). Because not all researchers in this area concur about the evidence for face cells, we must examine the observations that have been gathered to support the claim.

Some of the first claims (Bruce, Desimone, and Gross 1981; Desimone and Gross 1979; Gross, Rocha-Miranda, and Bender 1972) for face cells were derived from observations of cells responding selectively to faces, while failing to respond to a host of visual stimuli with comparable spatiotemporal patterns such as colors, textures, and so forth. Moreover, follow-up work provided extensive details about the factors that cause significant changes in neural response, as well as the factors that seem to have little effect. Both results are relevant to the selectivity argument, and the specifics are summarized in Table 4.5.

Though many authors take the evidence reviewed here as prima facie support for face-selective cells (Tovee and Cohen-Tovee 1993), others disagree, claiming instead that neurons (especially, neuronal populations) within IT are only *sensitive* to faces. Gross provides the strongest version of this alternative perspective: "There is actually no evidence that face-selective cells play a role in perceiving and remembering faces, whether as individual detectors or members of an ensemble. That is, both the 'grandmother' and 'ensemble' codes for faces are, at best, 'candidate' codes rather than biological ones" (p. 6).

If there are no face cells, then what are the properties of these cells? In a review paper, Gross and Rodman (1992) suggest that IT neurons exhibit the following properties:

1. Visual stimuli are solely responsible for eliciting a neural response.

2. The fovea is included within the receptive field of several neurons within IT, thereby suggesting that such cells are perceptually involved in gaze.

Table 4.5
Response Patterns by Putative "Face Cells" in the Inferior Temporal Cortex of the Rhesus Macaque

Stimulus or Featural Change	Neural Response–Unmodified	Neural Response–Weaker
Alter relative size of the face	X	
Real face, plastic face, video image of face	X	
Line drawing of face		X
Human versus nonhuman primate face	X*	
Coordinate arrangement of features (e.g., eyes, nose, mouth) rearranged		X
Low- and high-pass band filtered faces	X	
Color pattern of face altered or removed	X	
Decrease in contrast of face image	X	

The patterns described in this table were derived from the following references: Bruce, Desimone, and Gross 1981; Desimone et al. 1984; Perrett, Rolls, and Caan 1982; Perrett et al. 1984, 1985, 1988b; Rolls and Baylis 1986; Rolls, Baylis, and Leonard 1985; Yamane, Kaji, and Kawano 1988; Yamane et al. 1990.
*This comparison was between humans and rhesus macaque faces.

3. In contrast with cortical areas dedicated to lower-level visual processing, cells within IT generally have larger receptive fields.

4. Both halves of the visual space provide input to the receptive fields.

5. Only a small proportion of the neurons sampled respond specifically to faces and hands, whereas a majority respond to more general features of the object such as color, shape, and texture. Consequently, most neurons within IT are not considered feature detectors. Object recognition appears to occur by a population-level neural response.

6. Neurons that do respond exclusively to faces and hands, do so across a variety of transformations, including changes in color and size.

7. The neural response pattern of six-week-old infant rhesus is similar to that of adults. Neural patterns of response in adults are, however, unaffected by anesthetic (nitrous oxide), whereas infants exhibit a reduced response.

8. Experience with visual stimuli has a significant effect on the pattern of neural responses.

9. Neural response patterns can be instantiated by altering the subject's attention and the relevance of the task presented.

10. The response properties in IT neurons require input from striate cortex, V2, and V4. Considering this entire visual pathway, IT neurons have a larger re-

ceptive field, have greater response specificity with regard to stimulus features, are more susceptible to changes in attention, and show considerable ipsilateral effects.

11. Neurons in close proximity tend to produce similar patterns of response relative to spatially separate neurons. IT neurons appear to require a greater number of inputs for activation (average = 40) than neurons in striate cortex (average = 2–10).

In summary, research on face cells and the IT cortex has focused on three questions, and results allow for the following general conclusions. First, do face-specific cells exist? For most researchers, results suggest that cells within IT are highly responsive to faces, even when facial stimuli undergo quite extensive transformations (e.g., size, color, orientation). However, because these cells do respond to some nonface visual stimuli, they should be classified as "face-sensitive" cells rather than "face-specific." Moreover, whereas some researchers wish to talk about face-sensitive *cells,* others prefer to characterize the system in terms of *cell ensembles.* Second, are all or even most cells within IT sensitive to faces, and what other areas of the brain are involved in similar processes? Although a great deal of research has been conducted, this question cannot be resolved until more cells have been sampled, using additional manipulations of stimulus features. The discussion that follows provides information on other brain areas involved in face processing. Third, to what extent do face-sensitive cells play a role in behavior (e.g., recognizing and interacting with familiar individuals)? As the next section will reveal, there are cells that fire quite exclusively to particular facial expressions, independent of identity, and other cells that fire to individuals, independently of facial expression. These cells, together with those found in the amygdala, are likely to play a more significant role in communication than cells found within IT.

Thus far we have established that in the rhesus macaque brain there are areas that appear, minimally, to be highly responsive to faces. This ability is not yet communication. To explore the communicative aspects of the face, several research teams (Hasselmo, Rolls, and Baylis 1989; Perrett and Mistlin 1990; Perrett et al. 1984, 1988a, 1988b) have presented rhesus macaques with either still images of facial expressions produced by conspecifics or human subjects posing with different facial configurations. Before facial communication can occur, it is necessary for individuals to recognize that they are looking at each other. There are at least two relevant cues: head orientation and gaze direction. Perrett and colleagues (Harries and Perrett 1991; Perrett and Mistlin 1990; Perrett et al. 1988a), as well as Heywood and colleagues (Campbell et al. 1990; Heywood and Cowey 1992), have tested for neural sensitivity to these cues and have found a number of cells in the superior temporal sulcus that are responsive to head position (Figure 4.59), in addition to gaze direction, independently of head position. The latter result is particularly interesting because ablation of this area results in the failure to discriminate slides on the basis of where an animal is looking.

We have just learned that areas of the rhesus macaque's brain are sensitive to head orientation and gaze direction. Are there areas that are selectively responsive to different facial expressions? In one study, Perrett and colleagues have shown (Figure 4.60) that some cells are highly responsive to particular facial expressions, and the relative level of response is maintained across head orientations (e.g., straight-on versus profile). Similarly, Hasselmo, Rolls, and Baylis (1989) have shown that a small proportion of cells are responsive to particular facial expressions, independent of subject identity, whereas other cells are responsive to subject identity independent of expression; cells responding to expression tend to be located within the superior temporal gyrus, whereas cells responding to identity tend to be located within the inferior temporal gyrus.

In humans, results from brain-damaged patients, neuroimaging studies, and visual half-field experiments suggest that the right hemisphere is dominant with regard to face processing (reviewed in Tovee and Cohen-Tovee 1993). In rhesus, early reports indicated no asymmetry. For example, Overman and Doty (1982) conducted a study involving the presentation of chimeric faces (i.e., within-species chimeras, but no mixed-species chimeras) to humans and rhesus. They found that humans, but not rhesus, identified left-left chimeras as more similar to the original than right-right chimeras, but only when human chimeras were presented. Although the methods employed were sound, the data are partially[52] in conflict with single-unit recording studies (reported earlier) indicating relatively indiscriminate neural responses to nonhuman and human faces. More recent studies provide support for hemispheric asymmetries in face processing. Using split-brain rhesus monkeys, C. R. Hamilton and Vermeire (1988) showed a significant right-hemisphere bias for discriminating facial features and expressions of conspecifics; left-hemisphere biases have been demonstrated for nonface visual forms (see also Hopkins, Washburn, and Rumbaugh 1990). Moreover, though Perrett et al. (1988b) find slightly more face-sensitive cells in the left than in the right hemisphere, they also report that hemispheric asymmetries in processing faces

---

52. The conflict is only partial because the units of analysis differ considerably.

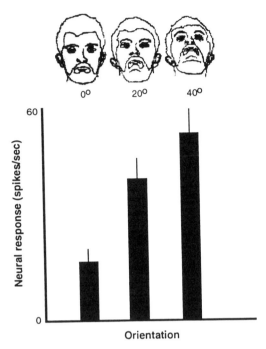

Figure 4.59
Changes in response strength (spike rate) of neurons within the superior temporal sulcus of the rhesus macaque as a function of human head orientation (redrawn from Perrett et al. 1985).

are eliminated by rotating the image to an inverted position. That is, as we will see for studies of human adults, when faces are presented upside down, hemisphere biases in perception tend to disappear. This effect appears to result from the fact that inverted faces are not processed as faces. Additional research must be carried out on nonhuman primates to ascertain which stimulus features are used during visual processing and categorization of faces, whether one hemisphere is dominant independently of stimulus properties, and whether the pattern of hemispheric dominance obtained for rhesus differs from other primates, especially arboreal species.

    To conclude this section, let me end with some of Leslie Brothers' recent work, which pushes the social complexity of presented stimuli to almost dizzying heights. Specifically, Brothers and her colleagues (Brothers, Ring, and Kling 1990; Brothers 1992; Brothers and Ring 1992, 1993) set out to explore whether cells within the stump-tailed macaque brain are specifically sensitive to socially

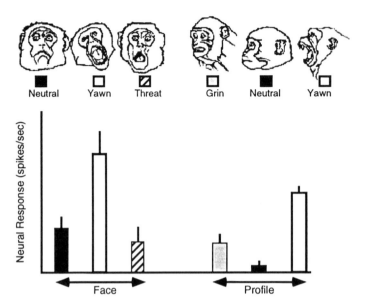

Figure 4.60
Response of neurons within the rhesus macaque superior temporal gyrus to different facial expressions, presented face-on and in profile. For the cells illustrated, there is an increased sensitivity to the threatening "yawn" display, independent of face orientation (redrawn from Perrett and Mistlin 1990).

mediated interactions, including vocal and nonvocal displays associated with different affective states (aggression, fear), and purposeful/goal-directed movement. Rather than presenting stills, Brothers and colleagues used a much more powerful system, storing digitized sequences of naturally occurring social interactions onto a laser disc system, and then retrieving them for subsequent presentations. Once cells are located that respond to particular sequences, the details of the sequence can be edited and analyzed to determine the salient eliciting features. Although the data set was limited to a relatively small sample of cells from only two adult females, several important results emerged. Specifically, cells that were highly responsive to affective displays and apparently goal-directed movement were located within the amygdala and the neighboring anterior temporal region. Second, cells could be classified into five general categories based on the relative selectivity/sensitivity of their response: (1) motion, (2) identity, (3) combined auditory-visual action, (4) eye contact, and (5) complex, heterogeneous featural set, including individuals with open mouths and commonly, mother-infant pairings. Third, the rostral entorhinal cortex provided the richest supply of responsive cells. This area receives significant input from the temporal cortex, thereby suggesting that the final output for responses to social stimuli involves processing of affective valence, individual identity, and motion. Although these studies only skim the surface, they provide an elegant entry into the neurobiology of social cognition, a topic I return to in chapter 8.

Humans   When humans produce facial expressions, a variety of muscles are put into gear, especially when contrasted with nonhuman primates (Table 4.6). There are *superficial muscles* that either constrict or pull on a given area of the face and *deep muscles* that, although generally involved in mastication, are incidentally activated when the jaw is involved in signaling, as occurs when the teeth are clenched shut. The superficial muscles are controlled by the facial nerve (VII), whereas the deep muscles are controlled by the trigeminal nerve (V).[53] Like that of other mammals, the lower part of the human face is under contralateral control. The upper part of the face, however, is under contralateral and ipsilateral control. Laterality of control is most dramatically seen in individuals suffering unilateral damage to the neural pathway connecting the facial nucleus with the motor region of the precentral gyrus of the neocortex. In these patients, the contralateral portion of the lower face is paralyzed. The strength of this facial asymmetry, however, is primarily observed in brain-damaged subjects, with little evidence of asymmetric control in normals (Hager and Ekman 1985).

An intriguing communicative deficit, called mimetic palsy, occurs in subjects with intact motor strips but, apparently, damage to areas often associated with Parkinson's disease (e.g., basal ganglia, thalamus, substantia nigra). In this syndrome, individuals can produce facial expressions on command, but fail to produce appropriate expressions spontaneously in response to jokes, simple conversations, and common greetings. The various kinds of deficits associated with facial expression suggest that what is at issue, from a neurobiological perspective, is the extent to which the expression is spontaneous or overlearned (Fridlund 1994). In this sense, studies of young infants or children might provide some insights, assuming that their expressions are likely to be more spontaneously generated.

In the literature on anuran advertisement calls and avian song that has been reviewed, I pointed out the consistent connection between calling behavior and changes in reproductive physiology. In particular, there are specific pathways

---

53. Interestingly, the facial nerve nuclei receive sensory input from the auditory nerve (VIII), thereby allowing facial expressions and movements to be generated in response to hearing particular sounds.

Table 4.6
The Primary Superficial Muscles Involved in Human Facial Expressions (Fridlund 1994)

Muscle	Location	Innervation*	Action
Scalp			
Frontalis	Forehead	Temporal	Wrinkles forehead and raises eyebrow
Occipitalis	Back of the head	Posterior auricular	Tightens the scalp
Temporoparietalis	Temple	Temporal	Elevates the ear
Ear			
Auricularis anterior	Anterior to ear	Temporal	Pulls auricle forward and up
Auricularis superior	Above ear	Temporal	Pulls auricle up
Auricularis posterior	Behind ear	Posterior auricular	Pulls auricle back
Eye			
Orbicularis oculi	Around the oribt	Temporal and zygomatic	Closes the eye
Corrugator	Deep to the orbicularis oculi	Temporal	Makes furrows between eyebrows
Nose			
Procerus	Bony bridge of nose	Buccal	Depresses eyebrows medially
Nasalis	Cartilaginous bridge and wing of nose	Buccal	Dilates nostrils
Depressor septi	Lateral to philtrum	Buccal	Constricts nostrils
Mouth			
Levator labii superioris	Upper lip	Buccal	Elevates upper lip
Levator labii superioris alaque nasi	Upper lip and side of nose	Buccal	Dilates nostrils and elevates upper lip
Levator anguli oris	Corner of mouth	Buccal	Lifts corner of mouth
Zygomaticus major	Cheek and mouth corner	Buccal	Lifts corner of mouth
Zygomaticus major	Cheek and mouth corner	Buccal	Elevates upper lip
Risorius	Cheek	Buccal	Draws corner of mouth laterally
Depressor labii inferioris	Lower lip	Mandibular	Depresses lower lip
Depressor anguli oris	Corner of mouth	Mandibular	Depresses corner of mouth
Mentalis	Chin	Mandibular	Wrinkles chin and protrudes lower lip
Orbicularis oris	Circumscribes mouth	Buccal	Closes, purses, and protrudes lips
Buccinator	Cheek	Buccal	Compresses cheek
Neck			
Platysma	Neck and chin	Cervical	Depresses mandible, mouth corner, and lower lip

*All of the innervation is from the eighth cranial nerve, except where noted.

connecting auditory, endocrine, and motor control regions, and when vocalizations are produced and perceived, changes in hormone titers arise. Recently Ekman and colleagues have provided intriguing evidence that when humans voluntarily generate facial expressions, this action induces changes in the autonomic nervous system (Ekman, Levenson, and Friesen 1983) as well as the central nervous system (Ekman, Davidson, and Friesen 1990). For example, in one study, subjects were asked to pull their eyebrows down while raising the upper eyelid, tightening the lower eyelid, narrowing and then pressing the lips together—the facial changes that underly the expression of anger. These motor actions, in contrast to those associated with the expression of happiness, were associated with significant increases in heart rate and skin conductance. In other words, simply going through the expressive actions, so to speak, generated quite explicit physiological changes. These kinds of effects have been demonstrated in several cultures and for several types of emotional expression (Ekman 1992; Levenson, Ekman, and Friesen 1990; Levenson et al. 1991).

Most of our understanding of the neurobiology of face processing in humans has been derived from brain-damaged patients (De Renzi, Scotti, and Spinnler 1968; Warrington and James 1967)—in particular, individuals with lesions located in the inferior occipital and temporal visual association cortices (Damasio, Damasio, and Van Hoesen 1982; Damasio, Tranel, and Damasio 1990; Tovee and Cohen-Tovee 1993; see Figure 4.61). Specifically, a number of clinical reports have described cases of individuals who fail to recognize familiar faces, but not familiar nonface visual stimuli. Moreover, many of these subjects were capable of recognizing the voices of familiar individuals, and at a covert level—revealed by galvanic skin response measures—were even capable of recognizing their faces (Tranel and Damasio 1985). Individuals with such deficits are known as prosopagnosics. More recent work using PET and MRI (Haxby et al. 1991; Sergent, Ohta, and MacDonald 1992) has illuminated the anatomy of face recognition in normal subjects, indicating extensive activation of Brodmann's areas 21, 22, 36, and 37. In addition, single-unit recordings and electrostimulation work indicate that areas 21 and 22 are responsive to face identity and facial expression (Ojemann, Ojemann, and Lettich 1992). Last, much of this research has converged on the conclusion that there is a clear dissociation between the neural architecture subserving face and nonface visual object processing (Sergent et al. 1992).

Although prosopagnosics provide an important study group for assessing how the brain recognizes faces, it is also clear, as in any analysis of brain-behavior relationships, that the damaged brain (naturally occurring or experimentally

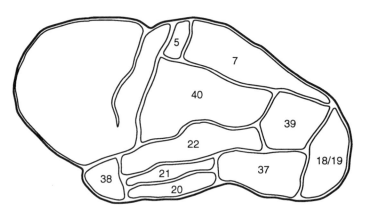

Figure 4.61
Schematic diagram of some of the primary areas associated with prosopagnosia in humans. These areas have been identified on the basis of brain-damaged patients and brain-imaging techniques (redrawn from Damasio, Tranel, and Damasio 1990).

perturbed) may not recognize faces in the same way as the normal brain. To address this problem, Sergent and colleagues (Sergent, Ohta, and MacDonald 1992; Sergent et al. 1992) have looked at face processing in normal subjects using PET. Using the subtraction technique, they found three areas that were specifically involved in face recognition: right medial lingual and fusiform gyri, right parahippocampal gyrus, and anterior cortex of the temporal lobes. Moreover, in four prosopagnosic patients that were studied on a comparable battery of tests, these areas were typically damaged, though other areas were also involved. Sergent and colleagues conclude that face recognition in normals and subjects identified as prosopagnosic involves a large proportion of the right hemisphere, from the occipital pole to the temporal pole. Interestingly, PET failed to show activation of the superior temporal sulcus and the lateral inferotemporal cortex. Recall that in single-unit recording studies of rhesus macaques, these were precisely the areas reported to be involved in face processing. Consequently, comparisons between humans and nonhuman primates should be made cautiously. Equally important, however, data from PET and other imaging techniques should not be taken as the final, definitive statement on areas of activation.

In a recent study by Tranel and Damasio (1993), an experiment was designed to determine whether face recognition in prosopagnosics could be reinstated by means of changing the affective valence of faces. Using a patient (BM) whose clinical history was well documented (bilateral damage to medial temporal lobe, as well as the interconnecting cortices of the anterior temporal and medial frontal areas), and whose inability to recognize familiar faces[54] was severe, the following test situation was established. First, BM was exposed on a daily basis to three nurses. One nurse, the "good guy," never tested BM, did nice things for him, and was very kind. The "bad guy" constantly ran BM in tedious neuropsychological tests, did not do nice things for him, and in fact, often tried to delay such things as meals and breaks in order to run the tests. The "neutral guy" tested BM on a reasonable schedule, but did not go out of his way to do anything special for him. After BM had been exposed to the three nurses, he was asked whether they were familiar, whether he recognized them. His answer was that they were unfamiliar. When skin conductance responses were obtained, however, there was clear evidence of recognition. In other words, familiarity was instantiated by means of increasing the affective valence of faces. When BM was tested one year later, the affective valence of each nurse had completely disappeared.

Most prosopagnosics have no trouble identifying faces as faces, distinguishing human from nonhuman faces, and identifying the expression, gender, and age of the face (Damasio, Damasio, and Tranel 1990; Tranel, Damasio, and Damasio 1988). However, patients with bilateral damage to the occipitotemporal region exhibit significant deficits in recognizing the expression, gender, and age of faces, but not in recognizing the stimuli as faces. Moreover, such patients can produce proper facial expressions, and comprehension of facial expressions is only impaired when the stimuli are presented statically, performing normally when an entire scene can be viewed dynamically (e.g., on a video monitor).

Cases of delusional misidentifications also appear to involve a connection between faces and emotional valence (Young 1992). Thus individuals with Capgras syndrome have delusions that their relatives have been supplanted by impostors. One reason for this deficit may be that they lack the ability to attribute the appropriate emotion to visually familiar stimuli. Thus impairments in facial expression and perception both appear to be intimately related to impairments in emotional processing of information. It is clear, however, that individuals who suffer from a face-recognition deficit may not suffer with regard to facial-expression processing (Parry et al. 1991). More succinctly: face-related deficits are often associated with impairments to anatomical regions of the brain involved in emotion, but there are numerous cases of dissociation between face recognition and processes involved in facial expression.

---

54. Patient BM is also impaired with regard to other nonface objects, events, and names. Concerning faces, he cannot remember the faces of family members, nor photographs of himself. He is severely amnesic as a result of herpes simplex encephalitis.

Neurobiological and neuropsychological research suggests that the right and left hemispheres are engaged in face processing, but that the right hemisphere may be dominant (reviewed in Tovee and Cohen-Tovee 1993). For example, some studies have indicated greater deficits in face recognition as a result of right-hemisphere damage (De Renzi, Scotti, and Spinnler 1968; Warrington and James 1967). Visual half-field presentation experiments reveal that face recognition is faster (Hillger and Koenig 1991) and more accurate (Klein, Moscovitch, and Vigna 1976) when images are presented to the left visual field, and this asymmetry disappears when inverted faces are presented (Carey and Diamond 1977; Leehey et al. 1978).

Turning to the more communicative potential of the face—its ability to convey information about emotional state by means of expressions—some studies have also revealed a strong right-hemisphere bias (Damasio 1994). Specifically, based on video observations of expressive faces, patients with neurological damage, and EMG and EEG recordings, several research teams have observed that the left side of the face starts the expression earlier than the right and tends to be more expressive. For the naked eye, this difference is most convincingly demonstrated by creating visual chimeras of "posed" faces (Levy et al. 1983; Sackheim, Gur, and Saucy 1978), where the right side of the face is paired with a mirror-reversed duplicate, and the same kind of pairing for the left side of the face. In these situations, both chimeras are perceptually different from the original, but the left-left chimera either appears more expressive than the right-right or appears to convey a different emotional expression. This finding suggests that for posed expressions, there is greater right hemisphere activation than left hemisphere activation. Interestingly, when facial expressions are produced spontaneously,[55] the side asymmetry disappears (Ekman, Hager, and Friesen 1981), suggesting bilateral hemispheric control of the expression.

In a more detailed neurobiological study by Gazzaniga and Smiley (1991), asymmetries in posed and spontaneous smiling were examined for three split-brain patients. Results showed that when the request to smile was lateralized to the left hemisphere, the right side started the expression significantly earlier than the left. In contrast, when the request to smile was issued to the right hemisphere, subjects were not able to respond. The authors suggest that the pattern of results

---

55. Note that posed facial expressions are controlled by the cortical pyramidal nervous system, whereas spontaneous expressions are controlled by the extrapyramidal system. It has been suggested that this difference in neural innervation leads to the symmetry of spontaneously produced expressions.

fits with the idea that the corpus callosum is critically involved in coordinating the two hemispheres during the production of voluntarily generated facial expressions; for spontaneous productions, either hemisphere can play a role. Moreover, the data conflict with the general claim that the right hemisphere is generally responsible for facial expression. Although split-brain data are far more convincing than behavioral data with regard to the underlying neurobiology, we should be somewhat cautious in our generalizations because analyses of lower facial movement were restricted to smiling. Recall that in Davidson's model, the left hemisphere is more specifically involved in positive/approach emotions, whereas the right hemisphere is more directly involved in negative/withdrawal. Thus it is possible that split-brains would also show a greater right-hemisphere bias were they requested to frown or show a fear expression. Nonetheless, the cautionary note voiced by Gazzaniga and Smiley should be taken seriously in future investigations: "[Our] data also emphasize the difficulty of making structure-function claims based on seemingly lateralized behavioral phenomena" (p. 242).

In summary, just as we saw earlier for spoken and signed language, the primate brain has evolved dedicated neural machinery both for recognizing faces and for processing facial expressions. At the periphery, we see evidence of considerable tinkering by natural selection: the primate face has undergone extensive remodeling with regard to the musculature controlling facial expression. Whereas the prosimian face is relatively unexpressive, the monkeys and apes tend to exhibit a quite significant range of expressions, culminating in the Marcel Marceau of expressiveness, modern humans. These general patterns are interrupted by striking interspecific differences, which emphasize the importance of considering communicative actions in the context of a species' socioecology. Thus, although the gorilla has access to the same sorts of facial musculature as does the chimpanzee, gorillas have a relatively impoverished repertoire of facial expressions. Before we can explain such species differences, we require far more information on the ecology of facial expressions and, in particular, the conditions favoring a diversity of expressions as opposed to a more limited set, or of signaling more extensively in a different sensory domain.

For humans, studies of brain-damaged patients and normals (using imaging techniques) have revealed areas that are heavily involved in face processing, including the decoding of identity and emotional expression. Although some have argued that there is a face module, current evidence is less convincing than it is for human language. For example, though subjects with damage to the occipito-temporal region commonly suffer from poor face recognition, they may also suffer from a more general visual recognition deficit. Similarly, whereas evidence for a left-hemisphere bias underlying language processing is relatively robust, support for a right-hemisphere bias for face processing is weaker, with several studies indicating bilateral control or activation. Nonetheless, the face appears to hold a special status in the cognitive neurosciences, and several research teams are actively working to uncover the neural pathways involved in recognizing faces and their expressive power.

5 Ontogenetic Design and Communication

Even learning can be innately guided, so that a creature "knows" in advance how to recognize when it should learn something, what object or individual it needs to focus on, which cues should be remembered, how the information ought to be stored, and when the data should be retrieved in the future. (J. L. Gould 1990, 84)

Development depends not only on the materials that have been inherited from parents—that is, the genes and other materials in the sperm and egg—but also on the particular temperature, humidity, nutrition, smells, sights, and sounds (including what we call education) that impinge on the developing organism. Even if I knew the complete molecular specification of every gene in an organism, I could not predict what that organism would be. Of course, the difference between lions and lambs is almost entirely a consequence of the difference in genes between them. But variations among individuals within species are a unique consequence of both genes and the developmental environment in a constant interaction. Moreover, curiously enough, even if I knew the genes of a developing organism and the complete sequence of its environments, I could not specify the organism. (Lewontin 1992, 26)

5.1 Introduction

Some organisms are born with the essential mechanisms for responding appropriately to biologically meaningful stimuli in the environment. For others, appropriate responses emerge gradually over time, shaped in part by maturation and experience, some of which occurs in utero. It is not uncommon for such accounts to generate a set of false conclusions. In the first case, one inappropriate conclusion is that simply because a behavior emerges in the absence of experience (e.g., practice), it can no longer be modified by experience. In the second case, researchers often state, inaccurately, that because a process unfolds gradually, genetic factors are unimportant or perhaps less important than environmental ones. Genetic factors determine the mode of responsiveness to experience, and are precisely the ones that often lead to a gradual unfolding of states during development.

Figure 5.1 illustrates the potential complexities that one must tackle in assessing an organism's ontogeny, including the development of its brain, body, and behavioral repertoire. The importance of the structure laid out in this chapter is that it emphasizes the complexities involved in building an organism or trait and shows that such complexity involves both deterministic and random factors. Recognizing the impact of random effects ("developmental noise") on both genetic and environmental processes serves to buttress, in part, the point made in the statement by Lewontin (1992) that complete knowledge of an individual's genome or environmental surroundings would not guarantee absolute predictive

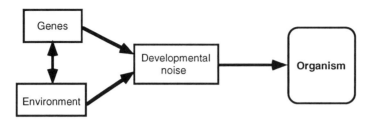

Figure 5.1
Some of the primary factors involved in the development of an organism. Arrows refer to the potential direction of effects.

power with respect to the end product. For most developmentalists, however, absolute predictive power is not the ultimate research goal. In fact, as I will illustrate, it is satisfactory, as well as scientifically interesting, to conduct experiments that tease apart how and when genetic and environmental factors influence the developmental process even if it is not possible to state precisely the relative contribution of these factors, including the various interaction terms. The view I hold in this book, therefore, is that genetic and environmental contributions to communicative abilities can and should be addressed experimentally, and that in terms of learning, we must allow for the possibility of a significant genetic component, or, as Peter Marler (1989) has put it, "the instinct to learn":

How do genetic mechanisms make learning possible? How do they provide appropriate substrates for the pervasive ontogenetic plasticity that so many behaviour patterns display, opening the way for social transmission of behavioural traits, and for the striking variations in the behaviour of individuals, populations and species that result?   (Marler 1991a, 43–44)

This view is not, as some would argue, nativism run wild or fanatic biological determinism. Rather, it is a theoretical position that acknowledges the importance of genetically specified mechanisms that guide learning.

   In thinking about the ontogeny of a structure or system such as communication, one can ask a variety of questions, clearly formulated by Bateson (1991, 21):

1. Does the activity appear at a particular stage in development?

2. Are individual differences in the activity due to genetic differences?

3. Was the activity adapted to its present function by the Darwinian process of evolution?

4. Once present, are the frequency and form of the activity unchanged by learning?

5. Is the activity shared by all members of the species?

6. Does the activity develop without previous opportunities for learning?

7. Does the activity have the characteristics of an organized behavioral system?

These questions will be examined in detail over the course of this chapter, and for some of the case studies, answers to all of the questions can and will be provided. Such completeness is, of course, quite exciting. Incompleteness is also exciting because for many of these issues, the requisite experimental and observational paradigms are already in place, primed for implementation.

An important concept for this chapter is the idea of canalization, formalized by the developmental biologist Waddington (1957) and discussed more recently by Gottlieb (1991a, 1991b; 1992). To illustrate the core of this notion, imagine a ball rolling down a narrow mountain path (Figure 5.2). In one situation, which I describe as strongly canalized, the ball follows a fairly direct course to the end of the path, with relatively few deviations along the way. Though the ball experiences changes in speed, spatial location, and so on, it maintains its trajectory to the designated end point. In contrast, the second situation represents weak canalization. Here, experiences that are encountered cause more substantial deviations and, at one point, either force the ball off its path or force an alternative trajectory and, thus, end point; the ball falling off the path can be likened to an organism that fails to develop and perhaps dies prematurely. Considering real-world organisms, the notion of canalization tell us that during the course of development, individuals will encounter a variety of experiences that have the potential to throw them off of their species-typical trajectory. Both within and between species, there will be considerable variation in the extent to which a given developmental process is strongly or weakly canalized. And, presumably, the degree of canalization will be highly correlated with the survival consequences of deviations from the species-typical trajectory. Importantly, however, deviations from the main path do not necessarily represent development abnormalities, but rather may indicate an ontogenetic alternative. An example of a strongly canalized trait is human language. The literature, reviewed in this section, shows that even quite dramatic perturbations to the mental and sensory abilities (e.g., deafness, blindness) of the individual fail to block language acquisition (Locke 1993; Pinker 1994a). What else could give us more confidence that language is critical for human survival?

Most of the developmental systems that we shall be investigating in this chapter are embedded within a social context. As such, there are opportunities for young to learn from other group members. To set the stage for the kinds of social experiences that can shape the developmental process, we need to be clear about

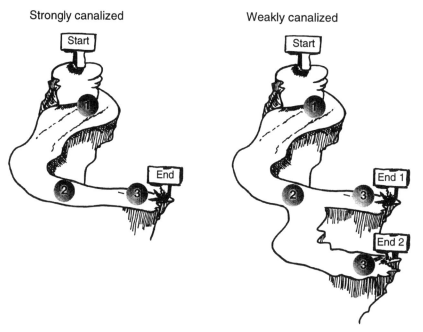

Figure 5.2
A schematic illustration of Waddington's (1957) concept of canalization. The left path illustrates the developmental trajectory (i.e., the rolling ball) of a system that is strongly canalized, showing that there are few deviations from the main path and a specified end point. In contrast, the right canal illustrates a weakly canalized system, where perturbations that arise during development cause significant deviations from the central trajectory. The ball may fly off the path (e.g., a developmental abnormality), or it may reach a different end point, which may or may not represent abnormality.

the mechanisms that underlie socially mediated changes. Operational definitions are provided in Box 5.1, with a more loosely structured summary provided in the following text.

In the case studies discussed, the forum for social encounters ranges from the general dynamics of group life to intensive caretaking interactions between parents with their young, with a few potential instances of teaching. In such situations, young have the opportunity to learn from more experienced members of their species. The methodological challenge is to provide an accurate depiction of the type of experience obtained by the developing individual and, more importantly, to show that particular kinds of experience are causally related to the acquisition of a specific behavior. Under laboratory conditions, of course, one can readily control both the nature and timing of experience, as in Galef's work

Box 5.1
Some of the More Significant Social and Nonsocial Learning Mechanisms

A considerable amount of research on animal intelligence has focused on the problem of learning, including the kinds of problems that individuals can solve on their own, in addition to the kinds of problems that individuals solve as a result of social interactions with group members. Two problems have plagued this area of research: (1) consensus over operational definitions and (2) the necessary and sufficient conditions for demonstrating, unambiguously, that a particular type of social learning has occurred. To facilitate discussion and interpretation of the case studies, I provide the following list of terms used in this line of research, together with a set of definitions that are operationally coherent. The emphasis is on mechanisms that can lead to *similarity* in the expression of a trait; some of these mechanisms involve learning and some do not. I draw extensively from the conceptual frameworks of papers by (1) Galef (1988), (2) Whiten and Ham (1992), and (3) Caro and Hauser (1992). **A** and **B** represent either two populations or two individuals.

Term	Definition	Ref #
I. Similarity via nonsocial processes		
Convergence	Natural selection has caused **B** to resemble **A** (e.g., subsong in birds and babbling in human infants).	1
Common descent	**B** resembles **A** because of evolutionary descent from a common ancestor (e.g., opposable thumb in Primates).	1
Mimicry	Natural selection has selected for a resemblance between members of one population **A** and a second population **B**, thereby allowing one population to exploit the other.	1
Individual learning	**A** and **B** independently acquire the same behavior due to trial-and-error learning, having confronted comparable socioecological problems.	1,2
II. Similarity via social mimetic processes **A. Social influence**		
Contagion or social facilitation	An innate behavior that is released in **A** as a function of **B**'s performance.	1,2
Exposure	As a result of **A**'s association with **B**, both are exposed to comparable environments and thus acquire comparable behaviors.	1,2
Social support	**A** is more likely to learn **B**'s behavior because **B**'s performance induces a comparable motivational state in **A**.	1,2
Matched dependent learning	**A**'s behavior converges on the form of **B**'s behavior as a result of reinforcement contingencies.	1
B. Social learning		
Stimulus/local enhancement	**A**'s attention is drawn to a location or object as a function of **B**'s behavior.	1,2
Observational conditioning	**A** learns from **B** which objects deserve its attention, and the response produced by **A** is contingent upon **B**'s.	1,2
Goal emulation	**A** produces the same end product as **B** following observation of **B**'s act, but the form of **A**'s behavior differs from **B**'s.	1
Imitation	**A** learns some aspect of the form of a behavior performed by **B** that is novel to **A**'s repertoire, and is capable of expressing the behavior in the absence of **B**.	1,2
Teaching	**A** (teacher) modifies its behavior only in the presence of a naive observer **B**, at some cost or at least without obtaining an immediate benefit for itself. **A**'s behavior thereby encourages or provides **B** with experience, or sets an example for **B**. As a result, **B** acquires knowledge or learns a skill earlier in life or more rapidly or efficiently than it might otherwise do, or that it would not learn at all.	3

on the ontogeny of social foraging in rats or Marler's research on the development of passerine song (discussed in section 5.2).

Under natural conditions it will be significantly more difficult both to identify the kinds of experience obtained by the developing individuals and to show that this experience has directly influenced the pattern of behavior observed. As a result, most studies of supposed "cultural transmission" or social learning in the wild have been challenged (Galef 1992), except in the few cases where detailed ontogenetic studies have also been conducted in the laboratory. One of the most exquisite examples of the latter is the study of avian song dialects (M. C. Baker and Cunningham 1985; Balaban 1988b, 1988c; Baptista 1975; Baptista and Morton 1982; Kroodsma et al. 1984; Marle 1970a; Marler and Tamura 1962; Nelson 1992; Payne, Payne, and Doehlert 1988).

To illustrate the kinds of problems that arise in attempting to isolate whether social learning has occurred, and if so, which form of social learning (Box 5.1), consider J. Fisher and Hinde's (1949) classic observations of milk bottle opening in the blue tit. In the late 1940s, observations in the London area suggested that blue tits had solved a major problem: individuals were able to peck through and remove the aluminum foil covering milk bottles and skim off the cream from the surface. Because a large population of individuals had acquired this ability quite rapidly, Fisher and Hinde concluded their study by suggesting that the transmission process must have occurred by means of some sort of observational learning, such as imitation; the argument for imitation was based on the intuition that the fidelity of the copy and the speed of transmission are superior under imitative learning relative to all other social learning mechanisms. To directly test the hypothesis that milk bottle opening spread through imitation, Sherry and Galef (1984, 1990) conducted a series of experiments with black-capped chickadees. One group of individuals, the demonstrators, knew how to remove the foil from milk bottles and skim the cream. Two other groups were naive. One of the naive groups watched demonstrators perform and were then presented with a foil-capped milk bottle. The second group of naive individuals were simply presented with an open milk bottle next to the foil cap. Both groups readily learned to remove the foil from the bottle and skim the cream. These results show that the acquisition of at least this particular behavior *can* emerge in the absence of social interactions and, more specifically, the absence of imitation. Population-level homogeneity in a trait can arise by means of several different mechanisms, including a combination of social learning and trial-and-error acquisition. The main point: we cannot assume that the transmission mechanism is a social one simply because all individuals in a population behave in the same way. This criticism applies directly to the fascinating observations of potato washing and wheat mincing in the Japanese macaque (Nishida 1987) and to the extensive studies of tool use in chimpanzees (for review, see Galef 1992; McGrew 1992; Tomasello 1990).

Armed with the requisite theoretical background, I now turn to the first case study, the ontogeny of avian song. Of the communication studies discussed in this chapter, avian song is *the* success story in developmental research. It shows how thinking hard about the developing organism can lead to a variety of fascinating problems including interspecific differences in brain plasticity and the fitness consequences of acquiring a large or small repertoire.

5.2 Mating Signals: Birds

5.2.1 Avian Song

General

The young males continue practicing, or, as the bird-catchers say, recording, for ten or eleven months. Their first essays show hardly a rudiment of the future song; but as they grow older we can perceive what they are aiming at. . . . Nestlings which have learnt the song of a distinct species, as with the canary-birds educated in the Tyrol, teach and transmit their new song to their offspring. The slight natural differences of song in the same species inhabiting different districts may be appositely compared, as Barrington remarks, "to provincial dialects"; and the songs of allied, though distinct species may be compared with the languages of distinct races of man. (Darwin 1871, 55–56).

In chapter 4, I reviewed evidence for the neuroanatomical substrates underlying birdsong and pointed out that in most oscine species, at least in temperate regions, males sing and females are silent. In this section, we examine the ontogenetic processes leading to the emergence of adult song, including a discussion of behavioral, sensory, and neurobiological factors. As a reminder (see Figure 5.3), though the oscines are a relatively homogeneous group with regard to the relative importance of auditory experience in song acquisition, auditory experience is significantly less important in other avian species, including the closely related suboscines. Further, some oscines acquire and then maintain a single song throughout life, whereas others acquire new songs seasonally, or in the most extreme case, acquire the songs of other species in a seemingly limitless Pavarottian performance. The literature reviewed in this section attempts to provide a synthesis of explanations for why some birds learn their songs and others do not (i.e., what is the functional significance of experientially modifiable song as opposed to song that is fully formed at birth—or soon after—and unmodifiable by experience?), and for those that do, why some species are limited to a single

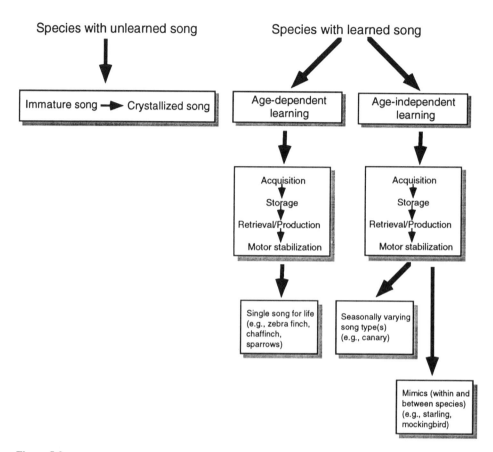

Figure 5.3
A general summary of interspecific variation in avian song development, emphasizing the relative plasticity of the system with regard to changes in structure over time.

song whereas others continue to learn new songs throughout life. At the core of this analysis will be an assessment of how different forms of experience, at different developmental periods, lead to structural changes in the brain that underlie both the production and perception of birdsong. Due to the range of issues discussed, therefore, we will often jump from mechanistic problems (e.g., what is the neural circuitry guiding song learning?) to functional problems (e.g., what is the adaptive significance of large vocal repertoires?).

Some of the earliest hints of vocal learning in the songbirds emerged from observations of geographic variation in song structure, or song dialects. Avian dialects, like human linguistic dialects, are vocal variations that can be observed across geographically distinctive[1] populations of a species (M. C. Baker and Cunningham 1985). Because dialects, by definition, occur within a species, it is presumed that much of the variation in song structure across populations is the result of learning (Lemon 1975).[2]

Research on the white-crowned sparrow represents one of the first (Marler 1970b; Marler and Tamura 1962) and most carefully explored cases of avian dialects (M. C. Baker and Mewaldt 1978; Baptista 1975; Baptista and Morton 1982; Kroodsma et al. 1984; Petrinovich, Patterson, and Baptista 1981). Observations in the San Francisco Bay area revealed significant acoustic variation in song structure (Figure 5.4), with discrete song clusters identifiable by means of differences in note morphology and note ordering within song. Playback experiments revealed that such variation was perceptually meaningful, evidenced by the fact that territorial males responded more aggressively to foreign dialects than to dialects of neighbors. Moreover, genetic studies of this species (M. C. Baker and Mewaldt 1978; Kroodsma et al. 1984) and other closely related oscines (Balaban 1988b, 1988c) suggested that dialects may represent barriers to dispersal. Specifically, when individuals leave their birthplace to either set up a territory or find a mate, they tend to search for breeding areas where individuals sing the dialect they experienced during development. One important consequence of this searching strategy is that dispersal distance is limited, and thus, so is the geographic range of acoustic variation. The proposal that dialects function in this way has yet to be investigated systematically and awaits detailed analyses of the interaction between genotype, dialect, and mating patterns.

Given the significance of dialect variation in oscine communities, Marler and several other ornithologists set out to assess the precise mechanism by which birds learn their song. In particular, research focused on identifying the specificity of the acquisition mechanism (i.e., what kind of input is accepted/acceptable) and the consequences of varying the timing and quality of exposure. The general approach is to capture individuals during the nestling phase of development, hand

1. Like human dialects, geographical distances separating avian dialects need not be great. Baptista (personal communication) recounts an extraordinary case of two white-crowned sparrow dialects, one present at the front door of the California Academy of Sciences, the other at the back door!

2. Absolute confirmation of the significance of learning requires detailed genetic studies of the populations under study. Given the possibility that small migrant populations can separate from larger populations, genetic drift could readily arise and drive at least some of the variation in song structure.

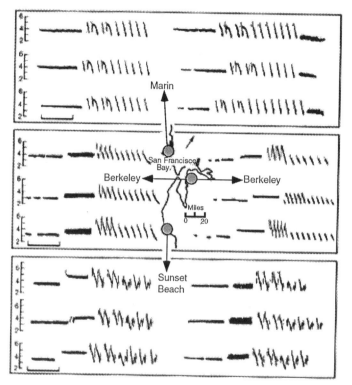

Figure 5.4
Dialects of the white-crowned sparrow from the San Francisco Bay area (redrawn from Marler 1970b).

rear them in acoustic isolation, and then selectively present them with auditory material via tape recorder and loudspeaker (Figure 5.5).

Results from some of the earliest tape tutoring experiments revealed an extremely narrow window of exposure, which was originally called the *critical period*[3] for song acquisition. For example, Marler's results on the white-crowned sparrow indicated that individuals exposed to species-typical song between 10 and 50 days subsequently produced a song whose acoustic features closely matched those presented during tape tutoring. Exposure outside of this period led to the development of an abnormal song. In addition to an experiential window

3. The notion of a *sensitive period* has been substituted for the critical period. This semantic change resulted from the fact that few developmental processes have rigid cutoff periods, beyond which experience plays no role.

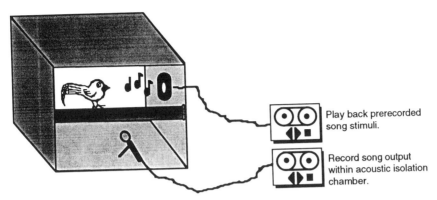

Figure 5.5
A typical tape tutoring setup for quantifying avian song ontogeny. Subject sits perched in an acoustic isolation chamber and is presented with song material from a mounted loudspeaker. All song productions are then recorded onto audiotape for subsequent comparison with stimuli presented earlier.

of opportunity, these experiments, along with others conducted at about the same time on other closely related oscines, uncovered several other factors relevant to song ontogeny:

1. The sensory acquisition phase is generally distinct from the motor production phase. Consequently, when birds begin to sing in their first potential mating season, they must access stored representations that were put in place 100 or more days before. This type of acquisition process has been called memory-based learning (Marler 1991c).

2. Following the sensory acquisition phase, and prior to song crystallization in adulthood, individuals experience a phase known as subsong. Subsong, like babbling in human infants, consists of relatively amorphous song elements whose features bear no relationship to those presented during taped tutoring. These only appear with the onset of the second stage, plastic song.

3. The developmental phase immediately preceding song crystallization—plastic song—is characterized by an overproduction of song syllables; many of these syllables have been heard during tutoring, but some represent improvisations on a theme. Gradually, most of these syllables are discarded as the individual emerges with its crystallized song or songs. Overproduction provides strong evidence that much more acoustic material is memorized than is evidenced by the structure of crystallized song.

4. The developing avian brain is *not* an acoustic sponge that accepts all song material for use in subsequent production. Rather, there appears to be a blueprint (Thorpe 1959) that guides young birds to select the conspecific song from among different songs in the environment. The innate template, in contrast, appears to provide a reference for the feedback control of song development in the absence of an external model (Konishi 1965a, 1965b, 1985).

5. In species where mimicry of heterospecific song is uncommon (e.g., white-crowned sparrows), the auditory template is rigid. Acceptable acoustic input is restricted to conspecific song, except under special circumstances.

6. Several changes in the species-typical auditory environment can lead to the production of abnormal song: (a) withholding all song input (i.e., complete acoustic isolation) prior to and during the sensitive period; (b) deafening during any phase of song development, and in particular during the transition between subsong and song crystallization when individuals appear to require auditory feedback to fine-tune the specific features of their song; (c) tape or live tutoring with heterospecific song, especially when conspecific song is withheld.

With these general findings presented, let us now look at some specific problems in song development and a few exemplary case studies. In particular, I will review recent behavioral and neurobiological studies and show how some of the results force important emendations to the earlier account of song ontogeny.

Instincts to Sing and the Comparative Method   Imagine that the young songbird is one of Waddington's balls rolling down a canal. The problem we are interested in addressing in this section is the extent to which the environment can displace the ball from its course, leading to different developmental trajectories and therefore, in the case of oscines, different song morphologies. The most powerful approach to this problem is a comparative one, involving detailed analyses of closely related species. In this light, studies by Marler and colleagues on song and swamp sparrows are illuminating and will be discussed in detail.

Song and swamp sparrows are closely related species, living in comparable habitats and exhibiting identical mating patterns. They differ, however, in song structure (Figure 5.6), particularly repertoire size, number of note types (phonological variation), and sequential ordering of notes within song (syntax).[4] For

---

4. As used here, and in the birdsong literature, *syntax* refers to a rule-based system for ordering notes or syllables within a song. The temporal organization of elements within song is not meaningful, in the sense that word order within a sentence *is* meaningful. Rather, temporal organization encodes information about species, dialect, and individual identity.

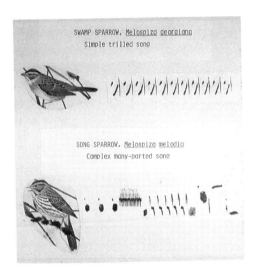

example, whereas swamp sparrow songs are constructed out of a set of six notes, song sparrow song consists of a syntactically complex arrangement of alternating notes and trills (Balaban 1988a; Clark, Marler, and Beeman 1987; Marler and Pickert 1984). These acoustic differences are sufficient for species recognition. Marler and colleagues set out to understand whether there are preferences to learn the wild-type song, at what age such preferences emerge, and what kinds of experience are necessary for producing a species-typical song.

Given natural variation in song structure between species, Marler and colleagues attempted to tease apart how rearing condition affects song ontogeny. When individuals were raised in isolation but tape tutored with conspecific song, interspecific differences obtained under natural conditions were maintained (Figure 5.6). More strikingly, interspecific differences in song structure were even maintained when individuals were deafened early on[5] or raised in acoustic isolation (Figures 5.6 and 5.7). For example, like wild song sparrows, those raised in acoustic isolation also have more notes and trills per song than do swamp sparrows. Similarly, swamp sparrows reared in the wild or in acoustic isolation have larger repertoires than do song sparrows (Figure 5.7). Among deafened individuals, though one can find songs from each species that are indistinguishable, some evidence of species distinctiveness is nonetheless maintained. These results show

5. The issue here is only partially resolved because studies have yet to be conducted involving deafening of individuals at the same physiological, developmental, or chronological age.

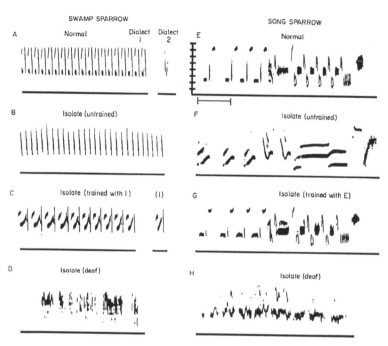

Figure 5.6
The effects of rearing condition on the songs of swamp and song sparrows. In the wild (*A, E*), song sparrows produce a significantly more complex song than swamp sparrows. These species differences are generally maintained when individuals are raised in acoustic isolation (*B, F*), in isolation but tutored with conspecific song (*C, G*), and when raised in isolation but deafened (*D, H*). However, as pointed out by Marler and Sherman (1985), there is a significant loss of specificity as one progresses from the natural song to an isolate song and last, an early deafened song. In panel *E*, frequency markers (1 kHz intervals) and time scale (0.5 sec interval) are indicated (redrawn from Marler and Nelson 1992).

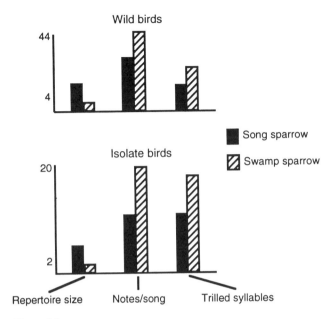

Figure 5.7
Differences in song structure between song sparrows (black bars) and swamp sparrows (striped bars) under natural conditions (wild birds), and raised in isolation (isolate birds) (redrawn from Marler and Sherman 1985).

that in the absence of song input—either from conspecifics or heterospecifics—species differences in song morphology are preserved (Marler and Sherman 1985). As one drifts further and further away from the species-typical acoustic environment, however, there is a significant loss of species-specific song structure.

Though there are important similarities in the structure of isolate song for both species of sparrow, there are also important differences. For example, whereas the number of trills per song in isolate and wild song was similar for song sparrows, the swamp sparrow's isolate song consisted of significantly more trills than did their normal song. In addition, when tutored with synthetic two-part songs, swamp sparrows, but not song sparrows, showed an increase in the number of such songs during crystallization. Overall, then, the stucture of swamp sparrow song is more susceptible to changes in the acoustic environment than is the song sparrow's song.

How an individual selects a song model is probabilistic and depends on such factors as the pattern of auditory exposure, age, and social interactions. To investigate the process of song selection, Marler and Peters (1989) conducted standard tape tutoring experiments, using a mixture of song material from swamp and song sparrows. Results showed that each species exhibited a preference to acquire its own species-typical song over the equally presented heterospecific song. For example, even though swamp sparrows were exposed to the same number of swamp and song sparrow songs during ontogeny, they consistently produced swamp sparrow song in the crystallization phase. Song sparrows also showed a preference for conspecific song, but produced some swamp sparrow song in adulthood, again emphasizing comparative differences in the plasticity or rigidity of the auditory template.[6] In addition, whereas the key unit for learning in swamp sparrows is the syllable, in song sparrows both syllabic structure and syntactic structure are important. The latter is clearly demonstrated in Figure 5.8 by the relationship between syntactic structure of tape tutor material, age at exposure, and syntax of crystallized song: the syntax of the song material heard early on in the sensitive period is more likely to be reproduced in adult song than material heard later on.

There are at least two reasons why the demonstrated preferences for conspecific song should not be taken as evidence that individuals—either song or swamp sparrows—lack the capability to accept heterospecific song. First, the auditory system of most oscine species is similar (Dooling 1980, 1982, 1992; see chapter 4). Consequently, song and swamp sparrows can readily discriminate the significant acoustic features of each other's songs (Okanoya and Dooling 1988). Second, if conspecific song is withheld during the sensitive period for acquisition, and heterospecific song presented instead, heterospecific song will be learned. For example, when song sparrows are exposed to swamp sparrow song, but not song sparrow song, they will learn swamp sparrow song. Interestingly, such heterospecific reproductions are often cast in the syntax of song sparrow song. These results suggest that the processes guiding song selection are, in part, similar to the processes guiding imprinting (see Box 5.2): innate biases guide the young individual in the direction of species-specific stimuli, but in the absence of such material, preferences toward other stimuli can be instantiated.

Thus far we have seen that there are significant differences in the structure of swamp sparrow and song sparrow songs, and that such species-specific differences are maintained under a variety of conditions. A point raised earlier was

---

6. In nature, swamp and song sparrows are often sympatric, living in geographic proximity. However, recordings indicate that they never produce heterospecific song. Thus, in contrast to the wild, laboratory conditions appear to create an opportunity to learn, and subsequently reproduce, heterospecific song material.

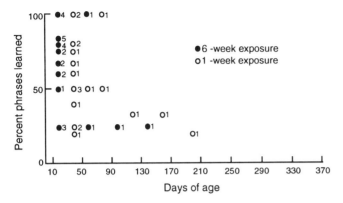

Figure 5.8
The relationship between early and late exposure to song on song sparrow reproduction of the syntax experienced. Songs acquired during the early phases of the sensitive period are more likely to be reproduced with the original phrasing of the model. Those acquired later in development are more likely to be different, including recombinations of specific phrases (redrawn from Marler 1987).

that songbirds appear to require auditory feedback to monitor the structure of the song produced—early deafening often wipes out species-typical phonology. If genetic differences contribute significantly to species-specificity in song morphology, then one might expect such specificity to be retained even after the young bird has been deafened. When swamp and song sparrows are deafened, their songs are extremely variable and amorphous. This observation parallels those obained for such species as the zebra finch, chaffinch, and white-crowned sparrow (Nottebohm 1968; Price 1979). However, amid such variability, quantitative analysis of swamp and song sparrow song also revealed that following deafening, several species-typical features were preserved. These observations are comparable to those obtained for the Oregon junco, black-headed grosbeak, and canary (Konishi 1964, 1965a; Marler and Waser 1977). Deafening experiments therefore reveal that in the absence of auditory feedback, some species produce completely amorphous song material whereas other species produce songs that are acoustically variable but consist of conspecific details. The latter provides strong evidence for a genetically based central motor program for song output.

Memory: Song Storage and Access   What we have observed thus far is that for the young oscine, acoustic input is perceived, stored, and used later in development to reproduce the species-typical song. The question we must now address is how the avian brain stores communicatively salient information for subsequent usage. An excellent example of how the timing and quality of exposure affects

Box 5.2
Avian Imprinting and Song Learning: Similar Developmental Mechanisms?

One of Konrad Lorenz's (1937) most significant contributions to ethology was his discovery of the processes guiding avian imprinting. At the time, Lorenz claimed that imprinting represented a fundamentally different learning process because it was irreversible—once the young individual had imprinted on an object, introducing a new object failed to elicit the typical following response. In this sense, imprinting is like song learning in close-ended or age-dependent learners: once song crystallizes, subsequent exposure to other song types fails to result in structural modifications in song output. These are quite general claims. Although imprinting is not strictly communication (i.e., the to-be-imprinted object [e.g., hen] is not signaling to the young chick "follow me"), the imprinting process represents the first phase in attending to what is ultimately[7] the relevant communicative source of information (conspecific/mate). Let us look at some of the specifics so that we can properly evaluate the ontogenetic mechanisms underlying song learning and imprinting (this literature is reviewed in Bateson 1991; Bolhuis 1991; M. H. Johnson and Morton 1991; ten Cate 1994).

In a recent analysis of the literature, Bateson (1991) proposed a three-stage system for the imprinting process. In the first stage, the chick must *analyze* the object as a potential imprinting/following target. The second stage involves *recognition* of the object, including its salient features. The third stage *executes* or controls the following response by means of a representation of the familiar object. Although the terminology differs, these three stages (or perhaps more conservatively, "processes") are also involved in song learning. The young bird must analyze the properties of sounds in the environment, distinguishing between song and other noise in the environment. Moreover, and as appears to be the case for imprinting, there is a predisposition ("template") to learn (recognize) particular songs, though the rigidity of this predisposition varies considerably between species. Once song is recognized, it is represented in the brain and serves as a template for motor output and perceptual matching. As Marler (1991b) puts it, "When conspecific stimuli are presented, it is as though the bird suddenly becomes attentive, and a brief time window is opened during which the stimulus cluster in view becomes more salient, more likely to be memorized, and probably destined to be used later for guiding song development." And like neurobiological studies of birdsong, there appear to be dedicated neural systems underlying imprinting (reviewed in Horn 1991).

These similarities should not, however, override some of the significant differences between song learning and imprinting. For example, the nature and timing of experience will, under most circumstances, be quite different: exposure to song may occur prior to birth (i.e., in the egg) whereas exposure to potential stimuli for imprinting will not occur until after birth and until the visual system is capable of resolving the relevant stimulus features;[8] though no one has yet provided evidence of song learning in the egg or nestling stage, acoustic experience clearly occurs, and its effects on song development require more formal analysis. Whereas song learning requires specific motor output based on a stored representation, imprinting does not. Last, although the imprinting process can certainly be reversed—depending upon the timing and nature of the stimulus input—it appears significantly less plastic than many song learning systems.

7. Under natural conditions, chicks will imprint upon their mothers, who will, in turn, communicate with them about biologically salient events in the environment.

8. These differences in the timing of sensory exposure need to be taken into account in studies involving prenatal sensory manipulations. Thus, for example, Lickliter and colleagues (Lickliter 1990; Lickliter and Hellwell 1992; Lickliter and Virkar 1989) provided young precocial birds (e.g., bobwhite quail) with early (prior to hatching) auditory exposure as well as early visual exposure. Results showed that experimentally induced visual stimulation before hatching interfered with subsequent responsiveness to the maternal call, while enhanced prenatal auditory experience facilitated species-specific visual responsiveness. Whereas prenatal auditory exposure clearly occurs, prenatal visual exposure does not, except under fairly atypical conditions (e.g., the shell is punctured).

not only what is stored, but also how it is stored and then accessed for subsequent reproduction, is Hultsch and Todt's studies of nightingales (reviewed in Hultsch 1993). In contrast to the oscines discussed in the previous sections, where males have one or a few song types within their repertoires, nightingales have an immense repertoire of song types, and this truly pushes the computational problems of storage and retrieval. Why, from a functional perspective, nightingales have evolved such virtuosity, remains somewhat of a mystery. I do, however, return to this problem, and some speculations, in chapter 6.

Male nightingales have repertoires of approximately 200 song types and often switch between song types during singing bouts. Rules for the sequence of song types within a bout appear to be probabilistic, with the order of songs produced exhibiting a nonrandom pattern, derived from a hierarchical organization. Evidence for such organization is based on tutoring experiments. Specifically, individuals acquired "packages" of song types, defined as a serial cluster of acoustically discrete songs whose order and typology could be directly traced to the tutoring regime. In addition, neither the frequency with which song packages were repeatedly presented during tutoring, nor the length of the song type string within a package, had a significant effect on the fidelity or structure of the individual's acquisition[9] (Hultsch and Todt 1988, 1989). However, in a tutoring experiment where one group was presented with species-typical song bouts (i.e., *versatile:* no song type repeated within a bout) and species-atypical song bouts (i.e., *repetitive:* identical song type repeated within a bout), and a second group was only presented with the species-typical bouts, results showed that song types from the repetitive segment were produced more often than those acquired from the versatile segment (Hultsch 1991). Last, when individuals were tutored with songs produced at either a *normal* rate of delivery (i.e., 4 sec between successive songs), *dense* (i.e., 1 sec between successive songs), or *spaced* (i.e., 10 sec between successive songs), subsequent performances indicated that individuals were under capacity and time constraints. Specifically, song packages were not significantly larger for individuals tutored on the densely organized than on the normally organized sequence (capacity effect). And, song packages from individuals tutored on the spaced regime were significantly smaller than those recorded from individuals tutored on the normal regime (time effect). The latter results suggest that as soon as the memory system begins registering incoming sound,

---

9. Of further note, nightingales can reproduce precise copies of the song heard and in the serial order they were tutored with, after as few as twenty repetitions of the string; this is comparable to some of the best performances by sparrows and European blackbirds.

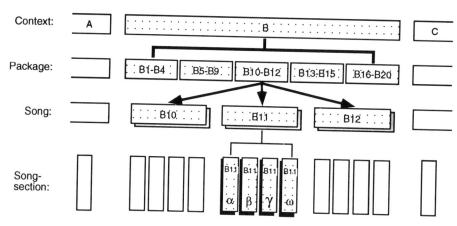

Figure 5.9
Informational components and memory representations of the nightingale's song system. Components were derived on the basis of selective tutoring regimes during ontogeny and subsequent song performance in adulthood (redrawn from Hultsch 1993).

there is a time window that parses the stream of information, and this allows for the segmentation of song types into song packages. The general structure of the nightingale's song system and memory storage is illustrated in Figure 5.9.

The work reviewed thus far suggests that for most oscine species studied, song ontogeny is generally characterized by an initial sensory acquisition phase followed, often after a relatively long period of time, by a motor output phase that *refers* to the stored representation. This developmental scenario is, in many ways, ecologically imposed, for the young bird hears the relevant song material while he is in the nest, or nearby, but then must wait until the following spring to use the material for acquiring a territory and attracting a mate. In species where acquisition is restricted to a sensitive period, the material acquired enters a storage template and forms the basis for subsequent reproductions. It appears, based on deafening studies (Bottjer and Arnold 1984; Konishi 1965b; Nottebohm 1968; Price 1979; Todt and Hultsch 1982), that song production in adulthood is possible in the absence of auditory feedback once the motor pattern has crystallized, or come close to it. This finding suggests that a central motor program may guide song output. In contrast to species where song acquisition is restricted to a sensitive period, more open-ended learners such as the canaries appear to incur significant motor output costs when deafened. Specifically, deafening, which causes drastic song deficits in infancy (Marler and Waser 1977), also causes substantial deterioration of song structure in adulthood. Thus feedback is critically involved in guiding the motor output.

Although the differential consequences of deafening on closed- versus open-ended learners make intuitive sense, most of the studies discussed here examined the short-term consequences of deafening on song production, following sensory acquisition. Some exceptions include Konishi's (1965a) study of an adult white-crowned sparrow who, 16 months after deafening, showed only minor changes in song morphology, with similar results obtained by Price (1979) for a three-year-old zebra finch. To address this general empirical gap, Nordeen and Nordeen (1992) replicated the early experiments of Price (1979) on the zebra finch, but recorded song output for 16 weeks postoperatively. In contrast to the lack of song change observed by Price within the first few weeks, Nordeen and Nordeen reported quite significant changes, including the insertion of novel syllables and marked deterioration in the actual structure of maintained syllables. These changes are summarized in Figure 5.10.

The importance of these findings is that they challenge the often-held assumption that song, once acquired, is firmly stored in the bird's brain, allowing the individual to sing on automatic pilot, in the absence of auditory feedback (Figure 5.11). Although birds clearly access stored motor commands and sensory

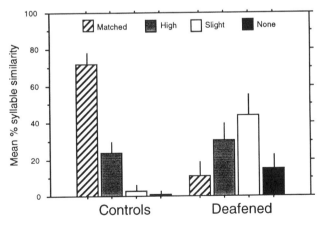

Figure 5.10
The effects of deafening on the structure of zebra finch song. The *y*-axis plots the mean proportion of syllables that, when compared across consecutive renditions in controls or between pre- and postdeafening in experimental subjects, were rated to be matched (striped bars), highly similar (stippled bars), slightly similar (white bars), or not similar/novel (black bars) (data extracted from Table 2 in Nordeen and Nordeen 1992).

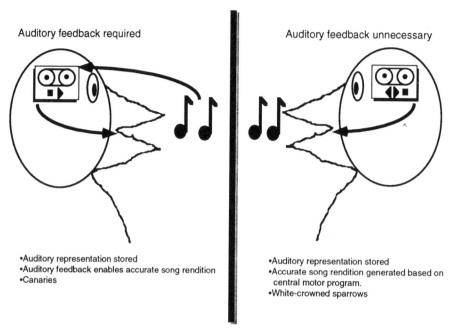

Figure 5.11
A schematic characterization of the importance of auditory feedback on the maintenance of song output in adulthood. The drawing on the right may characterize only a small subset of species, since even some close-ended learners that only acquire a single song (e.g., zebra finch) appear to require feedback for long-term upkeep of song structure.

representations, they also appear to require feedback from the output, feedback that is used to fine-tune the structure of their song. It seems, therefore, that this fine-tuning occurs in species that only sing one song and in those that either sing multiple songs at once or change songs seasonally. These general issues are captured in Figure 5.11 and will require more formal testing to confirm.

Neurobiological Changes during Song Ontogeny   In chapter 4, I provided an overview of how the production and perception of song in adult birds is orchestrated by a set of dedicated neural pathways and endocrinological changes. Two observations are of direct relevance to the following discussion. First, we saw that in most adult songbirds, changes in testosterone levels are causally related to seasonal changes in singing behavior. As the mating season approaches, testosterone levels rise, as do rates of singing. In addition, castration tends to eliminate singing in males, and implants of testosterone can reinstate singing in castrates, increase

levels of singing in intact males, and induce singing in typically silent females. Second, for open-ended learners such as canaries, it has been hypothesized that seasonal changes in song structure are associated with the birth and death of neurons.[10] This finding is striking because it provides evidence of neuronal plasticity in the adult brain. What has yet to be documented empirically is that such plasticity is causally related to song learning. This kind of evidence is critical, especially since cell birth occurs in the brains of nonpasserine adults where vocal learning is absent.

Given these observations, we now turn to a discussion of how the young oscine brain changes during sensory acquisition and sensorimotor learning (reviewed in Bottjer and Johnson 1992; Nordeen and Nordeen 1990). An important point to keep in mind here is that learning commonly occurs in songbirds in the absence of a vocal model. Specifically, as song production is initiated, vocal output is adjusted with reference to the innate template. In contrast, when acoustic models are presented and copied or imitated, then a different form of learning has occurred. Both types of learning are likely to involve changes in brain state and structure.

When the newly hatched songbird begins to perceive, process, and store acoustic input, the brain must undergo changes. Later on in development, the individual

---

10. Recall from chapter 4 that, as a result of Gahr's (1990) findings, we should treat such evidence cautiously because different staining techniques may reveal different attributes of the nervous system, critically related to how we assess such things as brain volume and changes in neural circuitry.

refers to this stored information during the process of song production. Once again, the brain must undergo changes in order to guide song output. Studies of the neurobiology of song development have been conducted on a number of species, including canaries; swamp, song, and white-crowned sparrows; and zebra finches. As illustrated in Figure 5.12, observations indicate that several nuclei are directly involved in the process of sensory acquisition and sensorimotor learning. Based on neuroanatomy and lesioning work, the primary circuit for song acquisition appears to involve area X, the dorsolateral nucleus of the anterior thalamus (DLM), and the lateral magnocellular nucleus of the anterior neurostriatum (I-MAN). To explore the specifics of this circuit and its developmental formation, I discuss in detail work on the zebra finch song system (Bottjer 1991). Recall that zebra finches are age-dependent learners, and thus many of the neurobiological patterns and processes observed may be restricted to this class of songbird. I point out significant interspecific differences when they are relevant to the discussion.

The sensitive period for song development in zebra finches (Bohner 1990; Eales 1985; Immelman 1969) is 20–40 days of age posthatching (Figure 5.13). During

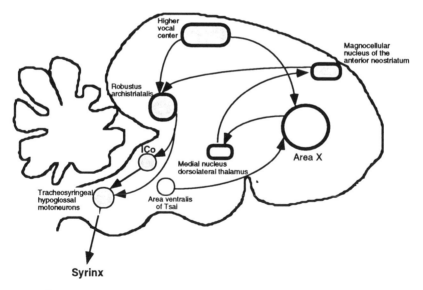

Figure 5.12
Neural circuitry involved in song production and perception, emphasizing in particular the pathways involved in song ontogeny. Nuclei with sex hormone receptors are stippled. Nuclei illustrated with a thick line are those believed to play a critical role in song acquisition. *ICo:* Intercollicular nucleus of the tectum.

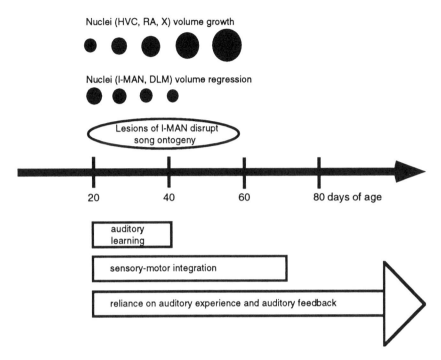

Figure 5.13
Changes in neural circuitry during song acquisition in the zebra finch. The solid circles in the upper portion of the figure are only schematic and do not represent specific changes in volume. Volume growth in HVc and area X is primarily due to the integration of new neurons, whereas the growth in RA is due to an increase in spacing between neurons. Volume decrease in l-MAN is primarily due to normal cell death (redrawn from Bottjer 1991).

this period, individuals show a preference for conspecific song, especially for the father's song. As a result, the structure of adult song, produced at approximately 90–100 days, is similar to the father's song. Zebra finches that have been cross-fostered to the closely related Bengalese finch will, however, produce Bengalese song in adulthood (Immelman 1969), emphasizing the relative plasticity of the auditory template. In contrast to several other songbirds, the sensory acquisition and sensorimotor learning phases overlap in the zebra finch, with early song production occurring at around 25 days.

We saw earlier that auditory experience is critical for proper song development. The gonadal hormones are also critical. If a male zebra finch is castrated, or if the effects of sex hormones are blocked pharmacologically, abnormal vocal behavior emerges during the course of development. For example, blocking hormonal action early on and then administering hormonal therapy in adulthood can result in extremely delayed emergence of stereotyped song. Conversely, administering abnormally high levels of testosterone during the sensitive period results in stereotyped song production, but the song lacks species-typical syllabic and phrasing structure. In swamp sparrows and song sparrows, castrated males with no circulating testosterone show song learning and normal progression through the early phases of song development. Testosterone does, however, appear to play a significant role in song crystallization: individuals lacking sufficient circulating levels of testosterone never produce crystallized song (Marler, Peters, et al. 1988). Interestingly, significant levels of estradiol were measured in castrated males, and in a parallel study by Marler, Peters, and Wingfield (1987), results suggested that males with high levels of estradiol learned song more effectively than males with low levels of estradiol. Work by Schlinger and Arnold (Schlinger 1994; Schlinger and Arnold 1991, 1992) on zebra finches indicates that the estrogen present in male plasma is synthesized in the telencephalon. Together, then, results from zebra finches and sparrows suggest that the development of a species-typical song is influenced by the quantity and timing of release of gonadal hormones. Moreover, as suggested by Bottjer (Bottjer 1991; Korsia and Bottjer 1991), the requisite plasticity for song ontogeny (e.g., which song models are accepted and when) appears to depend on the maintenance of low steroid levels.

A number of nuclei within the song control system show significant androgen concentrations. Given the consequences of changes in the endocrine environment for song development, several research teams have explored the functional and morphological changes that arise at the neural level. Most of the nuclei that have been examined experience significant changes in volume[11] over the course of development (Figure 5.13). Specifically, HVc, RA, and area X show an increase in volume, whereas l-MAN exhibits a decrease. Comparable changes occur in the swamp sparrow, and these correspond to the period of sensory acquisition (Figure 5.14). As pointed out previously, l-MAN, DLM, and area X appear to be crucial neural foci for song development in the zebra finch. Lesioning l-MAN, area X, or the axonal projections from DLM to l-MAN during the sensitive phase of song development results in abnormal song output. In contrast, when these regions are lesioned in adult birds, there are no detectable differences in singing behavior or song structure. These results, together with documented changes in neural composition (Figure 5.14), indicate that the area X–DLM–l-MAN circuit

---

11. Changes in neuron density could be the result of one of three factors (or a combination of all three): changes in the rate of neurogenesis, cell death, or differential migration.

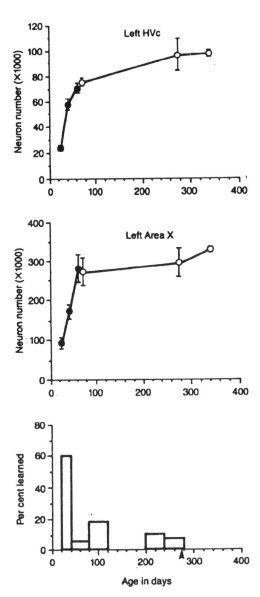

Figure 5.14
Neuronal changes associated with song ontogeny in the swamp sparrow. The top and middle panels show temporal changes in the number of neurons (mean plus-or-minus standard error of the mean) for HVc and area X. The bottom panel shows the results of tape tutoring experiments, indicating the relationship between percentage of song material learned and age. The most significant changes in neuronal number and song material acquired occur up to 120 days posthatching. When song production commences at approximately 275 days (indicated by an arrow in the bottom panel), there is an asymptote in the number of neurons added to HVc and area X (redrawn from Nordeen, Marler, and Nordeen 1989).

undergoes functional changes during development. Though the precise function of this circuit in the adult bird is unclear, some recent work suggests that area X may play a significant role in the adult bird's perception of song structure.

I have repeatedly stated that the particular pattern of acoustic experience obtained during the sensitive period of vocal development has a direct effect on the structure of the song produced in adulthood. I have also stated, based largely on the studies by Marler and colleagues, that the young oscine enters the world with a strong instinct to sing. What we have not yet addressed is the possibility that much of the structure of conspecific song is preencoded in the brain, and that various processes operate on this structure to select a specific song type. As Marler and Nelson (1992) hypothesize, "The songbird brain has extensive fore-knowledge about song structure, and the song is learned, not by instruction, but by selection operating on preexisting circuitry" (p. 415). This "neuroselection" hypothesis (for a general account of this perspective, especially as it applies to issues in the neurosciences, see Changeux, Heidman, and Patte 1984) has thus been proposed as an alternative to two other models of song learning, and the key components of these models are summarized in the following list (Marler, in prep):

1. *Sensorimotor or memory-based learning:* The first phase of the sensorimotor process involves memorization of song elements by means of instruction (live or tape tutoring). The second phase consists of storage song elements. The final phase involves retrieval and then reproduction of songs that have been memorized by instruction.

2. *Action-based learning:* The action-based learning model is actually a combination of the first and third. In phase one, there is memorization of song elements by means of instruction, followed by phase two, where there is storage of song elements. Phase three involves overproduction and then selective attrition of memorized and invented songs.

3. *Neuroselective learning:* The first phase of neuroselective learning involves memorization of song elements by means of selective attrition of predetermined or innate neural circuits. The elements that have *survived* the attrition process are then stored in phase two. Phase three is identical to that described in action-based learning, specifically, overproduction and selective attrition of memorized and invented songs.

Recall that swamp sparrows tend to produce a phonologically and syntactically simple song, especially in contrast with the song sparrow. Figure 5.15 illustrates

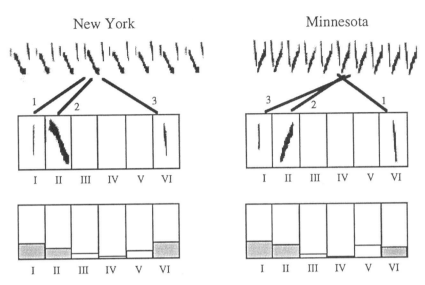

Figure 5.15
The six basic note types (I–VI) of the swamp sparrow song. Though the proportion of note types per song (histograms in lower panels) is consistent across populations (New York, Minnesota), there is geographic variation in the ordering of notes within songs (syntax). Note types without values are zeros (redrawn from Marler and Nelson 1992).

the six note types recorded for swamp sparrow song and shows the distribution of note types per song for two dialects, one from New York and the other from Minnesota. Swamp sparrows, like many other songbirds, overproduce song variants during ontogeny. Various processes may then function to weed out all but one of the variants, leading to the individual's song signature. An instructional view of song development would claim that overproduction of song types is due to the acoustics of the learning environment: what the bird produces prior to song crystallization reflects what he heard during the sensory acquisition phase. In contrast, a neuroselection hypothesis would claim that all the song variants were already in place, built into the neural substrate subserving song production. Though there is considerable within-species variability in song structure, Marler and colleagues have argued that for several species, including the swamp sparrow, there are an impressive number of acoustic universals. For example, as mentioned previously, swamp and song sparrows maintain a number of species-typical song features even when the normal auditory environment is perturbed during development. In the indigo bunting, though considerable individual and population-level variability has been observed, the song repertoire seems to be consistently constructed out of approximately 100 different syllables (Margoliash, Staicer, and Inoue 1991; Payne, Payne, and Doehlert 1988).

The neuroselection hypothesis for song learning is a relatively recent proposal, and much work remains before it can be properly pitted against the instructional hypothesis (see data from Nelson and Marler 1994, presented in the next section). Nonetheless, it is clear that in many species, song repertoires are not limitless with regard to acoustic variation, but rather appear to be constrained to a set of acoustic features—what Marler at least would like to consider to be universal species-specific features. As discussed in the following section, countersinging with the same song type as a neighbor or intruder may represent one mechanism by which song variants are fine-tuned, favoring some featural structures over others. Why, from a functional perspective, birds countersing, remains somewhat of a puzzle, though some have speculated that it provides the most potent stimulus in territorial interactions.

In summary, song development is accompanied by a number of significant changes in steroid levels, as well as changes in the structure and function of relevant neural circuitry. Moreover, the timing of such changes is fundamental, suggesting that the songbird brain may attain and ultimately surpass an optimal period for learning. Two questions require further attention. First, what is the extent to which sensitive periods for song development can be altered by different hormonal titrations, selective acoustic regimes, and altered neural pathways? Second, what are the specific effects on song structure of changes in the neural, endocrine, and social environments?

Social Effects on Song Ontogeny The standard technique for studies of song ontogeny is to present young birds with a preselected set of tape-recorded stimuli and then quantify the relationship between this input and the individual's output. The beauty of this approach is, of course, that one has complete control over the timing, quantity, and quality of the input. The disadvantage is that it fails to capture at least one facet of the natural context for song development: young are deprived of the opportunity to interact socially with their song tutors.[12] Song development in the wild is generally a social affair (see elegant review by Nelson, in press). One of the first explicit explorations of the effects of social factors on

---

12. The importance of social context for vocal learning has also recently been demonstrated in a nonsongbird, the mallard (*Anas platyrhynchos*), where ducklings reared in social groups are significantly more sensitive to temporal variation in maternal alarm calls than are individuals reared alone (D. B. Miller 1994). This result, which provides evidence of learning to recognize vocalizations, also highlights the importance of attempting to recreate naturalistic contexts in laboratory settings.

song ontogeny, and in particular, the rigidity of at least some species' auditory template, was Baptista and Petrinovich's (1984, 1986) experiments with the white-crowned sparrow. Recall that results obtained from Marler and colleagues suggested that white-crowns have a sensitive period that occurs between 10 and 50 days of age. If the species-typical song is not experienced during this period, individuals acquire an abnormal song; the template only "accepts" as input songs that fall within the white-crown range of variation.

Petrinovich and Baptista housed a young white-crowned sparrow with a strawberry finch (*Amandava*) and found that it not only acquired the finch song, but also acquired song several days after the supposed termination of the sensitive period; compared with a white-crowned sparrow, the strawberry finch is much smaller and has a strikingly different plumage. Subsequent experiments were conducted with live white-crown tutors, rather than tape tutors. In one of these experiments, individuals were isolated until 50 days of age, and then exposed to live tutors. In a second experiment, individuals were tutored with one dialect for the first 50 days, and then with a different dialect for another 50 days; the control involved tape tutoring from 50 to 90 days of age. Tutoring regimes therefore ended prior to plastic song. Experiments revealed that the sensitive period for song acquisition was extended beyond the 50-day period with social but not tape tutoring; all subjects failed to learn song after 100 days of age.

In a final experiment, Petrinovich and Baptista (1987) exposed white-crowned sparrows from the Nuttal's sub-species to two different live conspecific tutors, one before and one after 50 days of age. In contrast to the previous study, however, social tutoring in the second phase continued until song crystallized (200+ days). Results indicated that the sensitive phase was extended under these social conditions.

These results suggest two important modifications to the original template hypothesis. First, if oscines like white-crowned sparrows have an auditory template that guides vocal ontogeny, the template does not rigidly specify which songs are accepted as salient auditory input. Second, social tutoring appears to extend the sensitive phase beyond that demonstrated for tape tutoring.

The Petrinovich and Baptista results require three additional comments. First, the effects of social tutoring on the timing of song development may not generalize to all species, especially given the fact that in an experiment by Marler and Peters on swamp sparrows, there were no differences in the sensitive periods of birds tutored with tape or live birds. The causal factors underlying such putative species differences require further investigation. Second, social tutoring appears to extend the sensitive period for song acquisition beyond that observed under tape tutoring regimes. The mechanism underlying this change in the period of acquisition has yet to be specified, either in terms of behavioral effects or in terms of neural effects. Third, although it seems intuitively correct to conclude that social tutoring has different effects on song ontogeny than tape tutoring, we should tread cautiously, especially in light of Nelson's (in press) recent critique. Specifically, although social tutoring appeared to extend the sensitive phase for song learning in white-crowned sparrows, the experimental tutoring regimes were different in the two comparison groups. Tape tutoring was provided up to 100 days, whereas social tutoring was provided up to approximately 250 days. Tape tutoring thus stopped prior to the period associated with plastic song, whereas social tutoring stopped after song crystallization, thereby allowing for countersinging and song matching of pupil with tutor. As Nelson suggests, had tape tutoring been continued until the spring, when song crystallization occurs, evidence of song learning after 50 days would most likely have been observed.

Slater, Jones, and ten Cate (1993) have taken a slightly different approach to assessing the impact of social tutoring on song development, asking whether social deprivation causes significant changes in some of the temporal milestones of zebra finch song development. Males deprived of social tutors in the sensitive phase are more likely to produce elements of their father's song but show little evidence of learning from tutors encountered in adulthood (approximately 100–120 days). Males raised by females in a group context, but deprived of a male tutor until 120 days, learn little from their tutors thereafter, but the structure of their song is quite normal; moreover, as reported for black-capped chickadees (Nowicki 1989), the songs of group-reared zebra finches converge in structure. Zebra finches reared by both parents, but without siblings, and then deprived of song during the sensitive phase, show learning from both the father and from the tutors encountered after 120 days. Last, males reared alone by their mother and then housed singly until they were placed with singing adults at 120 days produced highly abnormal song, with some evidence of learning between 120 and 140 days. Overall, results from these experiments suggest that the social environment has a significant effect on the song material learned. However, depriving the young bird of social and acoustic experience does not lead to delayed learning abilities in adulthood.

In the Petrinovich and Baptista studies, the mechanism by which social tutors influenced song ontogeny was not quantified. Findings by M. J. West and King (1988) suggest one possible mechanism, at least for cowbirds. Because of their special life histories, cowbirds develop in a fairly unique auditory environment, hearing the songs and calls of other species. When they reach adulthood and

begin to sing, they are eventually exposed to an appropriate audience—potentially receptive females. During these encounters, males sing and females listen.
But females do more than listen. When they hear a song rendition that they like,
they give the equivalent of a human wink, what West and King describe as a
wingstroke display. Upon seeing this, males then selectively sing the song that
elicited the display. Moreover, the song associated with the display is highly
correlated with female sexual receptivity, illustrated by the strong copulatory
responses elicited by wingstroke-associated song but not by song produced either
before or after the display (Figure 5.16).

   To further explore the importance of social interactions on song development—in particular, the possibility that selection-based learning rather than instruction (see the subsection "Social Effects on Song Category" in section 4.2.1)
guides song acquisition—Nelson and Marler (1994) conducted laboratory experiments with two subspecies of the northern Californian white-crowned sparrow,
Zonotrichia leucophrys nuttalli and *Z. leucophrys oriantha; nuttalli* is a sedentary
species, whereas *oriantha* is migratory. The experiment involved tutoring subjects with multiple song types during the first few months posthatching. In the
early spring, an experimental group of birds was tutored with a song that matched
one produced in their overproduced plastic song repertoires. Control subjects
heard a novel song type that was never produced or perceived by them. This
design generated the following predictions: If selection processes guide the emergence of song dialects, then (1) experimental subjects should sing the matched
song rather than the nonmatched one, and (2) control subjects should sing one

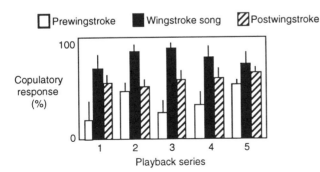

Figure 5.16
The influence of male cowbird song on female copulatory responses. A song was recorded from a
male prior to the female's wingstroke display (white bars), during the wingstroke display (black bars),
and after the wingstroke display (striped bars). These songs were then played back to females, and
the proportion of copulatory responses (*y*-axis) recorded (redrawn from M. J. West and King 1988).

of the songs that was memorized and produced during the plastic song phase. In contrast, if an instructional mechanism is dominant, then controls should acquire and then produce the novel song type. Results provide strong support for selection-based learning, with experimentals singing the matched song and controls singing a song produced during the plastic phase of development. Thus, in the white-crowned sparrow, selection of song type is dominated by auditory mechanisms, whereas in the brown-headed cowbird, visual mechanisms dominate.

What Makes Nonoscines Different from Oscines?

The name of this suboscine group is intriguing, for these birds are known more by what they are not and by what they have not achieved than by what they are themselves! They are *beneath* the songbirds, and are considered to be the more evolutionarily *primitive* of the two groups. (Kroodsma 1988, 157)

The problem that I have been developing since chapter 4 is that songbirds have neural pathways that are dedicated to song production and perception, that such pathways undergo developmental change, and that damage to these pathways can result in imperfect song acquisition, development, and production in adulthood. If these generalizations apply to all songbirds, then one might expect nonsongbirds to have extremely different brains, designed, for example, to produce species-typical vocalizations in the absence of auditory experience. This account is, of course, a mechanistic one, and fails to reveal *why,* from a functional perspective, oscines and nonoscine birds should differ with respect to song ontogeny. In other words, what is the adaptive significance of experientially guided song development, and what selection pressures caused the oscines to take one evolutionary branch of the phylogenetic tree whereas the nonoscines took a different branch? I will return to these questions in chapter 6, but the reader should be warned: no convincing scenario has yet been offered for the difference between oscines and nonoscines, in terms of either fundamental ecological or social factors that could have led to differential selection pressures. What emerges, however, from a comprehensive phylogenetic analysis, is that the oscines are extremely speciose relative to other birds, and the possibility of speciation is highest when there is plasticity in the system, especially the system that is most prominently involved in mating isolation mechanisms: song. An experientially guided system of vocal acquisition and production is more susceptible to change than one that is fixed or immune to experience.

Kroodsma and Konishi (1991) provided the first test of the hypothesis that oscines and suboscines differ with respect to the neural architecture subserving song output. They studied the eastern phoebe (*Sayornis phoebe*), a suboscine species that shows little geographical variation in song structure and that develops normal song when reared in acoustic isolation. In the first phase of their work, they showed that when phoebes were deafened prior to the age when singing commences, the structure of their song did not differ significantly from that produced by intact wild or laboratory-reared birds. Thus auditory feedback does not appear to be necessary for normal song development in this species. In the second research phase, phoebe brains were compared to zebra finch brains (Figure 5.17). Results suggested that in contrast to the oscines, phoebes lacked a number of critical song nuclei (e.g., HVc, RA, and MAN), at least with respect to the

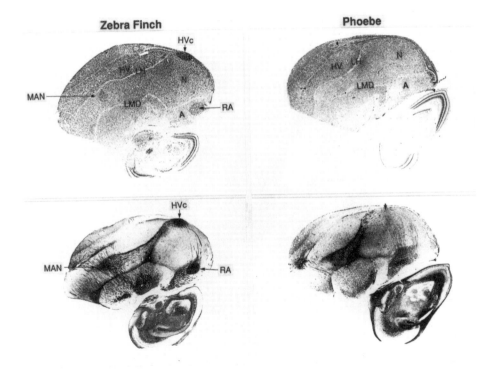

Figure 5.17
Brain sections (parasagittal) of a male zebra finch and a male eastern phoebe. The top row reveals cell clusters by means of thionin staining. The bottom row reveals axons by silver staining. The eastern phoebe appears to lack the primary song nuclei (HVc, RA, MAN; see chapter 4) evident in the zebra finch brain. Both species, however, show more coarse-grained similarity with regard to such areas as the hyperstriatum (HV), lamina hyperstriatica (HL), and lamina medullaris dorsalis (LMD) (from Kroodsma and Konishi 1991).

staining techniques employed. It is believed that such nuclei, together with their respective afferents and efferents, may play an important role in auditory feedback, as well as in song acquisition.

The data on phoebes are encouraging with regard to the importance of HVc, RA, and MAN in experientially guided song acquisition and production. However, given the significant variation in brain structure across the oscines, it will now be important to provide additional evidence that most of the suboscines lack the neural architecture that is presumed to underlie song acquisition and production in species that learn their song.

5.3   Survival Signals: Squirrels and Primates

Species that are vulnerable to predation are typically equipped with a suite of antipredator tactics, varying from subtle freezing behavior to elaborate vocal and visual signals that function to inform group members or the actual predator. The type of antipredator response exhibited by a given species represents a complex interplay between constraints imposed by the sensory system available for detection, the structure of the ecological environment, and the social organization. This section explores some of the ontogenetic processes and patterns that lead to the emergence of species-typical antipredator behavior.

5.3.1   Ground Squirrel Alarms

Ground squirrels and snakes have been engaged in an epic predator-prey relationship, dating back some 10 million years (Coss et al. 1993; Poran and Coss 1990) and involving both vocal and visual displays. What makes the squirrel-snake

relationship interesting, in contrast to other sorts of predator-prey interactions, is that snakes, unlike most mammalian predators, do not chase after their prey— they are sit-and-wait predators. Thus squirrels that detect snakes first have the opportunity (the luxury!) to explore the situation in some detail, and thereby make an informed decision with regard to the most appropriate type of antipredator behavior for the current context (e.g., proximity to the burrow, proximity to vulnerable pups, etc.). In this section, I focus on the visual antipredator signals of ground squirrels, leaving the vocal signals for chapter 6 where the adaptive consequences and design features of avian and mammalian alarm calls are discussed.

The California ground squirrel (*Spermophilus beecheyi*) is frequently preyed upon by nonvenomous Pacific gopher snakes (*Pituophis melanoleucus catenifer*) and venomous northern Pacific rattlesnakes (*Croatalus viridis oreganus*), but these snakes prey mostly on pups in the burrow. Adults are less vulnerable because they are able to retaliate by biting snakes; such attacks are usually preceded by prolonged staring, tail flagging, and dirt throwing. In some populations, moreover, adults have quite high resistance to snake venom (Poran, Coss, and Benjamini 1987); pups also have defenses against snake venom, but because they are smaller than adults and have had less exposure to the venom (i.e., have less immunity), they are more vulnerable. The different harassment techniques used by the squirrel often result in a response by the snake, and such responses tend to be species-specific (e.g., particular patterns of hissing, rattling, and striking). Among the two primary species of snake investigated, the rattlesnake is the more dangerous.

When pups first emerge from their burrow at approximately 45 days, they act like adults in the context of snakes: they investigate and produce vigorous tail flagging. In fact, young pups may even engage in such antipredator behavior upon encountering secondary cues of snakes, such as curled leaves, sticks, and rounded stones (Owings and Coss 1977; Poran and Coss 1990), or snakelike objects such as sticks placed next to the burrow (Coss et al. 1993). Moreover, the tail-flagging display employed by young pups is functionally meaningful as indicated by the fact that adults seeing this display come rushing in to protect them (Owings and Coss 1977; Poran and Coss 1990). Pups, either lab-born or wild-caught, that have experienced a snake in a simulated burrow will subsequently approach with caution and also throw dirt (Coss and Owings 1978). This observation provides some evidence of an associative memory mechanism enabling a connection between microhabitat variation and sources of danger. Such memory is experienced based but derived from innate predispositions to associate snakes with certain locations.

In a test of pup discrimination ability (Coss 1991), the antipredator behavior of snake-naive individuals was observed in the context of confronting a caged (1) gopher snake and (2) guinea pig. Results (Figure 5.18) indicate that from 41 days of age (approximately one day after pups begin to show visually guided behavior—eye opening occurs at 30–39 days of age), pups spend less time in proximity and show stronger tail-elevation and startle responses to the gopher snake than they do to the guinea pig. Interestingly (and serendipitously!), the startle response of 41-day-old pups was as strong to the textured strip on the wall of the testing chamber as it was to the gopher snake and guinea pig. This result suggests that the pups' earliest response to objects in the world is somewhat crude and thereby results in significantly more false alarms (chapter 3) than occur

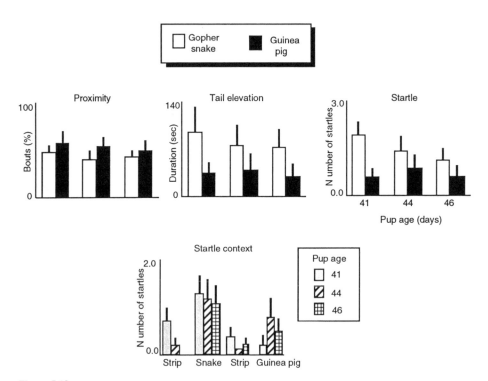

Figure 5.18
Developmental changes in the behavior of squirrel pups in response to gopher snakes (white bars), guinea pigs (black bars), and a textured strip affixed next to the animal test chamber. Histograms represent means and standard errors (redrawn from Coss 1991).

a few days later (i.e., by 44 days of age, there is a considerable decrease in the pups' startle response to the textured strip).

To explore the specific ontogenetic processes involved in antipredator responses to snakes, Poran and Coss (1990) experimentally observed snake confrontations by 60–73-day-old pups alone and in the presence of their mothers. Results indicated that for both rattlesnakes and gopher snakes, pups showed an overall reduction in response when in the presence of their mother. Specifically, they spent less time in proximity to the snakes, faced them less, and threw less dirt. Interestingly, for measures of both proximity and facing orientation, pups with and without their mothers showed the same patterns of response to the two snake species as adults. For substrate throwing, however, pups with and without their mothers threw less at rattlesnakes than gopher snakes, whereas adults showed the reverse pattern. When these pups were tested two years later, they behaved like adults. Last, based on radioimmunoassays of the serum-to-venom binding, pups showed a 74% increase from 14 to 30 days of age, and then exhibited adult levels. Thus, before visually guided behavior develops (at 40 days of age), pups have a sufficient defense mechanism. These measures are of absolute levels, however. Because pups are smaller, the relative ratio of serum to body weight makes them extremely vulnerable to attack by snakes.

Coss et al. (1993) examined the relationship between geographic variation in snake pressure (i.e., the relative selective pressures exerted by snakes as a function of their differential distribution in particular parts of northern California) and differences in pup antipredator behavior. Analyses revealed that venom resistance was approximately 60% lower in populations where snakes were rare as opposed to populations where they were abundant. Experiments showed that lab-reared pups were more aggressive to gopher snakes, whereas wild-borns (from snake-abundant areas) were equally aggressive to both snakes. However, adults from the snake-abundant area treated the rattlesnake more aggressively, a result that may reflect experience with snake encounters. In a population of wild-caught squirrels living in a recently colonized area with few rattlesnakes, rattlesnakes and gopher snakes were treated similarly, whereas in populations where rattlesnakes have been absent for long periods of time (reduced selection pressure), gopher snakes were treated more aggressively. One interpretation of this finding, suggested by Coss and colleagues, is that in the absence of discrimination problems (i.e., only one type of dangerous snake), the level of aggression is shifted toward the most threatening species. Without understanding the costs and benefits of such antipredator attacks, it is unclear whether alterations of aggression are beneficial.

5.3.2 Vervet Monkey Alarm Calls

The first comprehensive analysis of a wild primate's vocal repertoire was Struhsaker's (1967) studies of East African vervet monkeys. His account suggested more than 30 discrete call types, with several calls restricted to immatures. As reviewed in chapter 2, Struhsaker noted that in response to different predator classes, such as large cats (e.g., leopard, cheetah), birds of prey (e.g., martial eagle, crowned hawk eagle), and snakes (e.g., pythons, mambas), vervets produced acoustically distinct alarm calls and responded with different, behaviorally adaptive escape responses. Playback experiments by Seyfarth, Cheney, and Marler (1980a, 1980b) further confirmed the distinctiveness of the alarm call system and provided evidence of a semantic system of communication, with each alarm call type conveying information about the type of predator encountered (see also chapters 6 and 7). These careful behavioral and acoustical descriptions provided the foundation for looking at the ontogeny of the vocal repertoire.

Unlike research on captive primates (see the subsection "Ontogeny of Production"), where social situations are typically set up to evoke particular calls in the repertoire (Herzog and Hopf 1983; Winter et al. 1973), research on wild primates, and vervet monkeys more specifically, depends upon opportunistic recordings. Using spontaneously produced vocalizations by immature vervets, Seyfarth and Cheney (1986) analyzed the acoustic morphology of eagle alarm calls. Only a small sample of recordings were obtained, all from individuals over the age of four months. The structure of these calls was, however, similar to that of calls produced by adults, except for differences in the fundamental frequency, a feature that can be readily explained by age-dependent differences in the weight of the glottis. Recordings of immature alarm calls to snakes and leopards (Hauser, unpublished data) further confirm Seyfarth and Cheney's findings, suggesting that experience plays a relatively insignificant role in shaping the acoustic morphology of vervet alarm calls. (See section 5.4.1 for evidence that primate call structure undergoes little change ontogenetically; reviewed in Seyfarth and Cheney, in press.)

To determine when vervet monkeys begin to use alarm calls in the appropriate context, Seyfarth, Cheney, and Hauser (Caro and Hauser 1992; Hauser 1989; Seyfarth and Cheney 1980, 1986) analyzed observations of naturally occurring predator encounters and infants' vocal responses. In general, results indicate that up until the age of two to three years, immatures produce alarm calls to both predatory and nonpredatory species. With increasing age, the number of species eliciting alarm calls diminishes to the point where alarms are only given in response to predators—that is, species that have been observed preying on vervets. This refinement in category breadth is most elegantly illustrated by Seyfarth and Cheney's analyses of eagle alarm calls. Results show (Figure 5.19) that immatures often produce eagle alarm calls to nonpredatory species such as vultures. Interestingly, eagle alarm calls are restricted to objects in the air, and especially avian species, suggesting considerable canalization with regard to the process of categorizing predators. The same spatial restrictions are also observed for leopard and snake alarm calls, with several amusing cases of infants giving snake alarm calls to falling branches in a bush and eagle alarm calls to falling leaves (Cheney and Seyfarth 1990).

Seyfarth and Cheney interpret their developmental results as providing evidence that infants make classification "mistakes," producing alarm calls to inappropriate objects. Although this is a valid interpretation, there is an alternative that views the infant as a more active participant in the developmental process (see Hauser 1993b; Owings 1994; Owings and Coss 1977). Specifically, rather than making mistakes (e.g., giving an eagle alarm call to a vulture), infants may be using alarm calls to ask questions (e.g., "Is that thing in the air something I should give an eagle alarm call to?"). Whichever interpretation turns out to be correct, it appears that infants may obtain direct feedback about their vocalizations from other events in the environment, both acoustical and social. Thus, Seyfarth and Cheney (1986) and Hauser (Caro and Hauser 1992) have shown that when infants produce alarm calls, adults often follow with the same type of alarm call, especially if a vervet predator is detected. In addition to alarm calling, adults also respond with predator-specific escape behavior. Although the adults' response may not provide evidence of pedagogy, in the sense of intentional instruction (Cheney and Seyfarth 1990), it may nonetheless be functionally instructive[13] (Caro and Hauser 1992; see definition in Box 5.1). That is, when an infant's alarm call is followed by an adult's, the infant obtains feedback on its vocal behavior. And such feedback occurs even if the adults' alarm calls were not designed for this function (e.g., the adults' calls may be designed to amplify the severity of the impending danger). In fact, there is some evidence to suggest that such feedback has a positive effect on the infants' subsequent alarm call behavior. Thus, Hauser (discussed in Caro and Hauser 1992) observed that when infants produce an alarm call to the correct predator (e.g., eagle alarm call to a martial eagle) and adults follow with the same type of alarm call, such infants are more likely to

13. A more thorough discussion of pedagogy and instruction is provided in chapter 7.

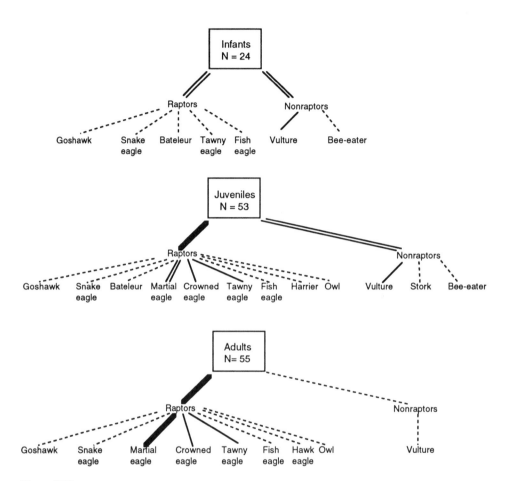

Figure 5.19
Developmental changes in the target of vervet monkey eagle alarm calls. Infants: <1 year old; Juveniles: 1–4 years old; Adults: >4 years old. Dashed lines: <5 alarms; thin solid lines: 6–10 alarms; double lines: 11–15 alarms; thick solid lines: >15 alarms (redrawn from Seyfarth and Cheney 1986).

produce the correct alarm in the subsequent encounter than infants who produce calls to species that do not prey on vervets. Perhaps more intriguing, but difficult to interpret because of small sample sizes, were observations of aggressive actions by mothers toward infants producing the inappropriate alarm call. For example, one infant produced a leopard-sounding alarm call to a mongoose (i.e., a nonpredator), and as a result, the mother ran up into a tree. Upon seeing the mongoose, the mother slapped her infant. Although one cannot conclude that this incident was causally related to the infant's subsequent alarm call accuracy, it suggests one potential mechanism for guiding the infant toward adult usage—punishment!

The ontogeny of the vervet's alarm call system is arguably the best case we have of nonhuman primate vocal development under natural conditions (Seyfarth and Cheney, in press). However, there are at least four pieces of data missing, and these pieces are critical if we are to determine the relative importance of learning in this ontogenetic system. First, given the observation that immature vervets produce alarm calls to the same general kinds of stimuli as adults (e.g., eagle alarm calls to "things" in the air), it appears that they are born with an innate category for "contexts-that-elicit-eagle-sounding-alarm-calls"; and the same is likely to be true for the other predator categories. What is unclear, however, is how experience fine-tunes this category. Thus, in the tree-diagram presented in Figure 5.19, the data indicate that immatures (pooled across all subjects) produce eagle alarm calls to many more species than do adults. But how many infants do this, and if some produce alarm calls to a narrower range of stimuli, what factors account for such between-individual differences? Second, and this will be difficult to assess under field conditions, how often do immatures, as opposed to adults, see predators and *fail* to give alarm calls? In a signal detection theory sense, immatures may call to a broader range of species because they have a lower criterion (beta value) for responding—they are "trigger happy"—but may also miss many opportunities to call as well (low sensitivity). Third, when adult alarm calls follow (temporally) those produced by immatures, we need to obtain a deeper understanding of how this sequence might benefit vocal development. For instance, is the developmental time course for fine-tuning the specific predator category (i.e., bird of prey, snake, terrestrial carnivore) delayed if the frequency of second alarm calls by adults is diminished? Fourth, are there populations of vervet monkeys that produce alarm calls that would be unintelligible to a geographically separate population? That is, is there evidence of alarm call dialects? If so, this would provide some of the strongest support for the importance of learning during vocal development.

5.4 Social Signals: Primates

5.4.1 Nonhuman Primate Vocalizations

Unlike insects, frogs, and precocial birds, most primates are relatively helpless at birth and depend on parental care (e.g., nutritionally and motorically) for an extended period of time (Altmann 1980; Hauser 1994; Pereira and Fairbanks 1994). One might expect, therefore, that experience would play a fundamental role in shaping the developmental trajectory to an adult form of communication. Paradoxically, studies of nonhuman primate vocal development have revealed extremely little evidence that experientially guided learning is involved in the production of species-typical calls (Cheney and Seyfarth 1990; Hauser and Marler 1992; Snowdon 1990; Todt, Goedeking, and Symmes 1988). To appreciate the significance of this paradox, it is important to keep several issues in mind. First, it is not the case that vocal learning is absent in all nonhuman mammals. There is now increasing evidence, for example, that acoustic experience directly influences the ontogeny of the dolphin's signature whistle and that in adulthood individuals exhibit the capacity to mimic the whistles of group members (Sayigh et al. 1990; Smolker, Mann, and Smuts 1993; Tyack 1993). Second, the common lore in the literature is that of "monkey see, monkey do" and, consequently, that most nonhuman primate behavior is learned. Although learning certainly guides much of their social behavior, there has been no convincing demonstration of imitation in either the vocal or nonvocal domain (Whiten and Ham 1992), with the exception of a few cases in captive chimpanzees (Tomasello, Savage-Rumbaugh, and Kruger 1993) and human-reared orangutans (Russon and Galdikas 1993); these exceptions are described further on in the chapter. Third, when one speaks of the role of experience in vocal development, it is necessary to distinguish three different acquisition systems:[14] (1) production of the appropriate acoustic morphology, (2) acquiring the ability to use calls in socioecologically correct contexts, and (3) comprehending the meaning or function of calls within the repertoire. The story presented in this section reveals that for nonhuman primates, experience plays a minimal role in call production, but a more significant role in the ontogeny of call usage and comprehension.

Ontogeny of Production   As the study of birdsong illustrates, there are several ways to document experientially guided development of call morphology (re-

---

14. These distinctions have generally not been made for nonprimate animals. Rather, most of the research has focused on the role of experience in producing the species-typical call morphology.

viewed in Nelson, in press). Among the nonexperimental approaches, one can look for dialects and cases of interspecific mimicry. Experimental approaches include acoustic isolation, deafening to remove auditory feedback, disruption of neural pathways, tape and social tutoring, and cross-fostering. As the following review indicates, relatively few of these approaches have been implemented with primates.

In the late 1960s and early 1970s, several German primatologists, led by Winter and Ploog (Talmage-Riggs et al. 1972; Winter et al. 1973; Winter, Ploog, and Latta 1966), initiated a long-term project on the vocal communication of the squirrel monkey, focusing in particular on its ontogeny. Many of the observations and experiments were guided by research on songbirds. Following quite detailed quantification of the acoustic features of the adult vocal repertoire (Schott 1975), studies were conducted to assess how the repertoire unfolds ontogenetically. In one study (Winter et al. 1973), newborn infants were raised under three different conditions: (1) raised by a muted mother until two weeks of age and then raised in isolation, (2) separated from hearing mothers at either two days or 19 days, and (3) deafened by cochleaectomy at four days. The vocal development of these infants was no different from that of control infants raised under normal conditions. Results from three additional studies further support the relative insignificance of acoustic experience and auditory feedback in vocal development: (1) experimentally deafened adults fail to show significant deficits in call production (Talmage-Riggs et al. 1972); (2) changes in individual "signatures" of the isolation peep reveal the importance of maturational effects (Lieblich et al. 1980);

and (3) subspecific differences in call morphology are maintained (Newman and Symmes 1982).

While the results on squirrel monkeys are interesting, they leave open the possibility that captivity itself creates conditions that are at odds with vocal learning, or that squirrel monkeys simply represent a taxonomic outlier among Primates. To this end, let us return to research on wild vervet monkeys. Vervets in Amboseli produce four acoustically distinct grunts, each grunt type associated with a different social context. In contrast with alarm calls, grunts are frequently produced by young infants. Detailed acoustical analyses (Seyfarth and Cheney 1984, 1986) revealed that grunts produced by one- to eight-week-old infants were different from those produced by adults. Although different acoustic features appeared to change at different developmental rates, the adult form was not observed until the age of 2–3 years. These data, like those presented by Gouzoules and Gouzoules (1989) for pigtail macaques (*Macaca nemestrina*) and by Snowdon and colleagues (Elowson, Sweet, and Snowdon 1992; Roush and Snowdon 1994; Snowdon and Elowson 1992) for pygmy marmosets (*Cebuella pygmaea*) and cotton-top tamarins (*Saguinus oedipus oedipus*), are insufficient to distinguish between ontogenetic changes that are guided by maturational factors and those that are guided by experiential factors.

Following up on Seyfarth and Cheney's (1986) work on vervet monkey vocal development, Hauser (1989) analyzed the ontogeny of the vervet monkey's ''intergroup wrr,'' a call given during aggressive encounters between neighboring groups. Wrrs given by adults are trilled vocalizations—discrete pulses of energy separated by significant, though brief, periods of silence. Infants from birth to three months produce trill-like calls (''lost wrrs'') in the context of separation that are acoustically quite similar to the adult wrr, especially in terms of the characteristic pulses separated by brief periods of silence. From three to ten months, infants do not produce trilled vocalizations in any context. After this period, trilled vocalizations emerge and are produced in the context of intergroup encounters. The acoustic morphology of these intergroup wrrs is not, however, like that of adults. Specifically, the pulses of energy are relatively fused, with virtually no detectable energy gaps in the waveform.

The vervet's developmental sequence is puzzling for at least two reasons. First, when trill-like calls disappear at the age of three months, it is not because the eliciting context has disappeared—infants continue to experience apparently stressful separations from their mothers for at least the first year of life, and probably longer. Second, the temporal quality of the trill produced by three-month-olds is much more adultlike than is the trill produced by ten-month-olds. That is, although three-month-old infants produce wrrs in the context of separation, rather than intergroup encounters, their calls are characterized by discrete pulses of energy followed by silence. The wrrs of ten-month-olds lack this temporal structure.

Comparing the ontogeny of the wrr with other vocalizations in the repertoire (Figure 5.20) provides some insights into the factors leading to morphological changes in call structure. When the lost wrr drops out of the repertoire at approximately three months, new vocalizations gradually appear within the adult repertoire. These new vocalizations require different articulatory routines[15] from those needed to produce either lost or intergroup wrrs. It is possible, therefore, that the introduction of new vocalizations with new articulatory gestures causes interference with the production of wrrs. This interference, if it occurs, could result in a poor rendition of the intergroup wrr by ten-month-old infants, a rendition that must be improved gradually over time. Based on the pattern observed, Hauser (1989) suggested that like humans infants (Locke 1993; Moskowitz 1970), vervet infants sometimes undergo a process of vocal regression. A similar phenomenon has been described in the callitrichids by Snowdon and his colleagues (Elowson and Snowdon 1994; Elowson, Sweet, and Snowdon 1992; Snowdon 1990; Snowdon and Elowson 1992). An alternative interpretation of this pattern of vocal production is that the calls produced by young infants are only coincidentally similar to those produced by adults: the two signals have completely different functions (i.e., "recruit parental care" versus "announce aggressive intergroup encounter") and just happen to share a similar acoustic morphology. When immatures reach the age at which they are involved in intergroup interactions, they have to learn the new structure-function association *de novo*. Testing between these two alternatives will require, minimally, a better understanding of (1) the articulatory requirements underlying the production of different calls in the repertoire and (2) the specific function/meaning of the immature and adult calls.

To more carefully assess the plasticity of the pygmy marmoset's vocal repertoire, Elowson and Snowdon (1994) conducted experiments involving manipulation of the social environment. Initially, unfamiliar groups (infants, juveniles,

---

15. As pointed out in chapter 4, relatively little is known about the articulatory gestures associated with vocal production in nonhuman primates. However, the lost and intergroup wrrs are the only trilled calls in the repertoire and appear to be produced by either rapid tongue movement or rapid velar movement.

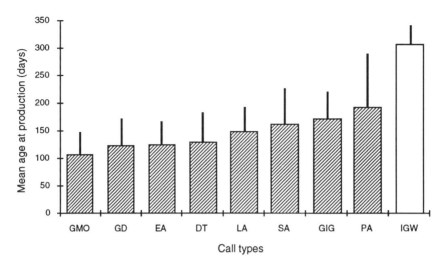

Figure 5.20
Differences in the average age (days) at first production of vervet monkey vocalizations. GMO: grunt given during movement into an open area; GD: grunt given to a dominant; EA: eagle alarm call; DT: dominance threat; LA: leopard alarm call; SA: snake alarm call; GIG: grunt given during an intergroup encounter; PA: primate alarm call; IGW: intergroup wrr. Standard deviations shown (redrawn from data presented in Hauser 1989).

subadults, and adults) exhibited significant between-group differences in the duration of the "trill" vocalization. As time in acoustical contact increased, group differences in duration vanished. Though spectral features also changed, they did so in a parallel fashion for both groups. That is, individuals of all ages exhibited a general increase in peak frequency and bandwidth over time. Elowson and Snowdon argue that these changes are similar to those described by Nowicki (1989) for the black-capped chickadee's D-note. In this situation, newly constituted flocks show acoustic convergence over a period of several weeks. Mitani and Brandt (1994) have recently reported a similar finding for chimpanzee pant-hoots, a long-distance vocalization. These results are taken as evidence of experientially guided changes in call structure.

Although acoustic isolation experiments clearly provide a useful tool for the ornithologist, they are problematic for highly social organisms such as primates. Isolating an individual from its group eliminates not only relevant acoustic experience, but an entire suite of experiences that are relevant to normal development. Consequently, abnormal vocal ontogeny potentially reflects the negative consequences of two sources of deprivation, social and acoustical. A more appropriate test for nonhuman primates, therefore, is cross-fostering[16] among compatible species.

Masataka and Fujita (1989) published the first results of a cross-fostering experiment involving rhesus macaques and Japanese macaques. Both species produce a repertoire of at least 20–30 different call types, and there is quite extensive overlap with regard to the acoustic structure of the calls (Gouzoules, Gouzoules, and Marler 1984; S. Green 1975b; Hauser and Marler 1993a; Rowell and Hinde 1962). For example, both species produce coo vocalizations in the context of affiliation and screams in the context of submission. However, there are also differences between the two species, both in terms of fine-grained acoustic structure and the frequency with which particular call types are used in a given context. Masataka and Fujita focused on coos given in the context of food. For adult coos, there was no overlap in the fundamental frequency between rhesus and Japanese macaques. Thus, if acoustic experience influences the ontogenetic production of these calls, then rhesus infants raised by Japanese macaque mothers should produce coos that more closely resemble those of their cross-fostered parents than those of their biological, but physically and acoustically removed, parents. Results showed that for both rhesus and Japanese macaques, infants produced coos that more closely resembled those of their cross-fostered parent.

Masataka and Fujita's results provided the first suggestive evidence that acoustic experience directly influences the ontogeny of call morphology. Unfortunately, an identical study by Owren and colleagues (1992, 1993), which was ongoing at the time, has failed to provide support for the conclusions reached by Masataka and Fujita. In the Owren et al. studies, a slightly larger sample of cross-fostered subjects was observed, and more comprehensive acoustical analyses were performed on the data set. Concerning the latter, analyses of adult food coos failed to reveal a statistically significant difference between rhesus and Japanese macaques. As a result, each species fails to provide a distinctive model in the cross-fostering paradigm. Species' differences in call usage were observed, however, together with some evidence that infants used calls in accordance with their foster parents' usage.

In sum, the available data on vocal production provide only weak support for the role of experience in modifying call structure. Although call structure changes ontogenetically, no study has provided convincing evidence that acoustic experience is causally related to such changes. The few reported cases of dialect

---

16. Cross-fostering is probably the most powerful approach for a number of species, including birds. It is not, however, suitable for problems related to learning preferences.

variation in red-chested mustached tamarins (*Saguinus l. labiatus;* Maeda and Masataka 1987; Masataka 1988) and chimpanzees (Mitani et al. 1992) have not yet been sufficiently well studied. Thus, for example, in the tamarin case it is not clear whether we are dealing with sub-specific differences or true populational differences. And in the chimpanzee study, though the pant-hoot vocalization of two geographically separated populations differed statistically, there was considerable overlap in call morphology, weakening the support for dialects. The conflicting results from cross-fostering experiments need to be resolved through further tests. Such investigation is important for two reasons. First, the conflicting evidence may be due to methodological differences between the studies, or to differences in call structure across macaque populations. Although dialect differences among macaque populations have not been convincingly documented,[17] if such differences exist, they may very well explain why one study revealed the effects of learning and the other study failed to do so. Second, rhesus and Japanese macaques may not provide us with the most ideal species for testing in a cross-fostering paradigm. Although interspecies compatibility is important for rearing, it is also important to pick species with significantly different vocal repertoires. Macaques tend to have similar vocal repertoires.

Ontogeny of Usage and Comprehension   In addition to Seyfarth and Cheney's work on alarm call development in vervet monkeys, studies of both New World and Old World monkeys have provided evidence that experience influences the ontogeny of correct usage and comprehension (reviewed in Seyfarth and Cheney, in press). Thus recent work by Roush and Snowdon (1994) on cotton-top tamarins shows that juveniles and subadults more frequently produce food-associated calls to nonfood items than do adults. In pigtail macaques, it takes several years before individuals produce screams in the appropriate context, with females acquiring this ability earlier than males (Gouzoules and Gouzoules 1989, 1990). In vervet monkeys, correct usage of intergroup wrrs develops over a period of approximately two years, shaped in part by experience (Hauser 1989). Thus young vervets often produce intergroup wrrs when there is no apparent threat from a neighboring group. Infants living in groups with high intergroup encounters—and therefore higher exposure to hearing intergroup encounter calls—produce wrrs in the appropriate context at an earlier age than infants living in groups with lower encounter rates.

17. S. Green (1975b) reported evidence of dialects in Japanese macaque coo calls, but this result was not substantiated by subsequent studies (Green, personal communication).

Child-language-acquisition research indicates that comprehension of word meaning commonly precedes correct usage (reviewed in Locke 1993; see next section for some subtleties in this developmental pattern). Two points should be kept in mind with regard to this generalization. First, production sometimes precedes comprehension as, for example, when children produce rote output that they do not understand, but have, perhaps, learned to produce in the appropriate context. At the age of about two years, my daughter went through a brief phase when she said, "No way José," whenever she wanted my attention. I don't believe she understood the wit of her expression. Second, when comprehension does not precede production, this sequence may not reflect a fundamental difference in the cognitive processes underlying these systems. Rather, the temporal separation between these two systems may reflect the fact that comprehension is an easier task because it uses redundancies in the signal, and human language has many, both acoustical and visual.

As with other aspects of research on primate communication, several investigations of vocal ontogeny have attempted to document whether immatures respond appropriately to species-typical calls at an earlier age than they produce them. Although researchers have provided careful accounts of developmental changes in usage, studies of vervet monkeys provide the only relevant data for assessing ontogenetic differences in the timing of comprehension and usage. In Seyfarth and Cheney's work on alarm calls, playback experiments were conducted to determine when infants produced an adult response to eagle alarm calls. Several interesting results emerged. First, consistent adult responses (i.e., scan up in the air, run for cover under a bush) were not observed until 6–7 months. Second, when infants failed to produce an adult response, they either produced an inappropriate response (e.g., running out into an open area) or ran to their mother; the latter is not adultlike but is appropriate in the sense that it secures safety. Third, based on analyses of film records from their playbacks, Seyfarth and Cheney found that infants who produced an adult response were most likely to look at an adult before they responded. This suggests that infants may learn about the appropriate escape response by first observing what adults do (obtaining cues from them), and then following their lead. In sum, results on alarm calls, together with those obtained for wrrs (Hauser 1989), suggest that comprehension of call function precedes correct usage, at least for vervet monkeys.

The data on nonhuman primate vocal development that I have reviewed in this section indicate that experientially guided learning plays a relatively insignificant role in the ontogeny of call morphology but plays a more important role in guiding correct usage and comprehension. Although the insignificant contribution of learning in vocal production is puzzling, we should not lose sight of the fact that such studies are only in their infancy. We know almost nothing about vocal development in the great apes, but at least one study suggests the possibility of dialect variation (Mitani et al. 1992),[18] and a second study (Mitani and Brandt 1994) suggests the possibility of call plasticity in the context of dynamically changing social relationships. In addition, although experience plays a role in call usage and comprehension, we should be equally impressed by the importance of innately structured categories for responding. Specifically, and here I focus primarily on vervet monkeys and macaques, immatures are born with a general understanding of call usage. Vervet infants, for example, do not just randomly produce alarm calls to all moving objects. Rather, they produce alarm calls to objects that fall within a loosely defined set of categories, such as "dangerous thing in the air" and "dangerous thing on the ground that is either big and four-legged (e.g., cats) or small and no-legged (e.g., snakes)." We now need a more thorough understanding of how, precisely, experience fine-tunes these general categories.

5.4.2   Human Spoken Language

Even if someone was uninterested in humans per se, they might still benefit from the study of children's acquisition of language—as children appear to be the only things (either living or nonliving) capable of performing this task. As such, computer scientists interested in language might study children for much the same reason that Leonardo da Vinci, who was interested in building a flying machine, chose to study birds.   (Bloom 1994a, 5)

The previous sections touched on several issues relevant to the study of child language acquisition, including the observation that comprehension of language often precedes production, and that language acquisition is a robust, highly canalized developmental process. Here, we take an in-depth look at the factors that guide and constrain the human child along the way to acquiring language, both spoken and signed. Because auditory experience is possible prior to birth, I begin by reviewing some of the exciting work on fetal language learning. I then turn to a discussion of the earliest phases of language acquisition, aimed at understanding how the child gains a foothold in the linguistic community. I then conclude with

---

18. As briefly mentioned at the end of the subsection "Ontogeny of Production," although Mitani et al. show significant differences between two geographically isolated chimpanzee populations, the differences are between mean values of six acoustical parameters. Inspection of the variation around the mean shows, however, that there is considerable overlap between the two populations. Thus, until perceptual experiments are conducted, it remains unclear whether there is sufficient distinctiveness between populations to create dialect boundaries.

a summary of how children make their transition from inarticulate to articulate members of the human community, capable of oration and reciting Shakespearean sonnets. Some of the more detailed work on the child's developing comprehension of word meaning and reference is provided in chapter 7.

Fetal Acoustics   We tend to think that birth marks the onset of experientially mediated developmental processes, especially cognitive processes. But just as neurobiologists such as Greenough and Alcantara (1993) and Diamond (1988) have illustrated the importance of early experience on the prenatal development of the nervous system,[19] so have developmental psychobiologists working with precocial birds (Gottlieb 1975, 1980; Lickliter 1990; Lickliter and Hellwell 1992) illustrated the importance of early fetal experience on the development of sensory systems (Turkewitz 1991, 1993).

In the warmth and comfort of the womb, what does the human fetus hear, and does such early auditory exposure influence subsequent language acquisition? To address the first part of this question, researchers have measured the acoustics of the intrauterine environment. Results indicate that the fetus is surrounded by considerable noise from blood flow, but that low-frequency sounds from the external environment can pass through the mother's stomach and, depending upon amplitude, rise above the background noise. In addition, by placing a microphone on the neck of a fetal sheep while simultaneously playing meaningful and non-meaningful speech through a speaker placed on the abdomen of the maternal ewe, results revealed that the recorded sound exhibited a high degree of intelligibility, with voicing contrasts well preserved (Griffiths et al. 1994). Most significantly, studies by Querleu and colleagues (Querleu, Renard, and Crepin 1981; Querleu, Renard, and Versyp 1985) indicate that the fundamental frequency of the maternal voice (approximately 200–250 Hertz), but not higher spectral features (e.g., formants), exhibits a substantially higher intensity (24 dB) than intrauterine noise. Consequently, the fetal (6–7 months) hearing system should be able to pick out some aspects of the mother's voice from the background. Does it?

19. From a neural perspective, there are at least three categories of experience-relevant changes that can arise during the course of development (Diamond 1988; Greenough and Alcantara 1993): (1) experience-independent: neural connections are formed in the absence of behavioral or sensory experience; overproduction of connections occurs, followed by attrition, but selection of a final set is not contingent upon experience; (2) experience-expectant: initial overproduction arises in the absence of experience, but the final set of connections results from external experience with the environment; and (3) experience-dependent: changes in synaptic number or processing efficiency occur during early development and into adulthood and are entirely due to experience (e.g., learning, memory).

To explore what the developing fetus learns from its acoustic environment, playback experiments have been conducted by attaching speakers to the mother's abdomen and using changes in fetal heart rate as a dependent measure of perceptual learning. In one study (Lecanuet et al. 1986), 40-week-old fetuses showed a decrease in cardiac response as a function of repeated exposure to noise (<800 Hz at approximately 100 dB). This finding suggests low-level learning by habituation. In a similarly designed experiment (Lecanuet, Granier-Deferre, and Busnel 1989), research showed that following habituation to one sound (e.g., "babi"), presentation of a new sound (e.g., "bibi") resulted in cardiac deceleration. These results indicate that the fetus is capable of perceptually discriminating between sounds. What is puzzling about these results is that the primary acoustic difference between stimuli such as "babi" and "bibi" lies in their formant structures. But, as pointed out, formants are generally filtered by the maternal womb and are thus relatively inaccessible to the fetus as a distinguishing feature. Thus differences in the fundamental frequency between a and i must have been significant in the stimuli presented and sufficient for the fetus to detect. Studies like these need to be replicated using a wider range of stimuli so that the precise nature of the perceptual response can be uncovered.

Early Postnatal Perception   If the fetus is fed relevant acoustic experience, what effect does this have on subsequent language development? Does the neonate show any preferences for particular kinds of sounds, such as the mother's voice, the native language? Regarding maternal voice, several studies have shown (using a wide assortment of techniques such as nonnutritive sucking and orienting responses) that within the first few hours to days of life, neonates exhibit a stronger preference for their mother's voice than for nonspeech, the voice of other women, and even the father's voice (reviewed in Locke 1993). Such preferences make good intuitive and evolutionary sense: the fetus has had greater exposure to the mother's voice than to any other voice, and to insure appropriate attachment, early recognition is highly adaptive[20]—a kind of human imprinting.

Although most of the important acoustic differences between human languages are the result of differences in formant structure, some languages can be differen-

---

20. It is interesting to note that although several studies of nonhuman animals, including birds, bats, and nonhuman primates, have provided convincing evidence that mothers recognize the voice of their offspring, evidence for offspring recognition of maternal voice has not been established. And yet, it would be relatively easy to outfit a pregnant mother with a speaker and provide the infant with differential auditory experience. One could then test for pre- and postnatal preferences for maternal as opposed to nonmaternal voices.

tiated on the basis of other spectrotemporal features, including prosodic patterns that arise from changes in the fundamental frequency. These lower-frequency features are the only ones that are likely to provide the developing fetus with a language-specific environment. To test this possibility, Mehler and colleagues (1988) conducted an experiment with four-day-olds to assess whether they exhibited preferences for either French or Russian. One group of infants had French-speaking mothers, whereas a second group had non-French-speaking mothers. Infants in the first group, but not the second, showed a clear preference for French. This preference was not evident in infants who were less than 12 hours old. The latter result is puzzling: if fetal exposure is significant (i.e., causally related to language preference), why does the preference require any time postpartum to kick in? An answer to this question, and related ones, will require a more complete understanding of the constraints imposed by the gradual maturation of the brain and the peripheral organs that it guides. In addition, we need to understand how much information the infant requires (pre- and postpartum) to learn the details of the native language. We will explore some of these issues momentarily, but first let us complete our tour of the infant's early perceptual abilities.

The speech as special debate provided the ideal forum for bringing young infants to the forefront of the subject pool and, more specifically, for exploring the topography of the controversy regarding special versus general mechanisms underlying speech processing.[21] One of the first experimental tests in this area focused on categorical perception (see section 7.3.2 for a detailed discussion of this topic). Using a high-amplitude sucking technique (a computer records the rate at which infants suck on a nonnutritive nipple), studies showed that infants as young as one month old could discriminate phonemes at category boundaries, but not those within a category, and this ability held for contrasts within and outside the native language (Aslin et al. 1981; Eimas et al. 1971). These results,

---

21. In the first rounds of this debate over specialized mechanisms, speech and nonspeech signals were contrasted, as were the abilities of adult humans and nonhuman animals. If speech and non-speech signals are processed differently, then speech consists of special features. If they are processed similarly, however, then two interpretations are possible. On the one hand, perceptual mechanisms may have evolved for processing speech, but some nonspeech signals can mimic the relevant acoustic details, thereby leading to similar patterns of perception. Alternatively, the mechanism is quite general and reponsible for processing a broad array of acoustic signals. Turning to nonhuman animals, if they too process speech in the same way as human adults, then such a similarity would suggest that the relevant perceptual mechanisms did not evolve for speech, though special mechanisms may nevertheless operate in human listeners.

together with those accumulating on nonhuman animals, set the stage for a more precise characterization of the special features of speech.

An important extension to the work on infant categorical perception was Kuhl's research on the problem of equivalence classes. Here, the subject is asked to discriminate stimuli that differ on a variety of measures (i.e., in contrast to typical categorical perception tests involving differences in one domain such as voice onset time) and to report whether two stimuli are the same. The test involves a conditioning paradigm and asks the infant to respond with a headturn toward a visually interesting object (e.g., a bear banging a drum) when she detects a novel exemplar from the same vowel class (e.g., after repetition of an adult-produced /a/, respond to a child-produced /a/—nonsignificant variation on an age dimension); no headturn is expected if the exemplar is novel but from a different vowel class (e.g., /i/). Results from experiments using both distinctly different vowel classes (e.g., the schwa sound as in "pop" and /i/ as in "peep"), and ones from quite similar classes (e.g., schwa sound as in "cot" and /ĕ/ as in "cat"), have demonstrated that by the age of six months, infants perform with adultlike accuracies, correctly categorizing equivalence classes of English vowels (Kuhl 1979, 1983, 1989).

A subsequent set of experiments was then designed to assess whether speech categories are associated with prototypical exemplars. The framework for these experiments was prototype theory (see chapter 3), originally developed to handle problems of categorization in the conceptual domain—in particular, with input from the visual modality (Medin and Barsalou 1987; Rosch 1975; E. E. Smith and Medin 1981). For example, when the category *bird* is explored, most people consider *robins* to be more prototypical of the category than either *eagles, chickens,* or *penguins*. Prototypes, then, represent something like the central or most representative exemplar of a category, and this assignation depends on their frequency of occurrence or the dominance of specific featural properties. In the speech version of this problem, Kuhl and her colleagues (Grieser and Kuhl 1989; Kuhl 1991) started by creating a set of synthetic continua within an acoustic space.[22] Adults were then asked to rate whether a particular exemplar was a "good" or "poor" instance of the category—that is, whether it perceptually matched what the listener considered to be a prototypical example of the phonetic category. The result of this rating procedure was a plot of exemplars that differed acoustically from the prototype (e.g., see Figure 5.21); *increasing* distance from

---

22. The acoustic space was defined on the basis of the first two formant frequencies.

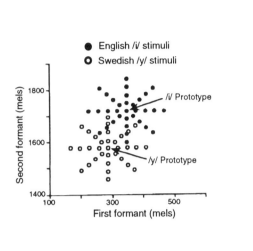

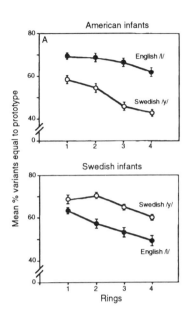

Figure 5.21
Exemplars of English /i/ and Swedish /y/ used to test for infant's perceptual biases. Each vowel set included a token that was considered exemplary with regard to its perceptual features; this token is labeled as the prototype and is located at the center of 32 other variants. Increasing distance from the prototype of each vowel (i.e., four rings around the prototype) corresponds with decreasing perceptual similarity to the prototype (redrawn from Kuhl et al. 1992).

the prototype (i.e., rings surrounding the prototype) corresponds to *decreasing* acoustic similarity. In tests of adults, using the vowel /i/ as in "peep," results indicated that prototypes were more similar (less discriminable) to other variants within the class than were nonprototypes. Or as Kuhl (1992) puts it," The prototype perceptually assimilates surrounding stimuli to a greater extent than is the case for nonprototypes. . . . The prototype appears to draw other stimuli toward it, effectively reducing the perceptual distance between the prototype and surrounding stimuli" (p. 452). This process has been called the "perceptual magnet effect" (Kuhl 1991).

   Given that phonetic prototypes have an organizing function in the perceptual world of adults, an additional set of experiments was conducted to assess whether such prototypes play a similar role in the perceptual world of infants, and if so, whether experience with native language modifies the nature of the prototype. Using the /i/ vowel and the ratings obtained from adults, Kohl found that infants were also guided by a magnet effect. Interestingly, macaques tested with the same stimuli failed to show the effect (i.e., prototype and nonprototype were treated similarly).[23]

How does the infant acquire its prototypes? To answer this question, Kuhl and colleagues (1992) conducted experiments with American and Swedish infants (six-month-olds), using the American English vowel /i/ and the Swedish vowel /y/. The paradigm was the same as in previous tests, except that half of the infants were tested with the prototype from their native language and half were tested with the prototype from the nonnative language. Results (Figure 5.21) showed that by six months infants perceived greater similarity between the native language prototype and its variants than they did for the nonnative language prototype and its variants. In other words, the perceptual magnet effect is influenced by experience with the native language. From these results, Kuhl (1992) argues that infants enter the world with innate knowledge about "natural psychophysical boundaries" (p. 454) that facilitate breaking up the continuous sound stream into functionally meaningful, though coarse-grained, categories. With linguistic experience, however, the categories and their prototypical exemplars are defined, leading to more fine-grained linguistic categories. Apparently, this process is already in progress by the age of six months. As discussed more fully in a later subsection, "Extracting Language from the Acoustic Environment," this line of reasoning conflicts with more traditional views of infant perception (Best 1993; Werker and Tees 1984; Werker and Lalonde 1988), suggesting that linguistic experience modifies phonetic perception before the infant can speak or properly process information about word meaning.

The perceptual studies reviewed thus far suggest that human neonates have quite impressive abilities to pick out, from the incoming acoustic stream, salient linguistic features. How does the immature brain perform this task? To address part of this problem[24]—specifically, where in the brain this task is performed— Dehaene-Lambertz and Dehaene (1994) recorded from the skull surface of two- to three-month-old infants[25] using high-density event-related potentials (ERPs). The task involved passive listening to CV (consonant-vowel) syllables varying in place of articulation. One presentation condition (standard) involved playing the

23. The failure here is of course interesting, especially in contrast to the results obtained with ma-caques and other nonhuman animals on tests of categorical perception. As Kuhl points out, tests of categorical perception and magnet effects are clearly tapping into different perceptual mechanisms.

24. Recall from chapter 3 that studies of brain imaging on human infants are only in their infancy, because of the problem of obtaining sufficiently high resolution with a relatively small brain.

25. The device used was a geodesic net, a kind of helmet outfitted with a large array of sensors for picking up electrical activity.

same CV string five times in succession, the other presentation (deviant) involved playing the same CV string four times followed by a fifth, different CV string. Dehaene-Lambertz and Dehaene predicted, based on neuropsychological studies of adults and behavioral studies of infants, that there should be (1) a decrease in ERP peak amplitude as a function of stimulus repetition, (2) an increase in ERP peak amplitude in response to discriminable novelty (i.e., to the change in CV), and (3) a left-hemisphere bias if the stimuli are processed linguistically. As indicated in Figure 5.22, infants showed a significant drop in amplitude for both Peak 1 and 2, especially after the first stimulus presentation. In addition, Peak 2 showed a significant increase in amplitude following the deviant fifth stimulus, but not in response to the standard stimulus. This increase in Peak 2 was particularly striking for recordings from the left hemisphere and was detected within 400 msec of the CV presentation. These results therefore provide general information about areas of brain activation during passive processing of linguistically salient information. They do not yet tell us *how* the neonate brain processes linguistic information.

First Utterances   The neonate is not a silent creature, but she is also not a linguistic creature, especially at the production end. Neonates and infants up to a few months of age cry, shriek, scream, coo, raspberry, and grunt (Figure 5.23). Such utterances are not referential, but they are meaningful: for the infant, they represent the only auditory vehicle for communication, and for the caretaker, they represent one of the most useful sources of information about the infant's emotional and motivational state. Put somewhat poetically, the earliest vocalizations provide a window into the infant's heart. Rapidly, however, this window opens up onto the mind. The goal then is to establish the relationship between the infant's first sounds and the subsequent emergence of full-blown speech.[26]

Some of the most detailed analyses of the child's earliest vocal utterances have focused on crying (also see chapter 7), motivated by applied and pure research questions (Gustafson, Green, and Cleland 1994; Lester and Boukydis 1992). At its onset, the child's cry represents a highly distinctive sound, recognized immediately by parents and nonparents, and serving the function of, minimally, eliciting an orienting response, and most often, eliciting care. Several research teams have assessed the structural design features of this vocalization, in addition to testing

---

26. One problem with the modern approach is that it often fails to acknowledge the immediate function of such early utterances. That is, vocalizations such as coos and grunts not only are acoustic bridges to the world of speech, but also have their own design features and their own functions; for an illustration of this kind of thinking, see McCune and Vihman (in press); McCune et al. (in press).

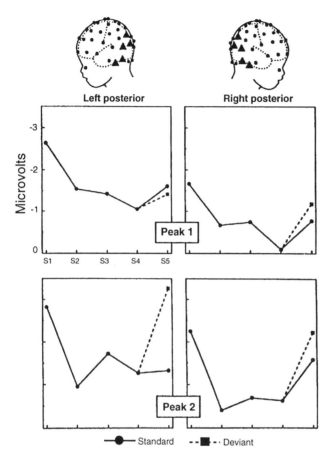

Figure 5.22
Changes in the ERP response (Peak 1 and Peak 2 voltages) during presentation of five syllabic stimuli (either "ba" or "ga"). High-density event-related potentials were obtained from the left and right posterior temporal lobes. *Standard* stimulus presentation (circle, solid line) refers to five identical repetitions of a syllable. The *deviant* stimulus (square, broken line) represents a sequence of four identical repetitions of a syllable followed by the presentation of a different syllable (redrawn from Dehaene-Lambertz and Dehaene 1994).

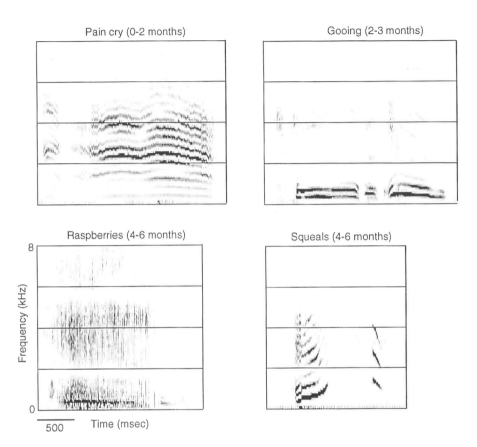

Figure 5.23
A sample of sounds produced by young infants. A relatively narrow filter (256 points, 22-kHz sample rate, 8 bits) was used to emphasize harmonic structure and for comparison with vocalizations produced by nonhumans.

for the perceptual salience of particular features. Cries are characterized by distinctive modulations of the fundamental frequency in addition to changes in the resonance frequencies. Cries of normal infants differ significantly from those of preterm or neurologically damaged infants, and such differences are medically diagnostic (Lester et al. 1991; Lester and Zeskind 1978). For example, preterms differ from normals on various measures of the fundamental frequency, the first two formants, and the overall amplitude envelope.

The earliest cries appear to be completely involuntary, driven by the limbic system. As cortical structures mature, however, cries can be voluntarily produced and used to manipulate the behavior of potential caretakers. Parents are clearly aware of changes in cry acoustics (Zeskind and Collins 1987; Zeskind et al. 1985) and often use such changes to guide their own parenting responses. As Lester and Boukydis (1992) point out, it seems clear that there has been strong selection pressures on the nature of the infant's cry and on the parents' perceptual sensitivities, since such factors serve the important function of regulating the parent-offspring relationship.[27]

During the phase associated with the production of involuntary cries, the child begins to produce a new set of sounds that includes the previously mentioned grunts, squeals, laughs, and so on. Many of these emerge at around three months, a period associated with descent of the larynx, a greater level of sound control at the level of the supralaryngeal cavity, and a significant increase in the ability to dictate the form and timing of motor output. How can we make sense of this broad array of sounds?

In a recent series of papers, Oller (1986) and Oller and Eilers (1988, 1992) have laid out a methodological framework for objectively analyzing the structure and function of prespeech utterances, including the transition to babbling and first words. This framework, sketched in Table 5.1, is significant because it provides a set of nonlinguistic, noncognitive criteria for characterizing the design features of the infant's first vocal utterances and, as such, provides the basis for interspecific comparisons. The criteria are derived from considering the acoustic and articulatory features that are necessary for the proper development of speech.

Stages are notoriously dangerous in developmental studies, but Ollers and Eilers are careful to define their periods broadly, allowing for significant variation between infants. As mentioned earlier, it is not until stage 2 that we begin to see the voluntary—intentional—use of vocalizations. That is, the voice becomes instrumental in achieving particular goals. It is at this stage as well that we begin to see the child play with her voice. As Oller and Eilers (1992) put it, "Squeals and growls are utilized in patterns that suggest the infant is exploring the ability to manipulate pitch. . . . Voluntary yells and whispers suggest exploration of the amplitude parameter. Raspberries show further exploration of the ability to produce articulated sounds, enlisting the lips and tongue to the task" (p. 183).

Babbling (Figure 5.24) is considered by most linguists to represent the first indication of speech precursors, and as such, it has received the most attention,

27. A question that we shall return to in chapter 6 is how parents, both human and nonhuman, can determine the relative honesty of the infant's cry, where honesty is defined in terms of the relationship between what the infant requests and what it actually needs (Trivers 1972).

Table 5.1
The Early Stages of Vocal Development in Human Infants (Oller and Eilers 1992)

Stage	Label	Age (months)	Description
1	Phonation	0–2	Quasi-vocalic sounds produced with normal phonation. The vocal tract is typically open, with minimal movement of the tongue and mandible.
2	Primitive articulation	1–4	Combination of quasi-vocalic and protoconsonantal utterances (i.e., primitive syllables) produced at the back of the throat. These sounds therefore indicate the first sign of normal phonation together with articulation.
3	Expansion	3–8	Full vocalic sounds (i.e., vowel precursors) produced repetitively. Squeals, growls, yells, whispers, raspberries, and grunts produced, indicating control of pitch, amplitude, and articulatory movement. Initiation of babbling, known as *marginal babbling,* characterized by protoconsonantal margins (i.e., sound that starts or terminates an utterance) and nucleus (resonant vowels); formant transitions for such utterances are not quite adultlike.
4	Canonical	5–12	First signs of clearly articulated reduplicated syllables such as "mama" and "dada," and later on (11–12 mos) variegated syllables such as "bada." Margins and nuclei are produced in the adult form. Infants will occasionally produce words of the native language, but without necessarily recognizing the fact that they have a specific meaning.

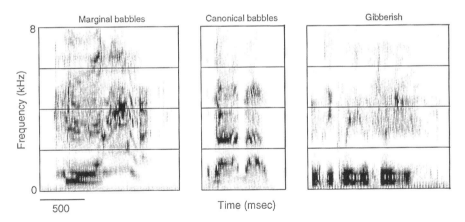

Figure 5.24
Sound spectrograms of infant babbling. A relatively wide-band filter (128 points, 22-kHz sampling rate, 8 bits) was used to emphasize the formant frequencies.

both theoretically and empirically. During babbling we begin to see repetitive motor actions (i.e., mandibular oscillations or cycles) that are associated with significant changes in formant frequencies, auditorily perceived as consonant-vowel strings (e.g., *repetitive babbling:* ba-ba; *variegated babbling:* ba-da). These repetitions, referred to as *frames* by MacNeilage and colleagues (Lindblom, Mac-Neilage, and Studdert-Kennedy, in prep; MacNeilage and Davis 1990, 1993), are perhaps the most fundamental articulatory unit in early sound production and may serve as the ultimate constraint on the kinds of sound produced. Let us look at some of the details of infant babbling.

Current research indicates that the sounds of babbling are confined to a relatively small set of vowels and consonants, suggesting, as J. Locke (1993) puts it, that there is "an irrepressible biological substrate . . . [a] phonetic substance of babble" (p. 179). Moreover, there appears to be a significant relationship between the frequency of occurrence of particular sounds and their occurrence in the child's repertoire of babbled sounds. This relationship is illustrated in Table 5.2 for the /m/ and /r/ sounds of American English; /m/ occurs in 97% of all languages, whereas /r/ occurs in only 5% (Ladefoged and Maddieson 1981).

The findings presented in Table 5.2, together with results from other studies (J. Locke 1983; MacNeilage and Davis 1990), have led developmental psycholinguists to suggest that the sounds of babbling are limited, that children babble in order to get a feeling (articulatorily and perceptually) for the range of relevant sounds in their acoustic environment, that the child's social world encourages such noise making, and that, ultimately, babbling forms the basis for the child's emerging capacity for speech. Can we conclude from these findings that babbling is necessary for the acquisition of speech—that babbling represents the crucial acoustic precursor to speech? Not quite, but some of the relevant data are avail-

Table 5.2
The Relationship between Frequency of Occurrence of a Phoneme and Its Ontogeny in Children's Vocal Utterances (modified from J. Locke and Pearson 1992)

Attribute	/m/	/r/
Frequency of occurrence in world's languages	Universal	Limited to a few languages
Variation across dialects	Low	High
Prominence in babbling	Prominent	Limited
Nature of behavior program	Closed	Open
Role of linguistic stimulation	Limited	Moderately important
Relative timing of appearance developmentally	Early	Late
Probability of disorder	Low	High

able for addressing this question. Specifically, if babbling represents a critical stage in child language acquisition, then children who do not babble (e.g., because of a tracheotomy, breathing tube, or some peripheral motor problem) should evidence significant delays in attaining relevant linguistic milestones. Oller and Eilers (1988; Oller et al. 1985) have compared the quality and quantity of babbling in deaf and hearing infants. Results show that in contrast to hearing infants, deaf infants begin babbling at a later age and produce a smaller proportion of cannonical syllables. This finding suggests that hearing oneself babble may represent an important feedback process in the acquisition of speech, comparable in kind to the process of feedback described for birds acquiring song.

In summary, and following the elegant discussion by J. Locke (1993), babbling appears to provide the normally developing child with auditory feedback about its capacity for vocal utterance. Moreover, babbling sets the infant up in a world that is ready—delighted—to respond because the sounds often resemble speech and therefore indicate a potential card-carrying member of the linguistic community.

Extracting Language from the Acoustic Environment   What kind of acoustic input is most likely to carry the child from prelinguistic to linguistic utterances? To address this question, I turn to a discussion of some of the more significant perceptual processes underlying child language acquisition and, in particular, the kinds of acoustic and visual input that the child potentially receives and uses for parsing the continuous flow of information into meaningful units of analysis (also see chapter 7). Although a considerable amount of research has been conducted in this area, I restrict the discussion to two topics. The first examines the role of affect, and in particular, prosodic characteristics of the speech environment (Fernald 1984, 1989, 1992a, 1992b; Fernald et al. 1989; J. Locke 1993). The second looks at developmental changes in perceptual biases for sounds and sound contrasts, within and outside of the native language (Best 1993, 1994; Best, McRoberts, and Sithole 1988; Kuhl 1989; Kuhl et al. 1992; Werker and Tees 1984; Werker and Lalonde 1988). These studies are relevant to an evolutionary and comparative perspective on communication because they directly assess which features of speech, as opposed to other sounds, are most salient to the preverbal infant. As such, they address the species-specificity of the human infant's perceptual apparatus.

When we speak to each other, not only do our voices convey specific information about objects and events in the world, including our thoughts about them, but they also convey information about our emotional or affective states (see chapter 7). If our voices encode information about emotional state, then such information could be used[28] by the developing infant to assess how individuals feel about the world. Additionally, infants may be able to use the emotive components of the voice to segment the acoustic stream into meaningful units. To examine these possibilities, developmental psycholinguists have looked at differences in the acoustic structure of speech directed toward preverbal infants as opposed to speech directed toward verbal children and adults. The earliest work in this area explored the possibility that when adults speak to preverbal infants, the structure of their speech is relatively more simple than adult-directed speech. One explanation for this modified form of speech, which has been called *motherese* (Snow and Ferguson 1977), is that by simplifying the structure of the child's input, language acquisition is facilitated. Although several studies have suggested that motherese helps the child learn language, other empirical studies and theoretical analyses have pointed out that not all language communities employ motherese, and in those communities where it is employed, it is not necessary for language acquisition. Moreover, if children were actually spoon-fed a Gerber's brand version of language, the input would be so extraordinarily impoverished and watered down that it is hard to imagine how they would develop the structural formalities of adult language (reviewed in J. Locke 1993; Pinker 1994a; Snow and Ferguson 1977).

More recent research in this area, especially by Fernald, Kuhl, and Papousek (Fernald 1984; Fernald 1992b; Fernald et al. 1989; Grieser and Kuhl 1988; Papousek, Papousek, and Haekel 1987), has focused on the actual form of the acoustic input to young preverbal infants, especially the relationship between voice pitch and communicative intent.[29] Based on a series of cross-linguistic analyses (Figure 5.25), there appear to be at least four different pitch contours (i.e., contours of the fundamental frequency), each associated with a different emotional state. *Approval* is conveyed by a contour that is heavily modulated, exhibiting a rapid and extensive rise and fall in the fundamental frequency. *Prohibition* is associated with stacatto-like utterances and only minor changes in the fundamental frequency, usually a brief rise then fall. *Attention,* like approval, is also associated with extensive frequency modulation, but the utterances are usually shorter in duration, and the rise to a frequency peak is faster. *Comfort* is expressed by relatively unmodulated contours that exhibit a frequency downsweep.

---

28. For exceptions and theoretical complications, see Gerken, Jusczyk, and Mandel (1994) and Fernald and McRoberts (1994).

29. The notion of intent is used loosely here, meaning the type of emotional state being conveyed by the mother to the child.

In addition to the pattern of results described in the preceding paragraph, two further observations are worth pointing out. First, the prosodic contours illustrated in Figure 5.25 are also found in tonal languages. This fact is significant because in such languages subtle changes in pitch are also used to convey differences in the *meaning* of a word. Second, mothers *and* fathers elevate the pitch of their voices when speaking to infants, as opposed to adults, and this change has been documented in six languages (Figure 5.26).

Fernald has used the results of such studies to argue for the adaptive significance of prosody in child language acquisition, as well as in the development and strength of the parent-offspring relationship. Specifically, she suggests that the pitch contours observed have been designed[30] to directly influence the infant's emotive state, causing the child to relax or become more vigilant in certain situations, and to either avoid or approach objects that may be unfamiliar. Moreover, by anchoring the message in the melody (Fernald 1989), there may be a facilitative effect on "pulling" the word out of the acoustic stream and causing it to be associated with an object or event. The logic of this claim, together with additional supporting evidence provided in other studies (Fernald and McRoberts 1994), is summarized in the following list (Fernald 1992b):

Stage 1: Intrinsic perceptual and effective salience

From the beginning, the infant is predisposed to respond differentially to certain prosodic characteristics of infant-directed speech. Certain maternal vocalizations function as unconditioned stimuli in alerting, soothing, pleasing, and alarming the infant.

Stage 2: Modulation of attention, arousal, and affect

Melodies of maternal speech become increasingly effective in directing infant attention and modulating infant arousal and emotion.

Stage 3: Communication of intention and emotion

Vocal and facial expressions give the infant initial access to the feelings and intentions of others. Stereotyped prosodic contours occurring in specific affective contexts come to function as the first regular sound-meaning correspondences for the infant.

---

30. As briefly discussed in the previous section on nonhuman primate calls and discussed in greater detail in chapters 7 and 8, the pitch contours associated with human emotional states may be evolutionarily quite ancient. Evidence from broad comparative analyses indicate that in birds and mammals (Hauser 1993a; Morton 1977, 1982), high-pitched vocalizations tend to be associated with either fear or affiliation, whereas low-pitched vocalizations tend to be associated with aggression. In the case of mammals, the similarity with humans represents a potential example of homology (primitive, derived), whereas in the case of birds, it most certainly represents a case of homoplasy (convergence).

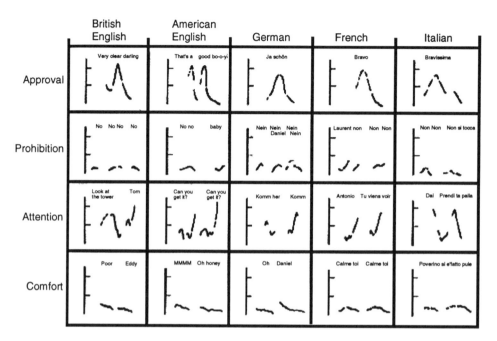

Figure 5.25
Differences in the fundamental frequency contour of infant-directed speech in four different emotive contexts and across five linguistic groups. The y-axis indicates frequency (kilohertz), and the x-axis indicates time (milliseconds) (redrawn from Fernald 1992a).

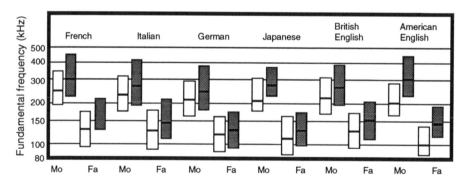

Figure 5.26
Differences in the fundamental frequency of mothers' (Mo) and fathers' (Fa) speech to adults (white bars) and infants (gray bars). The top of each bar represents the mean fundamental frequency maximum, the bottom represents the minimum, and the intersecting line represents the overall mean (redrawn from Fernald 1989).

Stage 4: Acoustic highlighting of words

Prosodic marking of focused words helps the infant to identify linguistic units within the stream of speech. Words begin to emerge from the melody.

The acoustic input described here is paralinguistic, serving to grab the infant's attention, mediate its affective state, and begin the process of deciphering bits of the linguistic code. We learned earlier that, even at the fetal and neonatal stages of development, a suite of phonemic distinctions are perceptually salient. Let us now explore how this initial ability matures and how the child becomes an expert with its own native language.

Everyone knows the feeling: you sit in a foreign language class attempting to learn some French, Spanish, or German, and it's as if you have to bang the information into your head in order to learn any of the words or pronunciations that your eloquent teacher throws at you. Why should learning a second language be so difficult for adults, especially when the young child seems to have an extraordinary talent with learning new languages? Or, more paradoxically, why should language be one of the very few domains where children appear to be better at learning than adults? The conceptual issue here is one that was raised earlier in reference to song acquisition in birds. It is the notion of a critical or sensitive period. Specifically, it seems that at some age, humans lose the ability to acquire a second language, or minimally, find it exceedingly difficult to do so. Even more subtly, several studies have demonstrated that adult listeners apparently do not perceive phonemic contrasts that are outside of their native language (e.g., native Japanese speakers fail to show a categorical discrimination/labeling function for the *ra-la* distinction; Miyawaki et al. 1975); it is only with quite extensive training that nonnative contrasts become perceptually distinctive (Lively et al. 1994; Pisoni et al. 1982; Werker and Logan 1985).[31] The interesting developmental question then, is at what age does the child go from an open language acquisition device to a closed one, and what are the factors that guide this process?

The first stabs at this problem were based on neuroanatomical and physiological data suggesting that at puberty the human brain generally loses a great deal of its

---

31. An interesting feature of the study by Lively and colleagues (1994) is that long-term retention of nonnative phonetic contrasts was obtained by means of a training program that emphasized the natural variation in phonetic structure. Specifically, rather than providing subjects with a limited set of training material, they provided them with a large range, apparently sufficient to capture differences between multiple speakers. Such variation has often been claimed to be essential for developing perceptual constancy in infancy (Kuhl 1983).

plasticity. Consequently, researchers predicted that prior to the onset of puberty individuals should be capable of making phonemic contrasts in all languages, but subsequently will be limited to contrasts within the native language (Lenneberg 1967). Werker and her colleagues (1981, 1983) have tested a range of subjects from six months to 12 years. In one set of experiments, contrasts between two variants of a /t/ sound in Hindi were presented to native Hindi adults and to English-speaking adults, infants, and children; this is an extremely subtle acoustic contrast and was therefore compared with a perceptually more robust voicing distinction (Figure 5.27). Results showed that 6–10-month-old infants had no problem discriminating nonnative sound contrasts, but after this age (and, presumably, somewhat before), discrimination rapidly deteriorated. There is, however, one exception to this pattern of experience-dependent change. Individuals who were exposed to Hindi early in life but subsequently matured into English-speaking adults maintained their ability to discriminate Hindi contrasts in adulthood. The implication here is that experience early in life can actually preserve some perceptual distinctions, even in the absence of relevant experience later in life.[32] Werker (1989) summarizes her research by concluding that there is "universal sensitivity in early infancy followed by only language-specific sensitivity beginning around ten to twelve months" (p. 59).

Werker's perspective suggests that the developmental cutoff for perceiving nonnative contrasts is quite rigidly specified at 10–12 months and that the perceptual changes associated with the discrimination of nonnative contrasts (i.e., phonology) are specifically coupled to changes in phonetic perception. This perspective, though supported by a considerable amount of data, is by no means supported by all of the data. For example, Best and her colleagues (1988) have tested infants from 6 to 14 months of age on a large number of nonnative contrasts, and though some of their results fit with the 10–12-month discriminative loss, other results do not. Thus, infants between 12 and 14 months, growing up in an English-speaking environment, can discriminate nonnative Zulu click contrasts that English-speaking adults perceive as nonspeech sounds (Best, McRoberts, and Sithole 1988). These results conflict with a familiarity hypothesis which suggests that infants lose the ability to discriminate contrasts that they have not experienced. However, infants in the 10–12-month-old category, and

---

32. Interestingly, when the same Hindi and Nthlakapmx contrasts are shortened, causing them to lose their linguistic distinctiveness but maintain their phonemic differences, adult listeners can readily make the perceptual discriminations (Werker and Tees 1984). Thus it is the language-processing ability which has been apparently lost, rather than the pure acoustic processing.

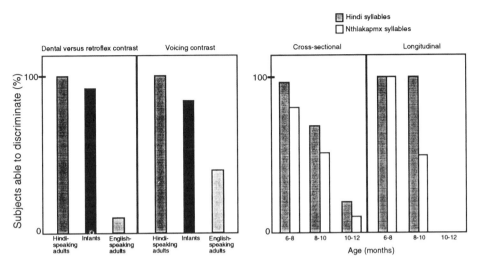

Figure 5.27
Differences in subjects' ability to perceive phonemic contrasts within and outside of their native language. The figure on the left shows results from subjects tested on a Hindi contrast involving a dental versus a retroflex /t/ sound, and two sounds that differed with regard to voicing; the infants were 6–8 months old. The figure on the right shows the results of a cross-sectional and longitudinal test on infants between 6 and 12 months. The Hindi syllables were the same as in the figure on the left, and the Nthlakapmx (a North American Indian language) syllables were a velar and uvular /k/ sound (redrawn from Werker et al. 1981; Werker and Tees 1984).

even older, are not able to discriminate all nonnative contrasts, and most importantly, actually fail to discriminate some contrasts that adults readily discriminate (Best 1993, 1994). As a working framework, Best offers a developmental analysis of the child's phonological change from a phonological universalist to a phonological specialist, presented in Table 5.3.

In general, the scenario proposed by Best shows some of the steps an infant takes as it moves along its path from a linguistically naive individual to a linguistically sophisticated one. Some of the mechanisms by which the child maneuvers along this trajectory are quite simple, as when the child builds a set of articulatory-phonetic associations based on experience with the native language. At this phase (II) the infant fails to recognize the linguistic import of the contrasts it distinguishes. More complicated, and still poorly explained, is the observation that infants discriminate native versus nonnative vowel contrasts at an earlier age than they do for consonant contrasts (Kuhl 1992); such differences have been attributed to constraints on memory and the distinctive functions of consonants and vowels in language. Gradually, and quite slowly, the child starts the process

Table 5.3
Changes in Speech Perception as a Function of Developmental Changes in Phonological Organization (modified from Best 1993)

Age	Phonological Phase
I. 0–6 months	Extraction of simple articulatory gestural information. • Language-universal phonetic details.
II. 6–10 months to 18–24 months	Recognition of native patterns of gestural information. • Language-specific phonetic categories.
III. 18–24 months to 5–6 years	Emergence of functional phonological contrasts. • Language-specific linguistic contrasts.
IV. 5–6 years to puberty (?)	Development of phonemic awareness. • Language-specific phonemic units.

of appreciating the discreteness of phonological contrasts and the extent to which they can be recombined into an even larger set of meaningful units. In its infancy, and clearly requiring more work, is the role of brain-state changes in the process of phonetic and phonological development (Mills, Coffey-Corina, and Neville 1993; Neville et al. 1993).

Explosion of the Lexicon   We now have the infant primed for becoming an articulate participant in the linguistic community. Coinciding with the perceptual tuning to native contrasts, as well as the decreasing sensitivity to nonnative contrasts, is the emergence of first words (L. Bloom 1973; R. Brown 1973). Production, however, generally follows comprehension by a couple of months (Huttenlocher and Smiley 1987), as children clearly evidence signs that they understand what people are talking about before they themselves can enter into the conversation. Most of the child's first words are names for specific individuals or commonly encountered objects. In addition, action verbs figure prominently in the lexicon as the child goes from grunting requests to productions such as "give" and "go." Thus, in Bretherton's (1988) analysis of 13-month-olds, the three most frequent word categories were parents, requests-refusals, and greetings. At 18 months, the child makes an important transition from single words and a quite limited lexicon to a lexical explosion and the first hint of word combinations or phrases (E. V. Clark 1993). In fact, from the age of 18 months to six years, the child reaches a lexicon of 9,000–14,000 words at a clip of approximately nine words per day (Carey 1978; Dromi 1987).

How does the child recognize a word from the apparently continuous stream of acoustic information in a person's utterance? Once a word is extracted, how is the word's referent or meaning grasped? And once word meaning and reference are extracted, how does the child work out the syntax that allows for words to be combined into phrases? In this section, I briefly discuss some relevant data on word recognition (the first question), and leave the primary discussion of meaning, reference, and syntax (the second and third questions) to chapter 7, where matters psychological are discussed in greater detail.

What cues are available to a child attempting to detect word boundaries? One study of 3-day-old infants suggests that timing events may provide at least some relevant cues to word boundaries. Thus Christophe and colleagues (1994) examined the hypothesis that infants should be able to discriminate between sequences that contain a word boundary from those that do not. Using a nonnutritive sucking approach for discrimination, they compared French bisyllabic stimuli from within words (e.g., **mati** as in mathe**mati**scien) with bisyllabic sounds between words (e.g., **mati** as in panora**ma ty**pique). Results support the position that babies discriminate between items that contain a word boundary from those that do not, though it is unclear from this study how babies perform this discrimination.

In cases of isolated naming, periods of silence surround the utterance. Thus, when a mother points to a toy truck and says "truck," the auditory gap on either side of this utterance is surely relevant, helping to pick out items in the environment. Also relevant is the fact that the mother points to an object. We know from several lines of research that at a relatively early age the child normally follows the gaze of other individuals[33] and also notices the significance of pointing gestures and head orientation—the idea of joint reference (Baldwin 1991). Visual cues are clearly insufficient to explain the child's ability to extract words, however, since children who are blind at birth reach the primary linguistic milestones at the same age as sighted children (Landau and Gleitman 1985). Thus, although pointing, nonverbal gestures, and eye gaze may facilitate word-object associations, they are clearly not necessary.

I previously discussed the role of prosody in parent-child speaking interactions, emphasizing the observation that such vocal modifications have a direct effect on the child's emotional state. Prosodic modifications also provide the child with a more structured speech environment and, as such, appear to facilitate word recognition. However, a structured speech input can only account for a small portion of the child's ability (Fernald and McRoberts 1994; Gerken, Jusczyk, and Mandel 1994). Consequently, a number of developmental psycholinguists have argued that the child enters the word-recognition task equipped with constraints

33. Young autistic children often fail to appreciate the importance of gaze in social interactions and are often delayed with regard to language-acquisition milestones (Baron-Cohen, Tager-Flusberg, and Cohen 1993).

that greatly reduce the number of potential hypotheses about word meaning.[34] Specifically, Markman and her colleagues (Markman 1990; Markman and Hutchinson 1984) have argued that word recognition emerges from a general constraint imposed by mechanisms of categorization, indicating that first words are often associated with *whole objects* and *taxonomically,* rather than *thematically,* related objects in a scene. For example, when a child hears a parent say, "Look at the lion in the cage," the most common first guess (hypothesis) is that "lion" refers to the entire creature, rather than to its mane, or tail, or paws. This is what developmental psycholinguists refer to as the whole object assumption. Moreover, though lions are often associated with cages (thematic), the child is more likely to say "lion" when it sees something that looks like a lion (e.g., a jaguar, mountain lion, or domestic cat) than when it sees cages. This bias, however, appears to be restricted to nouns, and more specifically, to things that are named (Macnamara 1982). Such early biases, which arise by 18 months, and are joined by a suite of conceptual riches (e.g., animate versus inanimate distinctions) form the basis for subsequent classification schemes (Carey 1985; Gelman, Coley, and Gottfried 1994; Keil 1994a, 1994b).

I now leave the normally developing child and turn to a discussion of language development in the deaf. Keep in mind, however, that we will return to language development in chapter 7, but in the context of parallel developments of the mind. Specifically, we will look at how children's understanding of other minds figures prominently in their use of language and how their use of language informs us of their cognitive powers—the relationship between language and thought (Jackendoff 1992, in press).

5.4.3   Human Sign Language

As discussed in chapter 4, most linguists consider human sign language to "count" as a bona fide natural language, equipped with all the essential design features. As Meier and Newport (1990) state, "The now-extensive literature on American Sign Language (ASL) and other sign languages clearly demonstrates

---

34. Note, as a prelude to issues discussed in chapter 7, that some of the constraints imputed by developmental psycholinguists are the same sorts of constraints required by the ethologist wishing to navigate around Quine's (1973) problem of referential opacity, or the indeterminacy of word meaning. Specifically, for every sound-context association, there would appear to be an infinite number of potential hypotheses regarding sound meaning, and yet all normal adult speakers of a language readily agree on the general meaning of words, as well as their referents. This universal agreement suggests that there are constraints on word meaning, and such constraints must serve the child well in acquiring a functional lexicon.

that there is neither an overall evolutionary advantage for speech nor an overall evolutionary disadvantage for sign'' (p. 2).[35] Given these attributes, it is reasonable to ask whether the process of acquisition parallels that seen for spoken language. If Chomsky is right—that language is based on an abstract nonspeech component—then the language organ should accept input from multiple sensory modalities. In this section the acquisition of signed language is described, focusing in particular on similarities and differences with the acquisition of spoken language.

The previous section showed that humans begin the path to spoken language in utero. After birth, the infant first produces a variety of nonlinguistic sounds, but then rapidly appears to transform these sounds into linguistic utterances that refer to objects and events in the world. Most infants move through these different phases of the language acquisition process at approximately the same pace—they attain linguistic milestones at the same age. A crucial question, therefore, is whether there are different maturational constraints on the visual-gesture modality, as opposed to the auditory-vocal modality, that would lead to different milestones for sign language acquisition.

Recent reviews of the literature suggest that, in general, children acquire sign and spoken language in much the same way and over the same general time course (Newport 1990, 1991; Newport and Meier 1985; Petitto 1988, 1993). However, data by Bonvillian and colleagues (Bonvillian et al. 1983a, 1983b; Folven and Bonvillian 1991) and by Meier and Newport (1990) suggest that in deaf children the one-sign and two-sign combination phase of development occurs earlier than the equivalent one- and two-word phase in hearing children.[36] They suggest that the difference in age at acquisition is approximately 1.5–2.5 months in favor

35. Although Meier and Newport never discuss the evolutionary aspects of spoken and signed languages, their statement is wrong in one sense: the auditory domain provides a selective advantage with regard to communication across visually obstructed channels, and this will certainly be the case for some of the primitive and ancestral signaling systems, such as warning calls.

36. A significant methodological issue for both the signing and spoken language acquisition work is the criterion by which the experimenter establishes what counts as a word or sign (see discussion in Vihman and McCune 1994). This is problematic at several levels. First, there are a number of studies that base their estimates of a child's developmental progress on reports by parents. Such reports are difficult to assess because parents may well overestimate what counts as a sign or word. Second, a number of studies are based on short videotaped records of children presented with an experimental task. This procedure may both bias the types of signs or words produced and, given the brevity of the sampling, underestimate repertoire size. The following represents a list of parameters used by several linguists to guide their assessment: The word or sign must (1) have a shape that resembles the adult form, (2) be uttered spontaneously, (3) be produced in an appropriate context, and (4) not be bound to a particular context.

of the deaf children, regardless of whether one considers the age at which the first word or sign was produced or the age at which the child acquires a lexicon of 10 signs/words. Although this difference appears small, given the life span of humans, it represents a potentially significant developmental difference if one considers the kinds of maturational changes that are ongoing during this period of the child's life. There is, however, no convincing evidence that more fully developed syntactical capabilities occur earlier in deaf than in hearing children. Meier and Newport argue that although both signed and spoken language are controlled by the same underlying mechanism, spoken language—specifically, the one-word phase—is relatively delayed because of difficulties associated with early processing in the auditory periphery.

Support for a peripheral constraint comes from published results showing earlier maturation of the visual cortex and the motor control system for hand movement than of the auditory cortex and vocal tract. In addition, the gestures used during sign represent motorically simpler motions than those used for speech, and parents often mold the signed gestures of their infants, whereas little or no molding can be performed on the spoken word. If the relative delay between signed and spoken language is due to the fact that the peripheral auditory system is unable to provide the necessary feedback, then the gap between the onset of correct production/usage and comprehension should be greater for signed than spoken language. This explanation would also work well with the observation that there are no differences between signed and spoken language with regard to later milestones (i.e., syntax) because auditory and visual processing are fully developed.

Other researchers working on language acquisition have been critical of the claims made by Meier and Newport for a timing difference between signed and spoken language. Specifically, Petitto (1988) has argued that Bonvillian and colleagues actually provide two different ages for the onset of first signs in deaf children. Signs that are similar to the adult form, but are not necessarily produced referentially, appear at 8.2 months. In contrast, signs that are consistently used to refer to objects and events are produced at 12.6 months. Meier and Newport favor the earlier date, which leads to the claim that first signs appear before first words. Petitto, however, claims that most researchers working in language acquisition treat nonreferential utterances as babbling and referential utterances as "true" signs or words. Consequently, the 8.2-month value corresponds precisely to the age at which infants produce syllabic babbling, and the 12.6-month value fits with the timing of first word production. Petitto therefore concludes that signed and spoken language acquisition follow the same developmental time course, including the attainment of key linguistic milestones. A more comprehensive data set is required to resolve the conflicting views regarding timing of first words and signs.

The question we must now address is whether the similarities observed thus far hold up when more specific processes in language acquisition are explored. Consider first some potentially significant differences between signed and spoken language: (1) Signed languages have a greater number of iconic elements, although this fact does not appear to influence the grammatical structure of either language. (2) Whereas most users of spoken language are exposed to the native language from birth—and even before—many users of signed language are not exposed to sign until weeks or even months after birth.[37] The remaining deaf children are born to hearing parents, and thus exposure to sign language is considerably delayed. This natural variation in exposure creates a beautiful little experiment of nature. This is unveiled by observations of children who, for example, invent their own gestural system of communication prior to explicit exposure to formal signing. And for those children who are not exposed to formal sign language until they are 12 years old or older, there are typically significant deficiencies in signing skills. These deficiencies, of course, provide some evidence of a sensitive period for language acquisition, which exists for both signed and spoken languages (Newport 1991). (3) Whereas most of the articulators for speech are relatively invisible to the infant, they are "transparent" in signed language. This difference is only important if, during the acquisition of spoken language, children use perceived articulatory gestures to guide their own vocal output (see section 7.3.5 for results on cross-modal perception of spoken language).

With these points in mind, let us look at two fascinating ontogenetic patterns: language development in normal infants exposed at birth to speech and sign, and the motoric gestures used by young infants prior to the emergence of fully crystallized adult signs. Concerning the former, Petitto studied a small number of bilingual and bimodal children aged 7–24 months. Analyses indicated that language-acquisition milestones were attained in both modalities at the same age as those communicating in one modality, and as children acquiring two spoken languages. Given the communicative dominance of acoustic over visual input for hearing children, one might have expected a timing bias in favor of spoken language. Since no biases could be detected, Petitto (1993) argues that "unitary

---

37. Moreover, whereas the auditory system can resolve a good deal of the speech stream in utero, the acuity of the visual system at birth is not great, and thus subtle differences in sign structure may not even be perceived by the young child.

timing constraints determine the acquisition of all linguistic milestones in both spoken and signed language'' (p. 368).

Turning to the early gestures used by young infants, Petitto and Marentette (1991) observed that deaf infants produced gestures that appeared to exhibit syllabic organization (Figure 5.28), were remarkably different from the manual movements performed by hearing infants, and perhaps most significantly, differed from the communicative but nonlinguistic gestures used by hearing and deaf infants (e.g., waving, pointing). For example, what Petitto and Marentette call *manual babbling* exhibited a rhythmic pattern and handedness distinct from other manual gestures (Petitto 1993). As with vocal babbling, infants exposed to signed languages from birth do not manually babble in any particular signed language. At approximately 12 months, however, language-specific phonetic units emerge, forming the basis of the native signing system. Last, at about 9–12 months, deaf and hearing children begin to use the pointing gesture in a communicatively meaningful way. From 12 to 18 months, however, pointing to people stops, though pointing to objects and locations continues. And at 12 months, both hearing and deaf children use naming instead of pointing to refer to people. In fact, deaf children even make the same kind of signing mistakes as hearing children make with words, as in the case where a child points to another individual to say ''you'' when in fact he means ''me.'' These observations provide strong evidence that deaf children, like hearing children, make a precise distinction between language and gesture, each communicative ''tool'' driven by different conceptual frameworks or theories (McNeil 1985; Petitto 1993).

Based on her own analyses, as well as those of others working in the same area, Petitto (1993) concludes that ''linguistically structured input—and not modality—is the critical factor required to begin and maintain very early language acquisition. . . . Language ontogeny begins through the complex interaction of three mechanisms: (1) general perceptual mechanism, (2) constraints on motor production, and crucially, (3) specific structural constraints that are especially tuned to particular aspects of linguistic input'' (p. 375). These conclusions lead to the prediction that over the course of human evolution, selection contributed to the design of a brain that could generate and perceive language in multiple modalities. Though most humans speak, everyone had—as a child—the capacity to sign.

Petitto's data on hand babbling and the logic of her arguments have been accepted by many researchers studying child language acquisition. However, important criticisms (Lindblom, MacNeilage, and Studdert-Kennedy, in prep) and contrary empirical findings (Meier and Willerman, in press) have also been raised,

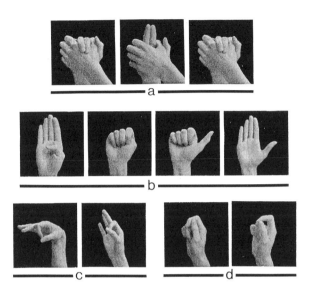

Figure 5.28
Single-frame snapshots of syllables (basic unit of gesture shown) produced by deaf infants during
manual babbling. First-row syllables: (*a*) two-handed handshape change; second-row syllables: (*b*)
single-hand, bisyllabic, handshape change showing four different configurations; third-row syllables:
(*c*) orientation change produced with wrist flexion; (*d*) handshape change as in *b*, but simpler, and
commonly produced by hearing infants (photograph provided by L. Petitto).

and these need to be treated seriously. First, vocal babbling is characterized by
a number of significant features including the following: (1) all normal infants
babble; (2) the onset of babbling follows a normal distribution that ranges from
18 to 48 weeks (mean = 31 weeks; SD = 6.5 weeks); (3) the structure of babbling
derives from an articulatory (mandibular) oscillation—called the syllable frame
(Lindblom, MacNeilage, and Studdert-Kennedy in prep); and (4) a limited repertoire of phones are common to the world's languages. Considering hand babbling,
we do not know whether *all* individuals do it, and for those that do, sample size
limitations preclude providing an estimate of the range of variation in its onset.
Second, though repetitive hand movements occur, we do not know if they are
similar to the repetitive articulatory actions (constonant-vowel iterations)
observed during vocal babbling; such articulatory oscillations appear to underlie
all the spoken languages of the world. Last, we do not know whether sign language universals exist in either infant or adult repertoires. Lindblom and colleagues (in prep) conclude by stating that there is "no convincing evidence that
sign babbling reveals a modality-free linguistic capacity. Instead . . . [there is] a

good deal of evidence for modality specificity, as would be expected if use begets form rather than the reverse, even in a type of language which probably did not evolve but was invented'' (p. 139). This is a strong claim. I would argue, contra Lindblom and colleagues, that the jury is out with regard to the salience of particular sensory modalities for language acquisition. Official reports from the jury will have to await the publication of a larger[38] and more detailed corpus of cross-cultural data on hand babbling, together with sophisticated analyses of gestures, both noncommunicative and communicative, from children who are normal, deaf, and blind, raised by parents with equally varied sensory worlds.

5.4.4   Facial and Gestural Expressions in Primates

In chapter 4, single-unit recording studies of rhesus macaques revealed, minimally, that in the inferior temporal cortex a number of cells are highly responsive to faces but not to nonface visual stimuli. Studies of humans, both brain-damaged and normals, indicated that the occipitotemporal region is often involved in face processing, including face recognition and facial expression. In both human and nonhuman primates, it appears that the right hemisphere is dominant for face processing. In this section, we turn to a discussion of studies that have focused on the ontogeny of face recognition and expression systems. I start with what little is known about nonhuman primates (again, restricted to rhesus macaques) and then turn to the more detailed work on humans.

Facial Expression and Face Recognition in Macaques   From a behavioral perspective, little is known about the ontogeny of facial expression and face recognition in nonhuman primates. Sackett (1966, 1970) conducted a series of experiments that were designed to assess the importance of social rearing conditions on macaque social development. Results showed that when slides of a threatening face were presented, infants reared in social isolation nevertheless responded with behaviorally appropriate withdrawal responses; neutral, or at least nonthreatening, faces failed to elicit such responses. Complementing these findings and the data reported in chapter 4 on the role of the superioral temporal sulcus of macaques in gaze detection and discrimination, Mendelson, Haith, and Goldman-Rakíc (1982) showed that macaques undergo a developmental change with regard to their responsiveness to gaze direction. Specifically, when presented with faces of indi-

---

38. These data are hard to come by, and even the most recent treatment by Meier and Willerman is only based on three deaf and two hearing children, thereby precluding the possibility of statistical inference testing.

viduals looking straight at them or to the side, one-week-old infants spend equally long periods of time looking at each gaze direction, whereas older infants tend to look longer at images of animals looking away from them. Said differently, older infants appear to recognize the importance of gaze direction and the fact that for most primate species, including macaques, staring represents an aggressive threat.

In a study by Rodman, Skelly, and Gross (1991), techniques similar to those used by Gross (1992) on adult macaques were used to assess whether young infant macaques process faces differently from nonfaces and, if so, whether the pattern of response differs from that exhibited by adults. In the first set of experiments, results revealed that extremely few neurons were responsive to visual stimuli in anesthetized 6-week-old infants (upper panel, Figure 5.29). In contrast, neural responses in alert infants were comparable to those exhibited by both anesthetized and alert adults. Superficially, at least, it seems counterintuitive that a process like face recognition, which from all accounts appears to involve higher cortical mechanisms, would be unaffected by anesthesia in adults but affected by anesthesia in infants. At present, there appears to be no clear explanation for these results.

Turning to a more specific age comparison, results showed that the magnitude of the neural response was greatly reduced in infants relative to both anesthetized and alert adults (lower panel, Figure 5.29). Interestingly, however, the stimulus specificities of the infant and adult responses were comparable. Thus both age classes showed selective responses to stimuli that varied as a function of shape, color, and apparent complexity. For example, a cell in one infant showed a strong response to the face of an adult macaque, but no response to the same orientation of an infant macaque face. In fact, even though anesthetized infants showed very poor responsiveness overall, those cells that responded did so with the same pattern of stimulus specificity as cells from alert infants and adults. Rodman and colleagues conclude their work by pointing out that some of the differences between infants and adults may be explained by the relative immaturity of the infant's brain, specifically the fact that connectivity within inferior temporal cortex requires several months from birth to reach adult maturity.

Production of Facial Expressions in Human Infants   We learned in chapter 4 that when adult humans and rhesus macaques produce facial expressions, the left side of their face is often[39] more expressive than the right. Studies of the neurobiology

---

39. Recall, however, that in the report by Gazzaniga and Smiley (1991), results from split-brain patients showed that the right side of the face started into a smile before the left.

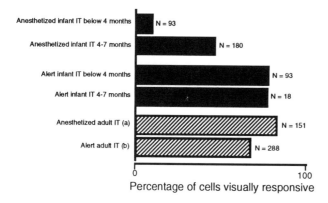

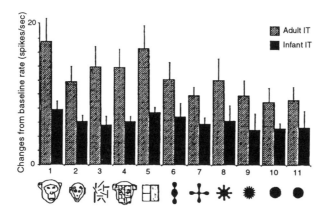

Figure 5.29
Upper: Differences in the proportion of visually responsive neurons within inferior temporal cortex (IT) for anesthetized and awake infant macaques (black bars) and adult macaques (striped bars). *Lower:* Average spike rates recorded from cells within IT of awake infants and adults in response to a set of visual stimuli; recordings were obtained from a mean of 18 cells (range = 13–26) in infants and a mean of 13 (range = 7–21) in adults (redrawn from Rodman, Gross, and Scaidhe 1993).

of human infants (K. R. Gibson and Petersen 1991; Molfese and Segalowitz 1988) have demonstrated that the right and left hemispheres of the brain mature at different rates. How does the differential maturation of the brain influence asymmetries in facial expression?

Best and Queen (1989) used photographs of infants (7–13 months old) spontaneously smiling and crying to create visual chimeras—mirror image pairings of two right and two left sides of the face (Figure 5.30). The test involved asking right-handed adults to judge the relative intensity of infant facial expressions from chimeric photographs. Results indicated that adults picked the *right* hemiface as the more intense. The selection of the right hemiface was strongest when judgments were restricted to the central features of the face. Data from other studies indicate that the mouth and eye regions receive contralateral as opposed to bilateral cortical input. When Best and Queen reanalyzed their data with regard to facial features, results showed that the mouth expressed the most significant rightward bias, implicating cortical mechanisms of control. In addition, the right hemiface bias was least strong in 7-month-olds, suggesting that cortical differentiation may occur gradually over time, beginning at about 7 months of age. Unlike the proposals of Davidson and colleagues (Davidson 1992, 1995; Davidson and Fox 1982; Fox and Davidson 1988; see chapter 7) on valence effects (i.e., different kinds of emotional expression and different patterns of brain asymmetries), no differences between smiling and crying were observed.

Best and Queen interpret their data by making two points. First, because the right hemisphere matures earlier than the left, it may cause inhibition on the left-side motor output, and this, in turn, may lead to greater expressiveness on the right side of the face. Although this is a reasonable hypothesis, it is unclear why the right hemisphere would inhibit the left side with regard to motor output. Second, because adults show a left-visual-field superiority for perception of emotions (Heller and Levy 1981; Levy et al. 1983), one explanation for the greater responsiveness of adults to infant facial expressions may be that the infant's right-face bias is being directly input into the more perceptive side of the adult brain. This is an interesting hypothesis that requires considerably more empirical attention.

Imitation of Facial Expression: A Passport into the World of Communication

Imitation is natural to man from childhood, one of his advantages over the lower animals being this, that he is the most imitative creature in the world, and learns at first by imitation.   (Aristotle 1941, 4486)

Smiling   Crying

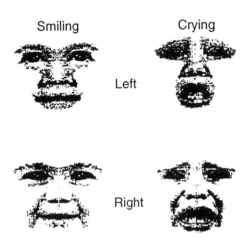

Left

Right

Figure 5.30
Chimeric infant faces emphasizing asymmetry in facial expression. Chimeras were created by pairing the left side of the face with a mirror-reversed left side, and the right side of the face with a mirror-reversed right side. Spontaneous smiling and crying expressions were used (redrawn from Best and Queen 1989).

Human parents often comment on their children's behavior by stating that they are acting like monkeys, copying the behavior of others, the old "Monkey see, monkey do." If only this phrase were accurate, the evolution of the capacity to imitate in humans would be more readily comprehended. To date, however, there is no evidence that monkeys can imitate one another, and only weak evidence that apes imitate (Visalberghi and Fragaszy 1991; Whiten and Ham 1992). In contrast, dolphins have an extensive imitative capacity in both the vocal and behavioral/gestural (e.g., display postures) domains (Reiss and McCowan 1993; Richards, Wolz, and Herman 1984; Sayigh et al. 1990; Smolker, Mann, and Smuts 1993), and at least in the vocal domain, so do songbirds. What, therefore, happened evolutionarily with regard to the neural hardware and software for imitation? If humans are imitators par excellence—*Homo imitans* in the words of Meltzoff (1988)—then how can our closest living relatives almost completely lack this capacity? Are we in the same theoretical boat as we appear to be for the evolution of human language—i.e., no obvious living precursors—or must we also attempt to break down imitation into its componential ingredients so that we can more carefully understand the evolution of this cognitive function? Alternatively, we might accept the possibility that the capacity to imitate evolved within the hominid line, leaving no trace in our primate cousins. To address these questions, which are of fundamental importance to understanding the developmental mechanisms underlying the transmission of species-typical behaviors, including culture (Galef 1992; McGrew 1992; Tomasello, Kruger, and Ratner 1993), I provide a synopsis of where things stand with regard to social learning in nonhuman animals, especially the monkeys and apes; most of this discussion will focus on the imitation of communicative gestures and expressions.[40] With this information laid out, I then turn to a synopsis of what we currently know about the infant's ability to imitate facial expressions.

Several field studies of Old World monkeys and apes (especially chimpanzees) have documented either the appearance and subsequent spread of a novel behavior or, more commonly, the observation of a behavior that is population-specific. In most of these cases, the authors have argued that such traits spread by imitation. The evidence for imitation as the one and only mechanism of social transmission is, however, weak. As emphasized by Sherry and Galef's (1984, 1990) experiments on milk bottle opening, homogeneity of expression within a population need not imply spread by imitation.

What about more controlled laboratory conditions? Perhaps the lack of convincing evidence for imitation in nonhuman primates is simply a reflection of the difficulties associated with field observations. This possibility can certainly be ruled out for Old and New World monkeys, where beautifully designed experiments have been conducted but have failed to reveal evidence of an imitative capacity (Visalberghi and Fragaszy 1991). For captive apes, especially chimpanzees and orangutans, the picture is somewhat less clear. In one study of captive chimpanzees, for example, individuals completely failed to imitate the demonstration of raking behavior by a human observer, whereas young human children readily did so. In a more recent study, however, enculturated chimpanzees (i.e., those raised by humans in a human environment) and human children, but not captive chimpanzees, showed significant evidence of imitation of actions on objects (Tomasello, Savage-Rumbaugh, and Kruger 1993). Similar results have been obtained for rehabilitated orangutans reintroduced into the wild (Russon and Galdikas 1993, 1995), and a description of one of these extraordinary imitative actions is described below:

cooks [at the campsite] used several techniques to make fires: They wet sticks with fuel kept in a metal container nearby, and they commonly scooped small amounts of fuel with a plastic cup also kept nearby; to start a new fire they often touched a burning stick to a fuel-soaked one; to make fires burn faster, they blew on them or fanned them with a round

---

40. Recall that in section 5.4.1, I concluded that there is no evidence of imitative vocal learning in nonhuman primates.

metal lid (held vertically in one hand and waved briskly back and forth horizontally toward the fire). . . . On entering the cooking area, Supinah [an orangutan] picked up a burning stick, blew on its burning end, and briefly bit gingerly at its hot tip. She next went to the metal container, removed the plastic cup and round metal lid sitting on top, scooped fuel from the container with the cup, and plunged the burning end of her stick into the fuel (this action was repeated several times since the burning tip was extinguished as a result of dipping it into the fuel). Next, she poured the fuel from her cup back into the container, placed the cup on the ground, picked up the container, poured new fuel from it into the cup, stopped when the cup overflowed, and put the container down. She retrieved her first stick, put it back into the cup of fuel, picked up the round metal lid, and fanned it repeatedly over the stick in the cup; in fanning, she held the metal lid in one hand, in vertical position, and waved it back and forth horizontally in front of the cup and stick. (Russon and Galdikas 1993, 151–153)

Last, in a project that was explicitly designed to assess the ontogeny of gestural use in chimpanzees (Tomasello et al. 1994), results showed that individuals failed to acquire the form of the gesture by means of imitation. Rather, individuals developed conventionalized styles of gesturing that appeared to be communicatively salient to other group members. Together, results on chimpanzees, and perhaps orangutans, suggest that certain features of the human environment are required for engaging the capacity to imitate in apes and, perhaps most importantly, to imitate a different species. In the absence of this environment, imitation fails to occur, including imitation of actions on objects and communicative gestures. Let us now turn to human imitation of facial expression and, in particular, review what is known about its ontogeny.

As discussed earlier, imitation of a motor sequence involves the coordination of perception and motor action. Based on the apparent complexity of this coordination event, Piaget (1962) suggested that motor imitation emerges quite late in development—at approximately one year of age. A major challenge to this claim has been put forward by Meltzoff and colleagues (reviewed in Meltzoff 1993) based on a series of studies, one involving early imitation of facial expression and a second involving cross-modal matching in the auditory and visual domain. I here provide a review of the facial imitation work because it raises extremely interesting comparative questions and because it provides novel methodological insights for those studying nonhuman animals; the cross-modal perception work is discussed in chapter 7.

The first challenge to the Piagetian perspective came with the publication of Meltzoff and Moore's (1977) experiments on 12- to 21-day-old infants. The paradigm for their experiment was to examine whether infant facial movement was simply the result of overall arousal, or whether, following the presentation of a specific facial gesture, infants could imitate the precise motor actions associated with this gesture. As illustrated in Figure 5.31, three facial expressions were presented to infants: tongue protrusion, mouth opening, and lip protrusion; the first and third are particularly interesting because both involve protrusion but with different articulators. Results showed that infants produced the specific motor action associated with the facial expression displayed.

Criticisms of these results rapidly emerged, raising important issues about the specificity of the response. Some, like those raised by Jacobson and Kagan (1979), were targeted at the eliciting features of the expression and subsequent imitation. Thus they claimed that, because tongue protrusion could be elicited in newborns by moving a pencil toward them, the imitative capacity documented had little to do with facial expression per se. However, as Meltzoff and others have argued, the fact that tongue protrusion can be elicited by a protruding pencil shows that the eliciting features—or in ethological terms, the sign stimuli or releasing mechanisms—can be captured by an object moving close to the face, rather than further away. Moreover, if protrusion in and of itself were the only relevant parameter, then one would have failed to predict a difference between tongue protrusion and lip protrusion.

Adding on to the power of the initial finding, subsequent experiments demonstrated that infants from a wide variety of cultures are capable of facial imitation, that the capacity appears in newborns (as young as 42 minutes old!), and that it can be elicited even when there is a delay between the initial facial display by the experimenter and the first opportunity to reproduce the gesture by the infant (Meltzoff 1993; Meltzoff and Moore 1983, 1989, 1994; Reissland 1988). Thus the general result of early facial imitation seems secure.[41]

To further assess the role of memory and individual identity in early facial imitation, Meltzoff and Moore (1994) conducted an additional series of experiments. Six-week-old infants ($n = 40$) were tested on three consecutive days with one of four facial states: no oral movement, mouth opening, tongue protrusion at midline, and tongue protrusion to the side; the last is a novel gesture, rarely performed spontaneously by infants. All gestures were presented for 15 sec followed by 15 sec of a neutral face, for a total of 90 sec of testing. The testing sequence is described in Table 5.4.

In trials 1, 3, and 5 subjects had the opportunity to show immediate imitation of the facial gesture displayed by the adult. In contrast, trials 2 and 4 were treated

---

41. Reader beware: this is a highly controversial area, and skeptics remain.

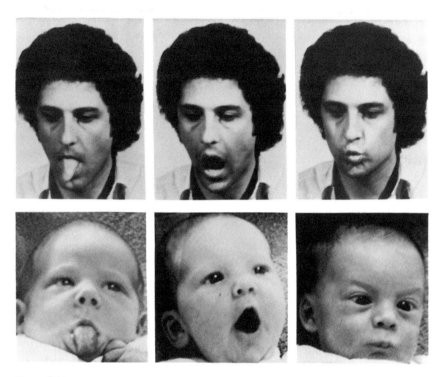

Figure 5.31
Three different facial expressions produced by an adult (top row) and subsequently imitated by a newborn infant (lower row) (Meltzoff and Moore 1977).

as memory trials because facial expressions by the infant were elicited not by an adult display, but by a passive face, implying that the infant was responding to the gesture performed on the previous day. In essence, because all infants received the passive face condition during the memory test, any difference between infants in facial expression was assumed to be the result of the previous test trial. Results showed that infants immediately imitated the expression presented, that after 24 hours they spontaneously produced the expression seen on the previous day (i.e., from memory), and that after repeated exposure and practice, several infants were capable of producing a high-fidelity reproduction of the novel tongue-protrusion-to-the-side gesture. In summary, therefore, infants appear to be able to store representations over a period of at least one day and, as suggested by Meltzoff and Moore, may use such early imitation to identify people they interact with. Thus, when a person who performed a tongue protrusion on the previous

Table 5.4
Experimental Design for Testing Facial Imitation in Human Infants (modified from Meltzoff and Moore 1994, 86)

Stimulus Information	Trials				
	1	2	3	4	5
Treatment group (adult display)					
No oral movement (NOM)	PF*	PF	PF	PF	PF
Mouth open (MO)	MO	PF	MO	PF	MO
Tongue protrusion–midline (TPm)	TPm	PF	TPm	PF	TPm
Tongue protrusion–side (TPs)	TPs	PF	TPs	PF	TPs
Trial duration (sec)	90	90	90	90	90
Day of test	1	2	2	3	3
Trial type†	II	M	II	M	II

*PF = passive face
†II = immediate imitation; M = memory

day reappears, "wearing" a neutral face, the infant's only tool for questioning the identity of this person is to "ask" by sticking out her tongue—as if to say, "Are you that tongue-protruding person?"

To determine whether the age of the model influences early infant imitation, Hanna and Meltzoff (1993) conducted a study using 14- to 18-month-old infants. The experiment (Figure 5.32) involved first training a 14-month-old infant to perform a specific motor act, and then second, to repeat this act in front of a naive age-matched infant. In one set of experiments, the opportunity for imitation occurred within five minutes of the trained infant's demonstration, and in a second series of experiments, the opportunity occurred two days after the demonstration; in the latter, the imitative event took place in a different social context from the demonstration. Infants successfully imitated their trained peers after short and long delays, and social context failed to have a significant effect on the quality of the imitative reproduction.

In an attempt to link early facial expression with early emotional sensation and perception—a form of intermodal interaction—Meltzoff has developed an interesting twist on the view developed by Ekman and colleagues (Ekman, Levenson, and Friesen 1983; Levenson et al. 1991; Levenson, Ekman, and Friesen 1990) for adult emotion (see chapter 7). Here is the argument in brief: Infants do not appear capable of reading the emotions of others from their facial expressions—of getting into their minds. But young infants can imitate the facial expressions of others, and such early imitation may occur in the absence of recognizing the corresponding or appropriate emotion underlying the expression. Once imitation

Figure 5.32
Method used to test for early imitation of a peer. In the first row, a knowledgeable ("expert") infant on the right demonstrates the use of a novel toy as a naive infant sits on his mother's lap and watches. In the second row, we see the same naive infant using a toy in the same way as the expert, but in his absence (photograph courtesy of Dr. A. Meltzoff).

occurs, however, this provides the requisite hook up for understanding—loosely stated—the emotion. As Meltzoff (1993) puts it, "Motor imitation of the facial acts provides a mechanism by which the 'invisible' emotional state is induced in the perceiver. . . . Imitation is more than a sensory-motor skill. It is a powerful mechanism for learning and interpersonal coordination in early infancy" (p. 484).

Face Recognition in Human Infants   The work described in the preceding subsection suggests that, at an extremely early age, human infants are interested in faces and what faces can do. But, do they recognize faces as faces, and how would we know? As in many of the other topics covered in this chapter, the fundamental problem in infant face recognition (see recent review by Chung and Thomson 1995) concerns the timing of recognition and the features used in the process. Two techniques have been used. The infant control procedure involves measuring the amount of time spent looking at either faces or nonfaces (e.g., scrambled or highly schematized faces). Maurer and Barrera (1981) found that infants under the age of four months looked longer at faces than nonfaces. In a

replication of this finding, M. H. Johnson and colleagues (1991) found no significant preference for faces at 5 weeks, a significant preference at 10 weeks, and
then no preference at 19 weeks. Although this result provides some support for
a face preference, the inverted U-shape of this developmental pattern is puzzling
and has yet to be explained, especially in terms of underlying neurocognitive
mechanisms.

The second approach involves presenting infants, lying on their backs, with
slowly moving schematized faces and then measuring head and eye tracking
(Goren, Sarty, and Wu 1975). When newborn infants (less than an hour old!)
were tested, they showed a significant preference (head and eye movement) for
a schematized face over a scrambled or blank face (Figure 5.33, top). However,
when comparably aged newborns were tested with a modified set of stimuli,
there were no differences in head tracking angle, and for eye tracking the face
configuration was only slightly different from the four other stimuli (Figure 5.33,
bottom). At this point, therefore, evidence of significant face tracking, over similarly structured facelike stimuli, is mixed.

One possible explanation for the relatively minor differences obtained for faces
and facelike stimuli (e.g., blotches placed in configurationally appropriate locations for the eyes, nose, and mouth) is that some nonface stimuli provide the
infant with features that are sufficiently attractive to elicit preferential visual
tracking. Kleiner's (1987, 1993) studies, for example, suggest that the preference
for looking at faces as opposed to nonfaces is reduced when face and nonface
patterns are equated for visibility (e.g., amplitude spectra and phase are equated).
Specifically, results indicate that infants under the age of two months respond
similarly to faces and abstract patterns, but by three months they use the specific
configurational features of upright faces to guide their preferences.

Two additional results, however, throw into question the suggestion that infants
are merely responding to low-level perceptual features. First, J. Morton's (1993)
analysis of the power spectra of a schematized face and a scrambled face (i.e.,
where all the face features are preserved, but their position on the paddle rearranged) indicates that there are differences. However, none of the differences
would provide an advantage (in terms of recruiting the infant's attention) for faces
over scrambled faces. Second, it has been demonstrated that newborns just a
few hours old can recognize their mother's face presented live or as a still image
(Bushnell, Sai, and Mullin 1989; Field et al. 1985; Walton, Bower, and Bower
1992). Evidence of recognition comes from a preferential looking paradigm, comparing the amount of time a newborn looks at its mother's face versus the face of
a strange adult female. Recently, Walton and Bower (1993) have used a computer

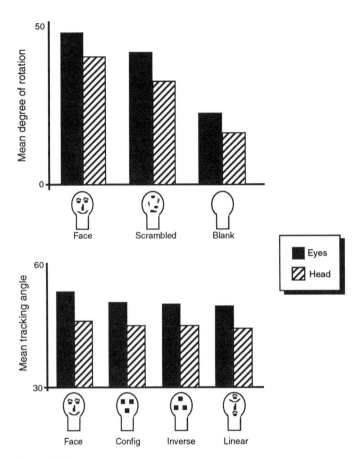

Figure 5.33
Top: Changes in newborn infants' eye and head turns in response to three stimuli. *Bottom:* Changes in newborn infants' eye and head tracking in response to four stimuli (redrawn from M. H. Johnson et al. 1991).

averaging technique[42] to assess whether newborns (8 to 78 hours old) use proto-types to guide their recognition of their mother's face. Based on results from a standard operant-sucking protocol, 14 of 16 newborns looked longer at composite faces they had seen compared to composite faces they had not seen. All of these results point to a recognition system that employs much more fine-grained fea-tures than implied by Kleiner's work. Nonetheless, this is an area of research that requires additional experiments aimed at providing a more complete descrip-tion of the properties of the recognition device at birth and the extent to which particular types of experience mold the output of the device.

Before we leave the problem of how a face-recognition system develops in humans, let me briefly touch upon a set of studies that have attempted to charac-terize some of the neural machinery underlying this problem. De Schonen and colleagues (1993) have tested infants between the ages of 4 and 9 months in a paradigm that is designed to assess whether hemispheric asymmetries are ob-served with regard to processing configural as opposed to local aspects of the face. As discussed in chapter 4, human adults tend to show a left-hemisphere bias for processing local aspects of the face (e.g., differences between individuals or species in terms of the shape of their eyes, nose, and mouth), whereas they show a right-hemisphere bias for processing the configural aspects of the face (e.g., the relative position of the eyes, nose, and mouth).

Results from a visual half-field presentation experiment indicate that from 12 weeks to 7 months, there is no hemisphere advantage for face recognition. By 4–6 months, however, infants show a faster ocular saccade to a photograph of their mother than to a photograph of a strange female, but only when the photo-graph is presented to the left visual field (i.e., right hemisphere). Interestingly, at four months, girls tend to show much less lateralization than boys in this particu-lar task. In a subsequent set of studies, de Schonen and colleagues set out to test whether hemispheric biases emerged when differences in configurational and local features were modified. Results indicate that by nine months, and potentially as early as four months, there is a right-hemisphere bias for discriminating configu-rational changes in the face and a left-hemisphere bias for discriminating local changes.

42. The computer averaging technique involved the following steps. First, images of individuals looking directly into the video camera with a neutral expression were digitized onto an Amiga 2000 computer. The sample included eight adult (ages 23–30 years) females (4 blondes and 4 brunettes) of north European descent. Composites were created (Digi-Paint software, NewTek, Inc.) by first aver-aging two blondes, then two brunettes, then averaging the average blonde with the average brunette. The end product was a sample of eight composite faces.

As pointed out in chapter 4, some researchers (Davidson 1992, 1995) hold that the right hemisphere is more dominant with regard to processing negative/withdrawal emotions, whereas the left hemisphere is more dominant with regard to processing positive/approach emotions. To date, however, no study has attempted to assess when such asymmetries emerge in the developing child and the extent to which particular experiences with emotional expressions—both perceived and produced—guide this ontogenetic process.

Studies with older children have demonstrated that, between the ages of 6 and 10 years, face processing improves as a result of specific experience and expertise with faces. This increase in expertise appears specific to faces, rather than to more general sophistication with regard to processing visual stimuli, because children at this age perform extremely well on discrimination tasks involving abstract visual images. To obtain a first approximation of how the developing brain influences face processing, Kolb, Wilson, and Taylor (1992) tested children between the ages of 6 and 15 years in their recognition and comprehension of facial expressions, and compared their results to those of normal adults and patients with frontal lobe damage, an area believed to be critically involved in emotion processing. Results showed that in a test involving facial chimeras (right-right pairs and left-left pairs), infants exhibited a bias to perceive the left-left chimera as more similar to the original by the age of 6–7 years. In a second test, involving a match-to-sample task with faces from Ekman's class of six standard emotional expressions (see chapter 7), and a second set of faces taken from *Life* magazine, results showed that the number of errors was high among 6–9-year-olds, and then gradually dropped to adult level from 10 to 15 years. Fifteen-year-olds showed adultlike error rates. A similar developmental pattern emerged in the third test where individuals were asked to pick the appropriate facial expression for an expressionless face in a cartoon, where the context was quite clearly depicted. Interestingly, young children (<10 years) showed error rates in the last two tasks that were comparable to those exhibited by patients with frontal lobe damage. Kolb, Wilson, and Taylor interpret these findings as evidence that the lack of comprehension of facial emotion may derive from an immature frontal lobe. Given comparative data on frontal lobe evolution, it would be fascinating to carry out similar experiments with nonhuman primates. My prediction would be that, although some apes and monkeys have frontal lobes that are, relative to body weight, the same size as or smaller than a ten-year-old child's, they nonetheless exhibit the capacity to discriminate among facial expressions. Preliminary support for this prediction comes from Dittrich's (1990) work with long-tailed macaques.

In summary, research on the ontogeny of face recognition suggests that human infants are born with an innate predisposition to respond to faces, to find them perceptually salient—interesting. The features that elicit such interest or attention are, however, quite simple. Consequently, visual objects that adults would reject as "faces" are apparently sufficient to grab the infant's attention and guide its visual tracking system. A brain that has such simple featural demands is well designed, however, because the most important thing for an infant to know is whether someone is looking at her, and most importantly, looking after her. Eyes are most salient, and infants can use quite simple feature detectors to pull eyes out of the visual array. With time, however, details of the face gain in importance, and experience with the world provides the requisite information, constrained by what the brain will accept as a "face."

6 Adaptive Design and Communication

We can really learn the truth about the evolution of signals best from the liars.   (Wickler 1968, 234)

6.1 Introduction

Imagine the following scenario. You are walking on the African savanna and you see a bird with an extremely long tail performing, over and over again, extraordinary leaps above the tall grass (Andersson 1991). From a mechanistic perspective, you might be curious about the energetic costs associated with leaping, and if you were a physiological ecologist, you might begin to design a series of experiments using a doubly labeled water technique to measure the metabolic costs (Vehrencamp, Bradbury, and Gibson 1989). From a functional perspective, however, you would be interested not only in the costs of the bird's acrobatics, but also in its benefits. And here, you would certainly begin to consider the costs of such a display in terms of survival (i.e., selected against by natural selection), as well as the potential benefits in terms of mating (i.e., selected for by sexual selection). In this chapter, I begin by reviewing some relevant issues in evolutionary biology as they relate to the study of animal communication; some of these issues were touched upon in chapter 2. Having laid out the theoretical turf, I then describe the results of studies that have focused on the adaptive consequences of communicative interactions. To follow the logic underlying these studies, it will be particularly important for the reader to keep the signal-cue distinction clearly in mind. For example, some studies show that females choose males on the basis of plumage coloration (Hill 1991) because variation in color maps on to variation in traits associated with male survivorship (e.g., quality of the territory). Plumage coloration in this case would be considered a cue. In contrast, other studies show that plumage coloration forms an integral part of territorial displays (Marchetti 1993; Metz and Weatherhead 1992), with particular wing movements used to reveal patches of color to potential competitors. Here, the display has been designed to *show off* specific portions of the plumage. Plumage coloration in this case would therefore be considered a crucial component of the signal.

When a trait exhibits complex design features, biologists commonly invoke selection as the most likely evolutionary mechanism (Darwin 1859; Dawkins 1986; G. C. Williams 1966). Although this view has long been an integral part of the evolutionary biologist's perspective on nonhuman animal behavior, it has only recently been integrated into discussions of psychological mechanisms and processes in humans and nonhumans, including language evolution (Pinker 1994a; Pinker and Bloom 1990), social exchange (Cosmides and Tooby 1992), deception

(Ristau 1991b), human aggression (Daly and Wilson 1988), and human mating and aesthetics (Buss 1994; Perrett, May, and Yoshikawa 1994). The selection mechanism invoked is, of course, natural selection. Although the advent of modern genetics has radically altered what we know about the molecular biology of inheritance (Gillespie 1991), the theoretical essence of natural selection theory remains largely untouched since Darwin's first formulation (G. C. Williams 1992).

Natural selection, as an evolutionary pressure, was not, however, sufficient to explain what Darwin (1871) considered to be exaggerated traits, morphological features, and behaviors that appeared deleterious with regard to survival. Darwin therefore developed the theory of sexual selection. Sexual selection, he argued, was responsible for maintaining traits in a population because, although deleterious with respect to survival, they conferred a mating advantage. Sexual selection theory, as originally conceived by Darwin, has undergone considerable change over the past one hundred or more years (Andersson 1994; Balmford and Read 1991; Bradbury and Andersson 1987; Cronin 1992; R. A. Fisher 1930; Harvey and Bradbury 1991; Kirkpatrick and Ryan 1991). Most of the changes, however, have focused on refining the mechanisms guiding female choice, the nature of intrasexual competition, and the coevolution of trait and preference. Box 6.1 reviews some of the primary mechanisms guiding mate choice and competition for access to mates, in addition to laying out a subset of scenarios for trait-preference coevolution.

Current discussions of sexual selection theory have focused quite explicitly on communication in both the visual and auditory domains. The issue that has stirred up the most excitement and, consequently, the most active empirical and theoretical research, is the notion of *honest signaling*. The problem, in a nutshell, is this: how do individuals determine whether a given signal accurately reflects the signaler's physical condition or ability to engage in a particular activity such as fighting, mating, or parental investment? Put somewhat differently: what information can be extracted from a signal that would enable perceivers to assess the signaler's genetic quality—its ability to survive and pass on genes that are functionally related to survival?[1] In a typically provocative series of papers, the evolutionary biologist Zahavi (1975, 1977, 1987) provided a solution. In brief, Zahavi

---

1. Throughout the rest of this chapter, I will be talking about the evolution of traits, such as survivorship, that appear to be genetically correlated with certain signals. It should be noted that only those traits that have a heritable component will be relevant to the kinds of selection pressures discussed. In this sense, the term *quality* will be used to refer to traits that correlate with survival and have a heritable component.

argued that signals are honest if and only if they are (1) costly to produce and maintain and (2) relatively more costly to individuals in poor condition than to individuals in good condition. If this idea is right, then individuals with *handicaps*—traits or displays that are costly in terms of survival—must be truly *super* individuals who have made it through a survival filter. As a result, so Zahavi argued, females should preferentially mate with handicapped males.

Although the notion of a handicap, as well as the more general idea of honesty being checked by signaling costs, was initially dismissed as implausible from an evolutionary perspective (i.e., it could not be formally modeled with population genetic techniques: Kirkpatrick 1986; Maynard Smith 1976), there is now considerable theoretical support (Grafen 1990; Grafen and Johnstone 1993; Hasson 1989; Johnstone and Grafen 1992, 1993; Maynard Smith 1991; Zahavi 1993) and empirical support (FitzGibbon and Fanshawe 1988; Hauser 1993b; Møller 1988b, 1993) for Zahavi's original formulation, though with some important emendations. For example, some of the newer models of communication (Johnstone and Grafen 1993) have realized that because of perceptual errors and differential costs of signaling, evolutionary stable systems must only be honest on average and can tolerate a certain level of cheating[2] (Caldwell 1986; Møller 1988a). As in other areas of behavioral ecology, however, quantifying the relevant costs and benefits of a particular signaling strategy has proved difficult for most studies. Consequently, and as the following discussion will reveal, there is a dire need to revamp current techniques for measuring the fitness payoffs of honest and deceptive communication strategies so that the theoretical models can be adequately tested (Collins 1993; Grafen 1990; Grafen and Johnstone 1993; Hasson 1989).

I present in Figure 6.1 a simple way to conceptualize the distinction between honest and dishonest signaling, and some of the parameters that are likely to figure into the calculation. The curves were not derived from a formal mathematical model, but are simply meant to illustrate some of the potential trade-offs and how one might go about thinking about the problem.

In one sense, the main point of the models in Figure 6.1 is methodological. Specifically, although the theoretical discussions by Zahavi and Grafen tell us that costs represent a check on signal honesty, they do not tell us how one distinguishes, empirically, between honest and dishonest signals based on the

---

2. Cheaters can also be held in check by costs associated with detection. Though few studies have documented retaliation or punishment, there is some evidence in honeybees (Ratnieks and Visscher 1989) and rhesus monkeys (Hauser 1992c), the first in the context of covert reproduction by workers, the second in the context of covert attempts to consume food.

Box 6.1
Sexual Selection Theory: Some Key Issues

For Darwin (1871), sexual selection had two important components: (1) intrasexual competition and (2) mate choice. As historical analyses have revealed (Cronin 1992), however, Darwin considered aggressive interactions between males as the primary form of intrasexual competition and female choice as the primary form of mate choice. Although in a wide variety of species, this characterization is certainly correct, more recent research has revealed species where males are choosy and where females compete among each other for access to mates (Gwynne 1991; Petrie 1983). The crucial point: the pattern of intra- and intersexual competition will be determined by the nature of the limiting resource (e.g., food, mates).

In an important review article, Kirkpatrick and Ryan (1991) provided a framework for understanding some of the key pressures imposed by sexual selection, especially the factors underlying the evolution of mating preferences. I reiterate these in the following outline of mechanisms of mating preferences (Kirkpatrick and Ryan 1991, 33):

I. Direct Selection

   A. Males provide resources to females or offspring

   B. Costs of searching for mates

   C. Selection against hybridization (i.e., members of two species interbreeding)

   D. Males differ in sperm fertility

   E. Pleiotropic effects of preference genes (multiple interacting consequences)

   F. Disease or parasite transmission

II. Indirect Selection of Preferences

   A. Runaway process (preference and trait coevolve)

   B. Good genes

      1. Host-parasite coevolutionary cycles

      2. Unconditionally advantageous mutations

      3. Unconditionally deleterious mutations

   C. Genetic epistasis and dominance

   D. Social system

   E. Mutation pressure on trait

III. Other Mechanisms

   A. Random genetic drift

   B. Group selection

   C. Mutation pressure on preference

To disentangle some of these alternative explanations, Ryan and Rand (1993b) provide an extremely simple and clear schematic illustration, which I reproduce on page 367:

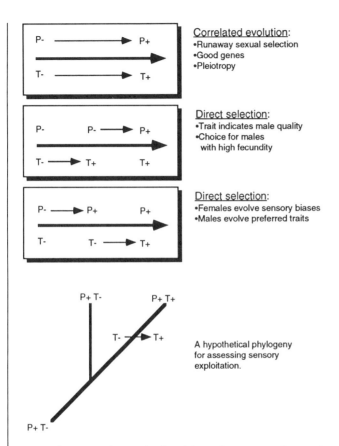

Correlated evolution:
•Runaway sexual selection
•Good genes
•Pleiotropy

Direct selection:
•Trait indicates male quality
•Choice for males
 with high fecundity

Direct selection:
•Females evolve sensory biases
•Males evolve preferred traits

A hypothetical phylogeny
for assessing sensory
exploitation.

In the figure P refers to the female's preference, T refers to the trait, a plus sign indicates presence, and a minus sign indicates absence. The thick arrow indicates evolutionary time, whereas the thinner arrows represent a state change, for example, from $P-$ to $P+$. Ryan and Rand propose three evolutionary scenarios. Under "correlated evolution," the preference and trait states change together, from absent to present. In the first "direct selection" model, there is initial selection on the trait and then selection on the preference for the trait change. For example, selection might favor males with large antlers over small antlers because of the competitive advantage they confer. Females initially show no preference for antler size, but because of the competitive advantage of large antlers, there is selection on females to mate with such males. In the second "direct selection" model, selection operates initially on the preference state and then on the trait state. For instance, imagine a predatory species that selectively preys on female frogs and, while moving through the environment, produces low-frequency signals. Selection should favor female frogs who are sensitive to such low-frequency signals over females whose hearing range is limited to higher-frequency sounds. As a result of such selection, a sensory bias will be established. Subsequently, there will be selection on males who produce low-frequency advertisement calls, since females will be maximally responsive to such signals.

Readers wishing more comprehensive information on sexual selection theory and the empirical data base should consult Andersson's (1994) thorough review.

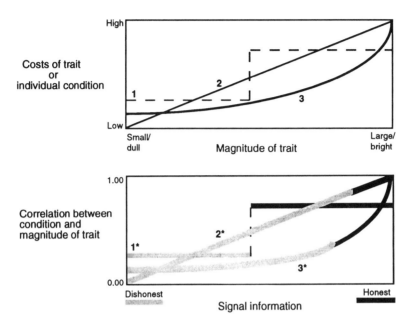

Figure 6.1
Models of honest signaling. In the *upper panel,* the *x*-axis represents the magnitude of the trait, upon which communicative interactions are decided; potential traits are plumage coloration (dull to bright) and size of sexual ornamentation (small to big). The *y*-axis represents either the cost of producing/ expressing the trait or the condition of the individual expressing the trait. The three curves indicate potential relationships. The *lower panel* provides some possible scenarios for signals lying on a continuum from honest to dishonest. As in the models by Zahavi and Grafen, honest signals are likely to be those where the correlation between condition and magnitude of the trait are high.

complex interactions between the magnitude of the trait, its associated costs, and the physical condition of the individual. As the lower panel of Figure 6.1 indicates, one can imagine several scenarios for how an individual might interpret the level of honesty in a signal. For example, in curve 2*, signals are not considered honest until the correlation between condition and the magnitude of the trait is fairly high. In contrast, curve 3* represents a situation where signals achieve honesty with a weaker correlation. All of these assessments will be significantly constrained by the resolving power of the sensory system, in addition to constraints (e.g., time with regard to energy budgets, vulnerability to predation such as occurs when females move within an open-area lek) on making decisions about potential mates and the fitness payoffs of incurring the costs associated with being choosy. Rather than work out the theoretical details of these issues based on our intuitions, let us turn to the empirical literature and review what researchers working in this area have done in light of the problem of signal honesty.

To review the main points then, I, like many others, have raised the notion of signal honesty as a central concept in current studies of signal function and design (Grafen 1990; Guilford and Dawkins 1991; Zahavi 1975, 1993). The paradigm is powerful because it provides a foundation for thinking about a wide array of communicative contexts, involving a broad spectrum of organisms. For example, as the empirical literature reviewed in this chapter will reveal, we can ask about how individuals distinguish between mates in good or poor physical condition, recognize when an apparently aggressive competitor is bluffing, and recognize when visual or auditory signals can be treated as reliable indicators of species identity. Moreover, we can ask not only about the adaptive consequences of the red-winged blackbird's song, but also about the functional significance of human crying and laughter, thereby allowing us to move freely within the taxonomic tree of life.

6.2   Mating Signals: Frogs, Birds, and Primates

6.2.1   Anuran Advertisement Calls

The two most common contexts for vocal communication in male anurans occur during competitive interactions with other males and in areas where vocalization is likely to attract sexually receptive females. In several species, large males experience higher reproductive success than small males (Arak 1988; Robertson 1986; Ryan 1985), and this is primarily due to their competitive abilities and female choice. In other species, large males fail to obtain higher reproductive success than small males, and this failure may be due to size-related assortative mating and sneaky mating by small satellite males who wait for incoming females on the periphery of breeding areas (Forester and Czarnowsky 1985; Gerhardt et al. 1987; Lopez and Narins 1991). Regarding design features, selection might favor male advertisement calls that provide relevant information about species identity (see chapter 4) and under some conditions, the male's ability to survive. In several species, there is a positive correlation between body size and the pitch of the male's advertisement call (Figure 6.2). Females, in contrast, should have a perceptual system that is designed to discriminate conspecifics from heterospecifics, and once this discrimination has been made, this system should enable them to choose the most genetically fit males among a potential pool of candidates. From an evolutionary perspective, mating with the wrong species is more

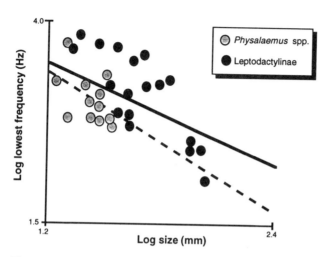

Figure 6.2
Relationship between body size and the lowest frequency of the advertisement call of species (each point represents a different species) within the genus *Physalaemus* and the family Leptodactylinae (redrawn from Ryan 1988).

costly to females than it is to males.[3] Consequently, one might expect the female's perceptual apparatus to be highly intolerant of signals that deviate from some prototypical exemplar of the species. Selection might first favor highly discriminating females in the context of species recognition, and then subsequently select for discriminating females in the context of mate choice. This process would lead to a situation where females drive, evolutionarily, the form of the male's advertisement call.

The differential selective pressures on anuran males and females have resulted in a number of fascinating evolutionary developments that highlight the importance of asking ultimate questions in light of proximate constraints. To maintain taxonomic and conceptual continuity with the discussion in section 4.2, I will describe results from the extensive field and laboratory studies of Ryan and colleagues (reviewed in Ryan 1988; Ryan and Rand 1993b) on the mating behavior of the Túngara frog (*Physalaemus pustulosus*) and cricket frog (*Acris crepitans*),

---

3. For those less familiar with the argument for differential costs to each sex, it is basically this: A cost asymmetry is set up initially by the differential investment in the ovum and sperm. This asymmetry continues as females are typically required to pay the costs of gestation and males are not. Such differences between males and females lead to a number of testable predictions with regard to mating strategies and patterns of parental investment. For reviews, see Clutton-Brock (1992) and Andersson (1994).

together with work by Gerhardt and colleagues (reviews in Gerhardt 1988b, 1992b) on several species of tree frogs (*Hyla*). Because these species exhibit an explosive lek mating system,[4] they are of significant interest to current studies of sexual selection theory (Box 6.1). In particular, they provide an opportunity to explore the selective pressures on female choice in the special case where males provide no direct benefits to females or their offspring. By mating with a male, females simply acquire sperm, with no bonus of real estate or other resources.

Túngara Frogs   The Túngara frog is a member of the family Leptodactylidae, a group that consists of approximately 30 species distributed throughout most of South and Central America. Though *P. pustulosus* has been more intensively studied than the others, all species appear to be characterized by a lek mating system, with males producing advertisement calls ("whines") that consist of a frequency downsweep and a number of harmonics (Figure 6.3). *P. pustulosus* and *petersi,* however, are also capable of adding a broad band "chuck" to the whine (Figure 6.4), a feature that contributes significantly to the attractiveness of *pustulosus* males to females (Ryan 1985). Moreover, this ability appears to be the result of an anatomical specialization—a fibrous mass associated with the larynx—that is absent in all other species within the genus (Ryan and Drewes 1990). Needless to say, though silent males are able to mate, vocal males have increased probability of mating success. Thus sexual selection will favor the production of loud, distinctive advertisement calls. Counteracting the pressure on males to call is natural selection favoring silence; for bats, the frog's call—especially the chuck component—represents a useful source of information with regard to prey localization and thus prey capture (Ryan and Tuttle 1987; Tuttle and Ryan 1981). But the chuck also augments attractiveness to females. This is a classic case of natural selection and sexual selection providing opposing pressures on trait expression.

The classic approach to studying mate choice in anurans is the two-speaker playback technique, involving the presentation of different stimuli to a female situated between the two speakers (reviewed in Gerhardt 1992a). Mate choice is derived from the female's phonotactic response, specifically whether she preferentially moves toward one speaker rather than the other. Using this approach, Ryan and colleagues have explored the preferences of female *pustulosus* and

4. Leks represent mating systems where females obtain no direct benefits from males (e.g., no resources or parental care). Thus female mate choice appears to be based solely on the male's genetic quality. Typically, there is considerable variation between males in mating success. A wide variety of species exhibit lek matings systems, including insects, fish, frogs, birds, and mammals.

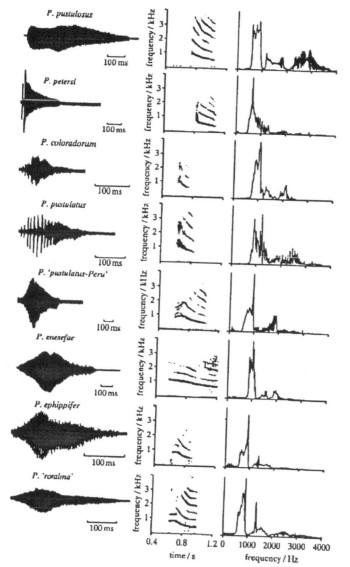

Figure 6.3
Variation in the advertisement calls of males within the genus *Physalaemus*. From right to left, the time-amplitude waveform, spectrogram, and power spectrum for each species are presented. For *pustulosus* and *petersi,* only the whine component is shown, though both species are capable of adding on a number of chucks (from Ryan and Rand 1993b). See Figure 6.4 for whine + chuck spectrograms.

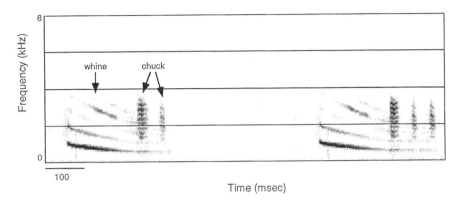

Figure 6.4
Spectrogram of the Túngara frog's advertisement call. The first call in the series consists of one whine and two chucks, whereas the second call in the series has a whine and three chucks. Frequency (kilohertz) on the *y*-axis and time (milliseconds) on the *x*-axis.

coloradorum, two species whose geographic ranges are nonoverlapping; recall that only *pustulosus* males produce a chuck. Neurophysiological studies of the basilar papilla reveal, however, that the best excitatory frequencies for both species are comparable (Ryan, Fox, et al. 1990), indicating that female *coloradorum* readily perceive the chuck of *pustulosus*. Results from one set of experiments (Figure 6.5) showed that each species preferred the calls of conspecifics over all heterospecific calls from the *Physalaemus* genus (Ryan and Rand 1993a, 1993c). However, when conspecific calls were modified, a number of interesting patterns emerged, providing insights into the evolutionary history of the female's sensory biases. For example, when the chucks of *pustulosus* were added to the whine of *coloradorum,* females preferred these calls over their species-typical call (i.e., the whine alone). Similarly, when call rate was modified in *pustulosus*, females preferred the double-pulsed call (i.e., like *coloradorum*) to the species-typical single-pulsed call.

Ryan and Rand (1993b) suggest that their experiments provide support for the notion of sensory exploitation. Specifically, in both species, females exhibit preferences for traits that do not yet exist in males (Figure 6.6). As such, males have the opportunity to evolve advertisement calls that more directly tap into the female's sensory biases. It would now be interesting to run additional experiments to determine if other acoustic parameters can alter a male's attractiveness. In particular, one would like to determine whether the female's sensory bias is more sensitive to some acoustic features than others, perhaps because of the neural topography of the papillae.

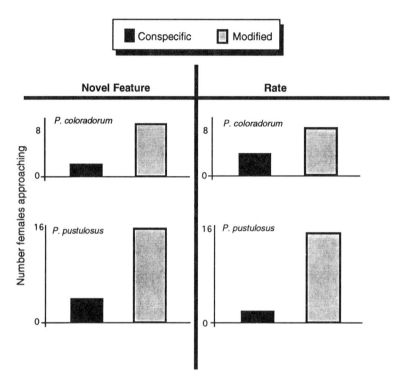

Figure 6.5
Results from female choice experiments using different species of *Physalaemus* and either the normal conspecific call (black bars) or a modified form of the call (stippled bars), comparing changes in call structure (adding a novel feature: left column) and call rate (right column). For *P. coloradorum,* novel chucks were added to the end of the whine, whereas for *P. pustulosus,* amplitude modulation was added (redrawn from Ryan and Rand 1993b).

Cricket Frogs   The genus *Acris* is a monophyletic group consisting of two species, *crepitans* and *gryllus*. Though it is a member of the family Hylidae, the phylogenetic relationship between the cricket frog and other hylids is unclear. Several studies have explored the sources of variation in the cricket frog's advertisement call, in particular, variation within *crepitans*. A significant finding is that although body size covaries with the dominant frequency of the male's call, body size fails to explain a majority of the variation in dominant frequency (Ryan, Crocroft, and Wilczynski 1990; Ryan and Wilczynski 1991). Rather, variation in frequency (see, for example, differences between an eastern and western population in Texas, illustrated in Figure 6.7) is primarily correlated with clinal variation

(longitude), habitat structure (environmental acoustics), and the pattern of social interactions (Nevo and Capranica 1985; Ryan, Crocroft, and Wilczynski 1990; Wagner 1992). For example, in one of the subspecies, *A. crepitans blanchardi*, observations and experiments revealed that dominant frequency was an unreliable indicator of male size (Wagner 1992). Specifically, though the size of the male may be indicated by the frequency of the first call in an aggressive interaction, subsequent calls can undergo quite substantial changes in frequency. As such, call frequency may more accurately reflect a male's willingness to engage in an aggressive competitive interaction than provide distant competitors with information about size. In summary then, there appears to be strong environmental selection on the design of cricket frog calls, selection that can lead to significant divergence in signal structure. What role does such variation play in mate selection by females?

Paralleling their research on the Túngara frog, Ryan, Wilczynski, and their colleagues have explored the relationship between the female's hearing organ (i.e., the tuning of the basilar papilla), the frequency of the male's call, and the phonotactic response elicited by playbacks of either conspecific (local versus foreign population) or heterospecific calls (Keddy-Hector, Wilczynski, and Ryan 1992; Ryan, Perrill, and Wilczynski 1992; Ryan and Wilczynski 1988; Wilczynski, Keddy-Hector, and Ryan 1992). Concerning sensory physiology, results indicate that the female's basilar papilla is tuned lower than the male's, that the population averages for male and female best excitatory frequencies are lower than the

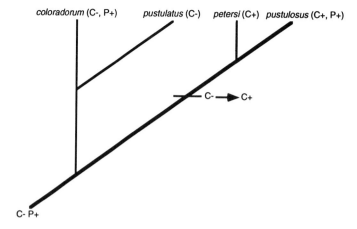

Figure 6.6
Phylogenetic relationship among species within the genus *Physalaemus*, showing changes in the expression of the chuck component of the male's call (absence = $C-$, presence = $C+$) and the female's sensory bias (tuning of the basilar papilla) or preference for the low-frequency chuck (absence = $P-$, presence is $P+$) (redrawn from Ryan 1990).

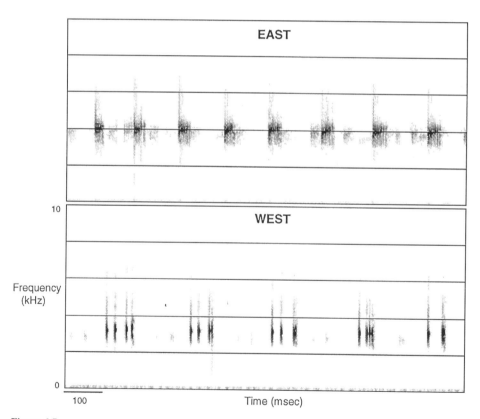

Figure 6.7
Advertisement calls of cricket frogs living in east (top) and west (bottom) Texas. Note the population difference in the dominant frequency (energy band with the greatest amplitude).

population average for male advertisement call frequencies, and that populational differences in tuning frequencies tend to covary with population differences in the frequency of the call. As such, females should prefer males from the local population who have the lowest call frequencies, and should prefer males from foreign populations if their call frequencies more precisely match the best excitatory frequency. Interestingly, this prediction also suggests that interpopulation gene flow should favor movement from forest to open habitat, or from east to west, since such geographical changes are associated with a general decrease in the tuning frequency of the basilar papilla.

To test for female preferences, Ryan and colleagues conducted playback experiments with females from three different populations, Austin and Bastrop, Texas, and Indianapolis, Indiana (Figure 6.8). In each playback, the female's phonotactic

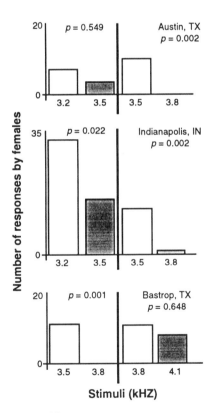

Figure 6.8
Response by female cricket frogs to playbacks of male advertisement calls. For each population, comparisons were made between a pair of stimuli, differing in frequency (white bar = lower frequency, gray bar = higher frequency) (redrawn from Ryan, Perrill, and Wilczynski 1992).

response to two calls, differing by 300 hertz, was assessed. Each population received two sets of stimulus comparisons. In all test situations, subjects preferentially approached the lower frequency call, though in two tests, the difference was not statistically significant; for two of these tests, preference for the lower frequency also represented preference for the foreign call. In addition to all these phonotactic responses, there was some evidence that females who preferred the lower-frequency calls were also larger in size than females who preferred the higher-frequency calls; this relationship was statistically significant only for the Indiana population.

Tree Frogs   The Hylidae are a species-rich group and, with regard to research on communication and mating, are perhaps the best studied family.[5] Within this family group, observations and experiments on the green tree frog (*Hyla cinerea*), gray tree frog (*Hyla versicolor*), and spring peeper (*Pseudacris crucifer*) excel in terms of detailed analyses of the relationship between acoustic structure of the male's call, the mechanisms guiding female choice, and the reproductive consequences of the male's call and female choice. Though several researchers have contributed to this knowledge, I focus on the work of Gerhardt and his colleagues (reviewed in Gerhardt 1988b, 1991, 1992a, 1992b). One reason for this focus is that results conflict with some of the patterns reported for the Túngara and cricket frog, especially in terms of the effects of males and female size, call frequency, and auditory tuning on mating success and female choice. Consequently, they provide a word of caution with respect to the generality of prior findings and, more importantly, stress the importance of comparative studies, even within a group of closely related organisms.

In a long-term study of the mating behavior of the green tree frog, results showed that a majority of males called and that silent satellite males were slightly smaller than vocal males. In contrast to several other reports, however, larger males of this species did not experience higher mating success than smaller males, nor was there evidence of size-assortative mating (e.g., large females did not mate preferentially with large males). Most importantly with regard to communication, mating success was not statistically correlated with the lowest frequency of the male's advertisement call, even though longer and heavier males produced lower-frequency calls than shorter and lighter males. The main point: we cannot assume that the negative correlation between body weight and pitch will hold across all

---

5. Unlike *Physalaemus*, however, the phylogenetic relationships among the Hylidae, in particular the tree frogs, are not well worked out, thereby limiting phylogenetic analyses.

anurans, and thus we must empirically derive this relationship for each new species (and possibly populations within species) studied.

As in other anuran species, the male's advertisement call exhibits variation in both temporal and spectral features of the signal. Most research on the perceptual mechanisms guiding both species recognition and mate choice has tended to focus on spectral variation, especially differences in the fundamental frequency. As discussed previously for the Túngara and cricket frog, this focus has led to important insights with regard to the coevolution of signals and perceptual systems. However, temporal features are also likely to be important in female mate choice, and given the data presented in chapter 4, we know that the anuran brain is designed to code for temporally significant components of the male's advertisement signal. The logic here, at least as articulated by Gerhardt and others, is that features such as call duration and call rate may provide a direct measure of the male's physical condition. This conclusion is valid because call production in general is energetically costly (Prestwich, Brugger, and Topping 1989; Wells and Taigen 1986), and males who are capable of producing either long calls or relatively shorter calls at high rates are likely to be in good condition (i.e., they can tolerate the costs). Of the several exquisitely detailed playback studies conducted by Gerhardt over the past 20 years, I focus on some of his more recent findings

(Gerhardt 1991, 1992b; Gerhardt et al. 1994), for they emphasize the necessity of considering the role of multiple acoustic features in call recognition, especially in female choice. As Gerhardt (1992b) states, "Different acoustic properties of communication signals potentially encode different kinds of biologically significant information" (p. 391).

In a detailed analysis (Figure 6.9) of acoustic variation in anuran (41 species, 96 populations) advertisement calls, Gerhardt (1991) found that some features were highly variable ("dynamic"), whereas other features were quite invariable ("static"); this terminological dichotomy is simply a shorthand for a conceptual continuum of possibilities. Static features, such as dominant frequency and pulse rate, tend to exhibit little variation either within or between males and consequently may turn out to be critical features in conveying information about identity, at either the individual, populational, or species level. In contrast, dynamic features such as call rate and call duration are variable, and much of the variability can be accounted for by contextual changes, as occurs during male-male competition and female approach. Given the potential range of variation, how do females sort out relevant from irrelevant features of the call so that they may approach and mate with males of high genetic quality?

To directly assess female preferences for a static feature (lowest frequency peak), and one that appears to correlate with body size in some species, Gerhardt conducted a two-speaker choice test, pitting a standard call (representing both the mean and median for the population; lower panel of Figure 6.10) against calls that were either lower or higher in frequency. Females showed a statistically significant preference for the standard call over calls that were either 200 Hertz lower or higher in frequency. However, females showed no preference when the standard call was compared with calls differing by only 100 Hertz. As Gerhardt points out, it is difficult to assess whether the lack of response is due to a lack of discriminability (i.e., the just noticeable difference) or to a lack of meaningful differences with regard to the approach response.

To examine the effects of dynamic features on female choice, Gerhardt tested green tree frogs in a two-speaker paradigm, pitting the standard call rate from a particular population against stimuli where call rate was increased either by 15% (i.e., within the range of natural variation) or by 100% (i.e., falling well outside the range of natural variation). As Figure 6.11 illustrates, females consistently preferred the stimuli played back at higher call rates. This pattern of preference for high rates, even supernormal levels, seems to fit with the general pattern of results obtained for other species—preference for exaggerated traits (Ryan and

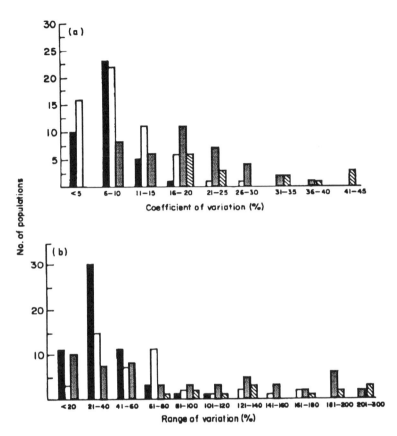

Figure 6.9
Frequency distribution of acoustic features for 96 populations of 41 species of anurans. The upper panel shows the coefficients of variation, and the lower panel shows the range of variation. The different patterns for each bar represent different acoustic features: black = dominant frequency, white = pulse rate, gray = call duration, striped = call rate (redrawn from Gerhardt 1991).

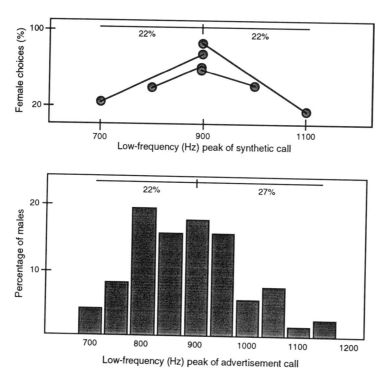

Figure 6.10
Upper panel shows the results of two-speaker playbacks with female green tree frogs (*Hyla cinerea*). Comparisons are between the synthetic versions of the standard call (900 Hertz) and calls that are either lower or higher in frequency. The lower panel represents the natural variation in the lowest frequency peak of the male's advertisement call (redrawn from Gerhardt 1991).

Keddy-Hector 1992). Yet to be demonstrated, however, is the relationship between a male's physical condition and call rate/duration, and whether the component of condition that is assessed by females is heritable.

The experiments discussed in preceding paragraphs show how manipulations of single features influence female choice, at least as defined by the phonotactic response. To explore the potential effects of multiple features, Gerhardt and colleagues conducted a series of experiments where at least two different features of the call were altered simultaneously. One of the more critical findings from gray and green tree frogs (Gerhardt 1987, 1988a), as well as the spring peeper (Doherty and Gerhardt 1984), is that female preferences for particular frequencies are highly sensitive to changes in intensity or sound-pressure level. Thus results

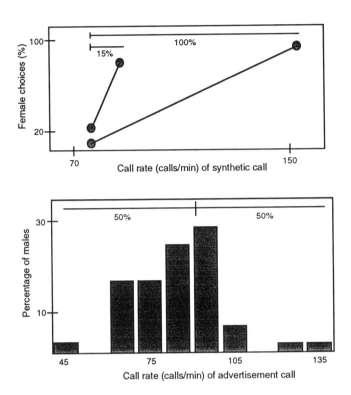

Figure 6.11
Upper panel shows the results of two-speaker playbacks with female green tree frogs (*Hyla cinerea*).
Comparisons are between the synthetic versions of the standard call rate (75 calls/minute) and calls
with either a lower or higher call rate. The lower panel represents the natural variation in call rate
of the male's advertisement call (redrawn from Gerhardt 1991).

from all three species have demonstrated that though females may selectively
approach relatively low-frequency, as opposed to high-frequency calls, even
small changes in intensity (3–6 dB) can cause this preference to shift. Such preference shifts presumably reflect the fact that the mechanisms underlying neural
tuning are often highly sensitive to changes in intensity.

How do females respond to calls where static and dynamic features have been
systematically altered? In one experiment with the gray tree frog (Gerhardt
1988a), a standard call of 20 pulses/sec was compared with an alternative of 30
pulses/sec, but with overall duration kept constant so that pulse number varied.
Females preferentially approached the standard call even when its intensity was
6 dB lower than the alternative; no preference was observed when intensity

dropped to a 12 dB difference. In contrast, when pulse number was held constant, thereby making the standard call 50% longer than the alternative, females preferentially approached the standard call even when an intensity difference of 12 dB was imposed. Although comparable experiments have produced more mixed results in terms of the weighting of static and dynamic call features (see references to unpublished data in Gerhardt 1991), it is clear that this approach to studying call recognition and female mate choice is both powerful and essential if we are to understand the mechanisms underlying phonotactic responses in frogs and, most likely, other species as well.

In several of the studies described in this section, it appears that females have perceptual strategies for distinguishing conspecifics from heterospecifics and for discriminating among males based on traits that are associated with survival. However, a number of playback experiments, especially those reported by Ryan and colleagues, have suggested that phylogenetically ancient biases may cause females to preferentially approach calls that deviate in some sense from the species-typical norm. As such, females risk approaching and thus mating with heterospecifics. To directly test whether females actually attempt to avoid heterospecifics, Gerhardt et al. (1994) designed and implemented a slightly modified version of the two-choice speaker test traditionally used in studies of anuran communication. Specifically, as schematically illustrated in Figure 6.12, rather than placing a gravid female midway between two speakers, the experimenters placed the test female at a starting point behind a barrier. Two speakers were placed on the opposite side of the barrier, out of sight from the female. One speaker sat directly behind the barrier and broadcast a heterospecific call, whereas the second speaker sat farther away and broadcast a conspecific call. The general idea here is beautifully simple. If females have been designed to avoid mating with heterospecifics, then in order to approach and mate with a conspecific in the Gerhardt design, the female must detour around the near speaker. The measure of avoidance is simply the distance from the near speaker at the point of hopping over the barrier.

Results clearly show that females (Hyla chrysoscelis, versicolor, gratiosa) do not avoid heterospecifics. In fact, several individuals passed right next to or over the speaker playing heterospecific calls, though no individuals showed directed head or body scanning toward this speaker. Additionally, none of the species exhibited a preference for approaching conspecific calls over heterospecific calls. As Gerhardt and colleagues point out, these results are surprising, especially in light of the fact that mating with the wrong species leads to significant reductions in viability, and those hybrids that survive are genetic cul-de-sacs. Two interpretations of the data are offered. First, though females may fail to avoid heterospe-

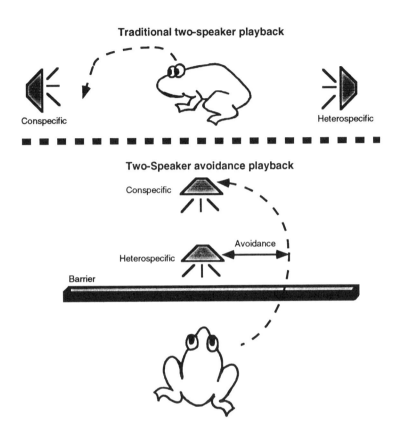

Traditional two-speaker playback

Conspecific   Heterospecific

Two-Speaker avoidance playback

Conspecific

Heterospecific   Avoidance

Barrier

Figure 6.12
Upper portion of figure shows the traditional setup for a phonotactic, female-choice playback experiment. Here, a gravid female is placed midway between two speakers, and approach behavior scored as a function of stimulus played. Thus one might compare approach to conspecific as opposed to heterospecific calls. In the lower portion of the figure, a different playback setup is presented, based on recent experiments by Gerhardt and colleagues (1994). Here, the aim of the experiment is to assess whether a gravid female placed on the opposite side of a barrier will avoid moving toward the speaker broadcasting heterospecific calls. Avoidance is measured by calculating how far away from the speaker the female moves in order to approach the alternative speaker playing conspecific calls.

cifics, interspecific matings may occur as the result of accidental contact. In some species, however, such contact may be reduced by the fact that individuals breed within spatially isolated species clusters. A second possible explanation for the lack of avoidance is that the particular features of the testing environment and setup failed to provide the necessary constraints on approach behavior. For example, in nature, females often approach an audible as well as a visible male. In the laboratory, visual input has been eliminated. Gerhardt and colleagues suggest that the role of male visibility on female approach could be tested under natural conditions. It could also potentially be tested in the lab, by using devocalized males accompanied by a hidden speaker broadcasting a call.

Some General Issues   As discussed in section 4.2, neurons within the papillae of female *P. pustulosus* and *A. crepitans* are tuned below the dominant frequency of the male's advertisement calls. As a result, the female's sensory preferences can ultimately push the form of the male's call to lower and lower frequencies,[6] exerting potentially strong directional selection. Given the female's sensory biases, it is important to consider the possibility that mate choice can lead to increasingly divergent character states and thus, ultimately, to speciation events (Lande 1981). To address this hypothesis, Ryan explored the relationship between species diversity and the complexity of the amphibian papilla[7] (Figure 6.13). The amphibian papilla rather than the basilar papilla was selected because the former shows significantly greater interspecific variation in morphology. Results indicate that in more advanced frogs, where the sensory epithelium is long and sensory constraints are relatively weak, there are far more species than in primitive frogs. As Ryan (1990) concludes, this finding is "consistent with the notion that sensory constraints influence the opportunity for speciation" (p. 164).

The theory of sensory exploitation is an attractive one because it explicitly emphasizes the coevolution of signaling and perceiving systems, acknowledges the importance of assessing issues of adaptive significance in light of evolutionary history, and provides important a priori predictions about the potential for speciation events. Over the past few years a handful of studies, other than those on frogs, have provided empirical tests of the predictions from the sensory exploitation hypothesis. Though phylogenetic information is less than ideal, and the neurophysiology of each sensory system has yet to be detailed (especially when

---

6. Limits on the frequency of the male's call will be largely established by the anatomy of the larynx.

7. Complexity was defined on the basis of measurements of four character states. As Ryan (1986, 1990) acknowledges, this is a limitation of the analysis.

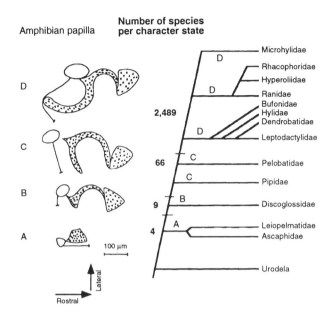

Figure 6.13
Variation in the complexity of the amphibian papilla (left) and species diversity. On the right is a phylogeny of the anurans, with values along the left edge indicating the number of species associated with a particular configuration of the amphibian papilla (redrawn from Ryan 1990).

compared with work on frogs), studies of water mites (Proctor 1993), swordtail fish (Basolo 1990), fiddler crabs (Christy 1988), anoline lizards (Fleishman 1992), and grackles (Searcy 1992) all seem to support the predictions of sensory exploitation. That is, individuals exhibit sensory biases that appear to predate the emergence of a sexually selected trait, and such biases are potentially responsible for the pattern of expression of such traits in the population.

One of the themes emerging from these studies is that playback experiments of anuran call recognition and mate choice, especially those conducted in the laboratory, often represent far cries from the natural environment. Some studies, for example, show a correlation between male size and vocal pitch, and then assume, based on previous studies, that females will preferentially mate with large males. Other studies conduct female preference tests and then, based on the results, assume that such preferences will bias reproductive success in the field. Studies by Ryan, Gerhardt, and their colleagues are exemplary in that they first demonstrated the relationship between male size and biased mating success, second showed that advertisement calls encoded relevant information about size, and last showed experimentally that information in the call influenced female preferences. Future research must heed some of the combined lessons emerging from field playbacks, as well as laboratory-based experiments that have attempted to simulate the natural conditions for call perception more carefully. This point is best summarized by Gerhardt (1988b, 475):

The conditions in which female anurans are tested with playbacks of synthetic calls are extremely simple compared with the conditions in nature. In most natural situations background noise is extremely high and its temporal and spectral structure, complex and variable. There are sound reflecting and scattering objects in the environment, and climatic factors such as wind and temperature gradients can also affect signal morphology. Natural lighting (moonlight) is seldom uniform in time and space, and temperature can easily vary by as much as 5°C from one part of a breeding site to another in temperate areas. Researchers usually attempt to eliminate or minimize all of these variables when testing females. Moreover, in a typical playback experiment, the female begins from a position midway between just two sound sources, and the SPLs [sound pressure levels] of the alternative stimuli are unusually equalized. Even in a relatively simple anuran chorus, there will be many conspecific males, and their calls will reach the female from different directions and with different SPLs.

Gerhardt's point: the real world is complicated! No surprise here. But, we should appreciate that this is the world in which selection operates to fine-tune vocal production and perception processes. Consequently, the design features of the anuran communication system represent one particular solution to the social and ecological problems associated with male advertisement and female choice.

6.2.2   Avian Advertisement Signals

Individual variation appears to be a universal phenomenon in oscine birdsong, providing an ample basis for natural selection to operate on, assuming that it has a heritable basis. (Marler 1991a, 61)

Chapters 4 and 5 described some of the mechanisms responsible for generating variation in the song structure of male songbirds and foreshadowed a point to be developed more carefully here between species that sing one song for life and those that either change songs seasonally or sing multiple song types within a season. The problem pursued in this section is to account for the functional consequences of repertoire variation, especially the kinds of selection pressures that might favor variation between and within species (Lambrechts 1992; Read and Weary 1992; Searcy 1992; Searcy and Andersson 1986). Figure 6.14 provides a schematic illustration of a possible evolutionary scenario for the emergence of song repertoire variation (Kroodsma 1982, 1988). The general idea here is that

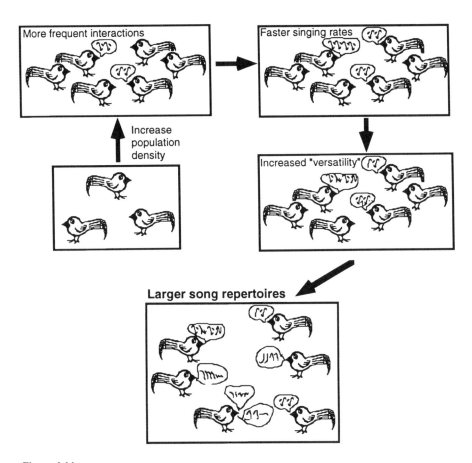

Figure 6.14
A possible scenario for how large repertoires emerge within avian species (modeled after the ideas presented in Kroodsma 1988).

increases in population density, which are likely to be correlated with increases in the frequency of competitive interactions, may lead to an increase in individual song output. Increases in song output may result in correlated increases in song variation, some of which may be selected for. The end product is that song repertoires may increase in size as a result of both male-male competition and female choice, females preferring to mate with males that are more vocally versatile and productive (in terms of pure output) (Catchpole 1986; Catchpole, Leisler, and Dittami 1984; Hiebert, Stoddard, and Arcese 1989; Howard 1974). Song output may therefore provide females with an accurate indication of a male's genetic quality, linked perhaps to his ability to survive, fend off parasites, or fight off intruders from his territory (Lambrechts 1992; Lemon, Monette, and Roff 1987; Mountjoy and Lemon 1991).

In parallel with the previous section on anurans, I focus here on a few selective cases where details of song structure and reproductive success have been well documented, especially those cases where there is some understanding of the neurobiology and ontogeny of song production and perception. Specifically, I describe the long-term studies of great tits (M. C. Baker et al. 1986; Krebs, Ashcroft, and Webber 1978; Lambrechts 1992; Lambrechts and Dhondt 1986; McGregor, Krebs, and Perrins 1981), red-winged blackbirds (Searcy and Yasukawa 1990; D. G. Smith and Reid 1979; Yasukawa 1981; Yasukawa et al. 1980), and song sparrows (Hiebert, Stoddard, and Arcese 1989; Searcy 1983, 1984; Searcy, MacArthur, and Yasukawa 1985; Searcy and Marler 1981; J. N. M. Smith 1988). To supplement this literature, and to provide a direct comparison with research on anurans, I discuss the visual and vocal displays of the sage grouse, a species that also exhibits a lek mating system (Bradbury et al. 1989; R. M. Gibson 1989; R. M. Gibson and Bradbury 1985; R. M. Gibson, Bradbury, and Vehrencamp 1991; Wiley 1973a, 1973b).

Song Repertoires and Mating Success   The history of research on the function of birdsong reveals a heavy emphasis on how males respond to song, but almost no effort has been invested into understanding how, and under what conditions, females respond to song. As Searcy (1992) points out, one reason for this asymmetry is that behavioral assays for sampling males are quite straightforward and consistent across studies, whereas assays for sampling females are problematic and are often difficult to replicate across studies. A number of recent experiments, however, have attempted to employ a combination of techniques and, in particular, to make use of the fact that in several species females respond to song by performing a copulation solicitation display. The reader should treat the studies reviewed here in light of the problem of different methodologies (Box 6.2) and, where relevant, should consider the feasibility of attempting alternative methods to measure female responsiveness to male song.

GREAT TITS   One of the longest-running field studies of avian breeding biology has been conducted on a population of great tits (*Parus major*) living in the woods surrounding Oxford, England (Hinde 1952; Perrins 1963). Male great tits typically sing a repertoire of one to eight song types (Lambrechts and Dhondt 1986; McGregor, Krebs, and Perrins 1981). In one experiment, Krebs, Ashcroft, and Webber (1978) showed that males with large repertoires were better able to defend their territories than males with small repertoires. Specifically, when resident males were removed and replaced with speakers broadcasting song repertoires differing in size, results showed that territories associated with small repertoires

Box 6.2
Playback Methods for Testing Responsiveness of Female Birds to Male Song (Searcy 1992).

A variety of methods have recently been employed to assess how females perceive and respond to variation in male song. Such methods differ in a number of ways. Searcy (1992) has provided a comprehensive review of this literature with an eye to showing the strengths and weaknesses of each technique. Each approach is evaluated on the basis of four factors: *practicality:* a judgment concerning whether the technique is easy to set up and relatively cost-effective; *generality:* an assessment of whether the technique can be used with a broad array of species, thereby providing useful comparative data; *sensitivity:* an evaluation of whether the technique is able to pull out differences in response (here, the crucial point is whether the response assay is sufficiently fine-grained to detect subtle differences among the test stimuli); and *interpretability:* assesses whether the approach permits analytical assessment of the function of song. Using these criteria, Searcy's review yielded the following summary table:

Method	Practicality	Generality	Sensitivity	Interpretability
Territorial playbacks	Good	Poor	Poor-good	Poor
Phonotaxis, laboratory	Fair	Fair	Good	Fair
Phonotaxis, field	Fair	Poor	Poor	Good
Solicitation assay	Fair	Good	Good	Good
Heart rate	Poor	Good?	?	Poor
Parental behavior	Fair	?	Good?	Good

Some of these method-evaluation pairings should be treated cautiously. In particular, techniques such as heart rate and parental behavior have been relatively infrequently used, especially in contrast with methods such as territorial playbacks. In general, however, this table serves as a useful set of guidelines for designing experiments on the reproductive function(s) of avian song.

were filled more quickly by floaters than territories associated with large repertoires. Although it is unclear from this study why large repertoires are more successful, the reported correlation suggests that territory quality is intimately connected with the ability to sing a large repertoire. This connection is further supported by studies showing a significant positive correlation between the quantity or quality of available food and song output (Cuthill and MacDonald 1990; Davies and Lundberg 1984; Searcy 1979), as well as recent experiments by Møller (1991) showing a statistically significant negative correlation between song output and parasite load (W. D. Hamilton 1982; W. D. Hamilton, Axelrod, and Tanese 1990; W. D. Hamilton and Zuk 1982).[8]

McGregor and colleagues (Krebs, Ashcroft, and van Orsdol 1981; McGregor and Krebs 1984a; McGregor, Krebs, and Perrins 1981) have tapped into the long-term data from the Oxford population to examine the relationship between repertoire size and several measures of mating success. As Figure 6.15 reveals, males with large repertoires were more likely to breed in multiple years, had more offspring survive and reproduce, obtained higher-quality nest sites, and fledged heavier young than males with small repertoires. These relationships all suggest that large repertoires result in reproductive advantages. In contrast, when a measure of lifetime reproductive success was calculated (Figure 6.16), males with intermediate repertoire sizes were more successful than males with either small or large repertoires. This pattern goes against the typical trend reported in the literature, where male mating success is positively correlated with the degree to which a sexually selected trait is exaggerated (e.g., brighter plumage, longer tails, larger antlers; Ryan and Keddy-Hector 1992). McGregor and colleagues suggest that the mating bias toward intermediately sized repertoires may result from the fact that large repertoires per se are only advantageous with regard to gaining access to nest boxes that have previously fledged large clutches. Additional analyses revealed that the heritability of song repertoire size was, however, low. Thus repertoire size is apparently not transmitted through evolutionary processes. This conclusion contradicts a theoretical model by Aoki (1989) arguing that sexual selection favors imitative learning of song variants and transmission of song repertoires from fathers to sons. Similar arguments have been offered for the evolution of speech in humans (Aoki and Feldman 1987, 1989; Cavalli-Sforza and Feldman 1983).

8. The idea here, put forward by W. D. Hamilton, is that parasite load should bring down a male's probability of survival; thus parasite-free individuals should be in better condition, and this fact should be revealed by means of differences in the expression of sexually selected traits such as song and plumage coloration—e.g., parasite-infested males will have duller plumage than parasite-free males.

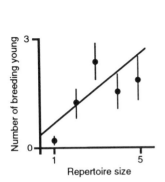
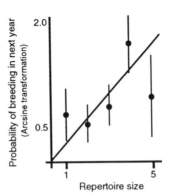

Figure 6.15
On the left, data showing a statistically significant correlation between repertoire size in the great tit and the number of young that survive to breed. On the right, data showing a statistically significant correlation between repertoire size and the probability of breeding in the next year (data arcsine transformed). Standard error bars shown (redrawn from McGregor, Krebs, and Perrins 1981).

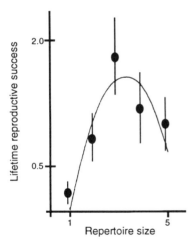

Figure 6.16
Data showing the relationship between lifetime reproductive success (measured as a weighted [by degrees of relatedness] sum of the total number of own and nondescendant kin surviving to breed) and repertoire size in great tits. Standard error bars shown (redrawn from McGregor, Krebs, and Perrins 1981).

A series of observations and experiments have focused on the possibility that particular features of the male's song are correlated with his ability to secure a territory and survive, thereby providing females with accurate information to guide their mating decisions (M. C. Baker et al. 1986; Lambrechts 1988; Lambrechts and Dhondt 1986; Weary et al. 1988). Although the experiments have been quite thorough, the emerging picture is unclear (summarized in Lambrechts 1992).

To explore potential acoustic correlates of a male's survival qualities, three features of the great tit's song were quantified: *strophe length* represents the number of phrases within a strophe or song type; *drift* is a measure of change in singing rate, with positive drift indicating a decrease in singing rate over the course of a strophe; and *repertoire size* is the number of acoustically different song types in the male's repertoire. All three measures were found to be significantly associated with reproductive success, thus providing additional support for the conclusions emerging from studies by McGregor, Krebs, and Perrins 1981). Lambrechts and Dhondt (1986) also found each of these measures of singing ability to be significantly correlated with dominance rank at winter feeding stations. Thus males with long strophes, negative drift, and large repertoires were higher ranking than males with short strophes, positive drift, and small repertoires. At this juncture, therefore, it appears that several song features may be highly correlated with a male's survival quality, thereby providing females with some of the requisite information for mate choice.

The acoustic characteristics of song could also be influenced by the motivational states of interacting males. If song features change with motivational state, then there may be insufficient information in the signal for females to assess a male's condition or survival prospects. To counteract this problem, Lambrechts and Dhondt (Lambrechts 1988; Lambrechts and Dhondt 1986) and Weary and colleagues (1988) conducted playback experiments involving manipulations of one of three different measures of singing performance. Results failed to reveal consistent responses across conditions, thereby leading Lambrechts (1992) to conclude that under these particular conditions, variation in song structure is apparently not related—at least reliably so—to a male's survivorship qualities. Moreover, when estradiol-implanted females were presented with songs differing in strophe length, there were no consistent preferences exhibited; strophe length, specifically, long versus short, was considered an accurate measure of a male's probability of survival. Overall then, variation in song structure among male great tits does not appear to map onto their ability to survive or acquire high-quality territories. As a result, if females can assess male genetic quality, they must base their reproductive decisions on factors other than song, or at least on acoustic features that differ from those originally manipulated.

RED-WINGED BLACKBIRDS   Like other passerines, the male red-winged blackbird uses its song in both intra- and intersexual competition. Males have repertoires of 2–8 song types. Acoustic differences among song types are not associated with differences in the socioecological context of production, and thus each song type appears to subserve the same function. In contrast to many other passerines, female red-winged blackbirds also sing, using one song type for what appears to be pair-bond maintenance and a second song type for aggressive conflicts with females (Beletsky 1983). In addition, the first song type is often given in response to the mate's song (Beletsky 1984) and consequently is somewhat analogous to vocal duetting among monogamous species (Farabaugh 1982; Thorpe 1972).

Prior to assessing the importance of male song in intersexual interactions, Ya-sukawa and Searcy (1985) examined the role of song in intrasexual competition, especially territorial defense. The specific aim of these studies was to test the beau geste hypothesis proposed by Krebs (1977). This hypothesis suggests that repertoire variation is deceptive, allowing territorial males to dupe nonterritorial males into perceiving that an area is densely populated when it is not. That is, extensive variation in song repertoire size should be perceived as an accurate indication of habitat saturation. As clearly articulated by Yasukawa and Searcy (1985), support for this hypothesis requires evidence that

1. the density of territorial males is positively correlated with the number of different song types.

2. non-territory-holders who are attempting to intrude and take over a territory avoid areas where male density is high.

3. non-territory-holders who are attempting to intrude and take over a territory avoid areas where the number of different song types is high.

In one experiment, Yasukawa (1981) removed territorial males and replaced them with a speaker broadcasting either a single song type or multiple song types. Results showed that the probability of intrusions by nonterritorial males was lower for multiple-song-type playbacks than for single-song-type playbacks; this outcome provides support for point 3. In a second study, Yasukawa and Searcy (1985) examined the relationship between male density, song repertoire size, ter-ritorial intrusions, and territory quality. In support of point 1, analyses revealed a positive correlation between the density of territorial males and the num-ber of different song types. No support, however, could be found for point 2.

Specifically, males did not avoid densely populated areas, even though they did avoid territories defended by males singing multiple song types. This finding suggests that in red-winged blackbirds, territorial intrusions are guided by factors other than song type and population density.

Given the results on intrasexual interactions, experiments with live and stuffed red-winged blackbirds were conducted to tease apart the relative contributions of male-male competition and female choice to song repertoire variation. Experiments in both New York and Pennsylvania revealed that males were more likely to switch song types in response to the stuffed female than in response to the stuffed male, and in general, males in both populations reduced switching rates in response to an unfamiliar male conspecific. Moreover, territorial males produced a greater variety of song types in response to a stuffed female than in response to a stuffed male (Figure 6.17). These results suggest that song type variation in the red-winged blackbird is primarily the result of intersexual selection. As Searcy (1992) has hypothesized, such variation may serve the function of maintaining the female's interest in the male, and this may be achieved by reducing the probability of sensory habituation. This idea dates back to Charles Hartshorn (1973) and his classic book *Born to Sing*.

SONG SPARROWS   Song sparrows are monogamous, with males defending territories and attracting mates by means of a species-typical song, each male producing between 5 and 13 different renditions or song types. To explore which features of the song are most significant to females with regard to eliciting copulatory display postures (see Box 6.2), Searcy and Marler (1981) conducted a series of experiments with wild females brought into the laboratory.[9] In all four experimental conditions, females preferred conspecific over heterospecific song (Figure 6.18). More specifically, females showed more copulatory displays in response to conspecific song than in response to either the closely related swamp sparrow's song or the more distantly related chaffinch song.[10] Females also preferred synthetic exemplars of song sparrow syllables arranged with the species-typical temporal pattern over synthetics comprised of a mixture of song sparrow and swamp sparrow syllables and temporal patterns. They responded more to large than small repertoires, and more to eventually varied song bouts (e.g., AAABBBCCC) than to immediately varied song bouts (ABCABCABC). And finally, sparrow

9. This test does not work with laboratory-reared birds. The reason for this failure is currently unclear.

10. Of further relevance here is a study by Searcy and Marler (1987) showing that females showed increasingly weaker responses (from greatest to weakest) to songs from normals, isolates, and deafened males.

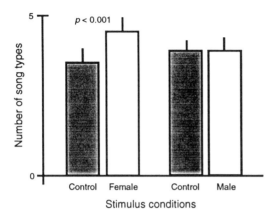

Figure 6.17
The number of song types produced by territorial red-winged blackbird males in response to a control
(a pole) and either a stuffed female or a stuffed male red-winged blackbird placed on top of a pole
(from Searcy and Yasukawa 1990).

species with small repertoires showed less of a response increment to increased
repertoire size than species with a large repertoire (Searcy and Marler 1984). In
sum, these results suggest that there is an elegant match between the structure of
the male's output and the female's perceptual and reproductive biases.[11] Several
questions remain unanswered by these studies, however. For example, do females merely prefer the larger of two repertoires, or are there boundaries to their
preferences, reflecting perhaps the natural mode or mean repertoire size in the
population? Additionally, does between-male variation in song structure or repertoire size map onto functionally meaningful differences in male quality, thereby
providing females with an accurate marker of mate quality? I now present results
from additional studies designed to answer each of these questions.

In Searcy and Marler's original studies, females showed a preference for males
with repertoires of four versus one song type. The population from which these
males were sampled produced between 5 and 13 song types each. To assess
whether females simply prefer the largest song repertoire or the repertoire size
most closely approximating the population's central tendency, Searcy (1984) presented captive estradiol-implanted females with repertoires of 4, 8, and 16 song

11. An additional set of experiments by Searcy, Marler, and Peters (1985) showed that females
responded more intensely to songs of isolate-reared conspecifics than to songs from normally reared
heterospecifics. These results show that though songs from isolates differ from those produced by
normals, there is sufficient information in these songs to elicit a preferential mating response.

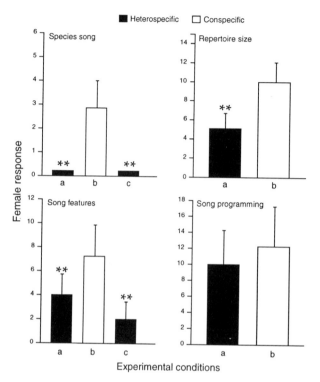

Figure 6.18
Responses of female song sparrows to playbacks of modified and unmodified playbacks of male song. *Upper left:* Females show a statistically significant (**p < 0.01) preference for conspecific song (*b*) over the songs of swamp sparrows (*a*) and chaffinches (*c*). *Upper right:* Females show a statistically significant preference for a repertoire size of four song types over a repertoire of one song type. *Lower left:* Females prefer (*b*) song sparrow syllables in a song sparrow temporal arrangement over (*a*) song sparrow syllables in a swamp sparrow temporal arrangement or (*c*) swamp sparrow syllables in a song sparrow temporal arrangement. *Lower right:* Females tended to prefer song bouts with (*b*) eventual variety (AAABBBCCC) over bouts with (*a*) immediate variety (ABCABCABC) (redrawn from Searcy and Marler 1981).

types. Results showed that in each pairing, females preferred the larger song repertoire, even though a repertoire of 16 song types exceeds the maximum observed in this population (i.e., 13). These data therefore indicate that females have a significant effect on male reproductive success, and do so by preferentially mating with males with large repertoires. However, based on two years of observation, field records failed to reveal a statistically significant correlation between repertoire size and two measures of reproductive success, the date of initial pair formation and the rapidity with which a second mate is acquired after the first is removed.

Given the observation that females produce more courtship displays in response to large repertoires than to small ones, Searcy, MacArthur, and Yasukawa (1985) set out to measure whether repertoire size was correlated with some aspect of male quality, such as territory size or the probability of survival. Analyses were carried out over a two-year period, focusing on a population of song sparrows in Dutchess County, New York. Based on a sample of approximately 50 individuals, there was no statistically significant association between repertoire size and any of the measured attributes of presumed male quality, including male age, wing length, weight, dominance status within a flock, territory size, or settlement date on the territory. As the authors acknowledge, it is possible that repertoire size correlates with some other measure of male genetic quality, but the factors measured are likely to be some of the most important. Searcy and colleagues therefore conclude that repertoire size does not provide females with an accurate indicator of male viability or condition, suggesting instead that female preference for large repertoires represents a ''phyletic holdover, with no real selective cost or benefit'' (p. 227). Although this explanation is possible, it is also conceivable that patterns of mate choice vary from year to year, and that in some years female choice is based on repertoire size and in others it is based on a different attribute or suite of attributes. This type of yearly variation, which may reflect the teeter-tottering pressures of sexual and natural selection, can only be detected with long-term field studies (see discussion by R. M. Gibson, Bradbury, and Vehrencamp 1991 in light of sage grouse mating patterns; reviewed in the next subsection).

To address the sampling problem raised above, Hiebert, Stoddard, and Arcese (1989) have recently analyzed data on the relationship between repertoire size and several measures of male viability in song sparrows, using life history data extending back to 1974 (reviewed in J. N. M. Smith 1988). Males with larger overall repertoires had longer tenures on their territories and higher annual and lifetime reproductive success than males with smaller repertoires. This relationship also held when repertoire size was evaluated on the basis of half-hour singing samples. The latter may represent a more meaningful unit of analysis from the female's perspective because it allows for a rapid assessment of a male's physical condition, at minimal costs (i.e., brief sampling by the female alleviates the costs associated with traveling to other males).

Visual and Vocal Signals of the Sage Grouse   The reproductive and social behavior of the sage grouse (*Centrocercus urophasianus*) has been observed for more than twenty years (Bradbury et al. 1989; Bradbury, Gibson, and Tsai 1986; Wiley 1973a, 1973b). Though controversy exists over the precise mechanisms driving female mate choice (i.e., which cues are used), researchers concur that this species exhibits a lek mating system. In general, the mating period is confined to approximately three months, with males appearing on a lek for a few hours each day, using vocal and visual displays (Figure 6.19) to attract potential mates. Females tend to approach and inspect leks for two or three days, and then mate. Because display areas occur in open habitats, females increase the risk of predation by approaching such areas. Therefore, the potential benefits obtained from assessing different males while they display (sexual selection for female mate choice) are at least partially offset by the costs associated with moving in an area where predation pressure is relatively high (natural selection on survival risks). All sage grouse studies have reported high variance in male mating success and tend to attribute such variation to biases in female choice. The key question, then, is how do females choose mates when there are no material goods to be had—no direct benefits?

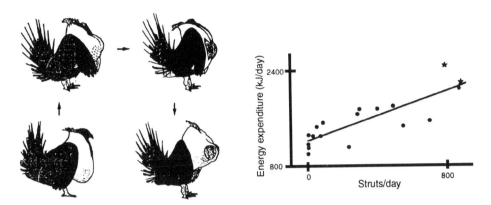

Figure 6.19
On the left, the sequence of motor actions exhibited by male sage grouse during their strut display (Hjorth 1970). On the right, data on the relationship between strut rate and energy expenditure (kilojoules per day) (redrawn from Vehrencamp, Bradbury, and Gibson 1989).

To determine which of a multitude of possible factors is responsible for the variation in mating success, R. M. Gibson, Bradbury, and Vehrencamp (1991) carried out a series of observations and experiments with a well-studied population of sage grouse in California. Analyses revealed that males producing strut displays at high rates and vocalizations with relatively long inter-pop intervals (i.e., an apparently salient acoustic feature of the call) achieved higher mating success than males producing struts at low rates and vocalizations with short inter-pop intervals (Figure 6.20).[12] These effects remained even when such factors as male morphology (wing length, tail length, body mass) and age were taken into account in a multivariate analysis. When the acoustic component of the display was played back through a speaker, females were more likely to approach on the day of the playback than on days preceding or following (i.e., nonplayback days). Although the acoustic signal may not have been designed to elicit approach (e.g., it may have been designed to enhance the potency of the visual display for females who are already nearby), it nonetheless appears attractive to females who have yet to approach a lek.

12. Note that in an earlier publication, Gibson and Bradbury (1985) reported that acoustic features other than inter-pop interval were statistically correlated with mating success. What the overall data set appears to reveal, then, is considerable yearly variation in the types of acoustic features correlated with mating success, though inter-pop interval represents a consistently repeatable measure of success.

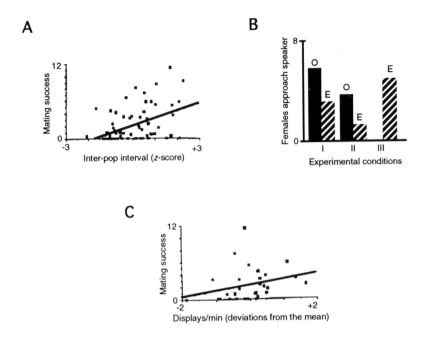

Figure 6.20
(A) Results of a statistically significant regression analysis of mean inter-pop interval and adjusted mating success (R. M. Gibson, Bradbury, and Vehrencamp 1991). (B) Results from playback experiments with the sage grouse's vocal display and the female's approach to the speaker. Female approach was assessed during three conditions: I: playback days; II: nonplayback days prior to a playback day; III: nonplayback days following a nonplayback day. O refers to the observed patterns of response, E to the expected patterns of response. (C) Data revealing the relationship between adjusted mating success and mean strut rate (number of displays per minute) when females were 50 meters away (redrawn from R. M. Gibson 1989).

The results presented here suggest that females use particular aspects of the male's communicative repertoire to make reproductive decisions. At least two questions remain, however. First, it is not clear from the data presented, nor from the conclusions drawn by Gibson and colleagues, why an acoustic feature such as inter-pop interval should be consistently associated with male mating success—that is, what does inter-pop interval reveal about the male's condition or ability to survive, or more specifically, the quality of his genes? For example, are longer inter-pop intervals more costly to produce than shorter intervals, and if so, are males in good physical condition more likely to produce them? Does the production of long inter-pop intervals represent a handicap, a costly signal in terms of its energetics and its effects on predation risk? Second, it is clear from the elegant analyses of Gibson and colleagues that females use several secondary strategies to guide their mating behavior. For example, in some years, females copy the mating patterns of other females or simply return to the same lek in which they mated in previous years. These strategies can statistically wash out the effects of particular male traits, thereby leading to highly skewed patterns of mating success. In the end, the picture that emerges is a bit murky: females appear to use different reproductive strategies from year to year, often using a combination of differently weighted features to determine who will be the lucky male in a given year. As Gibson and colleagues point out, such yearly variation may be the reason why some studies find a significant correlation between specific male traits and mating success whereas other studies do not. If observations are limited to a short period of time, as appears to be the case for most studies of lekking species, then researchers may fail to capture some of the more subtle factors causing females to shift their mating patterns each year, using particular male traits in one year and copying conspecifics in other years.

6.2.3   Primate Copulation Calls and Sexual Swellings

A recent synthesis of research on sexual selection theory (Andersson 1994) suggests that primates are poorly represented as a taxonomic group. However, primates pose some fascinating problems for sexual selection, and in this section I wish to explore some of these. To grab the reader's interest, consider the following laundry list of observations. Two of the most sexual species on earth are found within the primates: bonobos (*Pan paniscus*) and humans. These two species not only engage in sex for reproduction, but also engage in sex for nonreproductive purposes (e.g., the establishment and maintenance of social bonds) and do so in a wide variety of sexual positions with a wide variety of sexual partners (male-male, female-female, adult-juvenile). Among several primate species, individuals produce loud calls during copulation, and sometimes it's the male, sometimes the female, and sometimes both. During estrus, females in some species exhibit highly distinctive sexual swellings, and these visual signals appear to "flag" a period of sexual receptivity and thus tend to elicit intense male-male competition.

In this section, I will discuss research on primate copulation calls and sexual swellings. As will become evident, the patterns observed fall beautifully within current theoretical perspectives on the design and function of communicative signals (Hauser 1993b; Pagel 1994; Sillen-Tullberg and Møller 1993). However, much of the work is still observational, and experiments are sorely needed to assess whether these signals are causally related to individual variation in mating success.

Copulation Calls   In several nonhuman primate species, the male, female, or both members of the mating pair call *during* copulation. Such vocalizations are to be distinguished from those given by some nonhuman primates and several nonprimate species *prior to* or *following* copulation (McComb 1987). The former are commonly interpreted as courtship displays or signals, whereas the latter have often been interpreted as signals used in mate guarding. Calls given during copulation, in contrast, have been interpreted in at least six different ways, which I summarize as follows:

1. Females call to synchronize orgasm with the male (W. J. Hamilton and Arrowood 1978) or to facilitate proper endocrinological changes required for mating (Cheng 1992).[13] The idea here is that orgasm aids fertilization via an increase

13. Work by Cheng (1992) shows that in doves, the female's nest coo, used during courtship, actually sets off a series of neuroendocrinological changes that culminates in ovulation. Lesioning the requisite neural machinery or severing the relevant vocal production anatomy causes a failure in the endocrine response. Though this process is not strictly communication (i.e., self-stimulatory, akin to bat echolocation), it shows how vocal production can have a direct effect on the caller's endocrine system and thereby influence mating patterns.

in sperm transfer up the reproductive tract. This hypothesis would require coordination between the male and female, such that the female would signal that she is close to orgasm and the male would call just prior to ejaculation.

2. Females call to announce their reproductive state. Variation in call structure may covary with different stages of the reproductive cycle (Aich, Moos-Heilen, and Zimmerman 1990; Gust et al. 1990).

3. Females call to announce the presence of a male consort and to recruit aid from other females as a mechanism to reduce harassment (O'Connell and Cowlishaw 1994).

4. Females call when mating with low-ranking males as a mechanism for promoting male-male competition. As a result of generating a competitive arena, females can make more informed choices about male viability or physical condition (Cox and LeBoeuf 1977).

5. Females call to incite male-male competition, but at the level of sperm competition (Harcourt et al. 1981; O'Connell and Cowlishaw 1994).

6. Copulation calls are honest signals of an individual's genetic quality or a correlate, such as viability or physical condition. Honesty is derived, in part, from the costs associated with call production (Hauser 1993b).

W. J. Hamilton and Arrowood (1978) were the first to provide a theoretical assessment of copulation calls in primates, describing some of the general characteristics of calls given by chacma baboons (*Papio cynocephalus ursinus*), gibbons (*Hylobates* spp.), and humans. Although the data were insufficient to explicitly test any of the hypotheses put forward, it appeared that mating system might play a significant role. Specifically, they suggested that copulation calls, as sexually selected signals, might be more common in polygamous than monogamous primate species where the intensity of mating competition is higher.

Hauser (1990) analyzed the pattern of copulatory vocalizations in chimpanzees living in the Kibale Forest of Uganda. Chimpanzees are promiscuous and live in a fission-fusion society, where individuals form relatively small and ephemeral social groups. Hauser's observation period was short and unusual in that six females were simultaneously in estrus; chimpanzees typically cycle asynchronously. Observations revealed that both males and females vocalized during copulation, with males producing soft short pants and females producing loud, long screams; comparable vocalizations have been reported for chimpanzees in Gombe, Tanzania (Goodall 1986; Marler and Tenaza 1977). Females appeared to call more often than males, but this difference might have been the result of male calls being softer than those of females rather than in the rate of call production. Females called more often during matings with large and apparently high-ranking males than when mating with small and apparently low-ranking males. This pattern differs from those reported for elephant seals (Cox and LeBoeuf 1977) where females were more likely to call when small low-ranking males attempted to mate than when large high-ranking males attempted to mate. Cox and LeBoeuf interpreted their results in light of the hypothesis that females call to incite male-male competition, thereby providing an arena for female mate choice. This hypothesis cannot explain the pattern observed for chimpanzees. Instead, chimpanzees copulation calls may function to cement the social relationship or to facilitate reproductive coordination. Sample sizes were, however, far too limited to explicitly tease apart these alternative explanations.

O'Connell and Cowlishaw (1994) recently examined the function of female copulation calls in a wild population of chacma baboons, a polygamous species, where females call and males are silent during copulation. Females called in 97% of copulations, but only 35% of these were associated with ejaculation, and most calls occurred after ejaculation. Variation in call production could not be explained by predation risk or energetic costs of production. However, call duration was significantly correlated with the female's estrous cycle, reaching a maximum during the peak of sexual swelling. In addition, the longest calls occurred when matings were with adult males who ejaculated. O'Connell and Cowlishaw suggest that their data are consistent with the idea that the function of female copulation calls is to promote male-male competition at the behavioral and sperm level. Their data do not support the notion that calls reduce harassment from inferior males. Further work, in terms of physiological and fitness measures, will be required to substantiate the idea of copulation calls promoting sperm competition.

Among macaques, females call in some species, whereas males call in others (see Table 6.1). At present, it is unclear why such sex differences in calling exist, especially given coarse-grained similarities among macaques in their mating systems—all species studied to date are polygamous (Melnick and Pearl 1987). In rhesus monkeys, males call during copulation (Figure 6.21), whereas females are silent. Acoustic analyses of the male's call reveals that it is unique within the repertoire (i.e., the acoustic morphology of the copulation scream is distinctive) and that each male appears to have his own acoustic signature[14] (Hauser 1993b;

---

14. A discriminant function analysis revealed highly distinctive vocal signatures for each male's copulation call. However, like most other studies of vocal identity in nonhuman primates, Hauser's analyses are also limited to a small number of individuals. Consequently, it is currently unclear

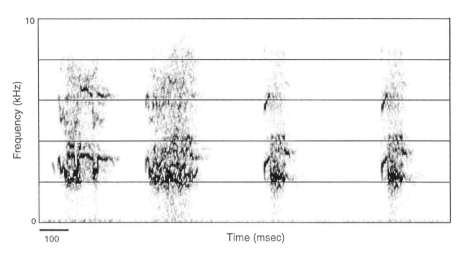

Figure 6.21
A spectrogram of a single copulation scream (four pulses separated by brief gaps of silence) produced by an adult male rhesus monkey.

Hauser and Marler 1993a). Thus, when a given male vocalizes during copulation, those out of view presumably know what the male has done, as well as his identity (i.e., there is sufficient acoustic information).

Based on several thousand copulations observed, involving more than 50 adult males and adult females, Hauser (1993d) reported a positive correlation between the number of females in estrus at the time and the proportion of mating males who called. In other words, when competition for estrous females was low, males were more likely to call during copulation. Why did males call? Hauser suggests that the rhesus monkey's copulation call may be an honest indicator (see Andersson 1994) of a male's viability and condition (i.e., good genes) and thus a signal that females might use in mate choice. Some support for this hypothesis comes from the costs and benefits associated with calling as opposed to being silent. Males who call during copulation are more likely than silent males to receive aggression. Thus vocal males incur a relatively greater cost than silent males. However, vocal males also achieve higher mating success than silent males, and the relationship between calling and reproductive success appears to be causal. Specifically, if one divides the mating season into temporal halves, results indicate

whether the notion of an acoustic signature will be preserved when a larger sample of males is assessed.

that males who call during the first half of the season experience a significant increase in mating success (total number of females mated) in the second half of the season, whereas silent males experience either no change or a slight decrease. These patterns suggest that vocal males are apparently able[15] (i.e., because they are in good physical condition) to tolerate the costs associated with calling, and as a result females prefer such males over those who are silent. Although this interpretation is plausible, there are at least three missing pieces of information and one cautionary note. First, male condition is assumed rather than measured. Ideally, one would obtain an independent measure of physical condition and then assess the actual costs of calling on condition. Second, the relatively higher mating success of vocal males is assumed to reflect female choice of vocal over silent males. This assumption requires more formal testing, perhaps using playback experiments to directly assess whether females prefer to mate with males who call over males who are silent and, perhaps more subtly, prefer to mate with males who produce calls of a particular structure. Third, no information is provided on heritability—whether vocal males pass on genes that are directly related to some kind of survival advantage. And last, the cautionary note: simply the fact that a signal is costly, and that females preferentially mate with individuals producing such signals, does not provide sufficient evidence for the idea that signals indicate good genes.

To determine the function of copulation calls at a more general level, I have collated socioecological information on a wide variety of species, and present the data in Table 6.1.

For males, none of the socioecological parameters measured account (statistically) for a significant proportion of the variation in calling across species. For example, whether or not a male calls does not appear to be influenced by sexual dimorphism in weight, length of the estrous cycle, presence or absence of a consort, or the pattern of intromission. Males living in multimale social groups were slightly more likely to call during copulation than males associated with different mating systems. Interspecific variation in female copulation calls was also not influenced by many of the same variables as assessed for males. However, females were more likely to call in species where sexual weight dimorphism was large than where it was small ($p < 0.02$). In addition, copulation calls by females were more common in the Old World monkeys and apes than in the New World monkeys and prosimians ($p < 0.001$). In sum, then, analyses appear to

15. Variation in calling patterns for this population could not be accounted for by differences in dominance rank.

Table 6.1
Copulation Calls in Nonhuman Primates and Some Relevant Social, Reproductive, and Ecological Variables

Genus	Species	Dimorph	Habit	Taxon	Mating Sys	Promisc	Cop Pattern	Consort	Estrus	Male Calls	Female Calls
Cheirogaleus	medius	1.00	W	P	D	Y	MI	N	30	N	N
Galago	senegalensis	1.14	W	P	D	Y	PI	N	•	Y	N
Lemur	catta	1.16	W	P	MM	Y	MI	Y	39	N	N
Lemur	macacao	1.00	W	P	MM	Y	MI	N	33	N	N
Nycticebus	coucang	1.08	R	P	D	Y	PI	N	40	Y	N
Propithecus	verreauxi	1.06	W	P	MM	Y	MI	N	•	N	N
Varecia	variegata	1.16	W	P	MM	Y	MI	N	30	N	N
Alouatta	paliatta	1.30	R	N	MM	Y	MI	N	16	N	N
Aotus	trivirgatus	0.92	R	N	MP	N	MI	Y	•	N	N
Ateles	belzebuth	1.07	R	N	MM	Y	PI	N	26	N	N
Ateles	fusciceps	0.98	R	N	MM	Y	PI	N	26	N	N
Ateles	geoffroyi	1.07	R	N	MM	Y	MI	•	26	N	N
Callicebus	moloch	1.05	R	N	MP	N	MI	Y	•	N	N
Callithrix	kuhli	1.09	R	N	MP	N	MI	Y	•	N	N
Cebuella	pygmaea	1.14	R	N	MP	N	MI	Y	•	N	N
Cebus	apella	1.36	R	N	MM	Y	MI	N	18	Y	Y
Cebus	nigrivittatus	1.26	R	N	MM	Y	MI	N	•	N	N
Cebus	olivaceus	1.26	R	N	MM	Y	MI	N	•	N	N
Leontopithecus	rosalia	1.02	R	N	MP	N	MI	Y	16	Y	Y
Saguinus	oedipus	0.88	R	N	MP	N	MI	Y	•	N	N
Saimiri	sciureus	1.29	R	N	MM	Y	MI	N	18	N	N
Cercocebus	albigena	1.41	R	O	MM	Y	MI	N	28	N	Y
Cercocebus	atys	1.85	R	O	MM	Y	MI	N	29	N	Y
Cercopithecus	aethiops	1.25	S	O	MM	Y	MI	N	33	N	N
Cercopithecus	cephus	1.41	R	O	OM	N	MI	•	•	N	N
Cercopithecus	talapoin	1.27	R	O	MM	Y	MI	N	36	Y	Y
Colobus	badius	1.81	R	O	MM	Y	MI	N	30	N	N
Macaca	arctoides	1.15	W	O	MM	Y	PI	N	29	Y	Y
Macaca	fascicularis	1.57	W	O	MM	Y	MI	Y	28	N	Y
Macaca	fuscata	1.18	W	O	MM	Y	MI	•	28	N	Y
Macaca	mulatta	1.09	W	O	MM	Y	MI	Y	29	Y	N
Macaca	radiata	1.58	W	O	MM	Y	MI	Y	28	Y	Y
Macaca	silenus	1.36	W	O	MM	Y	MI	N	•	Y	Y
Macaca	sylvanus	1.55	W	O	MM	Y	PI	Y	29	N	Y
Papio	anubis	2.00	S	O	MM	Y	MI	Y	31	N	Y
Papio	cynocephalus	1.80	S	O	MM	Y	MI	Y	31	Y	Y
Papio	ursinus	1.21	S	O	MM	Y	MI	Y	•	Y	Y
Presbytis	entellus	1.61	W	O	MM	Y	MI	N	22	N	N
Presbytis	johnii	1.23	W	O	OM	N	MI	N	22	N	N
Theropithecus	gelada	1.51	S	O	MM	Y	MI	Y	34	N	Y
Gorilla	gorilla	1.72	R	A	OM	Y	MI	N	28	Y	Y
Hylobates	hoolock	1.06	R	A	MP	N	MI	Y	28	Y	Y
Pan	paniscus	1.26	R	A	MM	Y	MI	Y	35	Y	Y
Pan	troglodytes	1.34	W	A	MM	Y	MI	Y	36	Y	Y
Pongo	pygmaeus	1.86	R	A	D	Y	PI	Y	30	N	N

Dimorph: Sexual dimorphism in body weight; values represent degree to which males weigh more than females (1.0 = no dimorphism; >1.0 = males are heavier); *Habit:* species-typical habitat; R = rain forest, S = savanna, W = woodland; *Taxon:* taxonomic affiliation; A = apes, N = New World monkeys, O = Old World monkeys, P = prosimians; *Mating Sys:* mating system; D = dispersed/solitary; MM = multimale; MP = monogamous pair/ polyandrous; OM = one male. *Promisc:* promiscuous (Y = yes, N = no); *Cop Pattern:* copulation pattern (data taken from Dixon 1991); MI = multiple (brief) intromissions; PI = single prolonged intromission; *Consort:* forms (Y = yes) or does not form (N = no) a mating consort; *Estrus:* values represent the average number of days that females are in estrus; *Male Calls:* male produces a call during copulation (Y = yes, N = no); *Female Calls:* female produces a call during copulation (Y = yes, N = no).

Data obtained from Dixon (1991), Hrdy and Whitten (1987), and a questionnaire to authors studying copulation calls specifically or vocal communication more generally.

support the hypothesis that for females, copulation calls function to incite male-male competition. The current data set is insufficient to explain the functional consequences of male calls or even the direct benefits to females of calling and inciting competition (e.g., by calling, do females obtain sufficient information about male competitive ability to choose the most viable male?).

Sexual Swellings

The word *estrus* comes from the Greek for gadfly, an insect that deposits its ova—eggs that are shortly to metamorphose into irritating larvae—in the skin of cattle. The image conveyed is of creatures driven to distraction by this temporary itching in their system. The phrase *females in estrus* thus implies the transformation of sedate foragers into active solicitors of male attention.   (Hrdy and Whitten 1987, 370)

In several primate species, the period of estrus is associated with a visually striking change in the area surrounding the female's perineum and, in particular, extensive swelling and reddening of the skin; in some species, the face also reddens (Dixon 1983). The size of the swelling typically reaches its maximum around ovulation. Two questions arise (Hrdy 1981): Why do some species exhibit swellings whereas others do not, and for those that do, what is the function of this apparent signal or cue? Phylogenetic analyses reveal that sexual swellings have evolved independently several times within the Primates, and in general, swellings are more common in species with multimale mating systems[16] than in species that are solitary-dispersed, monogamously pair-bonded, polyandrous, or one-male (Figure 6.22; Hrdy and Whitten 1987; Sillen-Tullberg and Møller 1993). In this section, I follow the pattern of discussion employed previously for copula-tion calls and first review some of the major hypotheses for the function of swell-ings. I then turn to the relatively limited empirical literature.

   With the exception of bonobos, female primates that exhibit sexual swellings tend to do so during periods of reproductive receptivity. Moreover, because swelling size tends to peak around ovulation, several researchers have argued that such dramatic changes may serve at least four functions:

1. As with copulation calls, female sexual swellings may incite male-male compe-tition, thereby allowing females to select the best male, either in terms of his

---

16. Some of the interesting exceptions are the species characterized by single-male breeding systems such as gelada and hamadryas baboons, drills, and mandrills. Though the breeding unit is unquestion-ably a single male and a cluster of females, these units form the core of a larger unit—a multimale society.

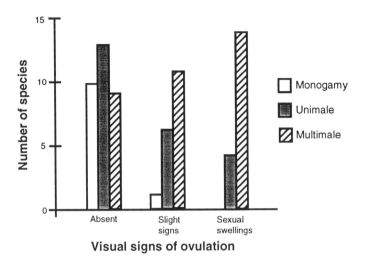

Figure 6.22
Distribution of visual signs of ovulation in nonhuman primates (data obtained from Hrdy and Whitten 1987; Sillen-Tullberg and Møller 1993).

fighting ability, sperm quality, or both (Clutton-Brock and Harvey 1976; Hrdy and Whitten 1987).

2. Swellings attract the attention of multiple males, thereby potentially confusing paternity; paternity uncertainty may subsequently reduce the risk of infanticide (Hrdy 1979, 1981) or increase the overall level of paternal care received (Taub 1980).

3. In contrast to hypothesis 2, W. J. Hamilton (1984) has argued that sexual swellings increase paternal certainty by precisely signaling the timing of ovulation. As a result, females can increase the probability of obtaining paternal care.

4. Swellings represent honest indicators of female condition or genetic quality (Grafen 1990; Zahavi 1975, 1977) and thus can be used by males to choose females who are likely to have high fitness (Pagel 1994). This hypothesis forces a reconsideration of the standard distinction within sexual selection theory between *female choice* and *male-male competition,* suggesting that we consider the potential role of *male choice* and *female-female competition.*

To properly assess these hypotheses, let us first address a conceptually more fundamental problem: are sexual swellings signals or cues? Recall from our previous discussion that signals can be turned ON and OFF, are associated with immediate costs of production, and are typically produced in response to a particular situation or event. Cues, in contrast, are permanently ON, are associated with initial (e.g., developmental) costs of production and maintenance, and are not given in response to a situation or event. Sexual swellings are not permanently ON. They are turned on, so to speak, when a specific physiological change occurs—a change that is correlated with a female's sexual receptivity. In this sense, therefore, sexual swellings are signals rather than cues, though the time course for turning the signal ON and OFF is significantly longer than for the signals that we have discussed thus far. Because sexual swellings are not permanently on, the cost of production is paid each time the female comes into estrus. Though no one to my knowledge has actually measured the costs of producing a swelling, the temporal pattern of costs is more in line with signals than cues. Last, though swellings may not occur in direct response to a particular situation or be directed at a particular individual, they are confined to the mating period. Taken together, then, current evidence suggests that sexual swellings are signals rather than cues. Now the more difficult problem: to whom are they signaling?

In all species that have been observed, males show considerable interest in females with sexual swellings, approaching them, sniffing and probing their vagina, and attempting to mate. In some species, such as yellow savanna baboons, male interest in females crests when swelling size reaches a maximum. Last, though several multimale species lack sexual swellings, swellings are completely absent in monogamous species, where presumably it is less important for females to provide males with information about receptivity.[17] These patterns strongly suggest that swellings have been designed to provide males with information about female receptivity and, potentially, the probability of achieving mating success (conception).

Discussions of the evolutionary origins and phylogeny of sexual swellings sometimes raise the possibility that estrus was lost in certain species and, more specifically, that indicators of ovulation have been lost—that is, some species exhibit *concealed ovulation* (Andelman 1987; Sillen-Tullberg and Møller 1993). First, as Hrdy (1988; Hrdy and Whitten 1987) has clearly articulated, it may be more appropriate to think about the conditions under which females change from temporally restricted periods of mating to either noncyclical or situation-dependent receptivity. Second, though females may fail to show *visual* indications

17. This discussion of course assumes that the system is rigidly monogamous. If extrapair copulations occur and if females are attempting to signal their willingness to engage in matings outside the pair bond, then swellings might be expected to evolve.

of ovulation, they may use other modalities. Acoustic (Aich, Moos-Heilen, and Zimmerman 1990; Gust et al. 1990) and olfactory (Ziegler et al. 1993) indicators of ovulation (perhaps more cautiously, "receptivity") have already been reported. Consequently, questions concerning the evolution of concealed ovulation must consider the possibility of signaling in multiple sensory modalities. And as this brief review clearly reveals, there is a dire need for field and laboratory primatologists to conduct more explicit empirical studies of the nature of sexual swellings, in terms of both the production and the perception of this signal.

6.3   Survival Signals: Insects, Birds, Squirrels, and Primates

6.3.1   Alarm Signals

General Introduction   A majority of species in the animal kingdom are subject to predation. Only a few species, such as elephants, lions, gorillas, chimpanzees, and whales appear immune. However, even these charismatic megafauna[18] are vulnerable under certain conditions. For example, chimpanzee infants are occasionally subject to leopard predation (Boesch 1991a), and in several parts of the world elephants, gorillas, and whales are, sadly, subject to human predation. Given the relative ubiquity of predation, therefore, one would expect strong selection for evolved mechanisms of warning and escape. This expectation is borne out by a vast empirical literature. What is less clear, however, is why any given individual should produce an alarm signal, for under most circumstances the signal appears altruistic and thus costly with regard to survival. In the first part of this section, therefore, I review some of the primary theoretical explanations for why individuals produce alarm signals, focusing quite specifically on the fitness consequences. I then turn to a discussion of design and show how the structure of alarm signals often represents a compromise between maximizing transmission of information to conspecifics and minimizing transmission of information to heterospecific predators.[19] I conclude with a description of studies on avian and mammalian alarm signals that allow us to disentangle some of the competing functional accounts for the evolution of alarm calls.

To set the stage, imagine a flock of birds. Two individuals, feeding together on the periphery, happen to spot a predatory eagle soaring above. One individual

---

18. Thanks to Kelly Stewart for this descriptive label.

19. Under some conditions, discussed later, alarm signals (e.g., mobbing calls) have been designed to maximize transmission to the predator.

gives an alarm call; the other flies for cover. In this scenario, callers are more likely to incur a cost (in terms of predator attack) than are silent individuals. Based on this economic asymmetry, one might expect selection to favor silent group members over alarm callers and thus expect alarm calling in general to be selected against over evolutionary time. We know, however, that alarm calling is extremely common, and consequently we are forced, as evolutionary biologists, to look for potential benefits to the caller and for costs to those who remain silent. Over the past twenty or so years, several functional hypotheses have been advanced. I lay these out in Table 6.2, borrowing directly from the extensive reviews by Curio (1978) and especially Klump and Shalter (1984).

Typical of several studies in behavioral ecology, quantifying the fitness consequences of alarm-signaling behavior has proved exceedingly difficult, particularly

Table 6.2
Functional Hypotheses of Alarm Calls (modified from Klump and Shalter 1984)

Function of Alarm Signals	Predictions	References
1. Individual selection		
To manipulate the actions of conspecifics	Genetic fitness change for alarm signaling (A) individual is greater than for silent (S) individual. The first to detect predator signals alarm. A majority of group members signal alarm. Fewer alarms by lone individuals.	(Charnov and Krebs 1975)
To facilitate group formation and cohesion, thereby increasing the probability of reaching cover	$A \leq S$ Individuals away from cover give alarms more often than those under cover. Individuals separated or peripheralized from the group are more vulnerable to predation.	(Dawkins 1976)
To silence group members, thereby decreasing the probability of predator detection	$A \leq S$ Alarm signals are primarily given when predator is nearby.	(Dawkins 1976; Maynard Smith 1965)
To reduce the probability of future attacks	$A \leq S$ Species with cohesive social groups and repeated interactions among group members should produce alarm signals more frequently than species with ephemeral grouping patterns. Resident group members should produce more alarm signals than migrants.	(Trivers 1971)

Table 6.2
(*continued*)

Function of Alarm Signals	Predictions	References
Alarm signals represent a form of mate investment	$A \leq S$ In polygynous species, males should produce more alarm signals than females. In monogamous species, rarer sex should produce more alarm signals. Regardless of mating system, unpaired individuals should signal least.	(Trivers 1972; G. C. Williams 1966; Witkin and Ficken 1979)
To invite predator pursuit	$A > S$ Predator is more likely to attack individual producing alarm signals.	(Smythe 1977)
To deter predator pursuit	$A > S$ or $A < S$ Predator is more likely to abandon attack on individuals who produce alarm signals. Only prey in good condition produce such costly signals.	(Caro 1994a; Hasson 1991; Tilson and Norton 1981; Zahavi 1987)
To confuse predator orientation when alone	$A > S$ Lone individuals are more likely to alarm than individuals in groups.	(Perrins 1968)
2. Kin selection To warn kin of varying degrees of relatedness	$A < S$ The rate or frequency of alarm signaling is influenced by the number of kin, the genetic relatedness and reproductive value of signaler and intended recipient. In species with female philopatry, females call more often than males. In species with male philopatry, males call more often than females.	(W. D. Hamilton 1963, 1964; Milinski 1978; Sherman 1977; G. C. Williams 1966)
To defend offspring	$A < S$ Alarm signals primarily produced when offspring are nearby. The rate and frequency of alarm signaling by parent(s) is influenced by offspring age and gender.	(W. D. Hamilton 1963, 1964; Trivers 1972, 1974; G. C. Williams 1966)

A = genetic fitness change for alarm-signaling individual in response to a predator; S = genetic fitness change for individual who fails to alarm in response to a predator. As in Klump and Shalter (1984), there can be positive, negative, and zero fitness change.

in the case of long-lived species. However, in several contexts the costs and benefits may be quite clear, as occurs when an alarm signal causes a predator to give up the hunt, or when the failure to alarm results in prey capture (i.e., death). More difficult are cases that rely on reciprocal relationships (Trivers 1971) where, commonly, the currency of exchange is different. Thus, for example, individual A might give an alarm call to a predator, thereby warning B who was not vigilant at the time. At a later date, individual B gives a food-associated call, thereby recruiting A to a valued resource. Here, A incurs the costs of alarm calling, and B benefits (i.e., escape from predation). In contrast, B incurs the costs associated with food calling (e.g., increased feeding competition), and A gains (i.e., access to food). Are the costs and benefits here of comparable magnitude to establish a functionally reciprocal relationship between A and B? These are the kinds of problems associated with assessing the adaptive significance of alarm calling. Rather than work out the answer theoretically, I hold off further discussion until the empirical literature has been reviewed. Let us now turn to what we know about the structural design features of alarm signals.

As discussed in chapters 2 and 3, Marler (1955) was the first to identify, from a comparative perspective, the kinds of selection pressures that would cause alarm signals to vary in structure. In particular, he demonstrated that there will be important trade-offs between detectability and localizability, and each of these will be further constrained by the nature of the acoustic environment (see chapter 3 for further details). Following up on Marler's early suggestions, Klump and Shalter's review suggests that there are at least five features influencing the detectability of alarm calls:[20]

1. Amplitude of the signal at the sound source

2. Attenuation characteristics of the environment

3. Signal-to-noise ratio at the perceiver

4. Discrimination ability of the perceiver against the background noise

5. Auditory sensitivity of the perceiver

To assess the relative importance of these factors, it is necessary to understand the production capabilities of the sender (i.e., the range of acoustic variation that can be achieved by the vocal apparatus), as well as the perceptual capacities of sender and perceiver. Importantly, the perceiver end includes both conspecifics

20. Note that Klump and Shalter's review focused on alarm calls, but many of the predictions concerning design and function apply equally to alarm signals produced in the visual domain.

(e.g., group members) and the targeted predator. For example, to demonstrate
that an alarm call enhances the probability of escape from predation, it is necessary to show that the acoustic properties of the call facilitate detection by conspecifics and minimize detection by predators. For this purpose, one would ideally
obtain audiograms of predator and prey, combined with playback experiments
aimed at assessing the degree to which predator and prey detect alarm calls
played back under natural conditions (Klump, Kretzschmar, and Curio 1986).
Although alarm call playbacks to free-living prey species have been conducted
quite extensively (see following subsections), most work on the predator's perception and response to alarm calls has been conducted under laboratory conditions; this limitation of course emerges from the logistical problem of setting up
a playback in time to confront a naturally (spontaneously) attacking predator.[21]
At present, experiments on predator localization of avian alarm calls provide
conflicting results. Whereas Shalter (1978) found that pygmy owls and goshawks
were as good at localizing alarm calls as they were at localizing other calls,
C. H. Brown (1982) found that red-tailed hawks and great-horned owls were less
accurate with alarm calls. These results may represent species differences in
sound localization or differences in the testing environment that have yet to be
characterized in terms of their relevance to localization.

The environment for transmission will impose its own effects on signal structure (reviewed in chapter 3), such that signals can change quite dramatically from
the point of release to the point of reception. And this statement holds for both
predator and prey. In fact, borrowing from some of the elegant work by McGregor
and Krebs (1984b) and Morton (1986) on distance cues in birdsong, if predators
have an auditory template of prey alarm calls, they may use this template to
gauge the degree of degradation in call morphology and, consequently, to gauge
distance to targeted prey. To my knowledge, this possibility has not yet been
investigated.

The final perceptual issue concerns localization. As discussed in greater detail
in chapters 3 and 4, results from birds and mammals indicate that, in general,
broadband noise is more readily localized than narrowband noise or pure tones.
Moreover, frequency-modulated (FM) calls are more readily localized than
non-FM signals, because FM increases the signal's bandwidth. If one looks at
the alarm calls of most mammals (Figure 6.23), for example, one finds that in
the absence of broadband noise, signals are typically frequency-modulated—the

---

21. One way around this problem might be to use a trained predator, as in the studies by Sherman
(1985) on ground squirrels.

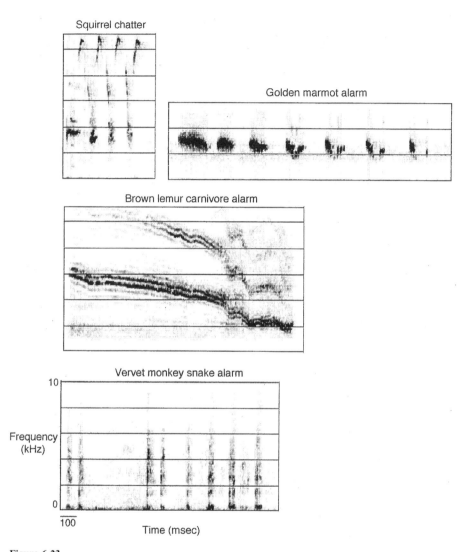

Figure 6.23
A sample of spectrograms of mammalian alarm calls, illustrating the importance of frequency band-width and frequency modulation for sound localization. The *y*-axis represents frequency (kilohertz), and the *x*-axis represents time (milliseconds).

classic chevron-shape—thereby emphasizing, once again, how selection has fine-tuned the design features of signals to meet some of the constraints imposed by the environment.

One last factor that may influence the structure of alarm calls is the composition of sympatric species assemblages. Specifically, one would predict that species living in the same environment, confronting the same kinds of predators, should have structurally similar alarm calls. Such similarity would enhance the probability of predator detection by increasing the number of vigilant eyes and, consequently, the number of useful warning systems. This prediction has been supported[22] for four sympatric squirrel species (Koeppl, Hoffmann, and Nadler 1978) and two sympatric primate species (Marler 1973), and similarity is also evident in the alarm pheromones of ants (Hölldobler 1977). Let us now turn to a discussion of the economics of alarm signaling, starting with avian species and ending with mammalian species.

Avian Alarm Calls Among avian species, young are often the most vulnerable to predation because they tend to be restricted to a particular location and their mobility is highly limited. Parents often respond to the threat of predation by giving alarm calls, mobbing the predator, or, in some cases, performing distraction displays in order to lure the predator away from the nest (Ristau 1991a). The degree to which a parent will engage in such antipredator responses will depend upon its own reproductive value (i.e., the average amount of expected reproductive success as a function of age and sex; R. A. Fisher 1930; G. C. Williams 1966), as well as its offspring's. In concert with such selective allocation of care will be differences among predators with regard to the preferred targets of predation. It is often claimed that predators prefer the young or infirm, but strong evidence for this claim is not overwhelming (though see, for example, data on predation pressure on cheetah cubs [Caro 1994a] and vervet monkey infants [Cheney and Seyfarth 1990]). Moreover, some predators may prefer young at different stages of development. These kinds of preferences might be expected to generate strong selective pressures on parents to exhibit differentiated forms and intensities of antipredator behavior.

To assess this possibility, T. L. Patterson, Petrinovich, and James (1980) carried out an experiment with wild white-crowned sparrows. In this population, individuals encounter three predators: western garter snake (*Thamnophis*

22. Although it is not explicitly tested, a recent report by Caro (1994b) on 13 species of African bovids suggests a fair amount of convergence in both visual and auditory signals used in antipredator defense.

elegans), kestrel (*Falco sparverius*), and scrub jay (*Aphelocoma coerulescens*). Observations of predation indicate the following rankings in terms of predator threat and developmental stage of the young white-crown:

Predator	Most Vulnerable Developmental Stage (most > least)
Snake	Nestlings > eggs = fledglings
Kestrel	Fledglings > eggs = nestlings
Scrub jay	Eggs = nestlings = fledglings

To assess the direct effects of these different predators on antipredator behavior, Patterson and colleagues presented parents with either a snake in a plexiglass box or a stuffed kestrel or scrub jay. As a control, they presented a stuffed Oregon junco (*Junco oreganus*), a nonpredator, comparable in size to an adult white-crowned sparrow. Results (Figure 6.24) provide support for the importance of both reproductive value and predator-specific threat. For example, snakes posed the greatest threat to parents with nestlings, and apparently as a result, parents produced the greatest number of alarm calls (''chinks'') at this developmental stage. The same kinds of effects were observed for the other predators. Added on to the predator-specific effect, alarm calling also changed as a result of offspring reproductive value. Thus, although snakes appeared to represent equivalent levels of threat to both eggs and fledglings, parental response to fledglings was higher; in terms of reproductive value, fledglings are worth more than eggs (i.e., they are closer to breeding age). Although these patterns provide a convincing fit to the reproductive-value–predator-threat model, some caution is necessary because of the relatively subjective measure of predator threat. As is the case in most studies of predator-prey interactions, extremely few predations are actually observed. In the Patterson et al. report, kestrels were observed taking adults twice and fledglings once, and on the basis of these data, the authors predicted that fledglings should be more vulnerable than either eggs or nestlings. At this stage, therefore, it is difficult to assess whether variation in antipredator response is primarily due to differences in offspring reproductive value or the degree of predator threat.

Given the consistent observation that alarm calls cause group members to respond with flight or engage in some other form of escape response, do individuals ever attempt to use such calls dishonestly by, for example, giving an alarm call when no predator is present and, as a result, causing group members to

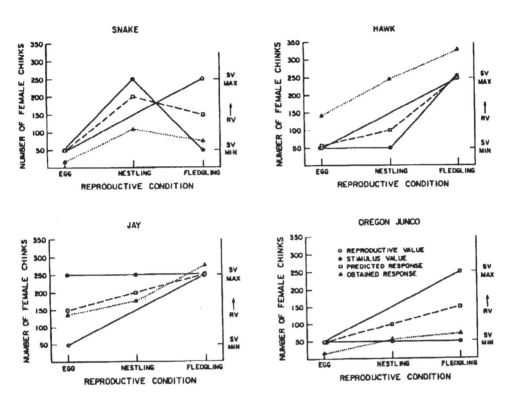

Figure 6.24
Changes in the alarm calling ("chinks") behavior of female white-crowned sparrows as a function of
predator type (snake, hawk [kestrel], scrub jay) and reproductive condition (egg, nestling, fledgling);
the Oregon junco is a control (nonpredator). Plotted along with data on alarm calls (triangles) is the
change in reproductive value (open circles), stimulus value (i.e., predator threat; closed circles), and
predicted alarm call response (squares) (from T. L. Patterson, Petrinovich, and James 1980).

abandon a valued resource? The answer is, apparently, yes, and Munn (1986a,
1986b) has provided one of the best examples of such functional deception.[23]

In a variety of birds and mammals, one observes cohesive groups comprised
of mixed-species assemblages. Although the functional advantages of such mixed-
species groups have not been resolved (see discussions in Cords 1990; FitzGibbon
1990), it seems clear that there are important fitness consequences in terms of
such factors as increased vigilance and food discovery. In the Peruvian rain

23. See chapter 7 for a more thorough discussion of the conceptual issues involved in studies of
animal deception.

forests, large mixed-species flocks of birds are abundant and subject to intense predation pressure. Following several years of observation, Munn noticed that certain species seemed to be responsible for finding food, whereas two other species (bluish-slate antshrike, *Thamnomanes schistogynus,* and white-winged shrike tanager, *Lanio versicolor*) were responsible for alarm calling to predators. The most fascinating observations centered on the context of interspecific competition over insect prey. When competition arose between the alarm-calling and food-finding species, the former often uttered an alarm call. As a result, the food-finding species looked up, giving the alarm-calling species an opportunity to snatch the prey. But Munn never observed a predator. So why was an alarm call given? Was it a case of deception, of the alarm caller duping the foraging competitor?

Munn (1986b) addressed this question in two ways. The first involved playbacks to the food-finding species to assess whether they responded differently to alarm calls given by the bluish-slate antshrike during "true" predator encounters and when no predator could be detected (i.e., by the observer); a nonalarm "rattle" call was played as a control. Based on his observations, Munn predicted that if alarm calls given during competition over insects are deceptive, then playbacks should elicit the same sorts of responses as alarm calls given during natural encounters with predators. Results presented in Figure 6.25 clearly support this prediction. Flock members consistently responded with alarm to both true and false alarm calls, but not to the rattle call. The implication here, then, is that when flock members hear alarm calls, there is nothing in the acoustic structure of the signal that gives away the signaler's honesty. If there is, and Munn has not yet conducted quantitative acoustic analyses, then the flock members have yet to make sense of the association between signal morphology and caller honesty.

Munn's second approach to the problem of deception was to provide more careful documentation of the context for putatively false alarm calls. Figure 6.26 presents data on the number of cases when the white-winged shrike tanager produced alarm calls in the absence of predators, but either in the presence or absence of a food competitor. Munn found that alarm calls were given most often when tanagers were directly competing over access to insect prey, but were generally not given when foraging alone. This finding provides additional support for the claim that false alarm calls are just that, and are used to increase the fitness gains of the signaler through access to valuable food resources. This system of false alarm calls works because of an asymmetry in the cost-benefit ratio: for food-finding flock members, the relative benefit of the insect is far less than the

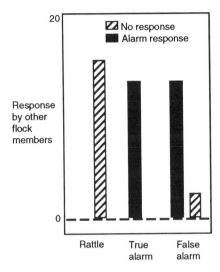

Figure 6.25
Response by flock members (i.e., those individuals in the food-finding species) to playbacks of alarm calls given by the bluish-slate antshrike (*Thamnomanes schistogynus*). True alarm calls were recorded during encounters with predators, whereas false alarm calls were recorded in the absence of predators and in the context of scramble competition over insect prey. The rattle, a nonalarm call, was used as a control (redrawn from Munn 1986b).

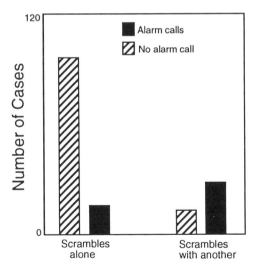

Figure 6.26
Alarm calls given by tanagers when foraging alone or in the presence of other flock members (redrawn from Munn 1986b).

costs of failing to scan for a predator upon hearing an alarm call. Thus antshrikes and tanagers can get away with their little trick because it never pays the food-finding species to challenge them. In terms of current discussions of the evolution of deception (Byrne and Whiten 1990, 1991; Cheney and Seyfarth 1990; Zahavi 1987, 1993) and, in particular, the selective pressures on cheating, Munn's results illustrate that if the relative costs to the deceived are low, then cheating can occur at relatively high rates.

There are some interesting pieces missing from Munn's elegant story: Do, for example, alarm-calling species ever try to trick members of their *own* species? What happens after a food-finding species has been deceived? Do they ever chase the alarm-calling species out of "anger"? Out of spite, do they attempt to conceal good food areas from the antshrikes and tanagers? How often can the alarm-calling species get away with such false calls? These are significantly more complicated questions, but ones within reach of empirical arms.

Mammalian Alarm Calls   In chapter 5, I reviewed current understanding of the ontogeny of antipredator defense systems in squirrels and vervet monkeys. I continue discussion of these two species here, but for squirrels, I shift from their tail-flagging display to their alarm call system. Thus, for both squirrels and vervet monkeys, we will compare and contrast some of the selective pressures that have led to the design features of their acoustic signaling systems.

Of several squirrel populations and species studied, it seems generally true that all adult individuals produce acoustically distinctive alarm calls to aerial and terrestrial predators (Figure 6.27). Differences exist, however, in the degree to which such calls are predator-specific. For example, the alarm calls of Belding's ground squirrels (*Spermophilus beldingi*) are produced in response to a much larger set of stimuli than are the alarms of the California ground squirrel (*Spermophilus beecheyi*). As Owings and his colleagues (Owings and Hennessy 1984; Owings and Leger 1980; Owings and Virginia 1978) have argued, however, such acoustic differences may have less to do with predator labeling (see discussion of vervet alarm calls in chapter 7) than with individual risk from a particular predator's attack. Thus a reasonable explanation for some of the variation in alarm call structure is that whistles are given in response to low-risk predator events, whereas chatters or trills (multiple note) are given to high-risk predator events.

As with most other mammalian species, the social structure of squirrel groups is characterized by a set of closely related females and genetically unrelated males. This pattern of within-group kinship results from the fact that females

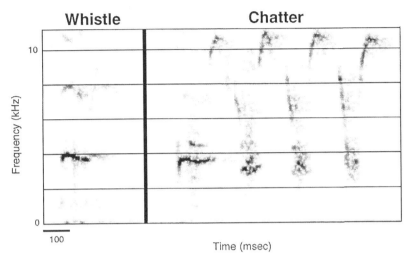

Figure 6.27
Spectrographic examples of an aerial ("whistle") and terrestrial ("chatter") predator alarm call given by California ground squirrels. The *y*-axis represents frequency (kilohertz) and the *x*-axis represents time (milliseconds).

remain in their natal groups for life, whereas males, upon reaching reproductive maturity, transfer into neighboring groups. Such demographic patterns may be causally related to the observed variation in alarm call production. Thus, if alarm calls are costly to produce but aid others in predator detection, then those individuals with the largest number of kin should produce alarm calls at the highest rates. Specifically, one would expect female squirrels to alarm more often than males, and for females with large kin groups to call more often than females with small kin groups. These are straightforward predictions from kin selection theory (Maynard Smith 1964).

Sherman's (1977, 1981, 1985) detailed study of Belding's ground squirrels provided the first test of these ideas. In his population near Tioga Pass, California, long-term demographic and behavioral data were available for several hundred marked individuals, in addition to observations of approximately 100 predator encounters; individuals at Tioga are preyed upon by badgers (*Taxidea taxus*), long-tailed weasels (*Mustela frenata*), coyotes (*Canis latrans*), dogs (*Canis familiaris*), goshawks (*Acipiter gentilis*), Cooper's hawks (*Acipiter cooperi*), prairie falcons (*Falco mexicanus*), peregrine falcons (*Falco peregrinus*), and golden eagles (*Aquila chrysaetos*).

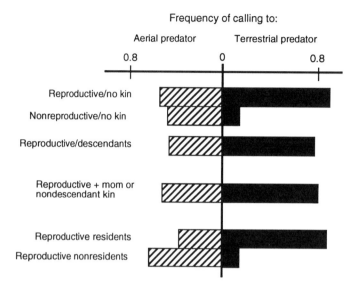

Figure 6.28
Differences in the frequency of alarm calling to aerial (striped) and ground (black) predators as a function of reproductive status and kinship (figure redrawn from Sherman 1985).

The first part of the study focused exclusively on ground predators. Results (Figure 6.28) showed that (1) females alarm called significantly more often than expected, (2) males called less than expected, (3) females called more than males, (4) adult females called more than juvenile females, and (5) adult females alarm-called as often in the presence of kin as in their absence. Sherman's conclusion is that his results provide strong support for the idea that alarm calling to ground predators is based on discriminative nepotism and thus is guided by kin selection. Although this interpretation is generally consistent with the data, the fact that the primary beneficiaries of alarm calling were offspring suggests that in some cases callers obtain a direct benefit—that alarm calling is a form of parental care.

To further evaluate the selective forces on ground squirrel alarm calls, Sherman conducted an additional set of observations and experiments, this time comparing individuals' responses to aerial and ground predators; observations were of natural encounters, whereas experiments involved presentation of a trained hawk. In contrast to the patterns obtained for ground predators, results revealed no sex differences in calling to aerial predators, and no effects of kinship (Figure 6.28). Consequently, kin selection does not appear to have played a significant role in aerial alarm calling.

To examine the costs and benefits of predator alarms, Sherman compared the consequences of calling for the caller. Results (Figure 6.29 indicated that squirrels had a much higher probability of escape when they called in response to aerial predators than when they called in response to ground predators. In addition, there were significant differences between alarm call types with regard to environmental context and the caller's behavior at the time of calling. Alarms to aerial predators were given when individuals were running and away from cover. In contrast, ground predator alarms were primarily given when the caller was close to its burrow and after it had stopped running. These socioecological factors further emphasize the different functions of these alarm calls.

In summary, data on squirrels (especially *beecheyi* and *beldingi*) show that kin and individual selection have played an important role in shaping the pattern of alarm calls to ground predators. Specifically, ground predator alarms increase the caller's vulnerability to attack, but such calls result in a genetic payoff by selectively alerting closely related kin, especially offspring.[24] In contrast, kinship plays no detectable role in the production of aerial predator alarms, and those who call reduce their vulnerability to attack. Thus, in those species that give aerial predator alarm calls,[25] individual selection appears to be the most important selective force. Let us now turn to vervet monkey alarm calls.

East African vervet monkeys, especially those observed in the Amboseli National Park, Kenya, are confronted by an impressive array of aerial and terrestrial predators, each class of predator (raptor, mammalian, carnivore, primate, snake) exhibiting importantly different hunting strategies. Empathize for the moment with the poor vervet monkey who confronts the following list (Cheney and Seyfarth 1981) of predators:[26] leopard (*Panthera pardus*), lion (*Panthera leo*), hyena (*Crocuta crocuta*), cheetah (*Acinonyx jubatus*), jackal (*Canis mesomelas*), baboon (*Papio cynocephalus*), python (*Python sebae*), cobra (*Naja* spp.), black mamba (*Dendroaspis polylepis*), green mamba (*Dendroaspis angustepis*), puff adder (*Bitis arientans*), martial eagle (*Polemaetus bellicosus*), tawny eagle

---

24. Studies of black-tailed prairie dogs also provide support for nepotistic alarm calling to ground predators, but this effect is revealed for both males and females (Hoogland 1983). Nepotism cannot, however, completely explain the pattern of calling in this species, evidenced by the fact that coterie members with no genetic relatives often produce alarm calls.

25. The thirteen-lined ground squirrel does not give aerial alarm calls (Schwagmeyer 1980).

26. Cheney and Seyfarth (1981) defined these predator types as follows: "'Confirmed' predators are animals known to prey on vervets in Amboseli or are known from other studies to prey on small monkeys. . . . 'Potential' predators are animals that are seldom observed to attack vervets, but prey regularly on species the size of vervets" (p. 29).

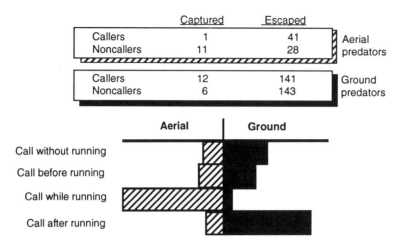

Figure 6.29
The top of this figure shows the relationship between capture rates by aerial and ground predators and whether or not individual squirrels called or remained silent. These differences were statistically significant. The bottom portion of the figure shows the relationship between types of escape responses and types of predators. Squirrels were significantly more likely to call while running in response to aerial predators and more likely to run after running in response to ground predators (redrawn from Sherman 1985).

(*Aquila rapax*), African hawk eagle (*Hieraetus spilogaster*), black-chested snake eagle (*Circaetus pectoralis*), and Verreaux's eagle owl (*Bubo lacteus*). As chapters 2 and 5 revealed, vervet monkeys have responded to such threats by evolving a suite of antipredator responses. In what follows, I provide further details on this system, paying particular attention to the relevant selective forces influencing alarm call production. In chapter 7, I explore how vervets make use of salient cues in their environment to facilitate early predator detection.

Like most mammalian species, vervets are characterized by female philopatry such that females remain in their natal groups throughout their lives, whereas males transfer to neighboring groups upon reaching reproductive maturity. The importance of this pattern of movement is that vervet groups are made up of genetically related females and typically unrelated adult males. Returning to the theoretical discussion at the beginning of section 6.3.1 and the previous discussion of squirrels, one would expect strong kin selection on females and strong individual selection on males. That is, alarm calling in females would be selected for its benefits to closely related group members, whereas individual selection would favor alarm calling in males if it benefits them directly in terms of survival. An additional complication, however, concerns the extent to which each predator class exerts differential selective pressures on particular age-sex classes. Thus, if alarm calling in vervets functions to warn vulnerable offspring, males and females might call at equal rates, or those with the most offspring might call most often. Although discussions of differential selective pressures tend to assess kin and individual selection separately, both pressures may be operative at the same time. Alarm calling will be favored both to protect relatives (kin selection) and to protect those who protect and care for relatives (individual selection).

To determine the factors causing individual differences in alarm calling, Cheney and Seyfarth first looked at differences between adult males, adult females, and juveniles during natural predator encounters. Overall, there were no significant differences between these age-sex classes in the frequency of producing first alarm calls.[27] Within age-sex classes, however, significant differences were uncovered. High-ranking adult males and females were more likely than low-ranking individuals to be the first to alarm, but this was not the case for juveniles, who failed to show a significant relationship. Data presented in Figure 6.30 indicate that in response to a stuffed leopard, high-ranking adult males and females spent more time alarm calling than did lower-ranking individuals; this relationship did not hold for juveniles. Differences between individuals could not be accounted for by differences in arousal, vigilance, or position in a group progression (i.e., front or back of group as it moved into a new area).

To evaluate the possible role of kin selection on observed differences in alarm calling, Cheney and Seyfarth examined several measures of reproductive success. For adult males, there was suggestive evidence that those individuals who sired more offspring (estimated from copulatory success)[28] produced more first alarm calls. For adult females, though there was no significant relationship between the number of first alarms and dominance rank or the number of offspring, there was an overall statistically significant relationship between the number of first alarms, dominance rank, and presence of closely related kin (Figure 6.31). Thus high-ranking females with kin tended to produce more first alarms than high-ranking females without close kin.

To evaluate the effects of different predatory styles and, potentially, different predatory "tastes," Cheney and Seyfarth explored the relationship between

27. First alarm calls were defined as the first detectable (by human observers) call to an identified predator.

28. Given that several studies of nonhuman primates, and nonprimates as well, have revealed that behavioral measures of mating success conflict with genetic data obtained from DNA fingerprinting (i.e., behavioral data can either underestimate or overestimate a male's actual genetic contribution), results on vervet male reproductive success should be treated cautiously.

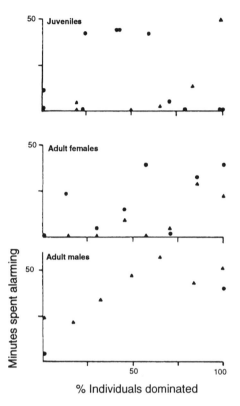

Figure 6.30
Alarm call response by vervet monkeys in Amboseli National Park, Kenya, to the presentation of a
stuffed leopard. Data show the number of minutes spent alarm calling to the leopard as a function of
age-sex class and dominance rank. The y-axis presents the number of minutes spent alarm calling,
and the x-axis shows the proportion of individuals dominated within the social group. The different
symbols represent two social groups (triangles = group A, circles = group C). Spearman rank
correlations generally showed a significant relationship between time spent calling and dominance
rank for adult males and adult females, but not for juveniles (figure redrawn from Cheney and Seyfarth
1981).

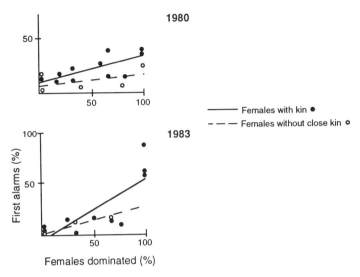

Figure 6.31
Natural observations of first alarm calls by female vervets in response to identified predators as a function of dominance rank (proportion of females dominated within the group) and kinship. Solid lines (slopes) and closed circles are for females with close kin. Dashed lines and open circles are for females without close kin (figure redrawn from Cheney and Seyfarth 1985a).

Table 6.3
Age-Sex Class Differences in Vervets' Vulnerability to Predation and Differences in Predator's Prey Preferences (Cheney and Seyfarth 1981)

1. Vulnerability to Different Predators

Age-Sex Class	Predator Vulnerability Rank	Alarm Call Rates Rank
Adult males	L > E > B	**L > E > B**
Adult females	L > E > B	**L > E > B**
Juveniles	B = E > L	**B > L > E**

2. Predator Preferences

Predator	Predator Preferences	Age-Sex Class Alarm Calling Rank
Leopard	M ≥ F ≥ J	**M ≥ F ≥ J**
Eagle	J > F > M	**F > M = J**
Baboon	J > F = M	**J > F > M**

L = Leopard, E = Eagle, B = Baboon; M = adult males, F = adult females, J = juveniles. Characters in bold represent statistically significant comparisons ($p < 0.05$).

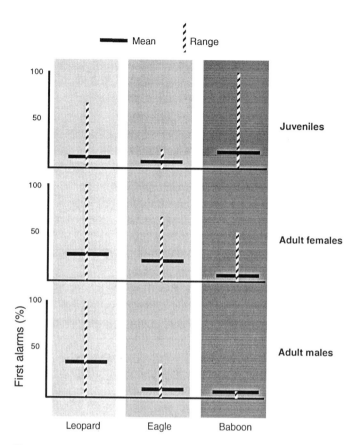

Figure 6.32
Proportion of first alarm calls given by different age-sex classes as a function of predator type. The solid black lines correspond to the means for each age-sex class by predator type comparison; the striped lines correspond to the range of variation (figure redrawn from Cheney and Seyfarth 1981).

age-sex class differences in alarm calling and predator type. Although too few observed predations were obtained to precisely assess whether particular predators have
favorite age-sex classes, it was nonetheless possible to estimate this relationship
by looking at the weight of each predatory species' typical prey item. For example, leopards prey on species with a mean weight of 20 kg. In an optimal foraging theory sense, adult male vervets, weighing in at approximately 5 kg, provide
the greatest yield for a leopard. In contrast, martial eagles typically prey on
species weighing an average of 1.2 kg, and thus adult female (1–3 kg) and immature ($\leq$ 1 kg) vervets are most appropriate (vulnerable). summarizing the results of this analysis, Cheney and Seyfarth presented the relationships given in
Table 6.3.

The data presented in Table 6.3 suggest that there are differences in vulnerability across age-sex classes and, potentially, different preferences across predator
types. As a result, one might expect strong individual selection on this population,
such that individuals would call most often in reaction to those predators to which
they are most vulnerable. Figure 6.32 provides some support for this prediction.
For example, juveniles produced fewer first alarms to leopards than did either
adult males or females. And, adult males, who appeared most vulnerable to leopards, produced more first alarms to leopards than to either eagles or baboons.
In contrast to these results, however, there were no statistically significant differences in the number of first alarms to martial eagles across age-sex classes.
In general, then, vulnerability and predator preferences can only account for
a relatively small proportion of the variation in alarm call production. This
finding, together with the data presented earlier, suggests that for vervets in
Amboseli, the pattern of alarm calling is determined by both individual and kin
selection.

6.3.2   Warning Colors

Distastefulness alone would be insufficient to protect a larva unless there were some
outward sign to indicate to its would-be destroyer that his contemplated prey would prove
a disgusting morsel, and so deter him from attack.   (Wallace 1867, lxxx)

Among several taxonomic groups, such as butterflies, terrestrial arthropods, sala-
manders, and anurans, individuals are both conspicuously colored and unpalatable to predators (see Figure 6.33). Furthermore, individuals in such species
typically form large aggregates, thereby maximizing saliency and potentially reducing predation. There is thus strong evidence for the color-palatability relationship (see review in Schuler and Roper 1992). At issue is the evolutionary
explanation for its occurrence and, as with alarm calls, the selection pressures

Figure 6.33
Warning colors in the European magpie moth (*Abraxas grossulariata*), in both the adult (top) and caterpillar (bottom) phase (from Owen 1980).

favoring an apparently altruistic act.[29] The traditional explanation, dating back at least as far as Alfred Russel Wallace (1867), the codiscoverer with Charles Darwin of natural selection theory, is that those species who have developed defenses to make them unpalatable will benefit by advertising such defenses. This general idea was formalized by Poulton (1890), who coined the term *aposematism*. Although aposematism is clearly not limited to a particular sensory modality (olfactory: Eisner and Grant 1981; visual: Grober 1988; auditory: Rowe, Coss, and Owings 1986), I focus here on coloration patterns, and thus visual processing systems.

Several explanations for the unpalatability-conspicuousness relationship have been offered, focusing in particular on the importance of kin selection and other

29. If this is not clear, consider for the moment the first individual in a group who obtained the conspicuous color genotype but not the unpalatability genotype. Here, conspicuousness would surely be a disadvantage to the individual, for he or she would be more vulnerable to predation.

selective forces in shaping the design features of the communicative component of this system. Before reviewing the alternatives, let me remind the reader of the signal-cue distinction. In the aposematic literature, warning colors are typically considered as signals. For example, in describing the function of warning colors, Guilford and Dawkins (1993) state, "All that the warning signal inherently conveys is that the animal is showing it is 'vulnerable' because it can afford easy detection by predators" (p. 401). Similarly, in Schuler and Roper's (1992) review of the literature, they state, "Insects and other animals (regardless of whether or not they are models in mimicry systems) might also use conspicuous color patterns as a visual signal of unpalatability" (pp. 112–113). Although this view, as well as those espoused by others working in this area, converges with my own sense of the function of warning colors, I would classify them as cues rather than signals: warning colors are permanently ON, and the costs associated with their expression are typically paid early in development, as a normal part of the organism's ontogeny. Although some may treat this distinction as merely semantic, I think there are important theoretical issues here, especially as they relate to the notion of a handicap. Specifically, Guilford and Dawkins (1993) have claimed, based on a review of the empirical literature and a conceptual model, that the handicap principle fails to account for the design features of warning colors. Though this claim has some validity, it may fail to capture the essence of the handicap principle (Zahavi 1987, 1991) because of the confusion regarding signals and cues. Before we can handle these particular theoretical problems, however, we must first review some of the key empirical evidence for warning colors, in addition to current propositions for their evolution.

In an elegant and clearly articulated paper, Guilford (1989a) provides a summary of three conceivable scenarios for how the relationship between conspicuous coloration and unpalatability evolved. I reproduce this summary in Table 6.4.

Guilford's review raises at least six critical issues (see also Schuler and Roper 1992). First, it forces one to think about the temporal sequence of evolutionary events: did conspicuous coloration evolve before, after, or with unpalatability? In cases where it evolved before unpalatability, conspicuously colored individuals will initially be at higher risk to predation than inconspicuously colored individuals. In contrast, where unpalatability evolved first, little may be gained because predators will still attack and potentially kill such individuals, but will not remember[30] this experience on subsequent encounters. Second, and related to the first

30. This claim makes the assumption that predators are more likely to remember conspicuously colored prey than inconspicuously colored prey. This claim is more completely evaluated further on.

Table 6.4
Factors Leading to the Evolution of Conspicuous Coloration and Unpalatability (modified from Guilford 1989a, Table 1, p. 8)

Evolutionary Scenario	Selection for Conspicuous Coloration	Selection for Unpalatability	Explanation for Conspicuous Coloration	Selective Conditions Favoring Conspicuous Coloration			
				Individual Selection	Kin Selection	Green-Beard Selection	Phenotypic Aggregation by Kin Grouping
1. Prior evolution of conspicuousness followed by unpalatability	Conspicuous coloration favored independent of prey palatability	Unpalatability favored only if prey are conspicuous	Not aposematic	Not considered	Not considered	Not considered	Not considered
2. Joint evolution of conspicuousness and unpalatability	Conspicuous coloration favored only if prey are unpalatable	Unpalatability favored only if prey are conspicuous	Aposematic	Yes (?)	Yes	Yes	Yes
3. Prior evolution of unpalatability followed by conspicuousness	Conspicuous coloration favored only if prey are unpalatable	Unpalatability favored independent of prey coloration	Aposematic / Not aposematic	Yes (?) / Not considered	No / Not considered	Yes / Not considered	Yes / Not considered

point, studies of aposematism not only must consider how prey aggregate and the relative frequency with which they encounter predators, but also must determine the limits of the predator's memory. For example, if a predator has just eaten a conspicuously colored and unpalatable prey item, is this experience sufficient with regard to future avoidance, and if not, how many encounters are required, over what time course, and are these factors influenced by degree of unpalatability (e.g., mildly noxious versus highly toxic)? And in terms of toxic prey, are the effects immediate or delayed, and if delayed, is the effect salient enough to enable predators to build the requisite association (Garcia and Koelling 1966; Kalat and Rozin 1973)? Third, although several studies have demonstrated that predators learn to avoid conspicuously colored and unpalatable prey, this finding does not necessarily mean that conspicuous coloration evolved as a mechanism to deter predator attack; for example, some coloration patterns may play a role in thermoregulation or mate attraction (see section 6.2). Fourth, unpalatability might be favored independently of coloration. This evolutionary route is possible if achieving unpalatability comes at no cost (e.g., an animal eats food that naturally provides toxic compounds linked to unpalatability) or if the costs incurred are outweighed by a benefit (e.g., by aggregating with other unpalatable individuals, predators learn to avoid those in spatial proximity). Fifth, the benefits of warning coloration may have more to do with favoring a local density of individuals sharing a common phenotype than with favoring a cluster of closely related kin (an explanation that invokes kin selection as the primary selective pressure). In the former case, common phenotypic characteristics arise among ancestrally related and unrelated individuals, and thus a mechanism other than kin selection must be invoked, such as Dawkins' (1976) *green-beard selection* or what Guilford has called *synergistic selection*. Under this type of selection regime,[31] which relaxes the requirement of kinship, it is possible for heterospecifics to evolve conspicuous coloration in the absence of unpalatability. Such mimics (see Box 6.3) would potentially boost the model's fitness by increasing abundance and, consequently, encounters with predators. Sixth, though there appears to be a group benefit associated with conspicuous coloration, grouping is not a requirement for the evolution of conspicuous coloration, as evidenced by Sillen-Tullberg's (1988) phylogenetic analyses of butterflies. With these issues in mind, let us now look at a few empirical studies of conspicuous coloration, especially those that have examined the effectiveness of coloration in predator defense.

---

31. Note that the different selection pressures on warning coloration need not be mutually exclusive.

Box 6.3
Mimicry

In the nineteenth century a British naturalist by the name of Bates and a German zoologist by the name of Müller made the observation that in Amazonian and African butterflies, several species often exhibit similar physical appearances in addition to similar palatabilities. In general, there were cases where the model species was unpalatable and the mimic was either palatable (Batesian mimicry) or unpalatable (Müllerian mimicry). Although these early classifications appeared reasonable, they have since been challenged for their simplicity. Specifically, similarity with regard to appearance and palatability is not an issue for the researcher to decide but rather an issue for the predator to decide. That is, whether something counts as a good mimic depends on whether the predator is fooled. And palatability is in the mouth of the beholder. As a result of these two issues, more recent work on mimicry has tended toward experimental analyses involving manipulations of prey coloration to test for the perceptual sensitivities of the predator, as well as manipulations of palatability. A useful review of experimental approaches to the problem of mimicry, especially warning coloration and palatability, is provided by Guilford (1989b). More complete reviews of antipredator mechanisms, including mimicry and crypsis, can be found in Endler (1991a) and D. L. Evans and Schmidt (1990).

The role of warning colors in naturally occurring predator-prey interactions has been looked at most carefully in the butterflies. In general, it appears that tropical butterfly species commonly exhibit warning colors, whereas temperate species do not. For several tropical species, such coloration is accompanied by unpalatability that is either the result of self-generated (i.e., synthesized) poisonous compounds or obtained by means of ingesting toxic plant compounds. For example, *Zygaena* moths synthesize hydrogen cyanide, and as a result, caterpillars are able to feed on plants containing cyanide. Other butterflies and moths with conspicuous coloration lack toxicity but are covered in spines, whereas some caterpillars take on the appearance of inedible objects, as in the case of *Apatele alni* which looks like avian fecal matter. Given the numerous observations of apparent antipredator devices, is there any evidence that such tricks actually work, that predators actually avoid conspicuously colored individuals over cryptic ones?

Sillen-Tullberg (1985a, 1985b) has conducted some elegant tests with great tits and zebra finches as predators and a species of lygaeid bug (*Lygaeus equestris*) as prey. In this lygaeid, there are two distinctive color morphs, one cryptically colored and one warningly colored; both morphs are equally unpalatable. In all of the experiments, at least some prey were rejected following handling, whereas other prey were apparently avoided, as evidenced by the lack of handling. The primary sources of variation in the data set were the predator's prior experience with the prey together with the prey's color morphology. Specifically, wild-caught

great tits rejected prey more often than hand-reared individuals, and the conspicuously colored morph was more likely to escape alive than the cryptically colored
morph (Figure 6.34). These results provide evidence that both direct effects induced by toxins and indirect effects induced by conspicuous coloration are involved in antipredator defense in lygaeid bugs and that both of these factors are
influenced by the predator's experience.

A variety of avian species also exhibit conspicuous coloration (R. R. Baker
and Parker 1979), and in several such species males are more brightly colored
than females, who are generally cryptic. As in butterflies and terrestrial arthropods, debate has arisen over the function of such coloration and, in particular, in
assessing whether brightly colored individuals (or species) are subject to greater
predation risks, whether females prefer brightly colored males over cryptically
colored ones, and whether coloration corresponds to palatability. A recent experiment by Götmark (1993) provides some important insights into this problem. Stuffed male and female chaffinches and pied flycatchers were placed in an
open area, and predations by sparrow hawks observed; in both species, males
are conspicuously colored, and females are cryptic. Results showed that male
chaffinches were attacked significantly more often than females, whereas for
flycatchers, females were attacked more often than males. These observations suggest that at least for chaffinches, conspicuous coloration represents a
cost with regard to survival. For flycatchers, on the other hand, conspicuous

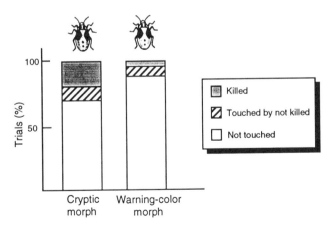

Figure 6.34
The results of a prey handling experiment, using hand-reared great tits and two naturally occurring
morphs of the species *Lygaeus equestris* (data from Sillen-Tullberg 1985a; figure redrawn from Alcock
1989, 351).

coloration does not appear to represent a cost. However, this relationship does not warrant the causal interpretation offered by Götmark (1993), who states that the "black-and-white plumage in pied flycatcher males reduces predation by sparrow hawks, a major predator of this species." Though coloration may not increase predation pressure, it is difficult to see how it would "reduce" it. Considerably more work on the visual system of hawks is necessary before this issue can be satisfactorily resolved.

A number of observations suggest that the advantages of conspicuous coloration are heightened by the effects of grouping. In essence, although a predator might remember a conspicuously colored unpalatable individual better than a cryptically colored individual, an aggregation of conspicuously colored individuals might be of even greater advantage to each group member. A critical test of this hypothesis would involve an experiment that measures discriminative aversion learning in a predator confronted with different prey group sizes (including solitary individuals), while controlling for ingestive/toxic effects. A recent study by Gagliardo and Guilford (1993), using the domestic chick, does just this. There were four prey conditions (Figure 6.35), where prey were either yellow or green food crumbs, and either palatable or unpalatable (treated with quinine and mustard). There were four conditions: (1) prey were visually encountered and ingested as singletons; (2) prey were visually encountered and ingested as aggregations; (3) prey were ingested as singletons but visually encountered as aggregates; and (4) prey were ingested as singletons, but when removed, a single crumb remained below, physically inaccessible.

In the first part of the experiments, focusing on acquisition of aversion to unpalatable prey, results showed that predators ingested more prey in the singly housed condition than in all other conditions (Figure 6.36). There were no statistically significant differences between the other conditions, suggesting that the protective consequences of aggregation are visual rather than ingestive. Moreover, when unpalatable prey were made palatable during the extinction phase of the experiment, predators ingested more of the previously unpalatable prey if they were solitary than in the other three conditions involving aggregation. When the time course for learning the association between coloration and palatability was examined, differences between conditions were revealed. Predators learned the discrimination between yellow and green crumbs more slowly when the prey were housed singly than when they were aggregated (conditions 2, 3, 4). Because birds learned more rapidly in the visual single condition than in the single condition, one possible mechanism for the enhanced learning effect lies in the visual reinforcement obtained following the ingestion of unpalatable foods. That is, in

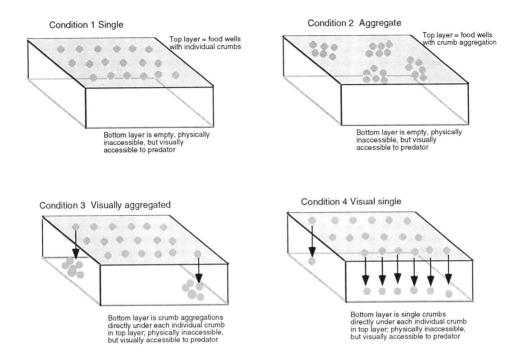

Figure 6.35
Four different conditions for assessing the effects of conspicuous coloration and gregariousness
(grouping) on prey capture success. Tests were conducted with domestic chickens as predators and
differently colored food crumbs as prey; crumbs were either palatable or unpalatable (treated with
quinine and mustard). *Condition 1–single:* 18 green and 18 yellow crumbs, housed singly. *Condition
2–aggregate:* aggregates of six same-color crumbs, for a total of 36 crumbs. *Condition 3–visual
aggregate:* 18 green and 18 yellow crumbs housed singly on top shelf with aggregates of six same-color
crumbs on the lower shelf; the lower shelf is visually, but not physically, accessible; only a few
examples are shown for the bottom shelf. *Condition 4–visual single:* 18 green and 18 yellow crumbs
housed singly on top shelf with the same array on the bottom shelf; only a few examples are shown
for the bottom shelf (schematic illustration of details provided in Gagliardo and Guilford 1993).

the visual single condition, as in the other aggregate conditions, ingestion of
unpalatable prey is associated with seeing other similarly looking prey in the
same area. In contrast, during ingestion of single prey, visual associational cues
are lacking. In summary, Gagliardo and Guilford's data provide empirical support
for the idea that selection will favor aggregations in conspicuously colored unpalatable prey. Moreover, such effects appear to fit well with the design features of
the predator's perceptual system, including its memory.

So where does this leave us with regard to the handicap principle? Guilford
and Dawkins (1993) argue that warning colors are not handicaps because it is

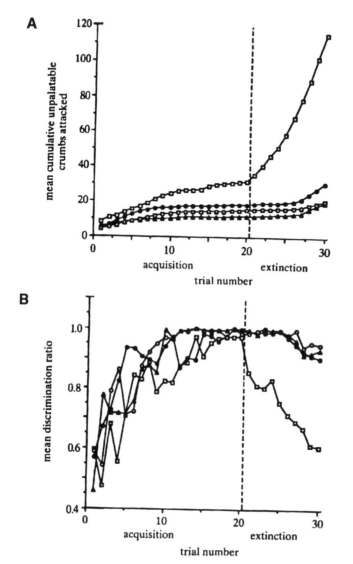

Figure 6.36
Results from experiments using domestic chicks as predators and different prey grouping patterns.
(*A*) Changes in the prey consumption during four test conditions, comparing the acquisition and
extinction phase when prey palatabilities are switched. Open squares = single condition, filled circles
= aggregate condition, open circles = visually aggregated condition, open squares = visual single
condition. See Figure 6.35 for details. (*B*) Data showing discrimination performance during the four
prey conditions; symbols as in *A* (redrawn from Figures 1 and 2 in Gagliardo and Guilford 1993).

hard to imagine how an individual would pay a cost that would in some sense take away from its palatability. In the same way, I would argue that it would be even harder to imagine how particular costs might take away from its conspicuous colors. To my knowledge, there is no evidence that individuals with warning colors undergo significant, temporal variation in coloration as a function of communicative interactions, though other factors, such as parasite load, might alter coloration. But, the reason why warning coloration does not fit with Zahavi's handicap principle is that only signals can be handicaps, not cues. And this is so because there must be a cost associated with production, the penalty imposed by this cost must be relative to the individual's current condition, and these factors must lead to variation in trait expression among individuals in the population. Warning colors are cues, permanently ON, imposing no costs of production, and yet clearly designed to be communicative to predators.

6.3.3 Food-Associated Signals

For socially living organisms, foraging typically involves a combination of cooperative and competitive interactions. Regarding cooperation, one often thinks of interactions between parents and offspring (Caro and Hauser 1992; Emlen 1992; McGrew and Feistner 1992), as when primate mothers allow their offspring to steal food or feed next to them (Altmann 1980; Boesch 1991b; Boesch and Boesch 1992; Hauser 1994) or when parents provide food to young who are immobile or lack the requisite foraging skills (Caro 1994a; Wilkinson 1987). Outside of the parent-offspring context, there is evidence of cooperative hunting (Boesch 1994; Packer, Scheel, and Pusey 1990; Scheel and Packer 1991; for a theoretical discussion, see Packer and Ruttan 1988) and of individuals within groups either taking advantage of the foraging skills of others (Barnard and Sibly 1981; Giraldeau, Caraco, and Valone 1994; Giraldeau and Lefebvre 1986) or forming coalitions with them to outcompete members of a neighboring group (Cheney 1992; Isbell 1991; Janson and van Schaik 1988; Wrangham 1980). A less commonly discussed form of resource-based cooperation involves the production of distinctive signals by individuals discovering food, which have often been labeled as *food calls*. As implicated for alarm calls, food calls would appear to be altruistic. Those who announce their discoveries are essentially inviting increased food competition and, consequently, potentially decreasing their own access to food. As pointed out for alarm calls, empirical quantification of the costs and benefits of food calls is required in order to demonstrate that such signals are altruistic.

A second issue raised by observations of food calls is their informational content. Do such signals convey specific information about the quality or quantity of food discovered, or do they represent a much more generic type of contact call? The specificity of this information is important (see discussion in chapter 7), for it may determine the kind of response elicited by other group members. For example, when an individual A is feeding in a food patch and hears a distant individual B calling (i.e., producing the call typically heard upon food discovery) from a different food patch, A functionally obtains additional information about the foraging environment without paying the usual traveling costs. In an optimal foraging sense, therefore, food calls may provide individuals with a cheap way of sampling the environment.

In this section, I first review two studies of food calls, one on birds and one on primates, and then, for comparative purposes, conclude with a brief synthesis of studies on other species. The focus is on the selection pressures and socioeco-logical contexts eliciting such signals. The informational content potentially conveyed and perceived is detailed in chapter 7.

House Sparrows Joining a foraging flock can be of benefit to individuals in at least two ways. By being a member of a group, predation risk is reduced (selfish herd effects or increased vigilance leading to early predator detection; W. D. Hamilton 1971), and the probability of successfully locating food is increased (Giraldeau and Lefebvre 1986). As several theoretical models and empirical tests have demonstrated, however, there are trade-offs between these two potential benefits, as well as conflicts of interest between current group members and those wishing to join. Added to these dimensions is the relative fluidity of the social group with regard to membership. Some groups exhibit a stable composition and tend to move as a cohesive unit. Other groups maintain a consistent composition but are characterized by temporally and spatially varying subgroups. In the latter situation, individuals will encounter food resources and be confronted with the choice of attracting others or feeding alone. Attracting others will undoubtedly increase feeding competition, but the level of competition will depend upon the divisibility of the food source. In addition, though feeding competition might increase, predation risk is likely to decrease as other individuals join the foraging flock. To explore these types of decision parameters, Elgar (1986) conducted an experimental study of individually marked house sparrows (*Passer domesticus*). House sparrows were of particular interest because observations indicated that upon food discovery, individuals often produced quite distinctive vocalizations known as *chirrups*.

A series of observations indicated that when solitary individuals ("pioneers") alighted and gave chirrups at high rates, other sparrows arrived and joined the

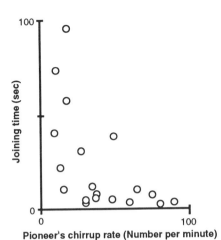

Figure 6.37
Data show the relationship between the rate (number per minute) at which pioneer house sparrows produced chirrup calls and the delay in arrival by other group members. Joining time (seconds) was calculated as the time elapsed from the arrival of the pioneer at an artificial feeder to the arrival of other group members (redrawn from Elgar 1986).

pioneer more rapidly than when chirrup rates were low (Figure 6.37). To test the attractive function of the chirrup call more carefully, Elgar conducted a playback experiment; sparrow arrivals were scored in response to a high chirrup rate and, as a control, an equally high rate of human whistles. Results showed that significantly more sparrows arrived in response to the chirrups than to the whistles.

Having established the attractive function of the chirrup call, Elgar then conducted a second experiment to assess how food divisibility affected a pioneer's vocal behavior. Five experimental conditions were presented: (1) no food, (2) ⅛ of a whole slice of bread, (3) ½ of a whole slice of bread, (4) ⅛ of a slice of bread in crumbs, and (5) birdseed. These conditions were designed on the basis of natural observations indicating that lone individuals could monopolize a whole piece of bread but not bread crumbs. Moreover, individuals were involved in more aggressive interactions at whole pieces of bread than at bread crumbs. As Figure 6.38 indicates, pioneers were more likely to give at least one chirrup call upon discovering divisible food sources than upon discovering nondivisible foods. Divisible foods also elicited a higher rate of chirrup calls than nondivisible foods. Last, chirrup rate was negatively correlated with flock size, with solitary individuals calling at almost three times the rate of individuals accompanied by one or more additional flock members.

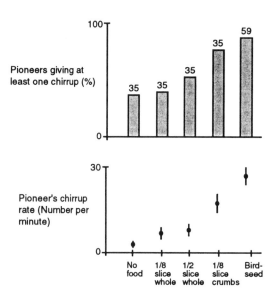

Figure 6.38
The effects of food divisibility on the pioneer's vocal behavior. The top panel indicates that pioneers were more likely to give at least one chirrup call when foods were divisible (crumbs and birdseed) than when they were not (whole slice). The bottom panel shows that pioneers also chirruped at a higher rate when foods were divisible. The numbers above the histograms in the top panel represent the number of pioneers observed (figure redrawn from Elgar 1986).

Data on house sparrows show that chirrup calls are used to attract conspecifics. Whether or not an individual chirrups, and at what rate, depends on food divisibility and the number of conspecifics associating in a flock with the discoverer or pioneer. Divisible resources are more likely to elicit chirrups than nondivisible resources because the pioneer can cofeed with those who join. Similar results have been obtained with chimpanzees (Hauser and Wrangham 1987; Hauser, Teixidor et al. 1993).[32] The ultimate function of chirrups may be to reduce predation risk or time spent vigilant for predators. However, such benefits are clearly outweighed in certain situations by significant costs, such as when the food item is nondivisible and when attracting other flock members results in increased levels of aggression.

---

32. One question raised by the chimpanzee results, which is equally applicable to the house sparrow data, is whether differences in the rate of food-associated calls may have more to do with the discoverer's perception of quantity than divisibility. Specifically, when a pioneer finds food, pieces of bread may actually appear to represent more food than a whole piece of bread. Experiments are needed to distinguish between these two possibilities.

Rhesus Monkeys   Rhesus monkeys living on the island of Cayo Santiago, Puerto Rico, produce five acoustically distinct vocalizations in the context of food (see chapter 7 for description of informational content). For this population, approximately 50% of the diet consists of provisioned monkey chow and the remaining proportion of the diet is made up of fruits (coconut), flowers, grass, and nutrient-rich soil. Although a significant proportion of the diet comes from provisioning, there is, nonetheless, significant competition for food. Moreover, some of the nonchow food items are relatively rare and, when discovered, are the source of intense aggressive competition.

Natural observations reveal that females produce food calls at a significantly higher rate than males, and this proportion generally holds true for all call types and all contexts (Figure 6.39). Because rhesus monkeys exhibit female philopatry and male transfer, this calling asymmetry makes sense in light of kin selection theory. Specifically, by announcing their food discoveries, individuals functionally recruit individuals to the food source. This hypothesis has been confirmed by preliminary playback experiments (Hauser, in prep), showing that individuals rapidly approach the speaker upon hearing food calls—in particular, those given to rare and highly preferred foods (i.e., warbles, harmonic arches, and chirps).

To further explore the effects of kinship on food calling, the relationship between food call rate and matriline size (number of kin whose degree of genetic relatedness was greater than or equal to 0.25) was examined. Figure 6.40 shows that females with large matrilines called at a significantly higher rate than females with small matrilines. Missing from these analyses is quantification of the direct benefits of calling to kin. For example, it was not possible to assess whether, by calling, kin actually obtained more food. Nor was it possible to determine that food calls are altruistic, because the relative costs associated with recruitment were not quantified relative to food quality or abundance. In chimpanzees, for example, food-associated calls such as the pant-hoot are given by individuals at relatively large food sources, implying that the costs of increased feeding competition may be negligible (Wrangham 1977). Nonetheless, and as described more completely in chapter 7, individual rhesus who found food but failed to call and were detected by other group members received more aggression than individuals who called upon discovery (Hauser 1992c; Hauser and Marler 1993b).

Synthesis of Studies on Food-Associated Signals   Table 6.5 provides a summary of current studies on food-associated signals. Although there is considerable variability in the potential information conveyed (see chapter 7), a number of similarities emerge with regard to the function of such signals. This result is somewhat

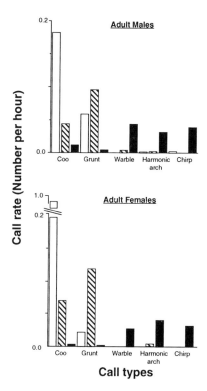

Figure 6.39
Differences in the rate of calling by adult male and female rhesus monkeys in the context of food. The *x*-axis represents the different call types, each bar representing the type of food context: white = individual waiting for access to provisioned monkey chow; striped = individual eating monkey chow; black = eating food other than chow (typically, highly preferred and rare food items). The *y*-axis is call rate (number of calls per hour) derived from focal samples of adult males (*n* = 20) and adult females (*n* = 20) (redrawn from Hauser and Marler 1993a).

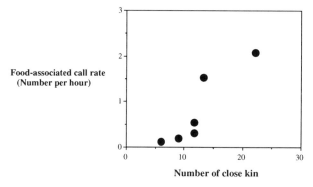

Figure 6.40
Differences in call rate (number of calls per hour) for focal females as a function of matriline size (i.e., the number of close kin, *r* ≥ 0.25) (redrawn from Hauser and Marler 1993a).

Table 6.5
Synthesis of Some of the Primary Studies of Food-Associated Signaling

Species	Signal Info	Signal Context	Function	References*
Leptothorax Solenopsis Atta Pogonomyrmex	Location of food (chemical signal)	Following immobilization of prey that are too large for transport by a single individual	To recruit workers to prey and obtain aid in transport	1
Apis spp.	Location and quality of food (visual and auditory signal)	Return to the hive following successful foraging trip	To provide information to nest members about location of food	2
Passer domesticus	Location of food (acoustic signal)	Following discovery of divisible food item	Recruit conspecifics to divisible food items	3
Corvus corax	Location of food (acoustic signal)	Following discovery of carcass	Recruit others to food; carcass defense	4
Gallus gallus	Food quality (acoustic signal)	Discovery of food by rooster	Announce discovery and attract mates	5
Hirundo pyrrhonota	??Shareable food (acoustic signal)	Foraging	?Recruit others to food	6
Saguinus oedipus	Food preference Food quality (acoustic signal)	Foraging	?Recruit others to food	7
Leontopithecus rosalia	Food preference Food quality (acoustic signal)	Foraging	?Recruit others to food	8
Ateles geoffroyi	Location of food (acoustic signal)	Food discovery	Recruit group members	9
Macaca fuscata	?Location of food (acoustic signal)	Food discovery	?Recruit group members	10
Macaca mulatta	Caller's hunger level and food quality (acoustic signal)	Forager anticipates access to food, discovery, and/or possession of food	Recruit group (?kin) members to food	11
Macaca sinica	Location of high-quality/rare food (acoustic signal)	Forager discovers high quality food	Recruit group members to food	12
Pan troglodytes	Arousal, food quantity, and divisibility (acoustic signal)	Food discovery and possession	Recruit community members to food source	13

*References: 1 (Hölldobler 1977, 1984; Hölldobler and Carlin 1987; Hölldobler, Obermayer, and Wilson 1992); 2 (Gould and Gould 1988; Kirchner and Towne 1994; Seeley 1992; von Frisch 1967a,b); 3 (Elgar 1986); 4 (Heinrich 1988; Heinrich and Marzluff 1991); 5 (Evans and Marler 1994; Gyger and Marler 1988; Marler, Duffy, and Pickert 1986a, 1986b; Marler, Karakashian, and Gyger 1991); 6 (C. R. Brown, Bomberger, Brown, and Shaffer 1991); 7 (Elowson, Tannenbaum, and Snowdon 1991); 8 (Benz 1993; Benz, French, and Leger 1990; Benz, Leger, and French 1992); 9 (Chapman and Lefebvre 1990); 10 (S. Green 1975a, 1975b; Owren et al. 1992); 11 (Hauser 1991, 1992c; Hauser and Marler 1993a, 1993b); 12 (Dittus 1984, 1988); 13 (A. Clark 1993; A. Clark and Wrangham 1993; Wrangham 1977).

surprising given the taxonomic diversity and the variety of signaling modalities. Thus, all food-associated signals appear to play a role in recruiting other group members to the food source, be this by directly calling at the source or by providing directions to others about food location. A further similarity concerns the relationship between call rate and some measure of food quality. Thus chickens, cotton-top tamarins, and golden lion tamarins call at high rates when the food item discovered is of high quality. In house sparrows and chimpanzees, high rates of food calling are associated with large and divisible food sources. Last, the calling behavior of many species appears to be guided by social factors, such as the composition of nearby animals or the advantages of forming larger groups. At this stage, however, few studies have quantified the fitness consequences of calling as opposed to remaining silent. Such studies are sorely needed if we are to do for food signals what has been done for either alarm or mating signals.

6.4   Social Signals: Birds and Primates

6.4.1   Dominance Signals and Cues

In sections 6.2 and 6.3 we discussed the outcome of competitive interactions for access to mates and food. In many of these interactions, especially those involved with mating, there are elaborate displays, designed to maximize saliency or attrac-

Box 6.4
Game Theory: The Conceptual Kernel

In the mid-1970s evolutionary biologists interested in the dynamics of aggressive competition started to apply economic game theory analysis to the problem (Maynard Smith 1974, 1979, 1982; Parker 1974). The goal was to determine which of several potential strategies was the most successful, over both the short and long term. The methodological trick was to determine how a particular set of parameters influenced the success of a given strategy and whether a specific set of strategies would be stable over time or open to invasion by a mutant strategy. If a stable solution could be derived, it was called an *evolutionary stable strategy, or ESS*.

The starting point for thinking about ESSes is to realize that the best strategy for an individual depends critically on what other individuals in the population are doing. Imagine two ravens, Bernd and Jay, competing for access to a freshly killed carcass. The carcass represents a potential benefit to each raven, and entering into a fight for access to the carcass represents a potential cost (Heinrich 1988; Heinrich and Marzluff 1991). In the simplest scenario, Bernd and Jay aggressively display at each other but do not fight. At some point, Jay runs away, leaving Bernd with the entire carcass. The interesting evolutionary question is, given Bernd's success, why don't all ravens act Bernd-like, threatening up to the level required for avoiding a fight and gaining access to the resource? The answer is simple. If the raven world was full of Bernds, it would leave open the possibility of a mutant raven, capable of escalating the level of display and thereby driving competitors away.

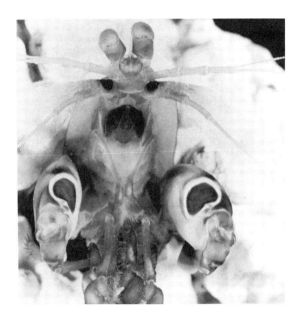

tiveness relative to other competitors. In this section we focus exclusively on communicative signals and cues used during social negotiations for resources other than food or mates. Although the ultimate outcome of such negotiations may very well be increased access to mates and higher-quality foods, the signals we will explore have more to do with working out the details of a social relationship, as occurs between territory owner and intruder, dominant and subordinate, or mother and offspring. Much of this discussion relies on logic or intuitions from game theory. I thus provide a terse summary of the key ingredients in Box 6.4.

Residents and Intruders: Stomatopods   Stomatopod crustaceans (Figure 6.41), also known as mantis shrimp, exhibit a diversity of mating systems (e.g., promiscuity, lifelong pair bonds). Following the juvenile molt, individuals attempt to obtain "homes"—cavities in the substrate. Because growth is indeterminate, with molts occurring once every few months, individuals frequently confront the problem of finding a new cavity suitable for their size. In general, nest cavities represent a limited and highly valued resource. Most remarkably, the stomatopod has evolved an extremely powerful claw (raptorial appendage) for defending its cavity and fighting off intruders.[33] When an intruder approaches a cavity, the

---

33. In one of the largest species, *Hemisquilla ensigera*, it has been reported that individuals can break the glass of an aquarium tank with a single strike of the claw.

Figure 6.41
A stomatopod intruder approaching the cavity of a resident, tucked away inside (photograph courtesy of Dr. R. Caldwell).

resident progresses through a series of displays, starting with a mild approach and escalating to the meral spread and strike; the latter is associated with heightened aggressiveness and involves spreading the claw and then striking the ground or intruder (Caldwell and Dingle 1975).

Over the past ten years, Caldwell and his colleagues (Adams and Caldwell 1990; Caldwell 1986, 1987; Caldwell and Dingle 1975; Steger and Caldwell 1983) have investigated the processes guiding competitive interactions over nest cavities in *Gonodactylus bredini*, a species of stomatopod found in the Caribbean. The story that has emerged is perhaps one of the best examples of the dynamics of resource competition, and it provides one of the most intriguing cases of functional deception (see chapter 7).

Because stomatopods show indeterminate growth, competitive interactions over nest cavities often involve residents and intruders that differ in size. To assess the effects of size asymmetries, Adams and Caldwell (1990) staged competitive interactions over cavities where residents were 15% larger, equal in size, or 15% smaller than intruders. Results (Figure 6.42) indicated that when residents were either larger or equal in size to intruders, they frequently threatened the

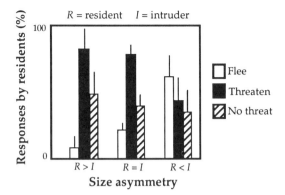

Figure 6.42
Responses by resident (*R*) stomatopods to conspecific intruders (*I*). Data show the relationship between relative size asymmetries (*R* is greater than [>], equal to [=], or less than [<] the size of *I*) and the type of behavior produced by *R* in response to *I*'s intrusion (redrawn from Adams and Caldwell 1990).

intruder and rarely fled. In contrast, when the intruder was larger, residents left the nest cavity on 50% of all challenges. In addition, when residents were larger or equal in size to intruders, they were more likely to retain their nest cavities than when they were smaller (Figure 6.43). Interestingly, although intruders were more likely to escalate their level of threat against residents who failed to threaten them, size per se had no effect on the proportion of intruders who escalated (Figure 6.44). What these data indicate is that size asymmetries play an important role in determining the outcome of competitive interactions over nest cavities. This is not a surprising result given the relatively common finding that larger animals outcompete smaller animals for access to valued resources. What is surprising, however, is how the effects of size interact with the effects of molting.

When a stomatopod enters the molting phase, its exoskeleton becomes soft. As a result, use of the claw is limited, since any strike received would cause significant damage. Based on this feature of stomatopod biology, one would predict that during the molt, individuals should abstain from producing the meral spread display since they cannot back up their display with an aggressive attack. In fact, Caldwell and colleagues (Adams and Caldwell 1990; Caldwell 1986; Steger and Caldwell 1983) have found just the opposite. Figure 6.45 indicates that during the intermolt (controls), individuals shift from low- to high-intensity displays as a function of changes in intruder pressure. Thus, when intruders probe for cavity availability, residents first approach the entrance, and then, if the intruder probes

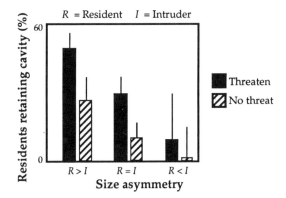

Figure 6.43
The effects of size asymmetries and aggressive responses on the proportion of stomatopod residents (*R*) retaining their cavities when confronted by conspecific intruders (*I*) (redrawn from Adams and Caldwell 1990).

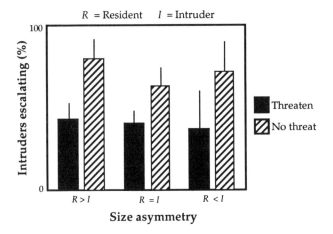

Figure 6.44
The proportion of stomatopod intruders escalating their aggressive responses toward residents based on size asymmetries (redrawn from Adams and Caldwell 1990).

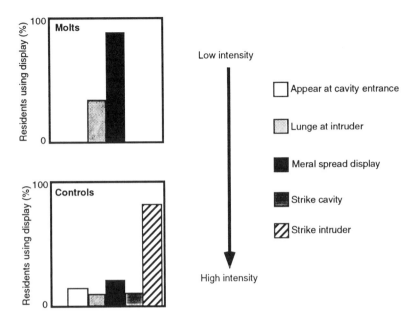

Figure 6.45
Changes in the resident's display behavior toward intruders as a function of molting state. The behavioral displays listed along the right side of the figure are arranged according to aggressive intensity, with the lowest-intensity display (i.e., appear at cavity entrance) shown at the top (redrawn from Steger and Caldwell 1983).

further, the resident gradually escalates the aggressive intensity of the display, culminating in a strike with the claw. As Figure 6.45 illustrates, however, the pattern of displays used during intermolt is not consistent with the pattern exhibited during the molt. Individuals in molt appear to skip the lowest-intensity display, jumping right into lunges and meral spreads. In fact, most residents in molt produce the meral spread display after the first probe by the intruder. Given the resident's inability to defend the message of the display (i.e., "Attack if intruder probes again"), it appears that stomatopods are actively falsifying information, or in the words of Caldwell and colleagues (Caldwell 1986; Steger and Caldwell 1983), residents are "bluffing."

Additional support for the inference that stomatopods actively falsify information about aggressive intent comes from observations of the frequency and timing of use of the meral spread display and its relation to the relative size of the intruder (Caldwell 1986). Figure 6.46 indicates that during the intermolt phase the meral spread and meral spread + strike displays (two of the more aggressive

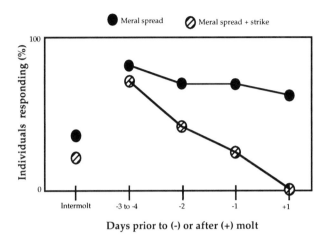

Figure 6.46
Use of the meral spread (black) and meral spread + strike (striped) displays by stomatopods in the intermolt phase (data on far left) and in the period immediately surrounding and including the molt (redrawn from Caldwell 1986).

displays) occur at a fairly low rate. However, three to four days prior to the molt, individuals produce these displays at high rates. What the resident appears to be signaling, therefore, is a heightened state of aggressiveness. Because the costs to an intruder of probing are high (possibly death following a strike from the claw), residents generally succeed in maintaining their cavities.

Size asymmetries between resident and intruder have a significant effect on the proportion of newly molted individuals using the meral spread as a bluff. Specifically, a greater proportion of new molts produce the meral spread display during encounters with same-sized intruders than during encounters with intruders who are 15% larger, or when residents are in the intermolt phase.

The stomatopod data are relevant to some of the issues raised earlier concerning honest signaling and deception. There are two points in particular: (1) The meral spread display is typically given by a cavity resident when challenged by an intruder over access to the cavity; most often, the cavity resident is in an intermolt phase and can vigorously back up this display with a potentially costly strike. (2) During the molt, individuals use the meral spread when challenged by intruders, but because of their soft exoskeleton, they are not able to back up the display with additional aggression. Consequently, when newly molted individuals produce a meral spread, they are actively falsifying the information conveyed (i.e., "If you continue to approach, I will physically attack you"). In general, the system works because the costs associated with continued probing (i.e., death or severe injury as a result of being struck by the claw) far surpass the benefits (i.e., gaining access to a cavity).

Badges of Status   Given the stomatopod's lifestyle, social interactions are typically one-on-ones, are generally quite short-lived, and are often not repeated. For group-living animals, social interactions may include one-on-many or many-on-one, can last for extended periods of time, and depending upon the stability of the group with regard to composition and size, will often be repeated several times. Among avian species, there are at least two situations where dominance-related interactions arise and form the centerpiece for establishing dominance relationships: winter-feeding flocks prior to the mating season and territorial interactions. The interesting question, comparable in kind to the questions raised in the section on mating, is, How do individuals assess the reliability of dominance-related signals (i.e., signals that are indicators of viability, strength, condition, etc.) so that dominance hierarchies can be established and maintained in a relatively stable state? Although a great deal of research has been done in this area, I focus here on visual characteristics that appear to covary with dominance and have thus been called *badges of status*. As with the topic of warning coloration, however, I think that the discussion of badges has also confused the signal-cue distinction, but let us look at the data first before we cast a vote.

Previously I alluded to observations of territorial interactions among red-winged blackbirds and argued that the motor action of revealing the red epaulet by pulling back the black wing feathers represents a signal, a display of aggressive intent (Metz and Weatherhead 1992). Support for this signal's function, and for the importance of the red splotch or badge, comes from studies that blackened the epaulet of some territorial males; these individuals subsequently lost their territories as a result of a significant increase in intruder pressure. Studies involving comparable manipulations of conspicuous coloration have generally failed to find evidence of disruptive territorial or mating patterns (reviewed in Rowher and Røskaft 1989). However, these other studies have been conducted on species whose badges are fixed in place and cannot be covered. Consequently, differences in the function of badges appear to map on to differences between signals and cues. Coverable badges represent communicative signals, whereas fixed badges represent communicative cues.

As a critical test of the badges-of-status hypothesis, Rowher and Røskaft (1989) conducted a field study with the yellow-headed blackbird (*Xanthocephalus xanthocephalus*). This species is a close relative of the red-winged blackbird, it has a similar mating system, and males are generally black except for the distinctively colored badges. The primary difference is that the yellow-headed blackbird has a fixed badge (yellow head) whereas the red-winged blackbird has a coverable badge. The experiment involved a complete blackening of the yellow badge and subsequently measuring changes in territorial interactions and harem size (i.e., a measure of mating success). There were no differences between these blackened individuals and controls in terms of mate attraction or territorial defense. The only significant effect was that approximately 40% of blackened males who obtained a territory following removal obtained a better quality territory than did controls.

Results on yellow-headed blackbirds stand in striking contrast to those obtained for red-winged blackbirds. Rowher and Røskaft offer an explanation. First, they suggest that much of the difference lies in the contrast between fixed and coverable badges, and their relationship to signals of subordinance. In red-winged blackbirds, covering the epaulet signals subordinance. Experimentally blackening the epaulet causes territorial males to experience high rates of intrusion, apparently because they are signaling submission or because they are failing to signal ownership. In yellow-headed blackbirds, blackening failed to produce a cue of subordinance, as evidenced by the lack of difference in intrusion rates for experimentals and controls. Second, they propose the notion of an arbitrary identity badge. Specifically, the color patch per se is not an indication of dominance but rather represents an arbitrary marker of fighting ability that subsequently facilitates recognition memory in others. In other words, as in the case of warning colors discussed earlier, conspicuous colors will be favored if they facilitate recognition of fighting ability and, consequently, avoidance of exceptionally good fighters. One potentially important consequence of this relationship is that novel badges can invade a population, thereby leading to significant morphological changes in the population over evolutionary time. Such invasion is contingent upon three criteria: (1) conspicuous badges must be heritable; (2) there must be repeated interactions between the same pair of individuals; and (3) the number of potentially interacting individuals must be large, such that individual recognition is difficult in the absence of distinctive markers or cues. Support for this idea comes from the fact that those blackened individuals who reclaimed higher-quality territories tended to be larger individuals and thus the winners of aggressive interactions.

In Harris sparrows (*Zonotrichia querula*), males exhibit considerable variation in plumage color (fixed badges), especially the degree of blackness covering their head and chest (Figure 6.47). Observations (Rowher 1977, 1982; Rowher and Rowher 1978) indicate that black males are more dominant than white or pale males, as evidenced by the outcome of competitive interactions over food and mates; black males are not larger than pale males. The intriguing problem raised by this system, as with blackbirds, is why subordinates don't attempt to cheat by growing black feathers. Wouldn't this deception provide the necessary boost for stepping up through the hierarchy? To explore these ideas, Rowher and Rowher (1978) conducted an experiment with a flock of Harris sparrows. In one condition, subordinates had their chests painted black. These dominant-looking individuals were attacked as often as they were before the test and failed to rise in status. The second condition involved pale subordinates who were injected with testosterone but whose plumage coloration was left intact. As known from

Figure 6.47
Schematic illustration of plumage variation in Harris sparrow (drawings from Krebs and Davies 1993).

previous work with testosterone, this hormone is commonly associated with heightened aggressiveness during communicative interactions (reviewed in Wingfield and Marler 1988). Results showed that testosterone-treated subjects were, as expected, more aggressive, but they failed to rise in rank. Last, condition three involved painting the subordinate black and injecting him with testosterone. In this case, subjects consistently rose in rank. This result has been interpreted as evidence that the Harris sparrow's black splotch represents a *signal* or badge of status (reviewed in Butcher and Rowher 1989). Such signals, if this is an appropriate descriptive term, are only functional if backed up by an appropriate level of aggressive motivation.

The reason for questioning the claim that fixed badges of status represent signals is related to their design features. Specifically, although fixed badges change in structure as a function of age, once acquired they are permanently ON, and there are no further production costs associated with their expression. As with warning colors, fixed badges are likely to vary as a function of condition and, as such, are likely to be useful indicators of the resources available to the individual. However, individuals cannot control the degree to which they express or show off their badges (in contrast to coverable badges) or when they do so. In addition, the actual function of badges may differ according to whether individuals are in the process of establishing dominance interactions or maintaining them. Most of the experimental studies of avian badges have been conducted prior to the establishment of dominance relationships.

One aspect of the badges-of-status hypothesis that needs to be worked out in greater detail is the extent to which differences in lighting conditions, as imposed by variation in the structure of the habitat, influence the use and salience of the cue or signal, as well as its perception. A nice example of how such effects might work has recently been provided by Marchetti (1993), based on observations and experiments of eight warbler species within the genus *Phylloscopus*. Marchetti's study was designed to test the notion of sensory drive (*sensu* Endler 1991b, 1992, 1993b), which predicts that the design features of a signal or cue will be shaped by environmental factors or constraints (also see chapter 3). Observational data revealed that species living in the darkest habitats had the highest reflectance at the feather tip, the greatest number of color patches, and the most elaborate territorial displays, designed to reveal the color patches. Outside of the context of territorial displays, these patches were concealed. Experimentally reducing color patches (applying green acrylic paint) in three consecutive years resulted in a significant decrease in territory area relative to controls and relative to individuals who either had their color patches enlarged (applying yellow paint) or had a novel patch added. The latter group experienced an increase in territory area relative to controls. Laboratory simulations of such experiments could be carried out with species such as the red-winged blackbird to determine whether lighting conditions influence the display, as well as its perception (measured behaviorally).

Birds are clearly not the only species to exhibit badges of status. Humans do too, and recent work falling within the domain of evolutionary psychology (Buss 1994; reviewed in Barkow, Cosmides, and Tooby 1992) has attempted to assess which features are perceived by others as signals of dominance and which as signals of attractiveness. In one study, C. F. Keating (1985) took advantage of an image manipulation package (Identification Kit) to create facial composites including mature and immature features. The perceptual task entailed rating each face with regard to dominance and attractiveness. For both men and women, results showed that mature facial features were associated with dominance. However, features associated with dominance had different effects on attractiveness for males and females. Males rated as highly mature were also rated as highly attractive. In contrast, females with relatively immature features were rated as more attractive. This finding, Keating argues, fits in with theories concerning mate choice and reproductive value, with younger women providing a higher probability of mating success than older women, and mature dominant men providing potentially greater resources (see also R. A. Fisher 1930).

Most recently, Perrett, May, and Yoshikawa (1994) examined which specific facial features are used by human observers to assess attractiveness. Previous studies, both theoretical and empirical, have suggested that people find the average face to be the most attractive (Langlois and Roggman 1990). The first step of the Perrett, May, and Yoshikawa study involved a rating of faces (both Caucasian and Asian women) along a scale ranging from 1 (very unattractive) to 7 (very attractive). Following this rating procedure, three composite sets were created. The first was obtained by averaging all of the faces in the original sample, yielding the *sample average face*. The second was obtained by averaging those faces that obtained the highest attractiveness scores, yielding the *average attractive face*. The third was derived from a 50% exaggeration of the shape differences in the average attractive face, yielding a *caricatured attractive face*. Both men and women, from both cultures, preferred the average attractive face to the sample average face, and preferred the caricatured attractive face to the average attractive face. Similar results were obtained for male faces, though the caricatured face did not enhance the preference. Together, these results provide evidence against the idea that attractive faces are average. Moreover, when one examines the particular features of the caricatures, attractiveness appears to be equated with high cheek bones, narrow jaws, and large eyes. The extent to which such features are associated with functionally significant reproductive parameters (e.g., fertility) has yet to be explored explicitly, although there are hints that physiological abnormalities (e.g., heightened exposure to steroids) may lead to exaggerated traits that deviate from the gender-typical norm (e.g., broad jaw in women).

Although the results presented by Perrett and colleagues argue against the average attractive face, they do not speak to the issue of symmetrical faces. Recently evolutionary biologists have developed an interest in the notion of fluctuating asymmetries (FA) in trait expression. Fluctuating asymmetry is said to occur when the normal or typical expression of a trait is bilaterally symmetrical and some individuals show small, random deviations from symmetry; unlike the data reported in chapter 4 on hemispheric asymmetries, FAs are not directionally biased (i.e., there is no population level bias in whether one side or the other is longer, brighter, etc.). In general, FAs occur during development, when some problem in the coding sequence results in random deviations in expression. Although geneticists have long been interested in this phenomenon (Palmer, Richard, and Strobeck 1986; Van Valen 1962), behavioral ecologists have provided a new twist on its role in behavior. Specifically, if FAs indicate developmental problems, then asymmetrical members of the population should receive fewer matings than symmetrical members—assuming, of course, that there is a heritable component to FA. Thus in several nonhuman animal species there appears to be evidence that females prefer mating with individuals showing significant symmetry in trait expression over individuals showing asymmetries (Møller and Hoglund 1991; Thornhill 1992). A beautiful example has been provided by Møller (1993), who showed that the nuclear disaster at Chernobyl caused significant tail asymmetries in a population of barn swallows and that those with the most asymmetrical tails had the lowest reproductive success as measured by laying date; surprisingly, the nuclear fallout did not have any effect (direct or indirect) on other morphological characteristics. Similar relationships between FA and mating success have been suggested for humans. Thus Thornhill and Gagestad (1993) have argued, based on measurements of facial asymmetry and questionnaire data on mating patterns, that individuals with more symmetrical faces have higher reproductive success. In contrast, Jones and Hill (1994) failed to find such relationships, using a much larger sample of individuals from a greater diversity of cultures.

Although the data summarized here suggest that a diversity of organisms, including scorpion flies and humans, exhibit asymmetries in trait expression, and such asymmetries may serve a functional role in aggressive combat and mating, several critical questions remain: (1) Are fluctuating asymmetries better explained by invoking natural or sexual selection? If sexual selection is the dominant pressure, than in a majority of polygynous species, one would expect females to show less FA than males. This prediction has found support in Møller's studies of barn swallows, but not in several other studies of insects and birds. (2) Are asymmetries actually associated with more significant costs than cases of symmetry, and if so, what is the relationship between cost and the degree of asymmetry? (3) How good is the perceptual system at detecting—recognizing—asymmetries in trait expression? In several studies of FA, the extent of the asymmetry seems extremely small. Although the sensory system may not be capable of picking up FA directly, there may be other factors associated with FA that are more readily detected (e.g., asymmetries in tail length might be revealed by distinctive patterns of flight). (4) Is the degree of asymmetry heritable? These questions are well in reach of empirical tests, and many researchers are actively working on these problems using both novel experimental methods and new modeling techniques (Enquist and Arak 1995; Johnstone 1995).

Care-Elicitation Cries   When Trivers (1972, 1974) published his two influential papers on parent-offspring relationships, he essentially turned around traditional ethological and psychological views, forcing researchers to take seriously the idea that there was a genetic war at hand, with offspring selected to get the most of their caring parents. The essence of Trivers' argument was to show that whereas

offspring are related to themselves by 1.0, parents are only related to their off-spring by 0.5 (the traditional measure of genetic relatedness, or *r*, can vary from 0.0 to 1.0). Consequently, offspring will tend to want more from their parents than their parents are willing to give. Moreover, for each unit of care allocated to current offspring, parents functionally reduce the available units for subsequent offspring. Once again, since offspring are related to themselves by 1.0 and, at best, only related to their future siblings by 0.5 (i.e., it will depend on whether the father is the same or different), they should continue to milk their parents for all that they are worth. Since Trivers' publications, there have been numerous empirical tests and several controversial theoretical challenges (reviewed in Clutton-Brock 1992; Mock and Forbes 1992), including discussions of the ef-fects of mixed parentage, current reproductive value of the parent, and clutch size on the allocation of parental care and parental investment.[34] Here I would like to focus on some of the communicative tools available to offspring to ma-nipulate their parents. In particular, I describe some work on avian begging calls and primate crying or fussing noises to showcase some of the interesting features of the mother-offspring relationship and to reintroduce Zahavi's handi-cap principle. I then use the results from this discussion to make a brief and speculative foray into the functional significance of human crying, that is, weep-ing with tears.

Prior to and during the weaning process, young appear to use different tactics in order to maximize the probability of obtaining care (Altmann 1980; Mock 1985). The primary types of care that young seem most intent on getting are access to food (provisioning in birds, milk in mammals), protection, and in mammals, transportation. And one of the primary mechanisms for obtaining care is vocal behavior, often accompanied by distress, that I will simply call crying. The ques-tion we need to address here is how parents, mothers in particular, discriminate between honest cries produced out of need and manipulatively dishonest ones produced out of gluttony. In general, there have been at least two dominant hypotheses regarding the function of such calls. Trivers' view, and the one sup-ported by many authors, is that cries are manipulative, designed to extract as

---

34. In Trivers' formulation, the distinction between parental care and parental investment is important, even though in practice it is difficult to measure empirically (see Clutton-Brock's thorough 1992 analysis of this problem). In general, the term *parental investment* has been reserved for cases where the parent's allocation of care to its offspring has a direct effect on future investments. *Parental care* tends to be used for cases where the concern is with investment in the current generation of offspring, independently of costs to subsequent offspring.

much care from the parent as possible without jeopardizing their lives.[35] An alternative view, recently developed by Godfray (1991), following in the modeling footsteps of Grafen (1990), suggests that crying may be an honest signal of need. For cries to be honest, they must be costly to produce, costs and benefits must vary as a function of need (e.g., those in greater need will pay greater signaling costs; Redondo and Castro 1992), and cryer and caretaker must be genetically related by descent.

R. M. Evans (1994) examined the relationship between offspring need and vocal behavior in American white pelicans. Young are born in an altricial state. As they increase in size and grow feathers, they also develop the requisite physiological mechanisms necessary to shiver and thereby control body temperature. Since control of temperature is critical to survival, one might expect young to produce care-eliciting cries as a function of their need to increase body temperature, especially if they are incapable of doing so on their own and require parental help. Under laboratory conditions, Evans manipulated brooding temperature and measured body temperature and call rate at three developmental stages: 1 day, 1 week, and 2 weeks. As Figure 6.48 clearly shows, young at all three developmental periods showed increases in call rate as body temperature decreased and, when subjected to rewarming, showed a decrease in call rate. The effect was most dramatic when chicks were only one day old. At this stage, chicks are incapable of shivering, lack insulating feathers, and thus cannot control their own body temperature. Another interesting feature of the pattern observed is that chicks begin calling at high rates even before there is a significant drop in body temperature. This suggests an anticipatory response which, intuitively, appears highly adaptive given the potential costs of experiencing a significant drop in temperature.

Are white pelican cries honest signals of need, as suggested by Godfray's theoretical position? The data presented provide strong support for the correspondence between calling and the need to modify body temperature. Specifically, call rate was highest when body temperature was lowest, or at least moving in the direction of a significant drop in temperature. Moreover, because shivering represents an involuntary response, driven by the individual's physiological system, and is directly observable by the parent, the close temporal relationship between shivering and calling provides further support for the honest signaling

---

35. For some relatively short-lived species, obtaining the most care possible, even at the cost of parental survival, might be advantageous. However, in long-lived species such as primates having a surviving parent may provide direct fitness benefits to the offspring.

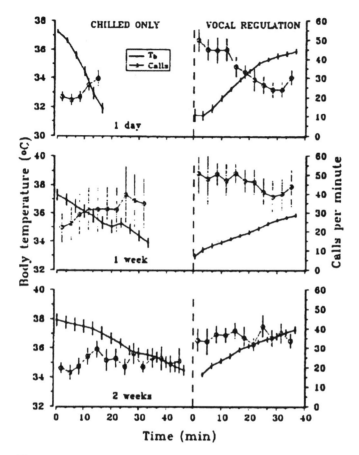

Figure 6.48
Calling behavior in young white pelicans as a function of changes in temperature. The left side of the figure shows changes in calling due to continuous lowering of body temperature ("chilled only"). The right side shows changes in calling as a result of rewarming. Means and standard deviations shown (from R. M. Evans 1994).

position. Missing from the analyses, however, is a direct assessment of the costs of calling and, in particular, the costs relative to condition. In addition, and as Evans points out, there is certainly room for dishonest signaling in this system. Although cycles of shivering and calling are quite synchronous, infants who started calling before they shivered would conceivably obtain care from their parents and, as a result, be in the position to increase body temperature without incurring the costs of shivering.

To evaluate the function of primate cries, Hauser (1993a) analyzed data from the Amboseli vervet monkey population. In this population, infants give several vocalizations during distress (Hauser 1989; Seyfarth and Cheney 1986; Struhsaker 1967), especially when they want their mother to carry them or want to nurse. During a two-year study, some infants survived past the first year of life, whereas some did not. To determine whether the communicative interactions of these two mother-infant groups differed, data on cry effectiveness and call rate were plotted for the first four months. The weaning process is apparent quite early on, but is most dramatically observed at around 3–4 months. Moreover, for the Amboseli population, there is a significant mortality peak at 4 months.

Figure 6.49 shows that during the first six weeks, infants in both groups call at relatively low rates, and their calls are highly effective at eliciting care. In the eighth week, infants in both groups increase call rate, and associated with this change is a decrease in call effectiveness. It is not until week 12 that one begins to observe a significant difference between these two groups. Specifically, in the surviving infant group, there is a substantial decrease in call rate and a corresponding increase in call effectiveness. This pattern continues until week 16. In contrast, for the infants who fail to survive their first year, call rate continues to increase, and call effectiveness continues to plummet.

How can we explain the pattern observed? Why don't all infants maintain a relatively low call rate, thereby maximizing call effectiveness and, presumably, the amount of care obtained? Do infants who call at high rates simply need more care, or are they trying to manipulate their mother, perhaps beyond her means? Answers to these questions are limited by the fact that Hauser did not have direct quantitative measures of infant condition or costs of crying. More qualitative measures, however, provide some insights. For example, the variation in vocal behavior could not be accounted for by subjective measures of infant condition (e.g., quality of hair, size, weight), maternal rank, predation pressure, or the availability of high-quality resources. If infants were signaling need, as revealed by call rate, then why didn't mothers respond to such requests? There are at least three possibilities. First, although call rate may represent an honest signal

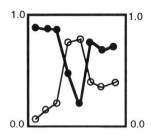

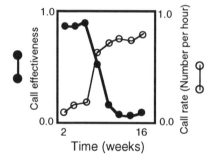

Time (weeks)

Figure 6.49
Changes in call effectiveness (*y*-axis on the left) and call rate (*y*-axis on right; rate = number of calls
per hour) over time (*x*-axis; weeks) in vervet monkeys. The offspring's fussing calls were considered
100% effective if, for each call associated with an attempt to nurse or be carried, the mother accepts;
0% effective means infants requests are always rejected. The top panel presents data for infants who
survived their first year. The lower panel is for infants who died before the end of their first year
(redrawn from Hauser 1993a).

of need, mothers may have simply lacked the requisite resources for allocating
more care to their offspring. Second, call rate may be honest, but mothers may
decide that the level of care required outweighs the potential benefits; for example, the current offspring may be a runt, and it would be better to hold off for
future offspring. Third, call rate is dishonest, and offspring are engaged in an
evolutionary arms race with their parents, attempting to acquire as much care as
possible. As with white pelicans, more formal tests of honest signaling in vervets
will require more quantitative measures of signaling costs and signaler condition.

What about human crying, specifically crying with tears? Humans are the only
species to use tearful crying as an expression of emotion,[36] and thus, as Darwin

---

36. Foreshadowing some of the literature reviewed in chapter 7, in particular the controversy between
Ekman and Fridlund, let me acknowledge here that tearful crying may be both an expression of
emotion and an expression that manipulates the behavior of others.

(1872) pointed out, this expression "must have been acquired [in humans] since the period when man branched off from the common progenitor of the genus *Homo* and of the non-weeping anthropomorphous apes" (p. 153). Tearful crying is associated with at least two corresponding emotional states, sadness and joy (even euphoria or hysteria). These observations raise two questions. First, why don't other species exhibit tearful crying as part of their species-typical communicative repertoire? Second, what is the function of tearful crying in humans? Let me start with the second question, and make the claim that tearful crying in humans is a handicap,[37] with a fundamental developmental origin. Early in life, tearful crying elicits almost immediate parental response. Over time, however, the child's increasing cognitive prowess allows it to use tearful crying to manipulate parental response—we all know about the child who falls off of a swing when no one is around and only begins crying when the parent leans out the window to ask if everything is okay. Thus, at some point, tearful crying becomes an extraordinary tool for manipulating others.

What makes tearful crying interesting, relative to all other human expressions, is that it leaves a permanent trace. If you have been crying, because you were either moved by hearing Bach's Double Violin Concerto, sad because of having lost someone close to you, or laughing hysterically because a comedian like Jim Carrey has contorted his mouth into some abnormal position, your face and eyes will be red, and unlike blushing, the red will stay for a while. Moreover, when you cry, your vision is temporarily out of commission. Last, crying can be extremely costly, as evidenced by the apparently debilitating effects of a good hard cry or laugh—just think of the English expression "I laughed so hard my stomach hurt."

These observations lead to the proposition that crying is a handicap, an honest signal of need. Crying is costly and difficult to fake. Actors require considerable training before they can cry and have an audience believe them; moreover, good Hollywood-type crying apparently involves getting into a particular emotional state, of thinking about something extremely sad. Because it reduces your visual acuity, it directly signals vulnerability and thus need. Humans have been designed to respond to those who cry, that is, to empathize. Although we can empathize with individuals of all ages, the origins of this response emerged, I suggest, in the context of parent-offspring relationships—in particular, situations involving infant vulnerability (see chapter 7 for further discussion of empathy).

So why don't nonhumans cry with tears? They certainly cry without tears, and they certainly experience many of the same emotions associated with tearful

37. I thank Richard Wrangham for contributing this insight and for clarifying my own thoughts.

crying in humans. And mammals, at least, have the requisite tear ducts, though perhaps they lack the requisite neural circuitry linking them to the limbic system. Even if such circuitry is lacking, we might ask why, or why such circuitry evolved in humans? I unfortunately don't have a good answer to these questions, but have one suggestion. Rather than think about the reason why nonhumans lack tearful crying, think about why humans do. The reason I would like to suggest is that, if we accept tearful crying as a handicap, it may be the only honest signal of emotion that we have. Although Ekman and his colleagues have argued that the Duchenne smile is an honest signal—a true smile—and is difficult to produce voluntarily, I would argue that it is under far greater control than is tearful crying. Rather than continue this discussion here, I will postpone it for chapter 7, where I unite some functional and psychological perspectives.

7 Psychological Design and Communication

7.1 Introduction

When my daughter Alexandra was about three years old, we were playing at the park on a hot sunny day when she asked if she could take off her shoes and go barefoot in the sand. I said that she couldn't because it was too hot and because there were likely to be sharp things that might hurt if stepped on. The following conversation transpired:

Alexandra: Daddy, I want to go play on the slide.

Marc: Okay, I will come with you.

Alexandra: I want to go alone.

Marc: Okay, I will be here if you need me.

Alexandra walked over to the slide, and then disappeared under it. Several minutes passed.

Marc: Alexandra, what are you up to?

Alexandra: Oh nothing, just playing.

Marc: Do you have your shoes on?

Alexandra: Yes. See.

Out from under the slide appeared her two arms, with her shoes covering her hands.

Marc: That's not where your shoes go.

Alexandra: Yes it is. I am wearing them.

All parents, and even nonparents, have wonderful stories like this from children. They are wonderful because they tell us about children's minds, what they think, what they believe, and what they want. In essence, we all seem to have, intuitively, the folk psychological view that language, even the child's language, provides a direct window onto the mind. One problem, of course, is that we might want to investigate the mind—the possibility of mind—of prelinguistic human infants and even nonhuman animals. How might we go about addressing this problem? Some would argue that the empirical study of mind is empirically out of reach and is better left to philosophers. I don't buy this view, nor do a large number of developmental cognitive psychologists (Astington 1994; Astington, Harris, and Olson 1988; Carey 1985; Keil 1994a; Meltzoff and Moore 1994) and philosophers. I do, however, feel that a number of more basic cognitive

mechanisms (e.g., perception of visual and auditory events, classification of stimuli into functionally meaningful categories) or abilities must be investigated first
before we can make proper sense of the ideas surrounding the study of mind
in both humans and nonhumans (Griffin 1992). In this chapter, therefore, I pick
up some of the issues that were touched on in our discussion of neurobiological
and ontogenetic design. This chapter thus constitutes a continuation of the theme
of proximate causation, focusing on putatively higher-level brain processes
than those discussed in chapters 4 and 5. For want of a better descriptor, the
literature discussed here is described under the problem of psychological
design—specifically, the mental processes guiding the production and perception of communication.

The material covered in this chapter has been divided along conceptual lines
rather than by case study or socioecological context. In section 7.2, I discuss
what is currently known about the information content of different communicative signals. The primary inroad into this problem is to think about the function
of a signal, what it has been designed to accomplish. By thinking about function,
it is possible to assess what the signaler is attempting to convey to the pool of
perceivers, and how the signal's design features have been selected to alter the
behavior of relevant perceivers. By adopting a comparative approach, we will
see that there is considerable interspecific variation in the quality and quantity
of information conveyed by communicative systems, and much of this variation
can be explained by interspecific differences in the socioecological contexts in
which communication is embedded.

In section 7.3, I explore the mechanisms underlying the perception of communicative signals, in particular the processes that cause individuals to classify signal
exemplars into functionally meaningful categories such as predator, food, enemy,
and potential mate. As stated throughout the previous chapters, variation in signal
structure will ultimately be constrained by perceptual systems. It is therefore
necessary to understand how signals are categorized by a receiver (e.g., what
features are used, what circumstances alter the categorization process) and
why intra- and interspecific differences exist in the extent to which within-
category variation is resolved or ignored.[1] To address these issues, I review
studies that have focused explicitly on how communicative signals are categorized. I also discuss more general research on categorization and concepts, particularly experiments designed to assess what individuals know not only about their
social world, but also about their physical world. Though this work is not specifi-

---

1. Recall once again the distinction made in chapter 3 between just noticeable differences and just
meaningful differences.

cally related to communication, the methodology employed is directly applicable to certain problems in communication. Section 7.3 concludes with a review of relevant literature on cross-modal perception. Here, for the first time in the book, we will enter the interesting conceptual and methodological arena of intersensory integration. In particular, we will see how visual and auditory components of a signal often interact, one modality tending to subsume a dominant role in the message conveyed and, consequently, the information extracted by perceivers.

Section 7.4 looks at the beliefs and intentions of the signaler as they relate to deceptive communication. I begin by reviewing key theoretical issues in the study of *functional deception,* cases where an individual experiences an increase in fitness as a result of producing a signal[2] that has been designed to manipulate the behavior of a perceiver. Such cases have been described as deceptive because the *behavior* of the signaler functionally dupes the targeted perceiver into performing a costly act or to miss out on an opportunity for acquiring beneficial resources; we have already come across some of these cases (e.g., stomatopods) in chapter 6. Following this discussion, I turn to potential cases of *intentional deception,* where the signaler's behavior is viewed as a window into the mind—in particular, a window into an individual's beliefs and intentions. Though studies of human children have made considerable progress in this area (reviewed in Astington, Harris, and Olson 1988; Baron-Cohen, Tager-Flusberg, and Cohen 1993; Whiten 1991), studies of nonhuman animals are only in their infancy (Byrne and Whiten 1990, 1991; Premack and Woodruff 1978; Ristau 1991b). Nonetheless, the few observations and experiments that exist are suggestive and can be more rigorously explored by parasitizing some of the methods developed by cognitive psychologists working on humans, especially human infants.

7.2   Conveying Information

Within the animal kingdom there is considerable interspecific variation in repertoire complexity. For example, the acoustic repertoires of most insect species are limited to a few functionally distinct acoustic signals, and much of the communicative burden lies in species and individual identification (Hoikkala, Kaneshiro, and Hoy 1994; reviewed in Bailey 1991). In contrast, nonhuman primate vocal repertoires are generally comprised of multiple call types, each call type is

2. As discussed in greater detail, deception can arise by active falsification of information (i.e., producing a signal that is deceptive) or by withholding information (i.e., suppressing a signal).

characterized by a variable set of exemplars (i.e., within-call type variation),[3] and variation in acoustic structure covaries with species and individual identification as well as with subtle changes in the call-eliciting context (Cheney and Seyfarth 1990; Snowdon 1990; Todt, Goedeking, and Symmes 1988). In this section, I examine inter- and intraspecific signal variation, focusing especially on the kind of information *potentially* available. I want to emphasize the notion of potentially available information because in several studies of nonhuman animal communication there has been a tendency to provide detailed quantification of signal structure (i.e., lists of univariate features describing signal morphology) in the absence of perceptual studies. The latter are of course critical if one wants to understand the significance of signal variation in the communication system. For example, several studies have shown that mothers can recognize their young based on distinctive acoustic signatures (birds: Beecher 1982; bats: Scherrer and Wilkinson 1993; humans: Gustafson, Green, and Tomic 1984; Gustafson, Green, and Cleland 1994), and yet, under certain conditions, information about individual identity may be relatively less important (alarm calling in primates: Cheney and Seyfarth 1988) or acoustically less salient than other features (birds: Beecher, Campbell, and Burt 1994). The material presented in this section therefore adopts the position that salient information is in the ear or eye of the beholder, and relative salience will vary as a function of significant changes in the ecological and social environment.

One way of conceptualizing the informational structure of communicative signals is in terms of a taxonomy with multiple branch points as seen in Figure 7.1. This schematic illustration is clearly not the only way to organize such information, but it provides a sketch of the material that will be presented. Most organisms use some form of communicative signal for species identification, and tend to have at least one signal for mating and one for survival. Depending upon the nature of the socioecological environment and the selection pressures favoring the design of increasingly specialized signals, these two branch points will diversify into increasingly fine-grained details of the communicative context. In terms

3. Some researchers have noted that the mammalian vocal system is relatively simple compared with birds. Thus, Seller (1983) writes, "Mammals use their hearing mainly in interspecific situations, such as location of predators or prey, and they emit relatively simple sounds. Birds communicate extensively by voice; they produce complex and varied sounds at much faster rates than mammals. They include species that have the most complex repertoires in the animal kingdom" (p. 94). There are several problems with this comment. First, as discussed in greater detail in this chapter, mammals, especially nonhuman primates, have acoustically variable vocal repertoires, and *most* of their calls are used in intraspecific situations rather than interspecific situations. Second, as pointed out in chapter 3, without some objective notion of acoustic complexity it makes little sense to talk about one species having a more complex repertoire than another.

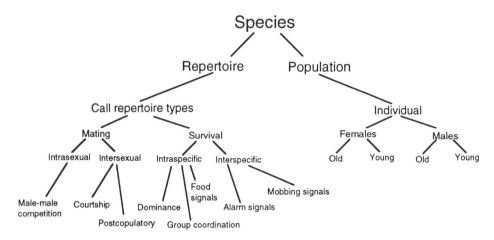

Figure 7.1
A taxonomy of communicative information. The right branch represents information about the identity of the caller, whereas the left branch represents information about behavioral context.

of identification, these might include differentiation by gender, age, and kinship, whereas contextual information might break up into different signals for different predators and types of food. Some of these identifiers may result from relatively rigid associations between the signal production mechanism and the resultant signal structure. For example, in fishes such as the midshipman (*Porichthys notatus;* Bass 1990) and the bicolor damselfish (*Pomacentrus partitus;* Myrberg, Ha, and Shamblott 1993) the peak frequency of the power spectrum (i.e., perceptually, the pitch of the voice) is significantly negatively correlated with body size, with obvious implications for dominance.[4] Moreover, in the damselfish, interindividual differences in the "chirp" vocalization can be largely accounted for by differences in the volume of the gas bladder.

Human language is perhaps the most extraordinary case of signal diversification in the animal kingdom, raising important questions about the selective advantage of expanding to literally limitless repertoires. In fact, many scholars in linguistics have commented on this point. Studdert-Kennedy (1986a) succinctly states the general view from linguistics:

If we compare language with other animal communication systems, we are struck by its breadth of reference. The signals of other animals form a closed set with specific, invariant

---

4. Recall from chapter 4 that in many anurans and some primates, there is also a negative correlation between body weight and vocal pitch; this relationship by no means characterizes all species, however.

meanings (Wilson 1975). The ultrasonic squeaks of a young lemming denote alarm; the swinging steps and lifted tail of the male baboon summon his troop to follow; the "song" of the male white-crowned sparrow informs his fellows of his species, sex, local origin, personal identity, and readiness to breed or fight. Even the elaborate "dance" of the honeybee merely conveys information about the direction, distance, and quality of a nectar trove. But language can convey information about many more matters than these. In fact, it is the peculiar property of language to set no limit on the meanings it can carry. (pp. 208–209)

Let us now review the evidence for the informational content of animal signals. This review process will allow us not only to describe, for example, the honeybee's dance language, but also to make a stab at understanding why the specificity or sophistication of this system stops where it does.[5]

7.2.1 Information about Affective State

The pitch of the voice bears some relation to certain states of feeling. . . . a person complaining of ill treatment . . . almost always speaks in a high-pitched voice. (Darwin 1872, 88)

All organisms experience changes in affective state as a result of direct confrontations with the environment and, in some species, expected or imagined confrontations with the environment. The affective states I am referring to include both motivational and emotional. In a majority of species, affective states are responsible for the production of communicative signals. In this section I review what is known about the relationship between changes in affective state and the structure of communicative signals. This connection is particularly rich from a comparative perspective because it allows for an unambiguous comparison of species as diverse as red-winged blackbirds, chimpanzees, and human infants. As a reminder for those who have forgotten (!), humans begin life "talking" about their affective states and do so through laughter, crying, and growling. Only later do such emotive outpourings develop into full-blown speech.

On the basis of comparative observations, Darwin (1872) argued that animal vocalizations are designed to convey explicit information about emotional state. His logic went as follows:

1. Dominant individuals are generally larger (i.e., weigh more) than subordinates.

2. The pitch of an individual's voice is negatively correlated with its weight. Thus large individuals will have relatively lower pitched voices than small individuals.

---

5. In chapter 2, I pointed out that the early ethologists considered repertoire size to be limited by memory constraints.

3. Dominants will have lower pitched voices than subordinates.

4. Because of the relationship between size, pitch, and dominance, aggressive vocalizations will be low in pitch and submissive vocalizations will be relatively high in pitch.

Morton (1977, 1982) was the first to provide a comprehensive treatment of Darwin's prediction. Using data on avian and mammalian vocalizations, Morton provided qualitative support for the relationship between pitch and motivational state. Moreover, he extended Darwin's logic to other motivational states and acoustic features, laying out a set of "motivation-structure (MS) rules." For example, he predicted that aggressive vocalizations would be broadband and noisy, whereas fearful/appeasement vocalizations would be narrowband and tonal. Further, in contexts where mixed or ambiguous motivational states arise (e.g., alarm/mobbing calls), the acoustic structure of the signal should reflect this ambiguity (e.g., a chevron-shaped signal that rises and falls in frequency). Some of the patterns derived from MS rules are schematically illustrated in Figure 7.2.

More quantitative but comparatively narrower tests of Morton's MS rules have recently been carried out for several primate species. Analyses of a male

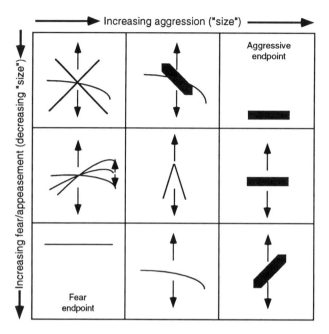

Figure 7.2
Some of the potential relationships between changes in motivational state and acoustic structure, emphasizing frequency (*y*-axis) and tonality/bandwidth. Thin lines represent tonal sounds; thick lines represent broadband noisy (atonal) sounds. Arrows indicate that frequency can shift up or down. The central panel represents a chevron-shaped sound, commonly associated with the context of alarm/ mobbing—a mixed motivational state (redrawn from Morton 1982).

chimpanzee's vocalizations,[6] briefly discussed in chapter 4, indicated that the pitch of aggressive barks was relatively lower than the pitch of submissive screams (Bauer 1987). In contrast to this case study, results from research on two macaque species (Gouzoules and Gouzoules 1989; Gouzoules, Gouzoules, and Marler 1984) and the ring-tailed lemur (Macedonia 1990) failed to provide support for the predictions from MS rules. For example, both rhesus and pigtail macaques produce five acoustically distinct screams in the context of submission, each scream type conveying information about the type of aggressive interaction encountered (e.g., attacks involving dominant kin versus nonkin, and with or without physical contact). However, whereas one species uses a relatively low-frequency, broadband, atonal scream in the context of physical attack from a

6. Analyses were derived from a single film sequence.

dominant, the other species uses a relatively high-frequency, narrowband tonal scream. That is, similar motivational states appear to be associated with different acoustic signals in closely related species.

In addition to tests of the relationship between motivational state and acoustic structure, other studies have explored the relationship between body size and pitch. As previously stated (see also chapters 4 and 6), large males produce lower-pitched advertisement calls than small males in a number of anurans (Ryan 1988; Ryan and Brenowitz 1985). This correlation, however, does not hold in all species, including other anurans (Arak 1988; Doherty and Gerhardt 1984; Gerhardt et al. 1987) and nonanuran vertebrates (e.g., red deer: McComb 1991; humans: Künzel 1989).

Though it is important to understand why exceptions arise, Morton's MS rules, like Darwin's earlier predictions, were intended to explain quite general and comparatively broad patterns of communication. Recently, Hauser (1993a) used a data set of several hundred vocalizations from 43 nonhuman primate species (23 genera, including the prosimians, New and Old World monkeys, and apes) to examine, first, the relationship between body weight and pitch (a between-species comparison) and, second, the relationship between motivational state and pitch (a within-species comparison). In general, body weight accounted for a significant proportion of the variation in vocal pitch among species. Thus the largest species within the order Primates (e.g., gorillas, chimpanzees) produce the lowest-pitched calls, whereas the smallest species (e.g., bush babies, marmosets) tend to produce the highest-pitched calls. Important exceptions to this general pattern were, however, observed, indicating that for some genera, body weight is clearly not the most important factor underlying the range of frequencies used. Within the genus *Macaca,* for example, differences in the species-typical habitat may account for a significant proportion of the variation in frequency. Forest-dwelling species (e.g., lion-tailed macaques) use a narrow range of low-frequency calls (150–1000 Hz), whereas species living in more open habitats (e.g., bonnet macaques) use a broader range of low- and high-frequency calls (150–15,000 Hz). This explanation must be treated cautiously, however, since for some species, recordings were obtained from a captive setting, whereas for other species, recordings were obtained in a natural setting (see chapter 3 for details on how the environment can exert specific pressures on the design features of the signal). Nonetheless, the macaques represent a beautiful test case because differences in the range of frequencies used are unlikely to be due to species-specific differences in the mechanisms or hardware underlying phonation; though see Schön Ybarra (in press) for a review of subtle variation in vocal-fold structure among primates.

Results from analyses of the relationship between motivational state and pitch among nonhuman primates also corroborated the prediction set forth by Darwin and Morton. Across species, low-frequency calls were generally given during aggressive interactions, and high-frequency calls were given during submission or fear. As with the body weight–frequency relationship, however, not all species provided support for the predicted relationship between frequency and motivational state, and the overall relationship between motivational state and tonality was not statistically significant. For example, although most species within the Old World monkeys produced low-frequency vocalizations in the context of aggression, this was not consistently the case for the prosimians, New World monkeys, and apes. In summary, therefore, the pitch of the voice appears to convey approximate information about body weight and motivational state in many nonhuman primate species. These two factors, however, do not account for all, or even most, of the variation in vocal pitch. To tackle the issue of motivational-structural rules head-on, we will require more accurate measurements of individual body weights within species together with quantitative assays of motivational state. Concerning the latter, it will be useful if future researchers attempt to combine behavioral measures of motivational state (e.g., facial expressions and body postures associated with a particular context) with physiological measures (e.g., changes in cortisol or heart rate); see discussion of human facial expressions in section 7.3.3 for an elegant example of how behavioral and physiological measures can be combined.

Although a number of linguists have provided elegant analyses of how speech-like structures emerge in the developing child (Locke 1993; Oller 1986; Oller and Eilers 1992; Vihman 1986), relatively fewer studies have been conducted on the purely emotive sounds produced by young infants, with the exception of research on infant cries (Lester and Boukydis 1992; Zeskind and Collins 1987; Zeskind et al. 1985). Recently, however, a handful of papers have emerged focusing on the structure and function of gruntlike sounds (McCune and Vihman, in press; McCune et al., in press) and laughter (Nwokah et al. 1993). In all of these studies, the goal has been to provide a taxonomy of signal variation and stereotypy under different conditions and to assess the perceptual salience of various acoustic parameters. Analyses of the child's emotive utterances are of considerable significance with regard to comparative analyses and are discussed in the following paragraphs.

Biomedical researchers have studied crying as an assay for screening healthy and unhealthy infants (Colton and Steinschneider 1980; Lester et al. 1991). Developmentalists, in contrast, have analyzed the structure and contexts associated with crying in order to understand how the child makes use of a limited communicative tool kit. As described in chapter 5, though infants, older children, and even adults cry, for the young preverbal infant crying represents one of the primary acoustic vehicles for transmitting information about affective state. Infant cries are also individually distinctive (J. A. Green and Gustafson 1983; Gustafson, Green, and Tomic 1984) and appear to be designed for long-range propagation (Gustafson, Green, and Cleland 1994).

From a functional perspective (chapter 6), it would make good sense for infant cries to convey at least some accurate information about changes in affective state and for caretakers to be sensitive to such covariation (Lester and Boukydis 1992; Lester and Zeskind 1978). Research has shown that infants produce at least two cry types, labeled *pain* and *hunger* cries. In general, parents are able to discriminate between these cries, and they show behavioral responses that are appropriate to the affective state conveyed. Infants may use cries to express other affective states, but thus far it has not been possible to assess whether the acoustic characteristics of these putative cry types are species-specific or represent individual or culture-specific changes due to subtle contextual influences.

In a wide variety of mammalian species, individuals produce communicative grunts. Several authors, especially those working on nonhuman primates (Cheney and Seyfarth 1982; Stewart and Harcourt 1994), have argued that such vocalizations are not merely reflexive sound emissions but rather intentional signals that convey information about the caller's social environment (e.g., dominance relationships, group movement, intergroup encounters). Now, in a stimulating set of empirical and theoretical papers, McCune and colleagues (McCune and Vihman, in press; McCune et al., in press) have followed the approach of primatologists in an analysis of human infant grunts. Grunts were identified in infants on the basis of three shared criteria: "(1) abrupt glottal onset, (2) lack of supraglottal constriction, and (3) short duration" (p. 5). Given these acousticomotoric parameters, three grunt types were observed based on contextual information. "Effort grunts" appeared to be generated simply on the basis of the infant's effort to obtain or do something, such as stand up or reach an object. "Attention grunts" occurred in the absence of movement or effort, and appeared to be in response to an attentional interaction; no communicative intent could be observed. "Communicative grunts" were distinguished from the other two on the basis of gestures, including eye gaze, apparently designed to attract someone's attention, usually the mother; such attention-seeking utterances were typically produced when the infant was in need of help or wanted to show something off. Based on these operational definitions, observations of five infants were collected between the ages of 9 and 16 months (Figure 7.3). In general (see summary, Figure 7.4), effort and attention grunts occupied a large proportion of the infant's early vocal repertoire. It was not until at least 13 months that communicative grunts appeared, and interestingly, there was a reasonably close correspondence between the onset of communicative grunts and the onset of referential word use.

The data presented in Figure 7.3 suggest that prior to the onset of communicative utterances, human infants produce a number of sounds that are mere expressions of changing affective states, including pain, hunger, and joy. Such sounds are also accompanied by facial expressions (see Lewis 1994). Eventually, however, some of these sounds are coopted for use in intentional communication, as occurs when infants use grunts to indicate their desires. Further support for the intentionality of such utterances comes from the observation that communicative grunts are commonly accompanied by nonvocal gestures, such as pointing to a desired object. Moreover, communicative grunts appear quite synchronously with the steep rise in the production of referential words, as well as what appear to be planned actions. A synthesis of this information, together with results on other communicative milestones, suggests an interesting ontogenetic system. I present this scenario (McCune and Vihman, in press) in Figure 7.4 and return to its implications later on in the chapter.

Research by Nwokah and colleagues (1993) has focused on the laughter of young children (three-year-olds) and, in particular, on documenting how changes in acoustic structure map on to changes in emotional state. Results suggest that children produce at least four different types of laughter: comment, chuckle, rhythmical, and squeal. The primary acoustic differences among these types were fundamental frequency contour, number of units, and unit duration. Such acoustic differences, in turn, appear to covary with the degree of excitement or arousal instantiated by the social context (in the current study, mother-child play). It will now be important to extend these sorts of analyses to younger children, different social contexts, different cultures, and perhaps, different species such as the great apes where laughter has been described (van Hoof 1972; for spectrograms, see Marler and Tenaza 1977). In addition, as laid out by McCune and colleagues for grunts, it will be interesting to establish when, developmentally, children begin to use laughs intentionally, as occurs when we snicker.

The literature reviewed thus far indicates that for nonhuman and human animals, acoustic signals encode some information about the signaler's affective state. Does the structure of the signal, however, have an effect on the affective state of relevant perceivers? McConnell (McConnell and Baylis 1985; McConnell 1990) has taken an interesting approach to this problem using cross-cultural data

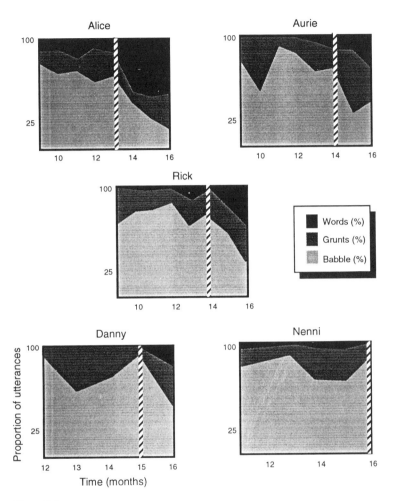

Figure 7.3
Variation in the onset of grunts, words, and babbles from six children ranging in age between 9 and 16 months. The striped vertical bar represents the age at which each child started producing communicative grunts (figure redrawn from McCune et al., in press).

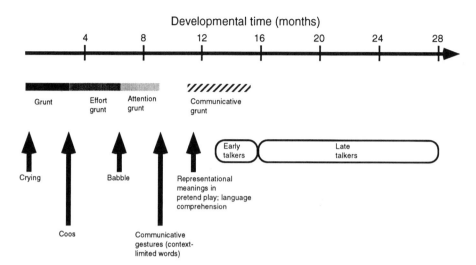

Figure 7.4
A developmental scenario proposed by McCune and colleagues (in press) for the onset and change (structural and functional) of grunts by human infants, relative to other communicative signals and milestones. Early talkers are considered to be children who have referential words, based primarily on recorded productions of two consonants. Late talkers are those who use words and sentences.

on the sounds produced by humans to mediate the movement of trained pets (primarily dogs and horses), in addition to developmental experiments with young dogs. The logic of her argument is that if signals are designed to convey information about the sender's affective state, then such signals should have an effect on the affective state and subsequent response of perceivers. An important theoretical motivation for this view, discussed in chapters 2 and 6, was that selection should favor manipulative signalers (Dawkins and Krebs 1978; Krebs and Dawkins 1984).

Based on interviews with animal trainers and acoustic analyses of their signals,[7] McConnell found a reasonably close correlation between the acoustic structure of sounds used to arrest movement and those used to increase activity. In general, when trainers wanted their subjects to move, they produced multinote signals that were short and frequency modulated. When they wanted their subjects to stop moving, they produced single notes that were relatively long in duration.

7. The sample was obtained from trainers speaking the following native languages: English, Spanish, Swedish, High German, Polish, Basque, Peruvian Quechua, Finnish, Chinese, Korean, Arabic, Farsi, Pakistani, Navajo, Shoshone, and Arapaho.

Table 7.1
Relationship between Whistle Acoustic Structure Produced by Shepherds and Intended Message for Shepherd's Dog (redrawn from McConnell and Baylis 1985)

	Multiple Notes				Single Notes						
	⌣	⌒⌒	- -	⌒⌐	⌒	⋀	⌐	⌣	⟋	∿	⌐
Fetch	.8	3	3	0	0	0	0	0	0	0	0
Stop	0	1	0	0	4	2	6	0	1	0	0
Go left	2	3	0	1	3	0	1	1	0	2	1
Go right	2	0	1	1	1	0	1	0	4	4	0

Cell values represent the number of signals that were assigned to the context on the left by the author (McConnell and Baylis 1985). The thick vertical line divides multiple- from single-note whistles.

Some examples are provided in Table 7.1 based on analyses of shepherds' whistles to dogs (McConnell and Baylis 1985).

In experiments with young dogs, McConnell (1990) trained one group of individuals to approach in response to a series of short, rising-frequency notes and stop-sit in response to a long, descending-frequency note; a second group was trained on the same signals, but with motor responses reversed. Results showed that the short, rising-frequency notes more readily caused individuals to approach than a single descending-frequency note. Although the long descending frequency caused more individuals to sit-stay, this result was not statistically significant.

The issues raised by McConnell's work are also directly relevant to studies of intraspecific communication, and in fact such studies have already sneaked into some of our earlier discussions. Specifically, in chapter 2, I reviewed the controversy between what some have called the traditional ethological view and the behavioral ecological view (Caryl 1979, 1982; Hinde 1981). To reiterate, much of the early ethological literature suggested that animals convey specific information about their emotional/motivational state (i.e., about the probability of initiating a particular behavioral response), and such information is used to select fitness-maximizing responses. Behavioral ecologists criticized this perspective, arguing instead that honest signaling of motivational state was evolutionarily unstable, open to cheaters who escalate dishonestly. Consequently, those working within the framework of behavioral ecology tended to argue that signals are designed to manipulate the behavior of the perceiver, and as such, are not informative with respect to motivational state.[8] These two views have made some peace with each other, based on both more recent theoretical models and empirical findings. In particular, although circumstances exist where individuals will be favored to conceal information about their motivations (e.g., Maynard Smith's "war of attrition" model makes this prediction for situations where the only cost incurred by individuals is time spent in an aggressive encounter), other studies suggest that accurate information about underlying motivations will be favored. Van Rhijn and Vodegel's (1980) model suggests, for example, that in species with individual recognition and repeated interactions among group members, truthful signaling about motivational state (subsequent behavior) will be selected for. Given these distinctions, let us look at a few studies where detailed analyses of the relationship between signal structure and motivational state have been explored.

One of the strengths of early ethological research was its attention to behavioral detail, especially fine-grained analyses of the motor actions involved in a display and the sequences of such displays used in communicative interactions. The general focus was on aggressive competition, especially in birds and fish. Most of the studies provided data on three components of the interaction: (1) type of display used by the actor, (2) response to the display by a reactor, and (3) follow-up behavior by the original actor. Thus, in a classic analysis format, there would be information on the probability of aggressive attack by the actor given that the reactor either attacks or flees. What is clear from these early studies is that *some* information about motivational state is conveyed, but as Caryl's (1979) reanalysis suggests, such information is not *absolutely* reliable. In other words, if A receives an aggressive threat display from B, A can not be sure that B will attack even if A holds his ground instead of running away.

An elegant and extremely thorough treatment of this issue was presented by Nelson (1984) based on observations of the pigeon guillemot (*Cephus columba*), a seabird. This species performs a variety of visual and vocal displays during territorial disputes (Figure 7.5). In general, territorial males consistently win aggres-

8. Recall, however, that the behavioral ecologists did not deny the existence of information exchange during communication. They suggested that individuals could obtain accurate information about such things as fighting ability by directly observing a competitor's resource holding potential (RHP). Attributes associated with RHP included body size and various ornamental traits such as antlers, tails, canines, etc. Moreover, in Krebs and Dawkins' (1984) revised framework, cooperative communicative interactions were seen as precisely those where information is exchanged by conspiratorial whispers.

Figure 7.5
The aggressive displays of the pigeon guillemot used during territorial disputes. (*a*) neck-stretch, (*b*) hunch-whistle, (*c*) bill-tuck, (*d*) trill, (*e*) duet trill, (*f*) trill-waggle (from Nelson 1984).

sive encounters against intruders, and most resolutions arise in the absence of fighting. In fact, based on more than 600 aggressive encounters, Nelson never observed an injury. The owner's use of the neck-stretch display was a highly effective threat as evidenced by the fact that intruders flew away in response to this display on 43% of all occurrences. In contrast, displays used by the owner failed to provide an accurate indication of the probability of attack: only 14% or less of all display types were followed by attack. Last, Nelson was able to use sophisticated statistical techniques to accurately predict the probability of the owner's subsequent behavior based on the spectrotemporal properties of the hunch-whistle signal. Specifically, owners were more likely to sit than attack when note duration or frequency increased; owners were more likely to move than attack when note duration decreased. These data indicate that contrary to Caryl's assertions, some of the pigeon guillemot's displays convey quite accurate information about subsequent behavior, information that is clearly used by individuals in selecting an appropriate response.

Countercalling, either in the form of one-on-one interactions or vocal choruses (reviewed in M. D. Greenfield 1994), is common among nonhuman animals and forms the basis of the human child's entry into the world of acknowledged communicative competence (i.e., being able to interact conversationally).[9] During such vocal duels, individuals may obtain information about their opponent's motivational state, and in nonhuman animals, such information is especially likely to be proferred when actor and reactor are territorial neighbors (i.e., familiar and engaged in frequent interactions).

In the European blackbird (*Turdus merula*), territorial males sing during defensive interactions with intruders and while attempting to attract mates. As noted for other species, much of the acoustic variation in song structure covaries with variation among individuals. However, as Dabelsteen and Pedersen (1990; Dabelsteen 1992) have documented, a considerable proportion of the variation is accounted for by changes in the motivational state[10] of the singer, changes that occur rapidly during vocal exchanges (Figure 7.6). Thus, at the start of a vocal exchange, males typically begin with what are called "low-intensity" songs. Once a response is elicited, the male rapidly moves into a "high-intensity" song. Thus, high-intensity songs appear to be indicative of an interaction that has escalated. The final level of intensity, and apparently, aggressive competition or threat, occurs when a rival male enters the resident's territory. At this point, males begin singing a "scrambled song." Dabelsteen's observations indicate that in the heat of a vocal duel, males rapidly alternate between these song types, and interject two other types as well. Sometimes, however, males match song type for song type; such song matching is common among oscine song birds and appears to provide some information about competitive ability (Kroodsma 1979; Todt 1981). Though the precise function of different patterns of vocal exchange is not yet clear, the European blackbird system provides a good example of a graded system of communication (see chapter 3), with song variants mapping onto changes in motivational state.

The studies on human infants reviewed earlier indicated that before they begin to communicate about objects and events in the external environment, infants convey information about their affective state. Human adults also communicate nonlinguistically about affective state, and one of the primary arenas for such communication occurs in infant-directed speech (see chapter 5). These signals, known in linguistic circles as the prosodic, paralinguistic, or melodic features of

9. In fact, one of the most striking hints that a child may be suffering from a developmental problem, as commonly occurs among young autistic children (Frith 1989), is that he or she fails to engage in conversational interactions, including the maintenance of vocal and visual attention.

10. Note that in Dabelsteen's work, as in much other research in this area, no independent measure of motivational state has been provided. Rather, motivational state is assumed on the basis of behavioral state.

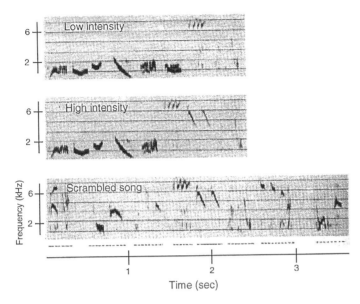

Figure 7.6
Three songs produced at different intensities by European blackbirds. Low-intensity songs are typically heard at the start of a song bout and are then followed by high-intensity songs, especially when the initial singer hears a vocal response from a neighbor. When a rival enters the singer's territory, he often switches to a scrambled song (highest intensity) (redrawn from Dabelsteen 1992).

language, not only provide the infant with direct information about its caregiver's emotional state (Fernald 1992a, 1992b), but also, in some circumstances, provide crucial information about the syntactical structure of language (Bahrick and Pickens 1988; Fernald and McRoberts 1994; Jusczyk et al. 1993; for some exceptions, see Gerken, Jusczyk, and Mandel 1994).

Adult speech to other mature listeners of the linguistic community also carries important information about the speaker's affective state. Studies of the acoustic correlates of affective state have, however, played a relatively minor role in the study of human communication (Goldbeck, Tolkmitt, and Scherer 1988; Hess, Scherer, and Kappas 1988; Scherer 1986; Scherer and Ekman 1982). Moreover, most of the research has been carried out with actors or public speakers rather than with naturally occurring conversations or utterances. Nonetheless, some general patterns have emerged, especially in terms of the relationship between relative pitch of the voice (fundamental frequency) and emotional state (Table 7.2). For example, Tartter (1980) has shown that smiling causes an increase in pitch, whereas frowning causes a decrease (Tartter and Braun 1994). Listeners

Table 7.2
Some of the Acoustic Correlates of Affective State (from Scherer 1986)

Emotion	Pitch Level	Pitch Range	Pitch Variability	Loudness	Tempo
Happiness/joy	High	?	Large	Loud	Fast
Confidence	High	?	?	Loud	Fast
Anger	High	Wide	Large	?	Fast
Fear	High	Wide	Large	?	Fast
Indifference	Low	Narrow	Small	?	Fast
Contempt	Low	Wide	?	Loud	Slow
Boredom	Low	Narrow	?	Soft	Slow
Grief/sadness	Low	Narrow	Small	Soft	Slow
Evaluation	?	?	?	Loud	?
Activation	High	Wide	?	Loud	Fast
Potency	?	?	?	Loud	?

find both acoustic changes to be perceptually meaningful (i.e., indicative of the speaker's emotional state).

The reported correlations between voice characteristics and emotional state suggest that a person's speech provides a window into what he or she feels. Given that one can voluntarily manipulate such acoustic features, however, it seems likely that listeners would use more than the voice to evaluate signal honesty. Indeed, Ekman's work on deception suggests this kind of multimodal approach to signal perception (see section 7.4).

Thus far, our discussion of the relationship between affective state and signal structure has focused on the auditory domain. Similar issues also arise within the visual domain and form the core of current debates over the function of facial expressions. One view, espoused by Ekman and his colleagues (Ekman 1973, 1992; Ekman and Friesen 1969, 1975; Ekman, Levenson, and Friesen 1983), suggests that human facial expressions reflect the signaler's emotional state. A number of such expressions are universal, based on significant physiological changes (Figure 7.7). The opposing view, defended by Fridlund (1994) and J. A. Russell (1994), is much more Dawkinsian-Krebsian (Dawkins and Krebs 1978; Krebs and Dawkins 1984), suggesting that facial expressions function to manipulate the emotional state and behavior of others and that the evidence for universality is weak. As in the animal literature, both views clearly share a piece of the explanatory pie.

For more than twenty years, Ekman has engaged in a rigorous research program designed both to quantify the neurophysiological and anatomical substrates

Figure 7.7
A sample of facial expressions from Duchenne's (in press) classic work. In frame 31, stimulation of
the zygomatic major muscle leads to what Ekman (1989) has called the Duchenne smile, or the smile
of enjoyment.

underlying facial expressions and to assess whether universal expressions exist,
and if so, why. A particularly attractive feature of this work is that it lays out a
set of methodologies that are quantitative and, further, that appear to be primed
for importing into studies of nonhuman animal expressions.

As I have pointed out, the most heated debate in the literature on facial expressions concerns the evidence for cross-cultural universals. Russell claims that
there is very little evidence for universals. Ekman claims that the evidence is
overwhelmingly in favor of universals and that Russell has simply misinterpreted
or misrepresented the existing data. The core of this debate emerges from differences in the definition of *universals* and the methodological approach that is best
suited for testing universality. Specifically, Russell argues that a universal facial
expression must show no variation in either production or perception across
cultures, whereas Ekman argues that some variation is to be expected but that

this in no way injures the claim for universality—a trait shared in the same basic form by most humans. Whereas Russell's detailed review suggests that there is little consensus on facial expression-emotion correlations, Ekman's reanalysis of the data presented by Russell provides a quite different picture, arguing strongly for universality. I present this reanalysis in graphical form in Figure 7.8.

In addition to the general findings already reported, a recent paper by Levenson and colleagues (1992) provides some of the strongest support for the universalist's position, based on behavioral and physiological analyses of expressions in Americans and Minangkabau. The Minangkabau are a group of people living in Western Sumatra. The society is Muslim, matrilineal, and agrarian. Most important for the current discussion, however, is the fact that overt displays of negative emotion are strongly discouraged.

To test for cross-cultural coherence within emotions, as well as consistent differences across emotions, Levenson et al. carried out a directed facial action task. In this task, subjects are required to move particular facial muscles based on instructions read by an experimenter; some subjects carried out these instructions while looking in the mirror (to obtain feedback), whereas others did so in the absence of mirrors. While carrying out the instructions, subjects were monitored for a host of physiological changes, including heart rate, finger temperature, skin conductance, finger pulse transmission and amplitude, and respiratory period and depth; these factors were measured in order to provide a comprehensive description of autonomic nervous system (ANS) reactivity.[11] Results, some of which are presented in Figure 7.9, generally indicated significant differences in ANS responses to different emotions and no significant difference within emotions across cultures. These data strongly suggest that ANS activation from facial expression is consistent cross-culturally.

In the description of the Minangkabau, I alluded to a societal convention regarding overt expression. Specifically, overt, public expressions of negative affect are strictly frowned upon. This social impact on communicative expression is reminiscent of Marler's (Marler, Dufty, and Pickert 1986a; Marler, Karakashian, and Gyger 1991) notion of an audience effect (see summary of work in the subsection "Referential Signals in the Domestic Chicken") and has been explicitly examined by Fridlund and colleagues (Chovil 1991; Fridlund et al. 1990; Fridlund, Kenworthy, and Jaffe 1992) with regard to the social dynamics of human facial expression. In particular, Fridlund's work challenges the facial-expression-as-

---

11. Previous research had demonstrated that such factors reliably predicted emotional differences among American subjects (Levenson, Ekman, and Friesen 1990; Levenson et al. 1991).

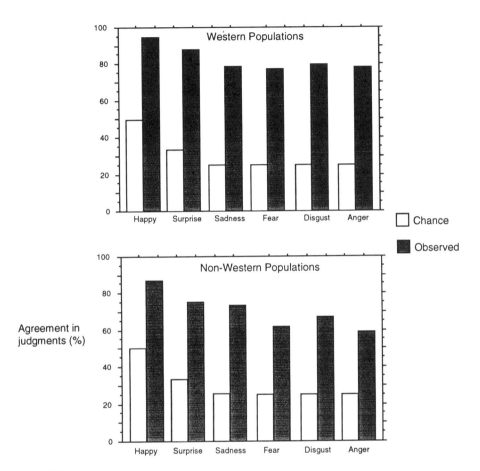

Figure 7.8
Plot of data tabulated by Ekman (1994) on cross-cultural agreement in facial expression, comparing Western (top) and non-Western (bottom) populations. Values expected by chance (white) were derived on the basis of the following logic. Happy is set at 50%, based on the fact that the choice is between either a positive or negative emotion. Surprise is set at 33.33%, based on the empirical finding that the expression is most often identified as either surprise, fear, or happiness. The four negative emotions, sadness, fear, disgust, and anger, are set at 25%, based on the probability of picking one of these four terms (for a more detailed description, see Ekman 1994, 272).

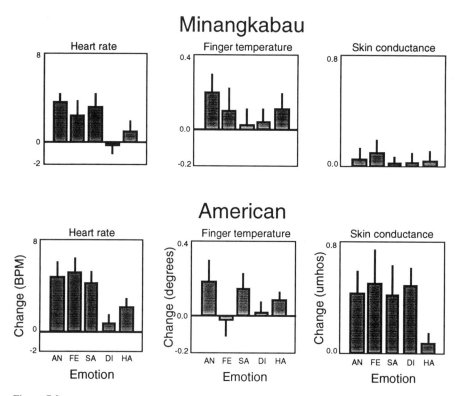

Figure 7.9
Data from Levenson and colleagues (1992) showing the similarity in autonomic response to different facial expressions. AN: anger; FE: fear; SA: sadness; DI: disgust; HA: happiness.

emotional-readout view (*sensu* Ekman) by making two points: first, all facial expressions are enacted *to* some social group, whether present or imagined; second, from a behavioral ecology perspective, such expressions have been designed to manipulate others rather than to provide them with explicit details of their affective states (i.e., expressions are highly deceptive). This view does not deny the relationship between emotion and facial expression. Rather, it argues that the relationship is more complex than previously entertained. What is the evidence for this alternative perspective?

To test the idea that all facial expressions are mediated by a social context, even when a person is physically alone (e.g., in a room that is visually and auditorily isolated), Fridlund and his colleagues (Fridlund et al. 1990; Fridlund, Kenworthy, and Jaffe 1992) have taken advantage of techniques in mental imagery designed by Kosslyn (reviewed in Kosslyn 1994). In one task, a subject arrives at the testing room either alone or with a close friend and is then placed in one of four conditions while viewing a comical videotape: (1) arrives alone and views tape alone; (2) arrives with friend and views tape alone but is told that friend is being tested on other sorts of psychological tests; (3) views tape alone but is told that friend is watching the same tape in another room; and (4) views tape with friend in same room. Results indicate a significant increase in the extent of smiling (quantified by EMG records) from conditions (1) to (4), although, interestingly, there were no differences between conditions (3) and (4);[12] the lack of difference between these two conditions suggests that imagined and actual presence of a friend *can* exert comparable effects on at least one form of facial expression, smiling.

From these kinds of results, Fridlund (1994) has argued that there are at least two problems with viewing facial expressions as simply mirrors onto the emotions. First, he claims that there is no agreed-upon definition of emotion, and thus the phenomenon to be explained is far too elusive for consensus, at least at present. In support of this claim, he quotes from a variety of sources, highlighting the conflicting perspectives. Second, and more critically, Fridlund argues that although current studies show a significant correlation between ANS change and facial expression (either through the directed facial action task or by means of recalling a particularly emotional event), they fail to tease apart two additional factors, believed to be critical to the face-emotion view. Specifically, physiological changes must be associated with something more than the subject's general state of arousal (e.g., increased heart rate associated with anger relative to sadness because the former requires greater energy) and must be the result of changes in emotion rather than in display function.

In summary, whereas Ekman explicitly argues in favor of the emotional content of facial expressions, maintaining that they have been designed to convey such information, Fridlund explicitly denounces this view, arguing instead that facial expressions are manipulative communication displays and can "be understood without recourse to emotions or emotion terms" (p. 186). My own reading of this literature is that the debate actually confuses levels of analysis. Whereas Ekman's work has generally focused on the *mechanisms* underlying facial expression (e.g., changes in physiology, brain state), Fridlund has considered the *function* of facial

12. Note that in Fridlund's (1994) review of this work he states that there was " a monotonic increase in smiling from" (p. 162) conditions 3 to 4. From his Figure 8.8 (p. 162), however, these two conditions show means and standard deviations that are virtually identical.

expression. As I have repeatedly pointed out in this book, both perspectives are important, but they are not mutually exclusive.

7.2.2   Information about the External Environment

Honeybees: A Puzzling Case?   One of the nice properties of human language is that if I am hungry and want the address of a good restaurant, I can stop someone on the street and ask for directions. If the person is familiar with the area, he or she might be able to say something like

There's this wonderful little French restaurant about 10 minutes away. Go to the first light and turn right, then walk about 10 blocks until you see a shoe store on the left side of the street. Turn left and walk down the short alley. The restaurant, called *La Vie et La Bouf,* has a red and white front, and a large picture glass window filled with bottles of Bordeaux, columns of salami, and baskets of baguettes.

These instructions provide explicit information about the location of the restaurant, how far away it is, and its quality—at least based on the experience of the person providing the information. Humans looking in at the nonhuman animal world would surely turn and say, "Beat that! No animal can communicate witl such expressive power." In my view, they would be partially wrong.

In the 1940s, Karl von Frisch initiated an experimentally rigorous research program designed to uncover the informational content of the dance system of honeybees, or what he later described as their dance language. To date, no other nonhuman animal communication system has come even close to matching the informational specificity of the honeybee's dance language. As I will describe in greater detail, von Frisch and those who have followed him (Dyer 1991; Dyer and Seeley 1989; Gould 1990; Gould and Gould 1988; Kirchner and Towne 1994; Seeley 1989, 1992) have demonstrated that within the dance system, information is conveyed about the distance to food, its location, and its quality, and the informational channel encompasses four sensory modalities: visual, auditory, olfactory, and tactile. Unfortunately, perhaps, the honeybee seems to be conversationally quite myopic, talking primarily about food. In this sense, they appear to be more limited than other nonhuman animals and certainly more limited than humans. More metaphorically, whereas bees are informational laser beams, humans are informational floodlights.

The notion of informational limitations or constraints on communication only makes sense in the context of the socioecological problems faced, and in this sense, perhaps the honeybee is no more limited than a human newborn without full-blown language—newborns, given their extraordinary altriciality, face the task of insuring parental care, and for this purpose, Shakespearean English would not buy them much. Let us take a more careful look at the kinds of information available in the honeybee dance system and how foragers make use of such information and regulate its flow.

Von Frisch focused his attention on two primary dance forms or types, a round dance used to convey information about nearby food sources and a waggle dance used to convey information about distant food sources. Appropriate for the early ethological zeitgeist, von Frisch provided an explanation for the origin of the dance, suggesting that in primitive bees, the dance emerged from a set of initial flight intention movements that subsequently became refined and stereotyped (i.e., ritualized; see discussion in chapter 2) into a clearly identifiable signal. The process of change from flight intention movements to signal required, evolutionarily, two crucial components. First, foragers must have been able to acquire, and then encode in the dance, information about the distance and direction of the flight to food. Von Frisch assumed that such information was derived from energy expenditure (rather than time in flight) and "reading" celestial patterns relative to the hive. Second, hive mates must have developed the perceptual capacity to attend and interpret motor patterns associated with the dance, and use such patterns to narrow their own foraging paths.

In a classic experiment, food was placed at different distances from the hive, and then the waggle dance of returning foragers was quantified. Results (Figure 7.10) indicated a negative correlation between the number of turns in the waggle dance and distance to the food source, such that distant foods were associated with relatively longer dances (fewer turns).

The data in Figure 7.10 show that distance information is accurately conveyed within the dance. But distance information is of limited use in the absence of information about location, relative to the hive. By placing food at different angles from the hive, von Frisch was able to determine that the orientation of the waggle portion of the dance (performed on the vertical surfaces within the hive), relative to the sun, provided foragers with the requisite information about food location (Figure 7.11). More specifically, direction, referenced to the sun's azimuth, is conveyed by the orientation of the waggle component of the dance relative to gravity.

To more precisely assess how foragers were calculating distance and direction, and relaying such information via their dance, von Frisch conducted two elegant experiments. In one, he located the hive at the bottom of a tower, thereby forcing foragers to fly with minimal displacement in the lateral plane. When these foragers returned to the hive, they performed round dances, suggesting that they were conveying information about a nearby resource. What makes this interesting is that the same amount of effort expended in a normal outward flight from the hive would have led to the production of a waggle dance of short duration.

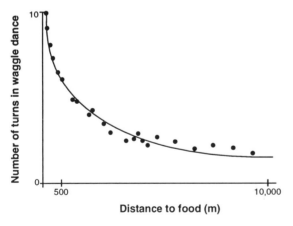

Figure 7.10
Results from one of von Frisch's experiments showing the relationship between the number of turns in the waggle dance and distance to the food source. As distance from the hive increased, the number of turns per dance decreased, resulting in dances of longer duration (redrawn from von Frisch 1950).

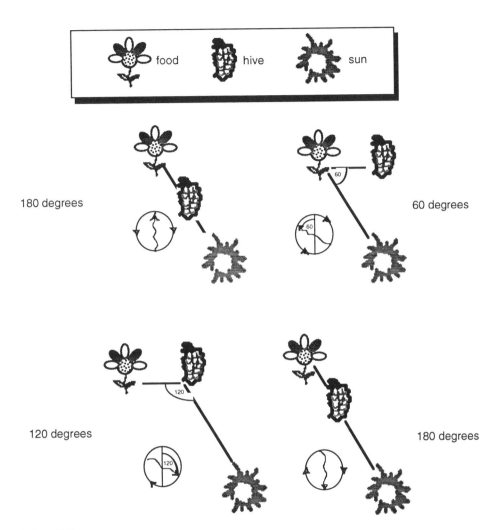

Figure 7.11
Experimental results indicating how a honeybee's waggle dance provides information about the location of food. Four experimental conditions were run, each involving the placement of food at different relative angles from the hive and sun. The schematic illustration of the waggle dance shows how bees communicate relative location (redrawn from von Frisch 1950).

In a second experiment, the hive was placed at the base of a vertically structured rock cliff, and the bees were trained to fly straight up. In contrast to the previous experiment, returning foragers, whose outward flight also lacked lateral movement, performed a waggle dance. Von Frisch suggested that the different outcomes could be explained by reference to the patterns of flight. Specifically, in the tower experiments, bees flew in a spiral with no reference to the sun[13] and, thus, no opportunity to calculate lateral displacement; consequently, a round dance was produced. In contrast, bees in the cliff experiment flew straight up while facing the cliff and kept the relative position of the sun constant. The relationship between flight pattern and sun position is what allowed the foragers to measure displacement and encode it in their dance.

An intriguing aspect of comparative studies of the honeybee dance language is the observation that species and geographical races appear to exhibit "dialects"[14] with regard to distance components of the dance. The common explanation for such dialects is that they reflect adaptive populational differences in foraging distance. Thus features of the dance are said to be highly correlated with flight range, thereby producing population-specific dialects with characteristic "slopes"; dance slopes are meant to represent the strength of the correlation between distance and particular features of the dance. In the first formal test of these ideas, Dyer and Seeley (1991) observed three Asian bee species (*Apis florea, A. cerana, A. dorsata*), analyzing the structure of their dance and the area of their foraging range. Though they found evidence of dance dialects, they failed to find a significant relationship between the structure of the dance and flight range. This result argues against the idea that dialects represent adaptively tuned

13. Note that considerable controversy (Dyer and Seeley 1989; Gould 1976; Gould and Towne 1987) exists over the precise mechanisms used by honeybees to navigate through their environment and find food. The essence of this controversy can be distilled to a disagreement over whether bees use mental/cognitive maps or a path integration mechanism. Some, like Gould (1986, 1990; Gould and Gould 1988), are convinced that honeybees use mental or cognitive maps both to convey relevant information to hive members and to respond appropriately to the information conveyed by the dance. Others, such as Dyer and Seeley (1991, 1989), are less convinced by the proposed evidence for cognitive maps and instead argue in favor of path integration. A recent study by Kirchner and Braun (1994), which involved experiments with bees flying in a wind tunnel, provides strong evidence for the path integration mechanism.

14. The term *dialect*, as used by bee biologists, differs slightly from the same term used by linguists and avian biologists for speech and birdsong, respectively. Specifically, as pointed out in chapter 5, dialects represent learned signal variants among members of a potentially interbreeding population of individuals. As far as we know, there is no evidence that bees learn dance dialects, and as reported, such dance variants are typically observed among closely related but different strains, subspecies, and species.

variants, suited to the specific ecological demands of the hive. Nonetheless, dance dialects exist, and their function requires explanation.

Seeley and his colleagues (Nieh 1993; Seeley 1992) have recently turned their attention to other kinds of bee displays that appear to serve the function of exchanging information regarding food sources. Specifically, experiments have explored the function of a display known as the *tremble dance*, first described by von Frisch[15] in 1923 (translated by Seeley 1992):

. . . a strange behavior by bees who have returned home from a sugar feeder or other goal. It is as if they had suddenly acquired the disease St. Vitus's dance (chorea). While they run about the combs in an irregular manner and with a slow tempo, their bodies, as a result of quivering movements of the legs, constantly make trembling movements forward and backward, and right and left. During this process they move about on four legs, with the forelegs, themselves trembling and shaking, held aloft approximately in the position in which a begging dog holds its forepaws.   (p. 375)

In general, this dance involves lateral shaking of the body while simultaneously rotating it around an axis of approximately 50 degrees every second; the dance lasts 30 minutes on average and involves travel throughout the brood nest area of the hive. Observations and experiments (Kirchner 1993b; Seeley 1992) reveal that this dance occurs primarily when a forager has returned from an area that is highly profitable with regard to nectar but encounters difficulties in terms of finding receptive food storers (Figure 7.12). Thus the primary cause of tremble dancing is increased search time to find bees who will unload nectar (Kirchner and Lindauer 1994). After the performance of a tremble dance, colony members respond by increasing nectar processing and decreasing recruitment to additional nectar sources.

Until quite recently, most work on the honeybee dance language focused on the correlation between visual features of the dance and the location, distance, and quality of food. A puzzle, of course, was how observer bees could decode the information in the dance under relatively poor conditions of visibility (i.e., dark nest cavities). Von Frisch (1967a) and Wenner (1962) both alluded to an auditory component of the dance, and even provided some evidence that the duration of the sound was proportional to the distance to food. In addition, a few studies suggested that vibrational stimulation may also be used in communicating about food (Frings and Little 1957; Markl 1983). Now, in an exciting series of studies, several researchers (Dreller and Kirchner 1993, 1994; Kirchner 1993a,

---

15. As Kirchner and Lindauer (1994) point out, von Frisch considered the tremble dance to be nonadaptive, a kind of bee neurosis.

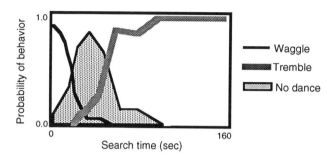

Figure 7.12
Data showing changes in the probability of performing the waggle dance, tremble dance, or no dance
as a function of search time (redrawn from Seeley 1992).

1993b; Kirchner, Lindauer, and Michelson 1988; Kirchner and Towne 1994; Michelsen 1992; Michelsen, Kirchner, and Lindauer 1986; Michelsen et al. 1992;
Towne 1985; Towne and Kirchner 1989) have provided some exquisitely detailed
observations on the airborne and vibratory signals accompanying the dance, in
addition to data on how nondancers perceive the potential information conveyed
and then act upon it by sending their own acoustic and vibratory signals.

In early reports, it was claimed that only two honeybee species—*Apis mellifera*
and *Apis cerana*—produced sounds while dancing; both these species are cavity
nesters (Towne 1985). More recently, however, Kirchner and Dreller have demonstrated that in a non–cavity nester, *Apis dorsata,* where dances are produced
in more open and well-lighted environments, sounds also accompany the visual
dance (Kirchner 1993a; Kirchner, Lindauer, and Michelsen 1988; Michelsen,
Kirchner, and Lindauer 1986; Michelsen et al. 1987). These near-field sounds are
readily picked up by Johnston's organ, a structure located in the bee's antenna
(Dreller and Kirchner 1994; Kirchner 1994). Interestingly, a closely related species, *Apis florea,* is also capable of perceiving near-field sounds, but fails to
produce them during the dance. This finding suggests a preadaptation, primed for
the evolution of an acoustic accompaniment to the visual dance.

Dancers are not the only ones who produce auditory signals. Followers—that
is, those that are tracking the movements of the dancer—also produce sounds
known as begging signals (Gould 1976; von Frisch 1967a), used to request food
samples from the returning forager.

When foragers return to the hive and dance, using either the round, waggle,
or tremble dance, they emit a corresponding sound. For the round and waggle
dance, the sound has a fundamental frequency of 200–300 Hz, a pulse duration

of 20 msec, and sound-pressure level of 73 dB measured at approximately 1–2 cm from the bee (Kirchner, Lindauer, and Michelsen 1988; Michelsen, Kirchner, and Lindauer 1986). The sounds emitted appear to provide information about direction, distance, and even profitability (Kirchner 1993a; Kirchner, Lindauer, and Michelsen 1988; Waddington and Kirchner 1992). In contrast, the sound of the tremble dance consists of a fundamental frequency of 350–450 Hz, an average duration of approximately 140 msec, and a sound pressure level of 80 dB recorded at 1 cm from the dancing bee. The function of these sounds is to arrest dancing in others and increase the rate of food storage.

The results presented thus far suggest that the airborne signals of the honeybee's dance are informative. What has not been addressed, however, is whether such signals are perceived as auditory or vibratory signals, and the extent to which the information conveyed is necessary, or even sufficient, with regard to localizing the food source. Studies of both the waggle (Michelsen, Kirchner, and Lindauer 1986) and round dance (Kirchner, Lindauer, and Michelsen 1988) indicate that the sound is transmitted in the air, with no detectable substrate vibrations. In contrast, analyses of the tremble dance reveal that information is potentially available by means of airborne sound, substrate vibrations, or tactile contact (Kirchner 1993b). To understand the perceptual significance of components of the dance language, Michelsen, Kirchner, and their colleagues have carried out some ingenious experiments, using a combination of operant conditioning (Dreller and Kirchner 1994; Kirchner, Dreller, and Towne 1991), analyses of mutants with shortened wings (Kirchner and Sommer 1992), and even the development and implementation of a robotic bee (Michelsen et al. 1992). Results from the robotic bee[16] experiments indicate that the wagging component appears dominant with regard to the information conveyed, but for the dance to be effective (i.e., in terms of recruiting hive members), the sound component must also be present. Michelsen and colleagues conclude that the auditory and visual features of the wagging dance provide redundant information about the distance and direction of food.

Much more could be said concerning the informational content of the honeybee's dance language. I hope the reader is convinced by what I perceive to be the punch line: these extremely small creatures, with their relatively simple nervous systems, are able to communicate quite explicit information about the

---

16. The beauty of this paradigm is that each component of the dance can be manipulated independently of all other components, thereby providing the opportunity to examine the relative contribution of visual, auditory, and vibratory features of the dance; possible olfactory effects could also be assessed.

distance, direction, and location of food. They also produce signals in other contexts, such as when newly emerged queens "toot" and young queens still confined to their cells respond by "quacking" (Michelsen et al. 1986). However, these signals, which are socially mediated, do not appear to be as specific, informationally, as those generated in the context of food. Honeybees are thus eligible for the *cordon bleu,* but would never make it onto a debating team or a quiz show such as "Jeopardy!"

More Referential than the Honeybee?

A dog cannot relate his autobiography; however eloquently he may bark, he cannot tell you that his parents were honest though poor.   (B. Russell 1961, 133)

The honeybee dance language is unquestionably precise and provides detailed information about the external environment. This claim would appear to fly in the face of historically ancient discussions of animal communication, reviewed in chapter 2 and section 7.2.1, favoring the view that animal signals *merely* reflect the signaler's affective state. That is, when an animal signals, it provides a direct readout of its emotional state (e.g., aggressive, fearful, distressed), but provides no information about the specific features of the eliciting context (W. J. Smith 1969, 1977). Is the honeybee case simply anomalous, a case of special design unworthy of mention with regard to the evolution of sophisticated communication systems? Or, is the honeybee system part of a cluster of systems that are capable of conveying information about specific external objects and events? Before describing the empirical studies that have addressed these questions, I must first attempt to clear away some of the terminological ambiguities that have accumulated within the past ten years or so. Such ambiguities are not uncommon when researchers working in quite unconnected disciplines attempt to cross-fertilize and share their respective fruits. In the case described in this section, the problem has been exacerbated by a lack of theoretical symmetry in the cross-fertilization process: biologists have attempted to use the language of psychologists and linguists to interpret the vocal behavior of nonhuman animals, but psychologists and linguists have often criticized such terminological usage. A more productive enterprise would have both sides working together, providing constructive criticism about theory, methods, and empirical findings (Roitblat, Herman, and Nachtigall 1993). I return to this more cooperative approach in chapter 8.

Within the ethological literature, there has been an extraordinary proliferation of terms associated with describing the characteristics of vocalizations associated with changes in the caller's affective and external state; similar sorts of confusions have arisen among researchers working on captive animals using either signed or artificial languages (Box 7.1). Thus, for affective state one reads about emotions, affect, and motivation, whereas for external state one reads about signals that are semantic, referential, representational, indexical, and symbolic (Cheney and Seyfarth 1990, 1992; Marler, Evans, and Hauser 1992; Roitblat, Harley, and Helweg 1993). Although various authors, working in fields such as psychology, linguistics, anthropology, and biology, provide different definitions and reasons for using these terms,[17] I think it is important to distinguish between emotional and motivational states on the one hand, and referential and representational signals on the other. Concerning the first distinction, I believe that what is generally meant by an emotional signal is that there is a correspondence or association between the structure of the signal emitted and the emotional state experienced by the signaler at the time of emission (this belief is directly analogous to the facial expression–emotion view discussed earlier). Thus human newborns produce acoustically differentiated cries associated with pain and anger (Lester and Boukydis 1992; Zeskind and Collins 1987), in much the same way that rhesus macaques have acoustically differentiated calls associated with fear (Gouzoules, Gouzoules, and Marler 1984; Hauser and Marler 1993a). In both the human and nonhuman primate case, one typically infers the emotional state underlying the acoustic signal from several external cues, including nonacoustic signals (e.g., facial expressions, body postures) and the social context that appeared responsible for eliciting the signal. In addition, several studies have taken such inferences one step further, tapping into physiological changes that covary with behavior, including heart rate, pupil dilation, and hormonal profiles (Bercovitch, Hauser, and Jones 1995; Berntson and Boysen 1989; Berntson et al. 1990; Berntson, Boysen, and Caccioppo 1992; Fernald 1984, 1992a, 1992b; Lester and Boukydis 1992). In contrast to emotional state, signals can also reflect changes in motivational state, a concept that is more akin to drive in the old ethological literature (Hinde 1959, 1960). Thus human infant cries can reflect the motivational state of hunger, and so can the calls of rhesus macaques (Hauser and Marler 1993a). And as with emotional states, it is also possible to obtain physiological measures of motivational state such as the amount of food consumption, glucose levels, and hormonal profiles associated with aggression.

The distinction between referential and representational is actually more difficult to articulate than the emotion-motivation distinction, because the former is

17. The various terms are often used interchangeably. Such usage has increased the level of confusion.

Box 7.1
A Brief Review of Language-Related Terminology in Studies of Captive Nonhuman Animals

More than twenty years ago, several research teams started training putatively smart animals (great apes, dolphins) and relatively not-so-smart animals (sea lions, parrots) to use sign language (Gardner and Gardner 1969; Gardner, Gardner, and Van Cantfort 1989), artificial languages (Herman, Richards, and Wolz 1984; Premack 1971, 1986; Premack and Premack 1983; Savage-Rumbaugh 1986; Savage-Rumbaugh et al. 1993; Schusterman et al. 1993) or, in one case, spoken language (Pepperberg 1987a, 1991). The goal of much of this research, both then and now, has been to assess whether nonhuman animals are capable of acquiring human language. Given this research agenda, it has been necessary for these researchers to provide quite explicit definitions of linguistic terminology, and equally explicit discussion of what they take as evidence of a given linguistic phenomenon. In cases where ASL was taught to chimpanzees, we have a fairly unambiguous definition of what counts as a sign for a given object or event, since this is defined by the structural position and movement of the hands relative to the body. What is at issue here is the intended meaning of the sign for the chimpanzee. In contrast, when "English-speaking" parrots use words to label objects in their environment, the actual sound-object pairing is not in question. Rather, it is the use that such a sound has for the parrot, the extent to which it applies to more than one object or event. The critical problem then is determining what counts as evidence for a particular linguistic capacity when the medium of expression and comprehension differ so dramatically between studies. The following table is a summary of the tools used by various researchers to assess language (and cognitive) abilities in nonhuman animals:

Linguistic Tool	Species	Reference
Formal ASL	Chimpanzee, gorilla	1
"Pidgin" ASL	Chimpanzee	2
"Pidgin sign English"	Orangutan	3
Arbitrary gestures and sounds	Dolphin, sea lion	4
Color/shape symbols	Chimpanzee	5
Lexigrams (abstract symbols)	Chimpanzee, bonobos	6
Comprehension of spoken English	Bonobo	7
Production and comprehension of spoken English	African gray parrot	8

References: 1 (Gardner and Gardner 1969; Gardner, Gardner, and Van Cantfort 1989; F. Patterson 1979, 1987); 2 (Terrace 1979); 3 (Miles 1978); 4 (Herman, Richards, and Wolz 1984; Schusterman et al. 1993); 5 (Premack 1971, 1986; Premack and Premack 1983); 6 (Rumbaugh 1977; Savage-Rumbaugh 1986); 7 (Savage-Rumbaugh et al. 1993); 8 (Pepperberg 1987a, 1991).

For a detailed but highly critical review of this literature, see Wallman (1992).

embedded within a much more challenging conceptual arena. Marler (1978), I believe, was the first ethologist to seriously entertain the possibility that nonhuman animal vocalizations might be capable of conveying information about the external environment, and in his discussion he labeled such signals *symbolic*. A symbolic signal, on Marler's account, would stand for an object or event in the environment, just as the golden arches stand[18] for McDonald's or, minimally, that fast-food chain serving burgers and fries. The difficulty with this claim is, of course, the part about "stand for." Over the years, Marler and those who have pursued his thinking have generally abandoned the term *symbolic* and instead have adopted one or more of the following terms and distinctions (the italicized words are the critical features of each sentence):

1. A signal stands for X in the sense that it has *semantic* status—it *means* that X has occurred, has been detected, etc. (Cheney and Seyfarth 1992; Seyfarth, Cheney, and Marler 1980a).

2. A signal stands for X in the sense that it *represents* X in the mind of the signaler and perceiver (Gouzoules, Gouzoules, and Marler 1984).

3. A signal stands for X in the sense that it *refers* to X—it is reliably associated with the occurrence of X (Marler, Evans, and Hauser 1992).

In item 1, signals are semantic insofar as they provide listeners with information about objects and events in the environment. Thus, as will be described more fully later, an alarm call given in response to a leopard would be semantically analogous to the human word *leopard* if it were given in response to leopards and not nonleopards, and if the alarm call itself (i.e., its acoustic structure alone) elicited behaviorally appropriate responses from those nearby. Moreover, it would truly be like the human word *leopard* if an acoustically similar call, produced in a different context, failed to elicit the kind of response elicited by the leopard call. Thus acoustic similarity is not what determines semantic similarity. Rather, semantic similarity is determined by the kinds of objects and events that are picked out by each sound. The philosopher[19] Quine (1973) described this property of terminological exchangeability or equity as *referential opacity*.

18. Granted: the golden arches may stand for the St. Louis gateway for some Americans, or for just plain old arches for those living in other parts of the world.

19. Critics of this line of research have often interpreted the claim about semantic communication to be synonymous with the notion of intentionality. It is not, and some of the problem in this area has come from a misreading of Dennett's philosophical interpretation of the vervet monkey's alarm call system. Briefly (see section 7.4), intentionality has to do with the notion of aboutness, the fact that certain mental states such as beliefs and desires are about other mental or nonmental states.

Signals are representational (item 2) if listeners create some kind of mental picture of the object or event eliciting the signal. Thus the production of an alarm call to a leopard evokes a mental image of the leopard, or something leopardlike, and as a result of this image, behaviorally adaptive responses can be initiated. The methodological trick here, then, is to find evidence of what is represented. Although studies of mental imagery in humans were initially met with skepticism, recent studies using brain imaging have gained considerable respect, thanks in large part to the work of Kosslyn and his colleagues (Kosslyn 1980, 1994; Kosslyn and Koenig 1992; see also Ishai and Sagi 1995; Miyashita 1995). Thus far, however, there is only limited evidence that nonhuman animals engage in mental imagery (e.g., mental rotation studies in baboons: Hopkins, Washburn, and Rumbaugh 1990; Vauclair, Fagot, and Hopkins 1993) and no evidence that I am aware of, that their communicative signals are capable of creating or evoking an image in the minds of listeners, or that signalers have an image in mind when they communicate. Although some cognitive scientists consider brain imaging studies to fall completely short of providing the requisite evidence for mental imagery, I disagree, side with Kosslyn and others on this point (see Miyashita 1995), and argue further that such imaging studies may provide the critical tool for assessing comparable issues in nonhuman animals.

A signal is referential (item 3) if it is reliably associated with objects and events in the world.[20] As a result of this association, listeners can accurately assess the range of potential contexts for signal emission. The breadth of this range depends, in part, on the specificity of the signal with regard to target objects and events. Thus, if a signal with a clearly defined acoustic morphology is produced only in the context of leopard confrontations, then the probability of accurately assessing the context underlying signal production must be nearly perfect, assuming of course that signalers are not attempting to use the strength of the association to deceive others (see section 7.4).

Within the last few years, most ethologists working on the informational content of nonhuman animal calls have favored the term *referential,* with one small addendum. Rather than adopt the purely linguistic meaning of the term, ethologists have coined the phrase *functionally referential* to describe the calls of some nonhuman animals, including several nonhuman primates and the domestic chicken (reviewed in Macedonia and Evans 1993; Marler, Evans, and Hauser

---

20. Human words can also refer to objects and events in imaginary worlds (unicorns, Star Trekkian Vulcans and Klingons) as well as the real world, an ability which is likely to be outside the mental grasp of nonhuman animals.

1992). The motivation for this terminological change was to make clear that non-human animal calls are not exactly like human words, but rather appear to function in the same way. Thus the acoustic structure of functionally referential signals provides listeners with sufficient information to determine the context underlying signal production. Although none of the functionally referential signals identified thus far have the expressive power of human words or, for that matter, the honeybee's dance language, such signals are more than mere expressions of emotional state. Let us take a look at how ethologists have mustered the support for this position.

Think back to when you were about eight years old, waiting around all day for a Thanksgiving meal.[21] Finally, you sit down, and a large golden brown bird is placed in front of you. You say, with extraordinary enthusiasm, "Yeah, turkey, turkey, turkey." You are served a healthy portion by your delighted grandmother. After finishing the first portion, you are served another. You clean your plate, and your grandmother pushes a third portion, at which point you exclaim, ready to burst, "Please, Grandma, no more turkey." Now, both sentences contain the word "turkey" and they were both said with slightly different emotional energy, the former with enthusiasm, the latter with exasperation. Nonetheless, "turkey" is "turkey," no matter. . . . The important point here is that the word "turkey" is characterized by a specific suite of spectral and temporal acoustic features, and it is these features that facilitate our ability to identify the referent of the utterance. And the sound-referent association is established because when speakers of the English language see a particular kind of bird, they consistently identify it as a turkey. If they did not, then labels in the market such as "Butterball Turkey" would not be of much help. This Thanksgiving day scene sets up the logic underlying studies of referential signaling in nonhuman animals.

Referential Signals in Nonhuman Primates   The first hint of a functionally referential system in nonhuman animals came from Struhsaker's (1967) observations of vervet monkey alarm calls. From a functional perspective, there are several ways to design an alarm call system (Klump and Shalter 1984; Marler 1955; see chapter 6). For example, in most avian species, one signal is used to warn other group members of an aerial predator's presence, and another signal is used to mob a predator on the ground. This type of system, in and of itself, would not work for vervet monkeys, especially those studied by Struhsaker in Amboseli National Park, Kenya. As reviewed in previous chapters, vervet monkeys in this area fall

---

21. For non-Americans: transform turkey into goose, ham, or even a chateaubriand!

prey to a diverse set of predatory species, including snakes, raptors, large cats, and other primates, both human and nonhuman. Associated with each predator class is a different style or strategy of hunting. Consequently, from the vervet's perspective, no spatial location in the environment guarantees safety from all predation, and in order for escape responses to be effective, they must be custom-made for the predator. Thus, if an eagle attacks, the safest response is to run under a dense bush. In contrast, if a leopard is sighted, the safest response is to run up into a tall tree, and not into a bush where the hunting leopard may be lurking.

Let's step back to the problem of design again. If vervet monkeys had an all-purpose alarm call, they would probably be an extinct species. But they are not extinct, do not have an all-purpose alarm call, and appear to be one of the most dominant nonhuman primate species in Africa, living in desert, savanna, woodland, and tropical rain forest environments. In fact, the vervet monkey's alarm call system is a beautiful illustration of how selection pressures might have favored signal diversification (see chapter 8 for a more in-depth analysis of this problem). For vervet monkeys, at least those living in Amboseli, an all-purpose alarm call would not work because it would not provide sufficient information about the kind of predator or about the kind of escape response that would be most adaptive. Although it may be difficult to determine why vervet monkeys evolved several different alarm call types (i.e., a question concerning origins), it is relatively easy to understand the selective advantage of having such a system: each alarm call type appears to allow listeners to select the most appropriate response strategy.

We have come full circle back to the problem of reference. The only way in which an alarm call would enable listeners to select an appropriate response is if the acoustic structure of the call, alone, provided sufficient information about the type of predator encountered—the call picks out or stands for a predator. A general solution to this kind of problem is for a sound to be paired with a context, in much the same way as a bell was paired to food in Pavlov's experiments. This solution need not require much in the way of high-level cognition. All that is required is consistent use of a call with a predator and a storage device that keeps the particular call-predator pairings separate. Is there evidence for this kind of association in vervet monkeys?

Struhsaker's careful observations clearly showed that vervet monkeys produced acoustically distinct and discrete alarm call types, and in response to hearing such calls or seeing the predator, individuals responded with behaviorally appropriate escape responses. From these observations, it was not possible to assess whether the context of alarm or the alarm call itself was responsible for guiding the type of escape response selected. As a result, Seyfarth, Cheney, and Marler (1980a) conducted a series of playback experiments (see Box 7.2 for procedural details) designed to reveal whether the acoustic properties of the alarm call system were sufficient to elicit behaviorally appropriate responses. Playbacks of alarm calls produced by adult males and females in response to large cats (i.e., leopards), eagles (i.e., martial eagles), and snakes (i.e., pythons and mambas) consistently elicited adaptive escape responses. Specifically, upon hearing alarm calls produced in response to leopards, vervets on the ground ran up into trees, and those already up in trees climbed higher; alarm calls produced in response to eagles caused individuals on the ground and in trees to look up and then run for cover under bushes; alarm calls associated with snakes caused individuals to stand bipedally and search the ground nearby. These were the exact responses observed during natural encounters (visual and auditory) with predators.

The playback results indicated that each alarm call type was sufficient to elicit a behaviorally appropriate response. But is this sufficient evidence for a referential signaling interpretation? Could one not still argue that each call is associated with a different level of fear and thus *merely* expresses the caller's emotional state? Several lines of evidence argue against this interpretation (Cheney and Seyfarth

Box 7.2
Field Playback Design for the Study of Call Meaning

Playback experiments have been used within the past ten years to study the meaning of animal vocalizations, a topic we have already touched on in discussing the habituation-dishabituation paradigm (chapter 3). Here, I wish to describe in somewhat greater detail how playback experiments of putatively referential signals are conducted outside of the laboratory, in the species-typical habitat or some close approximation. The following is written in cookbook form, with a more conceptually rich treatment in chapter 8.

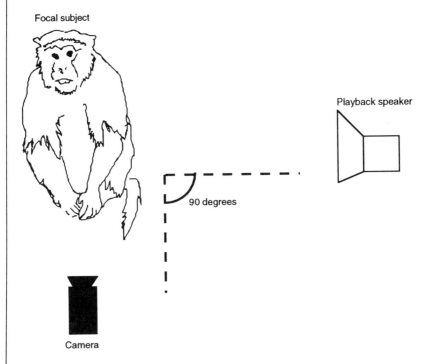

In step one, calls produced by known individuals, in clearly identified contexts, are tape-recorded onto audiocassette. Only if the recordings are clean—that is, free of background noise—are they included in the library of playback stimuli. Second, focal subjects in clear view are located. Third, a high-quality audio speaker is placed in dense vegetation, presumed to be out of sight from the target subject and from the individual whose call is being used for playback. The speaker is placed at a 90 degree angle from the focal subject and the human observer responsible for filming the trial (see illustration). Fourth, one person is responsible for playing back the alarm call, and a second person is responsible for recording the trial on film; the film record captures the subject's behavior for a designated period preceding and following playback. Fifth, a schematic map of the playback environment is sketched, indicating the location of nonfocal subjects, speaker-subject distance, and any ecologically relevant landmarks. Sixth, film records of each trial are analyzed blind, scoring the type of response elicited following the frame(s) in which the call was produced; blind scoring is achieved by marking the frame in which the call occurred, and then turning the sound off during editing. For video systems, several commercial systems available on PC platforms (e.g., Adobe Premiere) allow frame-by-frame analyses, accompanied by an audio waveform.

1992; Marler, Evans, and Hauser 1992). First, in the original experiments, Seyfarth, Cheney, and Marler selected call exemplars that varied—within alarm call
type—as a function of amplitude and duration; observations suggested that these
features covaried with changes in the caller's emotional state. Consequently, if
perceivers based their responses on the caller's emotional state, then changes in
amplitude or duration should have had a significant effect on the type of response
elicited. Results indicated that these two parameters might affect the vigor of the
response, but did not have an effect on the type of response exhibited; whether
the alarm call was loud and long or faint and short, vervets ran up into trees in
response to leopard calls, looked up for eagle calls, and stood bipedally and
scanned the ground for snake calls. Second, if the calls were simply an expression
of emotional state, then why would the behavioral responses be so specific? If
an alarm call expresses fear, even different levels of fear, then why do vervets
scan the ground nearby when they hear alarm calls for snakes and the sky when
they hear alarm calls for eagles? What are they looking for? These results, together with a second series of experiments described in section 7.3, make a
strictly emotion-based interpretation seem unlikely.

Referential Signals in the Domestic Chicken   A potential danger with studies of
nonhuman primates such as the vervet monkey is that because they are phylogenetically so close to humans, we are willing to impute sophisticated cognitive
machinery for the behavior observed, machinery that is perhaps less likely to
have evolved in species that are more distantly related. Thus, whereas vervet
monkey alarm calls appear to be strong candidates for a functionally referential
system, species with less mental sophistication are more likely to be driven by
raw affective states, leading to signals that merely convey information about their
emotions and motivations. If the logic here is correct, then surely the humble
domestic chicken would fall close to the bottom of the mental heap. The crucial
question then, and the one that has been pursued by Marler, Evans, and their
colleagues (reviewed in Evans and Marler, in press) for the past ten years, is
whether the calls of the domestic chicken, like the calls of vervet monkeys, are
functionally referential? If they are, then either the psychological mechanisms
underlying this communicative skill are less sophisticated then originally claimed
(see introduction to this section), or chickens have the same sorts of cognitive
abilities as do vervets. Though research on chickens has explored the nature of
food and alarm calling, I focus on the latter so that direct comparisons can be
made with the vervets' system; food calling is discussed in section 7.3.

The domestic chicken is a close relative of the jungle fowl. Because breeders
of ornamental fowl such as bantams have focused on the beauty of the domestic

chicken, rather than on egg production, the behavior of the ornamental fowl tends to be much closer to that of the wild type. This tendency is borne out in naturalistic studies of the vocal repertoire (Collias 1987; Gyger, Marler, and Pickert 1987), indicating that, like jungle fowl, domestic chickens have a large number of calls, including acoustically distinctive alarm calls to aerial and ground predators (Figure 7.13); moreover, chickens respond differently to each alarm call type, looking up and running for cover following aerial alarms and scanning horizontally following ground alarms. And, as with the vervet alarm call system, most of these calls tend to be produced to heterospecifics that represent direct threats. As the pie chart in Figure 7.14 indicates, however, the specificity of the chicken aerial alarm call system differs from that of vervet monkeys, with vervets producing fewer alarm calls to nonpredatory species. In a signal detection theory sense, chickens make more false alarm calls than vervets.

The pie charts presented in Figure 7.14 suggest that chickens, in contrast to vervets, use a relatively coarse-grained set of features during predator detection or classification. To more properly understand which features are used, Marler and Evans embarked on an elegant series of experiments, tapping into recent

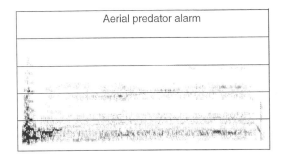

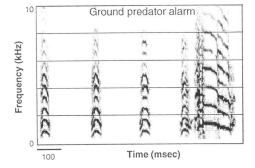

Figure 7.13
Spectrograms of aerial and ground predator alarm calls produced by domestic chickens. The *y*-axis represents frequency (kilohertz), and the *x*-axis represents time (milliseconds).

methodological breakthroughs in computer video technology. In one experiment, manipulated video scenes of real predators (*aerial*, red-shouldered hawk; *ground*, raccoon) were used to assess whether alarm calls were potentiated by the type of predator or its locomotory behavior (Figure 7.15). When hawks were flown either overhead or near the ground, aerial alarm calls were given more often than ground alarm calls. When raccoons were shown moving on the ground, only ground alarm calls were given. Interestingly, however, when raccoons were flown overhead, chickens produced more aerial than ground alarm calls. This either means that for chickens, all objects moving overhead are treated equally, deserving of aerial alarm calls, or that the system is not sufficiently flexible to permit new objects and events (i.e., raccoons flying in the air)[22] to be included or

22. One potential problem with the design of this experiment is that chickens are more likely to see a hawk at ground level than a raccoon floating in midair. More specifically, ''hawks on the ground'' is a possible category, whereas ''raccoon in the air'' is an impossible category. Parenthetically, one wonders how they might respond to a raccoon up in a tree.

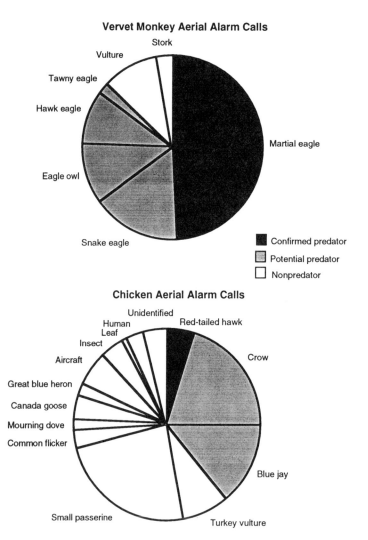

Figure 7.14
The proportion of aerial alarm calls produced to different objects by vervet monkeys (top panel) and domestic chickens (lower panel). Black areas correspond to confirmed aerial predators, speckled areas to potential predators, and white areas to nonpredators (redrawn from Evans and Marler, in press).

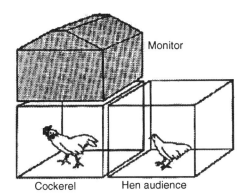

Monitor

Cockerel   Hen audience

Figure 7.15
Experimental setup for presenting computer animations or digitized video sequences to one individual (cockerel) while in the presence of a second (hen audience); note that the audience can also be presented via a second monitor (redrawn from Evans and Marler 1992).

excluded from particular call categories. Of course, the lack of flexibility in the system may be the *reason* why all objects moving overhead are classified as aerial predators. From a signal-detection theory perspective, the costs of false alarm calls might be low.

A second step toward refining the call-eliciting parameters of the chicken's alarm system involved manipulating the apparent size, speed, and shape of computer-generated hawk images. Results from the manipulation of size and speed (Evans, Macedonia, and Marler 1993) indicated that alarm calls were most likely to occur in response to hawklike shapes subtending an angle of 4° or more and traveling at a speed of 7.5 lengths/sec (Figure 7.16). Moreover, when the shape of the stimulus was varied from disklike (a round circle with no protrusions) to raptorlike (a body with wings, tail), alarm calls were primarily elicited by images with clearly discernible "wings."

Summarizing the results of their research on the factors eliciting aerial alarm calls, Evans and Marler (in press) state, "There is no reason to think that this is an exhaustive list. Our current estimate of the information content of aerial alarm calls [in chickens] is thus 'large, fast-moving, bird-shaped aerial object.'" Although this is a reasonable proposal, the reader should recall the potential dangers of inferring the informational content or meaning of a signal based solely on the contextual parameters eliciting the signal (i.e., Quine's referential opacity). And part of the danger stems from the fact that though "large, fast-moving, bird-shaped aerial object" elicits aerial alarm calls, chickens might be saying, "Run

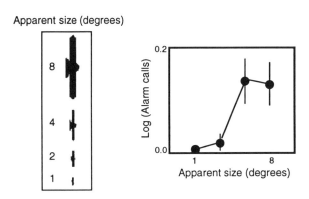

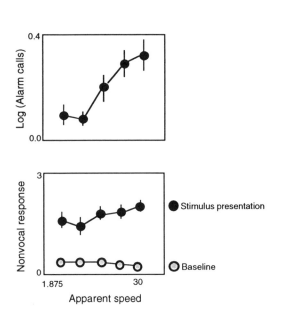

Figure 7.16
The upper panel shows the relationship between apparent speed and alarm call behavior, including vocal and nonvocal responses, and contrasted with a baseline period where no stimuli were presented. The lower panel shows the relationship between apparent size and alarm call rate (redrawn from Evans and Marler 1994).

for cover." Discriminating between these alternatives will require additional experiments, such as confronting a group of chickens already under cover with an aerial predator. Nonetheless, research conducted thus far does permit exclusion of at least one alternative explanation for the chicken alarm call system, that proposed by Owings and Hennessy (1984) for ground squirrel alarm calls. Specifically, though ground squirrels have two acoustically distinct alarm calls for aerial and terrestrial predators, these calls appear to convey information about urgency or the immediacy of danger, rather than about the type of predator encountered (see chapter 6). This explanation fails to account for the chicken system because small distant silhouettes of hawks consistently elicit aerial alarm calls, whereas large and apparently nearby images of a raccoon consistently elicit ground predator alarm calls. Moreover, whereas playbacks of terrestrial and aerial alarm calls elicited a higher proportion of head turning relative to control stimuli, only aerial alarm calls consistently elicited head rolling (looking up). As with vervet eagle alarm calls, when chickens look up they must be looking *for* something.

Other Studies of Referential Signaling   Since Seyfarth, Cheney, and Marler's (1980a) pioneering work on the vervet monkey's alarm call system, several other studies have focused on the problem of referential signaling in nonhuman animals, including other simian primates (e.g., rhesus macaques), prosimians (e.g., ringtailed lemurs), one nonprimate mammal (prairie dogs), and one avian species (i.e., domestic chickens). Moreover, the contexts for such signals range from predators to food to social relationships. A summary of studies is presented in Table 7.3, together with a set of relevant results. For completeness, I include a few studies that were designed to explore the possibility of referential signaling but either failed to provide conclusive evidence or found support for an alternative explanation concerning signal function.

The studies summarized in Table 7.3 have several features in common. Specifically, each of these studies was designed to show how a particular call or class of calls functions in the species-typical environment and to assess whether the primary message in the signal is *about* the caller's affective state or external environment. There is no question that for most of the calls investigated, a clear and exclusive external stimulus can be defined, such as food, predators, or social competitors. But the critical question, as raised for vervet monkey and chicken alarm calls is, What is the intended message, and how is it perceived? To this end, the studies differ with regard to the approach taken. Some studies provide observational details on the contexts eliciting the call, together with acoustic

Table 7.3
Synthesis of Research on Referential Signaling in Nonhuman Animals

Species	Call Context	Call Putatively Refers To?	Number of Call Types	Obs/Exp	Functionally Referential?	Reference
Chimpanzees	Discovering food	Food	2	O & E2	Maybe	1
Vervet monkeys	Avian and terrestrial predator encounters	Different predators	6	O & E1	Yes	2
Vervet monkeys	Intragroup dominance interactions, intergroup encounters	Different intra- and intergroup social relationships	4	O & E1	Yes	2
Rhesus macaques, pigtail macaques	Recruitment for agonistic aid	Different kinship relationships and the nature of the aggressive attack	5	O & E1	Yes	3
Rhesus macaques	Discovering and eating food	Food quality	5	O & E2	Yes	4
Toque macaques	Discovering and eating food	Food quality	1	O	Yes	5
Spider monkeys	Social interactions with group members	"Names" for group members	1+	O & E2	Maybe	6
Cotton-top tamarins	Discovering and eating food	Food quality	2	O & E2	No	7
Golden lion tamarins	Discovering and eating food	Food quality	1	O & E2	No	8
Ring-tailed lemurs	Avian and terrestrial predator encounters	Different predators	2	O & E1,2	Yes	9

Species	Context	Call Putatively Refers To	Number	Obs/Exp	Functionally Referential?	No.
Ruffed lemurs	Avian and terrestrial predator encounters	Different predators	2	O & E1,2	No	9
Ground squirrels	Avian and terrestrial predator encounters	Different predators	2	O & E1,2	No	10
Gunnison's prairie dog	Avian and terrestrial (human) predator encounters	Different predators, including different individuals within a predator class	3+	O & E2	Maybe	11
Domestic chickens	Avian and terrestrial predator encounters	Different predators	2	O & E1,2	Yes	12
Domestic chickens	Discovering and eating food	Food	1	O & E1,2	Yes	12

""Call Putatively Refers To?" represents information on what the authors of each study presumed to be the referent of the call. Studies vary considerably with regard to the precision with which such attributions can be made (see discussion). "Number of Call Types" represents an estimate. The accuracy of this estimate depends on the level of detail obtained from acoustic and behavioral analyses. For spider monkeys and Gunnison's prairie dogs, I have marked each call type with a plus sign because, if such calls refer to individuals (names for individuals), then the potential number of call types will be considerably larger. "Obs/Exp" provides data on the type of study conducted, either observational (O) or experimental (E). Under experimental, I have given two classifications. E1 represents playback experiments, and E2 represents studies manipulating the putative referent (e.g., food, predator). "Functionally Referential?" indicates whether the authors of the study concluded in favor of such an interpretation or not; those studies associated with a "Maybe" label represent situations that appear somewhat ambiguous. In some cases, terms other than "functionally referential" were used (e.g., representational, semantic), but for purposes of tabulation I have glossed over these distinctions.

References: 1 (A. P. Clark and Wrangham 1993; Hauser and Wrangham 1987; Hauser et al. 1993; Marler and Tenaza 1977; Wrangham 1977); 2 (Cheney and Seyfarth 1982, 1988, 1990; Seyfarth, Cheney, and Marler 1980b); 3 (Gouzoules and Gouzoules 1989; Gouzoules, Gouzoules, and Marler 1984); 4 (Hauser and Marler 1993a, 1993b); 5 (Dittus 1984, 1988); 6 (Masataka 1983); 7 (Elowson, Tannenbaum, and Snowdon 1991); 8 (Benz 1993; Benz, Leger, and French 1992); 9 (Macedonia 1990, 1991; Macedonia and Evans 1993; Pereira and Macedonia 1991); 10 (Owings and Hennessy 1984; Owings and Leger 1980; Owings and Virginia 1978; Sherman 1977; Sherman 1985); 11 (Slobodchikoff et al. 1991); 12 (Evans, Evans, and Marler 1994; Evans and Marler 1991, 1992, 1994; Gyger, Karakashian, and Marler 1986; Gyger, Marler, and Pickert 1987; Gyger and Marler 1988; Karakashian, Gyger, and Marler 1988; Marler, Dufty, and Pickert 1986a, 1986b; Marler, Karakashian, and Gyger 1991).

analyses of call structure. For example, Dittus's (1984) study of toque macaques shows that upon discovery of high-quality food, though in almost no other context (<5% of all calling situations with this call), individuals produce a highly distinctive call; individuals out of sight from the caller respond by running over. Evidence of referentiality is derived from the specificity of the context eliciting the call. Other studies provide the same sorts of observational evidence, but also carry out experiments to assess the perceptual salience and response specificity of the calls (i.e., playback experiments) or to more carefully determine the properties of the putative referent. For example, Macedonia's (Macedonia 1990, 1991; Macedonia and Evans 1993; Pereira and Macedonia 1991) work on lemur antipredator calls has combined playback experiments with presentation of both real predators (trained dogs) and model ones (hawk silhouettes). Although the general consensus among researchers working in this area is that nonhuman animals are capable of signaling about objects and events in their external environment, the precise meaning of such signals remains unclear. In the next section, I will return to this problem, but armed with a new set of methodological tools. Chapter 8 provides some final comments on this problem, along with a framework for how future work might profitably progress.

7.3 Categorizing Information

Animals, humans, and others, including birds, are able to categorize. . . .   (Herrnstein 1991, 385)

All living creatures, from cells to human populations, have been designed to classify stimuli in the environment into functionally meaningful categories. Thus the immune system must recognize foreign from familiar substances just as humans must recognize foreign from familiar cultures. At issue here are the mechanisms involved in categorization and the nature of the underlying representation. We know, for example, that in a wide variety of altricial species, young enter the world with predispositions to respond in particular ways to objects and events in their environment. With experience, however, the specificity of such responses changes, and these developmental patterns provide us with insights into the categorization process. Thus we learned in chapter 5 that young vervet monkeys only produce aerial alarm calls to objects flying in the air, but many of these objects (e.g., falling leaves) fail to elicit alarm calls from adults. Gradually, aerial predator alarm calls are fine-tuned, such that most calls are given to two species of raptors (i.e., objects deserving of aerial alarm calls because of their appetite for vervets).

Similarly, human infants are born with certain innate linguistic rules and have a tendency to overgeneralize (e.g., use of the past tense as in "I wented to the market") during the early years (Pinker 1995). Without any direct tutelage from expert speakers of the language, problems like overgeneralization disappear as the child's linguistic rules converge with the adult form. All of these developmental processes involve categorization—in particular, the generation of rules or propositions for gauging similarity between exemplars of a category or emerging concept.

For some species, categories are either fixed at birth or, once formed, undergo little to no modification as a result of experience. For other species, experience constantly shapes the categorization process and consequently modifies the underlying representation. Consider for the moment the concept or category[23] of "dead things." How does one recognize a dead thing from a living thing? This is, of course, a relevant question in the realm of communication, since one wants to be sure that attempts to communicate are with potentially responsive individuals. If you are an ant, it seems quite simple. Ed Wilson (1971) has demonstrated that when an ant dies, it emits a chemical known as oleic acid. If you place oleic acid on a live ant, colony members will drag the individual out of the nest area, apparently "kicking and screaming." For ants, therefore, the categorization process seems to be a trivial one: If oleic acid is detected, then the individual can be considered dead. No more, no less—a beautiful example of the ineffectiveness of experience in shaping the category. What about humans? How do we recognize dead things? Imagine you are walking down the street and see your friend Kim. You stop and chat with him for a moment, and then walk on. Five minutes later, you see Kim, laid out on the street. You rush over, put your hand to his heart, attempt to feel his breath, to wake him up, but to no avail. There are no signs of "life." Kim seems dead. You quickly call the police. They appear with an ambulance, and the medics say that Kim is in fact dead. This of course reinforces your experience with dead things, especially what you know about dead humans.[24] You now walk away from the incident, sad of course to have lost a good friend. All of a sudden, someone taps you on the shoulder and says, "Hey, why do you look so blue?" You turn around, and there is Kim. You do a double take and

23. For philosophers and cognitive psychologists reading this section, don't get upset yet! I am well aware of the distinction between category and concept (Carey 1985; Fodor 1994; Keil 1994b; Rey 1983; E. E. Smith, in press; E. E. Smith and Medin 1981), briefly discussed in chapter 3, and will return to it more thoroughly in a moment.

24. Needless to say, you may use a completely different set of criteria or inferences to assess (pick out) dead things in other organisms.

ask him what happened. He has no clue. You have a brief chat and then walk off. A few minutes later, there's Kim again, laid out on the ground. I think it is fair to say that most humans would not respond in the same way this second time around. One would conclude, not that Kim is dead, but that there is something wrong with him, that he is subject to fainting spells that wipe out all of his vital signs, or that he takes a drug with peculiar side effects. Okay. I think you have the picture. What makes us apparently more sophisticated than ants, let's say, is that experience does have an effect on how we represent things in the world. It is this notion of experience and representation that I would like to develop in this section by looking at a variety of communication systems and the ways in which organisms interact with their world.

To provide some structure to the following discussion, let me turn briefly to Herrnstein's (1991) attempt to develop a framework for assessing the mechanisms underlying categorization. Motivated in large part by his extensive operant work on pigeons, Herrnstein proposed a five-tiered categorization system (Figure 7.17). At the simplest level, *discrimination,* an individual's response to a stimulus, is dictated by perceptual features and constraints imposed by the sensory system. Thus a frugivore may classify objects as fruit based on their color. If the color falls within a particular wavelength of light corresponding to reddishness, then it's a fruit. The next level is categorization by *rote.* Here, the set of exemplars or stimuli that count as members of the category can be memorized and stored as a definable set because the set is finite. For example, a frugivore might have only ten fruiting trees within its territory. Assuming that there is no competition with other groups over which trees are or are not within the territory, these ten trees can be committed to memory. Level three is *open-ended categories.* In contrast to the previous level, here we have a category with a relatively large number of exemplars, so large that individuals cannot commit them to memory. The basis for classification would be some index of perceptual similarity, weighted perhaps by the frequency of encounters or experience. Thus, rather than using the rote category of ten fruiting trees within the territory, individuals may have a category that encompasses all potential fruiting trees; this would be important in cases of territorial expansions and contractions that led to changes in the number and position of trees. Fruiting trees would be recognized as such because all have little red fruits. The fourth level is *concepts,* defined on the basis of "fuzzy" perceptual features. The featural attributes are defined as fuzzy because no single perceptual parameter or set of parameters fully captures the exemplars that are treated similarly (i.e, behaviorally responded to in the same way) by the organism. For example, our frugivore might have a concept of tree,

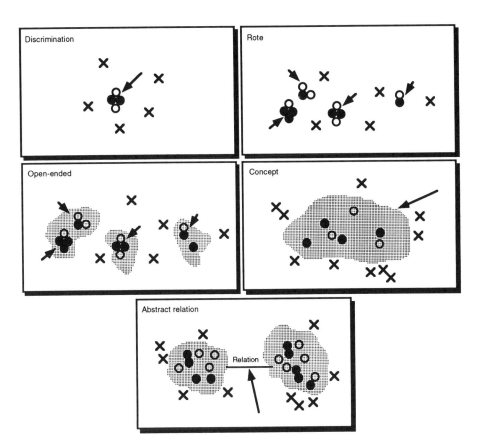

Figure 7.17
Five different levels of categorization as proposed by Herrnstein (1991). In each panel, filled circles refer to exemplars that are confirmed members of the category, whereas open circles represent exemplars for which generalization would be appropriate. Crosses represent exemplars for which generalization would be inappropriate. Arrows indicate the locus of control for the categorization process. Shaded areas represent boundaries for the category or concept. In "Abstract relation," the two shaded areas are connected, and the locus of control emerges from this relationship.

rather than just fruiting tree. Although one might be able to come up with a list of features that are useful for distinguishing trees from nontrees (e.g., trees have branches that stem from a trunk, but do not have eyes or flashing lights), it is hard to think of any particular set of features that are either necessary or sufficient. The final level is *abstract relations* involving specific connections or associations between two or more concepts. For example, frugivores might know about fruits *on* a tree as opposed to fruits *off* of a tree, or tree canopies that are *connected* as opposed to *disconnected*. At this level, perhaps even more so than at the level of concepts, simple perceptual features are insufficient for generating appropriate responses.

In reviewing the literature, Herrnstein claimed that, although work on operantly trained pigeons (Herrnstein, Loveland, and Cable 1976; Herrnstein et al. 1989) and on language-trained parrots (Pepperberg 1987a, 1987b) and chimpanzees (Matsuzawa 1985a, 1985b; Premack 1978, 1986) provided evidence of concepts, only humans show evidence of abstract relational concepts. "Beneficial as they are, abstract relations differentiate most sharply among animals. It has been noted that we see the largest gaps in comparative performance at the level of abstract relations. This observation is well grounded in the comparative data sampled here'' (Herrnstein 1991, p. 410). The studies reviewed below will allow me to make two points. First, I will show how research on natural communication provides an elegant empirical stage for looking at the process of categorization in the absence of formal training procedures. Second, I will present new data which suggest that Herrnstein's pessimism regarding abstract relations in nonhuman animals may have been premature. More specifically, and using the terminology of the preceding quote, absence of abstract relations may reflect "performance" differences but not differences in "ability." This distinction will be revealed by comparing two methodological approaches to the study of concepts in nonhuman animals and prelinguistic human infants.

7.3.1 Categorization of Predators

In section 7.2, the specificity of signaling systems was discussed. One argument offered for signal specificity within a categorical domain was selection pressure for differentiation. Thus, in the context of predator-prey relationships, if prey encounter predators with radically different hunting styles, then selection should favor acoustically different alarm calls. An extension of this argument is that if predation pressure is strong, selection should favor mechanisms that facilitate early detection and avoidance. Such selection should make good ecologists out of prey species. In the following section, I first discuss experiments that attempt to show how animals categorize different potential components of a signal. I then describe the results of studies designed to reveal what prey know about their predators, especially how they use both the animate and inanimate environments to increase the probability of predator detection.

In response to criticisms that vervet alarm calls were nothing like human words (e.g., see discussion in Bickerton 1990), Cheney and Seyfarth set out to provide a much more stringent test of their position. Using a modified form of the habituation-dishabituation technique, they asked whether vervet monkeys compare (i.e., categorize) vocalizations on the basis of their acoustic or semantic similarity. Recall (chapter 3) that this experimental paradigm relies explicitly on the fact that habituation to stimulus presentation is maintained unless the perceptual system detects a salient change in stimulation; detection results in dishabituation (i.e., response revival).

The first series of experiments compared two calls given during intergroup encounters, the *wrr* and *chutter*. Analyses revealed that these calls were acoustically different. Observations, however, showed that both calls were restricted to the context of intergroup encounters, though chutters tended to be produced during more aggressive interactions than wrrs. Thus, if vervet monkeys assess similarity on the basis of acoustic structure, habituation to wrrs should be followed by dishabituation to chutters, and vice versa.[25] In contrast, if similarity is determined by call meaning, then switching from wrrs to chutters should not alter the state of habituation. Cheney and Seyfarth found that when the same individual's (e.g., "Carlyle" in Figure 7.18) wrrs and chutters were used, habituation was maintained following a switch from wrrs to chutters, or the reverse. However, if exemplars from different individuals were played back, dishabituation was observed. For example, following habituation to Carlyle's wrr, there was a significant rebound in response to Austen's chutter. Cheney and Seyfarth offer an interesting functional explanation for these results. Repeatedly playing an individual's calls makes them unreliable: when Carlyle produces a series of wrrs in the absence of another group, she is crying wolf, and is thus ignored. However, just because Carlyle signals unreliably about neighboring groups doesn't mean that other individuals, such as Austen, are unreliable. This interpretation is certainly consistent with the data presented and is discussed in greater detail under the topic of deception (section 7.4).

25. Recall the point made by Tversky (1977), reiterated in chapter 3, that we should not necessarily expect symmetry even with entities that appear similar at some level.

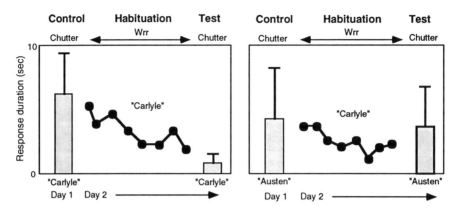

Figure 7.18
Results from habituation-dishabituation experiments on vervet monkey intergroup calls, with tests involving the same individual's calls (left panel) and different individual's calls (right panel) (redrawn from Cheney and Seyfarth 1988).

Using the same approach, Cheney and Seyfarth asked vervets about their eagle and leopard alarm calls. Here, as with the intergroup calls, there were significant differences in acoustic structure between eagle and leopard alarms, but coarse-grained contextual similarity—both calls are given during predator encounters. On a fine-grained level, however, the two alarm calls are associated with different types of threat and behavioral responses, differences that are far greater than those associated with wrrs and chutters. Playback results indicated a consistent revival of response following the habituation series, regardless of whether the two alarm call types were from the same or different individuals (Figure 7.19). Thus, although Amin may be unreliable about eagles, others do not perceive her to be unreliable about leopards. These results suggest that calls for eagles and leopards are not interchangeable but rather are semantically different.

The habituation-dishabituation technique is a potentially powerful tool for extracting meaning from a signal, but some of the results obtained by Cheney and Seyfarth are difficult to interpret, and I think some methodological changes might ameliorate the situation (see especially chapter 8). The interpretive problem stems from the fact that leopard and eagle calls are both acoustically different and semantically different. Thus habituation to one alarm call followed by disha-bituation to the other could be due to differences in either acoustics or seman-tics. Thus, unlike the comparison between wrrs and chutters, facing off acoustic differences with semantic similarity, the eagle-leopard comparison is more problematical.

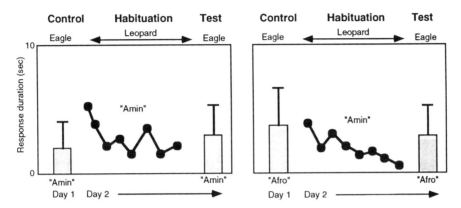

Figure 7.19
Results from habituation-dishabituation experiments on vervet monkey predator alarm calls, with tests involving the same individual's calls (left panel) and different individuals' calls (right panel) (redrawn from Cheney and Seyfarth 1988).

When an individual learns that exemplar *a* falls into category *A,* it often learns much more, because exemplars of a category often have multiple features or properties, and as one feature is learned, correlative features are also learned. Translate this abstract problem into a real-world problem, that of learning the species (exemplars) of predators (category) in the environment. For vervet monkeys, there are ground predators and aerial predators. Among the ground predators, there are big cats and snakes. Each of these predators behaves in a different way, and when they enter a vervet territory, the ecology of the environment changes as prey respond. Most notably, vervet monkeys shift from relatively quiet to noisy, producing their characteristic alarm calls. But they are not alone in the environment, and in fact there are other species that also shift from quiet to noisy. Consequently, associated with predator encounters is an orchestral composition of alarm calls. If vervets are good ecologists and have observed the association between predator encounters and the alarm calls of other vulnerable prey, then such calls can be used to increase the probability of early predator detection and subsequent escape. In this sense, the calls of other prey species represent signs (see chapter 1 for definition) from the vervet monkey's perspective.

Cheney and Seyfarth (1985a) set out to test this hypothesis with the Amboseli vervet population. In one experiment, they asked whether adult vervets recognize the alarm calls of the superb starling (*Spreo superbus*). Although starlings are preyed upon by species that never hunt vervets, they are also preyed upon by some of the same species, including both aerial and ground predators. Apparently in response to such predation pressure, starlings have also evolved different alarm calls, one for aerial predators and one for ground predators.

Results from playback experiments (Figure 7.20) indicate that extremely few subjects responded to starling song, whereas several subjects responded to the two starling alarm calls. More importantly, subjects responded by running up into a tree when they heard the starling ground predator alarm call, but looked up when they heard the aerial predator alarm call. These data suggest that for vervet monkeys in Amboseli, starling alarm calls provide an additional, valuable source of information about predator presence. As such, they are likely to contribute to the probability of survival.

Because superb starlings and vervet monkeys are not sympatric throughout Africa, recognition of starling calls must involve learning at some level. Hauser (1988a) examined the ontogeny of this ability in young vervet infants, recording between-group differences in exposure to starling alarm calls, as well as data on the age at which infants first recognize such calls as general alerting signals. Results revealed significant between-group differences in exposure to starling alarm calls, as well as in terms of the proportion of calls given to vervet-relevant predators. Differences in exposure were significantly related to the age at recogni-

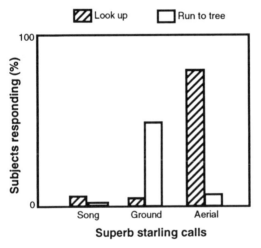

Figure 7.20
Vervet monkeys' response to superb starling vocalizations, including song, ground predator alarm calls, and aerial predator alarm calls. The *y*-axis represents the proportion of subjects responding by looking up (striped bars) or running up into a tree (open bars) (redrawn from Cheney and Seyfarth 1985a).

tion: as the rate of exposure to starling calls increased, the age at which infants recognized these calls as alerting signals decreased.

Extending their work on starling alarm call recognition, Cheney and Seyfarth (1985a) pushed the vervets to one more level of complexity and asked whether they recognized secondary cues associated with danger. In this case, the secondary cue was the mooing of cows brought into the park by the Masai. Because the Masai represent a threat to vervets in Amboseli and sometimes kill them, recognizing cow sounds would be beneficial as an advance warning. To test this hypothesis, playbacks of mooing cows were presented to vervets; the calls of wildebeests, a species that is not associated with danger, were played as a control. Vervets looked longer in the direction of the speaker following cow playbacks than following wildebeest playbacks (Figure 7.21), suggesting that the calls were discriminably different and the difference was salient.

Following up on the experimental approach laid out by Cheney and Seyfarth to assess what vervet monkeys know about their ecology, Hauser and Wrangham (1990) conducted experiments to determine whether birds and nonhuman primates recognized sounds made by their competitors and predators. Once again, this is a question about the recognition of signs. Because the experimental design was complicated, it is set out schematically in Figure 7.22. For each stimulus species, one representative call of each species was played back, attempting to use exemplars that were comparable in terms of duration and amplitude. Consequently, calls were not matched for context or meaning (e.g., the chimpanzee's call was a "pant-hoot" produced in the context of food discovery; the red colobus call was given during mating).

Quantitative data on scanning rates were obtained for all subject species, comparing the rates before and after the playback. Results (Figure 7.23) indicate that all species showed a statistically significant increase in scanning following eagle and chimpanzee calls, but only the frugivores showed a significant increase in response to the hornbill's calls. Turacos showed no significant change in scanning after colobus calls, but redtail and blue monkeys showed a significant decrease. Based on scanning data alone, it is not possible to assess whether subject species classify stimulus species as predators or competitors. The only conclusion that can be drawn from the data is that vigilance rates increase most dramatically in response to the two predators, only the frugivores respond to the hornbill (i.e., a competitor), and only the monkeys show a change in vigilance to the red colobus's call. The latter is consistent with the observation that these monkey species often form polyspecific associations, and one hypothesized function of such associations is to decrease predation pressure (Terborgh and Janson 1986; Waser

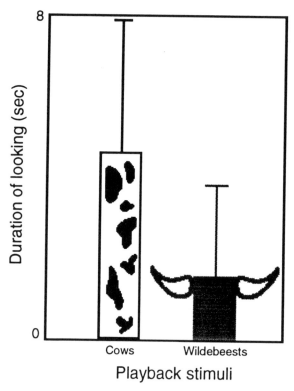

Figure 7.21
Vervet monkeys' response to playbacks of mooing by cows associated with Masai (threatening situation) and calls of wildebeest (nonthreatening situation). The y-axis represents the average amount of time (seconds) spent looking in the direction of the speaker following playback (redrawn from Cheney and Seyfarth 1985a).

1987); knowing that red colobus are nearby may enable redtails and blue monkeys to decrease their scanning rates with little cost.

More relevant data with regard to recognition of predator and competitor calls comes from observations of how subject species responded to the different stimulus species. In general, when individuals from all subject species heard the eagle's call, they responded by looking up and scanning, and then typically left the fruiting tree for up to 30 minutes. In contrast, although subjects also tended to leave the fruiting tree upon hearing the chimpanzee's call, they never looked up—all scanning was directed down toward the ground. Last, subjects never left the tree upon hearing the hornbill's calls.

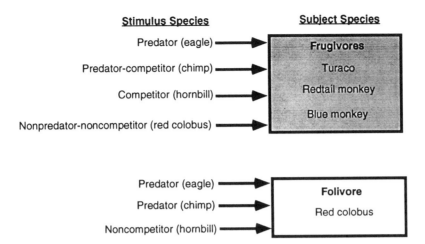

Figure 7.22
Playback experiment design (Hauser and Wrangham 1990) to assess whether each of four species recognizes the calls of its predators and competitors. For all subject species, the crowned hawk eagle is a predator, whereas the chimpanzee is a predator for all species but a competitor only for the three frugivores. The black-casqued hornbill is a competitor for all frugivores, but not the folivorous red colobus monkey. The red colobus is neither predator nor competitor for the three frugivores, but does move in polyspecific associations with the redtails and blue monkeys.

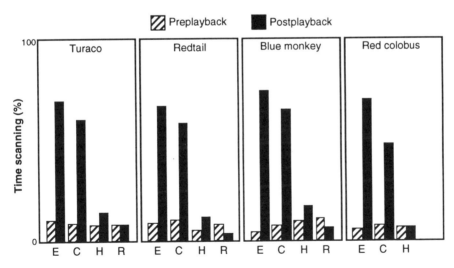

Figure 7.23
Preplayback (striped bars) and postplayback (black bars) scanning by three frugivores (great blue turaco, redtail monkey, blue monkey) and one folivore (red colobus) in response to calls by a predator (E, eagle), predator-competitor (C, chimpanzee), competitor (H, hornbill), and a nonpredator-noncompetitor (R, red colobus). The y-axis represents the proportion of time scanning in the pre- and postplayback period (redrawn from Hauser and Wrangham 1990).

Based on the four subject species tested, Hauser and Wrangham suggest that individuals can and do recognize the calls of their predators and competitors, and use this ability to guide their antipredator and foraging behavior. Future experiments will require significantly larger sample sizes to explore within-species variation in response, together with tests that assess how much of a predator's or competitor's repertoire is known by other species living in the same environment. My hunch would be that if predator and competitor avoidance is at a premium, selection should favor a recognition system that responds in a categorical fashion to the available acoustic variation: Is it a call from predator A's repertoire? If so, then move away. If not, stay put.

7.3.2 Categorical Perception of Vocal Signals

Categorization relies essentially on an assessment of similarity or equivalence (see chapter 3), in identifying whether two sounds, two colors, or two smells are the same or different. A classic approach to this problem is to use quite standard psychophysical techniques and theories to assess how (whether?) individuals take a stimulus continuum and divide it into perceptually meaningful categories. This process, known as *categorical perception* (CP), involves at least two fundamental steps. Harnad (1987) provides a succinct description of the methodological procedure and relevant phenomenological ingredients:

The method is to compare discrimination and identification performance. *Discrimination* requires a subject to tell apart stimuli presented in pairs (by indicating whether they are the same or different). *Identification* requires the subject to categorize individual stimuli using labels (to say, for example, whether they are /da/ or /ga/). A CP effect occurs when (1) a set of stimuli ranging along a physical continuum is given one label on one side of a category "boundary" and another label on the other side and (2) the subject can discriminate smaller physical differences between pairs of stimuli that straddle that boundary than between pairs that are within one category or the other. In other words, in CP there is a quantitative discontinuity in discrimination at the category boundaries of a physical continuum, as measured by a peak in discriminative acuity at the transition region for the identification of members of adjacent categories. (p. 3)

The most heated debates about categorical perception emerged in the 1970s when, following the discovery that several crucial phonemic distinctions were perceived categorically, a number of researchers claimed that the particular perceptual phenomenon observed were specific to speech—that categorical perception *shows* that speech is special. The theoretical framework for this interpretation was derived from the motor theory of speech (Liberman and Mattingly 1985; Liberman et al. 1967; Liberman et al. 1957). As discussed in chapter 4, this theory claims that speech perception, unlike perception of other acoustic or nonacoustic sensory stimuli, is special because it is anchored in the vocal production system. Categorical perception arises from the fact that the vocal tract creates discontinuities in the acoustic stream. Thus one can clearly generate (e.g., using sound synthesis techniques) an acoustic continuum from /ba/ to /ga/, but the articulatory gestures required to produce such sounds operate to constrain the output, thereby imposing acoustic discontinuities. The upshot of this theory, as well as the vast empirical literature that followed, was that speech researchers began to argue that the human auditory system is specialized—has been designed—to analyze the distinctive features of speech, and that this ability is learned.

To be clear, let us summarize the main claims from early work on categorical perception of speech:

1. Categorical perception is special to speech. Other, nonspeech, sensory stimuli (acoustic, visual) are not processed categorically.

2. Categorical perception is consistent with, and in fact predicted by, the motor theory of speech.

3. Categorical perception is based on learning the distinctive features of language.

Given these claims, there are at least four possible challenges to the theory. First, one might find particular phonemic distinctions within speech that are not processed categorically (Carney, Widin, and Viemeister 1977). Second, nonspeech sensory stimuli (auditory and other modalities) might be processed categorically (Bornstein 1987; Remez 1980; Remez et al. 1981). Third, human neonates may show evidence of categorical perception, thereby weakening the relative emphasis placed on learning for the establishment of phonemic categories (Eimas 1979; Eimas et al. 1971). Fourth, nonhuman animals may process human speech in a categorical fashion (Kuhl and Miller 1975, 1978), and might even process their own signals categorically (Ehret 1987b; Snowdon 1987). Let us now turn to a review of some of the most significant empirical challenges. Since I will focus primarily on the comparative data, readers interested in more comprehensive treatments of this topic, especially for humans, should see the excellent reviews in Harnad (1987) and Kuhl (1989); some of the work on human infants was already presented in chapter 5.

Significant challenges to the first two points have already been published. Thus there is evidence that certain phonemic distinctions such as /sh/ and /ch/ are *not* perceived categorically, and that several nonspeech acoustic stimuli (e.g., musical intervals, noise-buzz transitions) *are* perceived categorically (Burns and Ward 1978; Locke and Kellar 1973; Pastore 1987). Moreover, Bornstein (1987) reviews evidence of categorical perception for color categories. Regarding the third point, we learned in chapter 5 that by 3–4 months of age, human infants are sensitive to both between- and within-category variation in the speech stream. The former is relevant to the topic of categorical perception, and has typically been tested with a high-amplitude sucking technique (Eimas et al. 1971), whereas the latter is relevant to the issue of perceptual constancy or equivalence classes and has been tested with a discriminative head-turning response (Fodor, Garrett, and Brill 1975; Kuhl 1979). In general, infants appear to show the same sorts of category boundaries as adults. These results challenge two specific claims of the original formulation of categorical perception: (1) extensive learning[26] does not appear to be necessary for categorical perception, and (2) categorical boundaries are set up in the absence of vocal production (i.e., there is no support for the perception-production connection required by the motor theory).

Perhaps the hardest hits on the uniqueness claim for categorical perception came from studies of nonhuman animals. In the mid-1970s, Kuhl and colleagues initiated a series of experiments with chinchillas (Kuhl and Miller 1975, 1978) and then later with rhesus macaques (Kuhl and Padden 1982, 1983). The goal of these experiments was not to overthrow the idea that speech is processed by specialized mechanisms, since, after all, it must be the case that all communication systems, at some level, are processed by specialized structures. The interesting question is, Which aspects of speech require specialized mechanisms? In particular, does categorical perception require a specialized mechanism, and is this mechanism unique to humans?

In the first nonhuman animal experiments, chinchillas were trained in an avoidance-conditioning procedure and asked to discriminate among computer-generated exemplars of a voice-onset time[27] (VOT) continuum that ranged from /da/ to /ta/; in English speakers, VOT for stops (e.g., /da/) occurs at approximately 5–40 msec, whereas VOT for voiceless stops (e.g., /ta/) occurs at approximately 40–150 msec (Figure 7.24). The details of the initial training procedure

---

26. I want to emphasize the "extensive" component of this comment, for 3–4-month-olds, in one sense, have had experience with language, including some of the relevant phonemic contrasts.

27. In human speech, a number of important phonemes are distinguished on the basis of voice-onset time (VOT), an acoustic feature that results from a time difference between the energy burst before the start of laryngeal voicing and the energy burst that is released when a stop consonant such as /d/ or /t/ is produced. It is important to note that phonemes can often be distinguished by a suite of features, but in the case of such sounds as /da/ and /ta/, VOT is considered to be the most important.

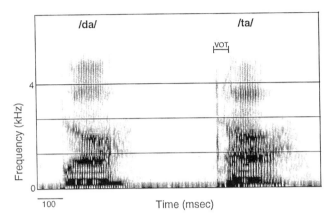

Figure 7.24
Sound spectrograms of an English speaker's production of /da/ and /ta/. Markers have been placed
showing the duration of the voice-onset time (VOT) for /ta/. VOT for /da/ is 5–40 msec (close to
5 msec in the example discussed in text), and for /ta/ it ranges from 40 to 150 msec (approximately
50 msec in this example).

are fundamental in such categorization problems. If subjects are provided with a
large number of exemplars from the category or categories of interest, then,
instead of revealing the processes involved in categorization, one may have set
up the process, including the featural boundaries of the category. Kuhl and Miller
trained chinchillas on two extreme cases, representing reasonable (based on human perceptual tests) end points for the /da/–/ta/ continuum: 0 msec for /da/ and
80 msec for /ta/. As such, the training procedure set up end points of the continuum to be tested. In the test, incorrect responses received a mild shock and
buzzer noise, whereas correct responses were rewarded with water. Once discrimination was close to perfect, a generalization procedure was started. In the
generalization session, half of the trials represented the two end points, and the
remaining half represented exemplars along the continuum from 0 to 80 msec; all
responses to novel stimuli, correct or incorrect, were rewarded. Results (Figure
7.25) provided clear evidence of a categorical boundary effect, with chinchillas
assigning the VOT boundary at 33.3 msec and humans at 35.2 msec. These values
do not differ statistically.

Following the success of their results with the /da/–/ta/ continuum, Kuhl and
Miller tested other contrasts (e.g., changes in place of articulation such as pa/
da/ga) and showed, as before, that chinchillas and humans appear to use similar
perceptual strategies, especially in terms of phonetic boundaries. In addition,

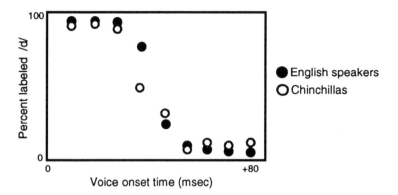

Figure 7.25
A test of categorical perception for the /da/–/ta/ continuum using English speakers (black solid circles) and chinchillas (open gray circles). The y-axis shows the proportion of exemplars that were responded to as /d/. The x-axis represents the differences in voice onset time (milliseconds) for the /da/–/ta/ continuum. Results from the experiment revealed a boundary value of 35.2 msec for humans and 33.3 msec for chinchillas (redrawn from Kuhl and Miller 1978).

results for both chinchillas and rhesus macaques (Kuhl and Padden 1982, 1983) have shown that individuals are most sensitive (measured in terms of psychophysical calculations of JNDs) to differences that are close to the phonetic boundaries and, apparently, less sensitive to differences within the phonetic category. Thus two nonhuman mammalian species show categorization and discrimination functions that are highly comparable to those exhibited by humans, at least for the phonetic contrasts explored. Moreover, recent work on three avian species—budgerigars, quail, and zebra finches—show that they too process human speech in a categorical fashion (Dooling 1989; Dooling, Best, and Brown 1995; Kluender, Diehl, and Killeen 1987). Taken together, these results suggest that the mechanisms underlying categorical perception are not specialized for speech. Rather, it appears that the process of categorizing and discriminating exemplars along an acoustic continuum is based on an evolutionarily ancient mechanism that either evolved in several different lineages independently (convergence) or perhaps evolved in a more limited number of lineages, then was maintained by selection over a long period of time.

Given the fact that nonhuman animals perceive speech categorically, is the underlying mechanism also recruited for categorization problems that arise during more ecologically natural or salient situations, such as occurs when individuals hear or see signals within their own species-typical repertoires? Studies of crickets[28] (Moiseff, Pollack, and Hoy 1978; Thorson, Weber, and Huber 1982), tree frogs (Gerhardt 1978a), bullfrogs (Capranica 1966), red-winged blackbirds (Beletsky, Chao, and Smith 1980), swamp sparrows (Nelson and Marler 1989), budgerigars (Dooling 1989), mice (Ehret 1992; Ehret and Haack 1981), pygmy marmosets (Snowdon 1987; Snowdon and Pola 1978), Goeldi's monkeys (Masataka 1983), and Japanese macaques (Hopp et al. 1992; May, Moody, and Stebbins 1989) have explored this problem and have provided relatively convincing evidence of categorical labeling functions (classifying stimuli as representative exemplars of one category but not another) but, in general, less convincing evidence of categorical discrimination functions (i.e., enhanced sensitivity at the category boundary). I review some of these results in the following paragraphs, focusing specifically on those studies that have explicitly tested for a categorical perception effect.[29]

In chapter 5 we discussed Marler's (Marler and Nelson 1992; Nelson and Marler 1994) notion of neuroselection during the acquisition of song. Some of the strongest evidence for this theory comes from the study of swamp sparrow song and, in particular, the apparent universality of note structure. Swamp sparrow song is composed of six note types, each with some degree of within-type variation. These notes differ along several acoustic dimensions, but temporal differences are highly significant. Most importantly for the current discussion, the positioning of notes within a song appears to vary geographically such that in New York individuals place note type 1 in the initial position and note type 6 in the terminal position. In contrast, Minnesota birds put note 1 in the terminal position and note 6 in the initial position. Playback studies show that the birds themselves are sensitive to this temporal distinction (Balaban 1988b,c).

To explore the possibility of categorical perception, Nelson and Marler (1989) first documented the natural variation in note duration for notes 1 and 6, and based on what appeared to be a bimodal distribution, they constructed playback stimuli that represented a continuum from note 1 to note 6 (Figure 7.26).

To test for a categorical boundary effect (discrimination function), Nelson and Marler used a habituation-dishabituation paradigm. Specifically, they repeatedly played one song series until subjects showed behavioral (i.e., an aggressive "wing

28. Hoy and colleagues (personal communication) have recently completed a set of experiments on crickets showing that they perceive the ultrasonic signals of predatory bats in a categorical fashion.

29. Many of the cited studies were not designed to test for categorical perception. Rather, experiments focused on the relationship between changes in acoustic morphology along a continuum and changes in behavioral or vocal responses.

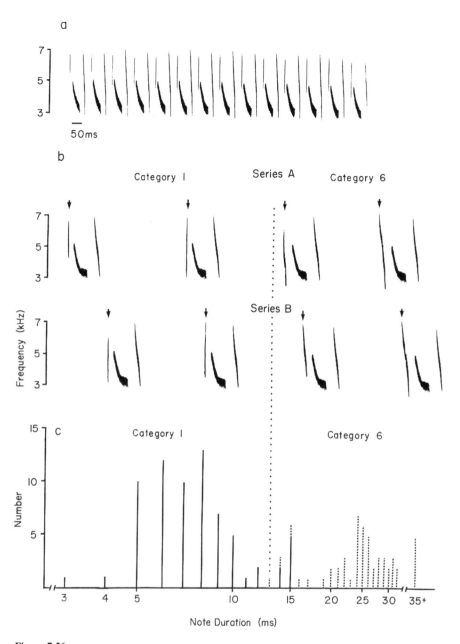

Figure 7.26
a. A sound spectrogram of a swamp sparrow song. b. The two tape series (A and B) used in playback experiments of categorical perception. c. Data showing the natural variation in note duration for category 1 notes (black bars) and category 6 notes (stippled bars) (from Nelson and Marler 1989).

wave" display by territorial males) habituation to 75% of the first block of trials,[30] and then played back a test series. All test series differed from the prior habituation series by an identical acoustic "amount" (i.e., equivalent acoustic distances of duration). Evidence of significant perceptual discrimination—meaningful differences—was derived from the degree of behavioral revival following habituation. Results (Figure 7.27) showed that the strength of the response was greatest when the habituation and test series straddled the between-note category boundary of 13 msec, with relatively weak responses observed to both within-category comparisons.

These results provide evidence that an acoustic continuum of note durations is naturally partitioned into two categories, representing fundamental units of the swamp sparrow's song. Although Nelson and Marler conclude that their results provide evidence of categorical perception, their data only speak to the discrimination-function component of this process. Experiments on labeling have yet to be conducted. Tests of labeling might profitably be pursued in this species using an operant classification procedure.

When mouse pups are separated from their mothers, they produce a distinctive ultrasound signal.[31] Under natural or even seminatural conditions, mothers approach their pups upon hearing the ultrasound. The pup's signal shows variability in both the time (duration: 20–120 msec) and spectral domains (frequency range: 40–90 kHz). Ehret and colleagues have conducted several categorical perception experiments, testing for both labeling and discrimination functions. The setup for these experiments was a two-speaker phonotactic design, comparable to those used in studies of anuran mate choice. The behavioral assay was movement by a lactating female toward one of the speakers.

Naturally produced ultrasounds result in maternal searching behavior, but a 20-kHz tone burst does not. Expected levels of movement were thus generated on the basis of the ratio of moves to the natural ultrasound as opposed to the 20-kHz burst. Based on this ratio, a comparison could be made with other frequencies, with differences in response providing the basis for establishing a labeling function—that is, which frequencies function (i.e., elicit searching) as pup ultrasounds. Results (Figure 7.28) indicate a significant boundary between 22.5 and 24 kHz (Ehret and Haack 1981). Thus ultrasound signals with a frequency

---

30. Note that this procedure is similar to the one employed by Cheney and Seyfarth, discussed previously.

31. There is no evidence that the pup's ultrasound conveys information about individual identity or that mothers recognize their pups by voice.

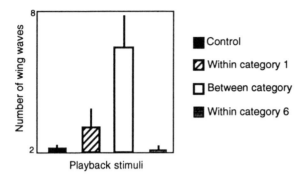

Figure 7.27
Swamp sparrow responses to four different playback series, using a habituation-dishabituation paradigm. The *y*-axis represents the number of wing wave displays given during the test (dishabituation) phase of the experiment, and thus indicates the degree of revival following habituation. *Control* (black): playback series is identical to the habituation series. *Within category 1* (striped): playback series represents a difference in duration of category 1 notes. *Between category* (white): playback series where differences in note duration straddle the 13 msec boundary (see Figure 7.26). *Within category 6* (stippled): playback series represents a difference in duration of category 6 notes (data from Nelson and Marler 1989).

bandwidth exceeding 23 kHz are preferred to the 20-kHz standard, and this contrast provided the relevant boundary conditions for setting up the discrimination test. The discrimination function clearly reveals that the best discrimination occurs when stimuli straddling the boundary are compared. Thus results provide strong evidence of categorical perception along a frequency-range continuum.

More recently, Ehret (1992) has used this procedure to examine the possibility of categorical perception for stimuli varying continuously along a temporal dimension—signal duration. Under natural conditions, ultrasounds vary from 30 to 120 msec (Sales and Smith 1978). For the labeling function, 50-kHz stimuli varying from 10 to 60 msec were played back. Results (Figure 7.29) indicate that for lactating females, signals that are 25 msec or shorter are labeled as different from the standard, whereas signals that are 30 msec or longer are labeled the same as the standard. More specifically, signals 25 msec or shorter resulted in response ratios that were close to the expected levels. As a result of these data, a discrimination test was set up. Lactating females showed poor within-category discrimination but good between-category discrimination; the latter was only evidenced, however, when the two comparison stimuli differed by at least 20–25 msec.

To date, therefore, the ultrasound signal of the mouse pup provides the only clear case of categorical perception of a natural call continuum—both labeling and discrimination functions have been tested and evidence provided with regard

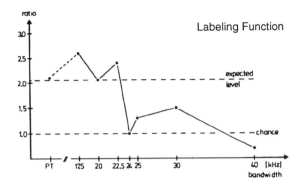

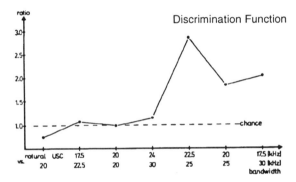

Figure 7.28
Data from labeling (*above*) and discrimination-function (*below*) tests on house mouse mothers in response to a pup's ultrasound signal. Stimuli varied along a continuum of noise bandwidths, centered at a frequency of 50 kHz. PT, pure tone of 60 kHz. Expected level indicates the response obtained to natural ultrasounds as opposed to 20-kHz tone bursts. The lower figure shows the boundary effect, with the strongest response occuring between 22.5 and 24.0 kHz (from Ehret 1987a).

to both temporal and spectral continua. From a functional perspective, categorical perception of ultrasound signals is highly adaptive, as it allows a lactating female to discriminate between her pup's calls and other biologically relevant sounds (e.g., postpartum and rough-handling pup sounds, defensive calls of adults). Similar functional arguments can be set up for other vocal stimuli (e.g., song sparrow note variation), and thus the lack of evidence may reflect methodological difficulties rather than the absence of requisite perceptual abilities.

In chapters 2 and 4, I discussed psychophysical and neurobiological studies of the Japanese macaque's vocal repertoire and, in particular, perceptual experiments conducted on their coo vocalization. Recall that Green's (1975a) field-work had established that coos represent an acoustically heterogeneous class of

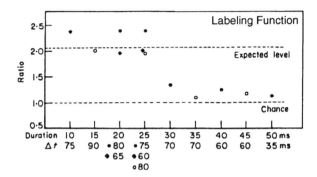

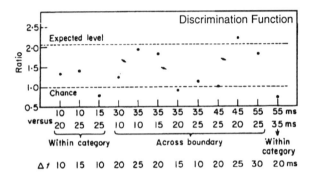

Figure 7.29
Categorical perception of a temporal continuum (duration) for the mouse pup's ultrasound. The upper figure shows a labeling function, and the lower figure shows a discrimination function. Arrows on the discrimination function reflect shifts in the duration difference between comparison stimuli (from Ehret 1992).

vocalizations, where within-class variation in the fundamental-frequency contour maps onto significant variation in the socioecological context eliciting the call. In particular, several perceptual experiments suggested that the location of the peak frequency in the fundamental-frequency contour was a crucial distinguishing feature, and that the left hemisphere of the brain was dominant in processing such information (Beecher et al. 1979; Heffner and Heffner 1984; M. R. Petersen et al. 1978; Zoloth et al. 1979).

To more carefully assess the relevant features for categorizing coos as ''smooth early highs'' (SEH) or ''smooth late highs'' (SLH), May, Moody, and Stebbins (1988) conducted several psychophysical experiments, using sound synthesis techniques to selectively manipulate specific features of the signal while holding others constant. The first set of results supported previous findings, indicating that individuals readily discriminate between SEH and SLH coos and that the strength of this discrimination is equivalent for natural and synthetic exemplars (Figure 7.30).

The subsequent experiments involved removing the frequency inflection but leaving amplitude modulation intact. Here, subjects failed to distinguish between exemplars. Similarly, when amplitude was held constant but the frequency inflection was shortened at the start or end, subjects again failed to discriminate. These results suggest that Japanese macaques may simply use a more coarse-grained assessment of the direction of frequency modulation to classify calls as either SEH or SLH. To test this hypothesis, frequency downsweeps were contrasted with frequency upsweeps (Figure 7.31). The results were extremely clear, showing that downsweeps are classified as SEH variants, whereas upsweeps are classified as SLH variants.

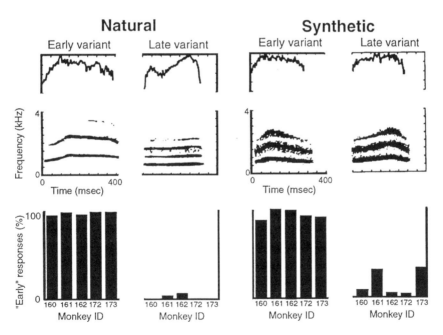

Figure 7.30
Performance by Japanese macaques with natural (*left panel*) and synthetic (*right panel*) versions of two coo types, SEH and SLH. The *y*-axis represents the proportion of trials where the exemplar was classified as an SEH coo (redrawn from May, Moody, and Stebbins 1988).

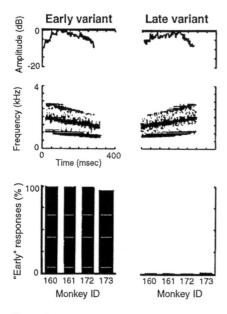

Figure 7.31
Discrimination performance by Japanese macaques on a task involving frequency downsweeps (*left panel*) and frequency upsweeps (*right panel*) (redrawn from May, Moody, and Stebbins 1988).

Together, these results emphasize the importance of frequency modulation in Japanese macaque perception and further strengthen the ubiquity of this acoustical feature in animal perception more generally. The next step was to set up a test of categorical perception, with an acoustic continuum running from synthesized SEH coos to synthesized SLH coos. In one testing procedure, subjects were asked to discriminate between two synthetic coos while simultaneously performing a discrimination task with natural coos ("high uncertainty procedure"). Results (Figure 7.32) showed a clear boundary effect, with all stimuli exhibiting a peak position before 150 msec labeled as SEH coos and all stimuli with peak positions above 150 msec labeled as SLH coos. Thus May and colleagues find support for a discrimination function, with enhanced sensitivity at the category boundary.

May and colleagues also used a "low-uncertainty procedure" (discriminate synthetic coos in the absence of a second discrimination task), and here they failed to find evidence of categorical perception. Moreover, a recent set of experiments by Hopp and colleagues (1992), which also used the low-uncertainty procedure, failed to find evidence of categorical perception of the SEH-SLH coo

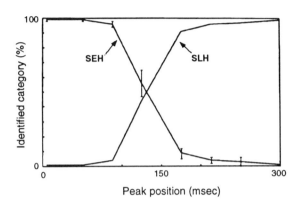

Figure 7.32
Identification functions by Japanese macaques (points represent the mean of four subjects) for a perceptual task involving exemplars within two coo categories, smooth early high (SEH) and smooth late high (SLH). Results show a category boundary at the peak position of approximately 125 milliseconds (redrawn from May, Moody, and Stebbins 1989).

continuum. In fact, not only did they fail to find the classic boundary effect, but they also found evidence that discrimination was superior at the end points. At least three factors may explain the discrepancy between these two studies. First, task demands differ between high- and low-uncertainty procedures. Second, the subjects used by May and colleagues had significant experience in other experiments, thereby potentially biasing the process of categorization; subjects were relatively naive in the study by Hopp and colleagues. Third, different sound synthesis techniques were employed, which may have set up different sorts of perceptually salient cues. Hopp and colleagues conclude that discrimination of coo exemplars in Japanese macaques does not appear to depend on categorical perception, but rather is guided by more general psychoacoustic processes (e.g., temporal discrimination). Additional experiments using different acoustic dimensions and training procedures will be required to resolve this conflict.

In summary, categorical perception of naturally communicative signals has only been convincingly demonstrated in house mice (Ehret 1987a, 1992; Ehret and Haack 1981). Other studies have either failed to show categorical perception, have only tested for a discrimination or labeling function but not both, or have provided conflicting data. Nonetheless, it is clear from the published literature that future research will most likely uncover further evidence of categorical perception. This expectation seems reasonable because there are several conditions under which such perceptual mechanisms are likely to be adaptive. For example, individuals are often confronted with signals that vary along a graded acoustic continuum, with representation along the continuum reflecting variation in the signaler's motivational state (e.g., Dabelsteen's work with European blackbirds; Figure 7.6). In such situations, perceivers are likely to be forced into a discrete response situation (e.g., flee or attack), thereby demanding categorical decisions. As several researchers (reviewed in Massaro 1987a) have warned, however, one must be careful to distinguish between categorical perception and categorical decisions. The former represents the perceptual process of interest, the latter, a presumably higher-level cortical process that works with or on the sensory information. The crucial point here, then, is that one's experiment must be designed in such a way that perception and decision are clearly differentiated. For example, tests where subjects are required to classify a stimulus as a member of category A or B by, let's say, pressing one of two buttons, will ultimately force a categorical process. Studies reviewed by Massaro (1987a) provide a number of solutions to this problem, such as setting up more than two response alternatives. In addition, Ehret (1987a, 320) provides an excellent methodological framework for approaching the problem of categorical perception in nonhuman animals, and I reiterate his recommendations here:

1. Inspect the call repertoire of a species together with behaviorally important sounds. Determine the discreteness between, gradations among, and overlap (along acoustic dimensions) of the acoustic stimuli.

2. Determine whether calls are species-, context-, or individual-specific, or whether they reflect motivational levels.

3. Describe natural response patterns to calls of an established specificity.

4. Conduct tests for categorical perception of a sound parameter with a known contribution to the specificity of a call.

In general, the first three steps have been carried out for a number of species. What remains for the future is step 4, the critical experimental tests of categorical perception.

7.3.3 Categorization of Faces and Facial Expressions in Nonhuman Primates

[Face recognition] is an effortless act performed as a matter of course and at a speed typical of an automatism. It is a function acquired very early during development and requiring no formal training, and it does not need to be cultivated or improved through special techniques as we feel no particular urge to achieve higher proficiency. It does not involve any thinking and it only occasionally fails us. There seem to be no limits to the number of faces that we can recognize and, as we make new acquaintances and as new actors or politicians become public figures, their faces are quickly learned and identified without obliterating the memories of already known faces. . . . Perceiving and recognizing faces is a private experience, uncommunicable to others, and we are not in the habit of sharing our impressions as to how we process the faces of people.   (Sergent and Signoret 1992, 55)

We learned in chapter 4 that face recognition in both human and nonhuman primates (at least macaques) involves regions of the brain that appear dedicated to the task. Although many would take issue with the notion of face cells or even face areas, it seems reasonable that some cells and some areas are more responsive to faces than to nonface visual stimuli. In section 7.2 we also learned that for humans inverted faces are often treated differently (i.e., are perceived differently) from upright faces, that certain transformations of the face have little effect on assessments of identity (e.g., size, orientation), that scrambling features (e.g., putting the eyes on the chin and the mouth on the forehead) is disruptive, and that facial expressions are made distinctive by moving particular features of the face. Such information is strikingly absent from the nonhuman primate literature. In this section, therefore, I review some of the experiments that have attempted to tackle the problem of face recognition and recognition of facial expressions in nonhuman primates.

    With the exception of work at the single-unit level, little research has focused on how nonhuman primates respond to and recognize different facial expressions. A recent paper by Dittrich (1990), however, provides an important start, using methods derived from research on humans. Using a captive colony of long-tailed macaques, young females (2–3 years old) were initially trained to press a button next to a facial expression called *slandering*. This expression is given as a threat and in an attempt to recruit support from others.[32] The presentation array consisted of four different facial expressions (all line drawings), including a second threatening display (*silent threatening*) and two submissive displays (*silent teeth baring* and *lip smacking*). Individuals acquired this discrimination within approximately 500–800 trials (i.e., presentations). Subsequently, Dittrich presented a series of manipulated images in order to assess the importance of particular features on the discrimination task. Confirming data from single-unit recordings (see

32. Dittrich states that this expression was selected "because it communicates an invitation to other individuals to interact, and so should not be frightening either for dominant or for subordinant group members." This assumption should be treated cautiously because it is presumably unclear whether the threatening component of this display is being directed toward the subject or toward someone else. Consequently, the expression may well be frightening, though I do not believe this fact influences the results obtained.

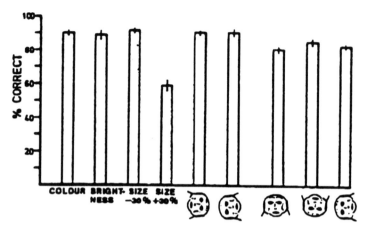

Figure 7.33
Results from a discrimination task using young female long-tailed macaques. The *y*-axis represents
the proportion of stimuli correctly discriminated, and the *x*-axis presents the types of featural changes
imposed (from Dittrich 1990).

chapter 4), results (Figure 7.33) showed that discrimination performance was
not significantly influenced by size, brightness, or color. Somewhat surprisingly,
perhaps, orientation also failed to influence discrimination, a result that contrasts
with one experiment on rhesus (Perrett et al. 1988b) and several studies of human
adults (Yin 1969). The contrasting results may reflect differences in experimental
design, a difference in how line drawings of faces (as opposed to photographs)
are perceived, or interspecific differences (rhesus versus long-tailed macaques).
    Dittrich was also able to explore the relative importance of particular features
of the face in recognition, as well as the potential effects of face asymmetries. A
number of studies on both humans and nonhuman primates have shown that
individuals pay particular attention to the eyes, especially when contrasted with
other facial features (C. K. Keating and Keating 1982; Kyes and Candland 1987).
Dittrich's results show, however, that long-tailed macaques perceive the outline
of the face as being most important, followed by the eyes and mouth. Moreover,
the right half of the face was perceptually more dominant than the left side. These
results conflict with studies by C. R. Hamilton (1977) and Overman and Doty
(1982) showing a lack of asymmetry in face recognition, in addition to studies by
Hauser (1993b) showing a left-side-of-the-face bias for rhesus macaque facial
expressions. At present, it is difficult to determine whether such differences are
the result of methodological details or species-specificity in the perception and
production of facial expressions. Although the latter is possible, it seems some-

what unlikely given that all of the studies focused on macaques, a genus within which variation in mating pattern and social organization tends to be small.

The discussion thus far has focused on how primates classify faces and facial expressions into functionally meaningful categories. But an entirely different set of problems and questions arise when one asks what the animal does with the information conveyed by a face or a facial expression. When an individual sees another giving a fear grimace, does it empathize, can it understand what the expression conveys and what it would feel like to be in that animal's state? Though few studies have attempted to tackle this issue quantitatively, an intriguing experiment on rhesus monkeys was carried out by R. E. Miller (1967, 1971), which I believe serves as an example of the kinds of experiments that could readily be carried out with other species and more fully developed with rhesus.

Miller's experiments were designed to assess whether both familiar and unfamiliar rhesus monkeys could use the facial expression of another to determine the appropriate behavioral response in a particular context. The following procedure was used (see Figure 7.34). Individuals were trained on a response panel that included two lights (cues) and two levers. When one light went on, pressing the appropriate lever resulted in shock avoidance. When the other light went on, pressing the lever resulted in a food reward. Once individuals were shaped up, one animal, say "Fred," was placed in a visually isolated chamber that included the same type of panel, but without levers (i.e., only lights). A second individual, "Sam," was placed in a different chamber and with a panel that lacked lights, but included levers. Sam could see Fred's face, however, because a video camera filmed Fred and the image was transmitted to a monitor in Sam's chamber. Relatively quickly, Sam learned to press the appropriate lever based, apparently, on Fred's facial expressions.[33] Miller's interpretation of his observations is that something like a fearful expression by Fred led Sam to pull the shock-avoidance lever, whereas a more neutral, nonfear face led Sam to pull the food lever. Most intriguingly, if an unfamiliar individual, also trained on the response panel, was put in Fred's place, Sam failed to press the appropriate lever.

These results suggest, minimally, that rhesus monkeys can read the emotional expressions (i.e., behavior) of familiar individuals and use such information to guide their own responses. It seems unlikely that they have simply been conditioned to respond in particular ways, for if such were the case, familiarity would

33. Since Miller does not provide details of the facial expression or, more generally, head movement, it is difficult to say whether the lever-pressing response was based on an expression of emotion or on some other motor action that, as a result of reinforcement contingencies, became associated with a reward.

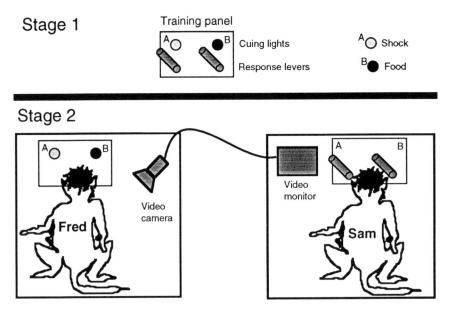

Figure 7.34
Setup for R. E. Miller's (1967, 1971) experiments on the recognition of facial expression in rhesus monkeys.

presumably be less important. Making the stronger claim that rhesus monkeys empathize—that they actually experience at some level what other's experience based solely on their observations—is a much more difficult claim to defend. As will be discussed more specifically in section 7.4, most of the research in this area shows quite elegantly that nonhuman animals are sensitive to subtle changes in behavior, but there is little evidence that nonhuman animals are sensitive to subtle changes in mental state. Nonetheless, the paradigm that Miller developed could readily be applied to a host of animal species, altering different features of the viewing environment. For example, the camera could be positioned in such a way that only the eyes are shown, or only the lower body. As a result, the researcher could quantify the necessary and sufficient features for "reading" the emotional state of others.

7.3.4   Categorization of the Inanimate and Animate World

Imagine the following scenario. You are a professional diver. You have just climbed the stairs to the 10-meter platform and are about to perform a backflip. To initiate the backflip, you must position your feet on the edge of the platform in such a way that you maintain balance and have enough support to flex and spring backward. For a professional diver, the backflip is a relatively trivial maneuver, and yet it requires substantial understanding of the notion of support. Less than the minimum amount of support and the diver will enter the water precipitously.

This example fits under the more general problem of studying how individuals come to understand the causal properties of the inanimate and animate world—of categorizing objects and events. Over the past ten years, developmental psychologists have investigated the ontogeny of this causal knowledge in extremely young human infants. Although the issues examined do not fall within the domain of communication, I will argue (also see Premack 1990) that the psychological mechanisms being tapped (e.g., issues of causality) are critically related to the emergence of a mentally rich communication system (e.g., issues of intentionality). As such they provide an important empirical step into some of the more cognitively challenging issues currently discussed by psychologists and biologists interested in the minds of other species, especially how communication provides a passport into their minds; these are the topics for section 7.4.

Good magicians surprise us because they appear to be violating physical laws, creating unexpected events such as chopping through a human body and placing each half on opposite sides of the room. And when we watch magicians, we indicate our surprise by looking intensively at the display, trying to find the trick. If increased looking is correlated with our detection of physically impossible events, then human infants who understand the mechanics of the physical event should also look longer. If, however, human infants lack such understanding, then possible and impossible events should elicit equally long looking times.

Figure 7.35 is an example of an experiment conducted by Spelke (1991) with infants under the age of 12 months. Note that the strength of this research comes less from one experiment than from the combination of several such experiments, each triangulating on a particular kind of knowledge of the world.

The expected control trials show a stationary ball in one of three positions (right to left): between two posts on the ground, on top of a table, and below a table. The test trials all involve motion. In the first, the ball is held above the posts (1), then a screen appears covering the posts, but not the ball. The ball is then dropped behind the screen. When the screen is taken away (2), the ball is seen between the posts—an expected position. In the second expected test, the ball is shown above the table (1), the screen appears, the ball drops, the screen is taken away and the ball is seen on the top of the table (2). The last test involving motion is the same as the previous test, with one exception: when the screen is

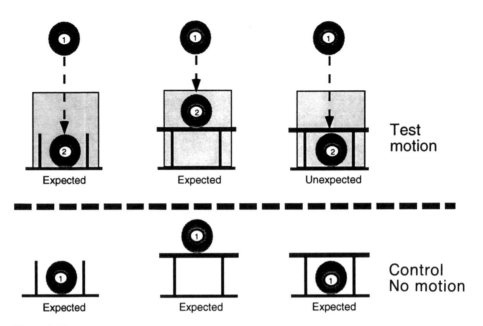

Figure 7.35
A test of infant's knowledge of falling objects and mass. The schematic drawings represent a ball (number inside indicates starting [1] and ending [2] positions), the ground (solid horizontal line), two posts emerging from the ground, and a table on top of the ground. The speckled rectangle represents an opaque screen. See text for further description.

removed, the ball is under the table (2). If infants respond to this event as adults do, then they should be surprised to see a ball passing through a table. Surprise is in fact what they seem to express, as evidenced by the fact that they look significantly longer at the unexpected event.

Consider a second experiment conducted by Kotovsky and Baillargeon (in press) designed to assess whether infants understand some of the crucial properties of propelled objects, especially those properties associated with collision phenomena (i.e., physical contact between one moving and one stationary object). In this experiment, infants (2.5 months old) were first habituated to a large cylinder rolling down a ramp and then stopping at the end of a track upon contact with a set of barriers. Subsequently, one group of infants was presented with either a possible or impossible event, and one group of infants was presented with a possible control event. In the possible test event, a toy bug on wheels was placed 10 cm from the ramp, and the cylinder was released, missing the bug and leaving the bug in place. In the impossible test event, the bug was placed at the base of the ramp, and the cylinder was released, striking the bug but leaving it in place. For the possible control events, a wall was stationed in front of the bug so that it could not be displaced.

When an adult observes one object move and strike a second, he or she expects the stationary object to be displaced forward, unless of course the stationary object weighs considerably more than the moving object. Infants in the test condition spent considerably longer looking at the impossible event than at the possible one. Infants in the control condition looked equally at the two events. Based on these findings, Kotovsky and Baillargeon infer that infants expect the cylinder to displace the toy bug and are surprised when it stays in place.

Given the inference that 2.5-month-olds know something about the properties of collision events, a second set of experiments was designed to assess whether they understand how the size of the moving object influences the trajectory of the stationary object (see Figure 7.36). In the habituation event, 6.5-month-old infants observed a medium-sized blue cylinder roll down a ramp, strike a toy bug, and then watch as the bug moved to the center of a scene of buildings. In the test events, they watched the same general sequence, but instead of the blue cylinder, either a larger yellow cylinder or a smaller orange cylinder rolled down and struck the bug; both of these cylinders caused the bug to roll farther than the blue cylinder, stopping at the end of the track. As a control, a second group of same-aged infants were habituated to a blue cylinder striking the bug and causing it to move to the end of the track. After watching the first blue cylinder sequence, adults expect the bug to move farther when struck by the larger yellow cylinder and to remain closer to the ramp when struck by the smaller orange cylinder. When watching the second blue cylinder sequence, adults are not surprised to see that both the large and small cylinders can cause the bug to move to the end of the track; they simply conclude that a size effect cannot be demonstrated with such a short track. Tests with infants indicated that they looked longer at the small cylinder than at the large one, whereas the control group looked equally at the two test events. From these experiments, Kotovsky and Baillargeon argue that by the age of 6.5 months, infants appear to understand the importance of size in collision events and show surprise when small objects propel a stationary object as far as relatively larger objects.[34]

Spelke (1991, 1994), Baillargeon (1994), and many others have conducted an impressive battery of such tests, providing us with a rich data set for how one might go about building a human child's mental data base (Mandler 1988, 1992)

34. Additional experiments were conducted with 5.5-month-olds, indicating that males but not females performed like 6.5-month-olds. This sex difference appears to be due to the fact that males fail to make the connection between size of the cylinder and expected travel distance.

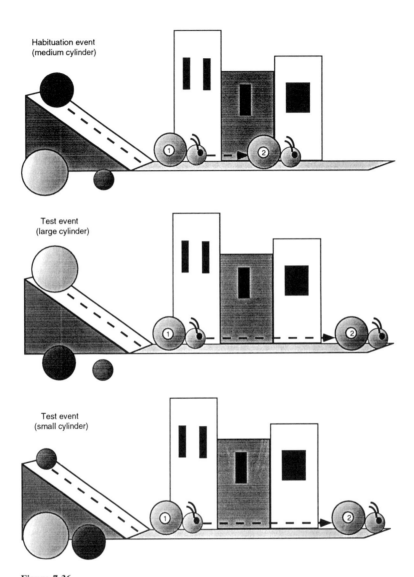

Figure 7.36
A test of collision events, modeled after Kotovsky and Baillargeon (in press). Infants are habituated to the top sequence, showing a medium-sized ball hitting a toy bug. Upon contact, the bug moves a specific distance. Infants are then shown a possible or impossible event. In the possible test (middle), a larger ball hits the bug and displaces it further than in the habituation test. In the impossible condition (bottom), a ball that is smaller than the medium-sized one used in the habituation strikes the bug and moves it as far as the largest ball. Looking time is used to assess infant's knowledge of these collision events.

or as Spelke (1994) states, the infant's "initial knowledge." What is clear from such work is that the human child enters the world with a relatively rich, innately specified understanding of causality, especially in terms of the physical world. In contrast, as I will illustrate subsequently, it takes the child much longer to figure out how causality guides events in the animate world, especially the world of socially interacting humans. Before detailing this line of research, let us first consider how the looking-time paradigm might work with nonhuman animals, and then explore a few potentially interesting situations and interactions that appear ripe for causal analysis.

Imagine the following encounters between two people, Harry and Betty, who know each other intimately: (1) Harry smiles at Betty; Betty smiles back and approaches Harry. (2) Harry frowns and furrows his eyebrows at Betty; Betty looks away and runs out of the room. (3) Harry frowns and furrows his eyebrows at Betty; Betty smiles, runs over, and embraces Harry. In these scenarios, there has been no contact between Harry and Betty, and yet we are likely to impute some causal connection between Harry's facial expression and Betty's response. The first two scenarios seem highly plausible; the third scenario seems unlikely, or at least less likely than the first two. The question of interest is, Do nonhuman animals understand that certain expressions, either facial or vocal, are causally related to certain types of responses and not others? And is there a way to distinguish between causally related effects and mere associations (established contingencies)? Cheney, Seyfarth, and Silk (1995) have attempted to address this question by setting up a series of possible and impossible vocal interactions among wild baboons, using a looking time assay as a measure of the baboon's understanding of causal relations.

Baboons grunt during social interactions (Cheney, Seyfarth, and Silk in press). Typically, dominant females grunt to subordinates while attempting to handle their infants. In response, subordinate females often give fear barks. Two play-back tapes were created to assess the possibility of causal reasoning about the relationship between call type and dominance rank (Cheney, Seyfarth, and Silk 1995). The first consisted of an impossible sequence involving a grunt by a low-ranking female followed by a higher-ranking female's fear bark. This sequence is impossible in that high-ranking females have never been observed[35] giving fear barks to subordinate females. The second involved the same sequence as the

---

35. Under field conditions, of course, it is impossible to say that particular types of events or sequences never occur. The best one can do is state, based on long and detailed observations, that such events have not been recorded and thus occur infrequently at best.

first, but the fear bark was followed by the grunt of an even higher-ranking female. The inclusion of this third call made the sequence possible, since fear barks are responded to (more conservatively, "followed"; see chapter 3) by grunts. Results (Figure 7.37) show that individuals looked significantly longer following presentation of impossible vocal sequences than possible ones.

Do these results show that baboons understand that only certain individuals can elicit (i.e., cause) fear barks? Cheney and colleagues suggest that their data permit exclusion of several alternative explanations. For example, overall energy is often an important parameter in eliciting responses during playbacks. However, the impossible sequence had less acoustic energy than the possible sequence, and yet the former elicited a far stronger response. Novelty of the sequence also fails to explain the data, since both sequences were novel. Moreover, high-ranking females do produce fear barks, and low-ranking females do produce grunts. But whether or not such particular calls are produced depends explicitly

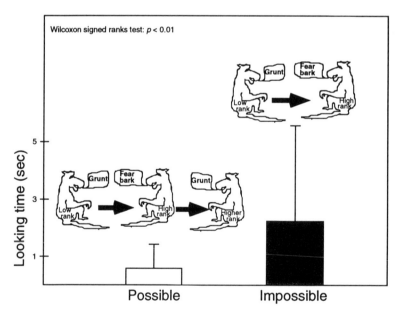

Figure 7.37
Results from playback experiments on wild baboons. *Impossible* vocal sequences consisted of a grunt from a low-ranking female, followed by a fear bark from a higher-ranking female, and terminating with a grunt from a higher-ranking female. *Possible* sequences consisted of a grunt from a low-ranking female, followed by a fear bark from a higher-ranking female, and then another grunt by an even higher ranking female. Response is the mean amount of time looking at the speaker (seconds); standard deviations shown (data obtained from Cheney, Seyfarth, and Silk 1995).

on the context, whether the interactant is higher or lower in rank. Last, and most important, the results cannot be explained by a simple set of associational rules, such as that grunts can be given to high- and low-ranking individuals, whereas fear barks are only given to higher-ranking individuals. If this were the case, then subjects should have responded as strongly to the first part of the possible se-quence as to the entire impossible sequence—they were identical. Rather, indi-viduals apparently listened to the entire possible sequence and judged the relevant vocal interaction as occurring between the second (fear bark) and third call (grunt). Thus it appears that baboons have some notion of cause-effect, though as Cheney, Seyfarth, and Silk point out, the precise nature of this relationship is still unclear.

Though experiments on baboon grunts were the first to use a looking-time procedure to assess what nonhuman animals know about causal relationships in a communicative context, other studies have begun to use the looking-time procedure to assess knowledge in noncommunicative contexts (e.g., numerosity: Hauser, MacNeilage, and Ware, in press) and several observational situations seem ripe for experimental treatment. Consider the following descriptions:

1. A chimpanzee infant prepares to crack open a palm nut. She places the nut on a stone anvil and then grabs a stick that is small and thin at one end, large and fat at the other. She grabs the fat end and prepares to strike with the thin end. Her mother approaches, grabs the stick, and reorients it in her hand so that she now bangs while holding the thin end and striking the nut with the more appropriate fat end.

2. A vervet monkey infant sees a mongoose and gives the referentially inappro-priate "leopard"-sounding alarm call. She and her mother dash up into the tree, an appropriate response to a leopard alarm call. The mother sees the mongoose, turns to her infant, and slaps her. Following this interaction, the infant was never observed giving a leopard alarm call for a mongoose.

3. A juvenile yellow-eyed junco picks up a mealworm, holding it at its center with one end sticking out of each side of the beak. An adult junco approaches, takes the mealworm from the juvenile's mouth and reorients it so that it can be slurped up like spaghetti; adults always pick up and eat mealworms while held in a vertical position (one end in the mouth, the other dangling below). Eventually, juveniles pick up mealworms in a vertical position.

All three cases (described in greater detail in Caro and Hauser 1992) share a common feature: In response to observing a particular behavior in animal A, animal B does something that appears to alter A's subsequent behavior (see Box 5.1). Two questions arise. First, can we be confident that B's behavior causally influenced A's behavior, or was it merely correlated with an observed change? Second, did B intend to change A's behavior? Did B actually have a plan to cause A to change, or at least, attempt to cause A to change? Both questions focus on the notion of causality. Unfortunately, the available data set is insufficient for addressing these questions in a rigorous fashion. There are, however, hints that nonhuman animals understand causal relationships (in addition to the evidence provided by Cheney and colleagues on baboon vocalizations) and use such understanding to guide their interactions with others. Premack (1986) and Premack and Dasser (1991), for example, describes experiments with the chimpanzee Sarah that were designed to assess relatively simple causal relationships. One task involved presenting Sarah with a series of three slides, where the first slide showed an object (e.g., a whole apple), the second showed the same object, but modified in a particular way (two apple halves), and the third showed a set of utensils (knife, glass of water) that may have been responsible for causing a change in the structure of the object in slide one to the structure in slide two. Sarah's job was to identify the utensil involved in causing a change in the object's properties. Sarah was highly successful (e.g., she picked the knife as the causal agent responsible for a change from whole to split apple), providing some evidence of causal reasoning.

Do nonhumans use their knowledge of causality in the inanimate world to work out problems of causality in the animate world? If so, do they understand that the rules governing causal relationships in the inanimate world can differ, quite fundamentally, from those governing relationships in the animate world? Thus, if a large heavy ball rolls toward a small stationary ball, the latter will move if contacted by the former. In contrast, when a dominant male rhesus monkey produces an open-mouth threat display, subordinates move away. Here, we clearly want to say that the dominant's threat *causes* subordinates to move off, but movement away is instantiated in the absence of direct contact. More concretely, causal interactions in the animate world often occur in the *absence* of physical contact. Causal interactions in the inanimate world often occur in the *presence* of physical contact. At present, no formal tests of this understanding have been carried out with nonhuman animals, but there are suggestive hints for nonhuman animals (see section 7.4) and slightly more encouraging data from human infants (Carey and Spelke 1994).

An enduring problem in the philosophy of mind and language is whether thought is possible without language (Jackendoff, in press). One would think that all of the exquisite data on preverbal human infants' cognitive abilities (see especially 7.4) ought to place this issue into exile. However, slippery philosophical discussions have managed to weasel out by arguing that even preverbal infants have a silent language of thought that, although primitive, guides their baby thoughts. Without entering that debate quite yet, one would think that even the slipperiest of philosophical moves couldn't handle data showing that nonhuman animals have complex thoughts. To start the ball rolling, let me mention some recent experiments on abstract relational concepts. Remember that even Herrnstein, who spent close to twenty years working on this problem, concluded (see opening quote in section 7.3) that nonhuman animals have concepts, but not abstract relational concepts; important supporting evidence was obtained from studies of pigeons that failed to understand the relational concept of inside versus outside (Herrnstein et al. 1989). I wish to report here on one study that challenges this view, for it will serve us well in the following section (see also Premack 1986).

Hauser, Kralik, and Botto (in prep) have recently explored the relational concepts of ON versus OFF in addition to CONNECTED versus DISCONNECTED, using a small New World monkey, the cotton-top tamarin (*Saguinus oedipus oedipus*) as a subject. However, rather than traditional operant techniques, a means-end relationship task developed by Willatts (1984, 1990) for human infants was employed. The task, schematically illustrated in Figure 7.38, involved presenting subjects with two physically separated pieces of cloth on a tray, the goal being to acquire a piece of food. In, for example, the ON versus OFF experiments, one piece of food was on the cloth, and one piece of food was off the cloth; pulling the cloth with food ON led to access, but not in the case of food OFF. Only one pull per session was permitted. In contrast to traditional tests of animal categorization, only a few training stimuli were presented. To assess the underlying concept, we then looked explicitly at the first novel probe trials, since responses to these conditions would be unpolluted by trial-and-error learning. Some of the stimuli for the ON-OFF and CONNECTEDNESS experiments are presented in Figure 7.38.

In general, results show that individuals quite readily move from the training set to novel probes. Thus subjects maintained close to 100% accuracy (i.e., always pulling the cloth with an accessible piece of food) even when such parameters as food distance, size, position, shape, and color were manipulated, in addition to alterations in the properties of the cloth, including cloth color and shape, gap size and shape, and the mechanical properties of the connection between cloths in the connectedness trials. This result provides evidence of abstract relational concepts in the absence of language, at least human language. These

ON versus OFF

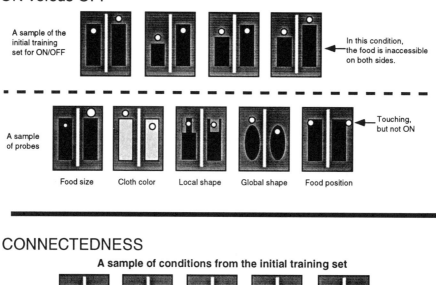

A sample of the initial training set for ON/OFF

In this condition, the food is inaccessible on both sides.

A sample of probes

Touching, but not ON

Food size | Cloth color | Local shape | Global shape | Food position

CONNECTEDNESS

A sample of conditions from the initial training set

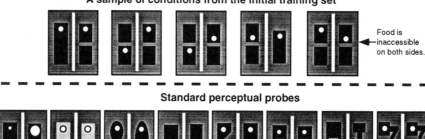

Food is inaccessible on both sides.

Standard perceptual probes

Food size | Cloth color | Cloth shape | Food position | Gap shape | Gap location | Gap location Cloth shape | Connection type

Mechanical efficiency probes

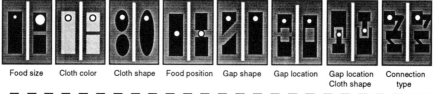

Sprinkled sand → ← Solid chain link

Chipped wood → ← Solid wood dowel

Figure 7.38
Stimuli used to test cotton-top tamarins on a mean-end relationship task involving the concepts of ON versus OFF (top row) and CONNECTED versus DISCONNECTED (bottom row). Note that, in the third presentation of the top row, the piece of food on the right is touching the cloth but is not on it, and is thus inaccessible; the same situation appears for connectedness in the second presentation from the right. For connectedness, this is an important condition, for it allows a distinction to be made between simply touching and a mechanical connection. The presentation in the fourth row shows a hole in the middle of the left cloth and a hole in the right cloth that creates two small gaps; pulling the right cloth is wrong (i.e., fails to bring food).

sorts of studies should set the stage for exploring the relationship between conceptual representations and communicative signals.

7.3.5 Cross-Modal Perception

The introduction to this book stated that the empirical data would be restricted to communication in the auditory and visual domains. Thus far, each of these modalities has been treated separately. However, a growing body of research in both neurobiology (Stein and Meredith 1993) and behavioral development (Kuhl and Meltzoff 1988; Kuhl, Williams, and Meltzoff 1991; Meltzoff 1993) has emphasized the importance of considering how information is integrated across sensory domains. This point is particularly relevant for mammalian species where auditory signals are often produced with a synchronously generated set of visual signals.[36] In this section, I summarize the results of research on intermodal transfer or exchange of information during communication in the visual and auditory domain. The cumulative data set is relatively small, but the significance of the results is large, and the potential for additional tests substantial.

Ventriloquists pull the proverbial wool over our eyes because we actually believe that, since the human performer's lips are still, the wooden dummy must be the one speaking. In more normal situations, however, the lips move during speech. In fact, deaf humans can become quite skilled at lip reading. What is perhaps most impressive about lip movement is that under certain conditions it appears to distort what is acoustically perceived. This sensory *fusion* between audition and vision is known as the *McGurk effect* (McGurk and MacDonald 1976). Here is how the phenomenon is instantiated in the perceiver. A speaker is filmed during the articulation of a syllable such as /ba/. The original sound track is supplanted by a syllable with a different point of articulation such as /ga/; the new syllable is dubbed onto the film record so that it is synchronized with the visual articulation. When subjects listen and watch this short clip, they characteristically report hearing a syllable that is articulatorily intermediate between /ga/ and /ba/, something like /da/. In fact, the phenomenon remains even when listeners are explicitly instructed to pay attention to what they hear rather than to what the speaker appears to be saying. In essence, it seems impossible for observers to turn off the McGurk effect. Several follow-up studies (Massaro 1987b) have since confirmed the findings of McGurk and MacDonald, and have

---

36. Other organisms, such as honeybees, also produce auditory and visual signals synchronously, but the output areas are different. In mammals, some visual and vocal signals emerge from the same anatomical area.

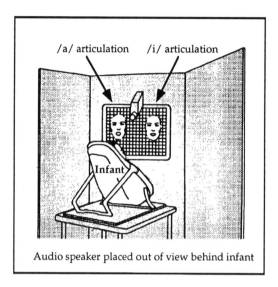

Figure 7.39
Setup used by Kuhl and Meltzoff (1982) to test for cross-modal perception in human infants.

even demonstrated that the phenomenon works when a female's voice is dubbed onto a film record of a male speaker, and vice versa (K. P. Green et al. 1991). Moreover, Fowler and Dekle (1991) have demonstrated that when a mismatch between spoken and written text is presented (e.g., listener hears /ba/ and sees the printed word /ga/), the McGurk effect is not obtained. In contrast, when untrained subjects[37] are required to feel in the mouth the articulation of /ba/ and hear /ga/, the McGurk effect is obtained. Based on these results, the authors argue that the McGurk effect is not derived from stored representations of relevant cues from multiple sensory modalities. Instead, the McGurk effect arises as a result of individuals directly perceiving the source (i.e., the articulation) of change in the environment (Browman and Goldstein 1986, 1989; Fowler 1989).

As Locke (1993) has put it, the McGurk effect is a beautifully powerful example of the "linguistic face" (p. 78). A crucial problem, therefore, is to determine when the linguistics of the face become meaningful to the developing child. Two approaches have been pursued, each tapping into a visual attention or looking-time paradigm. In one approach (Figure 7.39), a young infant (<4 months) is

37. As pointed out in chapter 1, the Tadoma system (Vivian 1966) is one in which touch is used by deaf-blind individuals to recover the linguistic information. The subjects tested by Fowler and Dekle were unfamiliar with this technique.

seated in a chair facing two video monitors. Directly behind the child is an audio speaker. The child is presented with two simultaneously running video clips, one showing a person articulating a vowel such as /i/ (as in "peep") and the other showing the same person articulating an /a/ (as in "pop"). Of critical importance for the visual presentation is two articulations that involve clearly differentiated gestures. Thus the /i/ vowel is produced by spreading/retraction of the lips, whereas the /a/ vowel is produced by lip opening in the vertical plane. Synchronized with the visual presentation is an auditory presentation from the speaker of one of the vowels (i.e., either /a/ or /i/ in the example described). The prediction is that if infants recognize the association between articulation and acoustic perception, they should look longer at the person articulating the sound played back. Kuhl and Meltzoff (1982, 1988) pioneered this approach, and their results support the predicted relationship: at the age of 18–20 weeks, infants look longer at the face articulating the vowel perceived acoustically. Moreover, control infants who were exposed to stimuli matched in duration and amplitude but devoid of formant structure[38] failed to exhibit attentional biases.[39] Thus the success of this cross-modal task depends critically on the spectral properties of the vowels presented.

To follow up on the claim that the infant's success at cross-modal matching depends upon the acoustic properties of speech per se, Kuhl, Williams, and Meltzoff (1991) conducted a methodologically comparable experiment using nonspeech stimuli. Following the logic underlying tests of categorical perception, these experiments were designed to assess which features of the auditory signal are necessary and sufficient for eliciting a match between sound and face. The first phase of the experiment involved testing adult subjects under three conditions:

1. *Auditory only:* Subjects were first presented with a tape recorded series of /i/ or /a/ vowels spoken by a woman, and were subsequently asked to match what they heard with a pure-tone frequency-generating device (i.e., they turned a dial to produce a matching frequency).

2. *Imagined only:* Subjects were shown a card with a printed word and asked to imagine the sound of the vowels. Subsequently, they were asked to match the imagined sound to a generated pure-tone frequency as in condition 1.

---

38. Recall that formants are critical with regard to the perceptual identification of vowels.

39. This result does not rule out the possibility that infants might be able to match auditory-visual events in nonspeech domains. In fact, some extremely interesting experiments by Spelke and her colleagues (Spelke 1979; Spelke, Born, and Chu 1983) have demonstrated that young infants are capable of understanding the synchronization between an apparently bouncing ball and the sound that it is likely to make upon contact with a surface (i.e., as opposed to when it is in the air).

3. *Visual only:* Subjects were presented with two side-by-side video clips, showing a woman articulating the vowel /a/ in one and the vowel /i/ in the other; the audio track was silent. Subsequently, they heard either a low- or high-frequency pure tone synchronized to the articulations in the video clips. The task was to identify which articulation best matched the tone heard.

For all three conditions, adults consistently matched high-frequency pure tones to /i/ and low-frequency pure tones to /a/.[40] In contrast to adults, infants tested in condition 3 did not appear to recognize a correspondence between pure tones and articulation. In fact, regardless of the tone played, infants showed a strong bias to look longer at the articulating /a/-face as opposed to the /i/-face.

A second approach to uncovering the developmental origins of the cross-modal nature of speech comes from research pioneered by Dodd (1979). In this paradigm, 10–16-week-old infants were presented with video clips of a person speaking. The only parameter varied in each video presentation was the degree to which the audio track was decoupled (i.e., out of synchrony) from the visual articulation. Results indicated that infants spent less time looking at presentations where the audio track was approximately 400 msec out of synchrony with the visual track, and interestingly, several infants actually looked away from the presentation. Like adults viewing poorly dubbed foreign movies, infants also appear sensitive to the synchrony between voice and face.

A more complicated version of Dodd's testing procedure has recently been developed by Pickens and colleagues (1994) to assess developmental changes in cross-modal perception in pre- and full-term infants ranging in age from three to seven months old. Here, rather than presenting a single face, two side-by-side faces were presented together with one audio track; the audio track was synchronized to one of the articulating faces. The articulation and corresponding sound was of a melodic speech passage such as "Jingle Bells." Results showed that 3- and 7-month-old full-term infants preferentially looked at the synchronized film and sound track, but 5-month-olds did not. Preterms failed to show any evidence of face-sound matching. Although it is difficult to explain the failure by 5-month-old full-terms to preferentially fixate on auditory-visual correspondences, these data underscore the importance of studying developmental problems across multiple periods (see also chapter 4 for comments on infant face recognition). The result of such sampling may be a more complete understanding of how different

40. Kuhl, Williams, and Meltzoff (1991) actually performed several other manipulations of nonspeech stimuli. Results from these experiments complicate some of the interpretations for adult cross-modal perception but are not entirely relevant to the current discussion.

components of the brain attend to particular features of perceptually salient events.

Though no study to date has attempted to look at cross-modal perception in nonhuman animals, I don't see significant methodological barriers. For example, I could imagine testing nonhuman primates on two versions of the McGurk effect, one with traditional human speech sounds and one with species-typical vocalizations. The first version would require training subjects in a standard operant procedure, including three response buttons. When a /ba/ is heard, subjects would press a green button. When a /ga/ is heard, subjects would press a red button. Last, when a /da/ is heard, subjects would press a white button. Once subjects achieved a high level of accuracy, test trials would begin. The crucial McGurk effect probe would involve the presentation of a person articulating /ba/ with a /ga/ sound track. If the McGurk effect is operative, subjects should press the white button. If not, they should either press the green button because the articulation dominates, or they should press the red button because the auditory signal dominates. The same kind of design could be used to test for cross-modal effects with the species-typical vocal repertoire and articulation, in addition to testing preverbal human infants.

Another interesting test population for the McGurk effect would be patients with different types of language aphasia. Because the phenomenon involves an integration of articulatory and perceptual cues, individuals with particular forms of brain damage might fail on the McGurk test (i.e., fail to experience sensory integration of the articulatory and auditory cues). Thinking, a priori, about which damaged areas are most likely to lead to failure depends on one's view of speech perception. Thus, in the direct realist or motor theory perspectives, perception involves, at some level, extraction of the articulation from the speech stream. If this occurs, one would expect the locus of control to be the primary area involved in perceiving the McGurk effect.

7.4 Mindful Manipulation of Information

The Earth is God's pinball machine and each quake, tidal wave, flash flood and volcanic eruption is the result of a TILT that occurs when God, cheating, tries to win a free game. (Robbins 1976, 243)

Many of the observations and experiments on categorization described in the previous section assess some aspects of human and nonhuman animal causal reasoning. Thus, in Cheney and Seyfarth's (1985a) work, experiments were

designed to reveal whether vervet monkeys understand the connection between cattle mooing and danger from the Masai. In Kuhl and Meltzoff's (1982, 1988) work on cross-modal perception, experiments were designed to assess whether infants recognize the connection between articulation and acoustic structure. From a philosophical perspective, one may wish to contest the claim that such cases represent evidence of causal reasoning, arguing instead that they merely represent an understanding of salient associations in the world. I think, however, that the cases described here, and in greater detail later, provide the organism with representations that set the stage for developing (ontogenetically and evolutionarily) a theory about causal relationships in the world (see Premack and Premack 1994 for a fascinating discussion of relevant issues). This section therefore begins by describing research on causal relationships in the social world—in particular, the intentional bases of communication, especially as they relate to deceptive signaling. Here, nonhuman animals appear to actively falsify or withhold information in order to cause other individuals to behave in a way that reduces their fitness. Although human deception may be one of the most exquisite expressions of our ability to causally manipulate belief systems (Ekman 1985), a number of studies of nonhuman animals have suggested that from a purely functional perspective, individuals have evolved communicative actions that have been designed to causally mislead others. In this sense, such actions are functionally deceptive (Lloyd 1984), providing fitness benefits to the deceiver and fitness costs to the deceived (see chapter 6). I will argue that, although our nonhuman animal relatives may only deceive each other in a functional sense,[41] such actions may have provided the requisite foundation from which a theory of other minds emerged. Following a discussion of the nonhuman animal data, I then review studies of how the human child's intentional system develops and becomes integrated into its communicative repertoire. This piece of the discussion will explore how children's early utterances inform us about the nature of their conceptual representations. I conclude by discussing observations and experiments on human adults, designed to reveal whether lies can be uncovered by means of the face, the voice, or both.

7.4.1   Functional Deception in Nonhuman Animals: Theoretical Issues

An organism deceives another when it misrepresents something, and, on the other hand, a deceived organism may not "know" reality, not being perfectly tuned, and be unable to distinguish bogus information from the real thing. Conscious thought permits human beings the luxury or burden of contemplating whether the factual truth can be known at all, but the distinction or difference may be quickly resolved in nature, for it can be tough on functional nihilists when the proof of the tiger is in the eating.   (Lloyd 1984, 48)

"I swear to tell the truth, the whole truth, and nothing but the truth." This bit of legalese conveys two interesting pieces of information relevant to the following discussion. First, by swearing to tell the truth, the person on the witness stand relays to the judge that he or she *intends* to be honest. That is, his or her responses to the lawyer's questioning will not be reflexive or unthoughtful, but will be *about* currently held beliefs. Second, by swearing to tell the truth, the witness claims that the answers provided will *accurately* represent his or her own observations of the situation. If the witness lies, then he or she has intentionally distorted the truth—has been deceptive by actively falsifying information. The legal trick, then, is to establish a course of questioning that will uncover the truth and discard the lies. To do so requires techniques that "catch" inconsistencies in statements and behavior, both present and past. Although there are currently no foolproof methods for telling liars apart from honest citizens, most of the available methods

41. See section 7.2 for more complete discussion of functional reference.

attempt to provide physiological, psychological, and behavioral measures that accompany (i.e., are correlated with) honesty, and then use these as a standard for comparison with conditions of potential dishonesty. Thus psychologists such as Ekman and Friesen (1978) have developed a quantitative scoring system for establishing the kinds of facial expressions that accompany deceptive acts, including an assessment of honest versus dishonest smiles (Ekman, Davidson, and Friesen 1990; Frank, Ekman, and Friesen 1993), and the detection of liars (Ekman 1985; Ekman and Friesen 1975; Ekman and O'Sullivan 1991; Ekman, O'Sullivan, et al. 1991). And in the context of criminal investigations, various physiological measures such as galvanic skin response, heart rate, and so forth are often called upon. In summary, therefore, there are several approaches to assessing deception in humans, and in general, the legal system assures us (!) that those who deceive are caught.

How would we know if a nonhuman animal was being deceptive? Why would nonhumans be deceptive? The second question can be answered quite easily. Like humans, nonhuman animals commonly engage in interactions that are competitive (see chapter 6), involving losers and winners. To increase the probability of winning, individuals have evolved strategies to outwit their opponents. Selection will clearly favor the individuals with the best strategies, and it is not difficult to imagine that dishonesty could pay off under certain conditions. Thus, if two individuals are competing over access to a valued food resource, it might benefit one individual to signal a level of fighting strength beyond what he can actually defend—the barroom wimp threatening the Schwarzenegger-like thug. The sense of dishonesty portrayed here is simply that there is a mismatch between the signal's typical message (i.e., "I can beat you up") and what the signaler is actually capable of doing—in the case described, the signaler would run away if the opponent pressed on with threats. An important component of the previous statement is "typical message." Although I pointed out in section 7.2 that current understanding of the meaning of nonhuman animal signals is limited, ethologists and behavioral ecologists have provided several comprehensive accounts of the frequency with which signals are given in particular socioecological contexts (e.g., Caryl 1979; Ryan 1985; Stokes 1962; Struhsaker 1967; von Frisch 1950). These kinds of data provide us with the requisite information for stating what the "typical message" is likely to be, as when a call with a specific acoustic structure is *only* heard in the context of predator encounters. Dishonest or deceptive signaling would occur when individuals take advantage of this association and produce the signal in a completely different context (see definition).

Animals can be deceptive in yet another way. They can withhold information about valuable resources or danger and thereby increase their fitness relative to others in the population. For example, in several avian and mammalian species, individuals call upon discovering food (reviewed in Hauser and Marler 1993a) and, as a result, recruit group members to the food source (see sections 6.3 and 7.2). Consequently, food-associated calls increase competition over food. But in some of these species, the probability of calling in the context of food is less than 100%, suggesting the possibility that individuals sometimes suppress their calls, thereby reducing feeding competition. Thus withholding information represents a second form of deception—one that is likely to be more prevalent than active falsification because it is relatively more difficult to detect cheaters.

How would one tell whether an individual failed to signal about the presence of food because of selfishness or because he or she made a mistake about the amount of food available to share? This question brings us directly back to the problem posed earlier: how can we recognize deception in nonhuman animals? To address this question, I turn now to a selective review of the available literature on deception in nonhuman animals. Because I wish to put off the possibility of *intentional deception* (i.e., deception based upon manipulation of belief states), keep the following definition of *functional deception* in mind when reading about the empirical studies described in subsequent sections:

FUNCTIONAL DECEPTION: For an animal's action to be considered functionally deceptive, the following conditions must hold: (1) There must exist a context C in which individuals typically (high probability) produce a signal S which causes other group members to respond with behavior B. (2) Under certain conditions, individuals produce S in a different context C′ which causes other group members to respond with behavior B and consequently enables the signaler to experience a relative increase in fitness (*active falsification*). (3) Under certain conditions, individuals fail to produce S in context C, consequently enabling them to experience a relative increase in fitness (*withholding information*). (4) The fitness increase comes from the fact that by being functionally deceptive, deceivers gain some benefit (e.g., access to resources) whereas those who are deceived obtain some cost or fail to gain a benefit.

In part 1, I envisage something like an alarm call (signal S) that is given when a predator has been spotted (context C). For part 2, an alarm call (signal S) is produced, but in the absence of a predator (context C′). Listeners nonetheless respond as if a predator had been detected by the alarm caller. For part 3, a predator is seen, but the subject fails to give an alarm call (signal S). Part 4 represents the fitness consequences of cases 2 or 3 in terms of costs and benefits to signalers and perceivers. Let us now turn to some concrete examples.

7.4.2   Empirical Evidence of Functional Deception in Nonhuman Animals

For deception to evolve, there has to be a certain level of behavioral plasticity in the system such that individuals do more than simply respond to a particular situation in a stereotyped fashion. Rather, they must coordinate their actions with subtle nuances in the socioecological conditions. The question is, then, Do individuals alter their behavior as a function of those who are around, those who are watching them, and the relative benefits of the resource at stake? If so, can such behavioral responses be explained by simple rules (e.g., "If alone, don't call. If with at least one other group member, then call"), or are more complex cognitive processes at play, suggesting perhaps, intentional manipulation of beliefs and desires? Some answers to these questions have recently been obtained for chickens and vervet monkeys by conducting experiments that explicitly manipulate the audience of the potential signaler.

In domestic chickens, males produce distinctive vocalizations upon discovering food, and as illustrated in Figure 7.40, call rate is positively correlated with food preference. Moreover, hens are more likely to approach males calling at high rates than males calling at low rates. As described in section 7.2, there is evidence that such calls are functionally referential, providing information about food, in addition to information about the caller's affective state. For the purposes of our

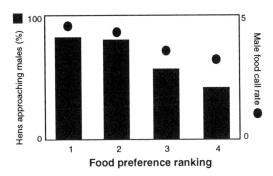

Figure 7.40
The relationship between male food call rate in the domestic chicken (right *y*-axis), food-preference (*x*-axis), and the proportion of females (i.e., hens) approaching the male (left *y*-axis) (redrawn from Marler, Dufty, and Pickert 1986a).

current discussion, what makes the chicken system interesting is that the production of food calls, as well as alarm calls, depends upon the composition of the audience. Consequently, individuals sometimes encounter food or a predator and remain silent. Is there any evidence that such silent discoverers are withholding information from particular group members, and if so, how would human observers find out? Let us first review some of the evidence for food calls, and then compare these results with those obtained for alarm calls.[42]

To address the issue of withholding information, Marler and colleagues (Marler, Dufty, and Pickert 1986a, 1986b; Marler, Karakashian, and Gyger 1991) embarked on a series of experiments designed to determine the precise factors eliciting food call production. In general, these experiments focused on the interaction between food quality and the composition of the social group, or audience. In the first set of experiments, cocks were presented with food under four social conditions set up in an adjacent cage: (1) familiar female, (2) strange female, (3) empty cage, and (4) adult male. As Figure 7.41 clearly reveals, target males called on almost every trial when a female, familiar or strange, was present in the adjacent cage. In striking contrast, males almost never called when another male was present. What makes the absence of calling in the presence of males all the more impressive is the fact that during trials with no audience, target males frequently called

---

42. Recall that the definitional conditions for deception require an unambiguous understanding of the signal. In human terms, the boy who cries "wolf" is detected as a liar because the signal "wolf" has an explicit meaning. If, after crying "wolf" repeatedly, the boy cried "banana," one would not necessarily accuse him—at least outright—of lying.

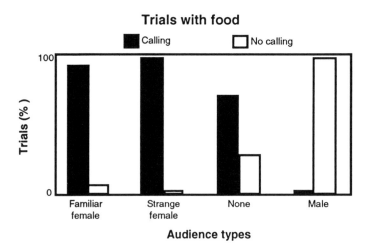

Figure 7.41
Audience effects on food calling in the domestic chicken. Results show the relationship between proportion of trials in which cocks call and the presence or absence of particular individuals in an adjacent cage (redrawn from Marler, Karakashian, and Gyger 1991).

during food presentations. Thus, although food presentation commonly elicits calling, males appear to suppress their vocal behavior in the presence of same-sex competitors.

In contrast to trials with food, males were generally less vocal during presentations of nonfood items. But again, the social composition of the audience played a significant role in mediating call production (Figure 4.42). Although target males rarely called with a male audience or an empty cage, they called on approximately 50% of trials with a familiar female and on approximately 25% of trials with a strange female. If we assume that the acoustic structures of calls given to food and nonfood items are similar,[43] then these data suggest that (1) males may use food calls in the absence of food to recruit potentially receptive females and (2) in the absence of an appropriate target audience, nonfood items fail to elicit such calls.

Although these data suggest that chickens withhold information about food under particular social conditions, there are, potentially, alternative explanations (W. J. Smith 1991). Specifically, some have argued that, because of the associa-

43. Quantitative acoustic analyses of calls given to food and nonfood items have not yet been conducted. As in Munn's (1986a,b) studies of alarm calling in birds (chapter 6), it will be necessary to show that the structure of calls given in food and nonfood contexts is the same, and that upon hearing such calls, hens respond similarly.

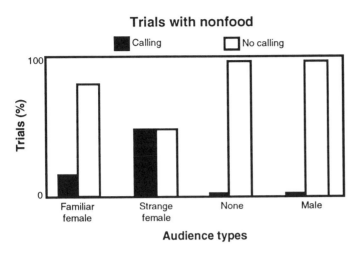

Figure 7.42
Audience effects on calling in response to nonfood items in the domestic chicken. Results show the relationship between proportion of trials in which cocks call and the presence or absence of particular individuals in an adjacent cage (redrawn from Marler, Karakashian, and Gyger 1991).

tion between calling in the context of food and approach by the hen, a more accurate functional description of the call is as a courtship signal, rather than a food signal. A second issue concerns the specificity of the audience effect, an issue raised earlier for the alarm call system. To disentangle these issues and resolve some of the problems, Evans and Marler (1994) designed a clever experiment, using the power of operant techniques. The first stage of the experiment involved training males to peck at a panel for food (see Figure 7.43); the males were alone. Subsequently, they were placed in the same operant setup, but with either an empty adjacent cage or a cage housing an unfamiliar female. If the function of the male's call is to increase mating opportunities, then the dispensation of food should have little effect on his call rate. If the presence of a hen increases general arousal, then we would expect to find corresponding changes in the operant task (e.g., faster and more accurate pressing).

Figure 7.44 summarizes the crucial results from the experiment. Prior to food dispensation (i.e., turning the pellet dispenser and its signal light on and making food available), the rate of food call emission is low, but the wing extension display—a significant component of the courtship display—is high, especially within the first 60 seconds of the trial. When food first becomes available, the rate of food calling increases dramatically, then drops off quite sharply. In striking

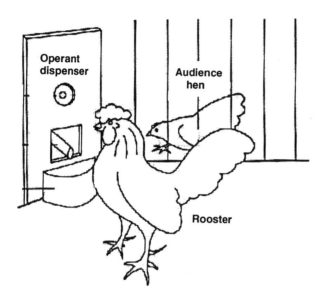

Figure 7.43
Schematic illustration of experimental design used by Evans and Marler (1994) to assess the relationship between food availability, calling, and courtship.

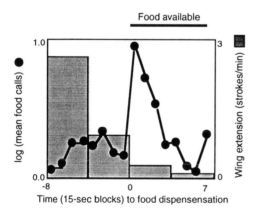

Figure 7.44
Changes in the rate of food call production (circles) and wing extension displays (speckled bars) by male chickens prior to and following food dispensation. The *x*-axis represents time to food dispensation (15-second blocks of time); at zero, the pellet dispenser was turned on and the males were able to gain access to food by pecking (redrawn from Evans and Marler 1994).

contrast, courtship is virtually absent during the period of food availability, and not because males are incapable of calling and displaying at the same time (i.e., the two displays are not mutually exclusive). Moreover, although males tested alone and in the presence of a hen showed similar food-calling behavior, males tested alone never produced the wing extension display or any other component of the courtship display.

Returning to the questions raised earlier, these experiments show a clear dissociation between food calling and courtship. Consequently, when males call to food, the call seems to be *about* food—it is, minimally, functionally referential. When hens hear such calls, they are attracted and approach, presumably because this action represents an opportunity to gain access to food. Because the call conveys information about the availability of food, the failure to call under particular social conditions (i.e., the presence of other males) certainly represents a case of withholding information in the purely functional sense.[44] Whether it represents withholding of information in the more *intentional* sense requires further experimentation, thought, and patience on the part of the reader. Before turning to the possibility of *intentional deception,* let us first look at two more cases of withholding information and one case of active falsification from a functional perspective.

Is the chicken's alarm-call system, like its food-call system, affected by the presence of an audience? Yes, but the details differ. Based on experiments involving manipulations of a live audience, Marler and colleagues (Gyger, Karakashian, and Marler 1986; Karakashian, Gyger, and Marler 1988) demonstrated that males produced more aerial alarm calls to a model hawk when (1) with a mate, nonmate, or male than when alone, and (2) with a hen than when alone or with a bobwhite quail. Rates of alarm calling were not, however, influenced by the sexual composition of the audience; this result contrasts with the conditions eliciting food calls.

To more precisely assess the stimulus characteristics of the audience with regard to alarm-call potentiation, Evans and Marler turned once again to the power of computer video technology. In their first experiment, they replicated the results obtained with a live audience, showing that a videotaped sequence of a hen was more effective at eliciting alarm calls than a video sequence of a bobwhite quail or empty cage. They then showed that alarm calls could be

---

44. The costs and benefits of this system have not yet been worked out in detail, and this issue is of course an important piece of the functional perspective on deception. For example, although males who find food and remain silent in the presence of other males benefit in terms of resource competition, the costs associated with detection by competitors (i.e., being caught with food but without having called) have not been assessed.

potentiated by the image of a hen without sound, by the sound of a hen without her image, and by the combination of image and sound. All three of these conditions were far more effective than an empty cage.

Individuals produced *more* alarm calls to conspecific audiences than to bobwhite quail audiences, but the presence of bobwhites was more effective than an empty cage. To assess the specificity of the audience effect more carefully, Marler and colleagues set up two experiments. In the first, the physical size of the audience was examined by comparing aerial alarm call rates by males when alone or in the presence of a mate, young chicks, or a female quail. Confirming previous findings, mates consistently elicited higher alarm call rates (Figure 7.45). More importantly, males called at higher rates in the presence of chicks than when they were alone or in the presence of a female quail. Thus size is not the only perceptual feature used by males to assess whether or not an alarm call should be produced.

In a second experiment, only adult domestic hens were used in the audience setting, but the strain was varied. This is a fascinating manipulation from the perspective of featural salience because the three selected strains are strikingly different. The Silky, for example, has long hairlike feathers and is black. Results (Figure 7.46) clearly indicate that all three strains are effective in potentiating alarm calls; there were no statistically significant differences between strains, though all three were significantly different from the empty cage.

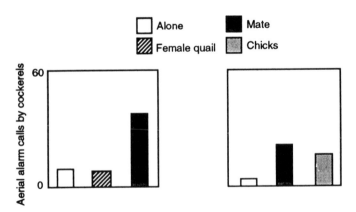

Figure 7.45
Rate of aerial alarm calls by cockerels when alone (white bar) or in the presence of a mate (densely speckled bar), chicks (loosely speckled bar), or a female quail (striped bar) (redrawn from Evans and Marler 1992).

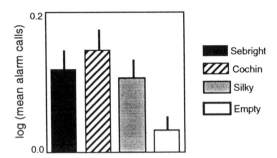

Figure 7.46
The effect of chicken strain on male aerial alarm calls. The Sebright (black bar) is the same strain as the test male. The y-axis shows the mean logarithm of alarm calls; standard deviations presented (redrawn from Evans and Marler 1992).

In summary, results from the domestic chicken indicate that the particular properties of an audience are relevant to the production of food and alarm calls. However, whereas the sex of the audience plays a significant role in the food-call system, it is less important for alarm calls, particularly aerial alarms.

A slightly different case of withholding information comes from rhesus monkeys. As discussed in chapter 6, rhesus on the island of Cayo Santiago, Puerto Rico, often vocalize when they either discover or eat food (Hauser and Marler 1993a). The vocalizations produced are acoustically variable (Figure 7.47), due in part to the eliciting contexts. Specifically, warbles, harmonic arches, and chirps are only given in the context of food—typically, the discovery of high-quality food (e.g., coconut). In contrast, coos and grunts are given in both food and nonfood contexts (e.g., mother-infant separation, dominant-subordinate interactions, group movement), and in general the quality of food eliciting these calls is lower than what has been observed for warbles, harmonic arches, and chirps. In addition to these contextual effects, observations reveal that adult females produce food-associated calls at higher rates than do adult males, and females with large matrilines call at higher rates than females with small matrilines.

Hauser and Marler used observational data to test for the effects of hunger level and food quality on calling behavior. As one measure of relative hunger, analyses of changes in food consumption were calculated during the day for a large sample of adult males and adult females. Results showed that the rate of food-call production (all call types pooled, but excluding coos) was highest when subjects were likely to be most hungry and declined to a relatively low rate when they were relatively satiated. This pattern suggests that call rate is correlated

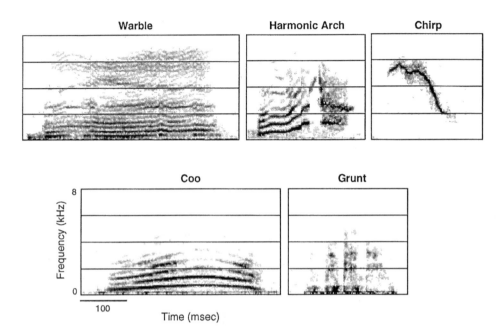

Figure 7.47
The five main call types used by rhesus monkeys during the discovery or consumption of food.
Frequency (kilohertz) plotted on the y-axis, time on the x-axis (milliseconds).

with hunger level. In contrast, call rate was not statistically associated with the type of food discovered, though call type was. These observational data suggest that call rate potentially encodes information about hunger level, whereas call type potentially encodes information about food quality.

The natural observations summarized in the preceding paragraphs and discussed in greater detail in chapter 6 suggest that rhesus monkeys do not always call when they discover food. To make a more precise determination of the factors that guide call production, Hauser (1992c, Hauser and Marler 1993b) conducted field experiments with the same population of rhesus studied under natural conditions. Three variables were directly assessed: the food discoverer's gender and sex, and the quality of the food discovered. The procedure for the experiment involved, first, locating a solitary individual (i.e., visually out of sight from group members)—the discoverer—and second, presenting this individual with either 15 pieces of chow or 15 pieces of coconut. Relative to chow, coconut is rare and of higher quality. The potential contribution of hunger level was assessed by testing some individuals early in the morning (before 7:00 A.M.) and others late in the afternoon (2:00–3:00 P.M.). On Cayo Santiago, chow is placed in the dispensers at approximately 7:30 A.M., and prior to this time, there is little foraging. Thus, because rhesus do not forage at night, individuals tested in the morning were considered maximally hungry. In contrast, by midafternoon individuals have completed approximately 85% of the foraging for the day and were thus considered relatively satiated.

When discoverers first sighted the food, they typically spent from a few seconds to a few minutes scanning their immediate environment. They then either called, approached silently, and fed, or approached, fed on a few pieces of food, and then called. Discoverers called in 45% of trials, and females called in significantly more trials than males. When discoverers called, they produced warbles, harmonic arches, and chirps in response to coconut, but never in response to chow; chow elicited coos and grunts. In addition, discoverers called at higher rates in morning tests with coconut than in afternoon tests with chow. In parallel with natural observations, then, these results indicate that hunger level covaries with call rate, whereas food quality covaries with the spectrotemporal structure of the call (i.e., call type). Moreover, although females call more often than males, individuals sometimes find food and simply fail to call. As with domestic chickens, rhesus are clearly withholding information from a functional perspective. What is missing from this story, and from most accounts of functional deception, is a description of both the specific costs and benefits to deceiver and deceived.

Several observations help us quantify the economics of functional deception in rhesus monkeys. As pointed out earlier, discoverers never beelined to the food source upon detecting it, but rather scanned the environment first. Although it is not possible to identify what or whom they were looking for, such scanning behavior at least suggests that discoverers assessed the situation before they approached the food source. A clue to the target of such scanning comes from observations following food discovery. Depending upon the density of animals nearby and whether or not the discoverer called, other group members quickly approached the food source and discoverer on 90% of all food trials. Upon detection, some discoverers were aggressively attacked and injured, whereas other discoverers either stayed and fed with group members or were peacefully supplanted by dominants but allowed to keep the food they had collected. Only one factor appeared to determine whether discoverers received aggression or escaped unharmed: whether or not they called when they found food (Figure 7.48). Independently of dominance rank, vocal discoverers received significantly less aggression than silent discoverers, and for females, those who called ate more food than those who were silent. In the few trials where discoverers were silent and

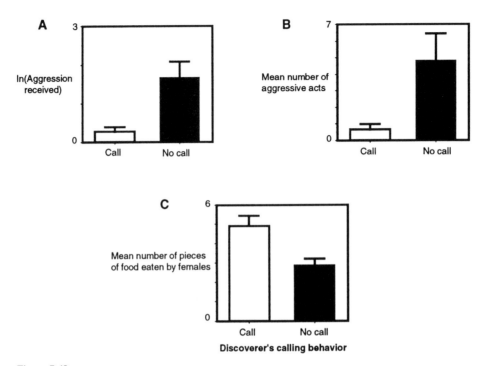

Figure 7.48
The consequences for discoverers of producing or failing to produce food-associated calls in rhesus monkeys. (A) The duration of aggressive events received as a function of calling behavior (white bars, call; black bars, no call); (B) the number of severe aggressive acts received as a function of calling behavior; (C) the amount of food eaten by female discoverers as a function calling behavior (redrawn from Hauser 1992c).

never detected, these individuals consumed more food than in any other trial. Last, individuals who failed to call under experimental conditions were observed calling under natural conditions. This observation suggests that the food-call system is relatively plastic and does not involve a population of callers and noncallers. It further suggests that individuals who withhold information about food are punished, in much the same way that retaliatory behavior has been observed in other organisms, including honeybees, birds, horses, and a variety of primates (recently reviewed by Clutton-Brock and Parker 1995).

Observations of targeted aggression suggest that prior to food consumption, discoverers may have been scanning for both allies and enemies. Depending upon the results of their scanning, individuals either call or remain silent. Are silent discoverers withholding information? Returning to the criteria laid out earlier,

(1) rhesus monkeys fail to signal in a context that commonly elicits a signal, and (2) as a result of their failure to signal, they sometimes experience an increase in fitness (i.e., those who are silent and remain undetected obtain more food than those who call and are detected). This sequence clearly shows some level of assessment, as opposed to a purely reflexive response. But does it imply that those who fail to signal are intentionally suppressing such information from competitors and that those who signal are intentionally communicating to others about the presence of food? Do rhesus monkeys actually understand the beliefs of others and the sequence of events that are likely to transpire when a call is either produced or withheld? I do not believe that the data presented thus far warrant such interpretations. But before we cast our votes, let us examine the notion of skepticism—a critical ingredient in any story on deception.

The notion of an evolutionary arms race discussed in chapter 6 tells us that in most competitive interactions there will be oscillations in the successes of each competitive strategy. Such oscillations are due in part to differential selection regimes operating to fine-tune each strategy under a set of constraints. In the context of deception, selection will often operate to make signalers ultimately sneaky and perceivers ultimately skeptical. What evidence is there for skepticism? In less loaded terminology, what evidence is there that individuals ignore, fail to respond to, or respond counter to the information content of the signal conveyed? Two experiments, one on honeybees and one on vervet monkeys, shed some light on this problem.

In section 7.2, I reviewed some of the evidence underlying the claim that honeybees use their dance to convey extraordinarily precise information about the location and quality of food in the environment. Gould (1990) wondered whether honeybees might ever treat such information skeptically, doubting the accuracy of the dancer's assessment of location or quality. Figure 7.49 provides a schematic illustration of the setup and experimental procedure.

The first experimental step involved taking a group of foragers away from their hive and allowing them to move from a station on land to a nearby boat on land; the boat contained high-quality pollen substitute. Gradually, Gould moved the boat farther and farther from the departure station, causing the foragers to travel greater distances. During these forays, the foragers were never allowed to return to the hive. The final position of the boat was at the center of a lake. Once the foragers had reliably visited the boat on the lake and returned from it, Gould allowed them to return to their hive and dance. When these foragers danced, however, virtually no recruits were observed at the boat. Thus, although the

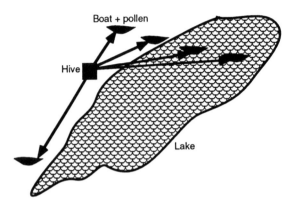

Figure 7.49
Experimental design for experiments on honeybee skepticism (redrawn from Gould 1990).

trained foragers were conveying information about the location of high-quality food, hive members appeared to ignore the information conveyed. Gould's working hypothesis at this point was that, although the foragers were saying, "There's great food out in the middle of the lake," hive members were responding, "I don't believe it. There has never been, and will likely never be, food out in the middle of the lake. I know, I fly around that area all the time." Thus skepticism is possible in honeybees because they have the ability to compare the information conveyed in a dance with their spatial map or representation of the environment (for reviews, see Dyer 1991; Gallistel 1992; Gould 1990; Gould and Gould 1988; Seeley 1989; for evidence against the notion of a cognitive map, see experiments by Kirchner and Braun 1994). In fact, counterstrategies to deceptive acts must generally rely on the ability to compare the current state of affairs with some standard or prototypical state. Honeybees appear to have this ability, at least at some level.

Numerous anecdotal observations of nonhuman primates have revealed that under certain conditions, individuals utter alarm calls in the absence of predators or produce intergroup encounter calls in the absence of another group. These calls have been interpreted as putative cases of deception (Byrne and Whiten 1990, 1991). Recall that a crucial feature of an evolutionary approach to deception is the cost involved in being detected as a cheater (Clutton-Brock and Parker 1995). For the cost to be significant, individuals must have the capability to treat information skeptically.

Cheney and Seyfarth's (1988, 1990, 1992) research on vervet monkeys repre-sents the only experimental attempt to address the problem of skepticism.[45] Inter-estingly, I have already described the data set, but disguised it in a somewhat different form. In discussing the issue of meaning in section 7.3, I presented the results from Cheney and Seyfarth's habituation-dishabituation experiments. These experiments were designed to provide evidence of meaning in the strong sense. Recall that in these experiments, one individual's calls were played back repeatedly until the intended perceiver habituated (i.e., showed a significant drop in response). Results showed that during the habituation phase, responsiveness decreased. In the dishabituation or test phase, responsiveness stayed low in response to hearing a call type with the same general meaning. Responsive-ness increased, however, in response to either a different individual's call or a call type with a different meaning. For example, following habituation to Leslie's intergroup wrr, individuals remained in a state of habituation if Leslie's intergroup chutter was played, but dishabituated if Borgia's chutter was played. Now rerun this description in your mind, and instead of thinking about perception and call meaning, think about why, under the conditions tested, individuals sometimes habituated and sometimes dishabituated. Cheney and Seyfarth's intriguing hypothesis is that habituation can be interpreted as a form of skepticism on the part of the perceiver. Thus, when Leslie continues to *report* via wrrs and chutters that a neighboring group is threatening but no group is in sight, she is unreliable. But simply because Leslie is unreliable about the presence of neighboring groups does not mean that Borgia, when she calls, is also unreliable. We are now veering extremely close to problems associated with the attribution of mental states. To end this discussion on a functional note, we can certainly conclude from Cheney and Seyfarth's work that in vervet monkeys individuals are treated as if they are unreliable if calls typically associated with particular environmental features are produced in the absence of such features. But can we conclude that listen-ers sometimes think that callers intend to mislead them into believing that a neighboring group is present when it is not? We have now entered the do-main of folk psychological attributions and, in particular, the topic of intentional deception.

45. Some of the cut-and-paste mate-choice studies, involving manipulations of sexually selected traits, in addition to research on badges of status in birds, might be considered as studies in skepticism. However, they do not directly attack the problem of a signal's information content or its perceptual salience.

7.4.3 Empirical Evidence of Intentional Deception in Nonhuman Animals

The data on rhesus monkeys provided evidence of withholding information from a functional perspective. Do they provide evidence of intentional deception, of individuals making decisions about the appropriate contexts for signal production as opposed to suppression? Although Hauser (1992c) originally described the rhesus data with terms that implied some level of conscious decision making, several empirical and conceptual issues have now forced an alternative, less sophisticated interpretation. One empirical challenge comes from an experiment conducted to assess the endocrine basis of vocal communication in rhesus monkeys (Bercovitch, Hauser, and Jones 1995). These experiments, described in the following paragraphs, suggest that physiological changes may largely dictate the conditions for call production and suppression, and that an intentional decision need not enter into the equation.

Studies of nonhuman primates, especially baboons and squirrel monkeys, have indicated that changes in cortisol levels are closely associated with changes in stress (Sapolsky 1992). An ecologically stressful context for most nonhuman primates occurs when predators are encountered. In such contexts, individuals commonly produce alarm calls. As a result, one might expect to find changes in cortisol levels during alarm-call production. To assess the relationship between cortisol level and alarm-calling behavior, experiments were conducted with captive female rhesus monkeys. In one group of females, cortisol levels were experimentally lowered through administration of metyrapone, a pharmacological drug that blocks the formation of cortisol in the blood; a second group of females received a vehicle. All females were exposed to three potentially threatening situations: (1) approach by personnel carrying a net; approach by personnel carrying the test female's (2) own infant or (3) an unrelated age-matched infant.

Results indicated that metyrapone-treated females, who were fully alert, were nonetheless silent when threatened by personnel carrying a net, whereas 50% of vehicle-treated females produced alarm calls. When threatened by personnel carrying an infant, vehicle-treated females produced a higher rate of alarm calls, as well as structurally more intense alarm calls, than metyrapone-treated females.

Cortisol levels in metyrapone-treated females were 20–30% lower than in vehicle-treated females. These results suggest, therefore, that below a particular physiological threshold, individuals fail to call. Returning to the problem of deception, rhesus monkeys who discover food and remain silent may have decided to intentionally withhold information from their competitors. Alternatively, silent discoverers may be under greater stress than vocal discoverers, and the failure to call may reflect a subthreshold cortisol level. At present, we have no way of distinguishing between these alternatives.

If the data on withholding information are problematic with regard to folk-psychological interpretations, what about studies of active falsification? Consider the following intriguing evidence. In a variety of species, including snakes (Burghardt 1991; Greene 1988) and birds (Gochfeld 1984), individuals have evolved mechanisms that apparently fake out potential predators. These mechanisms are expressed in the behavioral form of either playing dead or playing injured. Ristau (1991a) has spent the past few years conducting experiments designed to assess whether the injury-feigning displays of plovers are guided by relatively high-level intentional states. Although it appears clear that injury-feigning displays have been designed by selection to lead predators away from a breeding pair's nest, the specificity of the mental state underlying the display is unclear.[46] To investigate this problem, Ristau set up a series of experiments to examine how flexible the plover's (piping plover, *Charadruis melodus,* and Wilson's plover, *Charadruis wilsonia*) injury-feigning system is with respect to luring predators away from the nest. Keep the following distinctions in mind. At the simplest level, plovers may

---

46. The actual wording of this sentence is important. I am not facing off evolutionary design against mental state. Rather, I am suggesting that the evidence for design by natural selection is clear. In question is the nature of the underlying representation or mental state prior to and during the performance of the display.

simply be designed to lure predators away from the nest, so whenever they see a predator, their nervous system sparks a predetermined behavioral routine. No complex mentation required here. What is needed, however, is the ability to recognize predators from nonpredators (a nontrivial categorization problem) and to gauge the direction of movement by the predator in relation to the nest. To jump up to higher levels requires, so Ristau argues, a couple of key ingredients. Specifically, the plover must at least be capable of showing some flexibility in response based on changes in the particular parameters associated with the specific predator encounter. For example, individuals should be able to perform injury-feigning displays in all directions and at all distances from the nest, and should be sensitive to dynamic changes in the predator's behavior. More challenging, of course, is interest in showing that the plover goes through the following sort of thought process: "Yikes, I see a predator moving this way. That predator surely wants to eat my offspring. I don't want him to eat my offspring. I better pull out the old injury-feigning routine and lure the predator away because if I am successful, the predator will think that I am easy prey, will follow me, and will bypass the nest. And as soon as he gets close to me, licking his chops, I will put on the accelerators and get out of here." Okay, a bit elaborate, but pieces of this mental equation must be part of the plover's thoughts if we are to state, with confidence, that it intends to lure predators away by using the injury-feigning display. Enough speculation. Let's look at the plovers.

The specific display used by plovers is one in which they drop a wing down to the ground, as if it were broken, and then begin to perform an interesting version of a Monty Python silly walk, an extremely awkward-looking motor pattern that appears frenetic. From a human perspective, and presumably from a natural predator's as well, this display is highly salient. As soon as the predator approaches, licking his chops, the individual flies away.

Humans were employed as predators, and experiments were conducted by varying such things as the speed, direction, and identity of the subject. Results showed that a large majority of broken-wing displays were performed in highly visible locations and in such a way that they would lead predators who followed away from the nest; if displays were performed, for example, behind the predator, then this observation would hardly constitute evidence that the display was designed to attract attention. In order to perform such displays, individuals typically flew to a location closer to the predator instead of displaying in the location where the predator was first sighted; flight itself may help attract the predator's attention.

The next set of observations focused on the extent to which plovers monitor predator behavior. These observations are important because of the issue of behavioral plasticity. For example, if the first display fails to lure the predator away from the nest, a new tactic should be implemented. Observations revealed that plovers commonly looked back at predators as they were displaying, suggesting that they were monitoring the predator's actions. Moreover, in over 50% of encounters where predators failed to follow after the injury-feigning individual, subjects stopped displaying, reapproached the predator, and often displayed more vigorously. These observations appear to rule out the simplistic explanation that plovers automatically implement the display without any sensitivity to predatory behavior.

To more carefully assess whether plovers are sensitive to predator actions, experiments were conducted involving manipulation of gaze and relative threat. In the gaze experiments, human predators walked past looking either in the direction of the nest or away from it. Results showed that plovers were more likely to leave their nest and stay away when predators looked at the nest than when they looked away. To examine the effects of relative threat, one dangerous and one safe predator was established, each wearing distinctive clothing. The dangerous predator walked close to the nest and stared at it in a potentially threatening way. The safe predator simply walked past the nest at a significant distance, showing no detectable interest. Once these associations had been set up, the test trial commenced, and observations of the plover's response to the approach of each predator type was recorded. Results showed that plovers responded more strongly (e.g., leaving the nest) to the dangerous predator than to the safe predator in 25 of the 31 pairs tested. This discrimination is quite impressive given the fact that the putative associations (i.e., relative threat) between individuals were set up within two trials (i.e., approaches toward the nest area).

These observations indicate, minimally, that plovers are not operating on the basis of some fixed action pattern. There is clearly variation in the form and timing of the broken-wing display, and some of this variation can be explained by the predator's behavior. But is this variation in display performance anything more than the type of variation we saw earlier for chickens and their alarm calls or food calls? That is, can we say anything more about the mental state of the plover than that it modifies its behavior on the basis of specific rules,[47] rules that

---

47. Note that similar sorts of questions can and have been raised for alarm calling in chickens and vervet monkeys. For example, Dennett (1983, 1987) used the observation that vervets tend not to produce alarm calls when alone as evidence that their calls are intended to inform others. However,

are characterized by such factors as the predator's distance, direction, and speed of movement, and so forth? Consider the following sequential rules (set up as IF-THEN statements), and whether or not they would be sufficient to explain the behavior observed in the absence of folk psychological ascriptions:

1. Upon detecting a potential predator, estimate its trajectory, using both movement cues and head orientation.

2. IF the predator appears to be approaching the nest, THEN move in such a way as to attract its attention (e.g., fly up), and then perform the broken-wing display.

3. IF the predator does not follow, THEN move closer, and perform the display again.

4. IF the predator follows, THEN wait until it is far enough away from the nest, and then, as soon as it approaches, stop the display and fly away.

In my opinion, these rules are sufficient to explain the behavior of the plover. They are by no means simple, but they do not require attribution of intentional states. It is clear that plovers have been designed to lure predators away, and that this design results in repeated attempts that are sensitive to how predators respond during the interaction. But flexibility in and of itself does not seem to nail the lid on the intentional coffin, though it is a first step—an important ingredient. Moreover, the fact that plovers are sensitive to the direction of gaze is also not surprising. Gaze discrimination seems to be common among animals with visual systems that are capable of resolving changes in eye position. And it should be, especially among prey species whose lives depend on detecting where a predator is looking. Moreover, as Baron-Cohen (1995) has argued, though gaze discrimination is important for developing a theory of mind (see also Baron-Cohen, Tager-Flusberg, and Cohen 1993; Frith 1989; Gomez 1991), it is not necessary, as evidenced by the thoughts and behaviors of blind children. Last, the fact that plovers can discriminate between dangerous and safe individuals, even in a short period of time, is also not very impressive, nor does it seem to be relevant to the problem of assessing the content of the plover's mental state. Thus, although this is a critical treatment of the plover data, I think that Ristau is absolutely on the right track with regard to the kinds of manipulation that must be done if we are to obtain the requisite empirical data.

---

we can also invoke an extremely simple rule to explain their behavior, and one that does not stretch the imagination for a wide array of organisms: Remain silent when alone, but call when in the presence of at least one other group member.

Let us come back to chickens for a minute. We already discussed the results on withholding information, showing how alarm-call and food-call production appear sensitive to the caller's audience. But do chickens ever actively falsify information by, for example, giving a ground predator alarm call when a hawk is present, or giving a food call when no food is present? Although active falsification of predator presence has not yet been described, there is some evidence that chickens may call deceptively about food.

To set the stage, recall that food-call rate covaries with food preference, and that in response to hearing such calls, females typically approach males. Given this situation, males appear to be ideally situated to produce food calls even in the absence of food and, accordingly, to attract females that are potentially receptive (i.e., sexually) or keep their mates in close proximity. To examine the possibility of deceptive communication, Gyger and Marler (1988) observed the natural food-calling behavior of males in a free-ranging situation, focusing on the stimuli eliciting such calls in addition to the behavior of the male and female following call production. In 45% of all vocal utterances scored, no food item was detected by the observer. When a food item was identified and the male called, females approached in 53–86% of all cases; variation in female approach was closely associated with the type of food encountered by the male, with the highest level of approach occurring in the presence of insects and the lowest in the presence of mash. When males called in the absence of food, females only approached 29% of the time. Interestingly, males were more likely to approach females when they called in the absence of food than when they called in response to food.

If males are to get away with their little bouts of deceptive calling, then one would expect them to be sensitive to what the female can see with respect to the feeding environment; this is the point we developed in the previous discussion on plovers. Specifically, if a female can see that a male lacks food when he calls, then he will be caught red-handed. In contrast, if the female is sufficiently far away, then a call produced in the absence of food is not immediately detectable as deceptive. In this situation, females will have to approach in order to see what the male is calling about. Figure 7.50 shows the relationship between calling by the male in the presence and absence of food and his distance to a female. Overall, males were more likely to call in the presence of food when females were close and to call in the absence of food when females were far. Although these data do not demonstrate that males intend to deceive females with their calls, they at least show that the probability of successful deception is increased by the fact that females are relatively far away from calling males.

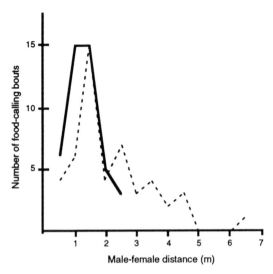

Figure 7.50
Food-calling behavior of male chickens in relationship to male-female distances. The solid line represents cases where food calls were produced in response to identifiable food items, and the dashed line represents cases where no food was detected by human observers (redrawn from Gyger and Marler 1988).

At least two key questions remain with regard to chicken deception. First, if food-associated calls given in the presence of food differ from those given in the absence of food, then a mechanism is in place for detecting deceptive calls. To investigate this question, Gyger and Marler looked at differences in call rate for food-present and food-absent calls. There were no differences. As the authors point out, this finding of course does not rule out the possibility that other acoustic features differ across calling contexts. Such analyses have yet to be conducted. Second, what are the fitness consequences of putatively deceptive as opposed to honest calling? For example, does the relationship between the male and female change depending upon some critical number of false as opposed to honest calls? Is the female less likely to mate if she is repeatedly deceived? Although these questions have not yet been addressed, the methodological tools seem to be in place. Until such questions have been addressed, however, this case provides only suggestive evidence of deception, and only at the functional level. The data are silent with regard to intentional deception.

Are there other studies of deceptive behavior in nonhuman animals that warrant the attributions of such mental states as beliefs, desires, and intentions. My own take is that there are large piles of terrifically entertaining anecdotes, especially for primates (Byrne and Whiten 1990), each suggesting the possibility of sophisticated mental analysis, but none providing the kind of evidence that seems necessary for such attributions.[48] In this sense, my own take may be more conservative than what developmentalists working on human children would attribute to their subjects (see the following section), but caution I believe is warranted and healthy at this stage, especially if it is buttressed by a keen interest in pursuing the problem experimentally. If such evidence is lacking in the domain of deception, what about other domains? Here, the results are mixed. Studies by Cheney and Seyfarth on captive macaques have suggested, using a perspective-taking task, that at least these monkey species appear to lack the ability to attribute mental states to others (Cheney and Seyfarth 1990). Studies by Povinelli and colleagues (Povinelli, Nelson, and Boysen 1990; Povinelli, Parks, and Novak 1991) have confirmed the conclusions of Cheney and Seyfarth for macaques, but have claimed that chimpanzees are capable of mental state attribution and, in particular, of distinguishing between individuals with or without particular knowledge. Some critics of this work (Heyes 1994) have argued that at present the battery of tests used to assess mental state attribution in nonhuman primates is simply insufficient to sway the decision one way or the other. For example, though Povinelli's experiments on chimpanzees are taken as evidence of mental state attribution, it required individuals several hundred trials to distinguish between knowledgeable and ignorant humans. Moreover, based on an impressive array of tests, Povinelli and Eddy (in press) showed that chimpanzees do not "understand seeing as a mental event" (p. 2). With such apparently lengthy training and the lack of appreciation that seeing-is-knowing, can one really argue for an *ability* to assess perspective, and the knowledge that comes with it? I think not, though I do find Povinelli's (1993) comparative approach and experimental creativity to be refreshing and a major advance over anecdotal reporting.

Before quitting, I would like to mention an experiment by Stammbach (1988) which provides interesting insights into how access to knowledge can transform the nature of social relationships that, at least for some primate societies, have often appeared quite rigid. This type of experiment seems ideal with regard to investigating questions about mental state attribution.

---

48. Although I find Dennett's intentional stance to be a useful heuristic for thinking about animals while observing them, I don't think that it provides the requisite ingredients for designing experiments to distinguish between a killjoy interpretation (automatic, reflexive responses) and a more folk-psychological interpretation.

Stammbach trained the lowest-ranking individual in each of three long-tailed macaque (*Macaca fascicularis*) groups to perform a lever-pressing task that provided access to a highly preferred food item (popcorn!); training occurred away from the social group so that others would not be able to solve the task, at least not immediately. Once the low-ranking individuals attained competence, they were put back with their social group, and the food-delivering device was affixed to the home cage. Initially, the trained individuals approached the device, pulled the levers, and waited for the food to emerge. Unfortunately, as soon as the higher-ranking individuals observed the food, they rushed over and snatched either all or most of it away. Gradually, the trained individuals stopped pulling the levers, and consequently no one obtained the food. The next set of events is the most fascinating. Eventually, the trained individuals went back to pulling, but this time the high-ranking individuals allowed them to eat some of the food. Moreover, their social relationships apparently changed in a subtle way, as evidenced by the fact that they received more grooming from the higher-ranking group members. Although the high-ranking individuals may have been sucking up to the skilled low-rankers, their general positions within the hierarchy did not change. Stammbach's interpretation, and one that I find compelling, is that the low-ranking individuals received "special" treatment based on their skills, but this was not sufficient to cause a rank change. Now one could create some complicated stimulus-response interpretations of these observations, but I believe that they would be far too convoluted and would fail to capture the subtle nature of the relationships observed. This kind of approach is well suited for more subtle experimental manipulations, such as changing the relative benefits of the resource to group members, in addition to altering group composition and individual knowledge of some of the reinforcement contingencies.

7.4.4   The Human Child's Discovery of Mind

It has often struck me as a curious fact that so many shades of expression are instantly recognized without any conscious process of analysis on our part.   (Darwin 1872, 359)

But if nobody spoke unless he had something to say, . . . the human race would very soon lose the use of speech.   (Maugham 1925, 38)

Language learning is not really something that the child does; it is something that happens to the child placed in an appropriate environment, much as the child's body grows and matures in a predetermined way when provided with appropriate nutrition and environmental stimulation.   (Chomsky 1988, 62–63)

The 2-year-old is clearly a mentalist and not a behaviorist. Indeed, it seems unlikely to us that there is ever a time when normal children are behaviorists. Even in infancy, children seem to have some notions, however vague, of internal psychological states. This [fact] is evidenced in very early "conversational" interaction and in facial imitation; almost from birth infants coordinate their actions with those of others in a sensitive and fine-grained way.   (Gopnik and Wellman 1994, 265)

What we learned in the previous section is that behavior might not provide the cleanest entry into the mind, though in several cases, it might be the only way in. Moreover, we learned that although animals *know how* to do many clever things, they may not *know that* they are doing clever things or know about the details underlying the clever thing or things. This distinction, between knowing how and knowing that, was made clear by the philosopher Ryle (1949), and it is crucially placed in the minds of most cognitive scientists who think that human adults clearly know how and know that, but nonhuman animals only know how. Whereas knowing that X relies on a representation (which in most discussions is a proposition of sorts), knowing how does not. Let us take a deeper look into this distinction, especially since many researchers working in the cognitive sciences think that young human infants are know-how-ers but not know-that-ers.

Chomsky's quote reminds one of the development-as-gardening metaphor: just add water and watch language grow. As language grows, however, there is the question of what the child knows about its use of the language. Thus, does the child simply know how to use the word "ball" whenever it sees something with the appropriate features, or does it know that "ball" refers to a particular object, that the utterance stands for that object? Similarly, but looking at the ontogeny of cognition more generally, Gopnik and Wellman suggest that humans are literally born with some notion of mental state, that they understand what might actually be *in* the minds of others. Importantly, they claim, evidence of thought, as well as thought about the thoughts of others, enters the child's repertoire of skills before language—that is, they know that others have minds (at least some psychological states) at a very early age. This view then strikes hard on those entrenched in the "language of thought" camp (the propositions in our minds are symbolic, languagelike). Although I side completely with the Gopnik-Wellman view on this particular point (for similar ideas, see reviews in Astington 1994; Astington, Harris, and Olson 1988; Hirschfeld and Gelman 1994; Mehler and Dupoux 1994; Weiskrantz 1988), it is nonetheless useful to challenge some of the findings on human infants, children, and adults by keeping the data on nonhumans firmly in mind. What I would therefore like to ask of the reader is that each time I say something like "the human infant or child *believes, wants, desires,* or *intends*," substitute your favorite nonhuman animal for human infant and child, and see whether you still buy the claim. Although some may feel that the simple *fact* that human adults *believe, want, desire,* and *intend* is sufficient evidence for arguing that human infants and children do as well, I think this type of thinking is misleading (see also Studdert-Kennedy 1986b). Without belaboring the point, consider what would happen if you attributed to autistic children a capacity for understanding the minds of others. You would generally be wrong. Autistics, who have actually provided developmentalists with some of the most telling results concerning social cognition (Baron-Cohen 1990, 1995; Baron-Cohen, Tager-Flusberg, and Cohen 1993; Frith 1989), commonly appear to lack a theory of mind. In the most severe cases, they don't understand that others have psychological states and that such psychological states form the basis for action. My main point: let us derive empirically what the child knows and thinks about rather than introspectively work it out. In this way, we will understand what they know, and will have gained such an understanding from an empirically solid foundation. This stance has been adopted by many developmental psychologists interested in cognition, and excellent reviews can be found in Astington (1994) and Mehler and Dupoux (1994).

The focus of this section is the child's theory of mind, how it develops and changes over time. In particular, I will review how developmentalists have probed the beliefs and desires of young children, and then work through some ideas on how the recognition of mental states, one's own and those of others, allows for some quite fancy linguistic maneuvers.

Earlier in this chapter, I talked about the potential links between causality and intentionality, and mentioned some of the fascinating work by Spelke, Baillargeon, and others. David Premack and his colleagues (Premack and Dasser 1991; Premack and Premack 1994) have now attempted to formalize the conceptual link between causality and intentionality (reviewed in Sperber, Premack, and Premack 1995) and to show how the emergence of this process sets the stage for thinking about thinking, especially thinking humans. Here is a synopsis of the key points:

For most researchers working in the area of cognitive development, the steps toward a theory of mind—of thinking of others as having psychological states— involve first, perception of an object or event, which then sets up a belief about that object or event, which in turn generates emotions, desires, intentions, and an outcome (Figure 7.51). Premack and Premack (1994) describe this process by making a distinction between things in the world that act/move as a result of external causes and things in the world that act/move as a result of invisible causes. In the latter case, human adults tend to invent causes for such actions, and such causes tend to be credited to the mind of the actor, to his or her psychology. The question then is, Do infants make similar attributions, and if so,

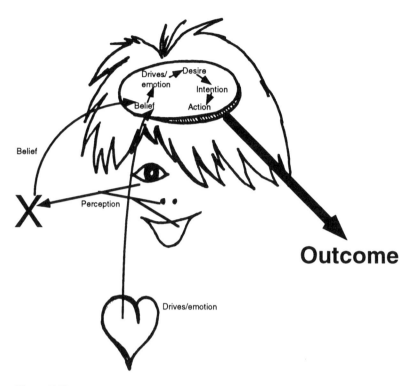

Figure 7.51
Schematic illustration of some of the key ingredients for a theory of mind (components adopted from Astington 1994).

at what age? Before we turn to what the child's language tells us about mental state attribution, let me briefly describe the Premacks' test of the preverbal infant's theory of mind. The reason for the mini-digression is that this experiment can easily be tested with nonhuman animals. Yes, that old saw again.

   Studies of motion perception indicate that human adults tend to perceive symmetrical objects as having no particular direction, whereas asymmetrical objects tend to have quite explicit directionality. Thus a circle has no specific direction, whereas the same circle with a triangle affixed to one side and the point sticking out appears to have a clear direction—moving with the point leading (Figure 7.52). The experiment with infants takes this distinction as a starting position and moves in a series of four steps. Here it is, formulated as a test of the preverbal infant's theory of mind:

Figure 7.52
On the left, a symmetrical (S) object (circle) with a shadow below. On the right, an asymmetrical (A) object (circle with triangular point) with a shadow below. Premack and Premack (1994) suggest that we judge the A-object as having an intended direction (movement to the right in this case), whereas the S-object lacks directionality.

1. If infants assume that asymmetric objects are directional with respect to motion, then when an asymmetric object appears on the screen they should look (anticipatorily) toward the most likely direction of movement (i.e., to the right in Figure 7.52). In contrast, there should be no bias in looking at symmetric objects.

2. To show that infants understand not only direction, but also the object's ''visual capacity,'' one could set up a barrier of sorts in the path of a moving asymmetric object. If the pointy end is the ''seeing'' end, then when moving in this direction, it should avoid the barrier. In contrast, if it is moving in the opposite direction (i.e., pointy end trailing), then it should be more likely to bump into the barrier. Again, if infants understand such distinctions, they might show surprise (look longer) when they see the asymmetric object bump into the barrier with its pointy end—with the end that supposedly can see.

3. To test for beliefs, the infant now watches the following animation. The asymmetric circle sits at the bottom of a hill, with its point directed toward the top. Repeatedly, it attempts to climb up the hill (i.e., it moves up and then slides down) to gain access to an apple on top. Before successfully reaching the top of the hill, the apple rolls down, past the asymmetric circle and into a black container, out of sight; the asymmetric circle appears to track this motion by turning its pointy end toward the apple as it rolls down the hill and into the black container. In one scenario, the asymmetric circle rolls down the hill and goes to the black container. In a second scenario, the asymmetric circle rolls down the hill and goes to a white container. If infants have a theory of mind, they should be

surprised to see the asymmetric circle go to the white container and not the black container. But to show such surprise, they must understand (a) that the asymmetric circle *wants* the apple, (b) that evidence of this *desire* is provided by its persistence in trying to climb the hill, and (c) that it *knows* where the apple is because its seeing end was properly oriented.

4. Following from step 3, we can move to a test of false beliefs by removing the asymmetric circle from the scene immediately after the apple rolls into the black container. While the asymmetric circle is away (the assumption here is that out of sight from the scene means that it lacks information about what is going on), we take the apple from the black container and place it in the white container. We now bring the asymmetric circle back. If infants have a theory of mind, then they should believe that the asymmetric circle believes that the apple is still in the black container. Consequently, if they see it searching in the white container they should be surprised, but not if they see it searching in the black container.

I think that this is an absolutely ingenious experiment, and as stated earlier, one that could readily be conducted on nonhuman animals. What I find surprising, however, is the pessimism expressed by Premack and Premack (1994) with regard to comparative issues, especially given their long involvement with such issues: "Though the complexity of most species consists in making automatic computations regarding items they perceive, human complexity consists in having evolved theories that seek to explain perceived items" (p. 217). Although it might be the case that nonhuman animals fail to show surprise at any step in this experimental sequence, this is an open empirical question. Given the recent successes of looking-time experiments with nonhuman animals (e.g., Cheney, Seyfarth, and Silk 1995; Hauser, MacNeilage, and Ware, in press), we should, minimally, hold off judgment on this issue. But enough said. Let us move on to a domain where humans do differ from nonhumans: how they use *their* language to communicate about *their* apparent thoughts.

Children are extraordinary imitators, and thus the simple utterance of a word need not imply knowledge of word meaning or its conceptual representation; this point is completely separate from, though by no means more trivial than, the problem of establishing when a word is a word rather than some other sound (Premack 1990; for a recent discussion and potential methodological solution, see Vihman and McCune 1994). Moreover, even when children produce a word in an appropriate context (i.e., correct usage), this event may not provide critical information on their knowledge because they may have simply learned, through observation, when to use the word to gain the desired response; several studies show that intelligibility and meaning are improved by nonlinguistic contextual information (Bard and Anderson 1994). At some point in development, this kind of behavioristic explanation fails to account for language use, and several developmentalists agree that at about two years, the language-as-vehicle-for-thought-expression begins to take shape.[49] Specifically, two-year-olds begin talking about their perceptions and their desires, in addition to showing an understanding of what others perceive and desire. Thus they appreciate that if Nancy wants coffee and sees that there is a freshly brewed pot of coffee, she will do something to get coffee. Thus, if Nancy says, "I want some coffee," and then looks around the room and sees the pot, she would be expected to walk over and grab a cup.

By three years, the child's mental ontology takes on a new form. The child begins to show an understanding of beliefs and representations of mental states more generally. But the earliest utterances and responses to questions indicate that the mental state is still anchored in a desire-perception psychology. Thus the youngest of three-year-olds treat beliefs as direct reflections of events in the world. It is only later that they realize that beliefs and the mental states that accompany them can be false, imagined, and readily altered in such a way that they deviate significantly from the "real" world. The classic test of the child's understanding of these psychological states is the false-belief test. A child is shown a bag of candy, presumed to be recognizable as such by its packaging (e.g., a bag of M&Ms). The child is then asked about its contents, and most children reply that it has candy inside. Now the experimenter opens the bag and shows that instead of candy, there are pencils. The next step is the critical one. The child is asked what someone else might think is in the bag. Before the age of about four years, children answer "pencils." This answer shows that their beliefs about the beliefs of others are dictated by real-world events: the bag contains pencils, and thus everyone ought to know about them. Moreover, if you then ask the child what he or she thought was originally in the bag, he or she will answer "pencils," suggesting that children fail to keep track of their prior beliefs and are highly influenced by the real state of affairs. Reinforcing this evidence for a failure to make accurate predictions about the beliefs of other individuals is the fact that three-year-olds make predictive mistakes in other related domains, including what other individuals will actually say (contrasted with

---

49. Although much of the early work in child language development focused on the identification of *dates* for passing particular linguistic milestones, and therefore placed an emphasis on the notion of a *modal child*, more recent work demonstrates considerable variability among children and emphasizes the importance of understanding the sources of variability (e.g., Fenson et al. 1994).

what they think—the candy-pencil false belief test) about an event that they have either failed to witness or only witnessed in pieces (i.e., fail to distinguish the strength of a belief or confidence in it) and the extent to which they distinguish between things that merely *appear* to be X and things that *really* are X. In the latter case, you might show a child what looks like a banana but turns out to be an eraser—you show the child that this banana-like object actually erases pencil from a piece of paper. Evidence for a failure to distinguish between appearance and reality comes from the fact that the three-year-old states that the object is a banana and that it looks like one. Last, this failure is not simply one that lies within the mental representation, but it also shows up in other representations, as demonstrated by Zaitchek's (1990) false-belief test using photographs—that is, falsely representing photographs.

A general consensus in the literature (for reviews, see Astington 1994; Gopnik and Wellman 1994) is that at approximately four years, the child begins to use words that reflect a more sophisticated understanding of own and other minds (e.g., "I *think, remember, believe,* that X", "I don't *think* she *believes* there's a tiger on the couch"). As Gopnik and Wellman have put it, children begin to use their representational abilities to predict and explain the behavior of other individuals and, in so doing, are in the process of developing a "theory" of other minds. Children at this stage pass false-belief tests that younger children fail and show an understanding of how specific kinds of knowledge influence beliefs and intentions. Thus, in a modified form of the candy-pencil test, four-year-olds know that if Fred sees Jane place cookies in the cookie jar, Fred will believe that cookies are kept in this cookie jar. If Fred wants cookies, he should go to the cookie jar. But, if Fred leaves the room and Jane moves the cookies to a shoe box, then when Fred returns, he should still search for cookies in the cookie jar, and not in the shoe box. These are the kinds of sophisticated expectations that older children set up. Such expectations allow the child to predict actions in the absence of direct observation.

An interesting twist on the false-belief paradigm has recently been published by Clements and Perner (1994), and their results return us to the distinction between knowing how and knowing that, or perhaps more appropriately, to the distinction between implicit and explicit understanding of false beliefs. Tests were conducted on children between the ages of two years, five months and four years, six months. Each child was videotaped during a session involving the presentation of a visual scene accompanied by a story. Like the cookie story, the stories presented by Clements and Perner also involved the placement and displacement of an object, with the relevant characters either seeing the object move or not, and then subsequently making a decision about where the object is located. The key difference between this study and others is that it uses the child's direction of gaze to assess implicit understanding of action as contrasted with explicit responses to the situation. Results show that children under the age of two years, ten months mistakenly look toward the real location of an object, rather than in the direction of where the actor is likely to move given that he or she did not see the displacement. Thus, returning to the Fred and Jane story, the young child looks toward the shoe box, not toward the cookie jar where Fred is most likely to go. Older children, in contrast, look toward the cookie jar. Surprisingly, however, when the older children are asked to state where Fred will go, almost half respond that he will go to the shoe box. These results show that by two years, 11 months, children have implicit understanding of false beliefs, but do not explicitly understand such false beliefs until much later. Although Clements and Perner are able to rule out a number of potential explanations for their data, they can only provide a speculative account of the pattern obtained. Specifically, they suggest that young children can only represent facts about the world, but cannot make judgments about them—a one fact–one representation stage of development. Gradually, however, the child can think about multiple representations, of comparing them, and of acting on them (e.g., giving a verbal response, a judgment). From a comparative perspective, the exciting piece of this story lies in the fact that quite young children generate expectations about the beliefs of others, and these are revealed by their gaze. Consequently, though we may be unable to elicit explicit responses from nonhuman animals, I would expect that we could present them with scenes that elicit directional eye movements and thereby reveal something about their implicit understanding of false beliefs.

Given the increasing sophistication of the child's mind with regard to representation, let me conclude this section with a more philosophical question: When does the child understand that its mind is the *source* of such representations? To address this question, Wellman and Hickling (1994) conducted a series of observations and experiments with children between the ages of 2.5 and 10 years. In part one, a large corpus of utterances produced by differently aged children was scanned for words associated with perceptual events or experiences (e.g., *see, look*), metaphors referring to the mind (e.g., "I see what you mean"), and nonmentalistic uses (e.g., using the word *see* to mean "find out"). Results showed that children use such perceptually based words as *see* and *look* toward the end of the second year, but metaphorical use of such words does not appear until the beginning of the fourth year.

The next part of the study asked children about their understanding of mental metaphors. Metaphors were selected from different domains, including personal statements ("My mind wandered") and sentences referring to mechanical objects ("The car is dead") and nature ("The wind howled"). In addition, children were asked to explain the phenomenon underlying a mental activity, such as making a mental image; as a control, they were asked to explain how a picture is made. Results (Figure 7.53) from the first part of this experiment showed that children understand metaphors for mechanical (cars) and natural (wind) events at an earlier age than for mental events (mind). Regarding the second part of the experiment, there were two significant findings. First, children from the age of six years on recognized the fact that information in a photograph is accessible to others, where the information in a mental image is not. Second, when asked about the mechanism underlying image generation, six-year-olds tended to give a personalized account of photographs, whereas older children tended to give a physical/mechanical description. Thus, six-year-olds described (Wellman and Hickling 1994) the photograph with statements like "The camera gets the idea of it (apple) and draws the picture"; older children described photographs with statements like "A little machine with gears makes it" (p. 1572). In contrast, personalized explanations of mental images ("Your mind draws it and puts it right in front of you"; p. 1572) only emerged as a consistent form of explanation by 8–10 years. In summary then, the child's language reveals that the mind and brain are initially (4–8 years old) viewed as entities that are critical to cognitive activities. Later, however, mind and brain are seen as relevant to a host of activities including thinking, eating, playing, and sleeping. Moreover, when older children describe the mind, they do so in reference to themselves or the thoughts of others.

7.4.5   The Human Adult's Capacity for Intentional Deception

In the preceding section we brought the child up to a quite sophisticated level of mentation and awareness of what he or she thinks, feels, desires, and intends. Some of the most convincing evidence for such awareness comes from studies of the linguistic child. When the child speaks, one hears the inner workings of its mind. And when the child listens to others speaking, it thinks that it hears the inner workings of their minds.

In this section, I return to a theme that was developed earlier in this chapter and in chapter 6: deception. Although several convincing cases of functional deception in nonhuman animals were provided, none of these cases provided sufficient support for intentional deception. But in humans, at least, we are all convinced that normal individuals are not only capable of being deceptive, but

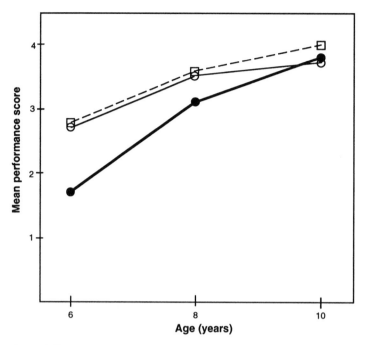

Figure 7.53
Performance scores on a metaphor task. The closed circles represent data from comprehension tests using mental metaphors (mind), the open circles are for mechanical metaphors (cars), and the squares are for natural metaphors (wind) (redrawn from Wellman and Hickling 1994).

in fact, extraordinarily good at it. The focus, therefore, will be on studies that have looked at some of the communicative correlates of deception and the extent to which perceptive audiences are aware of such factors, as evidenced by their ability to detect dishonesty. To put this topic in an evolutionary framework, recall from our earlier discussion that in nonhuman animals, at least, signal honesty is often preserved as a result of signal costs. Opposing this relationship, of course, will be strong selection on perceivers to recognize cues of dishonesty—selection for skepticism. In thinking about the cases discussed in this section, then, keep in mind the notion of signal costs and whether such economic thinking helps us understand the design of human signals. We will return to this interface between evolution and psychology in chapter 8.

One signal that we have already discussed in this chapter is smiling. Specifically, I referred briefly to the early work of Duchenne, together with the more recent work by Ekman and Friesen. Based on analyses of facial action, Ekman and his colleagues (Ekman, Friesen, and O'Sullivan 1988; Ekman, Davidson, and Friesen 1990) have proposed that there is not just one smile, but many smiles. This claim is based on a microanalysis of changes in facial musculature, differential activation of the right and left hemispheres, and reports of experienced emotion during smile production. To illustrate the distinctions, consider some recent findings on the Duchenne smile. Subjects were shown film clips that, based on prior analyses, elicited strong positive emotions (primarily, happiness) or negative emotions (e.g., fear, sadness, disgust, or pain). Subjects were tested alone and were told that the goal of the experiment was to obtain physiological data on their responses to films associated with positive and negative emotion. The reason for testing individuals alone was so that the emotional experience could be related to the expression in the absence of social interaction.[50] The first set of results showed that Duchenne smiles were primarily produced when positive-emotion films were viewed. The duration of Duchenne smiles was also longer than all other smiles,[51] and it was particularly long when given during positive-emotion films. When Duchenne smiles were produced, they were highly correlated with reports of amusement and happiness, and negatively correlated with feelings of anger and sadness. Interestingly, the other smiles were positively correlated with feelings of disgust. The second set of results examined the relationship between smile type and hemispheric activation, using EEG measures. Results showed that Duchenne smiles were associated with significant left-hemisphere anterior temporal and parietal activation, whereas the other smiles tended to show right-hemisphere activation. Referring back to the material in chapter 4, these results provide support for Davidson's (1992, 1995) ideas on hemispheric activation and emotional valence. Specifically, it appears that the left hemisphere is more commonly involved in positive/approach emotions, whereas the right hemisphere is involved in negative/withdrawal emotion. Data on smiles therefore suggests that at the production end there are robust expressions of honesty and dishonesty with regard to the verity of the emotion experienced.

Ekman and others have also looked at some additional visual and auditory correlates of dishonest communication (Table 7.4). For example, during conversations conveying truthful information, the pitch of the speaker's voice tends to be lower than during conversations conveying dishonest information. Although

---

50. Recall that one of Fridlund's counterarguments to this testing procedure is that even when individuals are tested alone, they have an imagined audience.

51. The "other" smiles were characterized by movement of the zygomatic major muscle, but no movement of the orbicularis oris, pars lateralis. The latter muscle is, definitionally, involved in the Duchenne smile.

Table 7.4
Visual and Auditory Cues to Deception

Visual Cues		Auditory Cues	
Variable	Change during Deception	Variable	Change during Deception
Blink rate	Higher	Length of utterance	Shorter
Pupil dilation	Increases	Content negativity	Higher
Self-manipulating gestures (e.g., rubbing)	Increases	Content relevance	Less
Changes in body position (fidgeting)	More frequent	Voice pitch	Higher
Rate of smiles other than Duchenne	Higher	Pauses between utterances	Longer
		Grammatical errors	Higher
		Slips of the tongue	Higher

Data obtained from the following sources: DePaulo 1994; DePaulo, Stone, and Lassiter 1985; Ekman 1985; Ekman, Friesen, and O'Sullivan 1988; Ekman, O'Sullivan, et al. 1991; Ekman and O'Sullivan 1991.

such differences are statistically significant, one potential problem is that they may not be discriminably different (e.g., in Ekman et al. [1991] a difference of 8.0 hertz is reported), at least not during conversational speech (i.e., as opposed to a psychophysical task where single utterances are presented and comparisons made under quite ideal conditions). Nonetheless, it is clear that a variety of cues to deception are potentially available to attentive listeners. The crucial question then is, How good are people at detecting lies, and what features do they use to make their judgments? To answer this question, let me now turn to a set of studies that have focused on the detection of liars.

Over the past few years, Americans have experienced an interesting cultural change. Rather than reading about legal debates in the papers and attempting to understand the subtleties of the law, they can now simply turn on the television and watch courtroom arguments ranging from Judge Wapner and "The People's Court" to Clarence Thomas and Anita Hill, and most recently, the infamous O. J. Simpson trial. What of course makes such television drama interesting is that people all over the country can now generate their own opinions about who is lying and who is not. But how good are we at detecting liars? In the traditional approach to this problem, people are asked to first describe something honestly and then to describe the same sort of situation dishonestly. For example, a person is asked to talk about a restaurant that he or she really likes (e.g., as if they were going to recommend it to someone) and then to provide the same sort of description but while imagining that they received food poisoning during their last visit. While such stories are being recounted, short film clips of the speaker are obtained, and these are shown to subjects who are unfamiliar with the storytellers. In study after study, it appears that people are quite inept at detecting lies. Surprisingly, this deficiency is even observed among people who have been explicitly trained to detect lies. In a study by Ekman and O'Sullivan (1991), secret service agents (U.S. Secret Service, CIA, FBI, National Security Agency), robbery investigators, judges, and psychiatrists were tested in a lie detection task and compared with college students. Only some of the secret service agents performed above chance, and of these, age, sex, and level of experience was not an accurate predictor of success at detecting lies.

One potential problem with studies like these is that subjects are required to make judgments about honesty without knowing the individual, and potentially, not caring about the situation. As Cosmides and Tooby (1992) have argued, being able to solve problems having to do with the logic of an argument is heavily influenced by the type of "story" being told. If the story is socially relevant, we tend to be better at detecting inconsistencies than if the stories are either irrelevant or fail to involve salient social interactions. In essence, our minds appear specifically designed to detect cheaters, rather than more general, nonsocially anchored inconsistencies. Although the logic and experiments by Cosmides, Tooby, and others (Gigerenzer and Hug 1992; see Cheng and Holyoak 1989 for a critical treatment of this work) are convincing, studies explicitly focusing on the issues of familiarity and the importance of the relationship have turned up negative evidence. Specifically, studies by McComack and Levine (1990) have shown that people involved in a romantic relationship do no better at detecting each other's deceit than at detecting the deceit of a stranger. In addition, individuals are no better at detecting lies from individuals of the same sex than individuals of the opposite sex. And last, although explicit training to detect lies from one individual improves performance for that individual, this ability does not generalize to other individuals.

Most of the studies reviewed thus far present analyses from grouped data, rather than exploring the successes of individuals in detecting lies. Moreover, they tend not to assess the relative contribution of visual as opposed to auditory expressions. To remedy these problems, Ekman, O'Sullivan, and colleagues (1991) have conducted analyses of the hit rates of individuals, looking at the independent and combined effects of face and voice measures. With the exception of hand movements (called "illustrators"), smiles (both Duchenne and masking) and changes in the pitch of the voice provided quite accurate information about whether the speaker was being honest or deceptive. However, a combination of smiling and voice pitch provided the highest hit rate with respect to detecting lies. Although these results generate evidence that people can distinguish truths from falsehoods, there is still a relatively high rate of incorrect attributions (i.e., 16%). As Ekman, O'Sullivan, et al. (1991) conclude, this level of error "would be an unacceptably *high* level if there were any serious consequences of such an identification. Rather than being used as signs of lying, recognition of changes in facial expression and voice might more judiciously be used to alert an interrogator to pursue a line of inquiry" (p. 133).

Can we conclude from these results that people are hopelessly lost in a sea of indistinguishable truths and lies? Perhaps, but some researchers think that current techniques for assessing sensitivity to dishonesty fail to capture implicit knowledge (DePaulo 1994). As a result, new methods must be developed for tapping into this knowledge, and this process may require using experiments that more directly probe issues that are of concern to the subject. And here is where the intuitions of evolutionarily minded psychologists appear to be helping. Specifically, by setting up situations that potentially affect the individual's survival and fitness prospects, we may be able to access what has appeared to be the inaccessible.

8 Comparative Communication: Future Directions

8.1 Introduction

Bioluminescent fireflies, dancing honeybees, snapping stomatopods, honking midshipmen, belching bullfrogs, singing blackbirds, echolocating horseshoe bats, alarm-calling vervets, fear-grimacing chimpanzees, and talking humans. Genes, neurons, brain modules, hormones, sensory organs, learning mechanisms, mental states, and sexual behavior. Yes, the world is replete with variation, and in this book I have attempted to grapple with it by, first, thinking about the functional design features of each system and, second, integrating different levels of analysis. But my job is not yet complete. This chapter attempts to tie up some loose ends, develop a number of theoretical and methodological ideas, and generate a set of suggestions for future research in what I hope will be a healthy and invigorated research program in comparative communication. In section 8.2, I work through a set of problems that I believe must be addressed before the payoffs of a comparative research program can be fully realized. In section 8.2.1 discussion centers on the neurosciences and suggests that there is a dire need to develop tools that allow for direct interspecific comparisons, while remaining sensitive to the ecological and social factors that both were and currently are responsible for the design features of each species' communicative repertoire. In section 8.2.2, I return to the problem of call meaning and sketch what I consider to be a feasible program of research, designed to elucidate the putative referents of nonhuman animal vocalizations. As a side benefit, this analysis also points the way to a more thorough treatment of the "units problem"—that is, how to determine the functionally meaningful units of communication for organisms without language. Section 8.2.3 explores the possibility that the preferential looking procedure, a method that originated in studies of human infants, is a candidate solution for comparative studies. Section 8.3 represents a thought experiment: thinking like an evolutionary engineer, I ask how one might design and then build communicating organisms, varying in their degree of sophistication. This thought experiment will be guided by the empirical observations reviewed in this book and by the conceptual tools of evolutionary theory. The end product will shed light on why some organisms communicate by means of simple reflexes, like mini on-off switches, whereas other organisms are born with an ability to convey their thoughts and feelings through melodic oratorios, descriptive prose, or highly rational and precise scientific treatises.

I conclude the chapter by suggesting that the study of communication, like the study of many other behavioral processes, will require a theoretical framework that is like an intricately constructed tapestry, weaving empirical threads from a wide diversity of species and integrating issues relevant to proximal mechanisms, developmental processes, functional design and adaptation, and phylogenetic patterns. And the only framework that can accommodate such goals is the one sketched by Tinbergen and emphasized throughout this book. It is not only interdisciplinary, but held together by one of the most powerful and all encompassing theories ever developed: Darwin's theory of natural selection.

8.2   Some Burning Issues

8.2.1   A Socioecologically Sensible Neuroscience

Imagine what the world would have been like for us humans if, rather than descending from the trees to live a terrestrial existence, we had stayed up in the canopy with some of our primate cousins—or even more radically, as the novelist Elaine Morgan has creatively embroidered, walked out of the ocean! Surely our brains would have evolved along a different trajectory and with a different neural architecture. And why should we have confidence in this claim? Because the animal kingdom provides us with case after case where the design features of the brain are well suited to the ecological problems confronted by the organism, and such results are due to selection pressures that nonrandomly eliminate failed solutions, while favoring successful ones. For those working on organisms living under natural conditions, facing real survival problems, it will come as no surprise that brains exhibit design features suited to solve socioecologically meaningful problems. However, it is just this kind of thinking that I believe has escaped many researchers working in the neurosciences (for similar comments on cognitive psychology, see the critical piece by Cosmides and Tooby 1994a) and is one of the greatest conceptual assets in the tool kit of neuroethologists, especially those with a strong evolutionary bent. In this section, I would therefore like to continue our discussion of the issues raised in chapter 4 by arguing that neuroscientists must hook up with ethologists in an attempt to create a more socioecologically rooted research agenda, driven by questions of design and the ability to process biologically meaningful stimuli. Conversely, ethologists working on the behavior of organisms living in their species-typical habitat must gain a better appreciation of the fact that an individual's ability to respond to events in the environment—to solve problems that are linked to fitness maximization—is heavily constrained by a package of neurobiological mechanisms that also have particular design features. To develop this argument, I would like to focus on the problem of auditory processing and, in particular, to explore some ideas about the perception and production of communication signals in nonhuman primates, especially rhesus monkeys. I have selected this problem and particular species for two reasons. First, unlike studies of frogs, birds, bats, and humans, we know extremely little about the neurobiology of acoustic processing in nonhuman primates, especially in terms of their vocalizations. Some studies of auditory processing in nonhuman primates have been conducted, but many of these have used artificial stimuli. These are not the kinds of signals that the primate brain was designed to process. Second, behavioral observations and experiments on rhesus provide suggestive hints that the neural architecture guiding their audiovisual communication system is complex and represents a potential precursor state along the evolutionary path to human language. Thus the prospects for a socioecologically sensible study of neural processing are extremely good.

To refresh the reader's memory of what I sketched in chapter 3, and what we learned about in greater detail in some of the later chapters, rhesus monkeys have an acoustically rich vocal repertoire, producing approximately 25–30 different call types in a broad range of situations, including the discovery of food, the initiation and maintenance of sexual and affiliative relationships, aggressively charged dominance interactions, and competitive intergroup encounters. To produce such calls, individuals must coordinate respiratory activity with muscular control of the larynx and articulators involved in configuring the supralaryngeal vocal tract. We know, primarily on the basis of behavioral and acoustic analyses, that rhesus monkeys are capable of producing communicative signals with an impressive range of fundamental frequencies (Hauser and Marler 1993a). It is impressive because, while weighing much more than a songbird and much less than a human adult male, rhesus can produce sounds in the range of both species. In addition to this acoustic potential, rhesus are also capable of filtering the sound source by means of articulatory gestures, involving, minimally, changes in the position of the lips, jaw, and teeth (Hauser, Evans, and Marler 1993; Hauser and Schön Ybarra 1994). An important acoustic consequence of supralaryngeal articulation is that vocalizations are characterized by time-varying resonance frequencies, providing an additional avenue for conveying meaningful information.

Perceptually, individuals must first localize the sound source and, based on the acoustic characteristics of the signal, identify the individual, his or her affective state, and the putative referent of the call. Psychophysical results have demonstrated that rhesus are highly sensitive to locational cues in the horizontal plane and have fine discriminative abilities in both the spectral and temporal domains (C. H. Brown et al. 1979, 1983). Field studies have revealed that at least some vocalizations within the repertoire are functionally referential, picking out—so to speak—salient objects and events in the environment (Gouzoules, Gouzoules, and Marler 1984; Hauser and Marler 1993a). Additionally, playback experiments have suggested that the left hemisphere of the brain is dominant in processing conspecific calls, whereas the right hemisphere is dominant in processing hetero-specific calls (Hauser and Andersson 1994). Last, single-unit recordings indicate that in the supratemporal plane there are neurons that appear to be highly responsive to rhesus monkey calls and significantly less responsive to pure tones and broadband noise (Rauschecker, Tian, and Hauser 1995). These brain-behavior relationships only skim the potential surface of what might be discovered empirically, and for the next few pages I will develop some ideas and experiments that will help fill in gaps in our current understanding. More specifically, by starting with current knowledge of how rhesus monkeys deploy their vocalizations in ecologically meaningful situations, we will be able to generate a research program that illuminates not only how the brain processes such communicative contexts, but also why it exhibits the design features uncovered, rather than some other possible set of design features.

To begin, consider the results of hemispheric asymmetry. In the Hauser and Andersson (1994) study we learned that adult rhesus monkeys show a preferential right-ear/left-hemisphere preference for listening to calls from within their repertoire, but evidence a left-ear/right-hemisphere bias when listening to a hetero-specific signal (i.e., the alarm call of a sympatric bird, the ruddy turnstone). This result leaves several questions wide open. First, what is the acoustic basis for this apparent perceptual bias? That is, what acoustic features of the rhesus monkey's repertoire anchor the hemispheric bias that drives the behavioral response? For example, although the turnstone's alarm call is acoustically different from all of the calls in the rhesus monkey's repertoire, it falls well within their spectrotemporal range of potential variation. In other words, the rhesus monkey's vocal apparatus could produce a call that is structurally like the turnstone's. What is therefore necessary, at a behavioral level, is to conduct experiments that determine which acoustic features are used by the perceptual system to place calls either within or outside of the general category of "conspecific vocalizations." And there are several approaches that I can think of. Under field conditions, run the same experiment as in Hauser and Andersson, but use calls from (1) closely related heterospecifics (e.g., other macaque species) or (2) synthesized exemplars of rhesus calls that selectively alter key features of the call. In the first case, we know that most macaques exhibit a striking degree of similarity in repertoire structure. Nonetheless, one would expect that at some level, each species is capable of, minimally, discriminating calls from its own repertoire from all others.

Such tests have never been formally conducted, but the preferential-head-turning response may provide a useful assay under field conditions. Specifically, if the conclusions from Hauser and Andersson hold, then playbacks of calls considered within the range of a species-typical rhesus call should be responded to by a preferential right-ear bias, whereas calls falling outside of this range should elicit a preferential left-ear bias. In the second case, sound synthesis techniques permit careful manipulations of signal structure to determine the necessary and sufficient features for call recognition. Thus one could take a call type within the repertoire and alter the values of putatively salient features such that they exceed species-typical values (i.e., the range of acoustical variation generated by conspecifics). As in the previous case, synthesized calls that exceed the species-typical range of variation would be expected to elicit a preferential left-ear bias, whereas those falling within this range would be expected to elicit a preferential right-ear bias.

The field experiments sketched in the preceding paragraph could readily be imported to the laboratory and run under more controlled conditions using standard psychophysical and neurophysiological techniques. At the neural level, there are several interesting possibilities. For example, single-unit techniques might be employed to assess whether the strength of the response to conspecific calls is greater for left- than right-hemisphere recordings, and even more subtly, to determine whether this asymmetry depends on either the affective or potentially semantic content of the signal. Recalling the points raised earlier in chapters 4 and 7, if Davidson (1992, 1995) is right about the relationship between emotional valence and human hemispheric biases, and if this relationship originated in an ancestral primate species (or earlier), then we might expect rhesus monkeys to show stronger right-hemisphere responses to calls associated with negative/withdrawal emotions, but greater left-hemisphere responses to calls associated with positive/approach emotions. Independently of hemispheric biases, and paralleling work conducted on face-sensitive neurons (see section 4.4), we might also ask which features of a call are either necessary or sufficient for a cell to fire. In the analyses by Rauschecker, Tian, and Hauser (1995), some extremely simple manipulations were performed to address this question. For example, a filter was applied to a coo vocalization to create two stimuli, one with only the lower harmonics present and the other with only the upper harmonics present (i.e., no fundamental frequency). For some cells, at least, there was a significantly stronger response to coos with only lower harmonics than to coos with only upper harmonics.

It is now possible to take the same procedure and run more fine-grained manipulations and perceptual tests. One experiment that I believe would prove highly instructive is a test of categorical perception at the neural level. Here's a quick run through one possible design. Assume, for purposes of discussion, that one can identify cells (cell ensembles) that are selectively responsive to certain calls in the repertoire—more colloquially, that there are certain neurons whose job it is to recognize particular call types. Using our knowledge of the rhesus repertoire, along with some of the intuitions raised by Ehret (1987b; see section 7.3), one would predict categorical perception in cases where functionally differentiated responses to two call-type categories are favored. Thus rhesus monkeys produce a coo vocalization when they find common, low-quality food and produce a harmonic arch when they find rare, high-quality food (see chapter 7). These calls, though structurally quite similar (see Figure 8.1) tend to be associated with different levels of competition and approach behavior in response to playbacks. Using sound synthesis, one can readily create a continuum of signal exemplars from coos to harmonic arches. The primary manipulation would involve increasing the level of frequency modulation in the terminal portion of the coo to attain the inverted-U frequency modulation that is characteristic of the harmonic arch. While recording from cells that respond strongly to prototypical coos, gradually shift the stimuli to calls that increasingly resemble a prototypical harmonic arch. As described in section 7.3, if the nervous system has been designed to process such signals categorically, then one should observe a discontinuity in firing rates between calls processed as coos and those processed as harmonic arches. If, in contrast, the nervous system processes the acoustic variation in a continuous fashion, then one should observe a gradual diminution in firing rates as one progresses from coos to harmonic arches.

The single-unit recording procedures described thus far for adults could also be conducted on infants. Given the recent success of Rodman and colleagues (Rodman, Skelly, and Gross 1991; Rodman, Gross, and Scaidhe 1993) using six-

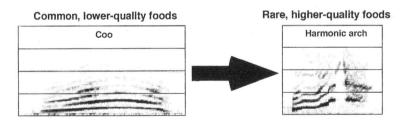

Figure 8.1
Sound spectrograms of a prototypical coo on the left and a prototypical harmonic arch on the right. Coos are given as a response to low-quality, common food, whereas harmonic arches are given for high-quality, rare food.

week old rhesus to test for face-sensitive cells, the prospects are promising for an equally intensive investigation into the ontogeny of putative call-recognition cells. Such studies are important on several counts. First, studies of rhesus monkeys indicate that experience plays, at best, a minor role in guiding the structure of their vocalizations over developmental time (Masataka and Fujita 1989; Owren et al. 1992; reviewed in Seyfarth and Cheney, in press; chapter 5). Changes that do occur ontogenetically are either maturationally driven (e.g., growth of the vocal tract) or represent the effects of acoustic experience on call usage and comprehension (Owren et al. 1993). Second, studies on closely related species such as vervet monkeys indicate that although experience fine-tunes call usage, infants are born with an innate predisposition to correctly use calls in coarsely defined contexts (e.g., "eagle"-sounding alarm calls for objects in the air). It will now be interesting to determine how the circuitry of the brain guides such changes and the extent to which neural wiring remains plastic (i.e., open to change), even after a critical period such as puberty. Given recent evidence in sexually mature pygmy marmosets (Elowson and Snowdon 1994) and chimpanzees (Mitani and Brandt 1994) of convergence in call structure as a function of changes in social conditions, there appears to be sufficient neurophysiological plasticity to support acoustical changes in the adult brain.

Let us return for a moment to the asymmetry results. The current hypothesis is that like humans, rhesus monkeys also show a left-hemisphere bias for processing species-typical vocalizations. Recall that in humans, this asymmetry can also be seen at the level of speech articulation (i.e., production). Thus, as Graves and colleagues (Graves, Goodglass, and Landis 1982; Graves, Strauss, and Wada 1990) have demonstrated, when humans speak, the right side of the mouth, relative to the left, starts the articulation earlier, opens wider, and if perturbed (e.g., by an injection of novocaine), shows greater deficits with regard to acoustic precision. These results provide us with a simple and readily testable prediction. In chapter 4, I presented data showing that adult rhesus monkeys exhibit a strong *left*-side-of-the-face bias during the production of facial expressions (Hauser 1993e). This finding suggests that the right hemisphere plays a dominant role in emotional expression. If rhesus vocalizations represent acoustic expressions of emotion, then one would expect articulatory biases to mirror those documented for silent facial expressions—that is, a left-side-of-the-face bias. However, if the dominant message in the signal is semantic, rather than emotive, then one would expect a right-side-of-the-face bias, as has been observed in humans.

Let me give a concrete example to illustrate this logic, and then summarize the results from some preliminary analyses. When rhesus monkeys are aggressively attacked by a group member, they produce either a silent fear grimace or a scream vocalization. Both the grimace and scream are produced by retracting the lips (i.e., the kinematics of the gesture are similar). Grimaces are associated with a left-side-of-the-face bias. If screams, like grimaces, are primarily emotive signals, then one would also expect a left-side-of-the-mouth bias during articulation. In contrast, if screams convey semantic information, as suggested by the work of Gouzoules, Gouzoules, and Marler (1984), then one would expect a right-side-of-the-mouth bias. Preliminary analyses reveal that some calls are associated with articulatory asymmetries (both timing and degree of mouth opening), whereas others are not. Specifically, coos, which are produced by rounding and protruding the lips, tend to be either symmetric or biased to the left. In contrast, screams and grunts are produced with a strong right-side-of-the-mouth bias. These data, though currently limited to a small sample of individuals and call exemplars, suggest that the asymmetries underlying at least some vocal signals may be quite different from those underlying visual signals. In particular, calls that convey semantic information appear to be primarily processed by the left hemisphere, whereas those that convey affective information are primarily processed by the right hemisphere. If supported by additional experiments, these results would indicate considerable similarity among humans and rhesus monkeys in how the two hemispheres of the brain process language-like stimuli. And, given the relatively close phylogenetic proximity between humans and monkeys, the identified similarity is likely to represent a case of homology, albeit one that is primitive rather than shared derived.

Before leaving the issue of hemispheric asymmetries, let us briefly return to a few developmental problems. Recall that in the playback studies, infants under the age of one year failed to show a perceptual bias in response to either conspecific or heterospecific calls. This result suggests that the development of the adult response is likely to depend upon both maturational factors associated with functional hemispheric differentiation and experiential factors that help the infant learn what each call in the repertoire means. These behavioral results raise some experimental possibilities, designed to reveal how experience fine-tunes the wiring of the nervous system. For example, several neurobiology laboratories conduct experiments with captive-born rhesus monkeys, reared alone but in visual and auditory contact with conspecifics. These individuals have certainly never heard, nor produced, a large number of vocalizations within their species-typical repertoire, because many vocalizations occur in contexts that would not arise under such laboratory conditions (e.g., intergroup encounters, group progressions into new areas, copulation, grooming). The limited experience of these individuals provides an ideal setup for addressing two fundamental questions: first, how does the brain process acoustic signals that are species-typical and yet have never been experienced? Second, once these calls have been acoustically experienced and then associated with a meaningful context, how does the brain change to accommodate this new and putatively meaningful information? Of interest, therefore, would be data from cellular responses to initially novel, then increasingly meaningful calls (e.g., pairing sounds with video clips of relevant contexts), in addition to tracer studies documenting potential changes in how the neural circuitry changes over time. Because tracer studies are necessarily invasive, developmental data can only be extracted from a cross-sectional design. However, owing to the increasing sophistication of neuroimaging techniques (e.g., see review by Posner and Raichle 1994), it will soon be possible to determine how the brain changes in single individuals, longitudinally.

The issue of necessary and sufficient conditions for causing cells to fire can also be approached from a somewhat different direction, which brings us back to the problem of intersensory integration raised briefly in chapter 7. Specifically, when rhesus monkeys call, they, like many other vertebrates, simultaneously produce both auditory and visual signals (i.e., multimodal displays).The visual signal is often complex, consisting of facial expressions and distinctive body postures (e.g., hair bristling, tail flicking, etc.). Recently, single-unit recordings by Brothers and colleagues (Brothers, Ring, and Kling 1990; Brothers and Ring 1992, 1993) have revealed a small number of cells within the amygdala of stump-tailed macaques that appeared responsive to stimuli consisting of both an auditory and visual component, as occurs, for example, when an animal is calling. Assuming that comparable cells will be uncovered within the amygdala of rhesus monkeys or several other cortical regions, several interesting research possibilities emerge. First, in order for such cells to fire, are both sensory modalities necessary or simply sufficient? Second, and most interestingly, do cells require a match between the visual and auditory display components? For example, if a cell fires in response to a video sequence of an individual producing a bark, does it also fire if presented with an individual who is visually articulating a bark but acoustically producing a scream vocalization instead?[1] Although there has been a considerable amount of work over the past ten or so years on intersensory integration and the extent to which any given sensory modality can compensate for deficits experienced by another (e.g., Lewkowicz and Lickliter 1994; Stein and Meredith 1993),

---

1. This kind of manipulation could readily be implemented with digital audiovisual techniques that permit different sounds to be synchronized with different visual displays.

only a small proportion of this work has focused on biologically meaningful signals—signals that the brain was designed to handle.

In summary, the main payoff of a socioecologically sensible neuroscience is that one asks not only *how* the brain processes information, but also *what* kinds of information it was designed to process and *why*. The conceptual and methodological ideas that I have discussed in this section and throughout other parts of the book would never have been conceived in a research program driven by the presentation of artificial stimuli. And this claim applies with equal force to studies of nonhumans and humans. Specifically, in research on nonhuman animals, failure to ask about socioecologically relevant problems would have caused us to miss out on the discovery of hemispheric asymmetries, neural plasticity guided by experience with particular acoustic stimuli, and the specificity of neuronal responses to signals that convey primarily semantic as opposed to affective information. Similarly, work on how the human visual system functions would have missed out on the fascinating status of faces, for both adults and very young infants. By focusing on problems that emerge in the organism's natural environment, we will not only enlighten our understanding of the brain's design, but most significantly, reveal why each species' brain exhibits both evolutionarily conservative and unique design features.

8.2.2   What Do Nonhuman Animal Vocalizations Mean? A Starter's Kit

Quine (1973) has claimed that we shall never really know the meaning of someone else's utterance. This is the problem of referential opacity. Nagel (1974) has claimed that we shall never really know what other animals think, or even perceive, because the computational machinery involved in their thoughts and their perceptions is so fundamentally different from our own. This is the problem of knowing another individual's mind. And several philosophers have argued that only individuals with language are capable of complex thought (reviewed in Jackendoff 1992, in press; Weiskrantz 1988). If these claims are correct, either singly or *in toto,* then surely those who study the semantics or referential properties of animal signals are involved in a research program that is destined to fail. As the reader must suspect by now, neither I nor several other ethologists studying animal communication are convinced that these arguments are absolutely correct. Nonetheless, I am convinced that if we are going to make progress in this area, it will require at least two major efforts: first, linguists and philosophers will need to work with (i.e., rather than against) behavioral biologists studying animal communication so that the critical issues underlying linguistic representations can be clearly articulated and explored in nonhuman organisms; second, and most

importantly, there is a dire need to develop new experimental techniques that
will uncover the fundamental units of communication, their relationship to one
another, and how they function in representing significant features of the organism's socioecology. In this section, I would like to explore some potential research paths by working through a few case studies that have been described in
previous sections. The context for this discussion will be a kind of play-by-play
handbook that will set out a research program which, I believe, is feasible and
designed to provide a more refined description of what nonhuman animal vocalizations mean. If this program succeeds, it will lay the foundation for a more
profound assessment of syntactic structure—at least, the possibility of syntactic
structure in nonhuman animals.

Let's start at square one: an ethologist's observations of a nonhuman animal's
vocal behavior. And to make this exercise more concrete, let us consider vervet
monkeys, since, for the problem at hand, our understanding of their vocal utterances is by far the most sophisticated to date. As Struhsaker so elegantly demonstrated in the late 1960s, the first step in our idealized research program is to map
out both the socioecology of the calling environment and the acoustic morphology
of the repertoire. Thus the vervet's repertoire can be captured by a multidimensional acoustic space. This space is, of course, only a rough estimate of the
potential range of acoustic and socioecological variation. It is rough, in part,
because changes in the ecology not only will influence the kinds of social relationships exhibited by a given species, but also will exert direct pressure on the form
of the signal produced (see chapter 3). Thus the topography of a species' vocal
behavior will represent the product of a series of compromises between constraints imposed by the nervous system, the peripheral organs involved in sound
generation (e.g., the vocal tract) and perception (e.g., outer and inner ear), the
structure of the habitat, and the selective advantage of signaling precisely to some
individuals and not others.

Methodologically associated with a description of the contexts in which particular calls are used is an assessment of how individuals respond to such calls.
Describing response patterns is not only important with regard to understanding
call function, but also provides the basis for establishing behavioral assays to be
used in perceptual playback experiments (see step three in this section) and for
generating proper inferences about the units of communication. Concerning
vervet monkeys, observations and experiments have shown that each alarm call
type elicits a highly specific response, with design features functionally suited to
the hunting strategy of the predator. As such, vervets have kindly provided the
ethologist with a beautifully packaged set of responses. In contrast, other call

systems, especially those associated with nonpredatory contexts, often lack clearly definable responses. And in this sense, some of the more interesting, socially inspired vocalizations produced by nonhuman animals are like the conversations exchanged among humans—words are passed from mouth to ear and are acknowledged by a subtle nod or, more commonly, sustained eye contact, expressing interest. Consequently, semantic analyses of an utterance depend on the ability to extract meaning from extremely subtle nonvocal behavioral responses. I will return to this problem later because it turns out to be the methodological linchpin for understanding what nonhuman animal vocalizations mean.

A brief parenthetical point: the process of describing the contexts and consequences of call production is by no means a trivial one, and yet it is often bypassed or treated superficially in current research on nonhuman animals. One reason for this neglect is that there has been a growing movement within behavioral biology to run experiments aimed at teasing apart a potentially complex array of correlated factors. Although such experiments are conceptually well grounded, their design and implementation rely on a strong descriptive foundation. Before moving to step two, therefore, I would like to propose a few emendations to the initial observational protocol that will enrich the descriptive process and thereby provide a more adequate foundation for the experimental phase of the research. In describing vocal signals, researchers classically provide information about socioecological parameters external to the caller. Such features of the calling environment are clearly important for establishing call function, but should also be supplemented by either indirect correlates or direct measures of the caller's affective state. Considering indirect measures first, animals often exhibit distinctive facial expressions and body postures during vocal signaling. Thus, while being attacked, an individual might grimace, crouch, or put its tail between its legs. These nonvocal features of the display, readily captured by video recordings, can provide independent measures of the level of fear or, more generally, the individual's emotional state. Similarly, a motivational state such as hunger can be qualitatively measured by monitoring the amount of food intake or, if chemical composition is known, quantifying nutrient intake. More directly, some situations may permit quantification of physiological changes known to be associated with changes in emotional or motivational state. For instance, urinary and fecal samples, commonly produced by animals in a state of fear, can be assayed for corticosteroids. More invasively, individuals can be captured, blood samples collected, and hormones assayed (see Sapolsky 1992 for an elegant example of field studies of neuroendocrinology in yellow baboons). Together, both direct and indirect measures can be effectively integrated into one's descriptive analyses of the acoustic correlates of affective state.

Step two in our research program is primarily statistical and involves analytically dissecting the pool of potential variables into those that account for the
greatest proportion of variation in signal structure. Thus we might find that a
vocalization with spectral parameters $S_1 \rightarrow S_n$ and temporal parameters $T_1 \rightarrow T_n$
is only given in the context of predator detection (i.e., researchers have only
heard the call produced by animals confronting a predator)—in particular, leopard
detection. One's first inference might be that the call means "leopard" or "There
is a leopard." As in Quine's case of "gavagai," however, interpretations or inferences of meaning are fragilely dependent upon the quality of the observational
database (a statistical sampling issue). Thus, when a white rabbit first runs by
and the native shouts "gavagai," the linguist's first inference is that gavagai
means rabbit.[2] However, there are several alternative inferences that cannot be
ruled out based on current observations: for example, white object, moving object, detached rabbit parts, and so on. If a second rabbit runs by that happens to
be gray, we can now at least rule out the inference that gavagai means white
object. And so on. In the vervet case, a call associated with seeing a leopard
could, in theory, mean different things depending upon both the caller's emotional
and mental state and the action and location of the predator. A limited observational sample might lead to a narrow interpretation of call meaning. Struhsaker's
original account was based on hundreds of hours of observations, including a
large number of predator encounters. Consistent with traditional interpretations
of warning signals (W. J. Smith 1969, 1977), however, the vervet's alarm calls
were originally considered to reflect changes in the signaler's emotional state,
providing no information whatsoever with regard to the putative external referent
or eliciting stimulus. The acoustic variability in the signal was attributed to differences in the level of threat imposed by each of the major predator types. Thus
leopards posed one kind of threat, eagles another, snakes yet another, and so on.

A critical insight into call meaning came several years later when Marler (1978,
1982) suggested that animal signals might be symbolic, mapping on to natural
categories that were functionally meaningful given the species-typical ecology.
This proposal acknowledged the importance of affective states in structuring animal signals, but further suggested that an organism's acoustic space was composed of both affective and referential components. Thus, although vervets may
experience different levels of fear when confronting a leopard as opposed to

2. This is a reasonable inference, especially when one considers the observation that human children
at the earliest stages of language acquisition first assign taxonomic content to words, and only later
assign thematic content; see discussion in chapter 7 and, especially, some of the writings of Ellen
Markman (1990; Markman and Hutchinson 1984).

an eagle, calls for these predators differ, acoustically, because of their external referents. The crucial test of this hypothesis was provided when Seyfarth, Cheney, and Marler (1980a) conducted their playback experiments with tape-recorded vervet alarm calls. In our model research program, this represents the third step. In the initial experiments, call parameters believed to be associated with changes in affective state (e.g., amplitude, call duration) were pitted against the overall spectrotemporal structure of the call, believed to convey information about predator type. Results showed that vervets responded to playbacks of alarm calls as if a predator had been detected—they ran up into trees for leopard alarm calls, scanned the sky or ran under a bush for eagle calls, and stood bi-pedally and scanned the ground below for snake alarms. In other words, the acoustic properties of each alarm call type appeared sufficient with regard to eliciting behaviorally appropriate responses. Moreover, the acoustic features as-sumed to be associated with changes in affective state failed to have a significant effect on the general response exhibited (i.e., whether leopard alarm calls were loud or soft, long or short, individuals responded by running up into a tree). These results led to the conclusion that vervet alarm calls were, rudimentarily, like human words in that they referred to external objects—specifically, different predatory species.

Before addressing the validity of this claim and describing the next experimen-tal step, let us first return to the problem of behavioral response assays. Excluding studies of functionally referential alarm calls (reviewed in Macedonia and Evans 1993), where highly distinctive antipredator responses are elicited, most playback experiments have employed a head-orienting response as an assay to examine call meaning. Specifically, when a call is played, the latency and amount of time spent looking toward the speaker are quantified using frame-by-frame analyses of film or video footage. For example, Gouzoules, Gouzoules, and Marler (1984) showed that in rhesus monkeys the latency to orient and the amount of time spent looking toward the speaker were significantly influenced by the type of recruitment scream played back. These results have been interpreted as evidence that each scream type provides rhesus with perceptually meaningful information about the nature of the agonistic relationship.

The orienting response assay is plagued, however, with problems that relate directly to the issue of just noticeable as opposed to just meaningful differences (see chapter 3). Specifically, latency (i.e., reaction time) is an extremely fragile measure even when it is extracted under highly controlled laboratory conditions. Though some field studies have obtained statistically significant differences in response latency across call conditions, it is unclear what such differences mean with regard to differences in call function or semantics. Latency to respond can be influenced by a host of factors, none of which may speak to the problem of call meaning. Thus, in the case of rhesus screams, why would one expect a "noisy scream" (given by a recipient of physical aggression) to be associated with a shorter response latency and a longer response duration than a "pulsed scream" (given by an individual following a nonphysical attack)? And, even if one can generate an a priori prediction (e.g., some calls are more urgent than others because of impending threat), what, precisely, do such temporal measures tell us about call meaning? They surely tell us that two or more calls are discriminable and that given the specifics of the perceptual context, one call is potentially more interesting (i.e., in terms of capturing the subject's attentional resources) than the others. But these experiments have been conducted to determine call meaning, to establish the putative referent of the rhesus monkey's scream. To contrast with vervets, when an individual scans the sky upon hearing an eagle alarm call, there is surely some kind of referent in mind—scanning in this case is not a random act of vigilance, but one that is targeted toward avian rather than terrestrial predators such as leopards and snakes. When rhesus look toward the speaker, although this action clearly shows that they are interested in the signal generated, *what,* precisely, are they interested in? Here is where the assay fails us. Can we do better? Yes, and as described in chapter 7, Cheney and Seyfarth were once again responsible for moving us closer to our goal, of doing what Dr. Doolittle apparently could do—talk to and with the animals. We have arrived at step four: the next generation of playback experiments.

In human language, words that sound the same (e.g., *right* and *write*) can have fundamentally different meanings. Conversely, words that sound fundamentally different (e.g., *car* and *automobile*) can have the same meaning—can be used interchangeably. Cheney and Seyfarth took advantage of a habituation-dishabituation technique to ask vervets whether calls that are acoustically dissimilar but given in contextually comparable situations are primarily evaluated on the basis of their acoustic or semantic properties. As described in chapter 7, vervets treated intergroup wrrs and chutters as similar—following habituation to wrrs, individuals failed to dishabituate to chutters, and vice versa. This pattern was only obtained, however, when the calls were from the same individual. When the methodologically identical experiment was run but with eagle and leopard alarm calls, dishabituation was consistently obtained independently of caller identity. Cheney and Seyfarth (1990, 1992) interpret these results as providing evidence that vervet calls are referential in a strong sense, in that similarity in meaning is derived from information about the putative referent, not acoustic similarity or dissimilarity.

As briefly mentioned in chapter 7, there are problems with the design of these experiments. Some of these problems can readily be fixed, whereas others may well be intractable. Let us discuss some of the remedies first. Consider the wrr-chutter playbacks. On day one, a single chutter from Carlyle is played. On day two, Carlyle's wrr is played eight times (habituation series), one exemplar per trial, each trial separated by approximately 20 minutes; in some sessions, different exemplars of Carlyle's wrrs are played, whereas in other sessions, the same exemplar is repeated. Following the habituation trials, a single exemplar of Carlyle's chutter is played, representing the test trial. For all trials, the response assay is the duration of time spent looking in the direction of the speaker.

I think this design can be improved by implementing the following, relatively minor changes. First, run all of the playback trials on one day. This procedure will provide greater uniformity with regard to ecological factors (e.g., changes in climate), group dynamics, and motivational state of the test subjects. Second, only present unique exemplars of a given call type. If the same call exemplar is used, then one cannot distinguish between habituation to *some sound* presented repeatedly, and habituation to *a call type* that is, inherently, characterized by a range of acoustic variation; the latter is what is desired for this kind of experiment, and many other playback designs as well (see discussions in McGregor 1992). If multiple call exemplars are not available, there is a partial solution: sound editing and synthesis to create new stimuli. Given modern digital technology (Beeman, 1996), it is now possible to take original call exemplars and make subtle modifications to spectrotemporal features of the signal. Such changes must, however, be guided by observed variation at the level of either the call type or the individual. Third, short intertrial intervals (i.e., the time between repeated playbacks of the same call type within the habituation series) are preferable. The logic behind this suggestion is that as intertrial interval increases, subjects are more likely to change motivational and attentional states, and such a change, in and of itself, can cause dishabituation. The problem with running shorter intertrial intervals, of course, is that it may represent an artificial situation. Intertrial interval will ultimately be constrained by the natural rate of call delivery, and this will depend upon the calling context. For example, alarm calls are often given in clusters—a massive bout of calls given within a short period of time, but then followed by a long gap of silence, subsequent bouts determined by the timing of predator encounters. In contrast, intergroup wrrs and chutters are often heard several times a day, almost every day. Such natural variation in call delivery must be considered at the design phase of the experiment. Fourth, instead of establishing a fixed number of habituation trials prior to testing, continue running habituation trials until the subject fails to respond (e.g., fails to show an orienting response toward the speaker) on two consecutive trials. There are a number of advantages to this approach. If there is no orienting response following two repeated playbacks, then one can be confident that the target subject has habituated. In contrast, if an individual fails to respond to a single presentation, there are several potential explanations (e.g., the call wasn't perceived, something else was perceptually more salient, etc.). Additionally, because all subjects enter the test phase following two consecutive trials with no response, one can more readily compare the magnitude of the dishabituating response across individuals. In the Cheney and Seyfarth design, only eight habituation trials are conducted, and thus each individual enters the test trial in a putatively different perceptual state: for some individuals, eight habituation trials will be sufficient to cause either no response or a highly diminished response, whereas for other individuals, additional trials will be necessary to bring the response levels down. By bringing response levels to zero, all subjects are likely to be in a more comparable attentional and motivational state. As a result, the test trial is more likely to extract what the researcher is after: a window into what the animal considers to be a perceptually meaningful difference between acoustic stimuli. Fifth, if the subject fails to respond on the test trial, run one last trial with a completely novel call type. The reason for doing this is to establish that the test subject has not simply habituated to playbacks in general. For example, if the subject has failed to respond to two consecutive playbacks of stimuli from call type A and a playback from call type B (test trial), then play back a final exemplar from call type C. If subjects have not habituated to the general playback setup, they should respond to call type C.

Now the potentially intractable problem: consider again the eagle-leopard alarm call playbacks. Here, habituation to one call type consistently leads to dishabituation to the other. Unfortunately, unlike wrrs and chutters, eagle and leopard alarm calls differ both acoustically and semantically. Thus one cannot determine whether vervets dishabituate because of a perceptually salient acoustic difference or because of a referentially meaningful difference. Moreover, such experiments do not allow us to distinguish between a call that means "leopard" and a call that means "Run up into a tree." I would like to explore two possible solutions to this latter problem.

Possible solution one: start by creating vervet groups comprised of two individuals. Each enclosure should have a few trees. In the first phase, remove one group member and place him or her in a holding cage that is acoustically and visually isolated from the home cage. Present a leopard to the remaining group member. If Cheney and Seyfarth's (1990) field observations hold, this individual should remain silent. If calls are produced, however, and if they mean "leopard," then one would expect to hear such calls when the individual is either on the ground or up in the trees. In contrast, if the call means "Get up into a tree," one would not expect to hear such calls unless alarm calling is an absolutely reflexive response or unless vervets call even when other group members are out of sight (i.e., they always consider other group members to be within earshot of their calls). The second phase is, I believe, somewhat more telling. Here, keep the group of two intact, and now, when subjects are either both on the ground or both up in the trees, present the leopard. If the call means "leopard," we would expect to hear such calls regardless of whether subjects were up in the trees or down on the ground. In contrast, if the call means "Run up into the trees," then one would not expect to hear such calls if both group members were up in the trees and could see each other. This experiment seems quite clear with regard to its interpretation, with one small hitch: the call could mean "Stay up in the trees." Returning to the problem of referential opacity, although we may never provide a completely precise account of what a vocalization means or refers to, we can conduct experiments that eliminate several reasonable inferences. In this sense, the study of call meaning is in the same boat as the study of concepts (see chapters 3 and 7): behavioral assays cannot definitively specify the content of the representation, but they can refine the potential boundary conditions. In the experiment described, it would be possible to eliminate the inference that the vervets' alarm call to a leopard is a label or descriptor of the object (i.e., a certain kind of predator), leaving the possibility that it is an instruction or command of sorts.

Possible solution two: a modified version of tests of language comprehension in human children. Like other areas of research on early infant cognition (see chapter 7), studies of language acquisition and particularly language comprehension have been forced to develop analytical techniques that rely on assays other than the subject's verbal report. A promising approach, especially because it can be imported into studies of nonhuman animal communication, is a paradigm pioneered by Hirsh-Patek and Golinkoff (1991). In the typical setup, a child is seated in a chair, facing two video monitors. Centered between each monitor is a light, a concealed speaker, and a video camera. When the child's attention is recruited to the light, a voice emerges from the speaker, producing a sentence that describes the event in one monitor but not the other. Thus, in monitor one the child might see a picture of Big Bird giving Cookie Monster a cookie, whereas in monitor two she would see Cookie Monster giving Big Bird a cookie.[3] The expectation is that if the child understands the relevance of syntactic structure, then, upon hearing "Big Bird gives Cookie Monster a cookie," there should be a preferential orienting response and fixation to monitor one. And this is precisely what Hirsh-Patek and Golinkoff find.

The following represents a nonhuman-animal version of the Hirsh-Patek–Golinkoff procedure, focusing on the vervet's alarm call system. Set up a situation where an individual can watch two side-by-side monitors with light, speaker, and video camera concealed as described in the preceding paragraph. In step one, present a video sequence on monitor one of a leopard moving around on the savanna; the leopard is the only animal present in the scene. While this image is presented, play back a tone, an arbitrary sound that is not associated with leopards. In the second step, present video footage on monitor two of vervets running up into a tree. Pair this sequence with a different arbitrary sound such as broadband noise. For step three, present both video sequences simultaneously, and combine these with yet a third arbitrary sound, different from the others. As in Hirsh-Patek and Golinkoff's design, these first steps (1) establish whether subjects show a visual preference for particular images and (2) inform subjects that visual stimuli can be presented either independently on each monitor or simultaneously, and that auditory stimuli are produced at the same time as the visual stimuli. Now comes the critical test phase: simultaneously present a leopard sequence on monitor one and the vervet running sequence on monitor two, but play back a leopard alarm call. If the call is a descriptor for leopard, then we would expect subjects to look preferentially at monitor one, whereas if it is a command for antipredator behavior, we would expect subjects to look at monitor two. If consistent results are obtained, there are several interesting and methodologically important probes that should be run. For example, in monitor two, one could show vervets running under a bush or scanning up in the air. Although these represent species-typical antipredator responses, they are inappropriate given the putative referent of the call (either "leopard" or "Run up into a tree"). Here, one would expect vervets to look preferentially at the leopard, since it is the best match to the sound. Other manipulations might involve presenting (1) images of species that look like leopards (e.g., cheetah) and might be considered

---

3. Hirsh-Patek and Golinkoff have implemented several important controls, such as presenting each visual stimulus alone with a vocal utterance that merely draws the child's attention to the monitor but does not describe the interaction between individuals in the stimulus (e.g., "Look at Big Bird!"). This procedure allows them to disentangle potential preferences prior to the test phase.

appropriate targets for their calls, as opposed to species that look quite different from leopards but nonetheless prey on them (e.g., snakes and raptors), and (2) scenes of vervets in different behavioral activities, some of which would be appropriate escape responses to leopards (e.g., climbing higher into a tree) and some of which would be inappropriate (e.g., running out into the open area). Again, although these experiments will not yield a definitive conclusion on call meaning,[4] they will narrow the range of possibilities.

The potential impasse reached at the perceptual level is not the end of our research program, however, for we have yet to push the production level, especially in terms of its impact on understanding the fundamental communicative units. Back to the vervet's repertoire. When Marler (1969a, 1970c, 1973) first started his comparative analyses of primate repertoires, vervets were considered to represent the ideal discrete case, whereas chimpanzees were considered relatively more graded (Marler 1976). The notion of discrete as opposed to graded repertoires makes the fundamental assumption that the relevant unit of analysis is clearly comprehended by the researcher (see discussion in chapters 2 and 3). Thus vervets were considered to exhibit a discrete call system because signals with well-defined acoustic boundaries could be readily discerned, using spectrographic techniques. This distinction is problematic on two counts. First, though acoustically bounded signals can be established, vervets often produce heterogeneous call bouts, each bout consisting of different call types (Hauser and Fowler 1991).[5] Consequently, although the call may represent one unit, a string of calls may represent a higher-order unit and one that makes the discrete-graded distinction more difficult to assess. Second, and more importantly, what we perceive as a discrete call type may not represent the fundamental unit from the vervets' perspective. To establish the vervet call unit or units, as we have established for humans (e.g., phonemes, words, phrases, etc.), and I believe to some extent songbirds (Cynx 1990; Margoliash 1983, 1987; Nowicki and Nelson 1990; Vu, Mazurek, and Kuo, 1994), we will require more sophisticated observations, experiments, and analyses than have been conducted to date. I now turn to a set of possible approaches that should help our search for functionally meaningful communicative units and their putative referents.

---

4. For example, vervets might preferentially look at the leopard when a leopard alarm call is played not because this is what the call means but because leopards are what typically trigger the call; vervet monkeys running up into a tree fail to trigger the call.

5. Note that a *bout* is considered to represent an acoustically glued unit, for individuals appear to generate such utterances during expiration and with relatively few pauses—the pauses that are inserted appear necessary to generate discretely separated calls within a string of calls.

To more fully understand the relationship between call morphology and changes in the caller's affective state and socioecological environment, experiments that manipulate these factors independently are required. Though such experiments have not been conducted with vervets, the first round of attempts has been completed for rhesus monkeys and domestic chickens. Specifically, as reviewed in chapters 6 and 7, Marler, Evans, and their colleagues have manipulated food and predator type, including a host of subtle features relevant to their categorization. Although most of the analyses have focused on the necessary and sufficient conditions for eliciting a call (i.e., presence/absence), studies of food calls have demonstrated that males increase call rate as a function of increasing food quality. What is difficult to discern, based on the analyses conducted thus far, is whether call rate maps onto variation in the caller's motivational state or whether call rate is an acoustic parameter that refers to food quality. Stated more transparently: when a chicken calls at a high rate it could be saying, "I am really *excited* about this food," or, "I have found some *high-quality food*," where the words in italics emphasize the main message in the signal. Although it is not yet possible to distinguish between these alternatives, the chicken system is ideally suited to carry such analyses further, especially since one can directly manipulate, through novel video technology, a host of features that are putatively relevant to their calling bahavior. Moreover, powerful signal analysis tools will permit more intricate quantification of the sources of acoustical variation.

Studies of rhesus monkeys have also experimentally manipulated the calling environment, and results presented in chapters 6 and 7 revealed that hungry individuals call at higher rates than do relatively satiated ones. Hunger level, however, does not appear to influence the type of call used during food discovery. Which call type is used depends on the type of food discovered. Thus, in contrast to the current state of affairs for chickens, the rhesus data suggest that when individuals call at high rates they *are* saying, "I am really *excited* about this food," but in order to convey information about the type of food discovered, they must alter the spectrotemporal features of the signal, producing, for example, harmonic arches for high-quality, rare food and grunts for low-quality, common food. And yet, even here, many ambiguities remain. We don't know, for instance, why rhesus monkeys have three acoustically different call types for high-quality, rare food items. Perhaps they are synonyms like *food*, *grub*, or *chow*. Alternatively, perhaps they reflect even more subtle variations in the affective dimension. Thus, in terms of high-quality, rare food, "warbles" may reflect relatively low-level excitement or motivation, whereas "chirps" may reflect relatively high-level excitement or motivation. As in statistical studies of avian communicative units, these call types clearly segregate acoustically, and yet we do not have a good handle on their meaning. It is now necessary to run additional experiments, where selective features of the calling context are manipulated, calls recorded, and detailed acoustic analyses performed, to determine the mapping between contextual and acoustical variation. Having carried out these steps, it will then be possible to return to the perceptual arena and conduct playback experiments to determine which components of the signal are meaningful.

A final approach to the study of call meaning is to carry out developmental studies, aimed at uncovering how subjects innately "label" or vocally respond to events in their environment and, further, how experience fine-tunes the process of categorization. For vervet monkeys, Seyfarth and Cheney's astute observations showed us that infants under the age of one year commonly produced eagle-sounding alarm calls to nonpredatory birds (i.e., species that pose no threat to vervets of any age). Although the class of objects eliciting such calls shrinks over time, certain features of the category remain stable. Thus vervet infants appear to be born with at least one predatory category that can be loosely specified as "dangerous things in the air." A suite of experiences (e.g., seeing group members respond both behaviorally and vocally) is then responsible for what appears to be an attrition of objects and events that elicit eagle alarm calls. In the field, it will be difficult to determine how the timing and type of experiences affects vocal development. In the laboratory, however, there are several options. For example, predatory species could be presented to infants while alone, visually and/or acoustically isolated from group members. This kind of setup would enable one to determine how visual and acoustic feedback from adults influences both the content of infant categories (e.g., "predatory birds of prey"), as well as the speed with which such categories are refined over developmental time. Of additional interest would be an experiment designed to assess whether there are innate constraints on the putative range of referents assigned to a particular call type. Thus the field data suggest that vervets only give eagle alarm calls to things in the air. The critical question here is whether they can be "convinced" that eagle alarm calls are also given to things on the ground. For example, mirroring experiments by Evans and Marler (in press) on chickens, would an infant vervet monkey give an eagle alarm call to a martial eagle on the ground? What about a leopard up in a tree? Both of these events occur naturally, but it is unclear how such spatial factors influence call production. Additionally, one could present an eagle on the ground or leopard up in the tree, while simultaneously playing back eagle and leopard alarm calls, respectively. Given such feedback, are infant vervet monkeys more likely to produce eagle alarm calls for eagles on the ground or up in the air? And if eagle alarm calls are played back during a presentation of a leopard up in the tree, can infants be convinced that this is the appropriate referent or match? Though these questions have not yet been addressed, they are certainly feasible.

I would like to conclude this section by suggesting how a more sophisticated understanding of the meaning of nonhuman animal vocalizations will shed light on both the adaptive significance of referentially precise signals and our understanding of the constraints on human language evolution, a topic that I will return to in section 8.3. Concerning the first, I strongly believe that the methodological tools that I have discussed, together with others that some of my colleagues are bound to think of, will bring us much closer to a quantitative analysis of call meaning. To repeat a primary theme of this section, though we may never know the specific meaning of a vocalization, the research program sketched here will, minimally, allow us to rule out several *possible* meanings. And it is in this area that an evolutionarily informed research program can help. By looking carefully at the socioecological problems faced by an organism, we are in a strong position to generate predictions about the design features of their signaling system. Thus, in the vervet monkey's case, knowing that this species is subjected to a variety of predators, each with its own hunting strategy, tells us that an all-purpose predator alarm call would fail miserably, because an all-purpose alarm call would simply signal that danger is nearby but nothing more. However, because eagles hunt from above and are adept at scooping vervets from trees, running up into a tree would be a bad move. Bad alarm-call system. Dead vervet. Thus, as Cosmides and Tooby (1994a,b) have carefully articulated in light of human cognitive adaptations, selection will favor increasingly specialized systems, wiping out the failed solutions and favoring those that lead to survival and reproduction. Such analyses must, however, be guided by an understanding of constraints—anatomical, physiological, developmental, and cognitive. Thus, although selection would presumably favor an alarm-call system that specified the distance and motivational state of the predator, the vervet's neural hardware and software may simply not be up to this task, at least for the moment. But selection is a tinkerer and will play with the goods provided. It will also capitalize on favorable mutations. Thus, as predators hone their prey-munching tactics, there may come a day (i.e., down the line, evolutionarily) when vervets produce an alarm call that specifies the predator's distance and indicates whether or not it is hunting around for a meal—especially a vervet meal.

A more thorough understanding of call meaning will also pay off handsomely with regard to current insights into the evolution of the human capacity for syntax. Specifically, we cannot even begin to make headway into a comparative analysis of syntax without completely understanding the units of each species' communication system as well as their meaning. Once established, however, it is then possible to determine whether units are recombined, whether they exhibit the property of generativity (see discussions by Corballis 1992, 1994; Bloom 1994b), and if so, what this buys the organism in terms of the richness of its communicative expressions. Black-capped chickadees, for example, produce a call that is made up of four notes, A, B, C, and D. Hailman and colleagues (Hailman, Ficken, and Ficken 1985, 1987) have demonstrated that this call exhibits combinatorial properties, where notes are strung together in a potentially infinite set of sequences, guided by certain constraints. Although there is some intuition that each note may mean something different, no formal analyses have been conducted. Without any sense of this system's semantics, however, there can be no real sense of its syntax. There are two further problems that must also be addressed. Even if each note means something different, the comparison with human language fails unless it can be demonstrated that each unique sequence of notes conveys some subtly different piece of information. Second, some of the statistical techniques that have been used to categorize notes as A, B, C, or D have yielded discrete note categories, whereas others have produced greater overlap or gradations among exemplars (Nowicki and Nelson 1990). If the notes are not discrete—and this question must be assessed at both the production and perception level—then it will be necessary to reevaluate the claim that this system is combinatorial.

8.2.3   A Comparative Method for All: Looking Time

In section 7.3.4, I described the preferential looking time technique used by developmental cognitive psychologists interested in assessing the conceptual representations of young preverbal infants. This technique is, in my opinion, ideally suited for studies in comparative communication. I want to repeat some of the issues raised previously because I am afraid that in a variety of circles, there is a surprising level of confusion about the purpose and validity of this technique.

Recall the fact that as normal human adults, we have developed an extraordinary number of expectations about the way things work in the world. Our expectations about physical properties of the world often differ from those that emerge in the context of living things, especially living humans. We expect, for example, that when coffee is poured into a mug, the mug will retain the coffee. We would be surprised to find the coffee sprinkling out from the bottom of the mug. That is, one expects coffee to remain inside the mug because mugs are containers and, as such, hold things in place. Similarly, after seeing a man and woman walking down the street arm in arm, we would not expect the woman to turn toward the man and slap him. However, in the absence of hearing their conversation or seeing their faces, the woman's reaction might not be totally unexpected. The main point here: we readily form expectations about the nature of particular events in the world, and when such expectations are violated, we are surprised and seek explanations for such violations. And one way that we do so is by allocating significant attentional energy to the observed event, attempting to obtain additional information about the causes of change.

No one would expect human infants to generate the same kinds of expectations that adults generate, and one would certainly anticipate an even greater difference between human adults and nonhuman animals, at least for some conceptual domains. In relative terms, however, the cognitive gulf between human infants and nonhuman animals might not be as great. If this were the case, it might shed light on what human children need to develop, and nonhuman animals would need to evolve, in order to implement some of the fancy cognitive skills that human adults recruit for understanding various facets of the animate and inanimate world.

When developmentalists use the preferential looking procedure, they take several steps to insure that the results illuminate the nature of the child's representation. There are at least three key points here. First, all of the objects and events involved in the experimental task are shown to the infant during either habituation or familiarization trials so that stimulus novelty can be ruled out as the primary factor guiding changes in visual attention. Second, different experimental conditions are set up in order to provide converging evidence relevant to a single conceptual representation. For example, Baillargeon (1991, 1994) and Spelke (1991, 1994) have conducted several experiments to evaluate Piaget's (1952, 1954, 1962) claims about the developmental timing of object permanence. With a Piagetian search task, children show evidence of object permanence toward the end of the first year. In contrast, results from looking time tasks reveal that 3–5-month-old children understand that objects placed out of sight continue to exist—that out of sight is still in the mind. These studies and several others indicate that it is a mistake to consider any single looking time experiment as definitive evidence for the nature of the child's representation. Rather, it is the combination of different experiments that provides confidence about the details of what children know (Baillargeon 1994; Spelke 1994). Third, all of the recent looking time experiments have adopted a cross-sectional approach, with the result that developmental changes in the details of the representation can be described and often accounted for by reference to changes in the child's direct and indirect experiences with the world (e.g., understanding the concept of *support* arises, in part, from the experience of placing objects on top of other objects and watching them either stay in place or fall). In sum, therefore, like several other researchers in developmental cognition, I find that the preferential looking procedure provides an elegant entry into the mind of the nonlinguistic organism. As a result, the door is open for comparative analyses. I now describe how researchers in cognitive psychology, animal behavior, and the neurosciences might profitably form an intellectual consortium with the goal of understanding how particular representations guide communicative interactions within the animal kingdom.

To set the empirical stage, consider the problem of individual recognition. As mentioned in sections 7.2 and 7.3, recognition of individuals occurs in both the auditory and visual domains for a wide range of species. Among primates, both the face and voice appear to convey meaningful information about identity. One question that has yet to be addressed in detail, however, is whether individuals are capable of matching face to voice, and if so, how this matching is accomplished. Bauer and Philip (1983) provide some experimental insights. They started by training a small group of chimpanzees to discriminate familiar faces and voices, and then to match face to voice. By reinforcing the chimps for correct choices, the experimenters were able to show a fairly high level of accuracy with regard to matching. With such training procedures, however, it is not completely clear which aspects of the process are naturally tied to the subject's cognitive tool kit. Moreover, because Bauer and Philip only varied a few dimensions of the face and voice, the mechanism underlying cross-modal matching is ambiguous, as is the nature of the representation. For example, it is unclear whether the chimpanzees were matching face to voice or, more simply, a few visual images with a few sounds. The preferential looking paradigm provides a potential way around these problems. One could, for example, present socially living chimpanzees (or any other species) with photographs of two group members; the two photographs would be presented in such a way that their spatial separation (e.g., on a monitor) permitted accurate measurement of the direction of eye gaze by test subjects. Synchronized with the presentation of faces, one would then play back a call from one of the two subjects. The researcher would measure differential visual fixation to each photograph.[6] If individuals recognize which face belongs to which

6. For relevant controls, see the previous discussion of a similar experiment on call semantics, modeled after the work of Hirsh-Patek and Golinkoff.

voice, then the preferential looking procedure would be expected to pick up this ability. To flesh out the sophistication of this ability, and to insure that subjects were in fact looking for faces and listening to voices (i.e., as opposed to simpler features), one could selectively alter various properties of the facial and vocal stimuli. For example, change the orientation, size, color, and expression of the face, using static photographs and digitized video footage. For vocalizations, present different call types by the same individual or use sound synthesis techniques to selectively manipulate features of a given call type. As one would expect from humans, if chimpanzees associate faces and voices from individuals, then it shouldn't matter whether the individual is viewed in profile or in full view, or whether the vocalization played is a food call or an alarm call.

If individuals correctly match face to voice, one could then present the test subject's own face and voice to assess whether *self-recognition* is possible: that is, match own face to own voice. This test might provide a nice companion technique to Gallup's mirror recognition procedure (see discussion in Gallup 1970, 1991; Povinelli 1993). Last, one could carry out all of the preceding steps, but do so while obtaining cellular recordings from the brain. In particular, given the observation that some cells within the temporal cortex and amygdala of macaques are polymodal (auditory and visual), one might find cell populations that are more responsive to matched than to nonmatched vocal-facial stimuli. As such, one might expect to find changes in firing rate as the subject alters his or her visual fixation from the inappropriate to the appropriate face-voice pairing. And jumping taxonomic study organisms, this kind of experiment is likely to be particularly telling in songbirds where there is already evidence of autogenous cells that respond selectively to the individual's own song, but not to songs from the same dialect, or even the bird's own song played backward (Margoliash 1983, 1987).

Now imagine a somewhat more complicated problem, one that has escaped firm empirical support in studies of nonhuman animals: the communication of intent (see discussion in chapter 7). In several experiments, researchers have demonstrated that at an early age, children respond appropriately to gestures used by adults to convey, minimally, their interest in an object (goal-oriented attention, sensitivity to gaze) and, somewhat more speculatively, their beliefs about the object and how they intend to interact with it (Baldwin 1991; Baldwin and Moses 1995; Campos 1983). Thus a child who sees an adult smile with outreached arms toward a novel object is likely to mind-read this action as "The adult wants the object." The fact that the child sets up such expectations based on the communicative information proffered by a smile and movement of the arms suggests that situations could be created that would violate such expectations. And the same might be expected of nonhuman animals.

Consider, therefore, the following experiment, designed for both humans and nonhuman primates.[7] In the first familiarization scenario, subjects view an individual who opens a refrigerator door, sees an apple (the only item inside), smiles, takes the apple out, and then shuts the door. The second familiarization scenario involves the same subject, but now she opens the refrigerator door, sees a rotting piece of cheese, produces a disgust face, and then closes the door. The test phase then involves several different scenarios. In one possible condition, the person opens the refrigerator and sees the apple on one shelf and the cheese on another. Looking at the apple, she smiles, pulls it out, and shuts the door. In the second possible condition, the refrigerator is opened, she sees both the apple and cheese, gives a disgust face while looking at the cheese, and then pulls out the apple. Last, the two impossible conditions. The refrigerator door is opened, the person sees an apple and a piece of rotting cheese, and while smiling and looking at the rotting cheese, pulls out the cheese. Alternatively, after giving the disgust face while looking at the apple, the person pulls out the apple. If, during familiarization, subjects have formed expectations based on communicative behaviors (facial expressions, nonfacial gestures), then they should look longer during the impossible conditions than during the possible conditions. Such expectations would, perhaps, not be upheld if a new person arrived who had never opened the refrigerator door. If the preferential looking patterns support such predictions, then this result would provide some evidence that subjects generate expectations about the intentional states of other individuals and that communicative expressions are used as reliable sources of information about such states.

One question that the previous experiment raises is how individuals, human or nonhuman, figure out which entities in the world *have* intentions. As briefly mentioned in chapter 7, Premack (1990), Premack and Premack (1994), and Leslie (1994) have suggested that a key entry into this problem is an understanding that objects that move on their own—that are self-propelled—are most likely driven by intentional states. And the most likely candidates are animate objects. Here is a simple experiment that might help us distinguish between organisms that do or do not understand the significance of objects that are self-propelled. In step one, present an individual with a two-chambered plexiglass box, divided by an opaque partition (see Figure 8.2); the partition has an opening in the middle, and

---

7. I am restricting this experiment to nonhuman primates because the test relies, minimally, upon an ability to attend to facial expressions. Though other organisms clearly do attend to faces, the face is unlikely to represent the same sort of communicative venue as it does for nonhuman primates, especially the Old World monkeys and apes.

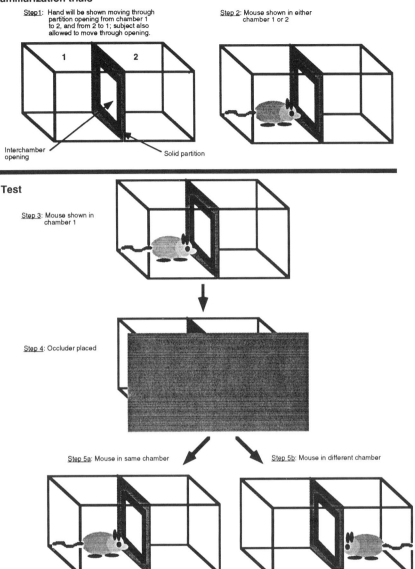

Familiarization trials

Step1: Hand will be shown moving through partition opening from chamber 1 to 2, and from 2 to 1; subject also allowed to move through opening.

Step 2: Mouse shown in either chamber 1 or 2

1   2

Interchamber opening

Solid partition

Test

Step 3: Mouse shown in chamber 1

Step 4: Occluder placed

Step 5a: Mouse in same chamber

Step 5b: Mouse in different chamber

Figure 8.2

Experimental design for test of an organism's understanding of the properties of self-propelled objects. In the familiarization phase (upper left), subjects are exposed to the two-chambered box, and specifically, made aware of the hole in the opaque partition. In the second phase, they see an object in either chamber 1 or chamber 2 but never see the object move through the hole in the partition. The test phase involves first showing the object in one compartment; the occluding screen is then lowered, and when subsequently raised, it reveals the object in the same compartment of the opposite one. Objects that are self-propelled (e.g., a mouse) are compared with objects that would require an external force to move (e.g., an apple). Looking time to each presentation is recorded.

subjects are made aware of this opening (e.g., putting a hand through it or allowing the subject to move back and forth between the two chambers). Following this familiarization phase, place an object in one chamber and score the subject's looking time. To start, one object should be inanimate (e.g., an apple), and one should be animate and mobile (e.g., a mouse). Subjects should see both objects in both chambers, but should not see objects moving back and forth (e.g., put up a screen when the objects are transferred between trials). The test phase is next. Place one of the objects in one chamber, bring down the occluding screen, and then lift it. In one trial the object remains on the same side, and in a second trial it appears on the opposite side. Now, if subjects understand the difference between objects that are or are not self-propelled, they should be surprised to see the apple in the opposite chamber from where it started, but should not be surprised to see the mouse in a different chamber. That is, although the mouse started in one chamber, it has the ability to move through the partition into the other chamber. The apple cannot, at least not on its own. Needless to say, surprise would be measured as an increase in looking time. If this experiment succeeds, it is then possible to move more deeply into the problem and assess what individuals know about the relevant properties of self-propelled objects in the world. For example, one might explore how subjects respond to a windup mouse that, in the familiarization phase, simply moves in a circle. Based on motion, which is nonetheless self-generated, one might not expect this "mouse" to move through the partition. Prior experience ought to suggest that it has restricted motion and thus should not move into the adjacent compartment. Pertinent to our earlier discussions of concepts and semantics, the proposed experiments will enable us to narrow the features that an organism uses to assign objects in the world to functionally meaningful categories such as self-propelled and animate. Because the technique can be used with any organism capable of attending to a visual scene, we are in a strong position to make comparisons across a taxonomically diverse group of species.

8.3   How to Build Communicating Organisms: Thinking Like an Evolutionary Engineer

Put on your engineering hat. Here is the problem: the first Model T Fords have just appeared on the scene, and you have been asked to design a device that will regulate traffic, that will allow vehicles moving in one direction to cross an intersection without smashing into vehicles coming from a different direction. As in any task, it is always best to break the general problem down into smaller problems, and to do so in the context of constraints on the system. As an engineer, you might first think about the most appropriate sensory modality for signaling to the oncoming traffic. The ecology of the situation is such that all modalities except vision are inappropriate. Thus, although you might entertain sending different sounds or smells for "stopping" and "going," you would quickly rule out these possibilities based on the fact that people often travel in cars with their windows up, thereby limiting the extent to which sounds or smells might effectively reach the targeted receiver. Visual signals represent the only feasible solution. Next problem: what kind of visual signal? How about an S for stop and a G for go? Bad idea. Bad because you are an engineer with foresight and quickly realize that in the future cities will be extremely international, with drivers from a diversity of linguistic groups. Thus, S and G would carry no inherent meaning to someone from Niger, Nicaragua, or Nepal, although clearly they could learn the meaning of such symbols. But there is a more severe problem. These letters are difficult to detect at a distance and, in particular, are difficult to discriminate: the upper portions of both letters are the same. The "Aha" feeling of discovery strikes, and you now realize that what is required are two unambiguous visual signals that can be accepted universally and can readily be discriminated from each other at a distance. The answer: two different colors. But which colors? Thinking a bit about human color vision, the way our color receptors work, limits the range of possibilities. The best solution is a set of colors that are opposites, and red-green is just about as good as one gets. So, you settle on these colors, and begin to build a device that will turn a RED light on when cars coming from one direction are meant to stop and will turn a GREEN light on when cars from a different direction are meant to go. The association between RED and stop is a sensible one, given that the evolutionary record provides us with case after case where red is paired with some kind of warning to stop: the poisonous areas of many insects, the red bellies of sticklebacks, the epaulettes of red-winged blackbirds, and so on.[8] Confident about the colors, you put in a timing mechanism so that when RED is on for one direction, GREEN is on for the other direction. You bring the device into your boss, feeling pleased about its design. You explain how it works; the boss is following, smiling. And then the boss frowns and says, "But, what happens when a car is about to cross the intersection, and all of a sudden, without any transition, there is this step function from a GREEN light

---

8. In some species, red is also an invitation to approach (e.g., the red inflatable pouch used by male frigate birds to attract females).

to a RED light?'' You scratch your head, feeling slightly defeated. But then, a new design solution emerges: build a light that can gradually fade from RED to GREEN and vice versa. Nope, another hopeless solution. A continuous, graded signal would nonetheless force the perceiver into a categorical decision, and people might very well differ in terms of where they place the categorical boundary between RED and GREEN. And if that were the case, it would lead to complete chaos. The solution then is to build in a third light, one that signals the fact that there will shortly be a transition from GREEN to RED. And place this light in between the GREEN and RED ones, and make it discriminably different. Again, thinking about color vision, you hit on YELLOW. We have arrived: the modern traffic light. It is a good solution, with design features suited to meet the task demands, sensitive to both the ecology of the situation and the sensory specializations, evolutionary history, and constraints of the participating organism.

The process of evolution by natural selection does not have such foresight, but it is a force that nonrandomly weeds out failed solutions, leaving the successful ones to thrive until a better solution arrives on the scene. And the process continues indefinitely,[9] molding systems with complex design and specializations that are well adapted to the current problems posed by the environment. Armed with this logic, let us look at some of the necessary and sufficient ingredients for creating an organism that can communicate with other members of its species. Although recent work in artificial life has addressed this problem and will undoubtedly contribute significant insights in the future, current investigations generally fail to think about the functional consequences of particular communicative solutions or pay insufficient attention to well-documented constraints on communication (for some exceptions, as well as general reviews of this literature, see M. G. Dyer, in press; MacLennan and Burghardt 1994; Todd and Miller 1991). The discussion that follows explicitly acknowledges constraints, uses the intuitions from natural selection theory to explore the functional design features of a system, and attempts to uncover how communication systems evolve from a set of primitives. We will start with a relatively simple system, and then gradually build in some complexity. This evolutionary cookbook approach will, I believe, shed some light on how normal human adults manage the fancy communicative moves they do (e.g., telling stories, reciting poetry, making fun of each other using witty repartees), and why fireflies, fish, frogs, birds, monkeys, apes, and newborn human infants fail.

---

9. Assuming, of course, that heritable variation can be maintained.

All communication systems require devices for sending signals and for receiving at least some components of the signal. Although many organisms have evolved send-receive systems in multiple modalities, most organisms have a dominant modality, one dictated by the ecological arena for communication. The sensory modalities employed in communication will constrain the content of the information conveyed and received. And as discussed most recently in the literature on human cognition, constraints on information acquisition can operate in a domain-general or domain-specific manner (see lucid analysis by Keil 1990). Domain-specific systems will be favored in cases where the potential range of solutions to a problem is far too great for a domain-general device to solve, and where failed solutions are costly with regard to survival and reproduction (see Cosmides and Tooby 1994b for a general treatment of this issue).[10] Domain-specific systems pick up on consistencies within the environment, bias the range of potential hypotheses or inferences for solving a problem, and generally limit the kinds of input that are accepted as relevant to the problem at hand.

To begin our thought experiment, consider fireflies, their bioluminescent flashes, and the ecological pressures on signal design and transmission (Lloyd 1984). The primary context for communication in fireflies is mating, specifically mate attraction. There are several species of fireflies, and each species has a unique flashing code. Some species have evolved the ability to mimic the flashing patterns of other species. This ability allows the mimetic species to dupe females from the nonmimetic species to approach. By approaching this seemingly honest signal *d'amour,* individuals from the nonmimetic species become delicious meals rather than mates.

The firefly system sets up the following engineering problems. The sender must be able to generate a visual flash that is discrete and can be spectrotemporally patterned into a species-typical signal. Given the ecology, the flash must be sufficiently bright to reach intended receivers. As with alarm calls, however, there are constraints on signal design, imposed by the need to avoid predators and

10. Note that for Cosmides, Tooby, Keil, and many others, domain-specific systems of knowledge are, generally, uniquely human. For example, Keil (1990) states: "There are by most accounts only a dozen or two of the Domain Specific Innate domains, corresponding to broad areas of cognition that have presumably been important to humans in the course of evolution and different enough from each other in terms of the natural laws describing them that different sets of constraints are optimal for learning each. These domains might include knowledge of spatial layout, naive mechanics, naive theories of persons or 'folk psychology,' aspects of language (e.g., syntax and phonology), moral judgments, and so on" (p. 145). As I discussed in chapter 7 and will continue to suggest over the next few pages, many of these domains are likely to have much deeper ancestral roots, certainly dating back at least as far as monkeys and apes.

preserve species distinctiveness, especially in areas where different firefly species coexist. For example, the longer each individual flash stays on, the more readily individuals can be detected by predators; the flash is like an arrow, pointing to the individual's precise location in space. On the perception side, individuals must be capable of detecting flashes, discriminating between conspecific and heterospecific flashes, and in the case of females, potentially using certain characteristics of the flash to make an assessment of male quality (e.g., viability).

Let us now take the firefly problem and break it down into a set of primitives—components that are likely to be critically involved in all communication systems. For communication to occur, signals must be reliably associated with certain events in the world. Thus the flash is consistently used by males to attract females and, as such, represents a signal that is reliably associated with the mating context. Fireflies know this signal; their knowledge is innately wired; and on a short time scale (e.g., within the individual's lifetime) experiences that challenge this association (e.g., a child using her flashlight to mimic the firefly flash) would not be accommodated into what individuals expect about the function of the flashing signal. In other words, fireflies know *that* the species-typical flash is associated with mating, but have no theory about *why* this association occurs, or why it should be true. And without some kind of theory, it is difficult to incorporate novel experiences into one's behavioral routines for responding. Frank Keil (1990) clearly articulates this distinction:

If an entity merely behaves in such a way that only certain inputs are "meaningful" to it, we might not want to attribute theory to it: at least without unpacking the meaning of "meaningful" in a way that does not circularly rely on theory. I am reluctant to grant spiders intuitive theories of the mechanisms of physical lattices like webs, even though their behavior displays a precise honoring of such principles. Similarly, cockroaches and other cognitively "simple" creatures also seem to pick out objects and follow their trajectories and the like, yet one cringes at calling them object theorists. They don't, I suspect, anticipate collisions for screened events and make inferences about numbers of objects involved based on seen and unseen trajectories; but in other respects, they do seem to think there are objects "out there."   (p. 152)

Summarizing, fireflies have evolved a system of communication that is characterized by a small number of signals; acquisition of the species-typical flashing code is innately specified with no evidence of learning; signals are processed by a brain that can discriminate species-typical signals from those that are not; and the result of this process is a reflexive response (i.e., approach the flashing code). The ability to discriminate among an array of signals is based, most simply, on some type of similarity analysis that rejects all signals deviating from a rigidly structured template. Last, though the firefly send-receive system operates like a beautifully choreographed light show, individuals do not know why their system works this way, and one would not expect selection to favor access to such knowledge in the future. Although an occasional predatory species may infiltrate their communicative exchanges, like a spy breaking the code, the bioluminescent flash represents a well-designed signal that solves the primary functional task: attract and find suitable mates.

What kinds of pressures might cause the firefly system to evolve a new suite of design features? So that we may incrementally build our thought experiment and provide the requisite engineering solutions, consider one aspect of this general question: specifically, what type of pressure would favor signal diversification and what kinds of neurocognitive machinery would be necessary to accommodate such requirements? Consider the possibility that females use the male's flash to find mates, but once they have located them, sit around waiting for the male to deliver more information about his genetic quality (e.g., his ability to survive, to produce high-quality offspring). To avoid losing the female once she has approached, males might begin producing additional flashes, and these might become increasingly elaborate and differentiated from the signals used to attract females. Consequently, selection (by means of female choice) would favor at least two different signals, one for attracting mates and one for strutting their stuff—for showing off their genes. In essence, selection will favor signal diversification when clearly differentiated contexts demand clearly differentiated signals to coordinate behavior.

As a species' repertoire diversifies, the neural machinery subserving signal production and perception must keep pace. On the production side, the system must be able to recognize the relevant contexts for using one signal rather than another and, once the context has been recognized, must be able to generate a signal with the appropriate morphology. Needless to say, as the number of signals increases, this task becomes computationally more difficult and places greater constraints on the motor control regions of the brain responsible for regulating the patterning of displays in the visual or auditory domains. It also raises the problem of how the neural hardware and software for producing signals develops ontogenetically. If the repertoire exhibits little variation in signal structure and there is only weak pressure to creatively generate novel variation, then an innately guided mechanism will be favored. All invertebrates and most vertebrates are characterized by repertoires with a small number of signals; variation in signal structure is, typically, closely tied to variation in some feature of the individual's anatomy (e.g., big animals producing vocalizations with a lower fundamental frequency); and learning plays no role in the ontogeny of the repertoire.

Selection to creatively generate novel variation in signal structure will arise in species with relatively complex social organizations.[11] Specifically, in group living organisms, and especially those where there are repeated interactions with the same individuals and where communication occurs in the context of competition and cooperation, there will be strong selection on the ability to encode subtle information about the dynamically changing nature of social relationships among group members. Recalling the arguments raised by Krebs and Dawkins (1984), whereas competitive interactions tend to be associated with displays that are exaggerated and highly visible (visually or acoustically), cooperative interactions are generally accompanied by more discrete conspiratorial whispers. A significant consequence of changes in the relationship between repertoire diversification and the relative complexity of group life is that signals become increasingly dissociated from what one can directly extract from simply looking at an individual's external features (e.g., its species, sex, age), reflecting instead a suite of hidden internal states, including emotions, motivations, and thoughts. A system like this would favor the implementation of learning mechanisms, either domain-general or domain-specific, but innately constrained with regard to the relevance or meaning of the incoming stream of sensory experiences.

Perceptually, repertoire diversification increases the complexity of the recognition task, thereby putting greater demands on the brain mechanisms guiding the process of categorization. Recall that for the early ethologists, nonhuman animals exhibited small repertoires precisely because of what they considered significant processing constraints on responding appropriately to a large number of meaningfully different signals (see chapter 2). The processing constraints appear at several levels. To clarify, consider a species with even just two signals, one used to attract mates and one to warn group members about predators. When the perceptual system detects a sound, it must first establish whether the signal is from a conspecific or a heterospecific. If the signal is from a conspecific, the listener can then move on to the next branch in the processing scheme, evaluating whether it is a mating or warning signal. Depending upon the species, there will then be further assessments, such as the identity of the signaler, whether the signal honestly reflects the individual's motivations or genetic quality, and so on. At each

---

11. In discussing song learning, Peter Marler has often raised the possibility of an instinct for innovation or creativeness. For songbirds, this instinct appears highly domain-specific. For humans, in contrast, the instinct for creativity (assuming it is present) appears more domain-general.

processing stage, the perceptual system must work through a similarity analysis based, presumably, on some neural representation of the species-typical morph, template, or prototype. The complexity of such analyses, together with the problem of storing representations, increases as the number of signals within the repertoire increases.

Let us bring these hypothetical ideas back to reality, and consider once again communication among East African vervet monkeys. As mentioned in section 8.2.2, selection has clearly favored repertoire diversification in the context of predator encounters. It has also favored diversification in the context of social interactions, including the production of acoustically differentiated calls during both within- and between-group dominance encounters. Surprisingly, perhaps, mechanisms underlying repertoire diversification have been engineered in the absence of sophisticated learning algorithms. Thus, unlike songbirds, vervets have not evolved the capacity to imitate (in any modality; Hauser 1988b, 1993f), and learning plays virtually no role in modifying signal structure, either during early development or later in life (chapter 5).

To fully capture the design problem that we have on our hands, let us focus on the alarm call system. Thus far, observations indicate that vervets produce acoustically discrete alarm calls to at least four predator classes (in capitals) and a set of primary target species (in bold subscript): LARGE CATS$_{leopard, cheetah}$; BIRDS OF PREY$_{martial eagle, crowned hawk eagle}$; SNAKES$_{mamba, python}$; PRIMATES$_{baboon, human}$. In terms of categorizing alarm call exemplars or predatory events, the problem is relatively easy, since each call type is acoustically discrete. Moreover, vervets are born with the ability to respond (i.e., vocally) appropriately to general predator categories (e.g., things up in the air, slithering things on the ground). Of interest, from both an evolutionary and constraints perspective, is the extent to which the structure of their alarm calls and the way in which they are perceived can be altered, either within an individual's lifetime or over several generations as a result of selection on heritable variation. Here are two intriguing problems that can be partially addressed with some observational appetizers. First, could the alarm-call system of vervets change if a new species began preying on them, and if so, what features of their calling behavior would change—could change? Second, what would happen if a population of vervets lost all predatory classes but one, thereby potentially reducing the range of hunting strategies confronted? Concerning the first question, I had the opportunity to observe a suspected predation by a species that had never been observed preying on vervets. Specifically, while driving through the study population established by Cheney and Seyfarth, I heard a chorus of alarm calls, suggesting that a leopard was nearby. The calls sounded slightly off, however. Rather than the intense, rapid-fire alarm calls typically associated with leopard sightings, these calls had a much slower temporal pattern of delivery, as would be perceived if the batteries of a tape recorder were run down during playback. When I arrived at the scene, the vervets were up in the trees, calling in the presence of a lion. Over the course of many years of research on this population, lions have never been observed preying on vervets and, given the potential payoffs, would not be expected to do so (i.e., given the energetic costs and benefits of hunting vervets, they provide minimal returns to a lion). What appears to have happened is that vervets added lions into the general category of large predatory cat and produced a call that was spectrally similar to the one used for leopards but was associated with a slower rate of delivery. These observations suggest that there is some flexibility in the system, both in terms of altering call structure and in classifying exemplars into a category with certain definitional features. Such changes may be observed over the course of an individual's lifetime or over the course of several generations.

The second question centers on the problem of how the structure and content of a category might change when certain representative exemplars no longer apply. More particularly, I have asked what happens to a predator alarm call system when all but one predator category is deleted experientially. The following situation provides a potential path into this problem. There are islands off the coast of Kenya where vervets have lived for many generations in the absence of land predators. Birds of prey, however, still pose a threat. Given that vervets have the *capacity* to produce different calls for predators with different hunting strategies, can they reallocate such cognitive reserves if pressures change. For example, could vervets encode more precise information about the species of birds of prey or their behavior, such as where they are, how fast they are moving, whether they are swooping or soaring, and so on? On a perceptual level, there is some evidence for such within-category differentiation based on recent experiments with wild western crows (Hauser and Caffrey 1995). Specifically, in the population studied, crows are subject to predation from three species of raptor, each with somewhat different hunting styles. When the calls of these raptors were played back, crows responded in different and apparently adaptive ways. Thus, when hearing the call of a horned owl, they approached and occasionally mobbed the speaker. In contrast, when they heard the call of a red-shouldered hawk, they exhibited either flight-intention movements or flew away from the speaker. These responses are appropriate given the hunting strategies of the raptors involved. The question we have been asking of the vervet's system is whether individuals can generate alarm calls capable of encoding subtle differences in raptor behavior, given that the pressure from other predatory species has been relaxed. Studies of island populations such as the ones described will help address this problem. Thus one could observe how this population naturally responds to the presence of raptors, quantifying such parameters as the relationship between call structure and raptor distance, flight pattern, and so forth. Additionally, one could experimentally investigate these relationships using either trained raptors or, more creatively, model airplanes "dressed up" as raptors. Thanks to recent advances in toy technology, there are now electric model airplanes that run on batteries and thus can be flown without making noise. More importantly, one can design such planes to look like raptors, allowing for a full range of visual features (e.g., feather patterning on the wings) to be selectively manipulated.

Species that have evolved a diversity of signals within their repertoire gain power with regard to the range of situations for exchanging information. But there are numerous design problems here, some of which we have touched upon in earlier sections. Specifically, what kind of information is encoded in the signal, how do individuals come to understand the meaning of the signal and the intent of the signaler, and what ecological problems have shaped the supporting cognitive architecture for signal production and perception? To address these problems, and conclude this section. I would like to have the reader think about nonhuman primates, the first hominids, the developing human infant, and normal human adults. And I would like to think about these organisms in light of the environmental problems that would have favored the emergence of a semantically rich repertoire, guided by a cognitively sophisticated brain, capable of reading the minds of other individuals and recognizing that what is in the mind is a collection of representations, beliefs, desires, and intentional propositions, some of which differ across individuals and some of which are the same, universal. Such representations can, however, form the basis for providing each individual with a model of the world, and one that establishes a foundation for anticipating communicative interactions and their potential outcomes.

We learned in section 8.2.2 that it is extremely difficult to establish the precise meaning of an utterance produced by an organism lacking full-blown language. And this statement holds as much for nonhuman animals as it holds for human infants, even children at the earliest stages of language acquisition. Nonetheless, the studies reviewed in this book have demonstrated that the calls of some primates and at least one bird convey information about objects and events in the environment, although at present it is difficult to say anything more precise about their meaning. The functional significance of a referential system is that individuals can understand what the caller is yapping about without having to see what is going on. Thus, when an individual hears an alarm call, an appropriate antipredator response can be initiated without having to observe the call-eliciting context. Similarly, when a food call is given, listeners obtain information about the availability of alternative food sources, and this can serve to guide their foraging decisions. These are statements about adaptive significance.

What ecological situations would select for a referential system, and what are the cognitive prerequisites? My hunch regarding the first part of this question is that referential signals will be favored in social species where individuals occasionally leave the core of their group and consequently have the opportunity to encounter objects or events that are relevant to other group members (e.g., food, predators, mating opportunities, dangerous neighbors). Clearly, several species can be characterized by this description, and yet not all species will evidence referential signals, or if they do, the capacity will be highly limited to a single context, and in this sense, we might not even want to talk about the capacity for referentiality, at least not in terms of the requisite psychological underpinnings. For example, the dance of the honeybee is clearly referential and even exhibits the property of displacement—of being able to refer to an object that is spatiotemporally removed from the present. However, even if functionally referential signals are restricted to one or a small number of contexts for a given species, such as the honeybee, this ability is nonetheless important, and it is one that might be enhanced over evolutionary time because of its adaptive consequences.

The ecological conditions favoring referential signals lead us directly to the cognitive prerequisites. In the literature on human language acquisition, reviewed in chapters 5 and 7, we learned that children enter the world with certain innate biases—constraints—that help narrow the range of possible inferences concerning word meaning (e.g., Markman's taxonomic-thematic distinction). They also have powerful learning devices (e.g., associative learning, working memory, and long-term storage), are sensitive to paralinguistic features of the language that help pick out a word's referent, and last, can readily take advantage of the adult's extraordinary desire to help the child acquire language competence—e.g., pointing to and staring at objects, slowly belaboring their pronunciation as they show the child the ''booooaaaaaatttt,'' and so on.

Using our knowledge of how children come to understand a word's referent provides some suggestions for how we might work out the prerequisites for referential signaling in nonhuman animals and the early hominids. Minimally, a referential system requires the ability to associate sound with some feature of the environment, and for this associative system to be powerful, the input device should be willing to accept almost any sound, even though there are likely to be design constraints set up by ecological factors and the organism's hearing ability. In fact, given the data on sign-language acquisition, we should probably restate the last sentence to read, "accept almost any sound or visual display." An associative mechanism must, however, be supplemented by a mechanism that facilitates the transfer of knowledge from one individual to another. Unlike humans (Baldwin and Moses 1995), nonhuman primates do not naturally point to pick out objects or events. Nonhuman primates do, however, have the ability to use gaze to assess where individuals are looking, though they may not have the capacity to understand that seeing is knowing (Gomez 1991; Povinelli and Eddy, in press; see following discussion). And in rhesus macaques, at least, we know that there are areas of the brain that are sensitive to the direction of gaze and that, if lesioned, cause severe deficits in the subject's ability to determine where someone is looking (see section 4.4). These observations suggest that if a vervet gives a leopard alarm call, group members who lack an understanding of the call's referent should be able to use gaze to determine the putative target. As in the case of children learning language, it appears that vervet infants may also enter the world with a taxonomic–whole object bias and subsequently, based on experience with actual predators, refine the featural properties associated with the category—thematic aspects of the utterance. The same sorts of arguments would, I believe, also apply to the first hominids. Given the ubiquity of associative learning mechanisms in the animal kingdom and the importance of eyes for most primate species, it is reasonable to assume that our human ancestors started off with a communication system capable of rudimentary referential signaling, which subsequently evolved into a system with a massive lexicon, supported by a recursive system that could combine entries in the lexicon into an infinite variety of meaningful utterances.[12]

Let us look more closely at this last step. What kind of mechanism might favor signal diversification—a lexical explosion—that is fully referential, or at least tilt a system toward it? From a design perspective, I would like to propose a mechanism that is necessary, though by no means sufficient: the capacity to imitate.

12. I am not committed to a particular order of events here, and I think that at present we do not yet have sufficient information to determine how various cognitive mechanisms subserving, for example, human language, unfolded over the course of evolution. Nonetheless, I am convinced that some of the fascinating speculations offered by Corballis (1992, 1994) and Bloom (1994b) on the relationship between a generative ability that originally evolved for numerosity as opposed to language will be resolved as further studies of preverbal children and nonhuman animals are conducted, focusing in particular on the ontogeny of their linguistic and arithmetic capacities (Hauser, MacNeilage, and Ware, in press; Wynn 1992; Xu and Carey, in press).

The comparative data reviewed in this book and in several other recent publications (e.g., Cheney and Seyfarth 1990; Tomasello, Kruger, and Ratner 1993; Whiten and Ham 1992) reveal that the human capacity to imitate is fundamentally different, if not uniquely different, from what has been observed in all other animals, including other primates. Thus, although songbirds are absolute virtuosos when it comes to imitating the songs of their own or another species, there is no evidence that this ability is employed in nonvocal domains, or even more dramatically, that it is employed for other, nonsong vocalizations. Nonhuman primates raised under natural conditions have provided no convincing evidence of imitation, although at least some apes raised by humans show a limited imitative capacity that is restricted to the nonvocal domain. Dolphins appear to be superb imitators, in both visual and auditory domains. Yet to be demonstrated, however, is the extent to which newborn dolphins can imitate, and if not, what kinds of experiences are crucially important for the ontogeny of this capacity.

Traditionally, imitation has been perceived as one of the most powerful social learning mechanisms because it enables individuals to accurately reproduce a particular motor action in the absence of a demonstrator, thereby facilitating the rapidity and fidelity with which information can be transmitted within a population.[13] Although this is an appropriate characterization, the ability to imitate also creates a situation that can introduce variation. Specifically, to imitate in the strong sense (see, for example, discussions in Galef 1988; Meltzoff 1988; Whiten and Ham 1992), individuals must be able to perceive a motor action, store this as a representation, and then repeat the motor act observed in the absence of the performer. Because the action must be performed on the basis of a stored representation, the accuracy of the copy will depend critically on the stability and resolution of the representation. If the representation is poor, the reproduction will also be poor, though it may very well be perceived by group members as a sufficient rendition of the act. Given that the transmission of behavioral routines often passes vertically, from parent to offspring, there is a strong possibility that variable reproductions of the imitated behavior would be passed on generationally, thereby setting up slightly new ways of doing things. Thus I would like to propose that the capacity to imitate represents a further design feature of a communication system that is referential and consists of a large number of referentially discrete signals. Imitation allows individuals to quickly pick up the meaning of a new signal and also provides a foundation for generating variation.

---

13. In fact, for some animal researchers, culture is possible only in organisms that exhibit an imitative capacity (Galef 1992).

The lack of an imitative capacity in nonhuman primates[14] severely constrains the expressive power of their communication system, including the size of the lexicon and the degree to which it can be modified as a function of changes in the environment. Based on some of Meltzoff and Kuhl's work on the newborn's ability to imitate facial expressions and the prosodic contours of human speech, it appears that humans are innately endowed with an imitative capacity. Although we may never be able to pinpoint when, during hominid evolution, imitation emerged, it must have represented a fundamental step in cognitive sophistication, one with far-reaching implications for our communicative interactions.

To cap this section, let us turn to the kinds of mental faculties that must be designed into a system that not only allows individuals to refer to objects and events in the environment, but also to understand why individuals utter such signals and what it reveals about their beliefs and desires. Once again, we can turn to research on child development, as well as to studies in comparative cognition. In the late 1970s, David Premack proposed, in his characteristically insightful way, that what makes the human adult mind a truly interesting one is that it has a theory of minds, its own and those of other human beings. Research on nonhuman primates, which is still in its infancy, provides only weak evidence that apes have a theory of mind, whereas experiments with Old World monkeys reveals that they do not (see review by Whiten 1993, 1994). For some researchers working on humans (e.g., Baron-Cohen 1995; Leslie 1994), infants are born with the requisite ingredients for a theory of mind in that they are sensitive to an object's goal directedness, are aware that self-propelled objects are driven by internal mechanisms, and are aware that an individual's direction of gaze or attention provides important hints as to what he or she thinks or feels. Over time, these initial abilities form the necessary support structure for understanding that shared attention means shared knowledge and that other individuals have beliefs and desires that guide their actions toward both animate and inanimate objects. For other researchers, these early abilities have more to do with knowing *that* than with knowing *why*—in other words, young infants have been designed with perceptual mechanisms that innately guide their actions or responses to physical events in the world. It is not until much later, around three or four years old, that they become aware of the reasons for such events—at this stage, they begin to develop a theory of mind and of the things in the world that have minds.

14. For those who might be concerned with the generality of this claim, I would be willing to narrow the argument and suggest instead that there is no evidence for an imitative capacity in the *vocal* domain in either monkeys or apes.

Regardless of the developmental timing of events, a brain that has been designed with a theory of mind is an awesomely powerful brain. Its power stems from the fact that it is now capable of creating a model of the world in the absence of direct experience with the myriad alternatives. Thus, a mind with a theory of mind represents belief-desire states, uses these to assess what others want and therefore are likely to do, and based on these assessments, generates a set of inferences that guide communicative interactions. And we see the unfolding of these events, at least in normal human children, through the use of their language. The child with a theory of mind has a richly textured language, joking, lying, conveying precise information about her feelings and how others feel about her, and so on. An organism without a theory of mind (e.g., many young autistic children, most nonhuman animals) will be frozen in an action-reaction state, with no understanding of why others feel the way they do or why they might have different beliefs or desires. They simply don't know how to care, though they behave as if they do.

We have reached the end of the engineer's assembly line. Many more design paths might have been pursued in this exercise. I have chosen those paths that have been trodden in this book. I hope this text provides the reader with a reasonable synthesis of how complexity can be built into a system of communication and how the design features of a system must be evaluated in terms of fundamental ecological problems and constraints that have been inherited over the course of evolution.

8.4   Final Remarks

Informational overflow. This is what most scientists surely feel as they enter the libraries of their respective institutions. There are journals that speak to highly specialized theoretical areas or to highly specialized taxonomic groups. There are journals that have an interdisciplinary flavor. There are journals designed to provide succinct reviews of hot topics and current controversies. And then there are specialized books, edited volumes, and highly polished popular accounts. How can anyone keep up? I raise this question because if Tinbergen's perspective is taken seriously, and this is the approach I have been advocating throughout this book, it will require us to keep up a bit more than we have. What I have attempted to argue in this book is that a complete analysis of communication requires a richly comparative perspective that grapples with how communicative systems work (neurobiological, developmental, and cognitive processes), why there are both similarities and differences in design across species, and what the functional consequences of such design features are for an organism's survival and reproduction. I have buttressed this argument by reviewing research in areas that commonly exist in the intellectual equivalent of solitary confinement. This is both devastating to the research vigor of comparative communication and, I believe, fully unnecessary. More to the point, by integrating levels of analysis and using the powerful intuitions derived from evolutionary theory, we will uncover new problems and new answers that ultimately will shed light on the diversity of communication systems in the animal kingdom. And how sweetly rewarding these discoveries will be.

References

Abbs, J. H. (1986). Invariance and variability in speech production: A distinction between linguistic intent and its neuromotor implementation. In J. S. Perkell and D. H. Klatt (Eds.), *Invariance and variability in speech production* (pp. 249–265). Hillsdale, NJ: Lawrence Erlbaum Associates.

Abbs, J. H., and Gracco, V. L. (1984). Control of complex motor gestures: Orofacial muscle responses to load perturbations of the lip during speech. *Journal of Neurophysiology, 51,* 705–723.

Adams, E. S., and Caldwell, R. L. (1990). Deceptive communication in asymmetric fights of the stomatopod crustacean *Gonodactylus bredini. Animal Behaviour, 39,* 706–716.

Aich, H., Moos-Heilen, R., and Zimmerman, E. (1990). Vocalizations of adult gelada baboons (*Theropithecus gelada*): Acoustic structure and behavioral context. *Folia primatologica, 55,* 109–132.

Aitken, P. G. (1981). Cortical control of conditioned and spontaneous vocal behavior in rhesus monkeys. *Brain and Language, 13,* 636–642.

Aitken, P. G., and Capranica, R. R. (1984). Auditory input to a vocal nucleus in the frog *Rana pipiens:* Hormonal and seasonal effects. *Experimental Brain Research, 57,* 33–39.

Aitken, L. M., Kudo, M., and Irvine, D. R. F. (1988). Connections of the primary auditory cortex of the common marmoset (*Callithrix jaccus*). *Journal of Comparative Neurology, 269,* 235–248.

Alcock, J. (1989). *Animal behavior, 4th ed.* Sunderland, MA: Sinauer Associates.

Alcock, J., and Sherman, P. (1994). The utility of the proximate-ultimate dichotomy in ethology. *Ethology, 96,* 58–62.

Allan, S. E., and Suthers, R. A. (1994). Lateralization and motor stereotypy of song production in the brown-headed cowbird. *Journal of Neurobiology, 25,* 1154–1166.

Allen, C., and Hauser, M. D. (1991). Concept attribution in nonhuman animals: Theoretical and methodological problems in ascribing complex mental processes. *Philosophy of Science, 58,* 221–240.

Allen, C., and Hauser, M. D. (1993). Communication and cognition: Is information the connection? *Yearbook of the Philosophy of Science, 2,* 81–91.

Altmann, J. (1980). *Baboon mothers and infants.* Cambridge, MA: Harvard University Press.

Alvarez-Buylla, A., Kirn, J. R., and Nottebohm, F. (1990). Birth of projection neurons in adult avian brain may be related to perceptual or motor learning. *Science, 249,* 1444–1446.

Alvarez-Buylla, A., Theelen, M., and Nottebohm, F. (1988). Birth of projection neurons in the higher vocal center of the canary forebrain before, during, and after song learning. *Proceedings of the National Academy of Sciences, 85,* 8722–8726.

Andelman, S. (1987). Evolution of concealed ovulation in vervet monkeys. (*Cercopithecus aethiops*). *American Naturalist, 129,* 785–799.

Andersson, M. (1971). Breeding behaviour of the long-tailed skua *Stercorarius longicaudus* (Vieillot). *Ornis Scandinavica, 2,* 35–54.

Andersson, M. (1982). Female choice selects for extreme tail length in a widowbird. *Nature, 299,* 818–820.

Andersson, M. (1994). *Sexual selection.* Princeton, NJ: Princeton University Press.

Andersson, S. (1991). Bowers on the savanna: Display courts and mate choice in a lekking widowbird. *Behavioral Ecology, 2,* 210–227.

Andrew, R. J. (1962). The origin and evolution of the calls and facial expressions of the primates. *Behaviour, 20,* 1–109.

Andrew, R. J. (1963). Evolution of facial expressions. *Science, 142,* 1034–1041.

Aoki, K. (1989). A sexual-selection model for the evolution of imitative learning of song in polygynous birds. *American Naturalist, 134,* 599–612.

Aoki, K., and Feldman, M. W. (1987). Toward a theory for the evolution of cultural communication: Coevolution of signal transmission and reception. *Proceedings of the National Academy of Sciences, 84,* 7164–7168.

Aoki, K., and Feldman, M. W. (1989). Pleiotropy and preadaptation in the evolution of human language capacity. *Theoretical Population Biology, 35,* 181–194.

Arak, A. (1988). Female mate selection in the natterjack toad: Active choice or passive attraction? *Behavioral Ecology and Sociobiology, 22,* 317–327.

Aristotle. (1941). *The basic works of Aristotle.* New York: Random House.

Arnnson, L. R., and Noble, G. K. (1945). The sexual behavior of Anura. 2: Neural mechanisms controlling mating in the leopard frog. *Bulletin of the American Museum of Natural History, 86,* 83–140.

Arnold, A. P. (1980). Quantitative analysis of sex differences in hormone accumulation in the zebra finch brain: Methodological and theoretical issues. *Journal of Comparative Neurology, 165,* 421–436.

Arnold, A. P. (1992). Developmental plasticity in neural circuits controlling birdsong: Sexual differentiation and the neural basis of learning. *Journal of Neurobiology, 23,* 1506–1528.

Arnold, A. P., Nottebohm, F., and Pfaff, D. W. (1976). Hormone concentrating cells in vocal control and other areas of the brain of the zebra finch (*Poephila guttata*). *Journal of Comparative Neurology, 165,* 487–512.

Aslin, R. N., Pisoni, D. B., Hennessey, B. L., and Perey, A. J. (1981). Discrimination of voice onset time by human infants: New findings and implications for the effects of early experience. *Child Development, 52,* 1135–1145.

Astington, J. W. (1994). *The child's discovery of the mind.* Cambridge, MA: Harvard University Press.

Astington, J. W., Harris, P. L., and Olson, D. R. (1988). *Developing theories of mind.* Cambridge, UK: Cambridge University Press.

Attneave, F. (1959). *Applications of information theory to psychology.* New York: Holt, Rinehart, Winston.

Au, W. L. (1993). *The sonar of dolphins.* Berlin: Springer-Verlag.

August, P. V., and Anderson, J. G. T. (1987). Mammal sounds and motivation-structural rules: A test of the hypothesis. *Journal of Mammology, 68,* 1–9.

Bahrick, L. E., and Pickens, J. N. (1988). Classification of bimodal English and Spanish language passages by infants. *Infant Behavior and Development, 11,* 277–296.

Bailey, W. J. (1991). *Acoustic behaviour of insects: An evolutionary perspective.* London: Chapman and Hall.

Baillargeon, R. (1994). A model of physical reasoning in infancy. In C. Rovee-Collier and L. Lipsitt (Eds.), *Advances in infancy research,* vol. 9 (pp. 112–145). Norwood, NJ: Ablex.

Baillargeon, R., and DeVos, J. (1991). Object permanence in young infants: Further evidence. *Child Development, 62,* 1227–1246.

Bakeman, R., and Gottman, J. M. (1986). *Observing interaction: An introduction to sequential analysis.* Cambridge, UK: Cambridge University Press.

Baken, R. J. (1987). *Clinical measurement of speech and voice.* Boston: College-Hill Press.

Baken, R. (1990). Irregularity of vocal period and amplitude: A first approach to the fractal analysis of voice. *Journal of Voice, 4,* 185–197.

Baker, M. C., and Cunningham, M. A. (1985). The biology of bird song dialects. *Behavioral and Brain Sciences, 8,* 85–133.

Baker, M. C., and Mewaldt, L. R. (1978). Song dialects as barriers to dispersal in white-crowned sparrows, *Zonotrichia leucophrys nuttali. Evolution, 32,* 712–722.

Baker, M. C., Bjerke, T. K., Lampe, H., and Espmark, Y. (1986). Sexual response of female great tits to variation in size of male's song repertoires. *American Naturalist, 128,* 491–498.

Baker, R. R., and Parker, G. A. (1979). The evolution of bird coloration. *Philosophical Transactions of the Royal Society of London, 287*, 63–130.

Balaban, E. (1988a). Bird song syntax: Learned intraspecific variation is meaningful. *Proceedings of the Royal Society, London, 85*, 3657–3660.

Balaban, E. (1988b). Cultural and genetic variation in swamp sparrows. (*Melospiza georgiana*): I. Song variation, genetic variation, and their relationship. *Behaviour, 105*, 250–291.

Balaban, E. (1988c). Cultural and genetic variation in swamp sparrows (*Melospiza georgiana*): II. Behavioral salience of geographic variants. *Behaviour, 105*, 292–322.

Balcombe, J. P., and Fenton, M. B. (1988). The communication role of echolocation calls in vespertilionid bats. In P. E. Nachtigall and P. W. B. Moore (Eds.), *Animal sonar: Processes and performances* (pp. 625–628). New York: Plenum Press.

Baldwin, D. A. (1991). Infant contribution to the achievement of joint reference. *Child Development, 62*, 129–154.

Baldwin, D. A., and Moses, L. J. (1995). Early understanding referential intent and attentional focus: Evidence from language and emotion. In C. Lewis and P. Mitchell (Eds.), *Children's early understanding of mind: Origins and development* (pp. 133–156). Hove, UK: Lawrence Erlbaum Associates.

Balmford, A., and Read, A. F. (1991). Testing alternative models of sexual selection through female choice. *Trends in Ecology and Evolution, 6(9)*, 274–276.

Baptista, L. F. (1975). Song dialects and demes in sedentary populations of the white-crowned sparrow (*Zonotrichia leucophrys nuttali*). *University of California Publications in Zoology, 105*, 1–52.

Baptista, L. F., and Morton, M. L. (1982). Geographical variation in song and mate selection in montane white-crowned sparrow. *Auk, 99*, 537–547.

Baptista, L. F., and Petrinovich, L. (1984). Social interaction, sensitive phases and the song template hypothesis in the white-crowned sparrow. *Animal Behaviour, 32*, 172–181.

Baptista, L. F., and Petrinovich, L. (1986). Song development in the white-crowned sparrow: Social factors and sex differences. *Animal Behaviour, 34*, 1359–1371.

Bard, E. G., and Anderson, A. H. (1994). The unintelligibility of speech to children: Effects of referent availability. *Journal of Child Language, 21*, 623–648.

Barkow, J., Cosmides, L., and Tooby, J. (1992). *The adapted mind.* Oxford, UK: Oxford University Press.

Barlow, G. W. (1977). Modal action patterns. In T. A. Sebeok (Ed.), *How animals communicate* (pp. 98–134). Bloomington: Indiana University Press.

Barnard, C. J., and Sibly, R. M. (1981). Producers and scroungers: A general model and its application to feeding flocks of house sparrows. *Animal Behaviour, 29*, 543–550.

Baron-Cohen, S. (1990). Autism: A specific cognitive disorder of mind-blindness. *International Review of Psychiatry, 2*, 81–90.

Baron-Cohen, S. (1995). *Mindblindness.* Cambridge, MA: Bradford Books/MIT Press.

Baron-Cohen, S., Tager-Flusberg, H., and Cohen, D. (1993). *Understanding other minds: Perspectives from autism.* Oxford, UK: Oxford University Press.

Basolo, A. L. (1990). Female preference predates the evolution of the sword in swordtails. *Science, 250*, 808–810.

Bass, A. H. (1990). Sounds from the intertidal zone: Vocalizing fish. *Bioscience, 40*, 249–267.

Bateson, P. (1966). The characteristics and context of imprinting. *Biological Reviews, 41*, 177–220.

Bateson, P. (1991). Are there principles of behavioural development? In P. Bateson (Ed.), *The development and integration of behaviour* (pp. 19–40). Cambridge, UK: Cambridge University Press.

Bauer, H. R. (1987). Frequency code: Orofacial correlates of fundamental frequency. *Phonetica, 44*, 173–191.

Bauer, H. R., and Philip, M. M. (1983). Facial and vocal individual recognition in the common chimpanzee. *The Psychological Record, 33,* 161–170.

Beecher, M. D. (1982). Signature systems and kin recognition. *American Zoologist, 22,* 477–490.

Beecher, M. D., Campbell, S. E., and Burt, J. M. (1994). Song perception in the song sparrow: Birds classify by song type but not by singer. *Animal Behaviour, 47,* 1343–1351.

Beecher, M. D., Petersen, M. R., Zoloth, S. R., Moody, D. B., and Stebbins, W. C. (1979). Perception of conspecific vocalizations by Japanese macaques. Evidence for selective attention and neural lateralization. *Brain, Behavior and Evolution, 16,* 443–460.

Beeman, K. (1992). *SIGNAL release notes.* Belmont, MA: Engineering Design.

Beeman, K. (1996). *SIGNAL, v 3.0.* Belmont, MA: Engineering Design.

Begun, D. R. (1992). Miocene fossil hominids and the chimp-human clade. *Science, 257,* 1929–1933.

Beil, R. G. (1962). Frequency analysis of vowels produced in a helium-rich atmosphere. *Journal of the Acoustical Society of America, 34,* 347–349.

Beletsky, L. D. (1983). Aggressive and pair-bond maintenance songs of female red-winged blackbirds (*Agelaius phoeniceus*). *Zeitschrift für Tierpsychologie, 62,* 47–54.

Beletsky, L. D. (1984). Intersexual song answering in red-winged blackbirds. *Canadian Journal of Zoology, 63,* 735–737.

Beletsky, L. D., Chao, S., and Smith, D. G. (1980). An investigation of sound-based species recognition in the red-winged blackbird (*Agelaius phoeniceus*). *Behaviour, 73,* 189–203.

Bellugi, U., Poizner, H., and Klima, E. S. (1983). Brain organization for language: Clues from sign aphasia. *Human Neurobiology, 2,* 155–170.

Bellugi, U., Poizner, H., and Klima, E. S. (1990). Mapping brain function for language: Evidence for sign language. In G. M. Edelman, W. E. Gall, and W. M. Cowan (Eds.), *Signal and sense* (pp. 521–543). New York: John Wiley & Sons.

Bellugi, U., and Studdert-Kennedy, M. (1980). *Signed and spoken languages: Biological constraints on linguistic form.* Berlin: Springer-Verlag.

Benz, J. J. (1993). Food-elicited vocalizations in golden lion tamarins: Design features for representational communication. *Animal Behaviour, 45,* 443–455.

Benz, J. J., French, J. A., and Leger, D. W. (1990). Sex differences in vocal structure in a callitrichid primate, *Leontopithecus rosalia. American Journal of Primatology, 21,* 257–264.

Benz, J. J., Leger, D. W., and French, J. A. (1992). The relation between food preference and food-elicited vocalizations in golden lion tamarins. (*Leontopithecus rosalia*). *Journal of Comparative Psychology, 106,* 142–149.

Bercovitch, F., Hauser, M. D., and Jones, J. (1995). The endocrine basis of alarm call production in rhesus macaques. *Animal Behaviour, 49,* 1703–1706.

Berntson, G. G., and Boysen, S. T. (1989). Specificity of the cardiac response to conspecific vocalizations in chimpanzees. *Behavioral Neuroscience, 103,* 235–243.

Berntson, G. G., Boysen, S. T., Bauer, H. R., and Torello, M. S. (1990). Conspecific screams and laughter: Cardiac and behavioral reactions of infant chimpanzees. *Developmental Psychobiology, 22,* 771–787.

Berntson, G. G., Boysen, S. T., and Cacioppo, J. T. (1992). Cardiac orienting and defensive responses: Potential origins in autonomic space. In B. Campbell, H. Hayne, and R. Richardson (Eds.), *Attention and information processing in infants and adults: Perspectives from human and animal research* (pp. 163–200). New York: Lawrence Erlbaum Associates.

Berntson, G. G., Boysen, S. T., and Torello, M. W. (1993). Vocal perception: Brain event-related potentials in a chimpanzee. *Developmental Psychobiology, 26,* 12–35.

Best, C. T. (1993). Emergence of language-specific constraints in perception of non-native speech: A window on early phonological development. In B. de Boysson-Bardies (Ed.), *Developmental neuro-cognition: Speech and face processing in the first year of life* (pp. 289–304). Netherlands: Kluwer Academic.

Best, C. T. (1994). Learning to perceive the sound patterns of English. In C. Rovee-Collier and L. P. Lipsitt (Eds.), *Advances in infancy research*, vol. 9 (pp. 217–304). Norwood, NJ: Ablex.

Best, C. T., McRoberts, G. W., and Sithole, N. M. (1988). The phonological basis of perceptual loss for non-native contrasts: Maintenance of discrimination among Zulu clicks by English-speaking adults and infants. *Journal of Experimental Psychology: Human Perception and Performance, 14,* 345–360.

Best, C. T., and Queen, H. F. (1989). Baby, it's in your smile: Right hemiface bias in infant emotional expressions. *Developmental Psychology, 25,* 264–276.

Biben, M. (1993). Recognition of order effects in squirrel monkey antiphonal call sequences. *American Journal of Primatology, 29,* 109–124.

Biben, M., Masataka, N., and Symmes, D. (1986). Temporal and structural analysis of affiliative vocal exchanges in squirrel monkeys (*Saimiri sciureus*). *Behaviour, 98,* 259–273.

Bickerton, D. (1981). *Roots of language.* Ann Arbor, MI: Karoma.

Bickerton, D. (1990). *Species and language.* Chicago: Chicago University Press.

Bickley, C., and Stevens, K. (1987). Effects of a vocal tract constriction on the glottal source: Data from voiced consonants. In T. Baer, C. Sasaki, and K. Harris (Eds.), *Laryngeal function in phonation and respiration* (pp. 239–253). Boston: College-Hill Press.

Bloedel, J. R. (1993). "Involvement in" versus "storage of." *Trends in Neurosciences, 16,* 451–452.

Bloom, L. (1973). *One word at a time: The use of single word utterances before syntax.* Cambridge, UK: Cambridge University Press.

Bloom, P. (1994a). Overview: Controversies in language acquisition. In P. Bloom (Ed.), *Language acquisition: Core readings* (pp. 5–48). Cambridge, MA: MIT Press.

Bloom, P. (1994b). Generativity within language and other cognitive domains. *Cognition, 51,* 177–189.

Boesch, C. (1991a). The effect of leopard predation on grouping patterns in forest chimpanzees. *Behaviour, 117,* 220–242.

Boesch, C. (1991b). Teaching among wild chimpanzees. *Animal Behaviour, 41,* 530–532.

Boesch, C. (1994). Cooperative hunting in wild chimpanzees. *Animal Behaviour, 48,* 653–667.

Boesch, C., and Boesch, H. (1992). Transmission aspects of tool use in wild chimpanzees. In T. Ingold and K. R. Gibson (Eds.), *Tools, language and intelligence: Evolutionary implications* (pp. 165–189) Oxford, UK: Oxford University Press.

Bohner, J. (1990). Early acquisition of song in the zebra finch. *Animal Behaviour, 39,* 369–374.

Bolhuis, J. J. (1991). Mechanisms of avian imprinting: a review. *Biological Reviews, 66,* 303–345.

Bonvillian, J., Orlansky, M. D., and Novack, L. L. (1983a). Developmental milestones: Sign language acquisition and motor development. *Child Development, 54,* 1435–1445.

Bonvillian, J., Orlansky, M. D., Novack, L. L., and Folven, R. J. (1983b). Early sign language acquisition and cognitive development. In J. Rogers and E. Sloboda (Eds.), *The acquisition of symbolic skills* (pp. 201–214). New York: Plenum Press.

Borden, G. (1979). An interpretation of research on feedback interruption in speech. *Brain and Language, 7,* 307–319.

Borden, G. J., and Harris, K. S. (1984). *Speech science primer: Physiology, acoustics, and perception of speech.* Baltimore: Williams & Wilkins.

Bornstein, M. H. (1987). Perceptual categories in vision and audition. In S. Harnad (Ed.), *Categorical perception* (pp. 287–300). Cambridge, UK: Cambridge University Press.

Bottjer, S. W. (1991). Neural and hormonal substrates for song learning in zebra finches. *Seminars in the Neurosciences, 3,* 481–488.

Bottjer, S. W., and Arnold, A. P. (1984). The role of feedback from the vocal organ: I. Maintenance of stereotypical vocalizations by adult zebra finches. *The Journal of Neuroscience, 4,* 2387–2396.

Bottjer, S. W., and Johnson, F. (1992). Matters of life and death in the songbird forebrain. *Journal of Neurobiology, 23,* 1172–1191.

Bottjer, S. W., Mieser, E. A., and Arnold, A. P. (1984). Forebrain lesions disrupt development but not maintenance of song in passerine birds. *Science, 224,* 901–903.

Bottjer, S. W., Schoonmaker, J. N., and Arnold, A. P. (1986). Auditory and hormonal stimulation interact to produce neural growth in adult canaries. *Journal of Neurobiology, 17,* 605–612.

Bradbury, J. W. (1985). Contrasts between insects and vertebrates in the evolution of male display, female choice, and lek mating. In B. Hölldobler and M. Lindauer (Eds.), *Experimental behavioral ecology and sociobiology* (pp. 273–289). Stuttgart, Germany: Gustav Fischer Verlag.

Bradbury, J. W., and Andersson, M. B. (1987). *Sexual selection: Testing the alternatives.* Berlin: Springer-Verlag.

Bradbury, J. W., Gibson, R. M., McCarthy, C. E., and Vehrencamp, S. L. (1989). Dispersion of displaying male sage grouse: II. The role of female dispersion. *Behavioral Ecology and Sociobiology, 24,* 15–24.

Bradbury, J. W., Gibson, R. M., and Tsai, I. M. (1986). Hotspots and the evolution of leks. *Animal Behaviour, 34,* 1694–1709.

Bradbury, J. W., and Vehrencamp, S. L. (1976). Social organization and foraging in emballonurid bats: I. Field studies. *Behavioral Ecology and Sociobiology, 1,* 337–381.

Bradshaw, J. L., and Rodgers, L. (1993). *The evolution of lateral asymmetries, language, tool use, and intellect.* San Diego: Academic Press.

Bregman, A. S. (1990). *Auditory scene analysis: The perceptual organization of sound.* Cambridge, MA: MIT Press/Bradford Books.

Brenowitz, E. A. (1982). The active space of red-winged blackbird song. *Journal of Comparative Physiology, 147,* 511–522.

Brenowitz, E. A. (1991). Altered perception of species-specific song by female birds after lesions of a forebrain nucleus. *Science, 251,* 303–305.

Brenowitz, E. A., and Arnold, A. P. (1986). Interspecific comparisons of the size of neural song control regions and song complexity in duetting birds: Evolutionary implications. *Journal of Neuroscience, 6,* 2875–2879.

Brenowitz, E. A., and Arnold, A. P. (1989). Accumulation of estrogen in a vocal control brain region of a duetting song bird. *Brain Research, 480,* 119–125.

Brenowitz, E. A., and Arnold, A. P. (1990). The effects of systemic androgen treatment on androgen accumulation in song control regions of the adult female canary brain. *Journal of Neurobiology, 21,* 837–843.

Brenowitz, E. A., Arnold, A. P., and Levin, R. N. (1985). Neural correlates of female song in tropical duetting birds. *Brain Research, 343,* 104–112.

Brenowitz, E. A., Rose, G., and Capranica, R. R. (1985). Neural correlates of temperature coupling in the vocal communication system of the gray treefrog (*Hyla versicolor*). *Brain Research, 359,* 364–367.

Bretherton, I. (1988). How to do things with one word: The ontogenesis of intentional message making in infancy. In M. D. Smith and J. L. Locke (Eds.), *The emergent lexicon: The child's development of a linguistic vocabulary* (pp. 225–262). New York: Academic Press.

Broca, P. (1861). Nouvelle observation d'aphémie produite par une lésion de la motie postérieure des deuxième et troisième circonvolutions frontales. *Bulletin de la Société d'Anatomique, Paris, 6,* 398–407.

Brothers, L. (1992). Perception of social acts in primates: Cognition and neurobiology. *Seminars in the Neurosciences, 4,* 409–414.

Brothers, L., and Ring, B. (1992). A neuroethological framework for the representation of minds. *Journal of Cognitive Neuroscience, 4,* 107–118.

Brothers, L., and Ring, B. (1993). Mesial temporal neurons in the macaque monkey with responses selective for aspects of social stimuli. *Behavioural Brain Research, 57,* 53–61.

Brothers, L., Ring, B., and Kling, A. (1990). Response of neurons in the macaque amygdala to complex social stimuli. *Behavioural Brain Research, 41,* 199–213.

Browman, C., and Goldstein, L. (1986). Towards an articulatory phonology. *Phonology Yearbook, 3,* 219–252.

Browman, C., and Goldstein, L. (1989). Articulatory gestures as phonological units. *Phonology, 6,* 201–251.

Brown, C. H. (1982). Ventriloquial and locatable vocalizations in birds. *Zeitschrift für Tierpsychologie, 59,* 338–350.

Brown, C. H. (1989a). The active space of blue monkey and grey-cheeked mangabey vocalizations. *Animal Behaviour, 37,* 1023–1034.

Brown, C. H. (1989b). The measurement of vocal amplitude and vocal radiation pattern in blue monkeys and grey-cheeked mangabeys. *Bioacoustics, 1,* 253–271.

Brown, C. H., Beecher, M. D., Moody, D. B., and Stebbins, W. C. (1979). Locatability of vocal signals in Old World monkeys: Design features for the communication of position. *Journal of Comparative and Physiological Psychology, 93,* 806–819.

Brown, C. H., and Gomez, R. (1992). Functional design features in primate vocal signals: The acoustic habitat and sound distortion. In T. Nishida, W. C. McGrew, P. Marler, M. Pickford, and F. de Waal (Eds.), *Topics in primatology,* vol. 1, *Human origins* (pp. 177–198). Tokyo: Tokyo University Press.

Brown, C. H., and May, B. J. (1990). Sound localization and binaural processes. In M. A. Berkley and W. C. Stebbins (Eds.), *Comparative perception,* vol. 1, *Basic mechanisms* (pp. 247–284). New York: John Wiley & Sons.

Brown, C. H., Schessler, T., Moody, D. B., and Stebbins, W. C. (1983). Vertical and horizontal sound localization in primates. *Journal of the Acoustical Society of America, 68,* 1804–1811.

Brown, C. H., and Waser, P. M. (1984). Hearing and communication in blue monkeys (*Cercopithecus mitis*). *Animal Behaviour, 32,* 66–75.

Brown, C. H., and Waser, P. (1988). Environmental influences on the structure of primate vocalizations. In D. Todt, P. Goedeking, and D. Symmes (Eds.), *Primate vocal communication* (pp. 51–68). Berlin: Springer-Verlag.

Brown, C. R., Bomberger Brown, M., and Shaffer, M. L. (1991). Food-sharing signals among socially foraging cliff swallows. *Animal Behaviour, 42,* 551–564.

Brown, R. (1973). *A first language: The early stages.* Cambridge, MA: Harvard University Press.

Brown, S. D., and Bottjer, S. W. (1993). Testosterone-induced changes in adult canary brain are reversible. *Journal of Neurobiology, 24,* 627–640.

Bruce, C. J., Desimone, R., and Gross, C. G. (1981). Visual properties of neurons in a polysensory area in superior temporal sulcus of the macaque. *Journal of Neurophysiology, 46,* 369–384.

Brugge, J. F., and Merzenich, M. M. (1973). Responses of neurons in auditory cortex of the macaque monkey to monaural and binaural stimuli. *Journal of Neurophysiology, 36,* 1138–1158.

Bruns, V. (1976a). Peripheral auditory tuning for fine frequency analysis by the CF-FM bat, *Rhinolophus ferrumequinum:* I. Mechanical specializations of the cochlea. *Journal of Comparative Physiology, 106,* 77–86.

Bruns, V. (1976b). Peripheral auditory tuning for fine frequency analysis by the CF-FM bat, *Rhinolophus ferrumequinum:* II. Frequency mapping in the cochlea. *Journal of Comparative Physiology, 106,* 87–97.

Brzoska, J., and Obert, H.-J. (1989). Acoustic signals influencing the hormone production of the testes in the grass frog. *Journal of Comparative Physiology, 140,* 25–29.

Bucher, L. T., Ryan, M. J., and Bartholomew, G. A. (1982). Oxygen consumption during resting, calling, and nest building in the frog *Physalaemus pustulosus. Physiological Zoology, 55,* 10–22.

Buck, J. R., and Tyack, P. L. (1993). A quantitative measure of similarity for *Tursiops truncatus* signature whistles. *Journal of the Acoustical Society of America, 94,* 2497–2506.

Burghardt, G. M. (1991). Cognitive ethology and critical anthropomorphism: A snake with two heads and hognose snakes that play dead. In C. A. Ristau (Ed.), *Cognitive ethology: The minds of other animals* (pp. 53–91). Hillsdale, NJ: Lawrence Erlbaum Associates.

Burns, E. M., and Ward, W. D. (1978). Categorical perception—phenomenon or epiphenomenon: Evidence from experiments in the perception of melodic musical intervals. *Journal of the Acoustical Society of America, 63,* 456–468.

Bushnell, I. W. R., Sai, F., and Mullin, J. T. (1989). Neonatal recognition of the mother's face. *British Journal of Developmental Psychology, 7,* 3–15.

Buss, D. M. (1989). Sex differences in human mate preferences: Evolutionary hypotheses tested in 37 cultures. *Behavioral and Brain Sciences, 12,* 1–24.

Butcher, G. S., and Rohwer, S. (1989). The evolution of conspicuous and distinctive coloration for communication in birds. In D. M. Power (Ed.), *Current ornithology,* vol. 6 (pp. 51–108). New York: Plenum Press.

Byrne, R., and Whiten, A. (1990). Tactical deception in primates: The 1990 database. *Primate Report, 27,* 1–101.

Byrne, R., and Whiten, A. (1991). Computation and mindreading in primate tactical deception. In A. Whiten (Ed.), *Natural theories of mind* (pp. 127–142). Cambridge, MA: Basil Blackwell.

Caldwell, R. L. (1986). The deceptive use of reputation by stomatopods. In R. W. Mitchell and N. S. Thompson (Eds.), *Deception: Perspectives on human and nonhuman deceit* (pp. 129–146). New York: SUNY Press.

Caldwell, R. L. (1987). Assessment strategies in stomatopods. *Bulletin of Marine Science, 41,* 135–150.

Caldwell, R. L., and Dingle, H. (1975). Ecology and evolution of agonistic behavior in the stomatopods. *Naturwissenshaften, 62,* 214–222.

Calvin, W. H., and Ojemann, G. A. (1994). *Conversations with Neil's brain.* Reading, MA: Addison-Wesley.

Campbell, C. B. G. (1988). Homology. In L. N. Irwin (Ed.), *Comparative neuroscience and neurobiology: Readings from the encyclopedia of neuroscience* (pp. 44–45). Boston: Birkhäuser.

Campbell, R., Heywood, C. A., Cowey, A., Regard, M., and Landis, T. (1990). Sensitivity to eye gaze with prosopagnosic patients and monkeys with superior temporal sulcus ablation. *Neuropsychologia, 28,* 1123–1142.

Campos, J. J. (1983). The importance of affective communication in social referencing. *Merrill-Palmer Quarterly, 29,* 83–87.

Canady, R. A., Kroodsma, D. E., and Nottebohm, F. (1984). Population differences in complexity of learned skill are correlated with the brain space involved. *Proceedings of the National Academy of Sciences, 81,* 6232–6234.

Cann, R. L. (1995). Mitochondrial DNA and human evolution. In J.-P. Changeux and J. Chavaillon (Eds.), *Origins of the human brain* (pp. 127–135). Oxford, UK: Oxford University Press.

Cann, R. L., Stoneking, M., and Wilson, A. C. (1987). Mitochondrial DNA and human evolution. *Nature, 325,* 31–36.

Caplan, D. (1987). *Neurolinguistics and linguistic aphasiology.* New York: McGraw-Hill.

Caplan, D. (1992). *Language: Structure, processing, and disorders.* Cambridge, MA: MIT Press.

Cappa, S. F., and Vignolo, L. A. (1988). Sex differences in the site of brain lesions underlying global aphasia. *Aphasiology, 2,* 259–264.

Capranica, R. R. (1965). *The evoked vocal response of the bullfrog.* Cambridge, MA: MIT Press.

Capranica, R. R. (1966). Vocal response of the bullfrog to natural and synthetic mating calls. *Journal of the Acoustical Society of America, 40,* 1131–1139.

Capranica, R. R. (1992). The untuning of the tuning curve: Is it time? *Seminars in the Neurosciences, 4,* 401–408.

Capranica, R. R., and Moffat, A. J. M. (1983). Neurobehavioral correlates of sound communication in anurans. In J. P. Ewert, R. R. Capranica, and D. J. Ingle (Eds.), *Advances in vertebrate neuroethology* (pp. 701–730). New York: Plenum Press.

Caramazza, A., Hillis, A., Leek, E. C., and Miozzo, M. (1994). The organization of lexical knowledge in the brain: Evidence from category and modality-specific deficits. In L. A. Hirschfield and S. A. Gelman (Eds.), *Mapping the mind: Domain specificity in cognition and culture* (pp. 68–84). Cambridge, UK: Cambridge University Press.

Carey, S. (1978). The child as word learner. In M. Halle, J. Bresnan, and A. Miller (Eds.), *Linguistic theory and psychological reality* (pp. 264–293). Cambridge, MA: MIT Press.

Carey, S. (1985). *Conceptual change in childhood.* Cambridge, MA: MIT Press.

Carey, S., and Diamond, R. (1977). From piecemeal to configurational representation of faces. *Science, 195,* 312–314.

Carey, S., and Spelke, E. (1994). Domain-specific knowledge and conceptual change. In L. Hirschfeld and S. Gelman, Eds. *Mapping the mind* (pp. 169–199). Cambridge, UK: Cambridge University Press.

Carney, A. E., Widin, G. P., and Viemeister, N. F. (1977). Noncategorical perception of stop consonants differing in VOT. *Journal of the Acoustical Society of America, 68,* 843–857.

Caro, T. M. (1994a). *Cheetahs of the serengeti.* Chicago: University of Chicago Press.

Caro, T. M. (1994b). Ungulate antipredator behaviour: Preliminary and comparative data from African bovids. *Behaviour, 128,* 189–228.

Caro, T. M., and Hauser, M. D. (1992). Is there teaching in nonhuman animals? *Quarterly Review of Biology, 67,* 151–174.

Cartmill, M. (1994). A critique of homology as a morphological concept. *American Journal of Physical Anthropology, 94,* 115–123.

Caryl, P. G. (1979). Communication by agonistic displays: What can game theory contribute to ethology? *Behaviour, 68,* 136–169.

Caryl, P. G. (1982). Animal signals: A reply to Hinde. *Animal Behaviour, 30,* 240–244.

Casey, R. M., and Gaunt, A. S. (1985). Theoretical models of the avian syrinx. *Journal of Theoretical Biology, 116,* 45–64.

Catchpole, C. K. (1986). Song repertoires and reproductive success in the great reed warbler *Acrocephalus arundinaceus. Behavioral Ecology and Sociobiology, 19,* 439–445.

Catchpole, C. K., Leisler, B., and Dittami, J. (1984). Differential responses to male song repertoires in female songbirds implanted with oestradiol. *Nature, 312,* 563–564.

Cavalli-Sforza, L. L., and Feldman, M. W. (1983). Paradox of the evolution of communication and of social interactivity. *Proceedings of the National Academy of Sciences, 80,* 2017–2021.

Changeux, J.-P., Heidman, T., and Patte, P. (1984). Learning by selection. In P. Marler and H. S. Terrace (Eds.), *The biology of learning* (pp. 115–133). Berlin: Springer-Verlag.

Chapman, C. A., and Lefebvre, L. (1990). Manipulating foraging group size: Spider monkey food calls at fruiting trees. *Animal Behaviour, 39,* 891–896.

Charif, R. A., Mitchell, S., and Clark, C. W. (1993). *Canary 1.1 user's manual.* Ithaca, NY: Cornell Laboratory of Ornithology.

Charnov, E. L., and Krebs, J. R. (1975). The evolution of alarm calls: Altruism or manipulation. *American Naturalist, 109,* 107–112.

Cheney, D. L. (1992). Intragroup cohesion and intergroup hostility: The relation between grooming distributions and intergroup competition among female primates. *Behavioral Ecology, 3,* 334–345.

Cheney, D. L., and Seyfarth, R. M. (1980). Vocal recognition in free-ranging vervet monkeys. *Animal Behaviour, 28,* 362–367.

Cheney, D. L., and Seyfarth, R. M. (1981). Selective forces affecting the predator alarm calls of vervet monkeys. *Behaviour, 76,* 25–61.

Cheney, D. L., and Seyfarth, R. M. (1982). How vervet monkeys perceive their grunts: Field playback experiments. *Animal Behaviour, 30,* 739–751.

Cheney, D. L., and Seyfarth, R. M. (1985a). Social and non-social knowledge in vervet monkeys. *Philosophical Transactions of the Royal Society of London, B, 308,* 187–201.

Cheney, D. L., and Seyfarth, R. M. (1985b). Vervet monkey alarm calls: Manipulation through shared information? *Behaviour, 94,* 150–166.

Cheney, D. L., and Seyfarth, R. M. (1988). Assessment of meaning and the detection of unreliable signals by vervet monkeys. *Animal Behaviour, 36,* 477–486.

Cheney, D. L., and Seyfarth, R. M. (1990). *How monkeys see the world: Inside the mind of another species.* Chicago: Chicago University Press.

Cheney, D. L., and Seyfarth, R. M. (1992). Meaning, reference, and intentionality in the natural vocalizations of monkeys. In T. Nishida, W. C., McGrew, P. Marler, M. Pickford, and F. de Waal (Eds.), *Topics in primatology,* vol. 1, *Human origins* (pp. 315–330). Tokyo: Tokyo University Press.

Cheney, D. L., Seyfarth, R. M., and Silk, J. (1995). The responses of female baboons (*Papio cynocephalus ursinus*) to anomalous social interactions: Evidence for causal reasoning? *Journal of Comparative Psychology, 109,* 134–141.

Cheney, D. L., Seyfarth, R. M., and Silk, J. (in press). The role of grunts in reconciling opponents and facilitating interactions among adult female baboons. *Animal Behaviour.*

Cheng, M.-F. (1992). For whom does the female dove coo? A case for the role of vocal self-stimulation. *Animal Behaviour, 43,* 1035–1044.

Cheng, P. W., and Holyoak, R. J. (1989). On the natural selection of reasoning theories. *Cognition, 33,* 295–313.

Cherry, C. (1957). *On human communication.* New York: Wiley Press.

Chevalier-Skolnikoff, S. (1973). Facial expression of emotion in nonhuman primates. In P. Ekman (Ed.), *Darwin and facial expression* (pp. 11–90). London: Academic Press.

Chiba, T., and Kajiyama, M. (1941). *The vowel: Its nature and structure.* Tokyo: Tokyo-Kaiseikan.

Chivers, D., and MacKinnon, J. (1977). On the behaviour of siamang after playback of their calls. *Primates, 18,* 943–948.

Chomsky, N. (1957). *Syntactic structures.* The Hague: Mouton.

Chomsky, N. (1965). *Aspects of the theory of syntax.* Cambridge, MA: MIT Press.

Chomsky, N. (1975). *Reflections on language.* New York: Pantheon Press.

Chomsky, N. (1986). *Knowledge of language: Its nature, origin, and use.* New York: Praeger.

Chomsky, N. (1988). *Language and problems of knowledge*. Cambridge, MA: MIT Press.

Chomsky, N. (1990). Language and mind. In D. H. Mellor (Ed.), *Ways of communicating* (pp. 56–80). Cambridge, UK: Cambridge University Press.

Chomsky, N., and Halle, M. (1968). *The sound patterns of English*. Cambridge, MA: MIT Press.

Chovil, N. (1991). Communicative functions of facial displays. *Journal of Nonverbal Behavior, 15*, 141–154.

Christophe, A., Dupoux, E., Bertoncini, J., and Mehler, J. (1994). Do infants perceive word boundaries? An empirical study of the bootstrapping of lexical acquisition. *Journal of the Acoustical Society of America, 95*, 1570–1580.

Christy, J. G. (1988). Pillar function in the fiddler crab *Uca beebi:* 1. Competitive courtship signalling. *Ethology, 78*, 113–128.

Chung, M.-S., and Thomson, D. M. (1995). Development of face recognition. *British Journal of Developmental Psychology, 86*, 55–87.

Clark, A. P. (1993). Rank differences in the production of vocalizations by wild chimpanzees as a function of social context. *American Journal of Primatology, 31*, 159–180.

Clark, A. P., and Wrangham, R. W. (1993). Acoustic analysis of wild chimpanzee hoots: Do Kibale Forest chimpanzees have an acoustically distinct food arrival pant hoot? *American Journal of Primatology, 31*, 99–110.

Clark, C., Marler, P., and Beeman, K. (1987). Quantitative analysis of animal vocal phonology: An application to swamp sparrow song. *Ethology, 76*, 101–115.

Clark, E. V. (1993). *The lexicon in acquisition*. New York: Cambridge University Press.

Clements, W. A., and Perner, J. (1994). Implicit understanding of belief. *Cognitive Development, 9*, 377–395.

Cleveland, J., and Snowdon, C. T. (1981). The complex vocal repertoire of the adult cotton-top tamarin, *Saguinus oedipus oedipus*. *Zeitschrift für Tierpsychologie, 58*, 231–270.

Clifton, R. K., Freyman, R. L., Litovsky, R. Y., and McCall, D. (1994). Listeners' expectations about echoes can raise or lower echo threshold. *Journal of the Acoustical Society of America, 95*, 1525–1533.

Clower, R. P., Nixdorf, B., and DeVoogd, T. J. (1989). Synaptic plasticity in the hypoglossal nucleus of female canaries: Structural correlates of season, hemisphere and testosterone treatment. *Behavioral and Neural Biology, 52*, 63–77.

Clutton-Brock, T. H. (1992). *The evolution of parental care*. Princeton, NJ: Princeton University Press.

Clutton-Brock, T. H., and Harvey, P. H. (1976). Evolutionary rules and primate societies. In P. P. G. Bateson and R. A. Hinde (Eds.), *Growing points in ethology* (pp. 195–237). Cambridge, UK: Cambridge University Press.

Clutton-Brock, T. H., and Parker, G. A. (1995). Punishment in animal societies. *Nature, 373*, 209–216.

Collias, N. E. (1987). The vocal repertoire of the red junglefowl: A spectrographic classification and the code of communication. *Condor, 69*, 510–524.

Collins, S. (1993). Is there only one type of male handicap? *Proceedings of the Royal Society, London, 252*, 193–197.

Colton, R. H., and Steinschneider, A. (1980). Acoustic relationships of infant cries to sudden infant death syndrome. In T. Murry and J. Murry (Eds.), *Infant communication: Cry and early Speech* (pp. 183–208). Houston: College-Hill Press.

Comrie, B. (1981). Language universals and linguistic typology: Syntax and morphology. Oxford, UK: Basil Blackwell.

Connine, C. M., and Clifton, C. (1987). Interactive use of lexical information in speech perception. *Journal of Experimental Psychology: Human Perception and Performance, 13,* 291–299.

Cooper, F. S., Delattre, P. C., Liberman, A. M., Borst, J. M., and Gerstman, L. J. (1952). Some experiments on the perception of synthetic speech sounds. *Journal of the Acoustical Society of America, 24,* 597–606.

Cooper, W., and Sorensen, J. (1981). *Fundamental frequency in sentence production.* Berlin: Springer-Verlag.

Corballis, M. C. (1992). On the evolution of language and generativity. *Cognition, 44,* 197–226.

Corballis, M. C. (1994). The generation of generativity: A response to Bloom. *Cognition, 51,* 191–198.

Cords, M. (1990). Vigilance and mixed-species association of some East African forest monkeys. *Behavioral Ecology and Sociobiology, 26,* 297–300.

Cosmides, L., and Tooby, J. (1992). Cognitive adaptations for social exchange. In J. Barkow, L. Cosmides, and J. Tooby (Eds.), *The adapted mind* (pp. 163–228). New York: Oxford University Press.

Cosmides, L., and Tooby, J. (1994a). Beyond intuition and instinct blindness: Toward an evolutionarily rigorous cognitive science. *Cognition, 50,* 41–77.

Cosmides, L., and Tooby, J. (1994b). Origins of domain specificity: The evolution of functional organization. In L. A. Hirschfield and S. A. Gelman (Eds.), *Mapping the mind: Domain specificity in cognition and culture* (pp. 85–116). Cambridge, UK: Cambridge University Press.

Coss, R. G. (1991). Context and animal behavior: III. The relationship between early development and evolutionary persistence of ground squirrel antisnake behavior. *Ecological Psychology, 3,* 277–315.

Coss, R. G., Guse, K. L., Poran, N. S., and Smith, D. G. (1993). Development of antisnake defenses in California ground squirrels (*Spermophilus beecheyi*): II. Microevolutionary effects of relaxed selection from rattlesnakes. *Behaviour, 124,* 137–164.

Coss, R. G., and Owings, D. H. (1978). Snake-directed behavior by snake naive and experienced California ground squirrels in a simulated burrow. *Zeitschrift für Tierpsychologie, 48,* 421–435.

Cox, C. R., and LeBoeuf, B. J. (1977). Female incitation of male competition: A mechanism in sexual selection. *American Naturalist, 111,* 317–335.

Crelin, E. (1987). *The human vocal tract.* New York: Vantage Press.

Cronin, H. (1992). *The ant and the peacock.* New York: Cambridge University Press.

Cronin, T. W., and Marshall, N. J. (1989). Multiple spectral classes of photoreceptors in the retinas of gonodactyloid stomatopod crustaceans. *Journal of Comparative Physiology, A, 166,* 261–275.

Crystal, D. (1987). *The Cambridge encyclopedia of language.* Cambridge, UK: Cambridge University Press.

Curio, E. (1978). The adaptive significance of avian mobbing: I. Teleonomic hypotheses and predictions. *Zeitschrift für Tierpsychologie, 48,* 175–183.

Curio, E. (1994). Causal and functional questions: How are they linked? *Animal Behaviour, 47,* 999–1021.

Cuthill, I. C., and MacDonald, W. A. (1990). Experimental manipulation of the dawn and dusk chorus in the blackbird (*Turdus merula*). *Behavioral Ecology and Sociobiology, 26,* 209–216.

Cynx, J. (1990). Experimental determination of a unit of song production in the zebra finch (*Taeniopygia guttata*). *Journal of Comparative Psychology, 104,* 3–10.

Cynx, J., and Nottebohm, F. (1992a). Role of gender, season, and familiarity in discrimination of conspecific song by zebra finches (*Taeniopygia guttata*). *Proceedings of the National Academy of Sciences, 89,* 1368–1371.

Cynx, J., and Nottebohm, F. (1992b). Testosterone facilitates some conspecific song discriminations in castrated zebra finches (*Taeniopygia guttata*). *Proceedings of the National Academy of Sciences, 89,* 1376–1378.

Cynx, J., Williams, H., and Nottebohm, F. (1992). Hemispheric differences in avian song discrimination. *Proceedings of the National Academy of Sciences, 89,* 1372–1375.

Dabelsteen, T. (1992). Interactive playback: A finely tuned response. In P. K. McGregor (Ed.), *Playback and studies of animal communication* (pp. 97–110). London: Plenum Press.

Dabelsteen, T., and Pedersen, S. B. (1990). Song and information about aggressive responses of blackbirds, *Turdus merula:* Evidence from interactive playback experiments with territory owners. *Animal Behaviour, 40,* 1158–1168.

Daly, M., and Wilson, M. (1988). *Homicide.* New York: Aldine de Gruyter.

Damasio, A. R. (1994). *Descartes' error.* Boston: Norton.

Damasio, A. R., and Damasio, H. (1992). Brain and language. *Scientific American, 267,* 88–109.

Damasio, A. R., Damasio, H., and Tranel, D. (1990). Impairments of visual recognition as clues to the processes of memory. In G. M. Edelman, W. E. Gall, and W. M. Cowan (Eds.), *Signal and sense: Local and global order in perceptual maps* (pp. 451–473). New York: Wiley-Liss.

Damasio, A. R., Damasio, H., and Van Hoesen, G. W. (1982). Prosopagnosia: Anatomical basis and neurobehavioural mechanism. *Neurology, 32,* 331–341.

Darwin, C. (1859). *On the origin of species.* London: John Murray.

Darwin, C. (1871). *The descent of man and selection in relation to sex.* London: John Murray.

Darwin, C. (1872). *The expression of the emotions in man and animals.* London: John Murray.

Davidson, R. J. (1992). Emotion and affective style: Hemispheric substrates. *Psychological Science, 3,* 39–43.

Davidson, R. J. (1995). Cerebral asymmetry, emotion, and affective style. In R. J. Davidson and K. Hugdahl (Eds.), *Brain asymmetry* (pp. 361–389). Cambridge, MA: MIT Press.

Davidson, R. J., and Fox, N. A. (1982). Asymmetrical brain activity discriminates between positive and negative affect stimuli in human infants. *Science, 218,* 1235–1237.

Davidson, R. J., and Hugdahl, K. (1995). *Brain asymmetry.* Cambridge, MA: MIT Press.

Davies, N. B., and Halliday, T. R. (1978). Deep croaks and fighting assessment in toads *Bufo bufo. Nature, 274,* 683–685.

Davies, N. B., and Lundberg, A. (1984). Food distribution and a variable mating system in the dunnock *Prunella modularis. Ibis, 127,* 100–110.

Dawkins, R. (1976). *The selfish gene.* Oxford, UK: Oxford University Press.

Dawkins, R. (1986). *The blind watchmaker.* New York: W. W. Norton.

Dawkins, R., and Krebs, J. R. (1978). Animal signals: Information or manipulation. In J. R. Krebs and N. B. Davies (Eds.), *Behavioural ecology* (pp. 282–309). Oxford, UK: Blackwell Scientific Publications.

Deacon, T. (1984). Connections of the inferior periarcuate area in the brain of Macaca fascicularis: An experimental and comparative investigation of language circuitry and its evolution. PhD diss., Harvard University, Cambridge, MA.

Deacon, T. (1990a). Fallacies of progression in theories of brain-size evolution. *International Journal of Primatology, 11,* 193–235.

Deacon, T. (1990b). Problems of ontogeny and phylogeny in brain-size evolution. *International Journal of Primatology, 11,* 237–282.

Deacon, T. W. (1991). The neural circuitry underlying primate calls and human language. In J. Wind (Ed.), *Language origins: A multidisciplinary approach* (pp. 131–172). Netherlands: Kluwer Academic.

Deacon, T. (1992). Brain-language co-evolution. In J. A. Hawkins (Ed.), *The evolution of human language* (pp. 629–705). Redwood City, CA: Addison-Wesley.

Deacon, T. (in press). Why a brain capable of language evolved only once: Prefrontal cortex and symbol learning. In B. Velichkovsky and D. M. Rumbaugh (Eds.), *Human by nature: Origins and destiny of language*. Hillsdale, NJ: Lawrence Erlbaum Associates.

Dear, S. P., Simmons, J. A., and Fritz, J. (1993). A possible neuronal basis for representation of acoustic scenes in auditory cortex of the big brown bat. *Nature, 364,* 620–623.

Dehaene-Lambertz, G., and Dehaene, S. (1994). Speed and cerebral correlates of syllable discrimination in infants. *Nature, 370,* 292–295.

deLacoste, M. C., Holloway, R. L., and Woodward, D. J. (1986). Sex differences in the fetal human corpus callosum. *Human Neurobiology, 5,* 93–96.

deLacoste-Utamsing, M. C., and Holloway, R. L. (1982). Sexual dimorphism in the human corpus callosum. *Science, 216,* 1431–1432.

Delattre, P. C., Liberman, A. M., and Cooper, F. S. (1955). Acoustic loci and transitional cues for consonants. *Journal of the Acoustical Society of America, 27,* 769–773.

Demonet, J.-F., Wise, R., and Frackowiak, R. (1993). Language functions explored in normal subjects by positron emission tomography: A critical review. *Human Brain Mapping, 1,* 39–47.

Denes, P. B., and Pinson, E. N. (1993). *The speech chain: The physics and biology of spoken language*. New York: W. H. Freeman.

Dennett, D. C. (1983). Intentional systems in cognitive ethology: The "Panglossian paradigm" defended. *Behavioral and Brain Sciences, 6,* 343–390.

Dennett, D. C. (1987). *The intentional stance*. Cambridge, MA: MIT Press.

DePaulo, B. M. (1994). Spotting lies: Can humans learn to do better? *Current Directions in Psychological Science, 3,* 83–86.

DePaulo, B. M., Stone, J. I., and Lassiter, G. D. (1985). Deceiving and detecting deceit. In B. R. Schlenker (Ed.), *The self and social life* (pp. 135–168). New York: McGraw-Hill.

De Renzi, E., Scotti, G., and Spinnler, M. (1968). The performance of patients with unilateral brain damage on face recognition tasks. *Cortex, 4,* 17–34.

de Schonen, S., Deruelle, C., Mancini, J., and Pascalis, O. (1993). Hemispheric differences in face processing and brain maturation. In B. de Boysson-Bardies, S. de Schonen, P. Jusczyk, P. MacNeilage, and J. Morton (Eds.), *Developmental neurocognition: Speech and face processing in the first year of life* (pp. 149–163). Dordrecht, The Netherlands: Kluwer Academic.

Desimone, R., Albright, T. D., Gross, C. G., and Bruce, C. (1984). Stimulus selective properties of inferior temporal neurons in the macaque. *Journal of Neuroscience, 4,* 2051–2062.

Desimone, R., and Gross, C. G. (1979). Visual areas in the temporal cortex of the macaque. *Brain Research, 178,* 363–380.

DeVoogd, T. J. (1991). Endocrine modulation of the development and adult function of the avian song system. *Psychoneuroendocrinology, 16,* 41–66.

DeVoogd, T. J. (1994). The neural basis for the acquisition and production of bird song. In J. A. Hogan and J. J. Bolhuis (Eds.), *Causal mechanisms in behavioural development* (pp. 114–165). Cambridge, UK: Cambridge University Press.

DeVoogd, T. J., Krebs, J. R., Healy, S. D., and Purvis, A. (1993). Relations between song repertoire size and the volume of brain nuclei related to song: Comparative evolutionary analyses amongst oscine birds. *Proceedings of the Royal Society, London, 254,* 75–82.

DeVoogd, T. J., Nixdorf, B., and Nottebohm, F. (1985). Formation of new synapses related to acquisition of a new behavior. *Brain Research, 329,* 304–308.

de Waal, F. (1982). *Chimpanzee politics*. New York: Harper & Row.

de Waal, F. (1989). *Peacemaking among primates*. Cambridge, MA: Harvard University Press.

Diamond, M. C. (1988). *Enriching heredity: The impact of the environment on the anatomy of the brain*. New York: Free Press.

Dittrich, W. (1990). Representation of faces in longtailed macaques (*Macaca fascicularis*). *Ethology, 85*, 265–278.

Dittus, W. (1988). An analysis of toque macaque cohesion calls from an ecological perspective. In D. Todt, P. Goedeking, and D. Symmes (Eds.), *Primate vocal communication* (pp. 31–50). Berlin: Springer-Verlag.

Dittus, W. (1984). Toque macaque food calls: Semantic communication concerning food distribution in the environment. *Animal Behaviour, 32*, 470–477.

Dixon, A. F. (1983). Observations on the evolution and behavioural significance of "sexual skin" in female primates. *Advances in the Study of Behaviour, 13*, 63–106.

Dixon, A. F. (1991). Sexual selection, natural selection and copulatory patterns in male primates. *Folia primatologica, 57*, 96–101.

Dodd, B. (1979). Lip reading in infants: Attention to speech presented in- and out-of-synchrony. *Cognitive Psychology, 11*, 478–484.

Doherty, J. A., and Gerhardt, H. C. (1984). Evolutionary and neurobiological implications of selective phonotaxis in the spring peeper (*Hyla crucifer*). *Animal Behaviour, 32*, 875–881.

Dooling, R. J. (1980). Behavior and psychophysics of hearing in birds. In A. N. Popper and R. R. Fay (Eds.), *Comparative studies of hearing in vertebrates* (pp. 261–288). Berlin: Springer-Verlag.

Dooling, R. J. (1982). Auditory perception in birds. In D. E. Kroodsma and E. H. Miller (Eds.), *Acoustic communication in birds*, vol. 1 (pp. 95–130). New York: Academic Press.

Dooling, R. J. (1989). Perception of complex, species-specific vocalizations by birds and humans. In R. J. Dooling and S. H. Hulse (Eds.), *The comparative psychology of audition: Perceiving complex sounds* (pp. 423–444). Hillsdale, NJ: Lawrence Erlbaum Associates.

Dooling, R. J. (1992). Hearing in birds. In D. B. Webster, R. F. Fay, and A. N. Popper (Eds.), *The evolutionary biology of hearing* (pp. 545–560). New York: Springer-Verlag.

Dooling, R. J., Best, C. T., and Brown, S. D. (1995). Discrimination of synthetic full-formant and sinewave /ra-la/ continua by budgerigars (*Melopsittacus undulatus*) and zebra finches (*Taeniopygia guttata*). *Journal of the Acoustical Society of America, 97*, 1839–1846.

Draper, M. H., Ladefoged, P., and Whitteridge, D. (1959). Respiratory muscles in speech. *Journal of Speech and Hearing Research, 2*, 16–27.

Dreller, C., and Kirchner, W. H. (1993). How honey bees perceive the information of the dance language. *Naturwissenshaften, 80*, 319–321.

Dreller, C., and Kirchner, W. H. (1994). Hearing in the Asian honeybees *Apis dorsata* and *Apis florea*. *Insectes Sociaux, 42*, 115–125.

Dretske, F. (1981). *Knowledge and the flow of information*. Cambridge, MA: MIT/Bradford Press.

Dromi, E. (1987). *Early lexical development*. Cambridge, MA: Cambridge University Press.

Dronkers, N., and Pinker, S. (in press). Language and the aphasias. In E. R. Kandel, J. H. Schwartz, and T. M. Jessell (Eds.), *Principles of neural science*, 4th ed. Norwalk, CT: Appleton & Lange.

Duchenne, B. (in press). *The mechanism of human facial expression or an electro-physiological analysis of the expression of the emotions*. New York: Cambridge University Press (original, 1862).

Duellman, W. E., and Trueb, L. (1986). *Biology of the amphibians*. New York: McGraw-Hill.

Dunbar, R. (1993). Coevolution of neocortical size, group size and language in humans. *Behavioral and Brain Sciences, 16(4)*, 681–735.

Dunia, R., and Narins, P. M. (1989a). Temporal integration in an anuran auditory nerve. *Hearing Research, 39*, 287–298.

Dunia, R., and Narins, P. M. (1989b). Temporal resolution in frog auditory-nerve fibers. *Journal of the Acoustical Society of America, 85,* 1630–1638.

Dusenbery, D. B. (1992). *Sensory ecology: How organisms acquire and respond to information.* New York: W. H. Freeman.

Dusser de Barenne, J. G., Garol, H. W., and McCulloch, W. S. (1941). The motor cortex of the chimpanzee. *Journal of Neurophysiology, 4,* 287–303.

Dyer, F. C. (1991). Comparative studies of dance communication: Analysis of phylogeny and function. In D. R. Smith (Ed.), *Diversity in the genus Apis* (pp. 177–198). Boulder, CO: Westview Press.

Dyer, F. C., and Seeley, T. D. (1989). On the evolution of the dance language. *American Naturalist, 133,* 580–590.

Dyer, F. C., and Seeley, T. D. (1991). Dance dialects and foraging range in three Asian honey bee species. *Behavioral Ecology and Sociobiology, 28,* 227–233.

Dyer, M. G. (in press). Toward the acquisition of language and the evolution of communication: A synthetic approach. In H. Roitblatt (Ed.), *Comparative approaches to cognitive science.* Cambridge, MA: MIT Press.

Eales, L. A. (1985). Song learning in zebra finches: Some effects of song model availability on what is learnt and when. *Animal Behaviour, 33,* 1293–1300.

Eggermont, J. J. (1988). Mechanisms of sound localization in anurans. In B. Fritzsch, M. J. Ryan, W. Wilczynski, T. E. Hetherington, and W. Walkowiak (Eds.), *The evolution of the amphibian auditory system* (pp. 307–336). New York: John Wiley & Sons.

Ehret, G. (1987a). Left hemisphere advantage in the mouse brain for recognizing ultrasonic calls. *Nature, 325,* 249–251.

Ehret, G. (1987b). Categorical perception of sound signals: Facts and hypotheses from animal studies. In S. Harnad (Ed.), *Categorical perception* (pp. 301–331). Cambridge, UK: Cambridge University Press.

Ehret, G. (1992). Categorical perception of mouse-pup ultrasounds in the temporal domain. *Animal Behaviour, 43,* 409–416.

Ehret, G., and Haack, B. (1981). Categorical perception of mouse pup ultrasounds by lactating females. *Naturwissenshaften, 68,* 208.

Eimas, P. D. (1979). Perceptual origins of the phoneme boundary effect and selective adaptation to speech: A signal detection theory analysis. *Journal of the Acoustical Society of America, 65,* 190–207.

Eimas, P. D., Siqueland, P., Jusczyk, P., and Vigorito, J. (1971). Speech perception in infants. *Science, 171,* 303–306.

Eisner, T., and Grant, R. P. (1981). Toxicity, odor aversion, and "olfactory aposematism." *Science, 213,* 476.

Ekman, P. (1973). *Darwin and facial expression.* London: Academic Press.

Ekman, P. (1982). Methods for measuring facial action. In K. R. Scherer and P. Ekman (Eds.), *Handbook of methods in nonverbal behavior research* (pp. 45–135). New York: Cambridge University Press.

Ekman, P. (1985). *Telling lies: Clues to deceit in the marketplace, marriage, and politics.* New York: W. W. Norton.

Ekman, P. (1989). The argument and evidence about universals in facial expressions of emotion. In H. Wagner and A. Manstead (Eds.), *Handbook of psychophysiology: The biological psychology of emotions and social processes* (pp. 143–164). London: John Wiley.

Ekman, P. (1992). Facial expressions of emotion: An old controversy and new findings. *Philosophical Transactions of the Royal Society of London, 335,* 63–70.

Ekman, P. (1994). Strong evidence for universals in facial expressions: A reply to Russell's mistaken critique. *Psychological Bulletin, 115,* 268–287.

Ekman, P., and Friesen, W. V. (1969). The repertoire of nonverbal behavior: Categories, origins, usage, and coding. *Semiotica, 1,* 49–98.

Ekman, P., Davidson, R. J., and Friesen, W. V. (1990). Emotional expression and brain physiology: II. The Duchenne smile. *Journal of Personality and Social Psychology, 58,* 342–353.

Ekman, P., and Friesen, W. V. (1975). *Unmasking the face.* Englewood Cliffs, NJ: Prentice-Hall.

Ekman, P., and Friesen, W. V. (1978). *The facial action coding system: A technique for the measurement of facial movement.* Palo Alto, CA: Consulting Psychologists Press.

Ekman, P., Friesen, W. V., and O'Sullivan, M. (1988). Smiles while lying. *Journal of Personality and Social Psychology, 54,* 414–420.

Ekman, P., Hager, J. C., and Friesen, W. V. (1981). The symmetry of emotional and deliberate facial actions. *Psychophysiology, 18,* 101–106.

Ekman, P., Levenson, R. W., and Friesen, W. V. (1983). Autonomic nervous system activity distinguishes among emotions. *Science, 218,* 1208–1210.

Ekman, P., and O'Sullivan, M. (1991). Who can catch a liar? *American Psychologist, 46,* 913–920.

Ekman, P., O'Sullivan, M., Friesen, W. V., and Scherer, K. R. (1991). Face, voice, and body in deceit. *Journal of Nonverbal Behavior, 15,* 125–135.

Elgar, M. A. (1986). House sparrows establish foraging flocks by giving chirrup calls if the resources are divisible. *Animal Behaviour, 34,* 169–174.

Elliott, L. L. (1994). Functional brain imaging and hearing. *Journal of the Acoustical Society of America, 96,* 1397–1408.

Elowson, A. M., and Snowdon, C. T. (1994). Pygmy marmosets, *Cebuella pygmaea,* modify vocal structure in response to changed social environment. *Animal Behaviour, 47,* 1267–1277.

Elowson, A. M., Sweet, C. S., and Snowdon, C. T. (1992). Ontogeny of trill and J-call vocalizations in the pygmy marmoset (*Cebuella pygmaea*). *Animal Behaviour, 43,* 703–716.

Elowson, A. M., Tannenbaum, P. L., and Snowdon, C. T. (1991). Food-associated calls correlate with food preferences in cotton-top tamarins. *Animal Behaviour, 42,* 931–937.

Emde, G. v. d., and Menne, D. (1989). Discrimination of insect wingbeat-frequencies by the bat *Rhinolophus ferrumequinum. Journal of Comparative Physiology, A, 164,* 663–671.

Emde, G. v. d., and Schnitzler, H.-U. (1990). Classification of insects by echolocating greater horseshoe bats. *Journal of Comparative Physiology, A, 167,* 437–450.

Emlen, S. T. (1992). Evolution of cooperative breeding in birds and mammals. In J. R. Krebs and N. B. Davies (Eds.), *Behavioural ecology* (pp. 301–337). Oxford, UK: Blackwell.

Endler, J. A. (1987). Predation, light intensity, and courtship behaviour in *Poecilia reticulata. Animal Behaviour, 35,* 1376–1385.

Endler, J. A. (1991a). Interactions between predators and prey. In J. R. Krebs and N. B. Davies (Eds.), *Behavioural ecology* (pp. 169–202). Oxford, UK: Blackwell Scientific.

Endler, J. A. (1991b). Variation in the appearance of guppy color patterns to guppies and their predators under different visual conditions. *Vision Research, 31,* 587–608.

Endler, J. A. (1992). Signals, signal conditions, and the direction of evolution. *American Naturalist, 139,* s125–s153.

Endler, J. A. (1993a). The color of light in forests and its implications. *Ecological Monographs, 63,* 1–27.

Endler, J. A. (1993b). Some general comments on the evolution and design of animal communication systems. *Proceedings of the Royal Society, London, 340,* 215–225.

Enquist, M., and Arak, A. (1995). Symmetry, beauty and evolution. *Nature, 372,* 169–172.

Espinoza-Varas, B., and Watson, C. S. (1989). Perception of complex auditory patterns by humans. In R. J. Dooling and S. H. Hulse (Eds.), *The comparative psychology of audition* (pp. 67–96). Hillsdale, NJ: Lawrence Erlbaum Associates.

Etcoff, N. L., and Magee, J. J. (1992). Categorical perception of facial expressions. *Cognition, 44,* 227–240.

Evans, C. S., Evans, L., and Marler, P. (1994). On the meaning of alarm calls: Functional reference in an avian vocal system. *Animal Behaviour, 45,* 23–38.

Evans, C. S., Macedonia, J. M., and Marler, P. (1993). Effects of apparent size and speed on the response of chickens, *Gallus gallus,* to computer-generated simulations of aerial predators. *Animal Behaviour, 46,* 1–11.

Evans, C. S., and Marler, P. (1991). On the use of video images as social stimuli in birds: Audience effects on alarm calling. *Animal Behaviour, 41,* 17–26.

Evans, C. S., and Marler, P. (1992). Female appearance as a factor in the responsiveness of male chickens during anti-predator behaviour and courtship. *Animal Behaviour, 43,* 137–145.

Evans, C. S., and Marler, P. (1994). Food-calling and audience effects in male chickens, *Gallus gallus:* Their relationships to food availability, courtship and social facilitation. *Animal Behaviour, 47,* 1159–1170.

Evans, C. S., and Marler, P. (in press). Language and animal communication: Parallels and contrasts. In H. Roitblatt. (Ed.) *Comparative approaches to cognitive science.* Cambridge, MA: MIT Press.

Evans, D. L., and Schmidt, J. O. (1990). *Insect defenses.* New York: State University of New York Press.

Evans, R. M. (1994). Cold-induced calling and shivering in young American white pelicans: Honest signalling of offspring need for warmth in a functionally integrated thermoregulatory system. *Behaviour, 129,* 13–34.

Fagen, R. M. (1978). Information measures: Statistical confidence limits and inference. *Journal of Theoretical Biology, 73,* 61–79.

Falk, D. (1991). 3.5 million years of hominid brain evolution. *Seminars in the Neurosciences, 3,* 409–416.

Falk, D. S., Brill, D., and Stork, D. (1986). *Seeing the light: Optics in nature, photography, color, vision, and holography.* New York: John Wiley & Sons.

Fant, G. (1960). *Acoustic theory of speech production.* The Hague: Mouton.

Fant, G. (1967). Auditory patterns of speech. In W. Wathen-Dunn (Ed.), *Models for the perception of speech and visual form* (pp. 111–125). Cambridge, MA: MIT Press.

Farabaugh, S. M. (1982). The ecological and social significance of duetting. In D. S. Kroodsma and E. H. Miller (Eds.), *Acoustic communication in birds,* vol. 2 (pp. 85–124). New York: Academic Press.

Fattu, J. M., and Suthers, R. A. (1981). Subglottic pressure and the control of phonation by the echolocating bat, *Eptesicus fuscus. Journal of Comparative Physiology, 143,* 465–475.

Fay, R. R. (1992). Structure and function in sound discrimination among vertebrates. In D. B. Webster, R. R. Fay, and A. N. Popper (Eds.), *The evolutionary biology of hearing* (pp. 229–266). New York: Springer-Verlag.

Fay, R. R., and Feng, A. S. (1987). Mechanisms for directional hearing among nonmammalian vertebrates. In W. Yost and G. Gourevitch (Eds.), *Directional hearing* (pp. 179–213). New York: Springer-Verlag.

Fay, R. R., and Popper, A. N. (1995). *Springer handbook of auditory research: Hearing by bats.* New York: Springer-Verlag.

Feduccia, A. (1980). *The age of birds.* Cambridge, MA: Harvard University Press.

Feng, A. S., Condon, C. J., and White, K. R. (1994). Stroboscopic hearing as a mechanism for prey detection in frequency-modulated bats? *Journal of the Acoustical Society of America, 95,* 2736–2744.

Feng, A. S., Gerhardt, H. C., and Capranica, R. R. (1976). Sound localization behavior of the green treefrog (*Hyla cinerea*) and the barking treefrog (*Hyla gratiosa*). *Journal of Comparative Physiology, 107,* 241–252.

Fenson, L., Dale, P. S., Reznick, J. S., Bates, E., Thal, D. J., and Pethick, S. J. (1994). Variability in early communicative development. *Monographs of the Society for Research in Child Development, 242,* 1–174.

Fernald, A. (1984). The perceptual and affective salience of mothers' speech to infants. In L. Feagans, C. Garvey, and R. Golinkoff (Eds.), *The origins and growth of communication* (pp. 5–29). Norwood, NJ: Ablex.

Fernald, A. (1989). Intonation and communicative intent in mothers' speech to infants: Is the melody the message? *Child Development, 60,* 1497–1510.

Fernald, A. (1992a). Meaningful melodies in mothers' speech to infants. In H. Papousek, U. Jürgens, and M. Papousek (Eds.), *Nonverbal vocal communication: Comparative and developmental approaches* (pp. 262–282). Cambridge, UK: Cambridge University Press.

Fernald, A. (1992b). Human maternal vocalizations to infants as biologically relevant signals: An evolutionary perspective. In J. Barkow, L. Cosmides, and J. Tooby (Eds.), *The adapted mind* (pp. 391–428). New York: Oxford University Press.

Fernald, A., and McRoberts, G. (1994). Prosodic bootstrapping: A critical analysis of the argument and the evidence. In J. L. Morgan and K. Demuth (Eds.), *Signal to syntax: Bootstrapping from speech to syntax in early acquisition* (pp. 110–138). Hillsdale, NJ: Lawrence Erlbaum Associates.

Fernald, A., Taeschner, T., Dunn, J., Papousek, M., Boysson-Bardies, B., and Fukui, I. (1989). A cross-language study of prosodic modifications in mothers' and fathers' speech to preverbal infants. *Journal of Child Language, 16,* 477–501.

Ficken, M. S., and Popp, J. W. (1992). Syntactical organization of the gargle vocalization of the black-capped chickadee, *Parus atricapillus. Ethology, 91,* 156–168.

Field, T. M., Cohen, D., Garcia, R., and Greenberg, R. (1985). Mother-stranger face discrimination by the newborn. *Infant Behavior and Development, 7,* 19–25.

Finlay, B. L., and Darlington, R. B. (1995). Linked regularities in the development and evolution of mammalian brains. *Science, 268,* 1578–1583.

Fischer, F. P., Köppl, C., and Manley, G. A. (1988). The basilar papilla of the barn owl, *Tyto alba:* A quantitative morphological SEM analysis. *Hearing Research, 34,* 87–101.

Fisher, J., and Hinde, R. A. (1949). The opening of milk bottles by birds. *British Birds, 42,* 347–357.

Fisher, R. A. (1930). *The genetical theory of natural selection.* Oxford, UK: Clarendon Press.

Fitch, W. T., and Hauser, M. D. (1995). Vocal production in nonhuman primates: Acoustics, physiology and functional constraints on honest advertisement. *American Journal of Primatology. 37,* 191–220.

FitzGibbon, C. D. (1989). A cost to individuals with reduced vigilance in groups of Thompson's gazelles hunted by cheetahs. *Animal Behaviour, 37,* 508–510.

FitzGibbon, C. D. (1990). Mixed-species grouping in Thomson's and Grant's gazelles: The antipredator benefits. *Animal Behaviour, 40,* 846–855.

FitzGibbon, C. D., and Fanshawe, J. W. (1988). Stotting in Thompson's gazelles: An honest signal of condition. *Behavioral Ecology and Sociobiology, 23,* 69–74.

Flanagan, J. L. (1963). *Speech analysis, synthesis and perception.* Berlin: Springer-Verlag.

Fleishman, L. (1992). The influence of the sensory system and the environment on motion patterns in the visual displays of anoline lizards and other vertebrates. *American Naturalist, 139,* S36–S61.

Fodor, J. (1983). *The modularity of mind*. Cambridge, MA: MIT Press.

Fodor, J. (1994). Concepts: A potboiler. *Cognition, 50*, 95–113.

Fodor, J. A., Garrett, M. F., and Brill, S. L. (1975). Pi Ka Pu: The perception of speech sounds by prelinguistic infants. *Perception and Psychophysics, 18*, 74–78.

Folkins, J. W. (1985). Issues in speech motor control and their relation to the speech of individuals with cleft palate. *Cleft Palate Journal, 22*, 106–122.

Folkins, J. W., and Zimmerman, G. N. (1981). Jaw-muscle activity during speech with the mandible fixed. *Journal of the Acoustical Society of America, 69*, 1441–1444.

Folven, R. J., and Bonvillian, J. (1991). The transition from nonreferential to referential language in children acquiring American Sign Language. *Developmental Psychology, 27*, 806–816.

Fooden, J. (1980). Classification and distribution of the living macaques. In D. Lindburg (Ed.), *The macaques* (pp. 1–42). New York: Van Nostrand.

Forester, D. C., and Czarnowsky, R. (1985). Sexual selection in the spring peeper *Hyla crucifer* (Anura: Hylidae): The role of the advertisement call. *Behaviour, 92*, 112–128.

Fowler, C. A. (1989). Real objects of speech perception. *Ecological Psychology, 1*, 145–160.

Fowler, C. A., and Dekle, D. J. (1991). Listening with eye and hand: Cross-modal contributions to speech perception. *Journal of Experimental Psychology: Human Perception and Performance, 17*, 816–828.

Fowler, C., Rubin, P., Remez, R. E., and Turvey, M. T. (1980). Implications for speech production of a general theory of action. In G. Butterworth (Ed.), *Language production*, vol. 1, *Speech and talk* (pp. 112–138). New York: Academic Press.

Fox, N. A., and Davidson, R. J. (1988). Patterns of brain electrical activity during facial expressions: signs of emotions in 10-month-old infants. *Developmental Psychology, 24*, 230–236.

Frank, M. G., Ekman, P., and Friesen, W. V. (1993). Behavioral markers and recognizability of the smile of enjoyment. *Journal of Personality and Social Psychology, 64*, 83–93.

Fridlund, A. J. (1994). *Human facial expressions: An evolutionary perspective*. New York: Academic Press.

Fridlund, A. J., Kenworthy, K. G., and Jaffey, A. K. (1992). Audience effects in affective imagery: Replication and extension to affective imagery. *Journal of Nonverbal Behavior, 16*, 191–212.

Fridlund, A. J., Sabini, J. P., Hedlund, L. E., Schaut, J. A., Shenker, J. I., and Knauer, M. J. (1990). Social determinants of facial expressions during affective imagery. *Journal of Nonverbal Behavior, 14*, 113–137.

Frings, H., and Little, F. (1957). Reactions of honey bees in the hive to simple sounds. *Science, 125*, 122.

Frith, U. (1989). *Autism: Explaining the enigma*. Oxford, UK: Blackwell Scientific.

Fromkin, V. A. (1973). Slips of the tongue. *Scientific American, 229*, 110–116.

Fuzessery, Z. M. (1988). Frequency tuning in the anuran central auditory system. In B. Fritzsch, M. J. Ryan, W. Wilczynski, T. E. Hetherington, and W. Walkowiak (Eds.), *The evolution of the amphibian auditory system* (pp. 253–273). New York: John Wiley & Sons.

Fuzessery, Z. M., and Pollak, G. D. (1984). Neural mechanisms of sound localization in an echolocating bat. *Science, 225*, 725–728.

Fuzessery, Z. M., Wenstrup, J. J., and Pollak, G. D. (1990). Determinants of horizontal sound location selectivity of binaurally excited neurons in the inferior colliculus of an isofrequency region of the mustache bat inferior colliculus. *Journal of Neurophysiology, 63*, 1128–1147.

Gagliardo, A., and Guilford, T. (1993). Why do warning-coloured prey live gregariously? *Proceedings of the Royal Society, London, 251*, 69–74.

Gahr, M. (1990). Delineation of a brain nucleus: Comparisons of cytochemical, hodological, and cytoarchitectural views of the song control nucleus HVC of the adult canary. *Journal of Comparative Neurology, 294,* 30–36.

Galaburda, A. (1995). Anatomic basis of cerebral dominance. In R. J. Davidson and K. Hugdahl (Eds.), *Brain asymmetry* (pp. 31–73). Cambridge, MA: MIT Press.

Galambos, R. (1942). Cochlear potentials elicited from bats by supersonic sounds. *Journal of the Acoustical Society of America, 14,* 41–49.

Galambos, R., and Griffin, D. R. (1942). Obstacle avoidance by flying bats. *Journal of Experimental Zoology, 89,* 475–490.

Galef, B. G., Jr. (1988). Imitation in animals: History, definitions, and interpretation of data from the psychological laboratory. In T. Zentall and B. G. Galef (Eds.), *Social learning: Psychological and biological perspectives* (pp. 3–28). Hillsdale, NJ: Lawrence Erlbaum Associates.

Galef, B. G., Jr. (1992). The question of animal culture. *Human Nature, 3,* 157–178.

Gallistel, C. R. (1992). *The organization of learning.* Cambridge, MA: MIT Press.

Gallup, G. G., Jr. (1970). Chimpanzees: Self-recognition. *Science, 167,* 86–87.

Gallup, G. G., Jr. (1991). Toward a comparative psychology of self-awareness: Species limitations and cognitive consequences. In G. R. Goethals and J. Strauss (Eds.), *The self: An interdisciplinary approach* (pp. 121–135). New York: Springer-Verlag.

Garcia, J., and Koelling, R. A. (1966). Relation of cue to consequence in avoidance learning. *Psychonomic Science, 4,* 123–147.

Gardner, R. A., and Gardner, B. T. (1969). Teaching sign language to a chimpanzee. *Science, 165,* 664–672.

Gardner, R. A., Gardner, B. T., and Van Cantfort, E. (1989). *Teaching sign language to chimpanzees.* Albany: State University of New York Press.

Gazzaniga, M., and Smiley, C. S. (1991). Hemispheric mechanisms controlling voluntary and spontaneous facial expressions. *Journal of Cognitive Neuroscience, 2,* 239–245.

Gaunt, A. S., and Gaunt, S. L. L. (1985). Syringeal structure and avian phonation. In R. F. Johnson (Ed.), *Current ornithology* (pp. 213–245). New York: Plenum Press.

Gelfer, C. E. (1987). A simultaneous physiological and acoustic study of fundamental frequency declination. PhD diss., CUNY, New York.

Gelfer, C. E., Harris, K. S., and Baer, T. (1987). Controlled variables in sentence intonation. In T. Baer, C. Sasaki, and K. Harris (Eds.), *Laryngeal function in phonation and respiration* (pp. 422–435). Boston: College-Hill.

Gelman, S. A., Coley, J. D., and Gottfried, G. M. (1994). Essentialist beliefs in children: The acquisition of concepts and theories. In L. A. Hirschfield and S. A. Gelman (Eds.), *Mapping the mind: Domain specificity in cognition and culture* (pp. 341–366). Cambridge, UK: Cambridge University Press.

Gerhardt, H. C. (1978a). Discrimination of intermediate sounds in a synthetic call continuum by female green treefrogs. *Science, 199,* 1089–1091.

Gerhardt, H. C. (1978b). Temperature coupling in the vocal communication of the gray treefrog, *Hyla versicolor, Science, 199,* 992–994.

Gerhardt, H. C. (1982). Sound pattern recognition in some North American tree-frogs (Anura: Hylidae): Implications for mate choice. *American Zoologist, 22,* 581–595.

Gerhardt, H. C. (1983). Communication and the environment. In T. R. Halliday and P. J. B. Slater (Eds.), *Animal behaviour,* vol. 2, *Communication* (pp. 82–113). Oxford, UK: Blackwell Scientific.

Gerhardt, H. C. (1987). Evolutionary and neurobiological implications of selective phonotaxis in the green treefrog (*Hyla cinerea*). *Animal Behaviour, 35,* 1479–1489.

Gerhardt, H. C. (1988a). Acoustic properties used in call recognition by frogs and toads. In B. Fritzsch, M. J. Ryan, W. Wilczynski, T. E. Hetherington, and W. Walkowiak (Eds.), *The evolution of the amphibian auditory system* (pp. 455–484). New York: John Wiley & Sons.

Gerhardt, H. C. (1988b). Acoustic communication in the gray treefrog, *Hyla versicolor:* Evolutionary and neurobiological implications. *Journal of Comparative Physiology, A, 162,* 261–278.

Gerhardt, H. C. (1991). Female mate choice in treefrogs: Static and dynamic acoustic criteria. *Animal Behaviour, 42,* 615–636.

Gerhardt, H. C. (1992a). Conducting playback experiments and interpreting their results. In P. K. McGregor (Eds.), *Playback and studies of animal communication* (pp. 59–77). New York: Plenum Press.

Gerhardt, H. C. (1992b). Multiple messages in acoustic signals. *Seminars in the Neurosciences, 4,* 391–400.

Gerhardt, H. C., Daniel, R. E., Perrill, S. A., and Schramm, S. (1987). Mating behaviour and male mating success in the green treefrog. *Animal Behaviour, 35,* 1490–1503.

Gerhardt, H. C., Dyson, M. L., Tanner, S. D., and Murphy, C. G. (1994). Female treefrogs do not avoid heterospecific calls as they approach conspecific calls: Implications for mechanisms of mate choice. *Animal Behaviour, 47,* 1323–1332.

Gerhardt, H. C., and Mudry, K. M. (1980). Temperature effects on frequency preferences and mating call frequencies in the green treefrog, *Hyla cinerea* (Anura: Hylidae). *Journal of Comparative Physiology, B, 137,* 1–6.

Gerken, L., Jusczyk, P. W., and Mandel, D. R. (1994). When prosody fails to cue syntactic structure: 9-month-olds' sensitivity to phonological versus syntactic phrases. *Cognition, 51,* 237–265.

Gibson, J. J. (1950). *The perception of the visual world.* Boston: Houghton Mifflin.

Gibson, J. J. (1966). *The senses considered as perceptual systems.* Boston: Houghton Mifflin.

Gibson, K. R., and Petersen, A. C. (1991). *Brain maturation and cognitive development: Comparative and cross-cultural perspectives.* New York: Aldine de Gruyter.

Gibson, R. M. (1989). Field playback of male display attracts females in lek breeding sage grouse. *Behavioral Ecology and Sociobiology, 24,* 439–443.

Gibson, R. M., and Bradbury, J. W. (1985). Sexual selection in lekking sage grouse: Phenotypic correlates of male mating success. *Behavioral Ecology and Sociobiology, 18,* 117–123.

Gibson, R. M., Bradbury, J. W., and Vehrencamp, S. L. (1991). Mate choice in lekking sage grouse revisited: The roles of vocal display, female site fidelity, and copying. *Behavioral Ecology, 2,* 165–180.

Gigerenzer, G., and Hug, K. (1992). Domain-specific reasoning: Social contracts, cheating and perspective change. *Cognition, 43,* 127–171.

Gillespie, J. H. (1991). *The causes of molecular evolution.* New York: Oxford University Press.

Giraldeau, L.-A., Caraco, T., and Valone, T. J. (1994). Social foraging: Individual learning and cultural transmission of innovations. *Behavioral Ecology, 5,* 35–43.

Giraldeau, L.-A., and Lefebvre, L. (1986). Exchangeable producer-scrounger roles in a captive flock of feral pigeons: A case for the skill pool effect. *Animal Behaviour, 34,* 797–803.

Gleich, O. (1994). Excitation patterns in the starling cochlea: A population study of primary auditory afferents. *Journal of the Acoustical Society of America, 95,* 401–409.

Glickstein, M. (1993). Motor skills but not cognitive tasks. *Trends in Neurosciences, 16,* 450–451.

Gochfeld, N. (1984). Antipredator behavior: Aggressive and distraction displays of shorebirds. In J. Burger and B. L. Olla (Eds.), *Shorebirds: Breeding behavior and population* (pp. 289–377). New York: Plenum Press.

Godard, R. (1991). Long-term memory of individual neighbors in a migratory songbird. *Nature, 350,* 220–229.

Godfray, H. C. J. (1991). Signalling of need by offspring to their parents. *Nature, 352*, 328–330.

Goedeking, P. (1988). Vocal play behavior in cotton-top tamarins. In D. Todt, P. Goedeking, and D. Symmes (Eds.), *Primate vocal communication* (pp. 133–141). Berlin: Springer-Verlag.

Goldbeck, T., Tolkmitt, F., and Scherer, K. R. (1988). Experimental studies on vocal affect communication. In K. R. Scherer (Ed.), *Facets of emotion* (pp. 119–138). Hillsdale, NJ: Lawrence Erlbaum Associates.

Goldman, S., and Nottebohm, F. (1983). Neuronal production, migration and differentiation in a vocal control nucleus of the adult female canary brain. *Proceedings of the National Academy of Sciences, 80*, 2390–2394.

Goldman-Rakic, P. (1987). Circuitry of primate prefrontal cortex and regulation of behavior by representational memory. In V. B. Mountcastlye, F. Plum, and S. R. Geiger (Eds.), *Handbook of physiology, Section 1: The nervous system* (pp. 373–418). Bethesda, MD: American Physiology Society.

Goller, F., and Suthers, R. A. (1995). Implications for lateralization of bird song from unilateral gating of bilateral motor patterns. *Nature, 373*, 63–65.

Gomez, J. C. (1991). Visual behaviour as a window for reading the mind of others in primates. In A. Whiten (Ed.), *Natural theories of mind* (pp. 195–208). Oxford, UK: Basil Blackwell.

Goodall, J. (1986). *The chimpanzees of Gombe: Patterns of behavior*. Cambridge, MA: Belknap Press of Harvard University Press.

Gopnik, A., and Wellman, H. M. (1994). The theory theory. In L. A. Hirschfeld and S. A. Gelman (Eds.), *Mapping the mind: Domain specificity in cognition and culture* (pp. 257–293). Cambridge, UK: Cambridge University Press.

Gopnik, M. (1990). Feature-blind grammar and dysphasia. *Nature, 275*, 344–346.

Gordon, B. (1990). Human language. In R. P. Kesner and D. S. Olton (Eds.), *Neurobiology of comparative cognition* (pp. 21–50). Hillsdale, NJ: Lawrence Erlbaum Associates.

Goren, C. C., Sarty, M., and Wu, P. Y. K. (1975). Visual following and pattern discrimination of face-like stimuli by newborn infants. *Pediatrics, 56*, 544–549.

Götmark, F. (1993). Conspicuous coloration in male birds is favoured by predation in some species and disfavored in others. *Proceedings of the Royal Society, London, 253*, 143–146.

Gottlieb, G. (1975). Development of species identification in ducklings: I. Nature of perceptual deficit caused by embryonic auditory deprivation. *Journal of Comparative and Physiological Psychology, 89*, 387–399.

Gottlieb, G. (1980). Development of species identification in ducklings: VI. Specific embryonic experience required to maintain species-typical perception in Peking ducklings. *Journal of Comparative and Physiological Psychology, 94*, 587–598.

Gottlieb, G. (1991a). Experiential canalization of behavioral development: Results. *Developmental Psychology, 27*, 35–39.

Gottlieb, G. (1991b). Experiential canalization of behavioral development: Theory. *Developmental Psychology, 27*, 4–13.

Gottlieb, G. (1992). *Individual development and evolution: The genesis of novel behavior*. New York: Oxford University Press.

Gould, J. L. (1976). The dance language controversy. *Quarterly Review of Biology, 51*, 211–244.

Gould, J. L. (1986). The landmark map of honey bees: Do insects have cognitive maps? *Science, 232*, 861–863.

Gould, J. L. (1990). Honey bee cognition. *Cognition, 37*, 83–103.

Gould, J. L., and Gould, C. G. (1988). *The honey bee*. New York: Freeman Press.

Gould, J. L., and Towne, W. F. (1987). Evolution of the dance language. *American Naturalist, 130*, 317–338.

Gould, S. J., and Lewontin, R. C. (1979). The spandrels of San Marco and the Panglossian program: A critique of the adaptationist programme. *Proceedings of the Royal Society, London, 205,* 281–288.

Gould, S. J., and Vrba, E. S. (1982). Exaptation—A missing term in the science of form. *Paleobiology, 8,* 4–15.

Gourevitch, G. (1987). Binaural hearing in land mammals. In W. A. Yost and G. Gourevitch (Eds.), *Directional hearing* (pp. 226–246). Berlin: Springer-Verlag.

Gouzoules, H., and Gouzoules, S. (1989). Design features and developmental modification in pigtail macaque. (*Macaca nemestrina*) agonistic screams. *Animal Behaviour, 37,* 383–401.

Gouzoules, H., and Gouzoules, S. (1990). Body size effects on the acoustic structure of pigtail macaque (*Macaca nemestrina*) screams. *Ethology, 85,* 324–334.

Gouzoules, S., Gouzoules, H., and Marler, P. (1984). Rhesus monkey (*Macaca mulatta*) screams: Representational signalling in the recruitment of agonistic aid. *Animal Behaviour, 32,* 182–193.

Gracco, V. L. (1988). Timing factors in the coordination of speech movements. *Journal of Neuroscience, 8,* 4628–4639.

Grafen, A. (1990). Biological signals as handicaps. *Journal of Theoretical Biology, 144,* 475–546.

Grafen, A. (1992). Modelling in behavioural ecology. In J. R. Krebs and N. B. Davies (Eds.), *Behavioural ecology* (pp. 5–31). Oxford, UK: Blackwell Scientific.

Grafen, A., and Johnstone, R. A. (1993). Why we need ESS signalling theory. *Proceedings of the Royal Society, London, 340,* 245–250.

Graves, R., Goodglass, H., and Landis, T. (1982). Mouth asymmetry during spontaneous speech. *Neuropsychologia, 20,* 371–381.

Graves, R., and Landis, T. (1985). Hemispheric control of speech expression in aphasia: A mouth asymmetry study. *Archives of Neurology, 42,* 249–251.

Graves, R., and Landis, T. (1990). Asymmetry in mouth opening during different speech tasks. *International Journal of Psychiatry, 25,* 179–189.

Graves, R. E., and Potter, S. M. (1988). Speaking from two sides of the mouth. *Visible Language, 22,* 129–137.

Graves, R., Strauss, E. H., and Wada, J. (1990). Mouth asymmetry during speech of epileptic patients who have undergone corotid amytal testing. *Neuropsychologia, 28,* 1117–1121.

Green, D. M. (1988). *Profile analysis: Auditory intensity discrimination.* New York: Oxford University Press.

Green, J. A., and Gustafson, G. E. (1983). Individual recognition of human infants on the basis of cries alone. *Developmental Psychobiology, 16,* 485–493.

Green, K. P., Kuhl, P. K., Meltzoff, A. N., and Stevens, E. B. (1991). Integrating speech information across talkers, gender, and sensory modality: Female faces and male voices in the McGurk effect. *Perception and Psychophysics, 50,* 524–536.

Green, S. (1975a). Variation of vocal pattern with social situation in the Japanese macaque (*Macaca fuscata*): A field study. In L. A. Rosenblum (Ed.), *Primate behavior,* vol. 4 (pp. 1–102). New York: Academic Press.

Green, S. (1975b). Dialects in Japanese monkeys: Vocal learning and cultural transmission of locale-specific behavior? *Zeitschrift für Tierpsychologie, 38,* 304–314.

Green, S., and Marler, P. (1979). The analysis of animal communication. In P. Marler and J. Vandenbergh (Eds.), *Social behavior and communication, Handbook of behavioral neurobiology,* vol. 3 (pp. 73–158). New York: Plenum Press.

Greene, H. W. (1988). Antipredator mechanisms in reptiles. *Biology of Reptiles, 16,* 1–152.

Greenewalt, C. H. (1968). *Bird song: Acoustics and physiology.* Washington, DC: Smithsonian Institution Press.

Greenfield, M. D. (1994). Cooperation and conflict in the evolution of signal interactions. *Annual Review of Ecology and Systematics, 25,* 97–126.

Greenough, W. T., and Alcantara, A. A. (1993). The roles of experience in different developmental information stage processes. In B. de Boysson-Bardies (Ed.), *Developmental neurocognition: Speech and face processing in the first year of life* (pp. 3–16). Netherlands: Kluwer Academic.

Grieser, D. L., and Kuhl, P. K. (1988). Maternal speech to infants in a tonal language: Support for universal prosodic features in motherese. *Developmental Psychology, 24,* 14–20.

Grieser, D. L., and Kuhl, P. K. (1989). Categorization of speech by infants: Support for speech-sound prototypes. *Developmental Psychology, 25,* 577–588.

Griffin, D. R. (1944). Echolocation by blind men and bats. *Science, 100,* 589–590.

Griffin, D. R. (1958). *Listening in the dark.* New Haven, CT: Yale University Press.

Griffin, D. R. (1967). Discriminative echolocation by bats. In R.-G. Busnel (Ed.), *Animal sonar systems* (pp. 273–311). Jouy-en-Josas, France: Laboratory Physiological Acoustics.

Griffin, D. R. (1992). *Animal minds.* Chicago: Chicago University Press.

Griffin, D. R., Dunning, D. C., Cahlander, D. A., and Webster, F. A. (1962). Correlated orientation sounds and ear movements of Horseshoe bats. *Nature, 196,* 1185–1186.

Griffin, D. R., and Galambos, R. (1941). The sensory basis of obstacle avoidance by flying bats. *Journal of Experimental Zoology, 86,* 481–506.

Griffiths, S. K., Brown, W. S., Jr., Gerhardt, K. J., Abrams, R. M., and Morris, R. J. (1994). The perception of speech sounds recorded with the uterus of a pregnant sheep. *Journal of the Acoustical Society of America, 96,* 2055–2064.

Grober, M. S. (1988). Brittle-star bioluminescence functions as an aposematic signal to deter crustacean predators. *Animal Behaviour, 36,* 493–501.

Gross, C. G. (1992). Representation of visual stimuli in inferior temporal cortex. *Philosophical Transactions of the Royal Society of London, 335,* 3–10.

Gross, C. G., Rocha-Miranda, C. E., and Bender, D. B. (1972). Visual properties of neurons in the inferotemporal cortex of the macaque. *Journal of Neurophysiology, 35,* 96–111.

Gross, C. G., and Rodman, H. R. (1992). Inferior temporal cortex: Neuronal properties and connections in adult and infant macaques. In R. Lent (Ed.), *The visual system from genesis to maturity* (pp. 245–266). Boston: Birkhauser.

Guilford, T. (1989a). The evolution of conspicuous coloration. *American Naturalist, 131,* S7–S21.

Guilford, T. (1989b). Studying warning signals in the laboratory. In R. J. Blanchard, P. Brain, D. C. Blanchard, and S. Parmigiani (Eds.), *Ethoexperimental approaches to the study of behavior* (pp. 87–103). Dordrecht, Germany: Kluwer Academic.

Guilford, T. (1990). The evolution of aposematism. In D. L. Evans and J. O. Schmidt (Eds.), *Insect defenses* (pp. 23–62). New York: State University of New York Press.

Guilford, T., and Dawkins, M. S. (1991). Receiver psychology and the evolution of animal signals. *Animal Behaviour, 42,* 1–14.

Guilford, T., and Dawkins, M. S. (1993). Are warning colors handicaps? *Evolution, 47,* 400–416.

Gulick, W. L. (1989). *Hearing: Physiological acoustics, neural coding, and psychoacoustics.* New York: Oxford University Press.

Gust, D., St. Andre, E., Minter, C., Gordon, T., and Gouzoules, H. (1990). Female copulatory vocalizations in a captive group of sooty magabeys (*Cercocebus torquatus atys*). *American Journal of Primatology, 20,* 196.

Gustafson, G. W., Green, J. A., and Cleland, J. W. (1994). Robustness of individual identity in the cries of human infants. *Developmental Psychobiology, 27,* 1–9.

Gustafson, G. W., Green, J. A., and Tomic, T. (1984). Acoustic correlates of individuality in the cries of human infants. *Developmental Psychobiology, 17,* 311–324.

Gwynne, D. T. (1991). Sexual competition among females: What causes courtship role reversal? *Trends in Ecology and Evolution, 6,* 118–121.

Gyger, M., Karakashian, S., and Marler, P. (1986). Avian alarm calling: Is there an audience effect? *Animal Behaviour, 34,* 1570–1572.

Gyger, M., and Marler, P. (1988). Food calling in the domestic fowl (*Gallus gallus*): The role of external referents and deception. *Animal Behaviour, 36,* 358–365.

Gyger, M., Marler, P., and Pickert, R. (1987). Semantics of an avian alarm calling system: The male domestic fowl, *Gallus domesticus. Behaviour, 102,* 15–40.

Habersetzer, J. (1981). Adaptive echolocation in the bat *Rhinopomae hardwickei,* a field study. *Journal of Comparative Physiology, 144,* 559–566.

Hager, J. C., and Ekman, P. (1985). The asymmetry of facial actions is inconsistent with models of hemispheric specialization. *Psychophysiology, 22,* 307–318.

Haglund, M. M., Ojemann, G. A., Lettich, E., Bellugi, U., and Corina, D. (1993). Dissociation of cortical and single unit activity in spoken and signed languages. *Brain and Language, 44,* 19–27.

Hailman, J. P. (1977). *Optical signals: Animal communication and light.* Bloomington: Indiana University Press.

Hailman, J. P., and Ficken, M. S. (1987). Combinatorial animal communication with computable syntax: Chick-a-dee calling qualifies as "language" by structural linguistics. *Animal Behaviour, 34,* 1899–1901.

Hailman, J. P., Ficken, M. S., and Ficken, R. W. (1985). The "chick-a-dee" calls of *Parus atricapillus:* A recombinant system of animal communication compared with written English. *Semiotica, 56,* 191–224.

Hailman, J. P., Ficken, M. S., and Ficken, R. W. (1987). Constraints on the structure of combinatorial "chick-a-dee" calls. *Ethology, 75,* 62–80.

Halliday, T. R., and Slater, P. J. B. (1983). *Animal behavior,* vol. 2, *Communication.* New York: W. H. Freeman.

Hamilton, C. R. (1977). An assessment of hemispheric specialization in monkeys. *Annals of the New York Academy of Sciences, 299,* 222–232.

Hamilton, C. R., and Vermeire, B. A. (1988). Complementary hemispheric specialisation in monkeys. *Science, 242,* 1691–1694.

Hamilton, W. D. (1963). The evolution of altruistic behavior. *American Naturalist, 97,* 354–356.

Hamilton, W. D. (1964). The genetical evolution of social behavior. *Journal of Theoretical Biology, 7,* 1–52.

Hamilton, W. D. (1971). Geometry for the selfish herd. *Journal of Theoretical Biology, 31,* 295–311.

Hamilton, W. D. (1982). Pathogens as causes of genetic diversity in their host populations. In R. M. Anderson and R. M. May (Eds.), *Population biology of infectious diseases* (pp. 269–296). Berlin: Springer-Verlag.

Hamilton, W. D., Axelrod, R., and Tanese, R. (1990). Sexual reproduction as an adaptation to resist parasites. *Proceedings of the National Academy of Sciences, 87,* 3566–3573.

Hamilton, W. D., and Zuk, M. (1982). Heritable true fitness and bright birds: A role for parasites? *Science, 218,* 384–387.

Hamilton, W. J., III. (1984). Significance of paternal investment by primates to the evolution of male-female associations. In D. M. Taub (Ed.), *Primate paternalism* (pp. 81–96). New York: Van Nostrand Reinhold.

Hamilton, W. J., III, and Arrowood, P. C. (1978). Copulatory vocalisations of chacma baboons (*Papio ursinus*), gibbons (*Hylobates hoolock*), and humans. *Science, 200,* 1405–1409.

Hanna, E., and Meltzoff, A. N. (1993). Peer imitation by toddlers in laboratory, home, and day-care contexts: Implications for social learning and memory. *Developmental Psychology, 29,* 701–710.

Harcourt, A. H., Harvey, P. H., Larson, S. G., and Short, R. V. (1981). Testes weight and breeding system in primates. *Nature, 293,* 55–57.

Harcourt, A. H., Stewart, K., and Hauser, M. D. (1993). The social use of vocalizations by gorillas: I. Social behaviour and vocal repertoire. *Behaviour, 124,* 89–122.

Harnad, S. (1987). *Categorical perception: The groundwork of cognition.* Cambridge, UK: Cambridge University Press.

Harries, M. H., and Perrett, D. I. (1991). Visual processing of faces in the temporal cortex: Physiological evidence for a modular organisation and possible anatomical correlates. *Journal of Cognitive Neuroscience, 3,* 9–24.

Hartley, D. J., and Suthers, R. A. (1988). The acoustics of the vocal tract in the horseshoe bat, *Rhinolophus hildebrandti. Journal of the Acoustical Society of America, 84,* 1201–1213.

Hartley, R. S., and Suthers, R. A. (1989). Airflow and pressure during canary song: Direct evidence for mini-breaths. *Journal of Comparative Physiology, A, 165,* 15–26.

Hartley, R. S., and Suthers, R. A. (1991). Lateralization of syringeal function during song production in the canary. *Journal of Comparative Physiology, A, 81,* 177–188.

Hartshorn, C. (1973). *Born to sing.* New York: Harper & Row.

Harvey, P. H., and Bradbury, J. W. (1991). Sexual selection. In J. R. Krebs and N. B. Davies (Eds.), *Behavioural ecology* (pp. 203–233). Cambridge, MA: Blackwell Scientific.

Harvey, P. H., and Pagel, M. D. (1991). *The comparative method in evolutionary biology.* Oxford, UK: Oxford University Press.

Hasselmo, M. E., Rolls, E. T., and Baylis, G. C. (1989). The role of expression and identity in the face-selective responses of neurons in the temporal visual cortex of the monkey. *Behavioural Brain Research, 32,* 203–218.

Hasson, O. (1989). Amplifiers and the handicap principle in sexual selection: A different emphasis. *Proceedings of the Royal Society, London, 235,* 383–406.

Hasson, O. (1991). Pursuit-deterrent signals: Communication between prey and predator. *Trends in Ecology and Evolution, 6,* 325–329.

Hausberger, M., Black, J. M., and Richard, J.-P. (1991). Bill opening and sound spectrum in barnacle goose loud calls: Individuals with "wide mouths" have higher pitched voices. *Animal Behaviour, 42,* 319–322.

Hauser, M. D. (1988a). How infant vervet monkeys learn to recognize starling alarm calls. *Behaviour, 105,* 187–201.

Hauser, M. D. (1988b). Invention and social transmission: New data from wild vervet monkeys. In R. W. Byrne and A. Whiten (Eds.), *Machiavellian intelligence: Social expertise and the evolution of intellect in monkeys, apes, and humans* (pp. 327–343). Oxford, UK: Oxford University Press.

Hauser, M. D. (1989). Ontogenetic changes in the comprehension and production of vervet monkey (*Cercopithecus aethiops*) vocalizations. *Journal of Comparative Psychology, 103,* 149–158.

Hauser, M. D. (1990). Do female chimpanzee copulation calls incite male-male competition? *Animal Behaviour, 39,* 596–597.

Hauser, M. D. (1991). Sources of acoustic variation in rhesus macaque vocalizations. *Ethology, 89,* 29–46.

Hauser, M. D. (1992a). Articulatory and social factors influence the acoustic structure of rhesus monkey vocalizations: A learned mode of production? *Journal of the Acoustical Society of America, 91,* 2175–2179.

Hauser, M. D. (1992b). A mechanism guiding conversational turn-taking in vervet monkeys and rhesus macaques. In T. Nishida, F. B. M. deWaal, W. McGrew, P. Marler, and M. Pickford (Eds.), *Topics in primatology*, vol. 1, *Human origins* (pp. 235–248). Tokyo. Tokyo University Press.

Hauser, M. D. (1992c). Costs of deception: Cheaters are punished in rhesus monkeys. *Proceedings of the National Academy of Sciences, 89*, 12137–12139.

Hauser, M. D. (1993a). Do vervet monkey infants cry wolf? *Animal Behaviour, 45*, 1242–1244.

Hauser, M. D. (1993b). The evolution of nonhuman primate vocalizations: Effects of phylogeny, body weight and motivational state. *American Naturalist, 142*, 528–542.

Hauser, M. D. (1993c). How monkeys feel about how they see the world. *Language and Communication, 14*, 31–36.

Hauser, M. D. (1993d). Rhesus monkey (*Macaca mulatta*) copulation calls: Honest signals for female choice? *Proceedings of the Royal Society, London, 254*, 93–96.

Hauser, M. D. (1993e). Right hemisphere dominance for the production of facial expression in monkeys. *Science, 261*, 475–477.

Hauser, M. D. (1993f). Social influences on the ontogeny of foraging behavior in wild vervet monkeys. *Journal of Comparative Psychology, 107*, 1–7.

Hauser, M. D. (1994). The transition to foraging independence in free-ranging vervet monkeys. In M. Mainardi and B. G. Galef (Eds.), *Ontogeny of social transmission of food preferences in mammals: Basic and applied research* (pp. 165–202). Reading, UK: Harwood Academic Press.

Hauser, M. D. (in press). Vocal communication in macaques: Causes of variation. In J. Fa and D. Lindburg (Eds.), *Evolutionary ecology and behaviour of macaques*. Cambridge, UK: Cambridge University Press.

Hauser, M. D., and Andersson, K. (1994). Left hemisphere dominance for processing vocalizations in adult, but not infant rhesus monkeys: Field experiments. *Proceedings of the National Academy of Sciences, 91*, 3946–3948.

Hauser, M. D., and Caffrey, C. (1995). Anti-predator response to raptor calls in wild crows. *Animal Behaviour, 48*, 1469–1471.

Hauser, M. D., Evans, C. S., and Marler, P. (1993). The role of articulation in the production of rhesus monkey (*Macaca mulatta*) vocalizations. *Animal Behaviour, 45*, 423–433.

Hauser, M. D., and Fowler, C. (1991). Declination in fundamental frequency is not unique to human speech: Evidence from nonhuman primates. *Journal of the Acoustical Society of America, 91*, 363–369.

Hauser, M. D., MacNeilage, P., and Ware, M. (in press). Numerical representations in primates: Perceptual or arithmetic? *Proceedings of the National Academy of Sciences*.

Hauser, M. D., and Marler, P. (1992). How do and should studies of animal communication affect interpretations of child development? In C. Ferguson, L. Menn, and C. Stoel-Gammon (Eds.), *Phonological development* (pp. 663–680). Baltimore: York Press.

Hauser, M. D., and Marler, P. (1993a). Food-associated calls in rhesus macaques (*Macaca mulatta*): I. Socioecological factors influencing call production. *Behavioral Ecology, 4*, 194–205.

Hauser, M. D., and Marler, P. (1993b). Food-associated calls in rhesus macaques (*Macaca mulatta*). II. Costs and benefits of call production and suppression. *Behavioral Ecology, 4*, 206–212.

Hauser, M. D., and Nelson, D. (1991). Intentional signaling in animal communication. *Trends in Ecology and Evolution, 6*, 186–189.

Hauser, M. D., and Schön Ybarra, M. (1994). The role of lip configuration in monkey vocalizations: Experiments using xylocaine as a nerve block. *Brain and Language, 46*, 232–244.

Hauser, M. D., Teixidor, P., Field, L., and Flaherty, R. (1993). Food-elicited calls in chimpanzees: Effects of food quantity and divisibility? *Animal Behaviour, 45*, 817–819.

Hauser, M. D., and Wrangham, R. W. (1987). Manipulation of food calls in captive chimpanzees: A preliminary report. *Folia primatologica, 48*, 24–35.

Hauser, M. D., and Wrangham, R. W. (1990). Recognition of predator and competitor calls in nonhuman primates and birds: A preliminary report. *Ethology, 86,* 116–130.

Hawkins, J. (1988). *Explaining language universals.* Oxford, UK: Basil Blackwell.

Haxby, J. V., Grady, C. L., Horwitz, B., Ungerleider, L. G., Mishkin, M., Carson, R. E., Hescovitch, P., Schapiro, M. B., and Rapoport, S. (1991). Dissociation of object and spatial visual processing pathways in human extrastriate cortex. *Proceedings of the National Academy of Sciences, 88,* 1621–1625.

Heffner, H. E., and Heffner, R. S. (1984). Temporal lobe lesions and perception of species-specific vocalizations by macaques. *Science, 226,* 75–76.

Heffner, H. E., and Heffner, R. S. (1990). Role of primate auditory cortex in hearing. In M. A. Berkely and W. C. Stebbins (Eds.), *Comparative perception,* vol. 2, *Complex signals* (pp. 136–159). New York: John Wiley & Sons.

Heiligenberg, W. (1986). Jamming avoidance responses. In T. H. Bullock and W. Heiligenberg (Eds.), *Electroreception* (pp. 613–649). New York: Wiley Press.

Heiligenberg, W. (1991). *Neural nets in electric fish.* Cambridge, MA: MIT Press.

Heinrich, B. (1988). Food sharing in the raven, *Corvus corax.* In C. Slobodchikoff (Ed.), *Ecology of social behavior* (pp. 286–311). New York: Academic Press.

Heinrich, B., and Marzluff, J. M. (1991). Do common ravens yell because they want to attract others? *Behavioral Ecology and Sociobiology, 28,* 13–21.

Heller, W., and Levy, J. (1981). Perception and expression of emotion in right-handers and left-handers. *Neuropsychologia, 19,* 263–272.

Hellige, J. (1993). *Hemispheric asymmetry. What's right and what's left?* Cambridge, MA: Harvard University Press.

Henson, O. W., Schuller, G., and Vater, M. (1985). A comparative study of the physiological properties of the inner ear in Doppler shift compensating bats (*Rhinolophus rouxi* and *Pteronotus parnelli*). *Journal of Comparative Physiology, 157,* 587–607.

Herman, L. M., Pack, A. A., and Palmer, M.-S. (1993). Representational and conceptual skills of dolphins. In H. L. Roitblat, L. M. Herman, and P. E. Nachtigall (Eds.), *Language and communication: Comparative perspectives* (pp. 403–442). Hillsdale, NJ: Lawrence Erlbaum Associates.

Herman, L. M., Richards, D. G., and Wolz, J. P. (1984). Comprehension of sentences by bottlenosed dolphins. *Cognition, 16,* 129–219.

Herrnstein, R. J. (1991). Levels of categorization. In G. M. Edelman, W. E. Gall, and W. M. Cowan (Eds.), *Signal and sense* (pp. 385–413). Somerset, NJ: Wiley-Liss.

Herrnstein, R. J., Loveland, D. H., and Cable, C. (1976). Natural concepts in pigeons. *Journal of Experimental Psychology (Animal Behavior), 2,* 285–311.

Herrnstein, R. J., Vaughan, W., Jr., Mumford, D. B., and Kosslyn, S. M. (1989). Teaching pigeons an abstract relational rule. *Perception and Psychophysics, 46,* 56–64.

Hersch, G. L. (1966). Bird voices and resonant tuning in helium air mixtures. PhD diss., University of California, Berkeley.

Herzog, M., and Hopf, S. (1983). Effects of species-specific vocalizations on the behavior of surrogate-reared squirrel monkeys. *Behaviour, 86,* 197–214.

Hess, U., Scherer, K. R., and Kappas, A. (1988). Multichannel communication of emotion: Synthetic signal production. In K. R. Scherer (Ed.), *Facets of emotion* (pp. 161–182). Hillsdale, NJ: Lawrence Erlbaum Associates.

Hetherington, T. E. (1994). The middle ear muscle of frogs does not modulate tympanic responses to sound. *Journal of the Acoustical Society of America, 95,* 2122–2125.

Heyes, C. (1993). Anecdotes, training, trapping and triangulating: Do animals attribute mental states? *Animal Behaviour, 46,* 177–188.

Heywood, C. A., and Cowey, A. (1992). The role of the "face-cell" area in the discrimination and recognition of faces in monkeys. *Philosophical Transactions of the Royal Society of London, 335,* 31–38.

Hiebert, S. M., Stoddard, P. K., and Arcese, P. (1989). Repertoire size, territory acquisition and reproductive success in the song sparrow. *Animal Behaviour, 37,* 266–273.

Hienz, R. D., and Brady, V. (1988). The acquisition of vowel discrimination by nonhuman primates. *Journal of the Acoustical Society of America, 84,* 186–194.

Hill, G. E. (1991). Plumage coloration is a sexually selected indicator of male quality. *Nature, 350,* 337–339.

Hillger, L. A., and Koenig, O. (1991). Separable mechanisms in face processing: Evidence from hemispheric specialization. *Journal of Cognitive Neuroscience, 3,* 42–58.

Hinde, R. A. (1952). The behaviour of the great tit (*Parus major*) and some related species. *Behaviour, 2,* 1–201.

Hinde, R. A. (1959). Unitary drives. *Animal Behaviour, 7,* 130–141.

Hinde, R. A. (1960). Energy models of motivation. *Symposia of the Society for Experimental Biology, 14,* 199–213.

Hinde, R. A. (1981). Animal signals: Ethological and games-theory approaches are not incompatible. *Animal Behaviour, 29,* 535–542.

Hinde, R. A. (1987). *Individuals, relationships and culture.* Cambridge, UK: Cambridge University Press.

Hirschfeld, L. A., and Gelman, S. A. (1994). *Mapping the mind: Domain specificity in cognition and culture.* Cambridge, UK: Cambridge University Press.

Hirsh-Patek, K., and Golinkoff, R. M. (1991). Language comprehension: A new look at some old themes. In N. A. Krasnegor, D. M. Rumbaugh, R. L. Schiefelbusch, and M. Studdert-Kennedy (Eds.), *Biological and behavioral determinants of language development* (pp. 301–321). Hillsdale, NJ: Lawrence Erlbaum Associates.

Hiscock, M., and Kinsbourne, M. (1995). Phylogeny and ontogeny of cerebral lateralization. In R. J. Davidson and K. Hugdahl (Eds.), *Brain asymmetry* (pp. 535–578). Cambridge, MA: MIT Press.

Hixon, T. (1973). Respiratory function in speech. In F. D. Minifie, T. J. Hixon, and F. Williams (Eds.), *Normal aspects of speech, hearing, and language* (pp. 297–356). Englewood Cliffs, NJ: Prentice-Hall.

Hjorth, I. (1970). Reproductive behavior in Tetraonidae. *Viltrevy, 7,* 184–196.

Hockett, C. F. (1960a). Logical considerations in the study of animal communication. In W. E. Lanyon and W. N. Tavolga (Eds.), *Animal sounds and communication* (pp. 392–430). Washington, DC: American Institute of Biological Sciences.

Hockett, C. F. (1960b). The origin of speech. *Scientific American, 203,* 88–96.

Hodos, W. (1988). Homoplasy. In L. N. Irwin (Eds.), *Comparative neuroscience and neurobiology: Readings from the encyclopedia of neuroscience* (p. 47). Boston: Birkhäuser.

Hodos, W., and Campbell, B. (1990). Evolutionary scales and comparative studies of animal cognition. In R. Kesner and D. Olton (Eds.), *Neurobiology of comparative cognition* (pp. 1–20). Hillsdale, NJ: Lawrence Erlbaum Associates.

Hoikkala, A., Kaneshiro, K. Y., and Hoy, R. R. (1994). Courtship songs of the picture-singed *Drosophila plantibia* subgroup species. *Animal Behaviour, 47,* 1363–1374.

Holder, M. D., Herman, L. M., and Kuczaj, S. (1993). A bottlenosed dolphin's responses to anomalous sequences expressed within an artificial gestural grammar. In H. L. Roitblat, L. M. Herman, and P. E. Nachtigall (Eds.), *Language and communication: Comparative perspectives* (pp. 443–454). Hillsdale, NJ: Lawrence Erlbaum Associates.

Hölldobler, B. (1977). Communication in social Hymenoptera. In T. A. Sebeok (Ed.), *How animals communicate* (pp. 418–471). Bloomington: Indiana University Press.

Hölldobler, B. (1984). Evolution of insect communication. In T. Lewis (Ed.), *Insect communication* (pp. 349–377). London: Academic Press.

Hölldobler, B., and Carlin, N. F. (1987). Anonymity and specificity in the chemical communication signals of social insects. *Journal of Comparative Physiology, A, 161*, 567–581.

Hölldobler, B., Obermayer, M., and Wilson, E. O. (1992). Communication in the primitive cryptobiotic ant *Prionopelta amabilis* (Hymenoptera: Formicidae). *Journal of Comparative Physiology, A, 170*, 9–16.

Holloway, R. L. (1983). Human brain evolution. *Canadian Journal of Anthropology, 3*, 215–230.

Holloway, R. L. (1995). Toward a synthetic theory of human brain evolution. In J.-P. Changeux and J. Chavaillon (Eds.), *Origins of the human brain* (pp. 42–54). Oxford, UK: Oxford University Press.

Hoogland, J. L. (1983). Nepotism and alarm calling in the black-tailed prairie dog (*Cynomys ludovicianus*). *Animal Behaviour, 31*, 472–479.

Hopkins, W. D., Washburn, D. A., and Rumbaugh, D. (1990). Processing of form stimuli presented unilaterally in humans, chimpanzees (*Pan troglodytes*) and monkeys (*Macaca mulatta*). *Behavioral Neuroscience, 104*, 577–582.

Hopp, S. L., Sinnott, J. M., Owren, M. J., and Petersen, M. R. (1992). Differential sensitivity of Japanese Macaques (*Macaca fuscata*) and Humans (*Homo sapiens*) to peak position along a synthetic coo call continuum. *Journal of Comparative Psychology, 106*, 128–136.

Hoptman, M. J., and Davidson, R. J. (1994). How and why do the two cerebral hemispheres interact? *Psychological Bulletin, 116*, 195–219.

Horn, A. G. (1992). Field experiments on the perception of song types by birds. In P. K. McGregor (Ed.), *Playback and studies of animal communication* (pp. 191–200). New York: Plenum Press.

Horn, G. (1991). Cerebral function and behaviour investigated through a study of filial imprinting. In P. Bateson (Ed.), *The development and integration of behaviour* (pp. 121–148). Cambridge, UK: Cambridge University Press.

Houde, A. E., and Endler, J. A. (1990). Correlated evolution of female mating preferences and male color patterns in the guppy, *Poecilia reticulata*. *Science, 248*, 1405–1408.

Houghton, P. (1993). Neanderthal supralaryngeal vocal tract. *American Journal of Physical Anthropology, 90*, 139–146.

Howard, R. D. (1974). The influence of sexual selection and inter-specific competition on mockingbird (*Mimus polyglottos*) song. *Evolution, 28*, 428–438.

Hoy, R. (1992). The evolution of hearing in insects as an adaptation to predation from bats. In D. B. Webster, R. F. Fay, and A. N. Popper (Eds.), *The evolutionary biology of hearing* (pp. 115–130). New York: Springer-Verlag.

Hrdy, S. B. (1979). Infanticide among animals: A review, classification, and examination of implications for the reproductive strategies of females. *Ethology and Sociobiology, 1*, 13–40.

Hrdy, S. B. (1981). *The woman that never evolved*. Cambridge, MA: Harvard University Press.

Hrdy, S. B. (1988). The primate origins of human sexuality. In G. Stevens and R. Bellig (Eds.), *The evolution of sex* (pp. 101–138). New York: Harper & Row.

Hrdy, S. B., and Whitten, P. L. (1987). Patterning of sexual activity. In B. B. Smuts, D. L. Cheney, R. M. Seyfarth, R. W. Wrangham, and T. T. Struhsaker (Eds.), *Primate societies* (pp. 370–384). Chicago: University of Chicago Press.

Hubel, D. H. (1988). *Eye, brain, and vision*. San Francisco: Freeman.

Hubel, D. H., and Wiesel, T. N. (1959). Receptive fields of single neurons in the cat's striate cortex. *Journal of Physiology, 148*, 374–391.

Huber, E. (1930a). Evolution of the facial musculature and cutaneous field of trigeminus, Part I. *Quarterly Review of Biology, 4,* 133–188.

Huber, E. (1930b). Evolution of the facial musculature and cutaneous field of trigeminus, Part II. *Quarterly Review of Biology, 4,* 389–437.

Huber, E. (1931). *Evolution of the facial musculature and facial expression.* Baltimore: Johns Hopkins University Press.

Hulse, S. H. (1989). Comparative psychology and pitch perception in songbirds. In R. J. Dooling and S. H. Hulse (Eds.), *Comparative psychology of audition: Perceiving complex sounds* (pp. 331–352). Hillsdale, NJ: Lawrence Erlbaum Associates.

Hultsch, H. (1991). Early experience can modify singing styles: evidence from experiments with nightingales, *Luscinia megarhynchos. Animal Behaviour, 42,* 883–889.

Hultsch, H. (1993). Tracing the memory mechanisms in the song acquisition of nightingales. *Netherlands Journal of Zoology, 43,* 155–171.

Hultsch, H., and Todt, D. (1988). Song acquisition and acquisition constraints in the nightingale, *Luscinia megarhynchos. Naturwissenshaften, 76,* 83–85.

Hultsch, H., and Todt, D. (1989). Memorization and reproduction of songs in nightingales (*Luscinia megarhynchos*): Evidence for package information. *Journal of Comparative Physiology, A, 165,* 197–203.

Hupfer, K., Jürgens, U., and Ploog, D. (1977). The effect of superior temporal lesions on the recognition of species-specific calls in the squirrel monkey. *Experimental Brain Research, 30,* 75–87.

Huttenlocher, J., and Smiley, P. (1987). Early word meanings: The case of object names. *Cognitive Psychology, 19,* 63–89.

Huxley, J. (1942). *Evolution: The modern synthesis.* London: Allen & Unwin.

Huxley, T. H. (1863). *Evidence as to man's place in nature.* New York: McGraw-Hill.

Ifune, C. K., Vermeire, B. A., and Hamilton, C. R. (1984). Hemispheric differences in split-brain monkeys viewing and responding to videotape recordings. *Behavioral and Neural Biology, 41,* 231–235.

Imig, T. J., Ruggero, M. A., Kitzes, L. M., Javel, E., and Brugge, J. F. (1977). Organization of auditory cortex in the owl monkey (*Aotus trivirgatus*). *Journal of Comparable Neurology, 171,* 111–128.

Immelman, K. (1969). Song development in the zebra finch and other estrilidid finches. In R. A. Hinde (Ed.), *Bird vocalizations* (pp. 61–74). Cambridge, UK: Cambridge University Press.

Isbell, L. C. (1991). Contest and scramble competition: Patterns of female aggression and ranging behavior among primates. *Behavioral Ecology, 2,* 143–155.

Ishai, A., and Sagi, D. (1995). Common mechanisms of visual imagery and perception. *Science, 268,* 1772–1774.

Isshiki, N. (1964). Regulatory mechanism of voice intensity variation. *Journal of Speech and Hearing, 7,* 17–29.

Izard, C. (1971). *The face of emotion.* New York: Meredity.

Jackendoff, R. (1992). *Languages of the mind.* Cambridge, MA: MIT Press.

Jackendoff, R. (in press). How language helps us think. *Pragmatics and cognition.*

Jacobson, S. W., and Kagan, J. (1979). Technical commentary on Meltzoff and Moore (1977). *Science, 205,* 215–217.

Janson, C. H., and van Schaik, C. P. (1988). Recognizing the many faces of primate food competition: Methods. *Behaviour, 105,* 165–186.

Jerison, H. (1973). *Evolution of the brain and intelligence.* New York: Academic Press.

Jerison, H., and Jerison, I. (1988). *Intelligence and evolutionary biology*. New York: Springer-Verlag.

Johnson, F., and Bottjer, S. W. (1993). Hormone-induced changes in identified cell populations of the higher vocal center in male canaries. *Journal of Neurobiology, 24*, 400–418.

Johnson, J. S., and Newport, E. L. (1991). Critical period effects on universal properties of language: The status of subjacency in the acquisition of a second language. *Cognition, 39*, 215–258.

Johnson, M. H., Dziurawiec, S., Ellis, H. D., and Morton, J. (1991). Newborns' preferential tracking of face-like stimuli and its subsequent decline. *Cognition, 40*, 1–19.

Johnson, M. H., and Morton, J. (1991). *Biology and cognitive development*. Cambridge, MA: Basil Blackwell.

Johnson-Laird, P. N. (1990). Introduction: What is communication? In D. H. Mellor (Ed.), *Ways of communicating* (pp. 1–13). Cambridge, UK: Cambridge University Press.

Johnstone, R. A. (1995). Female preference for symmetrical males as a by-product of selection for male recognition. *Nature, 372*, 172–175.

Johnstone, R. A., and Grafen, A. (1992). Error-prone signalling. *Proceedings of the Royal Society, London, 248*, 229–233.

Johnstone, R. A., and Grafen, A. (1993). Dishonesty and the handicap principle. *Animal Behaviour, 46*, 759–764.

Jones, D., and Hill, K. (1993). Criteria of physical attractiveness in five populations. *Human Nature, 4*, 271–296.

Jürgens, U. (1979). Vocalizations as an emotional indicator: A neuroethological study in the squirrel monkey. *Behaviour, 69*, 88–117.

Jürgens, U. (1990). Vocal communication in primates. In R. P. Kesner and D. S. Olton (Eds.), *Neurobiology of comparative cognition* (pp. 51–76). Hillsdale, NJ: Lawrence Erlbaum Associates.

Jürgens, U. (1992). Neurobiology of vocal communication. In H. Papoucek, U. Jürgens, and M. Papoucek (Eds.), *Nonverbal vocal communication: Comparative and developmental approaches* (pp. 31–42). Cambridge, UK: Cambridge University Press.

Jürgens, U., Maurus, M., Ploog, D., and Winter, P. (1967). Vocalization in the squirrel monkey (*Saimiri sciureus*) elicited by brain stimulation. *Experimental Brain Research, 4*, 114–117.

Jürgens, U., and Pratt, R. (1979). Cingular vocalization pathway: Squirrel monkey. *Experimental Brain Research, 10*, 532–554.

Jusczyk, P. W., Friederici, A. D., Wessels, J., Svenkerud, V. Y., and Jusczyk, A. M. (1993). Infants' sensitivity to the sound pattern of native language words. *Journal of Memory and Language, 32*, 402–420.

Kalat, J. W., and Rozin, P. (1973). Learned safety as a mechanism in long-delay taste-aversion learning in rats. *Journal of Comparative Physiological Psychology, 83*, 198–207.

Kanwal, J. S., Matsumura, S., Ohlemiller, K., and Suga, N. (1994). Analysis of acoustic elements and syntax in communication sounds emitted by mustached bats. *Journal of the Acoustical Society of America, 96*, 1229–1254.

Karakashian, S. J., Gyger, M., and Marler, P. (1988). Audience effects on alarm calling in chickens (*Gallus gallus*). *Journal of Comparative Psychology, 102*, 129–135.

Keating, C. K. (1985). Gender and the physiognomy of dominance and attractiveness. *Social Psychology Quarterly, 48*, 61–70.

Keating, C. K., and Keating, E. G. (1982). Visual scan patterns of rhesus monkeys viewing faces. *Perception, 1*, 395–416.

Keating, P. J., and Buhr, R. (1978). Fundamental frequency in the speech of infants and children. *Journal of the Acoustical Society of America, 63*, 567–571.

Keddy-Hector, A. C., Wilczynski, W., and Ryan, M. J. (1992). Call patterns and basilar papilla tuning in cricket frogs: II. Intrapopulation variation and allometry. *Brain, Behavior and Evolution, 39*, 238–246.

Keil, F. C. (1990). Constraints on constraints: Surveying the epigenetic landscape. *Cognitive Science, 14*, 135–168.

Keil, F. C. (1994a). The birth and nurturance of concepts by domains: The origins of concepts of living things. In L. A. Hirschfield and S. A. Gelman (Eds.), *Mapping the mind: Domain specificity in cognition and culture* (pp. 234–254). Cambridge, UK: Cambridge University Press.

Keil, F. C. (1994b). Explanation, association, and the acquisition of word meaning. *Lingua, 92*, 169–196.

Kelley, D. B. (1980). Auditory vocal nuclei in the frog brain concentrate sex hormones. *Science, 207*, 553–555.

Kelley, D. B., and Nottebohm, F. (1979). Projections of a telencephalic auditory nucleus—field L—in the canary. *Journal of Comparative Neurology, 183*, 455–470.

Kent, R. D., and Read, C. (1992). *The acoustic analysis of speech*. San Diego: Singular Publishing Group.

Kimura, D. (1961). Some effects of temporal-lobe damage on auditory perception. *Canadian Journal of Psychology, 15*, 156–165.

Kimura, D. (1992). Sex differences in the brain. *Scientific American, 267*, 118–125.

Kimura, D. (1993). *Neuromotor mechanisms in human communication*. Oxford, UK: Oxford University Press.

Kimura, D., and Hampson, E. (1994). Neural and hormonal mechanisms mediating sex differences in cognition. In P. A. Vernon (Eds.), *Biological approaches to the study of human intelligence* (pp. 114–146). Norwood, NJ: Ablex Press.

King, A. P., and West, M. J. (1983). Dissecting cowbird song potency: Assessing a song's geographic identity and relative appeal. *Zeitschrift für Tierpsychologie, 63*, 37–50.

Kirchner, W. H. (1993a). Acoustical communication in honeybees. *Apidologie, 24*, 297–307.

Kirchner, W. H. (1993b). Vibrational signals in the tremble dance of the honeybee, *Apis mellifera*. *Behavioral Ecology and Sociobiology, 33*, 169–172.

Kirchner, W. H. (1994). Hearing in honeybees: The mechanical response of the bee's antenna to near field sound. *Journal of Comparative Physiology, A, 667*, 1–5.

Kirchner, W. H., and Braun, U. (1994). Dancing honey bees indicate the location of food sources using path integration rather than cognitive maps. *Animal Behaviour, 48*, 1437–1441.

Kirchner, W. H., Dreller, C., and Towne, W. F. (1991). Hearing in honeybees: Operant conditioning and spontaneous reactions to airborne sounds. *Journal of Comparative Physiology, A, 168*, 85–89.

Kirchner, W. H., and Lindauer, M. (1994). The causes of the tremble dance of the honeybee, *Apis mellifera*. *Behavioral Ecology and Sociobiology, 35*, 303–308.

Kirchner, W. H., Lindauer, M., and Michelsen, A. (1988). Honeybee dance communication: Acoustical indication of direction in round dances. *Naturwissenshaften, 75*, 629–630.

Kirchner, W. H., and Sommer, K. (1992). The dance language of the honeybee mutant *dimunitive wings*. *Behavioral Ecology and Sociobiology, 30*, 181–184.

Kirchner, W. H., and Towne, W. F. (1994). The sensory basis of the honeybee's dance language. *Scientific American, 270*, 74–80.

Kirkpatrick, M. (1986). The handicap mechanism of sexual selection does not function. *American Naturalist, 127*, 222–240.

Kirkpatrick, M., and Ryan, M. J. (1991). The paradox of the lek and the evolution of mating preferences. *Nature, 350*, 33–38.

Kirn, J., Clower, R., Kroodsma, D., and DeVoogd, T. (1989). Song-related brain regions in the red-winged blackbird are affected by sex and season but not repertoire size. *Journal of Neurobiology, 20*, 139–163.

Kirzinger, A., and Jürgens, U. (1982). Cortical lesion effects and vocalization in the squirrel monkey. *Brain Research, 358*, 150–162.

Kitcher, P. (1986). *Vaulting ambitions*. New York: Academic Press.

Klatt, D. H., Stevens, K. N., and Mead, J. (1968). Studies of articulatory activity and airflow during speech. *Annals of the New York Academy of Sciences, 55*, 42–54.

Klein, D., Moscovitch, M., and Vigna, C. (1976). Perceptual asymmetries and attentional mechanisms in tachistoscopic recognition of words and faces. *Neuropsychologia, 14*, 44–66.

Kleiner, K. A. (1987). Amplitude and phase spectra as indices of infants' pattern preferences. *Infant Behaviour and Development, 10*, 45–59.

Kleiner, K. A. (1993). Specific vs. non-specific face recognition device. In B. de Boysson-Bardies, S. de Schonen, P. Jusczyk, P. MacNeilage, and J. Morton (Eds.), *Developmental neurocognition: Speech and face processing in the first year of life* (pp. 103–108). Dordrecht, The Netherlands: Kluwer Academic.

Klima, E. S., and Bellugi, U. (1979). *The signs of language*. Cambridge, MA: Harvard University Press.

Klima, E. S., Bellugi, U., and Poizner, H. (1988). Grammar and space in sign aphasiology. *Aphasiology, 2*, 319–327.

Kluender, K. R., Diehl, R. L., and Killeen, P. R. (1987). Japanese quail can learn phonetic categories. *Science, 237*, 1195–1197.

Klump, G., and Gerhardt, H. C. (1989). Sound localization in the barking treefrog. *Naturwissenschaften, 76*, 35–37.

Klump, G. M., Kretzschmar, E., and Curio, E. (1986). The hearing of an avian predator and its avian prey. *Behavioral Ecology and Sociobiology, 18*, 317–323.

Klump, G. M., and Shalter, M. D. (1984). Acoustic behaviour of birds and mammals in the predator context: I. Factors affecting the structure of alarm signals. II. The functional significance and evolution of alarm signals. *Zeitschrift für Tierpsychologie, 66*, 189–226.

Knudsen, E. I., Blasdel, G. G., and Konishi, M. (1979). Sound localization by the barn owl (*Tyto alba*) measured with the search coil technique. *Journal of Comparative Physiology, 133*, 1–11.

Knudsen, E. I., Esterly, S. D., and Knudsen, P. F. (1984). Monaural occlusion alters sound localization during a sensitive period in the barn owl. *Journal of Neuroscience, 4*, 1001–1011.

Koeppl, J. W., Hoffmann, R. S., and Nadler, C. F. (1978). Pattern analysis of acoustical behavior in four species of ground squirrels. *Journal of Mammalogy, 59*, 677–696.

Kolb, B., Wilson, B., and Taylor, L. (1992). Developmental changes in the recognition and comprehension of facial expression: Implications for frontal lobe function. *Brain and Cognition, 20*, 74–84.

Konishi, M. (1964). Effects of deafening on song development in two species of juncos. *Condor, 66*, 85–102.

Konishi, M. (1965a). The role of auditory feedback in the control of vocalization in the white-crowned Sparrow. *Zeitschrift für Tierpsychologie, 22*, 770–783.

Konishi, M. (1965b). Effects of deafening on song development in American robins and black-headed grosbeaks. *Zeitschrift für Tierpsychologie, 35*, 352–380.

Konishi, M. (1970). Comparative neurophysiological studies of hearing and vocalizations in song birds. *Zeitschrift für vergleichende Physiologie, 66*, 257–272.

Konishi, M. (1985). Birdsong: From behavior to neuron. *Annual Review of Neuroscience, 8*, 125–170.

Konishi, M. (1989). Birdsong for neurobiologists. *Neuron, 3*, 541–549.

Konishi, M. (1993). Listening with two ears. *Scientific American, 268,* 66–73.

Konishi, M. (1994a). An outline of recent advances in birdsong neurobiology. *Brain, Behavior and Evolution, 44,* 279–285.

Konishi, M. (1994b). Pattern generation in birdsong. *Current Opinion in Neurobiology, 4,* 827–831.

Konishi, M., Emlen, S. T., Ricklefs, R. E., and Wingfield, J. C. (1989). Contributions of bird studies to biology. *Science, 246,* 465–472.

Konorski, J. (1967). *Integrative activity of the brain.* Chicago: University of Chicago Press.

Korsia, S., and Bottjer, S. W. (1991). Chronic testosterone treatment impairs vocal learning in male zebra finches during a restricted period of development. *Journal of Neuroscience, 11,* 2362–2371.

Kössl, M., and Vater, M. (1985a). Evoked acoustic emissions and cochlear microphonics in the mustache bat, *Pteronotus parnelli. Hearing Research, 19,* 157–170.

Kössl, M., and Vater, M. (1985b). The frequency place map of the bat, *Pteronotus parnelli. Journal of Comparative Physiology, 157,* 687–697.

Kosslyn, S. M. (1980). *Image and mind.* Cambridge, MA: MIT Press.

Kosslyn, S. M. (1994). *Image and brain.* Cambridge, MA: MIT Press.

Kosslyn, S. M., and Koenig, O. (1992). *Wet mind: The new cognitive neuroscience.* New York: Free Press.

Kotovsky, L., and Baillargeon, R. (in press). Calibration-based reasoning about collision events in 11 month-old infants. *Cognition.*

Krebs, J. R. (1977). The significance of song repertoires: The beau geste hypothesis. *Animal Behaviour, 25,* 428–438.

Krebs, J. R. (1987). The evolution of animal signals. In C. Blakemore and S. Greenfield (Eds.), *Mindwaves: Thoughts on intelligence, identity and consciousness* (pp. 163–173). Oxford, UK: Basil Blackwell.

Krebs, J. R., Ashcroft, R., and Webber, M. (1978). Song repertoires and territory defense in the great tit. *Nature, 271,* 539–542.

Krebs, J. R., Ashcroft, R., and van Orsdol, K. (1981). Song matching in the great tit (*Parus major*). *Animal Behaviour, 29,* 918–923.

Krebs, J. R., and Davies, N. B. (1993). *Introduction to behavioural ecology.* Oxford, UK: Blackwell Scientific.

Krebs, J. R., and Dawkins, R. (1984). Animal signals: Mind-reading and manipulation. In J. R. Krebs and N. B. Davies (Eds.), *Behavioural ecology* (pp. 380–402). Sunderland, MA: Sinauer Associates.

Kroodsma, D. E. (1979). Vocal dueling among male marsh wrens: Evidence for ritualized expression of dominance/subordinance. *Auk, 96,* 506–515.

Kroodsma, D. E. (1982). Song repertoires: Problems in their definition and use. In D. E. Kroodsma, E. H. Miller, and H. Ouellet (Eds.), *Acoustic communication in birds,* vol. 2 (pp. 125–146). New York: Academic Press.

Kroodsma, D. E. (1988). Contrasting styles of song development and their consequences among passerine birds. In R. C. Bolles and M. D. Beecher (Eds.), *Evolution and learning* (pp. 157–184). Hillsdale, NJ: Lawrence Erlbaum Associates.

Kroodsma, D. E., Baker, M. C., Baptista, L. F., and Petrinovich, L. (1984). Vocal ''dialects'' in Nuttall's white-crowned sparrow. *Current Ornithology, 2,* 103–133.

Kroodsma, D. E., and Konishi, M. (1991). A suboscine bird (eastern phoebe, *Sayornis phoebe*) develops normal song without auditory feedback. *Animal Behaviour, 42,* 477–487.

Kuczaj, S. A., and Kirkpatrick, V. M. (1993). Similarities and differences in human and animal language research: Toward a comparative psychology of language. In H. L. Roitblat, L. M. Herman, and P. E. Nachtigall (Eds.), *Language and communication: Comparative perspective* (pp. 45–64). Hillsdale, NJ: Lawrence Erlbaum Associates.

Kuhl, P. K. (1979). Speech perception in early infancy: Perceptual constancy for spectrally dissimilar vowel categories. *Journal of the Acoustical Society of America, 66,* 1668–1679.

Kuhl, P. K. (1983). Perception of auditory equivalence classes for speech in early infancy. *Infant Behavior and Development, 6,* 263–285.

Kuhl, P. K. (1989). On babies, birds, modules, and mechanisms: A comparative approach to the acquisition of vocal communication. In R. J. Dooling and S. H. Hulse (Eds.), *The comparative psychology of audition* (pp. 379–422). Hillsdale, NJ: Lawrence Erlbaum Associates.

Kulh, P. K. (1991). Human adults and human infants show a "perceptual magnet effect" for the prototypes of speech categories; monkeys do not. *Perception and Psychophysics, 50,* 93–107.

Kuhl, P. K. (1992). Infants' perception and representation of speech: Development of a new theory. In J. Ohala, T. Neary, B. Derwing, M. Hodge, and G. Giebe (Eds.), *Proceedings of the international conference on spoken language processing* (pp. 449–456). Edmonton, Canada: University of Alberta Press.

Kuhl, P. K., and Meltzoff, A. N. (1982). The bimodal perception of speech in infancy. *Science, 218,* 1138–1141.

Kuhl, P. K., and Meltzoff, A. N. (1988). Speech as an intermodal object of perception. In A. Yonas (Ed.), *Perceptual development in infancy* (pp. 235–256). Hillsdale, NJ: Lawrence Erlbaum Associates.

Kuhl, P. K., and Miller, J. D. (1975). Speech perception by the chinchilla: Voiced-voiceless distinction in alveolar plosive consonants. *Science, 190,* 69–72.

Kuhl, P. K., and Miller, J. D. (1978). Speech perception by the chinchilla: Identification function for synthetic VOT stimuli. *Journal of the Acoustical Society of America, 63,* 905–917.

Kuhl, P. K., and Padden, D. M. (1982). Enhanced discriminability at the phonetic boundaries for the voicing feature in macaques. *Perception and Psychophysics, 32,* 542–550.

Kuhl, P. K., and Padden, D. M. (1983). Enhanced discriminability at the phonetic boundaries for the place feature in macaques. *Journal of the Acoustical Society of America, 73,* 1003–1010.

Kuhl, P. K., Williams, K. A., Lacerda, F., Stevens, K. N., and Lindblom, B. (1992). Linguistic experience alters phonetic perception in infants by 6 months of age. *Science, 255,* 606–608.

Kuhl, P. K., Williams, K. A., and Meltzoff, A. N. (1991). Cross-modal speech perception in adults and infants using nonspeech auditory stimuli. *Journal of Experimental Psychology: Human Perception and Performance, 17,* 829–840.

Künzel, H. J. (1989). How well does average fundamental frequency correlate with speaker height and weight? *Phonetica, 46,* 117–125.

Kyes, R. C., and Candland, D. K. (1987). Baboon (*Papio hamadryas*) visual preferences for regions of the face. *Journal of Comparative Psychology, 101,* 345–348.

Lack, D. (1966). *Population studies of birds.* Oxford, UK: Clarendon Press.

Lack, D. (1968). *Ecological adaptations for breeding in birds.* London: Methuen.

Ladefoged, P., and Broadbent, D. E. (1957). Information conveyed by vowels. *Journal of the Acoustical Society of America, 39,* 98–104.

Ladefoged, P., and Maddieson, I. (1981). *UCLA phonological segment inventory database: Data and index. no. 53.* UCLA.

Lambrechts, M. M. (1988). Great tit song output is determined both by motivation and by constraints in singing ability: A reply to Weary et al. *Animal Behaviour, 36,* 1244–1246.

Lambrechts, M. M. (1992). Male quality and playback in the great tit. In P. K. McGregor (Eds.), *Playback and studies of animal communication* (pp. 135–152). New York: Plenum Press.

Lambrechts, M. M., and Dhondt, A. A. (1986). Male quality, reproduction, and survival in the great tit (*Parus major*). *Behavioral Ecology and Sociobiology, 19*, 57–63.

Landau, B., and Gleitman, L. (1985). *Language and experience: Evidence from the blind child.* Cambridge, MA: Harvard University Press.

Lande, R. (1981). Models of speciation by sexual selection on polygenic characters. *Proceedings of the National Academy of Sciences, 78*, 3721–3725.

Lane, H., Boyes-Braem, P., and Bellugi, U. (1976). Preliminaries to a distinctive feature analysis of hand-shapes in American Sign Language. *Cognitive Psychology, 8*, 263–289.

Langlois, J. H., and Roggman, L. A. (1990). Attractive faces are only average. *Psychological Science, 1*, 115–121.

Larson, C. R., and Kistler, M. K. (1986). The relationship of periaqueductal gray neurons to vocalization and laryngeal EMG in the behaving monkey. *Experimental Brain Research, 63*, 596–606.

Larson, C. R., Ortega, J. D., and DeRosier, E. A. (1988). Studies on the relation of the midbrain periaqueductal gray, the larynx and vocalization in the awake monkey. In J. D. Newman (Ed.), *The physiological control of mammalian vocalizations* (pp. 43–65). New York: Plenum Press.

Lauder, G. V., Leroi, A. M., and Rose, M. R. (1993). Adaptations and history. *Trends in Ecology and Evolution, 8*(8), 294–297.

Lawrence, B. D., and Simmons, J. A. (1982). Echolocation in bats: The external ear and perception of the vertical positions of targets. *Science, 218*, 481–483.

Lecanuet, J.-P., Granier-Deferre, C., and Busnel, M.-C. (1989). Differential fetal auditory reactiveness as a function of stimulus characteristics and state. *Seminars in Perinatology, 13*, 421–429.

Lecanuet, J.-P., Granier-Deferre, C., Cohen, C., Le Houezec, R., and Busnel, M.-C. (1986). Fetal responses to acoustic stimulation depend on heart rate variability pattern, stimulus intensity and repetition. *Early Human Development, 13*, 269–283.

Leehey, S., Carey, S., Diamond, R., and Cahn, A. (1978). The upright and inverted faces: The right hemisphere knows the difference. *Cortex, 14*, 411–419.

Leiner, H. C., Leiner, A. L., and Dow, R. S. (1993). Cognitive and language functions of the human cerebellum. *Trends in Neurosciences, 16*, 444–447.

Lemon, R. E. (1975). How birds develop song dialects. *Condor, 77*, 385–406.

Lemon, R. E., Monette, S., and Roff, D. (1987). Song repertoires of American warblers (Parulinae): Honest advertisement or Assessment? *Ethology, 74*, 265–284.

Lenneberg, E. H. (1967). *Biological foundations of language.* New York: Wiley.

Leonard, M. L., and Fenton, M. B. (1984). Echolocation calls of *Euderma maculatum* (Chiroptera: Vespertilionidae): Use in orientation and communication. *Journal of Mammology, 65*, 122–126.

Leslie, A. M. (1994). ToMM, ToBy, and agency: Core architecture and domain specificity. In L. Hirschfeld and S. Gelman (Eds.), *Mapping the mind* (pp. 210–256), New York: Cambridge University Press.

Lester, B. M., and Boukydis, C. F. Z. (1992). No language but a cry. In H. Papousek, U. Jürgens, and M. Papousek (Eds.), *Nonverbal vocal communication: Comparative and developmental approaches* (pp. 145–173). Cambridge, UK: Cambridge University Press.

Lester, B. M., Corwin, M. J., Sepkoski, C., Seifer, R., Peucker, M., McGlaughlin, S., and Golub, H. L. (1991). Neurobehavioral syndromes in cocaine-exposed newborn infants. *Child Development, 62*, 694–705.

Lester, B. M., and Zeskind, P. S. (1978). Brazelton scale and physical size correlates of neonatal cry features. *Infant Behavior and Development, 1*, 393–402.

Levenson, R. W., Carstensen, L. L., Friesen, W. V., and Ekman, P. (1991). Emotion, physiology, and expression in old age. *Psychology and Aging, 6*, 28–35.

Levenson, R. W., Ekman, P., and Friesen, W. V. (1990). Voluntary facial action generates emotion-specific autonomic nervous activity. *Psychophysiology, 27,* 263–384.

Levenson, R. W., Ekman, P., Heider, K., and Friesen, W. V. (1992). Emotion and autonomic nervous system activity in the Minangkabau of West Sumatra. *Journal of Personality and Social Psychology, 62,* 972–988.

Levy, J., Heller, W., Banich, M. T., and Burton, L. A. (1983). Asymmetry of perception in free viewing of chimeric faces. *Brain and Cognition, 2,* 404–419.

Lewin, R. (1993). *Human evolution.* Boston: Blackwell Scientific.

Lewis, E. R., and Lombard, R. E. (1988). The amphibian inner ear. In B. Fritzsch, M. J. Ryan, W. Wilczynski, T. E. Hetherington, and W. Walkowiak (Eds.), *The evolution of the amphibian auditory system* (pp. 93–123). New York: John Wiley & Sons.

Lewis, M. (1994). Self-conscious emotions. *American Scientist, 83,* 68–78.

Lewkowicz, D. J., and Lickliter, R. (1994). *The development of intersensory perception: Comparative perspectives.* Hillsdale, NJ: Lawrence Erlbaum Associates.

Lewontin, R. C. (1992). *Biology as ideology: The doctrine of DNA.* New York: HarperCollins.

Leyton, A. S. F., and Sherrington, C. S. (1917). Observations on the excitable cortex of the chimpanzee, orangutan, and gorilla. *Journal of Experimental Physiology, 11,* 135–222.

Liberman, A. M. (1992). Plausibility, parsimony, and theories of speech. In J. Alegria, D. Holender, J. Junca de Morais, and M. Radeau (Eds.), *Analytic approaches to human cognition* (pp. 25–40). Dordrecht, Holland: Elsevier Science.

Liberman, A. M., Cooper, F. S., Shankweiler, D. P., and Studdert-Kennedy, M. (1967). Perception of the speech code. *Psychological Review, 74,* 431–461.

Liberman, A. M., Delattre, P. C., and Cooper, F. S. (1958). Some rules for the distinction between voiced and voiceless stops in initial position. *Language and Speech, 1,* 153–167.

Liberman, A. M., Harris, K. S., Hoffman, H. S., and Griffith, B. C. (1957). The discrimination of speech sounds within and across phoneme boundaries. *Journal of Experimental Psychology, 54,* 358–368.

Liberman, A. M., and Mattingly, I. G. (1985). The motor theory of speech perception revised. *Cognition, 21,* 1–36.

Lickliter, R. (1990). Premature visual stimulation accelerates intersensory functioning in bobwhite quail neonates. *Developmental Psychobiology, 23,* 15–27.

Lickliter, R., and Hellwell, T. B. (1992). Contextual determinants of auditory learning in bobwhite quail embryos and hatchlings. *Developmental Psychobiology, 25,* 17–31.

Lickliter, R., and Virkar, P. (1989). Intersensory functioning in bobwhite quail chicks: Early sensory dominance. *Developmental Psychobiology, 22,* 651–667.

Lieberman, M. R., and Lieberman, P. (1973). Olson's "projective verse" and the use of breath control as a structural element. *Language and Style, 5,* 287–298.

Lieberman, P. (1968). Primate vocalizations and human linguistic ability. *Journal of the Acoustical Society of America, 44,* 1574–1584.

Lieberman, P. (1984). *The biology and evolution of language.* Cambridge, MA: Harvard University Press.

Lieberman, P. (1985). The physiology of cry and speech in relation to linguistic behavior. In B. M. Lester and C. F. Z. Boukydis (Eds.), *Infant crying* (pp. 29–57). New York: Plenum Press.

Lieberman, P. (1991). *Uniquely human.* Cambridge, MA: Harvard University Press.

Lieberman, P. (1992). Could an autonomous syntax module have evolved? *Brain and Language, 43,* 768–774.

Lieberman, P. (1994). Functional tongues and Neanderthal vocal tract reconstruction: A reply to Houghton (1993). *American Journal of Physical Anthropology, 95*, 443–452.

Lieberman, P., and Blumstein, S. E. (1988). *Speech physiology, speech perception, and acoustic phonetics*. Cambridge, UK: Cambridge University Press.

Lieberman, P., Klatt, D. H., and Wilson, W. H. (1969). Vocal tract limitations on the vowel repertoires of rhesus monkeys and other nonhuman primates. *Science, 164*, 1185–1187.

Lieblich, A. K., Symmes, D., Newman, J. D., and Shapiro, M. (1980). Development of isolation peep in laboratory-bred squirrel monkeys. *Animal Behaviour, 28*, 1–9.

Lim, D. (1990). Representation of complex sounds in the peripheral auditory system of the green treefrog. PhD diss., Cornell University, Ithaca, NY.

Lindblom, B. (1990). On the communication process: Speaker-listener interaction and the development of speech. In K. Fraurud and U. Sundberg (Eds.), *AAC augmentative and alternative communication* (pp. 220–230). London: Williams & Wilkins.

Lindblom, B., Lubker, J., and Gay, T. (1979). Formant frequencies of some fixed-mandible vowels and a model of speech motor programming by predictive simulation. *Journal of Phonetics, 7*, 147–161.

Lindblom, B., MacNeilage, P., and Studdert-Kennedy, M. (in prep). *The evolution of speech*. Cambridge, MA: MIT Press.

Lindblom, B., and Maddieson, I. (1988). Phonetic universals in consonant systems. In L. M. Hyman and C. N. Li (Eds.), *Language, speech and mind* (pp. 62–78). London: Routledge.

Lindblom, B., and Studdert-Kennedy, M. (1967). On the role of formant transitions in vowel recognition. *Journal of the Acoustical Society of America, 42*, 830–843.

Littlejohn, M. J. (1977). Long-range acoustic communication in anurans: An integrated evolutionary approach. In D. H. Taylor and S. I. Guttman (Eds.), *The reproductive biology of amphibians* (pp. 263–294). Cambridge, UK: Cambridge University Press.

Lively, S. E., Pisoni, D. B., Yamada, R. A., Tohkura, Y., and Tamada, T. (1994). Training Japanese listeners to identify English /r/ and /l/: III. Long-term retention of new phonetic categories. *Journal of the Acoustical Society of America, 96*, 2076–2087.

Lloyd, J. E. (1984). On deception, a way of all flesh, and firefly signalling and systematics. In R. Dawkins and M. Ridley (Eds.), *Oxford surveys in evolutionary biology*, vol. 1 (pp. 48–54). New York: Oxford University Press.

Locke, J. L. (1983). *Phonological acquisition and change*. New York: Academic Press.

Locke, J. L. (1993). *The path to spoken language*. Cambridge, MA: Harvard University Press.

Locke, J. L., and Pearson, D. M. (1992). Vocal learning and the emergence of phonological capacity: A neurobiological approach. In C. Ferguson, L. Menn, and C. Stoel-Gammon (Eds.), *Phonological development: Models, research, implications* (pp. 91–130). Parkton, MD: York Press.

Locke, S., and Kellar, L. (1973). Categorical perception in a non-linguistic mode. *Cortex, 9*, 355–369.

Loomis, J., Poizner, H., Bellugi, U., Blakemore, A., and Hollerbach, J. (1983). Computer graphic modeling of American Sign Language. *Computer Graphics, 17*, 105–114.

Lopez, P. T., and Narins, P. M. (1991). Mate choice in the neotropical frog, *Eleutherodactylus coqui*. *Animal Behaviour, 41*, 757–772.

Lorenz, K. (1937). The companion in the bird's world. *Auk, 54*, 245–273.

Lorenz, K. (1966). Evolution of ritualization in the biological and cultural spheres. *Philosophical Transactions of the Royal Society of Britain, 251*, 273–284.

Lubker, J., and Gay, T. (1982). Anticipatory labial coarticulation: Experimental, biological and linguistic variables. *Journal of the Acoustical Society of America, 71*, 437–448.

Macedonia, J. M. (1986). Individuality in the contact call of the ring-tailed lemur (*Lemur catta*). *American Journal of Primatology, 11*, 163–179.

Macedonia, J. M. (1990). *Vocal communication and antipredator behavior in ring-tailed lemurs (Lemur catta)*. PhD diss., Duke University, Durham, NC.

Macedonia, J. M. (1991). What is communicated in the antipredator calls of lemurs: Evidence from playback experiments with ring-tailed and ruffed lemurs. *Ethology, 86,* 177–190.

Macedonia, J. M., and Evans, C. S. (1993). Variation among mammalian alarm call systems and the problem of meaning in animal signals. *Ethology, 93,* 177–197.

MacLennan, B. J., and Burghardt, G. M. (1994). Synthetic ethology and the evolution of cooperative communication. *Adaptive Behavior, 2,* 161–188.

Macnamara, J. (1982). *Names for things.* Cambridge, MA: MIT Press.

MacNeilage, P. F. (1991). The "postural origins" theory of primate neurobiology asymmetries. In N. Krasnegor, D. Rumbaugh, M. Studdert-Kennedy, and R. Schiefelbusch (Eds.), *Biological foundations of language development* (pp. 165–188). Hillsdale, NJ: Lawrence Erlbaum Associates.

MacNeilage, P. F. (in press). The sound pattern of the first spoken language. *Phonetica.*

MacNeilage, P. F., and Davis, B. (1990). Acquisition of speech production: Frames, then content. In M. Jeannerod (Ed.), *Attention and performance 13: Motor representation and control* (pp. 453–475). Hillsdale, NJ: Lawrence Erlbaum Associates.

MacNeilage, P. F., and Davis, B. L. (1993). Motor explanations of babbling and early speech patterns. In B. de Boysson-Bardies, S. de Schonen, P. Jusczyk, P. MacNeilage, and J. Morton (Eds.), *Developmental neurocognition: Speech and face processing in the first year of life* (pp. 235–247). Dordrecht, The Netherlands: Kluwer Academic.

MacNeilage, P. F., and DeClerk, J. L. (1969). On the motor control of coarticulation of CVC monosyllables. *Journal of the Acoustical Society of America, 45,* 1217–1233.

Macphail, E. (1987a). Intelligence: A comparative approach. In C. Blakemore and S. Greenfield (Eds.), *Mindwaves: Thoughts on intelligence, identity and consciousness* (pp. 177–194). Oxford, UK: Basil Blackwell.

Macphail, E. (1987b). The comparative psychology of intelligence. *Behavioral and Brain Sciences, 10,* 645–695.

Maddieson, I. (1984). *Patterns of sound.* Cambridge, UK: Cambridge University Press.

Maeda, T., and Masataka, N. (1987). Locale-specific vocal behaviour of the tamarin (*Saguinus I. labiatus*). *Ethology, 75,* 25–30.

Mandler, J. (1988). How to build a baby: I. On the development of an accessible representational system. *Cognitive Development, 3,* 113–136.

Mandler, J. (1992). How to build a baby: II. Conceptual primitives. *Psychological Review, 99,* 587–604.

Manley, G. A. (1990). *Peripheral hearing mechanisms in reptiles and birds.* Berlin: Springer-Verlag.

Manley, G. A., and Gleich, O. (1992). Evolution and specialization of function in the avian auditory periphery. In D. B. Webster, R. R. Fay, and A. N. Popper (Eds.), *The evolutionary biology of hearing* (pp. 561–580). Berlin: Springer-Verlag.

Maratsos, M., and Matheny, L. (1994). Language specificity and elasticity: Brain and clinical syndrome studies. *Annual Review of Psychology, 45,* 487–516.

Marchetti, K. (1993). Dark habitats and bright birds illustrate the role of the environment in species divergence. *Nature, 362,* 149–152.

Margoliash, D. (1983). Acoustic parameters underlying the responses of song-specific neurons in the white-crowned sparrow. *Journal of Neuroscience, 3,* 1039–1057.

Margoliash, D. (1987). Preference for autogenous song by auditory neurons in a song system nucleus of the white-crowned sparrow. *Journal of Neuroscience, 6,* 1643–1661.

Margoliash, D., Staicer, C. A., and Inoue, S. A. (1991). Stereotyped and plastic song in adult indigo buntings, *Passerina cyanea*. *Animal Behaviour, 42,* 367–388.

Markl, H. (1983). Vibrational communication. In F. Huber and H. Markl (Eds.), *Neuroethology and behavioural physiology* (pp. 332–353). Berlin: Springer-Verlag.

Markl, H. (1985). Manipulation, modulation, information, cognition: Some of the riddles of communication. In B. Hölldobler and M. Lindauer (Eds.), *Experimental behavioral ecology and sociobiology* (pp. 163–194). Stuttgart, Germany: Gustav Fischer Verlag.

Markman, E. M. (1990). Constraints children place on word meanings. *Cognitive Science, 14,* 57–77.

Markman, E. M., and Hutchinson, J. E. (1984). Children's sensitivity to constraints on word meaning: Taxonomic versus thematic relations. *Cognitive Psychology, 16,* 1–27.

Marler, P. (1955). Characteristics of some animal calls. *Nature, 176,* 6–7.

Marler, P. (1957). Specific distinctiveness in the communication signals of birds. *Behaviour, 11,* 13–39.

Marler, P. (1961). The logical analysis of animal communication. *Journal of Theoretical Biology, 1,* 295–317.

Marler, P. (1967). Animal communication signals. *Science, 157,* 769–774.

Marler, P. (1969a). Vocalizations of wild chimpanzees. *Recent Advances in Primatology, 1,* 94–100.

Marler, P. (1969b). Tonal quality of bird sounds. In R. A. Hinde (Ed.), *Bird vocalizations* (pp. 5–18). Cambridge, UK: Cambridge University Press.

Marler, P. (1970a). Birdsong and speech development: Could there be parallels? *American Scientist, 58,* 669–673.

Marler, P. (1970b). A comparative approach to vocal learning: Song development in white crowned sparrows. *Journal of Comparative Physiological Psychology Monographs, 71,* 1–25.

Marler, P. (1970c). Vocalizations of East African monkeys: I. Red colobus. *Folia primatologica, 13,* 81–91.

Marler, P. (1973). A comparison of vocalizations of red-tailed monkeys and blue monkeys, *Cercopithecus ascanius* and *C. mitis*, in Uganda. *Zeitschrift für Tierpsychologie, 33,* 223–247.

Marler, P. (1975). On the origin of speech from animal sounds. In J. F. Kavanagh and J. Cutting (Eds.), *The role of speech in language* (pp. 11–37). Cambridge, MA: MIT Press.

Marler, P. (1976). Social organization, communication and graded signals: The chimpanzee and the gorilla. In P. P. G. Bateson and R. A. Hinde (Eds.), *Growing points in ethology* (pp. 239–280). Cambridge, UK: Cambridge University Press.

Marler, P. (1977). The structure of animal communication sounds. In T. H. Bullock (Ed.), *Recognition of complex acoustic signals* (pp. 17–35). Berlin: Springer-Verlag.

Marler, P. (1978). Primate vocalizations: Affective or symbolic? In G. Bourne (Ed.), *Progress in ape research* (pp. 85–96). New York: Academic Press.

Marler, P. (1982). Avian and primate communication: The problem of natural categories. *Neuroscience and Biobehavioral Reviews, 6,* 87–94.

Marler, P. (1984). Song learning: Innate species differences in the learning process. In P. Marler and H. S. Terrace (Eds.), *The biology of learning* (pp. 289–309). Berlin: Springer-Verlag.

Marler, P. (1985). Representational vocal signals of primates. In B. Hölldobler and M. Lindauer (Eds.), *Experimental behavioral ecology and sociobiology* (pp. 211–221). Stuttgart, Germany: Gustav Fischer Verlag.

Marler, P. (1987). Sensitive periods and the roles of specific and general sensory stimulation in birdsong learning. In J. Rauschecker and P. Marler (Eds.), *Imprinting and cortical plasticity* (pp. 99–135). New York: Springer-Verlag.

Marler, P. (1989). Learning by instinct: Birdsong. *American Speech-Language Association, 89,* 75–79.

Marler, P. (1991a). Differences in behavioural development in closely related species: Birdsong. In P. Bateson (Ed.), *The development and integration of behaviour* (pp. 41–70). Cambridge, UK: Cambridge University Press.

Marler, P. (1991b). Song learning behavior: The interface with neuroethology. *Trends in Neurosciences, 14,* 199–206.

Marler, P. (1991c). The instinct to learn. In S. Carey and R. Gelman (Eds.), *The epigenesis of mind: Essays on biology and cognition* (pp. 212–231). Hillsdale, NJ: Lawrence Erlbaum Associates.

Marler, P. (1992a). Functions of arousal and emotion in primate communication: A semiotic approach. In T. Nishida, W. C. McGrew, P. Marler, M. Pickford, and F. B. M. deWaal (Eds.), *Topics in primatology,* vol. 1, *Human origins* (pp. 235–248). Tokyo: University of Tokyo Press.

Marler, P. (1992b). Introduction to "Communication: Behavior and neurobiology." *Seminars in the Neurosciences, 4,* 373–377.

Marler, P., Dufty, A., and Pickert, R. (1986a). Vocal communication in the domestic chicken: I. Does a sender communicate information about the quality of a food referent to a receiver? *Animal Behaviour, 34,* 188–193.

Marler, P., Dufty, A., and Pickert, R. (1986b). Vocal communication in the domestic chicken: II. Is a sender sensitive to the presence and nature of a receiver? *Animal Behaviour, 34,* 194–198.

Marler, P., Evans, C. S., and Hauser, M. D. (1992). Animal signals? Reference, motivation or both? In H. Papoucek, U. Jürgens, and M. Papoucek (Eds.), *Nonverbal vocal communication: Comparative and developmental approaches* (pp. 66–86). Cambridge, UK: Cambridge University Press.

Marler, P., Karakashian, S., and Gyger, M. (1991). Do animals have the option of withholding signals when communication is inappropriate? The audience effect. In C. Ristau (Ed.), *Cognitive ethology: The minds of other animals* (pp. 135–186). Hillsdale, NJ: Lawrence Erlbaum Associates.

Marler, P., and Nelson, D. A. (1992). Neuroselection and song learning in birds: Species universals in a culturally transmitted behavior. *Seminars in the Neurosciences, 4,* 415–423.

Marler, P., and Peters, S. (1989). Species differences in auditory responsiveness in early vocal learning. In R. J. Dooling and S. H. Hulse (Eds.), *The comparative psychology of audition: Perceiving complex sounds* (pp. 243–273). Hillsdale, NJ: Lawrence Erlbaum Associates.

Marler, P., Peters, S., Ball, G. F., Dufty, A. M., and Wingfield, J. C. (1988). The role of sex steroids in the acquisition and production of birdsong. *Nature, 336,* 770–772.

Marler, P., Peters, S., and Wingfield, J. (1987). Correlations between song acquisition, song production, and plasma levels of testosterone and estradiol in sparrows. *Journal of Neurobiology, 18,* 531–548.

Marler, P., and Pickert, R. (1984). Species-universal microstructure in the learned song of the swamp sparrow (*Melospiza georgiana*). *Animal Behaviour, 32,* 673–689.

Marler, P., and Sherman, V. (1985). Innate differences in singing behaviour in sparrows reared in isolation from adult conspecific song. *Animal Behaviour, 33,* 57–71.

Marler, P., and Tamura, M. (1962). Culturally transmitted patterns of vocal behavior in sparrows. *Science, 146,* 1483–1486.

Marler, P., and Tenaza, R. (1977). Communication in apes with special reference to vocalizations. In T. A. Sebeok (Ed.), *How animals communicate* (pp. 965–1033). Bloomington: Indiana University Press.

Marler, P., and Terrace, H. (1984). *The biology of learning.* Berlin: Springer-Verlag.

Marler, P., and Waser, M. S. (1977). Role of auditory feedback in canary song development. *Journal of Comparative Physiological Psychology, 91,* 8–16.

Marten, K., Quine, D. B., and Marler, P. (1977). Sound transmission and its significance for animal vocalization: 2. Tropical habitats. *Behavioral Ecology and Sociobiology, 2,* 291–302.

Martin, P., and Bateson, P. (1993). *Measuring behaviour. An introductory guide,* 2nd ed. Cambridge, UK: Cambridge University Press.

Martin, W. F. (1971). Mechanics of sound production in toads of the genus *Bufo:* Passive elements. *Journal of Experimental Zoology, 176,* 273–294.

Martin, W. F. (1972). Evolution of vocalizations in the genus *Bufo.* In W. F. Blair (Ed.), *Evolution in the genus* Bufo (pp. 279–309). Austin: University of Texas Press.

Masataka, N. (1983). Categorical responses to natural and synthesized alarm calls in Goeldi's monkeys (*Callimico goeldi*). *Primates, 24,* 40–51.

Masataka, N. (1988). The response of red-chested moustached tamarins to long calls from their natal and alien populations. *Animal Behaviour, 36,* 55–61.

Masataka, N., and Biben, M. (1987). Temporal rules regulating vocal exchanges of squirrel monkeys. *Behaviour, 101,* 311–319.

Masataka, N., and Fujita, K. (1989). Vocal learning of Japanese and rhesus monkeys. *Behaviour, 109,* 191–199.

Massaro, D. W. (1987a). Categorical partition: A fuzzy-logical model of categorization behavior. In S. Harnad (Ed.), *Categorical perception* (pp. 254–286). Cambridge, UK: Cambridge University Press.

Massaro, D. W. (1987b). *Speech perception by ear and eye: A paradigm for psychological inquiry.* Hillsdale, NJ: Lawrence Erlbaum Associates.

Masters, J. C. (1991). Loud calls of *Galago crassicaudatus* and *G. garnettii* and their relation to habitat structure. *Primates, 32,* 153–167.

Matsuzawa, T. (1985a). Colour naming and classification in a chimpanzee (*Pan troglodytes*). *Journal of Human Evolution, 14,* 283–291.

Matsuzawa, T. (1985b). Use of numbers by a chimpanzee. *Nature, 315,* 57–59.

Mattingly, I. G., and Studdert-Kennedy, M. (1991). *Modularity and the motor theory of speech perception.* Hillsdale, NJ: Lawrence Erlbaum Associates.

Maugham, W. S. (1925). *The painted veil.* New York: Penguin Books.

Maurer, D., and Barrera, M. (1981). Infants' perception of natural and distorted arrangements of a schematic face. *Child Development, 52,* 196–202.

May, B. J., Moody, D. B., and Stebbins, W. C. (1988). The significant features of Japanese monkey coo sounds: A psychophysical study. *Animal Behaviour, 36,* 1432–1444.

May, B., Moody, D. B., and Stebbins, W. C. (1989). Categorical perception of conspecific communication sounds by Japanese macaques, *Macaca fuscata. Journal of the Acoustical Society of America, 85,* 837–847.

Mayeux, R., and Kandel, E. R. (1991). Disorders of language: The aphasias. In E. R. Kandel, J. H. Schwartz, and T. M. Jessell (Eds.), *Principles of neural science,* 3rd ed. (pp. 839–851). Norwalk, CT: Appleton & Lange.

Maynard Smith, J. (1964). Group selection and kin selection. *Nature, 201,* 1145–1147.

Maynard Smith, J. (1965). The evolution of alarm calls. *American Naturalist, 99,* 59–63.

Maynard Smith, J. (1974). The theory of games and the evolution of animal conflicts. *Journal of Theoretical Biology, 47,* 209–221.

Maynard Smith, J. (1976). Sexual selection and the handicap principle. *Journal of Theoretical Biology, 57,* 239–242.

Maynard Smith, J. (1979). Game theory and the evolution of behaviour. *Proceedings of the Royal Society, London, B, 205,* 475–488.

Maynard Smith, J. (1982). *Evolution and the theory of games.* Cambridge, UK: Cambridge University Press.

Maynard Smith, J. (1991). Honest signalling: The Philip Sydney game. *Animal Behaviour, 42,* 1034–1035.

Maynard Smith, J. (1994). Must reliable signals always be costly? *Animal Behaviour, 47,* 1115–1120.

Mayr, E. (1982). How to carry out the adaptationist program. *American Naturalist, 121,* 324–334.

Mayr, E., and Provine, W. (1980). *The evolutionary synthesis.* Cambridge, MA: Harvard University Press.

Mazoyer, B. M., Tzourio, N., Frak, V., Syrota, A., Murayama, N., Levrier, O., Salamon, G., Dehaene, S., Cohen, L., and Mehler, J. (1993). The cortical representation of speech. *Journal of Cognitive Neuroscience, 5,* 467–479.

Mazzocchi, D., and Vignolo, L. A. (1979). Localisation of lesions in aphasias: Clinical-CT scan correlations in stroke patients. *Cortex, 15,* 627–654.

McCashland, J. S. (1987). Neuronal control of bird song production. *Journal of Neuroscience, 7,* 23–39.

McClintock, M. K. (1971). Menstrual synchrony and suppression. *Nature, 229,* 244–245.

McComack, S. A., and Levine, T. R. (1990). When lovers become leery: The relationship between suspiciousness and accuracy in detecting deception. *Communication Monographs, 57,* 219–230.

McComb, K. E. (1987). Roaring by red deer stags advances the date of oestrous in hinds. *Nature, 330,* 648–649.

McComb, K. E. (1991). Female choice for high roaring rates in red deer, *Cervus elaphus. Animal Behaviour, 41,* 79–88.

McConnell, P. B. (1990). Acoustic structure and receiver response in *Canis familiaris. Animal Behaviour, 39,* 897–904.

McConnell, P. B., and Baylis, J. R. (1985). Interspecific communication in cooperative herding: Acoustic and visual signals from shepherds and herding dogs. *Zeitschrift für Tierpsychologie, 67,* 302–328.

McCracken, G. F. (1987). Genetic structure of bat social groups. In M. B. Fenton, P. Racey, and J. M. V. Rayner (Eds.), *Recent advances in the study of bats* (pp. 281–299). Cambridge, UK: Cambridge University Press.

McCune, L., and Vihman, M. M. (in press). Grunt communication: A developmental phase in human infants with evolutionary significance. *Journal of Comparative Psychology.*

McCune, L., Vihman, M. M., Roug-Hellichius, L., Gogate, L., and Delery, D. (in press). Grunts—A gateway to language? *Behavioral and Brain Sciences.*

McGregor, P. K. (1992). *Playback and studies of animal communication.* New York: Plenum Press.

McGregor, P. K., and Krebs, J. R. (1984a). Song learning and deceptive mimicry. *Animal Behaviour, 32,* 280–287.

McGregor, P. K., and Krebs, J. R. (1984b). Sound degradation as a distance cue in great tits (*Parus major*) song. *Behavioral Ecology and Sociobiology, 16,* 49–56.

McGregor, P. K., and Krebs, J. R. (1989). Song learning in adult great tits (*Paurs major*): Effects of neighbours. *Behaviour, 108,* 139–159.

McGregor, P. K., Krebs, J. R., and Perrins, C. M. (1981). Song repertoires and lifetime reproductive success in the great tit (*Parus major*). *American Naturalist, 118,* 149–159.

McGrew, W. C. (1992). *Chimpanzee material culture.* Cambridge, UK: Cambridge University Press.

McGrew, W. C., and Feistner, A. T. C. (1992). Two nonhuman primate models for the evolution of human food sharing: Chimpanzees and callitrichids. In J. H. Barkow, L. Cosmides, and J. Tooby (Eds.), *The adapted mind* (pp. 229–249). New York: Oxford University Press.

McGurk, H., and MacDonald, J. (1976). Hearing lips and seeing voices. *Nature, 264,* 746–748.

McNeil, D. (1985). So you think gestures are nonverbal? *Psychological Review, 92*, 350–371.

Medin, D. L., and Barsalou, L. W. (1987). Categorization processes and categorical perception. In S. Harnad (Ed.), *Categorical perception: The groundwork of cognition* (pp. 455–490). New York: Cambridge University Press.

Medin, D. L., Goldstone, R. L., and Gentner, D. (1993). Respects for similarity. *Psychological Review, 100*, 254–278.

Megela-Simmons, A., Moss, C. F., and Daniel, K. M. (1985). Behavioral audiograms of the bullfrog (*Rana catesbeiana*) and the green tree frog (*Hyla cinerea*). *Journal of the Acoustical Society of America, 78*, 1236–1244.

Mehler, J., and Dupoux, E. (1994). *What infants know*. Cambridge, MA: Blackwell Scientific.

Mehler, J., Jusczyk, P., Lambertz, G., Halsted, N., Bertonici, J., and Amiel-Tison, C. (1988). A precursor of language acquisition in young infants. *Cognition, 29*, 143–178.

Meier, R. P., and Newport, E. L. (1990). Out of the hands of babes: On a possible sign advantage in language acquisition. *Language, 66*, 1–23.

Meier, R. P., and Willerman, R. (in press). Prelinguistic gesture in deaf and hearing children. In K. G. Emmorey and J. Reilly (Eds.), *Language, gesture, and space*. Hillsdale, NJ: Lawrence Erlbaum Associates.

Mellor, D. H. (1990). *Ways of communicating*. Cambridge, UK: Cambridge University Press.

Melnick, D. J., and Pearl, M. (1987). Cercopithecines in multimale groups: Genetic diversity and population structure. In B. B. Smuts, D. L. Cheney, R. M. Seyfarth, R. W. Wrangham, and T. T. Struhsaker (Eds.), *Primate societies* (pp. 121–148). Chicago: University of Chicago Press.

Meltzoff, A. N. (1988). The human infant as *Homo imitans*. In T. R. Zentall and B. G. Galef, Jr. (Eds.), *Social learning* (pp. 319–341). Hillsdale, NJ: Lawrence Erlbaum Associates.

Meltzoff, A. N. (1993). The centrality of motor coordination and proprioception in social and cognitive development: From shared actions to shared minds. In G. J. P. Savelsbergh (Ed.), *The development of coordination in infancy* (pp. 463–496). Amsterdam: North-Holland.

Meltzoff, A. N., and Moore, M. K. (1977). Imitation of facial and manual gestures by human neonates. *Science, 198*, 75–78.

Meltzoff, A. N., and Moore, M. K. (1983). Newborn infants imitate adult facial gestures. *Child Development, 54*, 702–709.

Meltzoff, A. N., and Moore, M. K. (1989). Imitation in newborn infants: Exploring the range of gestures imitated and the underlying mechanisms. *Developmental Psychology, 25*, 954–962.

Meltzoff, A. N., and Moore, M. K. (1994). Imitation, memory, and the representation of persons. *Infant Behavior and Development, 17*, 83–100.

Mendelson, M. J., Haith, M. M., and Goldman-Rakic, P. S. (1982). Face scanning and responsiveness to social cues in infant rhesus monkeys. *Developmental Psychology, 18*, 222–228.

Merzenich, M. M., and Brugge, J. F. (1973). Representation of the cochlear partition on the superior temporal plane of the macaque monkey. *Brain Research, 50*, 275–296.

Merzenich, M. M., Nelson, R. J., Kaas, J. H., Stryker, M. P., Jenkins, W. M., Zook, J. M., Cynader, M. S., and Schoppmann, A. (1987). Variability in hand surface representations in areas 3b and 1 in adult owl and squirrel monkeys. *Journal of Comparative Neurology, 258*, 281–296.

Merzenich, M. M., Recanzone, G., Jenkins, W. M., Allard, T. T., and Nudo, R. J. (1989). Cortical representational plasticity. In P. Rakic and W. Singer (Eds.), *Neurobiology of neocortex* (pp. 41–67). Chichester, UK: John Wiley and Sons.

Merzenich, M. M., and Schreiner, C. E. (1992). Mammalian auditory cortex—Some comparative observations. In D. B. Webster, R. F. Fay, and A. N. Popper (Eds.), *The evolutionary biology of hearing* (pp. 673–689). New York: Springer-Verlag.

Mesulam, M.-M., and Pandya, D. N. (1973). The projections of the medial geniculate complex with the sylvian fissure of the rhesus monkey. *Brain Research, 60,* 315–333.

Metz, K. J., and Weatherhead, P. J. (1992). Seeing red: Uncovering coverable badges in red-winged blackbirds. *Animal Behaviour, 43,* 223–229.

Michelsen, A. (1992). Hearing and sound communication in small animals: Evolutionary adaptations to the laws of physics. In D. B. Webster, R. F. Fay, and A. N. Popper (Eds.), *The evolutionary biology of hearing* (pp. 61–77). New York: Springer-Verlag.

Michelsen, A., Andersen, B. B., Storm, J., Kirchner, W. H., and Lindauer, M. (1992). How honey-bees perceive communication dances, studied by means of a mechanical model. *Behavioral Ecology and Sociobiology, 30,* 143–150.

Michelsen, A., Kirchner, W. H., Andersen, B. B., and Lindauer, M. (1986). The tooting and quacking vibration signals of honeybee queens: A quantitative analysis. *Journal of Comparative Physiology, A, 158,* 605–611.

Michelsen, A., Kirchner, W. H., and Lindauer, M. (1986). Sound and vibrational signals in the dance language of the honeybee, *Apis mellifera. Behavioral Ecology and Sociobiology, 18,* 207–212.

Michelsen, A., Towne, W. F., Kirchner, W. H., and Kryger, P. (1987). The acoustic near field of a dancing honeybee. *Journal of Comparative Physiology, A, 161,* 633–643.

Middlebrooks, J. C., and Green, D. M. (1991). Sound localization by human listeners. *Annual Review in Psychology, 42,* 135–159.

Middleton, F. A., and Strick, P. L. (1994). Anatomical evidence for cerebellar and basal ganglia involvement in higher cognitive function. *Science, 266,* 458–461.

Miles, L. (1978). Language acquisition in apes and children. In F. C. Peng (Ed.), *Sign language and language acquisition in man and ape: New dimensions in comparative pedolinguistics* (pp. 103–120). Boulder, CO: Westview Press.

Milinski, M. (1978). Kin selection and reproductive value. *Zeitschrift für Tierpsychologie, 47,* 328–329.

Miller, D. B. (1994). Social context affects the ontogeny of instinctive behaviour. *Animal Behaviour, 48,* 627–634.

Miller, L. A. (1988). Arctiid moth clicks can degrade the accuracy of range difference discrimination in echolocating big brown bats, *Eptesicus fuscus. Journal of Comparative Physiology, 168,* 571–579.

Miller, R. E. (1967). Experimental approaches to the physiological and behavioral concomitants of affective communication in rhesus monkeys. In S. A. Altmann (Ed.), *Social communication among primates* (pp. 125–134) Chicago: University of Chicago Press.

Miller, R. E. (1971). Experimental studies of communication in the monkey. In L. A. Rosenblum (Ed.), *Primate behavior,* vol. 2 (pp. 113–128). New York: Academic Press.

Mills, D. L., Coffey-Corina, S. A., and Neville, H. J. (1993). Language acquisition and cerebral specialization in 20-month-old infants. *Journal of Cognitive Neuroscience, 5,* 317–334.

Mitani, J. C. (1985). Responses of gibbons (*Hylobates muelleri*) to self, neighbor, and stranger duets. *International Journal of Primatology, 6,* 193–200.

Mitani, J. C., and Brandt, K. L. (1994). Social factors influence the acoustic variability in the long-distance calls of male chimpanzees. *Ethology, 96,* 233–252.

Mitani, J. C., Hasegawa, T., Gros-Louis, J., Marler, P., and Byrne, R. (1992). Dialects in wild chimpanzees? *American Journal of Primatology, 27,* 233–244.

Mitani, J. C., and Marler, P. (1989). A phonological analysis of male gibbon singing behavior. *Behaviour, 109,* 20–45.

Miyashita, Y. (1995). How the brain creates imagery: Projection to primary visual cortex. *Science, 268,* 1719–1720.

Miyawaki, K., Strange, W., Verbrugge, R., Liberman, A., Jenkins, J. J., and Fujimura, O. (1975). An effect of linguistic experience: The discrimination of [r] and [l] by native speakers of Japanese and English. *Perception and Psychophysics, 18,* 331–340.

Mock, D. W. (1985). Siblicidal brood reduction: The prey-size hypothesis. *American Naturalist, 125,* 327–343.

Mock, D. W., and Forbes, L. S. (1992). Parent-offspring conflict: A case of arrested development. *Trends in Ecology and Evolution, 7,* 409–413.

Moffat, A. J. M., and Capranica, R. R. (1976). Effects of temperature on the response properties of auditory nerve fibers in the American toad. *Journal of the Acoustical Society of America, 60,* S80.

Mogdans, J., Ostwald, J., and Schnitzler, H.-H. (1988). The role of pinna movement for the localization of vertical and horizontal wire obstacles in the greater horseshoe bat, *Rhinolopus ferrumequinum. Journal of the Acoustical Society of America, 84,* 1676–1679.

Moiseff, A., Pollack, G. S., and Hoy, R. R. (1978). Steering responses of flying crickets to sound and ultrasound: Mate attraction and predator avoidance. *Proceedings of the National Academy of Sciences, 75,* 4052–4056.

Molfese, D. L., and Segalowitz, S. J. (1988). *Brain lateralization in children: Developmental implications.* New York: Guilford Press.

Møller, A. P. (1988a). False alarm calls as a means of resource usurpation in the great tit, *Parus major. Ethology, 79,* 25–30.

Møller, A. P. (1988b). Female choice selects for male sexual tail ornaments in the monogamous swallow. *Nature, 332,* 640–642.

Møller, A. P. (1989). Viability costs of male tail ornaments in a swallow. *Nature, 339,* 132–135.

Møller, A. P. (1991). Parasite load reduces song output in a passerine bird. *Animal Behaviour, 41,* 723–730.

Møller, A. P. (1993). Morphology and sexual selection in the barn swallow *Hirundo rustica* in Chernobyl, Ukraine. *Proceedings of the Royal Society, London, 252,* 51–57.

Møller, A. P., and Hoglund, J. (1991). Patterns of fluctuating asymmetry in avian feather ornaments: Implications for models of sexual selection. *Proceedings of the Royal Society, London, 245,* 1–6.

Moody, D. B. (1994). Detection and discrimination of amplitude-modulated signals by macaque monkeys. *Journal of the Acoustical Society of America, 95,* 3499–3510.

Moody, D. B., May, B., Cole, D. M., and Stebbins, W. C. (1986). The role of frequency modulation in the perception of complex stimuli by primates. *Experimental Biology, 45,* 219–232.

Moody, D. B., and Stebbins, W. C. (1989). Salience of frequency modulation in primate communication. In R. J. Dooling and S. H. Hulse (Eds.), *The comparative psychology of audition* (pp. 353–378). Hillsdale, NJ: Lawrence Erlbaum Associates.

Moody, D. B., Stebbins, W. C., and May, B. J. (1990). Auditory perception of communication signals by Japanese monkeys. In W. C. Stebbins and M. A. Berkley (Eds.), *Comparative perception: Complex perception* (pp. 311–344). New York: John Wiley and Sons.

Moore, B. C. J. (1988). *An introduction to the psychology of hearing.* New York: Academic Press.

Morris, C. W. (1946). *Signs, language, and behavior.* New York: Prentice-Hall.

Morton, E. S. (1975). Ecological sources of selection on avian sounds. *American Naturalist, 109,* 17–34.

Morton, E. S. (1977). On the occurrence and significance of motivation-structural rules in some birds and mammal sounds. *American Naturalist, 111,* 855–869.

Morton, E. S. (1982). Grading, discreteness, redundancy, and motivational-structural rules. In D. Kroodsma and E. H. Miller (Eds.), *Acoustic communication in birds,* vol. 1 (pp. 183–212). New York: Academic Press.

Morton, E. S. (1986). Predictions from the ranging hypothesis for the evolution of long-distance signals in birds. *Behaviour, 83,* 66–86.

Morton, J. (1993). Mechanisms in infant face processing. In B. de Boysson-Bardies, S. de Schonen, P. Jusczyk, P. MacNeilage, and J. Morton (Eds.), *Developmental neurocognition: Speech and face processing in the first year of life* (pp. 93–102). Netherlands: Kluwer Academic Publishers.

Moskowitz, A. I. (1970). The two-year-old stage in the acquisition of English phonology. *Language, 46,* 426–441.

Moss, C. F., and Schnitzler, H.-U. (1995). Behavioral studies of auditory information processing. In R. R. Fay and A. N. Popper (Eds.), *Springer handbook of auditory research: Hearing by bats* (pp. 219–248). New York: Springer-Verlag.

Moss, C. F., and Zagaeski, M. (1994). Acoustic information available to bats using frequency-modulated sounds of the perception of insect prey. *Journal of the Acoustical Society of America, 95,* 2745–2756.

Mountjoy, D. J., and Lemon, R. E. (1991). Song as an attractant for male and female European starlings, and the influence of song complexity on their response. *Behavioral Ecology and Sociobiology, 28,* 97–100.

Moynihan, M. (1970). The control, suppression, decay, disappearance, and replacement of displays. *Journal of Theoretical Biology, 29,* 85–112.

Mullenix, J. W., and Pisoni, D. B. (1989). Speech perception: Analysis of biologically significant signals. In R. J. Dooling and S. H. Hulse (Eds.), *The comparative psychology of audition* (pp. 97–130). Hillsdale, NJ: Lawrence Erlbaum Associates.

Müller-Preuss, P. (1988). Neural correlates of audio-vocal behavior: Properties of the anterior limbic cortex and related areas. In J. D. Newman (Ed.), *The physiological control of mammalian vocalizations* (pp. 245–262). New York: Plenum Press.

Müller-Preuss, P., and Jürgens, U. (1976). Projections from the cingular vocalization area in the squirrel monkey. *Brain Research, 103,* 29–43.

Munn, C. A. (1986a). Birds that "cry wolf." *Nature, 319,* 143–145.

Munn, C. A. (1986b). The deceptive use of alarm calls by sentinel species in mixed species flocks of neotropical birds. In R. W. Mitchell and N. S. Thompson (Eds.), *Deception: Perspectives on human and nonhuman deceit* (pp. 169–175). Albany: State University of New York Press.

Myrberg, A. A., Jr., Ha, S. J., and Shamblott, M. J. (1993). The sounds of bicolor damselfish (*Pomacentrus partitus*): Predictors of body size and a spectral basis for individual recognition and assessment. *Journal of the Acoustical Society of America, 94,* 3067–3070.

Nagel, T. (1974). What is it like to be a bat? *Philosophical Review, 83,* 2–14.

Narins, P. M. (1982). Behavioral refractory period in neotropical treefrogs. *Journal of Comparative Physiology, 148,* 337–344.

Narins, P. M. (1992). Biological constraints on anuran acoustic communication: Auditory capabilities of naturally behaving animals. In D. B. Webster, R. F. Fay, and A. N. Popper (Eds.), *The evolutionary biology of hearing* (pp. 439–454). New York: Springer-Verlag.

Narins, P. M., and Capranica, R. R. (1976). Sexual differences in the auditor system of the treefrog, *Eleutherodactylus coqui. Science, 192,* 378–380.

Narins, P. M., Ehret, G., and Tautz, J. (1988). Accessory pathway for sound transfer in a neotropical frog. *Proceedings of the National Academy of Sciences, 85,* 1508–1512.

Narins, P. M., and Zelick, R. (1988). The effects of noise on auditory processing and behavior in amphibians. In B. Fritzsch, M. J. Ryan, W. Wilczynski, T. E. Hetherington, and W. Walkowiak (Eds.), *The evolution of the amphibian auditory system* (pp. 511–537). New York: John Wiley & Sons.

Neary, T. J. (1988). Forebrain auditory pathways in ranid frogs. In B. Fritzsch, M. J. Ryan, W. Wilczynski, T. E. Hetherington, and W. Walkowiak (Eds.), *The evolution of the amphibian auditory system* (pp. 233–252). New York: John Wiley & Sons.

Negus, V. E. (1929). *The mechanism of the larynx.* St. Louis: C. V. Mosby.

Negus, V. E. (1949). *The comparative anatomy and physiology of the larynx.* New York: Hafner.

Nelson, D. A. (1984). Communication of intentions in agonistic contexts by the pigeon guillemot, *Cepphus columba. Behaviour, 88,* 145–189.

Nelson, D. A. (1988). Feature weighting in species song recognition by the field sparrow (*Spizella pusilla*). *Behaviour, 106,* 158–182.

Nelson, D. A. (1992). Song overproduction, song matching and selective attrition during development. In P. K. McGregor (Ed.), *Playback and studies of animal communication* (pp. 121–134). New York: Plenum Press.

Nelson, D. A. (in press). Social interaction and sensitive phases for song learning: A critical review. In C. T. Snowdon and M. Hausberger (Eds.), *Social influences on vocal development.* Cambridge, UK: Cambridge University Press.

Nelson, D. A., and Marler, P. (1989). Categorical perception of a natural stimulus continuum: Birdsong. *Science, 244,* 976–978.

Nelson, D. A., and Marler, P. (1990). The perception of birdsong and an ecological concept of signal space. In W. C. Stebbins and M. A. Berkley (Eds.), *Comparative perception,* vol. 2, *Complex signals* (pp. 443–478). New York: John Wiley & Sons.

Nelson, D. A., and Marler, P. (1993). Innate recognition of song in white-crowned sparrows: A role in selective vocal learning? *Animal Behaviour, 46,* 806–808.

Nelson, D. A., and Marler, P. (1994). Selection-based learning in bird song development. *Proceedings of the National Academy of Sciences, 91,* 10498–10501.

Neuweiler, G., Bruns, V., and Schuller, G. (1980). Ears adapted for the detection of motion, or how echolocating bats have exploited the capacities of the mammalian auditory system. *Journal of the Acoustical Society of America, 68,* 741–753.

Neville, H. J., Coffey, S. A., Holcomb, P. J., and Tallal, P. (1993). The neurobiology of sensory and language processing in language-impaired children. *Journal of Cognitive Neurosciences, 5,* 235–253.

Nevo, E., and Capranica, R. R. (1985). Evolutionary origin of ethological reproductive isolation in cricket frogs, *Acris. Evolutionary Biology, 19,* 147–214.

Nevo, E., and Schneider, H. (1976). Mating call pattern of green toads in Israel and its ecological correlates. *Journal of Zoology, London, 178,* 133–145.

Newman, J. (1985). Squirrel monkey communication. In L. A. Rosenblum and C. L. Coe (Eds.), *Handbook of squirrel monkey research* (pp. 99–125). New York: Plenum Press.

Newman, J. D., and Symmes, D. (1982). Inheritance and experience in the acquisition of primate acoustic behavior. In C. T. Snowdon, C. H. Brown, and M. R. Petersen (Eds.), *Primate communication* (pp. 259–278). Cambridge, UK: Cambridge University Press.

Newman, J. D., and Wollberg, Z. (1973a). Multiple coding of species-specific vocalizations in the auditory cortex of squirrel monkeys. *Brain Research, 54,* 287–304.

Newman, J. D., and Wollberg, Z. (1973b). Responses of single neurons in the auditory cortex of squirrel monkeys to variants of a single call type. *Experimental Neurology, 40,* 821–824.

Newport, E. L. (1990). Maturational constraints on language learning. *Cognitive Science, 14,* 11–28.

Newport, E. L. (1991). Contrasting conceptions of the critical period for language. In S. Carey and R. Gelman (Eds.), *Epigenesis of mind: Essays on biology and cognition* (pp. 113–141). Hillsdale, NJ: Lawrence Erlbaum Associates.

Newport, E. L., and Meier, R. P. (1985). The acquisition of American Sign Language. In D. I. Slobin (Ed.), *The crosslinguistic study of language acquisition*, vol. 1, *The data* (pp. 881–938). Hillsdale, NJ: Lawrence Erlbaum Associates.

Nieh, J. C. (1993). The stop signal of the honey bees: Reconsidering its message. *Behavioral Ecology and Sociobiology, 33,* 51–56.

Nishida, T. (1987). Local traditions and cultural tradition. In B. B. Smuts, D. L. Cheney, R. M. Seyfarth, R. W. Wrangham, and T. T. Struhsaker (Eds.), *Primate societies* (pp. 462–474). Chicago: University of Chicago Press.

Noldus, L. P. J. J., van de Loo, E. L. H. M., and Timmers, P. H. A. (1989). Computers in behavioural research. *Nature, 341,* 767–768.

Nordeen, E. J., Marler, P., and Nordeen, K. W. (1989). Addition of song-related neurons in swamp sparrows coincides with memorization, not production, of learned song. *Journal of Neurobiology, 20,* 651–661.

Nordeen, E. J., and Nordeen, K. W. (1990). Neurogenesis and sensitive periods in avian song learning. *Trends in Neurosciences, 13,* 31–36.

Nordeen, K. W., and Nordeen, E. J. (1992). Auditory feedback is necessary for the maintenance of stereotyped song in adult zebra finches. *Behavioral and Neural Biology, 57,* 58–66.

Nottebohm, F. (1968). Auditory experience and song development in the chaffinch (*Fringila coelebs*). *Ibis, 110,* 549–568.

Nottebohm, F. (1976). Phonation in the orange-winged Amazon parrot, *Amazona azonica. Journal of Comparative Physiology, A, 108,* 157–170.

Nottebohm, F. (1981). A brain for all seasons: Cyclical anatomical changes in song control nuclei of the canary brain. *Science, 214,* 1368–1370.

Nottebohm, F. (1989). From bird song to neurogenesis. *Scientific American, 260,* 74–79.

Nottebohm, F., and Arnold, A. P. (1976). Sexual dimorphism in vocal control areas of the song bird brain. *Science, 194,* 211–213.

Nottebohm, F., Kelley, D. B., and Paton, J. A. (1982). Connections of vocal control nuclei in the canary telencephalon. *Journal of Comparative Neurology, 207,* 344–357.

Nottebohm, F., and Nottebohm, M. E. (1976). Left hypoglossal dominance in the control of canary and white-crowned sparrow song. *Journal of Comparative Physiology, A, 108,* 171–192.

Nottebohm, F., Stokes, T. M., and Leonard, C. M. (1976). Central control of song in the canary, *Serinus canarius. Journal of Comparative Neurology, 165,* 457–468.

Nowicki, S. (1987). Vocal tract resonances in oscine bird sound production: Evidence from birdsongs in a helium atmosphere. *Nature, 325,* 53–55.

Nowicki, S. (1989). Vocal plasticity in captive black-capped chickadees: The acoustic basis and rate of call convergence. *Animal Behaviour, 37,* 64–73.

Nowicki, S., and Capranica, R. R. (1986). Bilateral syringeal coupling during phonation of a songbird. *Journal of Neuroscience, 6,* 3593–3610.

Nowicki, S., and Marler, P. (1988). How do birds sing? *Music Perception, 5,* 391–426.

Nowicki, S., Mitani, J. C., Nelson, D. A., and Marler, P. (1989). The communicative significance of tonality in birdsong: Responses to songs produced in helium. *Bioacoustics, 2,* 35–46.

Nowicki, S., and Nelson, D. A. (1990). Defining natural categories in acoustic signals: Comparison of three methods applied to "chick-a-dee" call notes. *Ethology, 86,* 89–101.

Nowicki, S., Westneat, M., and Hoese, W. (1992). Birdsong: Motor function and the evolution of communication. *Seminars in the Neurosciences, 4,* 385–390.

Nwokah, E. E., Davies, P., Islam, A., Hsu, H.-C., and Fogel, A. (1993). Vocal affect in three-year-olds: A quantitative acoustic analysis of child laughter. *Journal of the Acoustical Society of America, 94,* 3076–3090.

O'Connell, S. M., and Cowlishaw, G. (1994). Infanticide avoidance, sperm competition and mate choice: The function of copulation calls in female baboons. *Animal Behaviour, 48,* 687–694.

Ogden, C. K., and Richards, I. A. (1923). *The meaning of meaning.* London: Routledge and Kegan Paul.

Ohala, J. J. (1983). Cross-language use of pitch: An ethological view. *Phonetica, 40,* 1–18.

Ohala, J. J. (1984). An ethological perspective on common cross-language utilization of Fø of voice. *Phonetica, 41,* 1–16.

Ohlemiller, K. K., Kanwal, J. S., Butman, J. A., and Suga, N. (1994). Stimulus design for auditory neuroethology: Synthesis and manipulation of complex communication sounds. *Auditory Neuroscience, 1,* 19–37.

Ojemann, J. G., Ojemann, G. A., and Lettich, E. (1992). Neuronal activity related to faces and matching in human right nondominant temporal cortex. *Brain, 115,* 1–13.

Okanoya, K., and Dooling, R. J. (1988). Obtaining acoustic similarity measures from animals: A method for species comparisons. *Journal of the Acoustical Society of America, 83,* 1690–1693.

Oller, D. K. (1986). Metaphonology and infant vocalizations. In B. Lindblom and R. Zetterström (Eds.), *Precursors of early speech* (pp. 21–36). New York: Stockton Press.

Oller, D. K., and Eilers, R. (1988). The role of audition in infant babbling. *Child Development, 59,* 441–449.

Oller, D. K., and Eilers, R. E. (1992). Development of vocal signaling in human infants: Toward a methodology for cross-species vocalization comparisons. In H. Paousek, U. Jürgens, and M. Paousek (Eds.), *Nonverbal vocal communication: Comparative and developmental approaches* (pp. 174–191). Cambridge, UK: Cambridge University Press.

Oller, D. K., Eilers, R., Bull, D., and Carney, A. (1985). Prespeech vocalizations of a deaf infant: A comparison with normal metaphonological development. *Journal of Speech and Hearing Research, 28,* 47–63.

Overman, W. H., and Doty, R. W. (1982). Hemispheric specialization displayed by man but not macaques for analysis of faces. *Neuropsychologia, 20,* 113–128.

Owen, D. H. (1980). *Camouflage and mimicry.* Chicago: University of Chicago Press.

Owings, D. H. (1994). How monkeys feel about the world: A review of "How monkeys see the world." *Language and Communication, 14,* 15–20.

Owings, D. H., and Coss, R. G. (1977). Snake mobbing by California ground squirrels: Adaptive variation and ontogeny. *Behaviour, 62,* 50–69.

Owings, D. H., and Hennessy, D. F. (1984). The importance of variation in sciurid visual and vocal communication. In J. O. Murie and G. R. Michener (Eds.), *The biology of ground-dwelling squirrels* (pp. 167–200). Lincoln: University of Nebraska Press.

Owings, D. H., and Leger, D. W. (1980). Chatter vocalizations of California ground squirrels: Predator- and social-role specificity. *Zeitschrift für Tierpsychologie, 54,* 163–184.

Owings, D. H., and Virginia, R. A. (1978). Alarm calls of California ground squirrels (*Spermophilus beecheyi*). *Zeitschrift für Tierpsychologie, 46,* 58–70.

Owren, M. J. (1990). Acoustic classification of alarm calls by vervet monkeys (*Cercopithecus aethiops*) and humans: I. Natural calls. *Journal of Comparative Psychology, 104,* 20–28.

Owren, M. J., and Bernacki, R. (1988). The acoustic features of vervet monkey (*Cercopithecus aethiops*) alarm calls. *Journal of the Acoustical Society of America, 83,* 1927–1935.

Owren, M. J., Dieter, J. A., Seyfarth, R. M., and Cheney, D. L. (1992). "Food" calls produced by adult female rhesus (*Macaca mulatta*) and Japanese (*M. fuscata*) macaques, their normally raised offspring, and offspring cross-fostered between species. *Behaviour, 120,* 218–231.

Owren, M. J., Dieter, J. A., Seyfarth, R. M., and Cheney, D. L. (1993). Vocalizations of rhesus (*Macaca mulatta*) and Japanese (*M. fuscata*) macaques cross-fostered between species show evidence of only limited modification. *Developmental Psychobiology, 26*, 389–406.

Packer, C., and Ruttan, L. (1988). The evolution of cooperative hunting. *American Naturalist, 132*, 159–198.

Packer, C., Scheel, D., and Pusey, A. E. (1990). Why lions form groups: Food is not enough. *American Naturalist, 136*, 1–19.

Pagel, M. (1994). The evolution of conspicuous oestrous advertisement in Old World monkeys. *Animal Behaviour, 47*, 1333–1341.

Palmer, A., Richard, D., and Strobeck, C. (1986). Fluctuating asymmetry: Measurement, analysis and patterns. *Annual Review of Ecology and Systematics, 17*, 391–421.

Pandya, D. P., Seltzer, B., and Barbas, H. (1988). Input-output organization of the primate cerebral cortex. In H. D. Steklis and J. Erwin (Eds.), *Comparative primate biology: Neurosciences* (pp. 39–80). New York: Liss.

Papousek, M., Papousek, H., and Haekel, M. (1987). Didactic adjustments in fathers' and mothers' speech to their three-month old infants. *Journal of Psycholinguistic Research, 16*, 491–516.

Parker, G. A. (1974). Assessment strategy and the evolution of fighting behaviour. *Journal of Theoretical Biology, 47*, 223–243.

Parry, F. M., Young, A. W., Saul, J. S. M., and Moss, A. (1991). Dissociable face processing impairments after brain injury. *Journal of Clinical and Experimental Neuropsychology, 13*, 534–547.

Pastore, R. E. (1987). Categorical perception: Some psychophysical models. In S. Harnad (Ed.), *Categorical perception* (pp. 29–52). New York: Cambridge University Press.

Patterson, F. (1979). Linguistic capabilities of a lowland gorilla. In R. L. Schiefelbusch and J. Hollis (Eds.), *Language intervention from ape to child* (pp. 325–356). Baltimore: University Park Press.

Patterson, F. (1987). *Koko's story*. New York: Scholastic Press.

Patterson, T. L., Petrinovich, L., and James, D. K. (1980). Reproductive value and appropriateness of response to predators by white-crowned sparrows. *Behavioral Ecology and Sociobiology, 7*, 227–231.

Payne, K. B., and Payne, R. S. (1985). Large scale changes over 19 years in songs of humpback whales in Bermuda. *Zeitschrift für Tierpsychologie, 68*, 89–114.

Payne, R. B., Payne, L. L., and Doehlert, S. M. (1988). Biological and cultural success of song memes in indigo buntings. *Ecology, 69*, 104–117.

Peacocke, C. (1992). *A study of concepts*. Cambridge, MA: MIT Press.

Peirce, A. (1985). A review of attempts to condition operantly alloprimate vocalizations. *Primates, 26*, 202–213.

Pepperberg, I. M. (1987a). Evidence for conceptual quantitative abilities in the African parrot: Labeling of cardinal sets. *Ethology, 75*, 37–61.

Pepperberg, I. M. (1987b). Acquisition of the same/different concept by an African Grey parrot (*Psittacus erithacus*): Learning with respect to categories of color, shape, and material. *Journal of Experimental Analysis of Behavior, 50*, 553–564.

Pepperberg, I. M. (1991). A communicative approach to animal cognition: A study of conceptual abilities of an African grey parrot. In C. A. Ristau (Ed.), *Cognitive ethology* (pp. 153–186). Hillsdale, NJ: Lawrence Erlbaum Associates.

Pereira, M., and Fairbanks, L. (1994). *Juvenile primates*. Oxford, UK: Oxford University Press.

Pereira, M. E., and Macedonia, J. M. (1991). Response urgency does not determine antipredator call selection by ring-tailed lemurs. *Animal Behaviour, 41*, 543–544.

Perkell, J. S., and Klatt, D. H. (1986). *Invariance and variability in speech processes*. Hillsdale, NJ: Lawrence Erlbaum Associates.

Perrett, D. I., May, K. A., and Yoshikawa, S. (1994). Facial shape and judgements of female attractiveness. *Nature, 368,* 239–242.

Perrett, D. I., and Mistlin, A. J. (1990). Perception of facial characteristics by monkeys. In W. C. Stebbins and M. A. Berkley (Eds.), *Comparative perception: Complex signals* (pp. 187–216). New York: John Wiley & Sons.

Perrett, D. I., Mistlin, A. J., Chitty, A. J., Harries, M., Newcombe, F., and de Haan, E. (1988a). Neuronal mechanisms of face perception and their pathology. In C. Kennard and F. Clifford Rose (Eds.), *Physiological aspects of clinical neuro-ophthalmology* (pp. 137–154). London: Chapman and Hall.

Perrett, D. I., Mistlin, A. J., Chitty, A. J., Smith, P. A., Potter, D. D., Broennimann, R., and Haries, M. (1988b). Specialized face processing and hemispheric asymmetry in man and monkey: Evidence from single unit and reaction time studies. *Behavioural Brain Research, 29,* 245–258.

Perrett, D. I., Rolls, E. T., and Caan, W. (1982). Visual neurones responsive to faces in the monkey temporal cortex. *Experimental Brain Research, 47,* 329–342.

Perrett, D. I., Smith, A. J., Potter, D. D., Mistlin, A. J., Head, A. S., Milner, A. D., and Jeeves, M. A. (1984). Neurones responsive to faces in the temporal cortex: Studies of functional organization, sensitivity to identity and relation to perception. *Human Neurobiology, 3,* 197–208.

Perrett, D. I., Smith, P. A. J., Potter, D. D., Mistlin, A. J., Head, A. S., Milner, A. D., and Jeeves, M. A. (1985). Visual cells in the temporal cortex sensitive to face view and gaze direction. *Proceedings of the Royal Society, London, 223,* 293–317.

Perrins, C. M. (1963). Population fluctuations and clutch size in the great tit (*Parus major*). *Journal of Animal Ecology, 34,* 601–647.

Perrins, C. M. (1968). The purpose of the high intensity alarm call in small passerines. *Ibis, 110,* 200–201.

Petersen, M. R., Beecher, M. D., Zoloth, S. R., Moody, D. B., and Stebbins, W. C. (1978). Neural lateralization of species-specific vocalizations by Japanese macaques. *Science, 202,* 324–326.

Petersen, S., Fox, P., Posner, M., Mintun, M., and Raichle, M. (1989). Positron emission tomographic studies of the cortical anatomy of single-word processing. *Nature, 331,* 585–589.

Petersen, S. E., Fox, P. T., Snyder, A., and Raichle, M. E. (1990). Activation of prestriate and frontal cortical activity by words and word-like stimuli. *Science, 249,* 1041–1044.

Petersen, M. R., and Jusczyk, P. (1984). On perceptual predispositions for human speech and monkey vocalizations. In P. Marler and H. S. Terrace (Eds.), *The biology of learning* (pp. 585–616). Berlin: Springer-Verlag.

Peterson, G. E., and Barney, H. L. (1952). Control methods used in the study of the identification of vowels. *Journal of the Acoustical Society of America, 24,* 175–184.

Petitto, L. A. (1988). "Language" in the pre-linguistic child. In F. S. Kessel (Eds.), *The development of language and language researchers* (pp. 187–221). Hillsdale, NJ: Lawrence Erlbaum Associates.

Petitto, L. A. (1993). On the ontogenetic requirements for early language acquisition. In B. de Boysson-Bardies, S. de Schonen, P. Jusczyk, P. MacNeilage, and J. Morton (Eds.), *Developmental neurocognition: Speech and face processing in the first year of life* (pp. 365–383). Dordrecht, The Netherlands: Kluwer Academic.

Petitto, L. A., and Marentette, P. (1991). Babbling in the manual mode: Evidence for the ontogeny of language. *Science, 251,* 1483–1496.

Petrie, M. (1983). Female moorhens compete for small fat males. *Science, 220,* 413–415.

Petrinovich, L., and Baptista, L. F. (1987). Song development in the white-crowned sparrow: Modification of learned song. *Animal Behaviour, 35,* 961–974.

Petrinovich, L., Patterson, T. L., and Baptista, L. F. (1981). Song dialects as barriers to dispersal: A re-evaluation. *Evolution, 35,* 180–188.

Pettigrew, J. D. (1988). Microbat vision and echolocation in an evolutionary context. In P. E. Nachtigall and P. W. B. Moore (Eds.), *Animal sonar: Processes and performances* (pp. 645–650). New York: Plenum Press.

Pfingst, B. E., and O'Connor, T. A. (1981). Characteristics of neurons in auditory cortex of monkeys performing a simple auditory task. *Journal of Neurophysiology, 45,* 16–34.

Piaget, J. (1952). *The origins of intelligence in children.* New York: International University Press.

Piaget, J. (1954). *The construction of reality in the child.* New York: Basic Books.

Piaget, J. (1962). *Play, dreams and imitation in childhood.* New York: Norton.

Piatelli-Palmarini, M. (1989). Evolution, selection, and cognition: From "learning" to parameter setting in biology and the study of language. *Cognition, 31,* 1–44.

Pickens, J., Field, T., Nwrocki, T., Martinez, A., Soutullo, D., and Gonzalez, J. (1994). Full-term and preterm infants' perception of face-voice synchrony. *Infant Behavior and Development, 17,* 447–456.

Piercy, J. E., and Daigle, G. A. (1991). Sound propagation in the open air. In C. M. Harris (Ed.), *Handbook of acoustical measurements and noise control* (pp. 1–26). New York: McGraw-Hill.

Pierrehumbert, J. (1979). The perception of fundamental frequency declination. *Journal of the Acoustical Society of America, 66,* 363–369.

Pilbeam, D. (1984). The descent of hominoids and hominids. *Scientific American, 11,* 84–96.

Pinker, S. (1994a). *The language instinct.* New York: William Morrow.

Pinker, S. (1994b). On language. *Journal of Cognitive Neuroscience, 6,* 92–97.

Pinker, S. (1995). Language acquisition. In L. R. Gleitman, M. Liberman, and D. N. Osherson (Eds.), *An invitation to cognitive science,* 2nd ed., vol. 1, *Language.* Cambridge, MA: MIT Press.

Pinker, S., and Bloom, P. (1990). Natural language and natural selection. *Behavioral and Brain Sciences, 13,* 707–786.

Pisoni, D. B., Aslin, R. N., Perey, A. J., and Hennessy, B. L. (1982). Some effects of laboratory training on identification and discrimination of voicing contrasts in stop consonants. *Journal of Experimental Psychology: Human Perception and Psychophysics, 8,* 297–314.

Pisoni, D. B., and Sawusch, J. R. (1975). Some stages of processing in speech perception. In A. Cohen and S. Nooteboom (Eds.), *Structure and process in speech perception* (pp. 111–132). Heidelberg, Germany: Springer-Verlag.

Ploog, D., Hopf, S., and Winter, P. (1967). Ontogenese des Verhaltens von Totenkopfaffen (*Saimiri sciureus*). *Psychologisches Forschung, 31,* 1–41.

Poeppel, D. (in press). A critical review of PET studies of language. *Brain and Language.*

Poizner, H., Klima, E. S., Bellugi, U., and Livingston, R. (1986). Motion analysis of grammatical processes in a visual-gestural language. In V. McCabe and G. Balzano (Eds.), *Event cognition* (pp. 155–174). Hillsdale, NJ: Lawrence Erlbaum.

Pola, Y., and Snowdon, C. T. (1975). The vocalizations of pygmy marmosets (*Cebuella pygmaea*). *Animal Behaviour, 23,* 826–842.

Pollak, G. D. (1992). Adaptations of basic structures and mechanisms in the cochlea and central auditory pathway of the mustache bat. In D. B. Webster, R. F. Fay, and A. N. Popper (Eds.), *The evolutionary biology of hearing* (pp. 751–778). New York: Springer-Verlag.

Pollak, G. D. (1993). Some comments on the proposed perception of phase and nanosecond time disparities by echolocating bats. *Journal of Comparative Physiology, A, 172,* 523–531.

Pollak, G. D., and Casseday, J. H. (1989). *The neural basis of echolocation in bats.* Berlin: Springer-Verlag.

Poor, H. V. (1988). *An introduction to signal detection and estimation.* New York: Springer-Verlag.

Poran, N. S., and Coss, R. G. (1990). Development of antisnake defenses in California ground squirrels (*Spermophilus beecheyi*): I. Behavioral and immunological relationships. *Behaviour, 112,* 222–245.

Poran, N. S., Coss, R. G., and Benjamini, E. (1987). Resistance of California ground squirrels (*Spermophilus beecheyi*) to the venom of northern Pacific rattlesnake (*Croatalus viridis oreganus*): A study of adaptive variation. *Toxicon, 25,* 767–777.

Posner, M. I., and Carr, T. H. (1992). Lexical access and the brain: Anatomical constraints on cognitive models of word recognition. *American Journal of Psychology, 105,* 1–26.

Posner, M. I., Petersen, S. E., Fox, P. T., and Raichle, M. E. (1988). Localization of cognitive operations in the human brain. *Science, 240,* 1627–1631.

Posner, M. I., and Raichle, M. E. (1994). *Images of mind.* New York: W. H. Freeman.

Poulton, E. B. (1890). *The colour of animals: Their meaning and use.* London: Kegan Paul.

Povinelli, D. J. (1993). Reconstructing the evolution of mind. *American Psychologist, 48,* 493–509.

Povinelli, D. J., and Eddy, T. J. (in press). What young chimpanzees know about seeing. *Monographs of the Society for Research in Child Development.*

Povinelli, D. J., Nelson, K. E., and Boysen, S. T. (1990). Inferences about guessing and knowing by chimpanzees (*Pan troglodytes*). *Journal of Comparative Psychology, 104,* 203–210.

Povinelli, D. J., Parks, K. A., and Novak, M. A. (1991). Do rhesus monkeys (*Macaca mulatta*) attribute knowledge and ignorance to others? *Journal of Comparative Psychology, 105,* 318–325.

Premack, D. (1971). Language in chimpanzees? *Science, 172,* 808–822.

Premack, D. (1978). On the abstractness of human concepts: Why it would be difficult to talk to a pigeon. In S. H. Hulse, H. Fowler, and W. K. Konig (Eds.), *Cognitive processes in animal behavior* (pp. 423–451). Hillsdale, NJ: Lawrence Erlbaum Associates.

Premack, D. (1986). *Gavagai! Or the future history of the animal language controversy.* Cambridge, MA: MIT Press.

Premack, D. (1990). Words: What are they, and do animals have them? *Cognition, 37,* 197–212.

Premack, D., and Dasser, V. (1991). Perceptual origins and conceptual evidence for theory of mind in apes and children. In A. Whiten (Ed.), *Natural theories of mind* (pp. 253–266). Oxford, UK: Basil Blackwell.

Premack, D., and Premack, A. (1983). *The mind of an ape.* New York: Norton.

Premack, D., and Premack, A. J. (1994). Origins of human social competence. In M. Gazzaniga (Ed.), *The cognitive neurosciences* (pp. 205–218). Cambridge, MA: MIT Press.

Premack, D., and Woodruff, G. (1978). Does the chimpanzee have a theory of mind? *Behavioral and Brain Sciences, 4,* 515–526.

Prestwich, K. N., Brugger, K. E., and Topping, M. (1989). Energy and communication in three species of hylid frogs: Power input, power output and efficiency. *Journal of Experimental Biology, 144,* 53–80.

Price, P. H. (1979). Developmental determinants of structure in zebra finch song. *Journal of Comparative Physiology and Psychology, 93,* 268–277.

Proctor, H. C. (1993). Sensory exploitation and the evolution of male mating behaviour: A cladistic test using water mites (Acari: *Parasitengona*). *Animal Behaviour, 44,* 745–752.

Prosen, C. A., Moody, D. B., Sommers, M. S., and Stebbins, W. C. (1990). Frequency discrimination in the monkey. *Journal of the Acoustical Society of America, 88,* 2152–2158.

Pye, J. D. (1983). Echolocation and countermeasures. In B. Lewis (Ed.), *Bioacoustics* (pp. 407–430). London: Academic Press.

Querleu, D., Renard, X., and Crepin, G. (1981). Perception auditive et réactive foetale aux stimulations sonores. *Journal de Gynecologie, Obstetrique et Biologie de Reproduction, 10,* 307–314.

Querleu, D., Renard, X., and Versyp, F. (1985). Vie sensorielle du foetus. In G. Levy and M. Tournaire (Eds.), *L'environment de la naissance* (pp. 114–136). Paris: Vigot.

Quine, W. V. (1973). On the reasons for the indeterminacy of translation. *Journal of Philosophy, 12,* 178–183.

Raichle, M. E. (1992). Mind and brain. *Scientific American, 267,* 48–57.

Raichle, M. E. (1994). Images of the mind: Studies with modern imaging techniques. *Annual Review of Psychology, 45,* 333–356.

Rakic, P. (1995). Evolution of neocortical parcellation: The perspective from experimental neuroembryology. In J.-P. Changeux and J. Chavaillon (Eds.), *Origins of the human brain* (pp. 84–100). Oxford, UK: Oxford University Press.

Rand, A. S. (1988). An overview of anuran acoustic communication. In B. Fritzsch, M. J. Ryan, W. Wilczynski, T. E. Hetherington, and W. Walkowiak (Eds.), *The evolution of the amphibian auditory system* (pp. 415–432). New York: John Wiley & Sons.

Ratnieks, F. L. W., and Visscher, P. K. (1989). Worker policing in the honeybee. *Nature, 342,* 796–797.

Rauschecker, J. P., Tian, B., and Hauser, M. D. (1995). Processing of complex sounds in the macaque nonprimary auditory cortex. *Science, 268,* 111–114.

Rauschecker, J. P., Tian, B., Pons, T., and Mishkin, M. (in press). Serial and parallel processing in macaque monkey auditory cortex. *Journal of Neurophysiology.*

Read, A., and Weary, D. (1992). The evolution of bird song: Comparative analyses. *Proceedings of the Royal Society, London, 338,* 165–187.

Redondo, T., and Castro, F. (1992). Signalling of nutritional need by magpie nestlings. *Ethology, 92,* 193–204.

Reeve, H. K., and Sherman, P. W. (1993). Adaptation and the goals of evolutionary research. *Quarterly Review of Biology, 68,* 1–32.

Reiss, D., and McCowan, B. (1993). Spontaneous vocal mimicry and production by bottlenose dolphins (*Tursiops truncatus*): Evidence for vocal learning. *Journal of Comparative Psychology, 107,* 301–312.

Reissland, N. (1988). Neonatal imitation in the first hour of life: Observations in rural Nepal. *Developmental Psychology, 24,* 464–469.

Remez, R. E. (1980). Susceptibility of a stop consonant to adaptation on a speech-nonspeech continuum: Further evidence against feature detectors in speech perception. *Perception and Psychophysics, 27,* 17–23.

Remez, R. E., Rubin, P. E., Pisoni, D. B., and Carrell, T. D. (1981). Speech perception without traditional speech cues. *Science, 212,* 947–950.

Rey, G. (1983). Concepts and stereotypes. *Cognition, 15,* 237–262.

Rheinländer, J., and Klump, G. (1988). Behavioral aspects of sound localization. In B. Fritzsch, M. J. Ryan, W. Wilczynski, T. E. Hetherington, and W. Walkowiak (Eds.), *The evolution of the amphibian auditory system* (pp. 297–306). New York: John Wiley & Sons.

Richards, D. G., Wolz, J. P., and Herman, L. M. (1984). Vocal mimicry of computer-generated sounds and vocal labeling of objects by a bottlenose dolphin, *Tursiops truncatus. Journal of Comparative Psychology, 98,* 10–28.

Ridley, M. (1983). *The explanation of organic diversity.* Oxford, UK: Clarendon Press.

Ristau, C. (1991a). Aspects of the cognitive ethology of an injury-feigning bird, the piping plover. In C. Ristau (Ed.), *Cognitive ethology: The minds of other animals* (pp. 91–126). Hillsdale, NJ: Lawrence Erlbaum Associates.

Ristau, C. (1991b). *Cognitive ethology: the minds of other animals.* Hillsdale, NJ: Lawrence Erlbaum Associates.

Robb, M. P., and Saxman, J. H. (1985). Developmental trends in vocal fundamental frequency of young children. *Journal of Speech and Hearing Research, 28*, 421–427.

Robbins, T. (1976). *Even cowgirls get the blues.* New York: Bantam Books.

Robert, D., and Hoy, R. R. (1994). Overhearing cricket love songs. *Natural History, 6*, 49–50.

Robertson, J. G. M. (1986). Female choice, male strategies and the role of vocalizations in the Australian frog *Uperoleia rugosa. Animal Behaviour, 34*, 773–784.

Robinson, J. G. (1984). Syntactic structures in the vocalizations of wedge-capped capuchin monkeys, *Cebus nigrivittatus. Behaviour, 90*, 46–79.

Rodman, H. R., Gross, C. G., and Scaidhe, S. P. (1993). Development of brain substrates for pattern recognition in primates: Physiological and connectional studies of inferior temporal cortex in infant monkeys. In B. de Boysson-Bardies, S. de Schonen, P. Jusczyk, P. MacNeilage, and J. Morton (Eds.), *Developmental neurocognition: Speech and face processing in the first year of life* (pp. 63–75). Dordrecht, The Netherlands: Kluwer Academic.

Rodman, H. R., Skelly, J. P., and Gross, C. G. (1991). Stimulus selectivity and state dependence of activity in inferior temporal cortex of infant monkeys. *Proceedings of the National Academy of Sciences, 88*, 7572–7575.

Roitblat, H. L., Harley, H. E., and Helweg, D. A. (1993). Cognitive processes in artificial language research. In H. L. Roitblat, L. M. Herman, and P. E. Nachtigall (Eds.), *Language and communication: Comparative perspective* (pp. 1–24). Hillsdale, NJ: Lawrence Erlbaum Associates.

Roitblat, H. L., Herman, L. M., and Nachtigall, P. E. (1993). *Language and communication: Comparative perspective.* Hillsdale, NJ: Lawrence Erlbaum Associates.

Rolls, E. T., and Baylis, G. C. (1986). Size and contrast have only small effects on the responses to faces of neurons in the cortex of the superior temporal sulcus of the monkey. *Experimental Brain Research, 65*, 38–48.

Rolls, E. T., Baylis, G. C., and Leonard, C. M. (1985). Role of low and high spatial frequencies in the face-selective responses of neurons in the cortex in the superior temporal sulcus in the monkey. *Vision Research, 25*, 1021–1035.

Römer, H. (1992). Ecological constraints for the evolution of hearing and sound communication in insects. In D. B. Webster, R. F. Fay, and A. N. Popper (Eds.), *The evolutionary biology of hearing* (pp. 79–94). New York: Springer-Verlag.

Römer, H., and Bailey, W. J. (1986). Insect hearing in the field: II. Male spacing behaviour and correlated acoustic cues in the bushcricket *Mygalopsis marki. Journal of Comparative Physiology, 159*, 627–638.

Rosch, E. (1975). Cognitive reference points. *Cognitive Psychology, 7*, 532–547.

Rose, G. J., Brenowitz, E. A., and Capranica, R. R. (1985). Species specificity and temperature dependency of temporal processing by the auditory midbrain of two species of treefrogs. *Journal of Comparative Physiology, A, 157*, 763–769.

Ross, L. S., Pollak, G. D., and Zook, J. M. (1988). Origin of ascending projections to an isofrequency region of the mustach bat's inferior colliculus. *Journal of Comparative Neurology, 270*, 488–505.

Roush, R. S., and Snowdon, C. T. (1994). Ontogeny of food-associated calls in cotton-top tamarins. *Animal Behaviour, 47*, 263–273.

Roverud, R. C., and Grinnell, A. D. (1985). Discrimination performance and echolocation signal integration requirements for target detection and distance determination in the CF/FM bat, *Noctilio albiventris. Journal of Comparative Physiology, A, 156*, 447–456.

Rowe, M. P., Coss, R. G., and Owings, D. H. (1986). Rattlesnake rattles and burrowing old hisses: A case of acoustic Batesian mimicry. *Ethology, 72*, 53–71.

Rowell, T. E., and Hinde, R. A. (1962). Vocal communication by the rhesus monkey (*Macaca mulatta*). *Symposium of the Zoological Society of London, 8*, 91–96.

Rowher, S. (1977). Status signaling in Harris' sparrows: Some experiments in deception. *Behaviour, 61*, 107–129.

Rowher, S. (1982). The evolution of reliable and unreliable badges of fighting ability. *American Zoologist, 22*, 531–546.

Rowher, S., and Rowher, F. C. (1978). Status signaling in Harris' sparrows: Experimental deceptions achieved. *Animal Behaviour, 26*, 1012–1022.

Rowher, S., and Røskaft, E. (1989). Results of dyeing male yellow-headed blackbirds solid black: Implications for the arbitrary identity badge hypothesis. *Behavioral Ecology and Sociobiology, 25*, 39–48.

Ruhlen, M. (1994). *The origin of language: Tracing the evolution of the mother tongue.* New York: John Wiley & Sons.

Rumbaugh, D. (1977). Language behavior of apes. In A. M. Schrier (Ed.), *Behavioral primatology* (pp. 105–138). Hillsdale, NJ: Lawrence Erlbaum Associates.

Russell, B. (1961). The uses of language. In R. E. Egner and L. E. Denonn (Eds.), *The basic writings of Bertrand Russell* (pp. 131–136). New York: Simon & Schuster.

Russell, J. A. (1994). Is there universal recognition of emotion from facial expression? A review of cross-cultural studies. *Psychological Bulletin, 115*, 102–141.

Russon, A. E., and Galdikas, B. M. F. (1993). Imitation in free-ranging rehabilitant orangutans (*Pongo pygmaeus*). *Journal of Comparative Psychology, 107*, 147–161.

Russon, A. E., and Galdikas, B. M. F. (1995). Constraints on great apes' imitation: Model and action selectivity in rehabilitant orangutan (*Pongo pygmaeus*) imitation. *Journal of Comparative Psychology, 109*, 5–17.

Ruvolo, M. (1991). Resolution of the African hominoid trichotomy by use of a mitochondrial gene sequence. *Proceedings of the National Academy of Sciences, 88*, 1570–1574.

Ryan, M. J. (1985). *The Túngara frog: A study in sexual selection and communication.* Chicago: University of Chicago Press.

Ryan, M. J. (1986). Neuroanatomy influences speciation rates among anurans. *Proceedings of the National Academy of Sciences, 83*, 1379–1382.

Ryan, M. J. (1988). Constraints and patterns in the evolution of anuran acoustic communication. In B. Fritzsch, M. J. Ryan, W. Wilczynski, T. E. Hetherington, and W. Walkowiak (Eds.), *The evolution of the amphibian auditory system* (pp. 637–678). New York: John Wiley & Sons.

Ryan, M. J. (1990). Sexual selection, sensory systems and sensory exploitation. *Oxford Surveys in Evolutionary Biology, 7*, 157–195.

Ryan, M. J., Bartholomew, G. A., and Rand, S. A. (1983). Energetics of reproduction in a neotropical frog, *Physalaemus pustulosus*. *Ecology, 64*, 1456–1462.

Ryan, M. J., and Brenowitz, E. A. (1985). The role of body size, phylogeny, and ambient noise in the evolution of bird song. *American Naturalist, 126*, 87–100.

Ryan, M. J., Crocroft, R. B., and Wilczynski, W. (1990). The role of environmental selection in intraspecific divergence of mate recognition signals in the cricket frog, *Acris crepitans*. *Evolution, 44*, 1869–1872.

Ryan, M. J., and Drewes, R. C. (1990). Vocal morphology of the *Physalaemus pustulosus* species group (Leptodatylidae). *Biological Journal of the Linnaean Society, 40*, 37–52.

Ryan, M. J., Fox, J. H., Wilczynski, W., and Rand, A. S. (1990). Sexual selection for sensory exploitation in the frog, *Physalaemus pustulosus*. *Nature, 343*, 66–67.

Ryan, M. J., and Keddy-Hector, A. (1992). Directional patterns of female mate choice and the role of sensory biases. *American Naturalist, 139*, S4–S35.

Ryan, M. J., Perrill, S. A., and Wilczynski, W. (1992). Auditory tuning and call frequency predict population-based mating preferences in the cricket frog, *Acris crepitans. American Naturalist, 139,* 1370–1383.

Ryan, M. J., and Rand, A. S. (1993a). Phylogenetic patterns of behavioral mate recognition systems in the *Physalaemus pustulosus* species group (Anura: Leptodactylidae): The role of ancestral and derived characters and sensory exploitation. In D. Lees and D. Edwards (Eds.), *Evolutionary patterns and processes* (pp. 251–267). London: Academic Press.

Ryan, M. J., and Rand, A. S. (1993b). Sexual selection and signal evolution: The ghost of biases past. *Proceedings of the Royal Society, London, B, 340,* 187–195.

Ryan, M. J., and Rand, A. S. (1993c). Species recognition and sexual selection as a unitary problem in animal communication. *Evolution, 47,* 647–657.

Ryan, M. J., and Tuttle, M. D. (1987). The role of prey-generated sound, vision, and echolocation in prey localization by the African bat *Carioderma cor. Journal of Comparative Physiology, 161,* 59–66.

Ryan, M. J., and Wilczynski, W. (1988). Coevolution of sender and receiver: Effect on local mate preference in cricket frogs. *Science, 240,* 1786–1788.

Ryan, M. J., and Wilczynski, W. (1991). Evolution of intraspecific variation in the advertisement call of a cricket frog (*Acris crepitans,* Hylidae). *Biological Journal of the Linnean Society, 44,* 249–271.

Ryle, G. (1949). *The concept of mind.* New York: Barnes and Noble.

Saban, R. (1995). Image of the human fossil brain: Endocranial casts and meningeal vessels in young and adult subjects. In J.-P. Changeux and J. Chavaillon (Eds.), *Origins of the human brain* (pp. 11–38). Oxford, UK: Oxford University Press.

Sackett, G. P. (1966). Monkeys reared in isolation with pictures as visual input: Evidence for an innate releasing mechanism. *Science, 154,* 1468–1473.

Sackett, G. P. (1970). Unlearned responses, differential rearing experiences, and the development of social attachments by rhesus monkeys. In L. A. Rosenblum (Ed.), *Primate behavior,* vol. 1 (pp. 81–98). New York: Academic Press.

Sackheim, H. A., Gur, R. C., and Saucy, M. C. (1978). Emotions are expressed more intensely on the left side of the face. *Science, 202,* 434–436.

Salasoo, A., and Pisoni, D. B. (1985). Sources of knowledge in spoken word identification. *Journal of Memory and Language, 24,* 210–231.

Sales, G. D., and Smith, J. C. (1978). Comparative studies of the ultrasonic calls of infant murid rodents. *Developmental Psychobiology, 11,* 595–619.

Sapolksy, R. M. (1992). Neuroendocrinology of the stress-response. In J. B. Becker, S. M. Breedlove, and D. Crews (Eds.), *Behavioral endocrinology* (pp. 287–324), Cambridge, MA: MIT Press.

Saunders, J. C., and Henry, W. J. (1989). The peripheral auditory system in birds: Structural and functional contributions to auditory perception. In R. J. Dooling and S. H. Hulse (Eds.), *The comparative psychology of audition* (pp. 35–66). Hillsdale, NJ: Lawrence Erlbaum Associates.

Savage-Rumbaugh, E. S. (1986). *Ape language: From conditioned response to symbol.* New York: Columbia University Press.

Savage-Rumbaugh, E. S., Murphy, J., Sevcik, R. A., Brakke, K. E., Williams, S. L., and Rumbaugh, D. M. (1993). Language comprehension in ape and child. *Monographs of the Society for Research in Child Development, 58,* 1–221.

Sawusch, J. R. (1986). Auditory and phonetic coding of speech. In E. C. Schwab and H. C. Nusbaum (Eds.), *Pattern recognition by humans and machines,* vol. 1, *Speech perception* (pp. 51–88). Orlando, FL: Academic Press.

Sayigh, L. S., Tyack, P. L., Wells, R. S., and Scott, M. D. (1990). Signature whistles of free-ranging bottlenose dolphins *Tursiops truncatus:* Stability and mother-offspring comparisons. *Behavioral Ecology and Sociobiology, 26,* 247–260.

Scharff, C., and Nottebohm, F. (1989). Lesions in area X affect song in juvenile but not adult male zebra finches. *Society for Neuroscience Abstracts, 15,* 618.

Scheel, D., and Packer, C. (1991). Group hunting behaviour of lions: A search for cooperation. *Animal Behaviour, 41,* 697–709.

Scherer, K. R. (1986). Vocal affect expression: A review and a model for future research. *Psychological Bulletin, 99,* 143–165.

Scherer, K. R., and Ekman, P. (1982). *Handbook of methods in nonverbal behavior research.* New York: Cambridge University Press.

Scherrer, J. A., and Wilkinson, G. S. (1993). Evening bat isolation calls provide evidence for heritable signatures. *Animal Behaviour, 46,* 847–860.

Schlinger, B. A. (1994). Estrogens and song: Products of the songbird brain. *BioScience, 44,* 605–612.

Schlinger, B. A., and Arnold, A. P. (1991). Brain is the major site of estrogen synthesis in a male songbird. *Proceedings of the National Academy of Sciences, 88,* 4191–4194.

Schlinger, B. A., and Arnold, A. P. (1992). Circulating estrogens in a male songbird originate in the brain. *Proceedings of the National Academy of Sciences, 89,* 7650–7653.

Schmid, E. (1978). Contribution to the morphology and histology of the vocal cords of Central European anurans (Amphibia). *Zoologische Jahrbücher: Abteilung für Anatomie und Ontogonie der Tiere, 99,* 133–150.

Schmidt, R. S. (1965). Larynx control and call production in frogs. *Copeia, 65,* 143–147.

Schmidt, R. S. (1966). Central mechanisms of frog calling. *Behaviour, 26,* 251–285.

Schmidt, R. S. (1968). Preoptic activation of frog mating behavior. *Behaviour, 30,* 239–257.

Schmidt, R. S. (1969). Preoptic activation of mating call orientation in female anurans. *Behaviour, 35,* 114–127.

Schneider, H. (1977). Acoustic behavior and physiology of vocalization in the European tree frog, *Hyla arborea* (L.). In D. H. Taylor and S. I. Guttman (Eds.), *The reproductive biology of amphibians* (pp. 295–335). New York: Plenum Press.

Schneider, H. (1988). Peripheral and central mechanisms of vocalization. In B. Fritzsch, M. J. Ryan, W. Wilczynski, T. E. Hetherington, and W. Walkowiak (Eds.), *The evolution of the amphibian auditory system* (pp. 537–558). New York: John Wiley & Sons.

Schnitzler, H.-U. (1970a). Comparison of the echolocation behavior in *Rhinolophus ferrumequinum* and *Chilonycteris rubiginosa. Bijdragen tot de Dierkunde, 40,* 77–80.

Schnitzler, H.-U. (1970b). Echoortung bei der Fledermaus *Chilonycteris rubiginosa. Zeitschrift für Vergleichende Physiologie, 68,* 25–39.

Schnitzler, H.-U., and Flieger, E. (1983). Detection of oscillating target movements by echolocation in the greater horseshoe bat. *Journal of Comparative Physiology, 153,* 385–391.

Schnitzler, H.-U., Kalko, E. K. V., Kaipf, I., and Grinnell, A. (1994). Fishing and echolocation behavior of the greater bulldog bat, *Noctillo leporinus,* in the field. *Behavioral Ecology and Sociobiology, 35,* 327–345.

Schnitzler, H.-U., Menne, D., Kober, R., and Heblich, K. (1983). The acoustical image of fluttering insects in echolocating bats. In F. Huber and H. Markl (Eds.), *Neuroethology and behavioral physiology: Roots and growing pains* (pp. 235–251). Berlin: Springer-Verlag.

Schön Ybarra, M. (1988). Morphological adaptations for loud phonation in the vocal organ of howling monkeys. *Primate Report, 22,* 19–24.

Schön Ybarra, M. (in press). A comparative approach to the nonhuman primate vocal tract: Implications for sound production. In E. Zimmerman, U. Jurgens, and J. Newman (Eds.), *Current topics in primate vocal communication.* New York: Plenum Press.

Schott, D. (1975). Quantitative analysis of the vocal repertoire of squirrel monkeys. *Zeitschrift für Tierpsychologie, 38,* 225–250.

Schuler, W., and Roper, T. J. (1992). Responses to warning coloration in avian predators. *Advances in the Study of Behavior, 21,* 111–146.

Schuller, G. (1979a). Coding of small sinusoidal frequency and amplitude modulations in the inferior colliculus of the CF-FM bat, *Rhinolophus ferrumequinum. Experimental Brain Research, 34,* 117–132.

Schuller, G. (1979b). Vocalization influences auditory processing in collicular neurons of the CF-FM bat, *Rhinolophus ferrumequinum. Journal of Comparative Physiology, 132,* 39–46.

Schusterman, R. J., Gisiner, R., Grimm, B. K., and Hanggi, E. B. (1993). Behavior control by exclusion and attempts at establishing semanticity in marine mammals using match-to-sample paradigms. In H. L. Roitblat, L. M. Herman, and P. E. Nachtigall (Eds.), *Language and communication: Comparative perspectives* (pp. 249–274). Hillsdale, NJ: Lawrence Erlbaum Associates.

Schwagmeyer, P. L. (1980). Alarm calling behavior of the thirteen-lined ground squirrel, *Spermophilus tridecemlineatus. Behavioral Ecology and Sociobiology, 7,* 195–200.

Searcy, W. A. (1979). Sexual selection and body size in male red-winged blackbirds. *Evolution, 33,* 649–661.

Searcy, W. A. (1983). Response to multiple song types in the male song sparrows and field sparrows. *Animal Behaviour, 31,* 948–949.

Searcy, W. A. (1984). Song repertoire size and female preferences in song sparrows. *Behavioral Ecology and Sociobiology, 14,* 281–286.

Searcy, W. A. (1992). Song repertoire and mate choice in birds. *American Zoologist, 32,* 71–80.

Searcy, W. A., and Andersson, M. (1986). Sexual selection and the evolution of song. *Annual Review of Ecology and Systematics, 17,* 507–533.

Searcy, W. A., and Brenowitz, E. A. (1988). Sexual differences in species recognition of avian song. *Nature, 332,* 152–154.

Searcy, W. A., MacArthur, P. D., and Yasukawa, K. (1985). Song repertoire size and male quality in song sparrows. *Condor, 87,* 222–228.

Searcy, W. A., and Marler, P. (1981). A test for responsiveness to song structure and programming in female sparrows. *Science, 213,* 926–928.

Searcy, W. A., and Marler, P. (1984). Interspecific differences in the response of female birds to song repertoires. *Zeitschrift für Tierpsychologie, 66,* 128–142.

Searcy, W. A., and Marler, P. (1987). Response of sparrows to songs of deaf and isolation-reared males: Further evidence of innate auditory templates. *Developmental Psychobiology, 20,* 509–519.

Searcy, W. A., Marler, P., and Peters, S. S. (1985). Songs of isolation-reared sparrows function in communication, but are significantly less effective than learned songs. *Behavioral Ecology and Sociobiology, 17,* 223–229.

Searcy, W. A., and Yasukawa, K. (1990). Use of the song repertoire in intersexual and intrasexual contexts by male red-winged blackbirds. *Behavioral Ecology and Sociobiology, 27,* 123–128.

Seeley, T. D. (1989). The honey bee colony as a superorganism. *American Scientist, 77,* 546–553.

Seeley, T. D. (1992). The tremble dance of the honey bee: Message and meanings. *Behavioral Ecology and Sociobiology, 31,* 375–383.

Seller, T. J. (1983). Control of sound production in birds. In B. Lewis (Ed.), *Bioacoustics: A comparative approach* (pp. 93–124). London: Academic Press.

Serafin, J. V., Moody, D. B., and Stebbins, W. C. (1982). Frequency selectivity of the monkey's auditory system: Psychophysical tuning curves. *Journal of the Acoustical Society of America, 71,* 1513–1518.

Sergent, J., Ohta, S., and MacDonald, B. (1992). Functional neuroanatomy of face and object processing: A positron emission tomography study. *Brain, 115*, 15–36.

Sergent, J., and Signoret, J.-L. (1992). Functional and anatomical decomposition of face processing: Evidence from prosopagnosia and PET study of normal subjects.

Sergent, J., Zuck, E., Terriah, S., and MacDonald, B. (1992). Positron emission tomography study of letter and object processing: Empirical findings and methodological considerations. *Cerebral Cortex, 2*, 68–80.

Seyfarth, R. M., and Cheney, D. L. (1980). The ontogeny of vervet monkey alarm-calling behavior: A preliminary report. *Zeitschrift für Tierpsychologie, 54*, 37–56.

Seyfarth, R. M., and Cheney, D. L. (1984). The acoustic features of vervet monkey grunts. *Journal of the Acoustical Society of America, 75*, 129–134.

Seyfarth, R. M., and Cheney, D. L. (1986). Vocal development in vervet monkeys. *Animal Behaviour, 34*, 1640–1658.

Seyfarth, R. M., and Cheney, D. L. (1990). The assessment by vervet monkeys of their own and another species' alarm calls. *Animal Behaviour, 40*, 754–764.

Seyfarth, R. M., and Cheney, D. L. (in press). Some general features of vocal development in nonhuman primates. In C. T. Snowdon and M. Hausberger (Eds.), *Social influences on vocal development*. Cambridge, UK: Cambridge University Press.

Seyfarth, R. M., Cheney, D. L., and Marler, P. (1980a). Monkey responses to three different alarm calls: Evidence of predator classification and semantic communication. *Science, 210*, 801–803.

Seyfarth, R. M., Cheney, D. L., and Marler, P. (1980b). Vervet monkey alarm calls: Semantic communication in a free-ranging primate. *Animal Behaviour, 28*, 1070–1094.

Shalter, M. D. (1978). Localisation of passerine seet and mobbing calls by goshawks and pygmy owls. *Zeitschrift für Tierpsychologie, 46*, 260–267.

Shannon, C. E., and Weaver, W. (1949). *The mathematical theory of communication*. Urbana: University of Illinois.

Shaywitz, B. A., Shaywitz, S. E., Pugh, K. R., Constable, R. T., Skularski, P., Fulbright, R. K., Bronen, R. A., Fletcher, J. M., Shankweiler, D. P., Katz, L., and Gore, J. C. (1995). Sex differences in the functional organization of the brain for language. *Nature, 373*, 607–609.

Sherman, P. W. (1977). Nepotism and the evolution of alarm calls. *Science, 197*, 1246–1253.

Sherman, P. W. (1981). Kinship, demography, and Belding's ground squirrel nepotism. *Behavioral Ecology and Sociobiology, 8*, 251–259.

Sherman, P. W. (1985). Alarm calls of Belding's ground squirrels to aerial predators: Nepotism or self-preservation. *Behavioral Ecology and Sociobiology, 17*, 313–323.

Sherry, D. F., and Galef, B. G., Jr. (1984). Cultural transmission without imitation: Milk bottle opening by birds. *Animal Behaviour, 32*, 937–938.

Sherry, D. F., and Galef, B. G., Jr. (1990). Social learning without imitation: More about milk bottle opening by birds. *Animal Behaviour, 40*, 987–989.

Sillen-Tullberg, B. (1985a). Higher survival of an aposematic than of a cryptic form of a distasteful bug. *Oecologia, 67*, 411–415.

Sillen-Tullberg, B. (1985b). The significance of coloration per se, independent of background, for predator avoidance of aposematic prey. *Animal Behaviour, 33*, 1382–1384.

Sillen-Tullberg, B. (1988). Evolution of gregariousness in aposematic butterfly larvae: A phylogenetic analysis. *Evolution, 42*, 293–305.

Sillen-Tullberg, B., and Møller, A. P. (1993). The relationship between concealed ovulation and mating systems in anthropoid primates: A phylogenetic analysis. *American Naturalist, 141*, 1–25.

Simmons, J. A. (1971). Echolocation in bats: Signal processing of echoes for target range. *Science, 171*, 925–928.

Simmons, J. A. (1973). The resolution of target range by echolocating bats. *Journal of the Acoustical Society of America, 54*, 157–173.

Simmons, J. A. (1979). Perception of echo phase information in bat sonar. *Science, 204*, 1336–1338.

Simmons, J. A. (1989). A view of the world through the bat's ear: The formation of acoustic images in echolocation. *Cognition, 33*, 155–199.

Simmons, J. A., Ferragamo, M., Moss, C. F., Stevenson, S. B., and Altes, R. A. (1990). Discrimination of jittered sonar echoes by the echolocating bat, *Eptesicus fuscus:* The shape of target images in echolocation. *Journal of Comparative Physiology, A, 167*, 589–616.

Simmons, J. A., Howell, D. J., and Suga, N. (1975). Information content of bat sonar echoes. *American Scientist, 63*, 204–215.

Simmons, J. A., Kick, S. A., Lawrence, B. D., Hale, C., and Escudie, B. (1983). Acuity of horizontal angle discrimination by the echolocating bat, *Eptesicus fuscus. Journal of Comparative Physiology, 153*, 321–330.

Simmons, J. A., Saillant, P. A., Dear, S. P., and Ferragamo, M. J. (1994). Auditory dimensions of acoustic images in echolocation. In R. R. Fay and A. N. Popper (Eds.), *Handbook of auditory research: Hearing by bats* (pp. 89–125). New York: Springer-Verlag.

Simpson, H. B., and Vicario, D. S. (1991). Early estrogen treatment alone causes female zebra finches to produce learned, male-like vocalizations. *Journal of Neurobiology, 22*, 755–776.

Sinnott, J. M. (1989a). Internal cognitive structures guide birdsong perception. In R. J. Dooling and S. H. Hulse (Eds.), *Comparative psychology of audition: Perceiving complex sounds* (pp. 447–464). Hillsdale, NJ: Lawrence Erlbaum Associates.

Sinnott, J. M. (1989b). Detection and discrimination of synthetic English vowels by Old World monkeys (*Cercopithecus, Macaca*) and humans. *Journal of the Acoustical Society of America, 86*, 557–565.

Sinnott, J. M., Petersen, M. R., and Hopp, S. L. (1985). Frequency and intensity discrimination in humans and monkeys. *Journal of the Acoustical Society of America, 78*, 1977–1985.

Slater, P. J. B. (1973). Describing sequences of behavior. In P. P. G. Bateson and P. H. Klopfer (Eds.), *Perspectives in ethology*, vol. 1 (pp. 131–153). New York: Plenum Press.

Slater, P. J. B. (1983). The study of communication. In T. R. Halliday and P. J. B. Slater (Eds.), *Animal behaviour*, vol. 2, *Communication* (pp. 9–42). Oxford, UK: Blackwell Scientific.

Slater, P. J. B., Jones, A., and ten Cate, C. (1993). Can lack of experience delay the end of the sensitive phase for song learning? *Netherlands Journal of Zoology, 43*, 80–90.

Slobodchikoff, C. N., Kiriazis, J., Fischer, C., and Creef, E. (1991). Semantic information distinguishing individual predators in the alarm calls of Gunnison's prairie dogs. *Animal Behaviour, 42*, 713–719.

Smith, C. A., Konishi, M., and Schull, N. (1985). Structure of the barn owl's (*Tyto alba*) inner ear. *Hearing Research, 17*, 237–247.

Smith, D. G., and Reid, F. A. (1979). Roles of the song repertoire in red-winged blackbirds. *Behavioral Ecology and Sociobiology, 5*, 279–290.

Smith, E. E. (in press). Concepts and categorization. In D. Osherson and E. E. Smith (Eds.), *Invitation to cognitive science*, vol. 3, *Thinking*. Cambridge, MA: MIT Press.

Smith, E. E., and Medin, D. L. (1981). *Categories and concepts*. Cambridge, MA: Harvard University Press.

Smith, J. N. M. (1988). Determinants of lifetime reproductive success in the song sparrow. In T. H. Clutton-Brock (Eds.), *Reproductive success* (pp. 154–172). Chicago: Chicago University Press.

Smith, W. J. (1969). Messages of vertebrate communication. *Science, 165*, 145–150.

Smith, W. J. (1977). *The behavior of communicating.* Cambridge, MA: Harvard University Press.

Smith, W. J. (1991). Animal communication and the study of cognition. In C. A. Ristau (Ed.), *Cognitive ethology: The minds of other animals.* (pp. 209–230). Hillsdale, NJ: Lawrence Erlbaum Associates.

Smolker, R., Mann, J., and Smuts, B. B. (1993). Use of signature whistles during separation and reunions by wild bottlenose dolphin mothers and infants. *Behavioral Ecology and Sociobiology, 33,* 393–402.

Smythe, N. (1977). The function of mammalian alarm advertising: Social signals or pursuit invitation? *American Naturalist, 111,* 191–194.

Snow, C. E., and Ferguson, C. A. (1977). *Talking to children: Language input and acquisition.* Cambridge, UK: Cambridge University Press.

Snowdon, C. T. (1979). Response of nonhuman animals to speech and to species-specific sounds. *Brain, Behavior and Evolution, 16,* 409–429.

Snowdon, C. T. (1982). Linguistic and psycholinguistic approaches to primate communication. In C. T. Snowdon, C. H. Brown, and M. R. Petersen (Eds.), *Primate communication* (pp. 212–238). New York: Cambridge University Press.

Snowdon, C. T. (1987). A naturalistic view of categorical perception. In S. Harnad (Ed.), *Categorical perception* (pp. 332–354). Cambridge, UK: Cambridge University Press.

Snowdon, C. T. (1988). Communication as social interaction: Its importance in ontogeny and adult behavior. In D. Todt, P. Goedeking, and D. Symmes (Eds.), *Primate vocal communication* (pp. 108–122). Berlin: Springer-Verlag.

Snowdon, C. T. (1990). Language capacities of nonhuman animals. *Yearbook of Physical Anthropology, 33,* 215–243.

Snowdon, C. T., and Cleveland, J. (1984). "Conversations" among pygmy marmosets? *American Journal of Primatology, 7,* 15–20.

Snowdon, C. T., and Elowson, A. M. (1992). Ontogeny of primate vocal communication. In T. Nishida, F. B. M. deWaal, W. McGrew, P. Marler, and M. Pickford (Eds.), *Topics in primatology,* vol. 1, *Human origins* (pp. 279–290). Tokyo: Tokyo University Press.

Snowdon, C. T., and Pola, Y. V. (1978). Interspecific and intraspecific responses to synthesized pygmy marmoset vocalizations. *Animal Behaviour, 26,* 192–206.

Sohrabji, F., Nordeen, E. J., and Nordeen, K. W. (1990). Selective impairment of song learning following lesions of a forebrain nucleus in juvenile zebra finches. *Neural Behavior and Biology, 53,* 51–63.

Sommers, M. S., Moody, D. B., Prosen, C. A., and Stebbins, W. C. (1992). Formant frequency discrimination by Japanese macaques (*Macaca fuscata*). *Journal of the Acoustical Society of America, 91,* 3499–3510.

Spelke, E. S. (1979). Perceiving bimodally specified events in infancy. *Developmental Psychology, 15,* 626–636.

Spelke, E. S. (1991). Physical knowledge in infancy: Reflections on Piaget's theory. In S. Carey and R. Gelman (Eds.), *The epigenesis of mind: Essays on biology and cognition* (pp. 37–61). Hillsdale, NJ: Lawrence Erlbaum Associates.

Spelke, E. S. (1994). Initial knowledge: Six suggestions. *Cognition, 50,* 431–445.

Spelke, E. S., Born, W. S., and Chu, F. (1983). Perception of moving, sounding objects by 4-month-old infants. *Perception, 12,* 719–732.

Sperber, D., Premack, D., and Premack, A. (1995). *Causal cognition.* Oxford: Oxford University Press.

Stammbach, E. (1988). Group responses to specially skilled individuals in a *Macaca fascicularis.* *Behaviour, 107,* 241–266.

Stebbins, W. C., and Sommers, M. S. (1992). Evolution, perception and the comparative method. In D. B. Webster, R. F. Fay, and A. N. Popper (Eds.), *The evolutionary biology of hearing* (pp. 211–228). New York: Springer-Verlag.

Steger, R., and Caldwell, R. L. (1983). Intraspecific deception by bluffing: A defense strategy of newly molted stomatopods (Arthropods: Crustacea). *Science, 221,* 558–560.

Stein, B. E., and Meredith, M. A. (1993). *The merging of the senses.* Cambridge, MA: MIT Press/Bradford Books.

Stein, R. D. (1968). Modulation in bird sounds. *Auk, 85,* 229–243.

Stetson, R. H. (1951). *Motor phonetics: A study of speech movements in action.* Amsterdam: North-Holland.

Stevens, K. N. (1972). Quantal nature of speech. In E. E. David and P. B. Denes (Eds.), *Human communication: A unified view* (pp. 67–84). New York: McGraw Hill.

Stevens, K. N., and Halle, M. (1967). Remarks on analysis by synthesis and distinctive features. In W. Wathen-Dunn (Ed.), *Models for the perception of speech and visual form* (pp. 88–102). Cambridge, MA: MIT Press.

Stewart, K. J., and Harcourt, A. H. (1994). Gorillas' vocalizations during rest periods: Signals of impending departure? *Behaviour, 130,* 29–40.

Stiebler, I. B., and Narins, P. M. (1990). Temperature-dependence on auditory nerve response properties in the frog. *Hearing Research, 46,* 63–82.

Stokes, A. W. (1962). Agonistic behaviour among blue tits at a winter feeding station. *Behaviour, 19,* 118–138.

Struhsaker, T. T. (1967). Auditory communication among vervet monkeys (*Cercopithecus aethiops*). In S. A. Altmann (Ed.), *Social communication among primates* (pp. 281–324). Chicago: Chicago University Press.

Studdert-Kennedy, M. (1981). The beginnings of speech. In K. Immelman, G. W. Barlow, L. Petrinovich, and M. Main (Eds.), *Behavioral development* (pp. 533–564). Cambridge, UK: Cambridge University Press.

Studdert-Kennedy, M. (1986a). Some developments in research on language behavior. In N. J. Smelser and D. R. Gerstein (Eds.), *Behavioral and social science: Fifty years of discovery* (pp. 208–248). Washington, DC: National Academy Press.

Studdert-Kennedy, M. (1986b). Sources of variability in early speech development. In J. S. Perkell and D. H. Klatt (Eds.), *Invariance and variability in speech processes* (pp. 58–76). Hillsdale, NJ: Lawrence Erlbaum Associates.

Studdert-Kennedy, M. (1991). Language development from an evolutionary perspective. In N. A. Krasnegor, D. M. Rumbaugh, R. L. Schiefelbusch, and M. Studdert-Kennedy (Eds.), *Biological and behavioral determinants of language development* (pp. 5–28). Hillsdale, NJ: Lawrence Erlbaum Associates.

Suga, N. (1988). Auditory neuroethology and speech processing: Complex sound processing by combination-sensitive neurons. In G. M. Edelman, W. E. Gall, and W. M. Cowan (Eds.), *Auditory function* (pp. 679–720). New York: Wiley Liss Press.

Suga, N., Niwa, H., Taniguchi, I., and Margoliash, D. (1987). The personalized auditory cortex of the mustached bat: Adaptation for echolocation. *Journal of Neurophysiology, 58,* 643–654.

Sundberg, J. (1987). *The science of the singing voice.* DeKalb, IL: Northern Illinois University Press.

Surlykke, A., and Miller, L. A. (1985). The influence of arctiid moth clicks on bat echolocation: Jamming or warning? *Journal of Comparative Physiology, A, 156,* 831–843.

Surlykke, A., Miller, L. A., Møhl, B., Andersen, B. B., Christensen-Dalsgaard, J., and Jorgensen, M. B. (1993). Echolocation in two very small bats from Thailand: *Craseonycteris thonglongyai* and *Myotis siligorensis. Behavioral Ecology and Sociobiology, 33,* 1–12.

Suthers, R. A. (1988). The production of echolocation signals by bats and birds. In P. E. Nachtigall and P. W. B. Moore (Eds.), *Animal sonar: Processes and performance* (pp. 23–45). New York: Plenum Press.

Suthers, R. A. (1990). Contributions to birdsong from the left and right sides of the intact syrinx. *Nature, 347,* 473–477.

Suthers, R. A., and Fattu, J. M. (1973). Mechanisms of sound production in echolocating bats. *American Zoologist, 13,* 1215–1226.

Suthers, R. A., and Fattu, J. M. (1982). Selective laryngeal neuroanatomy and control of phonation by the echolocating bat, *Eptesicus. Journal of Comparative Physiology, 145,* 529–537.

Suthers, R. A., Goller, F., and Hartley, R. S. (1994). Motor dynamics of song production by mimic thrushes. *Journal of Neurobiology, 25,* 917–936.

Suthers, R. A., Hartley, D. J., and Wenstrup, J. J. (1988). The acoustic role of tracheal chambers and nasal cavities in the production of sonar pulses by the horseshoe bat, *Rhilophus hildebrandti. Journal of Comparative Physiology, A, 162,* 799–813.

Suthers, R. A., and Hector, D. H. (1982). Mechanism for the production of echolocating clicks by the grey swiftlet, *Collocalia spodiopygia. Journal of Comparative Physiology, A, 148,* 457–470.

Suthers, R. A., and Hector, D. H. (1985). The physiology of vocalization by the echolocating oilbird, *Steatornis caripensis. Journal of Comparative Physiology, A, 156,* 243–266.

Suthers, R. A., and Hector, D. H. (1988). Individual variation in vocal tract resonance may assist oilbirds in recognizing echoes of their own sonar clicks. In P. E. Nachtigall and P. W. B. Moore (Eds.), *Animal sonar: Processes and performances* (pp. 87–91). New York: Plenum Press.

Sutton, D. (1979). Mechanisms underlying vocal control in nonhuman primates. In H. Steklis and M. Raleigh (Eds.), *Neurobiology of social communication in primates* (pp. 45–68). New York: Academic Press.

Symmes, D., and Biben, M. (1988). Conversational vocal exchanges in squirrel monkeys. In D. Todt, P. Goedeking, and D. Symmes (Eds.), *Primate vocal communication* (pp. 123–132). Berlin: Springer-Verlag.

Talmage-Riggs, G., Winter, P., Ploog, W., and Mayer, W. (1972). Effect of deafening on the vocal behavior of the squirrel monkey (*Saimiri sciureus*). *Folia primatologica, 17,* 404–420.

Tartter, V. C. (1980). Happy talk: Perceptual and acoustic effects of smiling. *Perception and Psychophysics, 27,* 24–27.

Tartter, V. C., and Braun, D. (1994). Hearing smiles and frowns in normal and whisper registers. *Journal of the Acoustical Society of America, 96,* 2101–2107.

Taub, D. M. (1980). Female choice and mating strategies among wild Barbary macaques (*Macaca sylvanus*). In D. G. Lindburg (Ed.), *The macaques: Studies in ecology, behavior and evolution* (pp. 13–45). New York: Van Nostrand Reinhold.

Teleki, G. (1973). *The predatory behavior of wild chimpanzees.* Lewisburg, PA: Bucknell University Press.

ten Cate, C. (1994). Perceptual mechanisms in imprinting and song learning in birds. In J. A. Hogan and J. J. Bolhuis (Eds.), *Causal mechanisms of behavioural development* (pp. 89–115). Cambridge, UK: Cambridge University Press.

Terborgh, J., and Janson, C. (1986). The socioecology of primate groups. *Annual Review of Ecology and Systematics, 17,* 111–135.

Terrace, H. S. (1979). *Nim.* New York: Knopf.

Thomas, S. P. (1987). The physiology of bat flight. In M. B. Fenton, P. Racey, and J. M. V. Rayner (Eds.), *Recent advances in the study of bats* (pp. 75–99). Cambridge, UK: Cambridge University Press.

Thomas, S. P., and Suthers, R. A. (1972). The physiology and energetics of bat flight. *Journal of Experimental Biology, 57*, 317–335.

Thornhill, R. (1992). Female preference of the pheromone of males with low fluctuating asymmetry in the Japanese scorpionfly (*Panorpa japonica*). *Behavioral Ecology, 3*, 277–283.

Thornhill, R., and Gagestad, S. W. (1993). Human facial beauty: Averageness, symmetry, and parasite resistance. *Human Nature, 4*, 237–270.

Thorpe, W. H. (1959). Talking birds and the mode of action of the vocal apparatus of birds. *Proceedings of the Zoological Society of London, 132*, 441–455.

Thorpe, W. H. (1972). Duetting and antiphonal song in birds. *Behaviour, 18*, 1–197.

Thorson, J., Weber, T., and Huber, F. (1982). Auditory behavior of the cricket: II. Simplicity of calling-song recognition in *Gryllus,* and anomalous phonotaxis at abnormal carrier frequency. *Journal of Comparative Physiology, 146*, 361–378.

Tilson, R. L., and Norton, P. M. (1981). Alarm duetting and pursuit deterence in an African antelope. *American Naturalist, 118*, 455–462.

Tinbergen, N. (1952). Derived activities: Their causation, biological significance, origin and emancipation during evolution. *Quarterly Review of Biology, 27*, 1–32.

Tinbergen, N. (1953). *The herring gull's world.* London: Collins.

Titze, I. R. (1976). On the mechanisms of the vocal fold vibration. *Journal of the Acoustical Society of America, 60*, 1366–1980.

Titze, I. R. (1989). On the relation between sub-glottal pressure and fundamental frequency in phonation. *Journal of the Acoustical Society of America, 85*, 901–906.

Todd, P. M., and Miller, G. F. (1991). Exploring adaptive agency: II. Simulating the evolution of associative learning. In J. A. Meyer, H. L. Roitblat, and S. W. Wilson (Eds.), *From animals to animats* (pp. 132–141). Cambridge, MA: MIT Press.

Todt, D. (1981). On the functions of vocal matching: Effect of counter-replies on song post choice and singing. *Zeitschrift für Tierpsychologie, 57*, 73–93.

Todt, D., Goedeking, P., and Symmes, D. (1988). *Primate vocal communication.* Berlin: Springer-Verlag.

Tomasello, M. (1990). Cultural transmission in the tool use and communication signalling of chimpanzees. In S. Parker and K. Gibson (Eds.), *Language and intelligence in monkeys and apes* (pp. 274–311). New York: Cambridge University Press.

Tomasello, M., Call, J., Nagell, K., Olguin, R., and Carpenter, M. (1994). The learning and use of gestural signals by young chimpanzees: A trans-generational study. *Primates, 35*, 137–154.

Tomasello, M., Kruger, A., and Ratner, H. (1993). Cultural learning. *Behavioral and Brain Sciences, 16*, 495–552.

Tomasello, M., Savage-Rumbaugh, E. S., and Kruger, A. (1993). Imitative learning of actions on objects by children, chimpanzees, and enculturated chimpanzees. *Child Development, 64*, 1688–1706.

Tovee, M. J. (1995). What are faces for? *Current Biology, 5*, 480–482.

Tovee, M. J., and Cohen-Tovee, E. M. (1993). The neural substrate of face processing models: A review. *Cognitive Neuropsychology, 10*, 505–528.

Towne, W. F. (1985). Acoustic and visual cues in the dances of four honey bee species. *Behavioral Ecology and Sociobiology, 16*, 185–187.

Towne, W. F., and Kirchner, W. H. (1989). Hearing in honey bees: Detection of air-particle oscillations. *Science, 244*, 686–688.

Tranel, D., and Damasio, A. R. (1985). Knowledge without awareness: An autonomic index of recognition of prosapagnosics. *Science, 228*, 1453–1454.

Tranel, D., and Damasio, A. R. (1993). The covert learning of affective valence does not require structures in hippocampal system of amydala. *Journal of Cognitive Neuroscience, 5,* 79–88.

Tranel, D., Damasio, A. R., and Damasio, H. (1988). Intact recognition of facial expression, gender, and age in patients with impaired recognition of face identity. *Neurology, 38,* 690–696.

Trivers, R. L. (1971). The evolution of reciprocal altruism. *Quarterly Review of Biology, 46,* 35–57.

Trivers, R. L. (1972). Parental investment and sexual selection. In B. Campbell (Ed.), *Sexual selection and the descent of man* (pp. 136–179). Chicago: Aldine Press.

Trivers, R. L. (1974). Parent-offspring conflict. *American Zoologist, 14,* 249–264.

Turkewitz, G. (1991). Perinatal influences on the development of hemispheric specialization and complex information processing. In M. J. S. Weiss and P. R. Zelazo (Eds.), *Newborn attention, biological constraints and the influence of experience* (pp. 443–465). Norwood, NJ: Ablex.

Turkewitz, G. (1993). The origins of differential hemispheric strategies for information processing in the relationships between voice and face perception. In B. de Boysson-Bardies, S. de Schonen, P. Jusczyk, P. MacNeilage, and J. Morton (Eds.), *Developmental neurocognition: Speech and face processing in the first year of life* (pp. 165–170). Dordrecht, The Netherlands: Kluwer Academic.

Tuttle, M. D., and Ryan, M. J. (1981). Bat predation and the evolution of frog vocalizations in the neotropics. *Science, 214,* 677–678.

Tversky, A. (1977). Features of similarity. *Psychological Review, 84,* 327–352.

Tyack, P. L. (1993). Animal language research needs a broader comparative evolutionary framework. In H. L. Roitblat, L. M. Herman, and P. E. Nachtigall (Eds.), *Language and communication* (pp. 115–152). Hillsdale, NJ: Lawrence Erlbaum Associates.

Urano, A., and Gorbman, A. (1981). Effects of pituitary hormonal treatment on responsiveness of anterior preoptic neurons in male leopard frogs, *Rana pipiens. Journal of Comparative Physiology, A, 141,* 163–172.

van Hoof, J. A. R. A. M. (1972). A comparative approach to the phylogeny of laughter and smiling. In R. A. Hinde (Ed.), *Nonverbal communication* (pp. 12–53). Cambridge, UK: Cambridge University Press.

van Hooff, J. A. R. A. M. (1982). Categories and sequences of behavior: Methods of description and analysis. In K. R. Scherer and P. Ekman (Eds.), *Handbook of methods in nonverbal behavior research* (pp. 362–439). Cambridge, UK: Cambridge University Press.

van Rhijn, J. G., and Vodegel, R. (1980). Being honest about one's intentions: An evolutionary stable strategy for animal conflicts. *Journal of Theoretical Biology, 85,* 623–641.

Van Valen, L. (1962). A study of fluctuating asymmetry. *Evolution, 16,* 125–142.

Vauclair, J., Fagot, J., and Hopkins, W. (1993). Mental images in baboons when the visual input is directed to the left cerebral hemisphere. *Psychological Science, 4,* 99–103.

Vehrencamp, S. L., Bradbury, J. W., and Gibson, R. M. (1989). The energetic cost of display in male sage grouse. *Animal Behaviour, 38,* 885–896.

Vicario, D. S. (1991). Contributions of syringeal muscles to respiration and vocalization in the zebra finch. *Journal of Neurobiology, 22,* 63–73.

Vihman, M. M. (1986). Individual differences in babbling and early speech: Predicting to age three. In B. Lindblom and R. Zetterström (Eds.), *Precursors of early speech* (pp. 95–112). New York: Stockton Press.

Vihman, M. M., and McCune, L. (1994). When is a word a word? *Journal of Child Language, 21,* 517–542.

Visalberghi, E., and Fragaszy, D. (1991). Do monkeys ape? In S. T. Parker and K. R. Gibson (Eds.), *"Language" and intelligence in monkeys and apes* (pp. 247–273). Cambridge, UK: Cambridge University Press.

Vivian, R. (1966). The Tadoma method: A tactual approach to speech and speech reading. *Volta Review, 68,* 733–737.

von Frisch, K. (1950). *Bees: Their vision, chemical senses and language.* Ithaca, NY: Cornell University Press.

von Frisch, K. (1967a). *The dance language and orientation of bees.* Cambridge, MA: Belknap Press of Harvard University Press.

von Frisch, K. (1967b). Honeybees: Do they use direction and distance information provided by their dancers? *Science, 158,* 1073–1076.

Vu, E. T., Mazurek, M. E., and Kuo, Y. (1994). Identification of a forebrain motor programming network for the learned song of zebra finches. *Journal of Neuroscience, 14,* 6924–6934.

Waddington, C. H. (1957). *The strategy of the genes.* London: Allen and Unwin.

Waddington, K. D., and Kirchner, W. H. (1992). Acoustical and behavioral correlates of profitability of food sources in honey bee round dances. *Ethology, 92,* 1–6.

Wagner, W. E., Jr. (1992). Deceptive or honest signalling of fighting ability? A test of alternative hypotheses for the function of changes in call dominant frequency by male cricket frogs. *Animal Behaviour, 44,* 449–462.

Walkowiak, W. (1988). Neuroethology of anuran call recognition. In B. Fritzsch, M. J. Ryan, W. Wilczynski, T. E. Hetherington, and W. Walkowiak (Eds.), *The evolution of the amphibian auditory system* (pp. 485–510). New York: John Wiley & Sons.

Wallace, A. R. (1867). Insect defenses. *Proceedings of the Entomological Society, 3,* lxxx–lxxxi.

Wallman, J. (1992). *Aping language.* New York: Cambridge University Press.

Walton, G. E., Bower, N. J. A., and Bower, T. G. R. (1992). Recognition of familiar faces by newborns. *Infant Behavior and Development, 15,* 265–269.

Walton, G. E., and Bower, T. G. R. (1993). Newborns form "prototypes" in less than 1 minute. *Psychological Science, 4,* 203–205.

Warrington, E. K., and James, M. (1967). An experimental investigation of facial recognition in patients with unilateral cerebral lesions. *Cortex, 3,* 317–326.

Waser, P. M. (1977a). Experimental playbacks show vocal mediation of avoidance on a forest monkey. *Nature, 255,* 56–58.

Waser, P. M. (1977b). Sound localization by monkeys: A field experiment. *Behavioral Ecology and Sociobiology, 2,* 427–431.

Waser, P. (1987). Interactions among primate species. In B. B. Smuts, D. L. Cheney, R. M. Seyfarth, R. W. Wrangham, and T. T. Struhsaker (Eds.), *Primate societies* (pp. 210–226). Chicago: Chicago University Press.

Waser, P. M., and Brown, C. H. (1984). Is there a "sound window" for primate communication? *Behavioral Ecology and Sociobiology, 15,* 73–76.

Weary, D. M., Krebs, J. R., Eddyshaw, R., McGregor, P. K., and Horn, A. (1988). Decline in song output by great tits: Exhaustion or motivation? *Animal Behaviour, 36,* 1242–1244.

Weary, D. M., Lemon, R. E., and Perreault, S. (1992). Song repertoires do not hinder neighbour-stranger discrimination. *Behavioral Ecology and Sociobiology, 31,* 441–447.

Weber, E. (1976). Die Veränderung der Paarungs und Revierrufe von *Hyla arborea savignyi* AUDOUIN (Anura) nach Ausschaltung von Kehlkopfmuskeln. *Bonn Zoologische Beitreitung, 27,* 87–97.

Webster, F. A., and Durlach, N. I. (1963). *Echolocation system of the bat.* No. 41-G-3. Massachusetts Institute of Technology.

Weinberger, N., and Diamond, D. (1988). Dynamic modulation of the auditory system by associative learning. In G. M. Edelman, W. E. Gall, and W. M. Cowan (Eds.), *Auditory function— Neurobiological bases of hearing* (pp. 485–512). New York: John Wiley and Sons.

Weiskrantz, L. (1988). *Thought without language*. Oxford, UK: Oxford University Press.

Wellman, H. M., and Hickling, A. K. (1994). The mind's "I": Children's conception of the mind as an active agent. *Child Development, 65,* 1564–1580.

Wells, K. D., and Taigen, T. L. (1986). The effect of social interactions on calling energetics in the gray tree frog (*Hyla versicolor*). *Behavioral Ecology and Sociobiology, 19,* 9–18.

Wenner, A. M. (1962). Sound production during the waggle dance of the honey bee. *Animal Behaviour, 10,* 79–95.

Werker, J. F. (1989). Becoming a native listener. *American Scientist, 77,* 54–59.

Werker, J. F., Gilbert, J. H. V., Humphrey, K., and Tees, R. C. (1981). Developmental aspects of cross-language speech perception. *Child Development, 52,* 349–353.

Werker, J. F., and Lalonde, C. E. (1988). Cross-language speech perception: Initial capabilities and developmental change. *Developmental Psychology, 24,* 672–683.

Werker, J. F., and Logan, J. S. (1985). Cross-language evidence for three factors in speech perception. *Perception and Psychophysics, 37,* 35–44.

Werker, J. F., and Tees, R. C. (1983). Developmental changes across childhood in the perception of non-active speech sounds. *Canadian Journal of Psychology, 37,* 278–286.

Werker, J. F., and Tees, R. C. (1984). Cross-language speech perception: Evidence for perceptual reorganization during the first year of life. *Infant Behavior and Development, 7,* 49–63.

Wernicke, C. (1874). *Der aphasische symptomenkomplex*. Breslau: Cohn and Weigert.

West, C., and Zimmerman, D. H. (1982). Conversation analysis. In K. R. Scherer and P. Ekman (Eds.), *Handbook of methods in nonverbal behavior research* (pp. 506–541), Cambridge, UK: Cambridge University Press.

West, M. J., and King, A. P. (1988). Female visual displays affect the development of male song in the cowbird. *Nature, 334,* 244–246.

Westneat, M. W., Long, J. H. J., Hoese, W., and Nowicki, S. (1993). Kinematics of birdsong: Functional correlation of cranial movements and acoustic features in sparrows. *Journal of Experimental Biology, 59,* 112–135.

Whitehead, J. M. (1987). Vocally mediated reciprocity between neighbouring groups of mantled howling monkeys, *Alouatta palliata palliata*. *Animal Behaviour, 35,* 1615–1627.

Whiten, A. (1991). *Natural theories of mind*. Cambridge, MA: Basil Blackwell.

Whiten, A. (1993). Evolving a theory of mind: The nature of non-verbal mentalism in other primates. In S. Baron-Cohen, H. Tager Flusberg, and D. J. Cohen (Eds.), *Understanding other minds* (pp. 367–396). Oxford, UK: Oxford University Press.

Whiten, A. (1994). Grades of mindreading. In C. Lewis and P. Mitchell (Eds.), *Children's early understanding of mind: Origins and development* (pp. 47–70). Hove, UK: Lawrence Erlbaum Associates.

Whiten, A., and Ham, R. (1992). On the nature and evolution of imitation in the animal kingdom: Reappraisal of a century of research. In P. J. B. Slater, J. S. Rosenblatt, C. Beer, and M. Milinski (Eds.), *Advances in the study of behavior* (pp. 239–283). New York: Academic Press.

Wickler, W. (1968). *Mimicry in plants and animals*. New York: McGraw-Hill.

Wilczynski, W. (1988). Brainstem auditory pathways in anuran amphibians. In B. Fritzsch, M. J. Ryan, W. Wilczynski, T. E. Hetherington, and W. Walkowiak (Eds.), *The evolution of the amphibian auditory system* (pp. 209–232). New York: John Wiley & Sons.

Wilczynski, W., Allison, J. D., and Marler, C. A. (1993). Sensory pathways linking social and environmental cues to endocrine control regions of amphibian forebrains. *Brain, Behavior and Evolution, 42,* 1–12.

Wilczynski, W., Keddy-Hector, A. C., and Ryan, M. J. (1992). Call patterns and basilar papilla tuning in cricket frogs: I. Differences among populations and between sexes. *Brain, Behavior and Evolution, 39*, 229–237.

Wilczynski, W., Zakon, H. H., and Brenowitz, E. A. (1984). Acoustic communication in spring peepers: Call characteristics and neurophysiological aspects. *Journal of Comparative Physiology, B, 155*, 577–584.

Wild, J. M. (1993). Descending projections of the songbird nucleus robustus archistriatalis. *Journal of Comparative Neurology, 338*, 225–241.

Wild, J. M. (1994). The auditory-vocal-respiratory axis in birds. *Brain Behavior and Evolution, 44*, 192–209.

Wiley, R. H. (1973a). The strut display of male sage grouse: A "fixed" action pattern. *Behaviour, 47*, 129–152.

Wiley, R. H. (1973b). Territoriality and non-random mating in sage grouse (*Centrocercus urophasianus*). *Animal Behaviour Monographs, 6*, 85–169.

Wiley, R. H. (1994). Errors, exaggeration, and deception in animal communication. In L. Real (Ed.), *Behavioral mechanisms in ecology* (pp. 121–143). Chicago: University of Chicago Press.

Wiley, R. H., and Richards, D. G. (1978). Physical constraints on acoustic communication in the atmosphere: Implications for the evolution of animal vocalizations. *Behavioral Ecology and Sociobiology, 3*, 69–94.

Wiley, R. H., and Richards, D. G. (1982). Adaptations for acoustic communication in birds: Sound propagation and signal detection. In D. E. Kroodsma and E. H. Miller (Eds.), *Acoustic communication in birds*, vol. 1 (pp. 131–181). New York: Academic Press.

Wilkinson, G. S. (1987). Altruism and cooperation in bats. In M. B. Fenton, P. Racey, and J. M. V. Rayner (Eds.), *Recent advances in the study of bats* (pp. 299–323). Cambridge, UK: Cambridge University Press.

Wilkinson, G. S. (in press). Information transfer in bats. *Symposium of the Zoological Society of London, 67*.

Will, U., And Fritsch, B. (1988). The eighth nerve of amphibians. In B. Fritsch, M. J. Ryan, W. Wilczynski, T. E. Hetherington, and W. Walkowiak (Eds.), *The evolution of the amphibian auditory system* (pp. 159–183). New York: John Wiley & Sons.

Willatts, P. (1984). The Stage-IV infant's solution of problems requiring the use of supports. *Infant Behavior and Development, 7*, 125–134.

Willatts, P. (1990). Development of problem-solving strategies in infancy. In D. F. Bjorklund (Ed.), *Children's strategies: Contemporary views of cognitive development* (pp. 143–182). Hillsdale, NJ: Lawrence Erlbaum Associates.

Williams, G. C. (1966). *Adaptation and natural selection.* Princeton, NJ: Princeton University Press.

Williams, G. C. (1992). *Natural selection: Domains, levels, and challenges.* Princeton, NJ: Princeton University Press.

Williams, H., Crane, L. A., Hale, T. K., Esposito, M. A., and Nottebohm, F. (1992). Right-side dominance for song control in the zebra finch. *Journal of Neurobiology, 23*, 1006–1020.

Williams, H., and Nottebohm, F. (1985). Auditory responses in avian vocal motor neurons: A motor theory for song perception in birds. *Science, 229*, 279–282.

Williams, J. M., and Slater, P. J. B. (1990). Modelling bird song dialects: The influence of repertoire size and number of neighbours. *Journal of Theoretical Biology, 145*, 487–496.

Wilson, E. O. (1971). *Insect societies.* Cambridge, MA: Harvard University Press.

Wilson, E. O. (1975). *Sociobiology.* Cambridge, MA: Harvard University Press.

Wingfield, J. C., and Marler, P. (1988). Endocrine basis of communication in reproduction and aggression. In E. Knobil and J. Neill (Eds.), *The physiology of reproduction* (pp. 1647–1677). New York: Raven Press.

Winter, P. (1969). Dialects in squirrel monkeys: Vocalizations of the Roman arch type. *Folia primatologica, 10*, 216–229.

Winter, P., Handley, P., Ploog, W., and Schott, D. (1973). Ontogeny of squirrel monkey calls under normal conditions and under acoustic isolation. *Behaviour, 47*, 230–239.

Winter, P., Ploog, D., and Latta, J. (1966). Vocal repertoire of the squirrel monkey (*Saimiri sciureus*), its analysis and significance. *Experimental Brain Research, 1*, 359–384.

Witkin, S. R., and Ficken, M. S. (1979). Chickadee alarm calls: Does mate investment pay dividends? *Animal Behaviour, 27*, 1275–1276.

Wrangham, R. W. (1977). Feeding behaviour of chimpanzees in Gombe National Park, Tanzania. In T. H. Clutton-Brock (Ed.), *Primate ecology: Studies of feeding and ranging behaviour in lemurs, monkeys and apes* (pp. 504–538). London: Academic Press.

Wrangham, R. W. (1980). An ecological model of female-bonded primate groups. *Behaviour, 75*, 262–300.

Wynn, K. (1992). Addition and subtraction by human infants. *Nature, 358*, 749–750.

Xu, F., and Carey, S. (in press). Infants' metaphysics: The case of numerical identity. *Cognitive Psychology*.

Xu, J., Gooler, D. M., and Feng, A. S. (1994). Single neurons in the frog inferior colliculus exhibit directional-dependent frequency selectivity to isointensity tone bursts. *Journal of the Acoustical Society of America, 95*, 2160–2170.

Yaeger, D. D., and Hoy, R. R. (1986). The cyclopean ear: A new sense for the praying mantis. *Science, 231*, 727–729.

Yamane, S., Kaji, S., and Kawano, K. (1988). What facial features activate face neurons in the inferotemporal cortex of the monkey? *Experimental Brain Research, 73*, 209–214.

Yamane, S., Komatsu, H., Kaji, S., and Kawano, K. (1990). Neural activity in the inferotemporal cortex of monkeys during a face discrimination task. In E. Iwai and M. Mishkin (Eds.), *Vision, memory and the temporal lobe* (pp. 133–172). New York: Elsevier.

Yasukawa, K. (1981). Song repertoires in the red-winged blackbird (*Agelaius phoeniceus*): A test of the beau geste hypothesis. *Animal Behaviour, 29*, 114–125.

Yasukawa, K., and Searcy, W. A. (1985). Song repertoires and density assessment in red-winged blackbirds: Further tests of the beau geste hypothesis. *Behavioral Ecology and Sociobiology, 16*, 171–175.

Yin, R. U. (1969). Looking at upside down faces. *Journal of Experimental Psychology, 81*, 141–148.

Yost, W. A. (1980). Man as mammal: Psychoacoustics. In A. N. Popper and R. F. Fay (Eds.), *Comparative studies of hearing in vertebrates* (pp. 399–420). Berlin: Springer-Verlag.

Yost, W. A., and Gourevitch, G. (1987). *Directional hearing*. Berlin: Springer-Verlag.

Yost, W. A., and Hafter, E. R. (1987). Lateralization. In W. A. Yost and G. Gourevitch (Eds.), *Directional hearing* (pp. 49–84). New York: Springer-Verlag.

Young, A. W. (1992). Face recognition impairments. *Philosophical Transactions of the Royal Society of London, 335*, 47–54.

Zahavi, A. (1975). Mate selection: A selection for a handicap. *Journal of Theoretical Biology, 53*, 205–214.

Zahavi, A. (1977). The cost of honesty (further remarks on the handicap principle). *Journal of Theoretical Biology, 67*, 603–605.

Zahavi, A. (1987). The theory of signal selection and some of its implications. In V. P. Delfino (Ed.), *International symposium of biological evolution* (pp. 305–327). Bari, Italy: Adriatica Editrice.

Zahavi, A. (1991). On the definition of sexual selection, Fisher's model, and the evolution of waste and of signals in general. *Animal Behaviour, 42,* 501–503.

Zahavi, A. (1993). The fallacy of conventional signalling. *Proceedings of the Royal Society, London, 340,* 227–230.

Zaitchek, D. (1990). When representations conflict with reality: The preschooler's problem with false beliefs and "false" photographs. *Cognition, 35,* 41–68.

Zakon, H. H., and Wilczynski, W. (1988). The physiology of the anuran eighth nerve. In B. Fritzsch, M. J. Ryan, W. Wilczynski, T. E. Hetherington, and W. Walkowiak (Eds.), *The evolution of the amphibian auditory system* (pp. 125–155). New York: John Wiley & Sons.

Zattore, R. J., Evans, A. C., Meyer, E., and Gjedde, A. (1992). Lateralization of phonetic and pitch discrimination in speech processing. *Science, 256,* 846–849.

Zeskind, P. S., and Collins, V. (1987). Pitch of infant crying and caregiver responses in a natural setting. *Infant Behavior and Development, 10,* 501–504.

Zeskind, P. S., Sale, J., Maio, M. L., Huntington, L., and Weiseman, J. R. (1985). Adult perceptions of pain and hunger cries: A synchrony of arousal. *Child Development, 56,* 549–554.

Ziegler, T. E., Epple, G., Snowdon, C. T., Porter, T. A., Belcher, A. M., and Kuderling, I. (1993). Detection of the chemical signals of ovulation in the cotton-top tamarin. *Animal Behaviour, 45,* 313–322.

Zimmerman, E. (1981). First record of ultrasound in two prosimian species. *Naturwissenshaften, 68,* 531.

Zimmerman, E. (1985). The vocal repertoire of the adult Senegal bushbaby (*Galago senegalensis senegalensis*). *Behaviour, 94,* 212–233.

Zoloth, S. R., Petersen, M. R., Beecher, M. D., Green, S., Marler, P., Moody, D. B., and Stebbins, W. C. (1979). Species-specific perceptual processing of vocal sounds by monkeys. *Science, 204,* 870–872.

Zook, J. M., and Casseday, J. H. (1982). Origin of ascending projection to inferior colliculus in the mustache bat, *Pteronotus parnelli. Journal of Comparative Neurology, 207,* 14–28.

Zook, J. M., and Leake, P. A. (1989). Correlation of cochlear morphology specializations with frequency representation in the cochlear nucleus and superior olive of the mustache bat, *Pteronotus parnelli. Journal of Comparative Neurology, 261,* 347–361.

Index